Stanislaw Ulam: Sets, Numbers, and Universes

Mathematicians of Our Time

Gian-Carlo Rota, series editor

1. Charles Loewner: Theory of Continuous Groups
   notes by Harley Flanders and Murray H. Protter

2. Oscar Zariski: Collected Papers
   Volume I
   Foundations of Algebraic Geometry and Resolution of Singularities
   edited by H. Hironaka and D. Mumford

3. Collected Papers of Hans Rademacher
   Volume I
   edited by Emil Grosswald

4. Collected Papers of Hans Rademacher
   Volume II
   edited by Emil Grosswald

5. Paul Erdös: The Art of Counting
   edited by Joel Spencer

6. Oscar Zariski: Collected Papers
   Volume II
   Holomorphic Functions and Linear Systems
   edited by M. Artin and D. Mumford

7. George Pólya: Collected Papers
   Volume I
   Singularities of Analytic Functions
   edited by R. P. Boas

8. George Pólya: Collected Papers
   Volume II
   Location of Zeros
   edited by R. P. Boas

9. Stanislaw Ulam: Sets, Numbers, and Universes
   edited by W. A. Beyer, J. Mycielski, and G.-C. Rota

Stanislaw Ulam

Sets, Numbers, and Universes

Selected Works

edited by
W. A. Beyer, J. Mycielski, and G.-C. Rota

The MIT Press
Cambridge, Massachusetts, and London, England

This book was printed on Finch Title 93
and bound in Columbia Millbank Vellum MBV-4632
by The Colonial Press Inc.
in the United States of America

Library of Congress Cataloging in Publication Data

Ulam, Stanislaw M.
Stanislaw Ulam: sets, numbers, and universes.

(Mathematicians of our time, v. 9)
Bibliography: p.
1. Mathematics—Collected works. 2. Ulam, Stanislaw M.—Bibliography I. Series.
QA3.U4 510′.8 73-21680
ISBN 0-262-02108-0

Stanslaw Ulam. Sketch by Z. Menkès, ca. 1940

Contents
(Bracketed numbers are from the Bibliography)

Part I
Mathematics

Part II
Computations, Games, and Numbers

Part III
Mathematical Problems

Commentaries

Preface

Stanislaw Marcin Ulam was born in Lwow, Poland. He received his M.Sc. at the Polytechnic Institute there in 1932 and his Ph.D. in 1933. At the invitation of John von Neumann he came to the Institute for Advanced Study in Princeton in late 1935. He was a member of the Society of Fellows at Harvard from 1936 to 1940. After two years at the University of Wisconsin he joined the Los Alamos Scientific Laboratory in New Mexico where he remained until 1967, occasionally spending time at Harvard, MIT, University of Southern California, California at La Jolla, Colorado at Boulder, and IBM. In 1967 he became Professor of Mathematics at the University of Colorado, but continued to visit and consult at Los Alamos.

Ulam's career and interests have been unusually broad for a mathematician and it is not easy to survey his work. The commentaries by other mathematicians assembled at the end of this volume attest to Ulam's impact on the development of mathematics and other sciences since his first paper in 1929. In a brief introduction, one can do little more than mention some of the areas of pure and applied mathematics, technology, computation, physics, astronomy, and biology to which he has contributed. In addition to direct contributions to these fields, he has posed concise problems whose solutions or attempts at solution have advanced these fields. Some of these problems are discussed in his book *A Collection of Mathematical Problems* which is included in this volume.

Ulam's early mathematical work was in set theory, an area fundamental to modern mathematics. Perhaps his most influential contribution here was [3] in which he relates measure theory to general set theory and proves, among other things, that no countably additive measure function $m(A)$ exists, defined for all subsets of a set E of cardinality $\aleph_1$, which vanishes for all subsets consisting of a single point, and for which $m(E) = 1$. With Schreier he made important contributions to group theory and with Borsuk he commenced the development of the notion of ϵ-mappings; from this, he in later years (with D. Hyers) laid the foundation for concepts of structural stability. Łomnicki and Ulam developed measure-theoretic foundations of probability theory, which were prior to and independent of Kolmogorov's book on the same subject.

A fundamental problem in science is to connect microscopic and macroscopic descriptions of matter. An important aspect of this is the ergodic hypothesis, which asserts roughly that the time averages of functions of material states almost always equal the corresponding phase averages. The time averages will usually, in contrast to the space averages, be inscrutable. G. D. Birkhoff showed in 1931 that the ergodic hypothesis can be replaced by the hypothesis that the only sets invariant under a given transformation are of measure 0 or 1. This contribution, however, fails to establish the relevance of the ergodic hypothesis to statistical mechanics. Oxtoby and Ulam in their long paper of 1941 [24] succeeded in showing that the ergodic hypothesis holds in general except for an exceptional set of the first category. Whether the transformations of statistical mechanics are in the exceptional set is not known. Nevertheless the ergodic transformations are dense in the space of all transformations which they considered. Thus Oxtoby and Ulam came closer than anyone (at least until the last few years) in showing the relevance of the ergodic hypothesis to statistical mechanics. For a more complete discussion, see the commentaries by Oxtoby and Smale.

Ulam (with Hawkins and Everett) developed in the years 1944–48 the theory of multiplicative systems (now called branching processes). This work was motivated by applications for the atomic project at Los Alamos.

Ulam has long been concerned with the possibility that other mathematical structures might better describe our space-time on the submicroscopic level than does Euclidean geometry. In developing suitable alternative structures, it is necessary to take account of the Lorentz group. In [86] Everett and Ulam develop a theory of vector spaces over a *p*-adic field and an associated Lorentz group. Beltrametti discusses the relation of their paper to modern physics in more detail in the commentary.

In the area of technology Ulam holds, with Everett, a patent for propelling very large space vehicles by a series of small external nuclear explosions. This idea has developed into what is now called the Orion project. He seems to have been the first to propose extracting gravitational energy from planets to propel space vehicles. This idea is now being used in the "flyby" missions to the outer planets and will provide part of the energy for the first spacecraft to travel beyond the planets. Ulam holds, with Teller, the patent disclosure for the first thermonuclear weapon.

From the earliest days of electronic computers, Ulam has been active in their application to mathematical and physical problems. He proposed (with Fermi and von Neumann) and developed Monte Carlo techniques as a means of computing solutions to probabilistic problems, and extended the applications to nonprobabilistic problems. With Fermi and Pasta [85] he studied by computer the time evolution of a model of a nonlinear vibrating string. To the surprise of the investigators the system did not evolve as one might have predicted; i.e., the energy did not equipartition itself among the modes. Instead, the system tended to return to its initial state. Ulam participated in writing the first computer program to play a game of chess. He made use of computers to study heuristically problems in number theory.

The discovery that biological organisms are encoded by discrete finite sequences over an alphabet of four letters was of great interest to Ulam, as it would be to any mathematician. Biological organisms, usually seemingly very complex, would seem to require very long codes. With Schrandt, Ulam studied the possibilities of simple recursively defined codes giving rise to complex objects. Thus it might be possible that short codes could define objects that might appear to be complex. With Beyer, Smith, and Stein, Ulam studied applications of the concept of distance between finite sequences as a means of reconstructing the evolutionary history of biological organisms.

The present volume contains reprints of Ulam's major papers in pure mathematics and his studies of the applications of computers to nonlinear computation in game theory and physics. The editors hope to prepare a future volume covering his contributions to biology, fluid mechanics, nuclear physics, astronomy, and pattern recognition.

The editors thank the Graphic Arts Group of the Los Alamos Scientific Laboratory for making the negatives for the plates, Margo Lang for typing the commentaries, and Barbara Hendry for preparing the bibliography of Ulam's works. This work and part of the work of the editors was done with support of the U.S. Atomic Energy Commission.

W. A. Beyer

Los Alamos, New Mexico
Nov. 9, 1973

Bibliography of Stanislaw Ulam

(Entries preceded by an asterisk are reprinted in this volume.)

*[1] *Remark on the generalised Bernstein's theorem,* Fund. Math., vol. 13 (1929), pp. 281–3

*[2] *Concerning functions of sets,* Fund. Math., vol. 14 (1929), pp. 231–3

*[3] *Zur Masstheorie in der allgemeinen Mengenlehre,* Fund. Math., vol. 16 (1930), pp. 140–50

*[4] (with K. Borsuk) *On symmetric products of topological spaces,* Bull. Amer. Math. Soc., vol. 37 (1931), pp. 875–82

*[5] (with J. Schreier) *Sur une propriété de la mesure de M. Lebesgue,* Comptes Rendus, vol. 192 (1931), pp. 539–42

*[6] (with K. Kuratowski) *Quelques propriétés topologiques du produit combinatoire,* Fund. Math., vol. 19 (1932), pp. 247–51

*[7] (with S. Mazur) *Sur les transformations isométriques d'espaces vectoriels, normés,* Comptes Rendus, vol. 194 (1932), pp. 946–8

*[8] *Zum Massbegriffe in Produkträumen,* International Congress of Mathematics (New Series), 4th, Zurich, 1932. Verhandlungen, vol. 2 (1932), pp. 118. Fussli

*[9] (with J. Schreier) *Sur le groupe des permutations de la suite des nombres naturels,* Comptes Rendus, vol. 197 (1933), pp. 737–8

*[10] (with J. Schreier) *Sur les transformations continues des sphères euclidiennes,* Comptes Rendus, vol. 197 (1933), pp. 967–8

*[11] (with K. Kuratowski) *Sur un coefficient lié aux transformations continues d'ensembles,* Fund. Math., vol. 20 (1933), pp. 244–53

*[12] (with J. Schreier) *Über die Permutationsgruppe der natürlichen Zahlenfolge,* Studia Math., vol. 4 (1933), pp. 134–41

*[13] (with K. Borsuk) *Über gewisse Invarianten der ϵ-Abbildungen,* Math. Ann., vol. 108 (1933), pp. 311–8

*[14] *Über gewisse Zerlegungen von Mengen,* Fund. Math., vol. 20 (1933), pp. 221–3

*[15] (with J. Schreier) *Eine Bemerkung über die Gruppe der topologischen Abbildungen der Kreislinie auf sich selbst,* Studia Math., vol. 5 (1934), pp. 155–9

*[16] (with Z. Łomnicki) *Sur la théorie de la mesure dans les espaces combinatories et son application au calcul des probabilitiés. I Variables indépendantes,* Fund. Math., vol. 23 (1934), pp. 237–78

*[17] (with J. Schreier) *Über topologische Abbildungen der euklidischen Sphären,* Fund. Math., vol. 23 (1934), pp. 102–18

*[18] (with H. Auerbach) *Sur le nombre de générateurs d'un groupe semisimple,* Comptes Rendus, vol. 201 (1935), pp. 117–9

*[19] (with J. Schreier) *Sur le nombre des générateurs d'un groupe topologique compact et connexe,* Fund. Math., vol. 24 (1935), pp. 302–4

*[20] (with H. Auerbach and S. Mazur) *Sur une propriété caractéristique de l'ellipsoïde,* Monat. Math., vol. 42 (1935), pp. 45–8

*[21] (with J. Schreier) *Über die Automorphismen der Permutationsgruppe der natürlichen Zahlenfolge,* Fund. Math., vol. 28 (1937), pp. 258–60

*[22] (with J. C. Oxtoby) *On the equivalence of any set of first category to a set of measure zero,* Fund. Math., vol. 31 (1938), pp. 201–6

*[23] (with J. C. Oxtoby) *On the existence of a measure invariant under a transformation,* Ann. Math., ser. 2, vol. 40 (1939), pp. 560–6

*[24] (with J. C. Oxtoby) *Measure-preserving homeomorphisms and metrical transitivity,* Ann. Math., ser. 2, vol. 42 (1941), pp. 874–920

*[25] *What is measure?,* Amer. Math. Monthly, vol. 50 (1943), pp. 597–602

[26] (with N. C. Metropolis and E. Teller) *Pressures developed by nuclear . . . ,* Los Alamos Scientific Lab., 1944, LA-161, Secret

*[27] (with D. H. Hyers) *On approximate isometries,* Bull. Amer. Math. Soc., vol. 51 (1945), pp. 288–92

*[28] (with C. J. Everett) *On ordered groups,* Trans. Amer. Math. Soc., vol. 57 (1945), pp. 208–16

[29] (with J. L. Tuck) *Possibility of initiating a thermonuclear reaction . . . ,* Los Alamos Scientific Lab., 1946, LA-560, Secret

*[30] (with C. J. Everett) *Projective algebra I.,* Amer. J. Math., vol. 68 (1946), pp. 77–88

[31] *Stefan Banach, 1892–1945,* Bull. Amer. Math. Soc., vol. 52 (1946), pp. 600–3

[32] (with D. Hawkins) *Theory of multiplicative processes,* Los Alamos Scientific Lab., 1944, LA-171

*[33] (with D. H. Hyers) *Approximate isometries of the space of continuous functions,* Ann. Math., ser. 2, vol. 49 (1947), pp. 285–9

[34] (with R. D. Richtmyer) *On the nuclear composition of the implosion-type bomb during the last stages of the explosion,* Los Alamos Scientific Lab., 1947, LAMS-559, Secret

*[35] (with C. J. Everett) *Multiplicative systems, I,* Proc. Nat. Acad. Sci., U.S.A., vol. 34 (1948), pp. 403–5

[36] (with C. J. Everett) *Multiplicative systems in several variables. Part I,* Los Alamos Scientific Lab., 1948, LA-683

[37] (with C. J. Everett) *Multiplicative systems in several variables. Part II,* Los Alamos Scientific Lab., 1948, LA-690

*[38] (with N. C. Metropolis) *The Monte Carlo method,* J. Amer. Statist. Assoc., vol. 44 (1949), pp. 335–41

[39] (with C. J. Everett) *Multiplicative systems in several variables. Part III,* Los Alamos Scientific Lab., 1948, LA-707

[40] (with G. F. Evans and J. von Neumann) *Outline of a method for the calculation of the progress of thermonuclear reactions,* Los Alamos Scientific Lab., 1949, LAMS-831, Secret

[41] (with E. Fermi and M. Planck) *Considerations of the thermonuclear reactions . . . ,* Los Alamos Scientific Lab., 1950, LA-1158, Secret

[42] (with C. J. Everett) *Ignition of a large mass of deuterium. I,* Los Alamos Scientific Lab., 1950, LA-1076, Secret

[43] (with J. L. Elliott, C. J. Everett, J. Houston, and F. A. Ulam) *Ignition of a large mass of deuterium. II,* Los Alamos Scientific Lab., 1950, LA-1124, Secret

[44] (with E. Teller) *On heterocatalytic detonations. I,* Los Alamos Scientific Lab., 1951, LAMS-1225, Secret

[45] *On the Monte Carlo method,* Symposium on Large-Scale Digital Calculating Machines, 2nd, Cambridge, Mass. (Sept. 13–16), 1949. Proc., pp. 207–12. Harvard University Press, 1951

*[46] (with D. H. Hyers) *Approximately convex functions,* Proc. Amer. Math. Soc., vol. 3 (1952), pp. 821–8

[47] (with W. Goad) *Observable effect . . . in thermonuclear reactions,* Los Alamos Scientific Lab., 1952, LAMS-1331, Secret

*[48] *Random processes and transformations,* International Congress of

Mathematicians, Cambridge, Mass. (Aug. 30–Sept. 6, 1950), Proc., vol. 2, pp. 264–75. Amer. Math. Soc., 1952

[49] (with C. J. Everett, E. D. Cashwell, and O. W. Rechard) *Monte Carlo calculation of the number of fissions per source neutron . . .*, Los Alamos Scientific Lab., 1953, LA-1592, Secret

*[50] (with N. C. Metropolis) *A property of randomness of an arithmetical function,* Amer. Math. Monthly, vol. 60 (1953), pp. 252–3

*[51] (with D. H. Hyers) *On the stability of differential expressions,* Math. Mag., vol. 28 (1954), pp. 59–64

[52] (with C. J. Everett) *On a method of propulsion of projectiles by means of external nuclear explosions,* Los Alamos Scientific Lab., 1955, LAMS-1955

*[53] (with V. L. Gardiner, R. Lazarus, and N. C. Metropolis) *On certain sequences of integers defined by sieves,* Math. Mag., vol. 29 (1956), pp. 117–22

[54] (with M. E. Battat et al.) *Some suggestions for experiments with nuclear explosions,* Los Alamos Scientific Lab., 1956, LAMS-2087, Secret

*[55] (with J. M. Kister, P. R. Stein, W. Walden, and M. B. Wells) *Experiments in chess,* J. Assoc. Comput. Mach., vol. 4 (1957), pp. 174–7

[56] (with P. R. Stein) *Experiments in chess on electronic computing machines,* Computers and Automation, vol. 6 (1957), pp. 14–8

*[57] *Infinite models in physics,* Symposia in Applied Mathematics, 7th, Polytechnic Institute, Brooklyn, April 14–15, 1955. Proc. Applied Probability, pp. 87–95. (Amer. Math. Soc. Symposia Appl. Math., vol. 7) McGraw-Hill, 1957

[58] *Marian Smoluchowski and the theory of probabilities in physics,* Amer. J. Phys., vol. 25 (1957), pp. 475–81

[59] *On some possibilities in the organization and use of computing machines,* International Business Machines, 1957 (IBM-RC-68)

[60] (with P. R. Stein) *Study of certain combinatorial problems through experiments on computing machines,* High-Speed Computer Conference, Baton Rouge, 1955. Proc., pp. 101–6. Louisiana State University, 1957

[61] *Book Review: The computer and the brain, by John von Neumann,* Sci. Amer., vol. 198, no. 6 (1958), pp. 128–30

[62] *John von Neumann, 1903–1957,* Bull. Amer. Math. Soc., vol. 64, no. 3 (1958), pt. 2, pp. 1–49

[63] *On the possibility of extracting energy from gravitational systems by navigating space vehicles,* Los Alamos Scientific Lab., 1958, LAMS-2219

[64] (with C. L. Longmire and F. Reines) *Some schemes for nuclear propulsion,* Los Alamos Scientific Lab., 1958, LAMS-2186

[65] *Bato and some suggestions for experiments with nuclear explosions,* Joint AEC Weapons Laboratory Symposium, University of California Radiation Laboratory, Livermore, Feb. 6–8, 1957. Proc., p. 54, Univ. of California Rad. Lab., 1959, UCRL-4893, Secret

*[66] (with J. R. Pasta) *Heuristic numerical work in some problems of hydrodynamics,* Math. Tables and Other Aids to Computation, vol. 13 (1959), pp. 1–12

[67] *Book Review: Funkcje rzeczywiste, by Roman Sikorski,* Bull. Amer. Math. Soc., vol. 65 (1959), pp. 305–6

[68] (with P. R. Stein and M. T. Menzel) *Quadratic transformations. Part I,* Los Alamos Scientific Lab., 1959, LA-2503

*[69] *Collection of Mathematical Problems.* Interscience, New York, 1960

[70] *Monte Carlo calculations in problems of mathematical physics,* Modern Mathematics for the Engineer, Edwin F. Beckenback, ed., 2nd ed. McGraw-Hill, 1961, pp. 261–81

*[71] *On some statistical properties of dynamical systems,* Berkeley Symposium on Mathematical Statistics and Probability, 4th, University of California, June 20–July 30, 1960. Proc., vol. 3, pp. 315–20, University of California Press, 1961

[72] *Electronic computers and scientific research,* Age of Electronics, Carl F. Overhage, ed., McGraw-Hill, 1962, pp. 95–108

[73] *On some mathematical problems connected with patterns of growth of figures,* Symposia in Applied Mathematics, 14th, New York, April 5–8, 1961. Proc., Mathematical Problems in the Biological Sciences, pp. 215–24. Amer. Math. Soc., Symposia Appl. Math., vol. 14, Amer. Math. Soc., 1962

[74] *Stability of many-body computations,* Symposia in Applied Mathematics, 13th, New York, April 14–15, 1960. Proc., Hydrodynamic

Instability, pp. 247–58. Amer. Math. Soc. Symposia Appl. Math., vol. 13, Amer. Math. Soc., 1962

[75] *Electronic computers and scientific research (pts. I and II)*, Computers and Automation, vol. 12, no. 8, pp. 20–4, pt. I; vol. 12, no. 9, pp. 35–40, pt. II; 1963

[76] *Nekotorykh statisticheskikh svoistvakh dinamicheskikh sistem,* Matematikai; Periodicheskii Sbornik Perevodov Ionostranykh Statei, vol. 7 (1963), pp. 137–42

*[77] *Some properties of certain non-linear transformations,* Conf. on Mathematical Models in Physical Sciences, University of Notre Dame, Apr. 15–7, 1962. Proc., pp. 85–95, Prentice-Hall, 1963

*[78] *Combinatorial analysis in infinite sets and some physical theories,* Rev. Soc. Ind. Appl. Math., vol. 6 (1964), pp. 343–55

[79] *Computers,* Sci. Amer., vol. 211, no. 3 (1964), pp. 203–16

[80] *Nereshennye Matematicheskie Zadachi,* Moscow, 1964

*[81] (with P. R. Stein) *Non-linear transformation studies on electronic computers,* Rozprawy Matematyczne, no. 39 (1964), pp. 1–66

[82] (with W. E. Walden) *Possibility of an accelerated process of collapse of stars in a very dense centre of a cluster or a galaxy,* Nature, vol. 201 (1964), p. 1202

*[83] (with M. L. Stein and M. B. Wells) *A visual display of some properties of the distribution of primes,* Amer. Math. Monthly, vol. 71 (1964), pp. 516–20

[84] *La Machine Créatrice?,* Rencontres Internationales de Genève "Le robot, la bête et l'homme." Proc., pp. 31–42. Histoire et Société d'Aujourd'hui, Editions de la Baconniere, Neuchâtel, Switzerland, 1966

*[85] (with E. Fermi and J. R. Pasta) *Studies of non linear problems,* Fermi, Enrico, Collected Papers, vol. 2, pp. 977–88, University of Chicago Press, 1965

*[86] (with C. J. Everett) *On some possibilities of generalizing the Lorentz group in the special relativity theory,* J. Combinatorial Theory, vol. 1 (1966), pp. 248–70

[87] *Thermonuclear devices,* Essays in Modern Physics, Robert E. Marshak and J. Warren Blaker, eds. Interscience, 1966

[88] *How to formulate mathematically problems of rate of evolution,*

Mathematical Challenges to the Neo-Darwinian Interpretation of Evolution, Symposium Held at the Wistar Institute of Anatomy and Biology, Apr. 25–26, 1966, P. S. Moorhead and M. Kaplan, eds. (Wistar Institute, Symposium Monograph no. 5). Wistar Institute, Philadelphia, 1967, pp. 21–3

*[89] (with M. L. Stein) *An observation on the distribution of primes,* Amer. Math. Monthly, vol. 74 (1967), p. 43

[90] (with R. G. Schrandt) *On recursively defined geometrical objects and patterns of growth,* Los Alamos Scientific Lab., 1967, LA-3762

*[91] (with G. H. Meisters) *On visual hulls of sets,* Proc. Nat. Acad. Sci., U.S.A., vol. 57 (1967), pp. 1172–4

[92] (with M. Kac) *Mathematics and logic; retrospect and prospects,* Britannica Perspectives, vol. 1, pp. 557–732. Encyclopaedia Britannica, Inc., 1968. Also issued as a book, Praeger, New York, 1968

*[93] (with W. A. Beyer) *Note on the visual hull of a set,* J. Combinatorial Theory, vol. 4 (1968), pp. 240–5

[94] *Computations on certain binary branching processes,* Computers in Mathematical Research, pp. 168–70, North Holland, 1968

*[95] (with P. Erdös) *On equations with sets as unknowns,* Proc. Nat. Acad. Sci., U.S.A., vol. 60 (1968), pp. 1189–95

[96] *Reminiscences from the Scottish Café,* Wiadom Mat., ser. 2, vol. 12 (1969), pp. 49–58

*[97] (with J. Mycielski) *On the pairing process and the notion of genealogical distance,* J. Combinatorial Theory, vol. 6 (1969), pp. 227–34

[98] (with W. A. Beyer and R. G. Schrandt) *Computer studies of some history-dependent random processes,* Los Alamos Scientific Lab., 1969, LA-4246

[99] (with C. J. Everett) *Entropy of interacting populations,* Los Alamos Scientific Lab., 1969, LA-4256

[100] The applicability of mathematics, in *The Mathematical Sciences: A Collection of Essays* (Committee for Support of Mathematics), MIT Press, 1969, pp. 1–6

[101] Foreword in *George Gamow: My World Line (An autobiography),* Viking, New York, 1970

[102] (with R. G. Schrandt) *Some elementary attempts at numerical model-*

ing of problems concerning rates of evolutionary processes, Los Alamos Scientific Lab., 1971, LA-4573-MS

*[103] (with P. Erdös) *Some probabilistic remarks on Fermat's last theorem,* Rocky Mountain J. Math., vol. 1 (1971), pp. 613–6

[104] (with W. A. Beyer and M. L. Stein) *The notion of complexity,* Los Alamos Scientific Lab., 1971, LA-4822

[105] (with W. A. Beyer, T. F. Smith, and M. L. Stein) *Metrics in biology, an introduction,* Los Alamos Scientific Lab., 1972, LA-4973

[106] *Some combinatorial problems studied experimentally on computing machines,* Applications of Number Theory to Numerical Analysis, Ed., S. K. Zaremba; Academic Press, 1972, pp. 1–10

[107] *Some ideas and prospects in biomathematics,* Annual Review of Biophysics and Bioengineering, vol. 1 (1972), pp. 277–91

[108] Gamow and mathematics, *Cosmology, Fusion & Other Matters,* George Gamow Memorial Volume, Colorado Associated Univ. Press, Boulder, 1972

Commentators

E. Beltrametti, University of Genoa
W. Beyer, Los Alamos Scientific Laboratory
K. Borsuk, Polish Academy of Science
W. Fuchs, Cornell University
T. Ganea, University of Washington
D. Hyers, University of Southern California
D. Knuth, Stanford University
J. Lindenstrauss, Hebrew University of Jerusalem
Z. Łomnicki, Codsall, England
D. Monk, University of Colorado
J. Mycielski, University of Colorado
J. Oxtoby, Bryn Mawr College
S. Smale, University of California, Berkeley
P. Stein, Los Alamos Scientific Laboratory
M. Wells, Los Alamos Scientific Laboratory

Part I

Mathematics

Extrait de »Fundamenta Mathematicae«, T. XIII.

Remark on the generalised Bernstein's theorem.

By

Stanisław Ulam (Lwów).

Bernstein's theorem, that from $2\mathfrak{m} = 2\mathfrak{n}$ follows $\mathfrak{m} = \mathfrak{n}$, for every pair of cardinal numbers was recently generalised in the following way [1]):

Let a set E be twice decomposed into two equivalent parts:

$$E = M + N = P + Q, \quad MN = 0 = PQ$$

and let φ and ψ be one-one correspondances between M and N and P and Q respectively, then the sets M and Q can be decomposed into *four* disjunctive subsets:

$$M = M_1 + M_2 + M_3 + M_4$$

$$Q = Q_1 + Q_2 + Q_3 + Q_4$$

such that:

$$Q_1 = \alpha_1(M_1), \quad Q_2 = \alpha_2(M_2), \quad Q_3 = \alpha_3(M_3), \quad Q_4 = \alpha_4(M_4),$$

where α_i $(i = 1 \dots 4)$ are four functions belonging to the group Γ created by combining the functions φ and ψ. (The first of them are:

$$0 = \text{Identity}, \quad \varphi, \ \psi, \ \varphi\psi, \ \psi\varphi, \ \varphi\psi\varphi, \ \psi\varphi\psi, \ \text{etc.})$$

I shall prove, that in this theorem the number *four* cannot be diminished [2]). I shall define indeed a set E, two subsets M and Q and two one-one correspondances between M and $E - M$ and Q

[1]) See C. Kuratowski, *Fund. Math.* VI, p. 240—243, D. König *Math. Ann.* 77, and *Fund. Math.* VIII. See also Banach and Tarski, *Fund. Math.* VI, where use is made of this theorem.

[2]) This problem was raised in the Seminary of Prof. C. Kuratowski.

For commentary to this paper [1], see p. 673.

and $E - Q$ respectively, so that there does not exist any decomposition of the sets M ane Q into *three* disjunctive subsets united by some functions belonging to the group Γ.

For the sake of better understanding I shall at first give an example proving the impossibility of a decomposition into *two* parts possessing the above stated properties.

The set E consists of 8 points (see fig. 1).

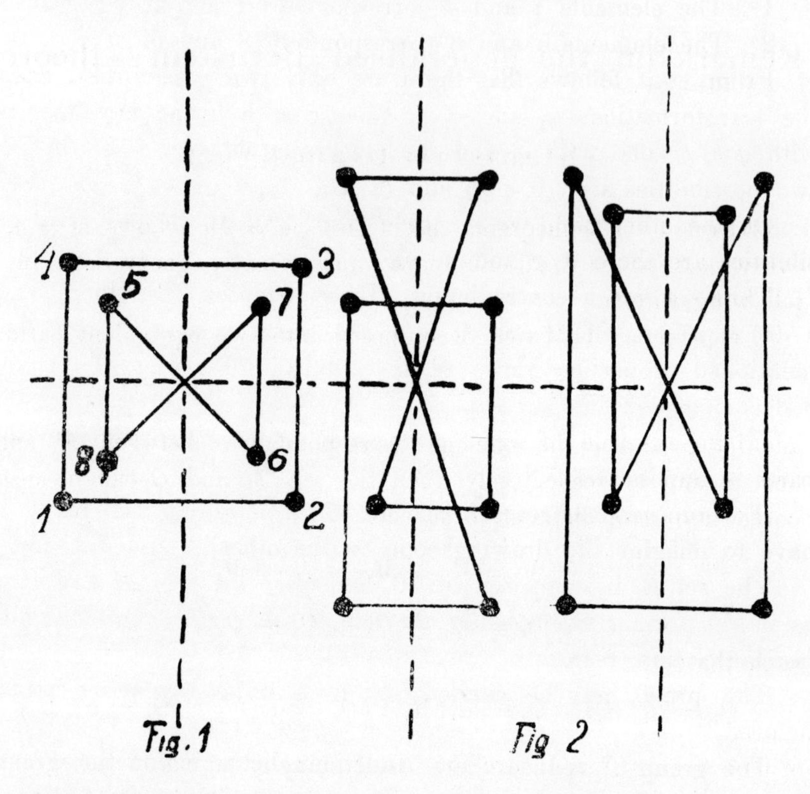

Fig. 1 Fig 2

M is the part of E on the left of the vertical line, Q below the horizontal one.

$$\varphi(1) = 2, \quad \varphi(4) = 3, \quad \varphi(5) = 6, \quad \varphi(8) = 7;$$

$$\psi(1) = 4, \quad \psi(2) = 3, \quad \psi(8) = 5, \quad \psi(6) = 7;$$

and inversely:

$$\varphi(2) = 1, \quad \varphi(3) = 4, \ldots, \quad \psi(7) = 6.$$

Two corresponding points are connected by a segment. In our case the group Γ reduces itself evidently to 4 functions:

$$0, \quad \varphi, \quad \psi, \quad \varphi\psi (= \psi\varphi)$$

Let us now suppose, there are existing two functions α_1 and α_2 of Γ and a decomposition of both M and Q into two disjunctive parts, such that:

$$M = M_1 + M_2, \quad Q = Q_1 + Q_2,$$

$$Q_1 = \alpha_1(M_1), \quad Q_2 = \alpha_2(M_2).$$

It follows that

1°: The elements 1 and 4 correspond to 1 and 2.

2°: The elements 5 and 8 correspond to 8 and 6.

From 1° it follows that there are only two possibilities: one of the transformations α_1 and α_2 is identic with 0 and the other one with $\varphi\psi$, *or* one with φ and the other one with ψ; i. e. the only two possibilities are: $(0, \varphi\psi)$ and (φ, ψ).

On the other hand we conclude from 2° that the only two possibilities are there $0, \varphi$ and $\psi, \varphi\psi$.

Thus we get a contradiction, if we suppose that there exists a decomposition of M and Q into *two* parts possessing the above mentioned properties. However, there exists in the case of the fig. 1 a decomposition into *three* parts.

In the example of fig. 2 even such a decomposition into *three* parts is impossible.

(For typographic reasons the set E is drawn in two parts. We have to imagine the drawings one *on* the other).

The set E is composed of 16 elements, the sets M and Q are as in the former example on the left of the vertical and below the horizontal line, resp.

The proof may be carried out in a quite analogous way as before.

The group Γ reduces here itself actually to 8 functions:

$$0, \quad \varphi, \quad \psi, \quad \psi\varphi, \quad \varphi\psi, \quad \varphi\psi\varphi, \quad \psi\varphi\psi, \quad \varphi\psi\varphi\psi (= \psi\varphi\psi\varphi).$$

Concerning functions of sets.

By

Stanisław Ulam (Lwów).

On considering the properties of functions defined on every subset of a certain space [1]) arises the problem of the existence of a function which satisfies the condition of „sustractivity“ i. e., that

$$F(A - B) = F(A) - F(B),$$

but does *not* satisfy the „infinite additivity“ i. e. the condition:

$$F(A_1 + A_2 + \ldots + A_n + \ldots) = F(A_1) + F(A_2) + \ldots + F(A_n) + \ldots$$

(the values of the function F are sets!)

I shall prove here the existence of such a function. In the proof Zermelo's axiom of choice will play an essential part.

The space (denoted by 1) on which the function is defined is the set of all natural numbers. We shall call two sets of natural numbers M and N „almost identical“ if the set $(M - N) + (N - M)$ is finite or vacuous.

M „almost contains“ N will mean, that M contains in the usual sense a set N_1 „almost identical“ with N.

It is easy to conclude, that if A is „almost identical“ with B and B with C, A is „almost identical“ with C. Every two finite sets are „almost identical“.

The class of all sets of natural numbers X may be ordered with the aid of Zermelo's axiom into a transfinite sequence $\mathcal{A}$.

We shall place now the sets of the sequence $\mathcal{A}$ into certain classes: $K_0, K_1, \ldots, K_\alpha \ldots$ in the following way:

[1]) See e. g. A. Tarski Ann. Soc. Pol. Math. VI. pp. 127 et 132.

For commentary to this paper [2], see p. 673.

(I) The class K_0 consists of all finite and the vacuous set.

(II) The class K_1 contains 1°: the first infinite set X of the sequence $\mathfrak{A}$, 2°: all the sets Y which „almost contain" X.

(It is obvious that $1 \varepsilon K_1$ i. e. the set of *all* natural numbers belongs to K_1),

(III) If η is *even* (all limit ordinal numbers are here to be considered as even) and > 0. K_η contains 1°: the first set X in the sequence $\mathfrak{A}$ which belongs to no K_ξ with $\xi < \eta$, 2°: all the sets „almost identical" with X, 3°: the vacuous set.

(IV) If η is *odd* > 1, K_η consists of all the sets which almost contain" the differences $A - B$, where A belongs to K_α, B to $K_{\eta-1}$ α is odd and $< \eta$

I shall prove now, that no set X can belong to two classes, one of which has an even index, the second an odd one.

For this purpose I shall prove that:

1) If α and η are odd and $\alpha < \eta$, then $K_\alpha \subset K_\eta$

2) If $X \varepsilon K_\alpha$ and $Y \varepsilon K_\alpha$ and α is odd, then $X . Y \neq 0$ and $X . Y \varepsilon K_\alpha$.

Ad 1). Let $X \varepsilon K_\alpha$. The set X is obviously a difference $X - 0$. 0 belongs to $K_{\eta-1}$ It follows from (IV) that the set X belongs also to K_η.

Ad 2). I shall prove this lemma by transfinite induction. The lemma holds obviously true for $\alpha = 1$.

Suppose it is true for all classes K_α, $\alpha < \eta$; we shall conclude that it is true for K_η.

Every set X of the class K_η „almost contains" a set of the form $X_1 - C$, where X_1 belongs to a class K_α with an odd index $\alpha < \eta$, and $C \varepsilon K_{\eta-1}$. Since K_α contains all sets „almost identical" with X_1, we may suppose that $X \supset X_1 - C$. Similarly $Y \supset Y_1 - D$, where Y_1 and D have an analogous meaning.

We shall prove, that $X_1 . Y_1 - (C + D) \neq 0$

By hypothesis: $X_1 . Y_1 \varepsilon K_\alpha$, on the other hand $(C + D) \varepsilon K_{\eta-1}$ for it is (by III) „almost identical" with either C or D. Now no set from the class $K_{\eta-1}$ can contain a set from the class K_α, $\alpha < \eta - 1$. [In this case it would belong by definition also to the class K_α!]

Thus $X . Y \supset X_1 . Y_1 - (C + D) \neq 0$, and as $X_1 . Y_1 \varepsilon K_\alpha$ and $(C + D) \varepsilon K_{\eta-1}$ it follows by (IV) that $X . Y \varepsilon K_\eta$.

The lemma 2) is proved. We shall conclude, that *no set X can figure in two classes, one with an even index, the second with an odd one.*

Suppose $X \varepsilon K_\xi \cdot K_\eta$, ξ even, η odd. By (IV) $(1-X) \varepsilon K_{\xi+1}$ (for $1 \varepsilon K_1$)

Thus, the two sets: X and $1-X$ belong to the classes K_η and $K_{\xi+1}$ resp. Now, by lemma 1) we have either $K_\eta \subset K_{\xi+1}$ or $K_{\xi+1} \subset K_\eta$. Hence the sets X and $1-X$ belong simultaneously either to $K_{\xi+1}$ or K_η.

Since:

$$(1-X) . X = 0,$$

we have a contradiction with lemma 2).

We may now define our function $F(X)$ as follows: let $X \varepsilon K_\alpha$; if α is odd: $F(X) = 1$ (= the whole space), if α is even: $F(X) = 0$.

In virtue of this definition, we have by lemma 2)

a) If $F(X) = 1 = F(Y)$, $F(X . Y) = 1$.

b) By (I): $F(0) = 0$.

c) By (II) and (IV): if $X \subset Y$ and $F(X) = 1$, then $F(Y) = 1$.

We shall prove, that:

$$F(X-Y) = F(X) - F(Y).$$

In this purpose we shall show at first:

d) $F(1-X) = 1 - F(X)$.

If $F(X) = 0$, $X \varepsilon K_\alpha$, α is even. By (IV): $(1-X) \varepsilon K_{\alpha+1}$ hence $F(1-X) = 1 = 1 - F(X)$.

If $F(X) = 1$, $F(1-X) = 0$, for otherwise the sets X and $1-X$ both belong to a class with an odd index (by lemma 1) which is a contradiction to lemma 2), since $X . (1-X) = 0$.

Now we shall prove that:

e) $F(X . Y) = F(X) . F(Y)$.

If $F(X) = 0$ we have by d) $F(1-X) = 1$ hence by c) $F(XY) = 0$ Thus to prove e) it remains to consider the case where $F(X) = 1 = F(Y)$. Now, by a) in this case we have $F(XY) = 1 = F(X) . F(Y)$.

By d) and e) the „sustractivity" is proved. On the other hand. the function does not satisfy the condition:

$$F(A_1 + A_2 + \dots A_n + \dots) = F(A_1) + F(A_2) + \dots F(A_n) + \dots$$

This is seen if we define A_n as the set composed of the one element n: $A_n = \{n\}$.

Zur Masstheorie in der allgemeinen Mengenlehre.

Von

Stanisław Ulam (Lwów).

Herr Banach stellte folgendes Problem, das als Verallgemeinerung des bekannten Lebesgue'schen Massproblems aufgefasst werden kann:

Gibt es eine, für alle Teilmengen des Intervalles (0, 1) erklärte reelle Funktion f, welche die folgenden Bedingungen erfüllt:

1°: Es gibt eine Menge Q, für welche $f(Q) > 0$ ist.

2°: Für einen einzelnen Punkt p ist stets $f(p) = 0$.

3°: Ist $\{A_n\}$ eine Folge von paarweise elementfremden Mengen, so gilt: $f(A_1 + A_2 + \ldots A_n + \ldots) = f(A_1) + f(A_2) + \ldots f(A_n) + \ldots$

Die Nichtexistenz einer solchen Funktion haben unter Voraussetzung der Richtigkeit der Kontinuumhypothese Banach und Kuratowski bewiesen [1]). Somit wurde das bekannte Vitali'sche Resultat verschärft, laut dessen es im Sinne von Lebesgue unmessbare Mengen gibt. [Die Lebesgueschen Bedingungen enthalten bekanntlich ausser den Bedingungen 1°... 3° auch das Postulat der Massgleichheit für kongruente Mengen].

Im Anschluss daran bewies Banach, unter Zugrundelegung der Cantorschen Hypothese: $2^{\aleph_\xi} = \aleph_{\xi+1}$, den folgenden, allgemeinen Satz [2]):

Ist f eine additive Funktion vom Typus $\aleph_\xi$, so ist $\aleph_\xi$ eine unerreichbare Kardinalzahl [3]); dabei soll eine Funktion f additiv mit

1) Fund. Math. T. XIV.

2) Fund. Math. T. XV.

3) Die Kardinalzahl $\aleph_\xi$ heisst unerreichbar, falls ξ eine Grenzzahl ist, $\aleph_\xi > \aleph_0$ ist und sich nicht als Summe von Kardinalzahlen die kleiner als $\aleph_\xi$ sind, so dass

For commentary to this paper [3], see p. 674.

der Mächtigkeit $\aleph_\xi$ heissen, wenn für elementfremde Mengen A_η stets

$$f\left(\sum_\eta A_\eta\right) = \sum_\eta f(A_\eta); \quad 0 \leqq \eta < \omega_\xi$$

gilt, wo ω_ξ die erste Ordinalzahl von der Mächtigkeit $\aleph_\xi$ bezeichnet; die Funktion f ist vom Typus $\aleph_\xi$, wenn sie mit jeder Mächtigkeit, die kleiner als $\aleph_\xi$ ist, additiv ist, nicht aber mit der Mächtigkeit $\aleph_\xi$ selbst.

In der vorliegenden Arbeit wird nun vor allem gezeigt:

(I): *In dem Satze von Banach ist die Hypothese $2^{\aleph_\xi} = \aleph_{\xi+1}$ vollkommen entbehrlich.*

(II): *Für das Banach-Kuratowski-sche Resultat genügt es nur die Richtigkeit der folgenden, schwächeren Hypothese vorauszusetzen:*

(I) Unter der Kardinalzahlen, die $\leqq \mathfrak{c}$ sind ($\mathfrak{c}$ bezeichnet die Mächtigkeit des Kontinuums) gibt es keine unerreichbare.

Es sei dem Verfasser gestattet Herrn Prof Kuratowski auch hier, für das Interesse, welches Er an dieser Arbeit nahm, sowohl wie für die wertvollen Ratschläge den wärmsten Dank aussprechen zu können.

§ 1.

Wir beweisen zunächst den folgenden grundlegenden

Satz (*A*): *Es sei Z eine Menge von der Mächtigkeit $\mathfrak{m}$, so dass es unter den Kardinalzahlen, die $\leqq \mathfrak{m}$ sind, keine unerreichbare gibt. Behauptung: $\mathfrak{m}$ ist „unmessbar", d. h. in der Menge Z lässt sich keine Funktion definieren, die die Bedingungen 1° — 3° erfüllen würde.*

(Aus diesem Satze folgt direkt die Richtigkeit der Behauptung II).

1. Beweis. Wenn die Mächtigkeit von Z $\aleph_0$ oder kleiner ist, ist die Behauptung trivial. Wir werden den Satz durch transfinite Induktion beweisen. Um die Methode deutlich hervortreten zu lassen, zeigen wir die Richtigkeit des Satzes im Falle wo die Mächtigkeit von Z $\aleph_1$ ist.

die Anzahl der Summanden auch kleiner als $\aleph_\xi$ ist, darstellen lässt. Es ist nicht bekannt, ob solche Kardinalzahlen überhaupt existieren. Vgl. Hausdorff. *Grundzüge der Mengenlehre*, p. 131, Leipzig 1914.

Die Unmöglichkeit der Angabe von f, des „Masses" in Z ergibt sich aus folgender Tatsache:

Es lässt sich eine unendliche Matrix von Teilmengen von Z konstruieren:

$$\begin{array}{l} A_1^1, A_2^1, \dots A_n^1 \dots A_\alpha^1 \dots \\ A_1^2, A_2^2, \dots A_n^2 \dots A_\alpha^2 \dots \\ \dots\dots\dots\dots\dots \\ A_1^n, A_2^n, \dots A_n^n \dots A_\alpha^n \dots \\ \dots\dots\dots\dots\dots \end{array}$$

für $n < \infty$, $\alpha < \Omega$, d. i. mit $\aleph_0$ Zeilen und $\aleph_1$ Spalten mit den Eigenschaften:

a) Je zwei Mengen derselben Zeile sind elementfremd, d. i. $A_{\alpha_1}^n \cdot A_{\alpha_2}^n = 0$, für jedes n, wenn nur $\alpha_1 \neq \alpha_2$.

b) Die Vereinigungsmenge jeder Spalte ergibt „fast" das ganze, d. i. mit Ausnahme von höchstens abzählbar vielen Elementen, Z. Anders gewendet: für jedes α ist $(Z - \sum_{n=1}^{\infty} A_\alpha^n)$ höchstens abzählbar.

Da die Existenz des Masses (der Funktion f) eine Invariante gegenüber ein-eindeutigen Transformationen der Menge Z ist, können wir voraussetzen, dass die Menge Z, mit deren Teilmengen die Matrix konstruiert worden ist, diejenige Menge Q ist, für welche laut der Bedingung 1° $f(Q) > 0$ ist. Da weiter alle abzählbaren Mengen nach 3° und 1° nur das Mass 0 haben können, so sieht man, dass die Vereinigungsmenge jeder Spalte eine Menge von positivem Masse sein muss: für jedes α, $f(\sum_{n=1}^{\infty} A_\alpha^n) > 0$.

Nun aber muss in jeder Spalte eine Menge hervortreten, die eine Menge enthält, für welche $f > 0$ ist.

Die Menge $\sum_{n=1}^{\infty} A_\alpha^n$ kann man nämlich in der Form darstellen:

$$\sum_{n=1}^{\infty} A_\alpha^n = A_\alpha^1 + [A_\alpha^2 - A_\alpha^1] + [A_\alpha^3 - (A_\alpha^1 + A_\alpha^2)] + \dots [A_\alpha^n - \sum_{k=1}^{n-1} A_\alpha^k] + \dots$$

wo offenbar alle Mengen elementfremd sind. Wenn die Vereinigungsmenge abzählbar vielen elementfremden Mengen eine Menge von positivem Masse ergibt, so muss nach der Bedingung 3 mindestens eine dieser Mengen ein positives Mass haben.

Da jedoch $A_\alpha^n - \sum_{k=1}^{n-1} A_\alpha^k \subset A_\alpha^n$, gibt es wirklich in jeder Spalte eine Menge, die eine Menge von positiven Masse enthält.

Unsere Matrix hat unabzählbar viele Spalten und nur $\aleph_0$ Zeilen Somit gibt es mindestens eine Zeile in der unabzählbar viele solcher Mengen hervortreten. Das ist aber unmöglich, weil laut b) die Mengen derselben Zeile elementfremd sind und höchstens abzählbar viele elementfremde Mengen von positivem Masse existieren können.

Es bleibt nur die Matrix wirklich zu konstruieren. Die Menge Z hat laut der Voraussetzung die Mächtigkeit $\aleph_1$. Ihre Elemente lassen sich also in eine transfinite Reihe bringen:

$$p_1, p_2, \ldots, p_\alpha, \ldots \qquad \alpha < \Omega.$$

Für jede Ordinalzahl $\alpha < \Omega$ bereiten wir uns eine Reihe von lauter verschiedenen natürlichen Zahlen, vom Typus α vor:

$$n_1^\alpha, n_2^\alpha, \ldots, n_\xi^\alpha, \ldots, \qquad \xi < \alpha.$$

Das Element p_α wird nun zugezählt den Mengen:

$$p_\alpha \in A_\xi^{n_\xi^\alpha}, \quad \text{für jedes } \xi < \alpha.$$

(Die Mengen A_ξ^n sind auf diese Weise Urbildmengen der Funktion, die jedem Paar der Ordinalzahlen ξ, α eine natürliche Zahl n_ξ^α zuordnet, so dass man die Zuordnung der α oder vielmehr der Elemente p_α hinschreiben kann: $p_\alpha \in A_\xi^n \equiv \{n = n_\xi^\alpha\}$ bei festem ξ).

Die oberen Indizes, bei einem gegebenem Elemente lauter verschiedene natürliche Zahlen, bestimmen die Zeilen. Jedes Element tritt daher nur höchstens einmal in jeder Zeile hervor; — die Mengen derselben Zeile sind daher elementfremd.

Damit ist die erste Eigenschaft unseres Schemas: a), bewiesen.

Was nun die zweite anbetrifft, so sieht man direkt aus der Einreihung der Elemente, dass sich in der α-ten Spalte alle Elemente, deren Nummer in der Reihe $> \alpha$ ist, befinden, d. i. alle mit Ausnahme von höchstens abzählbar vielen.

Die Matrix is konstruiert — der Satz ist für $\aleph_1$ bewiesen.

2. Wie man leicht einsieht, lässt sich für jede Menge Z von der beliebigen Mächtigkeit $\aleph_\xi$, wenn nur ξ keine Grenzzahl ist, eine analoge Matrix definieren; die Matrix hat dann $\aleph_{\xi-1}$ Zeilen und $\aleph_\xi$ Spalten. Je zwei Mengen derselben Zeile sind elementfremd, die

Vereinigungsmenge jeder Spalte ergibt Z mit Ausnahme von höchstens $\aleph_{\xi-1}$ Elementen.

Wir können voraussetzen, dass unsere Behauptung für Mengen von den Mächtigkeiten $< \aleph_\xi$ schon bewiesen ist. Wir wollen sie für $\aleph_\xi$ beweisen. Sei zunächst ξ keine Grenzzahl.

Wir unterscheiden die zwei einzig möglichen Fälle:

α) Bei jeder Zerlegung von Z in $\aleph_{\xi-1}$ Mengen, gibt es immer deren eine, die eine Teilmenge von positivem Masse enthält.

β) Es gibt mindestens eine Zerlegung von Z in $\aleph_{\xi-1}$ Mengen die alle, samt ihren Teilmengen dass Mass 0 haben [1]).

Wenn α zutrifft, definieren wir die oben genannte Matrix. Jede Spalte repräsentiert dann eine Zerlegung von Z in $\aleph_{\xi-1}$. Mengen. Es gibt also unter ihnen eine, die eine Teilmenge von positivem Masse enthält. Da es mehr Spalten, als Zeilen gibt, führt dies, wie wir schon wissen, zum Widerspruche.

Im zweiten Falle existiert eine Zerlegung von Z in $\aleph_{\xi-1}$ Mengen, die keine Teilmenge von positivem Masse enthalten.

Die Nichtexistenz des Masses ergibt sich hier aus der Voraussetzung, dass Mengen von Mächtigkeiten, die kleiner als $\aleph_\xi$ sind, nur dass Mass 0 haben können, und dem folgenden

Lemma 1. *Sei m eine unmessbare Kardinalzahl, n eine Kardinalzahl, die sich als Summe von m unmessbaren Kardinalzahlen darstellen lässt. Dann ist n auch unmessbar* [2]).

Das Lemma kann auch so ausgesprochen werden: Ist die Menge Z, in der eine Massfunktion f erklärt is, in m Mengen A_α zerlegt, wo m eine unmessbare Kardinalzahl ist, so ist für mindestens eine dieser Mengen $f(A_\alpha) \neq 0$.

In der Tat können die Summanden nur das Mass 0 haben, wir könnten sie als Elemente einer neuen Mannigfaltigkeit auffassen, die dann die unmessbare Mächtigkeit m haben würde. Ein Mass in n wird dann automatisch auch ein Mass in m definieren [3]), was unserer Voraussetzung widerspricht.

[1]) Die Möglichkeit des negativen Masses, als vollkommen symmetrisch zu α) brauchen wir nicht gesondert zu behandeln.

[2]) Die Beweismethode ist in der zit. Arbeit von Banach enthalten.

[3]) Es genügt einer Teilmenge der Mannigfaltigkeit, das Mass beizulegen, das die Vereinigungsmenge der Mengen, die den Elementen dieser Teilmenge entsprechen, hat.

Im Falle β) ist offenbar die Voraussetzung des Lemmas erfüllt nnd somit ist die Behauptung für β) erwiesen.

Auch im Falle, wo ξ eine erreichbare Grenzzahl ist, führt das gleiche Lemma zum Ziele.

Die Erreichbarkeit besagt eben eine Möglichkeit der Zerlegung von Z in $\aleph_\eta$ Mengen mit Mächtigkeiten $< \aleph_\xi$, so dass $\aleph_\eta < \aleph_\xi$ ist. Nach der Voraussetzung bei transfiniter Induktion sind alle diese Alephs unmessbar, so dass man wieder das Lemma anwenden kann. Damit is (A) in allen Teilen bewiesen

Bemerkung 1: Wir bemerken, dass wir zugleich durch Induktion die Behauptung erwiesen haben: wenn $\aleph_\xi$ unmessbar ist, so ist es auch $\aleph_{\xi+1}$.

Es dürfte vielleicht nicht ohne Interesse sein dem Beweise eine Tatsache zu entnehmen, die gewissermassen den Charakter eines abstrakten Überdeckungssatzes hat, der für die Mächtigkeit $\aleph_1$ lautet:

(B). *Werden alle Teilmengen einer Menge Z von der Mächtigkeit $\aleph_1$ in zwei Klassen M und N eingeteilt, so dass in M höchstens abzählbar viele elementfremde Mengen existieren, so gibt es in N abzählbar viele Mengen $\{A_n\}$ so dass $(Z - \sum_{n=1}^{\infty} A_n)$ höchstens abzählbar ist.*

Das folgt aus der Existenz der Matrix für $\aleph_1$: Da es in M nur abz. viele elementfremde Mengen gibt, muss in der Matrix eine solche Spalte existieren, die nur Mengen aus N enthält. (Im anderen Falle gäbe es in jeder Spalte eine Menge aus M. Dann würden diese Mengen, wie es sich zeigte, in unabzählbarer Anzahl in einer Zeile, also elementfremd, hervortreten, was unmöglich ist). Die Spalte ergibt aber die gewünschte abz. Anzahl von Mengen, die Z, mit Ausnahme von höchstens abz. v. Elementen, überdecken.

Ein entsprechender Satz gilt für die höheren Mächtigkeiten.

Es bleibt die Frage offen, ob der Überdeckungssatz in seiner Fassung für $\aleph_1$ vielleicht für alle Mächtigkeiten gültig ist.

Die Bedingung, dass f eine reelle Funktion sein soll wurde in dem Beweise nicht vollends verwertet.

Der Beweis würde fast unverändert verlaufen, wenn f nur an die Bedingung gebunden wäre, dass die Werte von f einen Zahlkörper durchlaufen, wo höchstens abzählbare Addition von „Null" verschiedener Elemente zulässig ist.

Für $\aleph_1$ folgt dies unmittelbar aus dem „Überdeckungssatze" wenn wir M als die Klasse aller Mengen mit nichtverschwindendem Masse deuten.

Der Satz besagt nämlich, dass es abz. viele andere Mengen, also solche vom Masse 0 gibt, die zusammen eine Menge vom Masse $\neq 0$ ergeben, was mit der Bedingung 3 im Widerspruche steht.

Bemerkung 2. *Wenn eine Kardinalzahl m unmessbar ist, so ist auch jede kleinere Kardinalzahl n unmessbar.*

Beweis: Wenn ein Mass in der Menge Z existiert, können wir in jeder Menge Y, die Z umfasst, die Massfunktion f folgendermassen definieren: eine Teilmenge Y' von Y enthält immer eine (ev. leere) Teilmenge $Z' = ZY'$ von Z Wir setzen einfach: $f(Y') = f(Z')$.

Man verifiziert leicht, dass die Bedingungen 1—3. von selbst erfüllt sind. Wir zeigten: ist n messbar, so ist es auch jede Kardinalzahl $m > n$, was natürlich mit der Bemerkung 2 äquivalent ist.

Nach leichter Überlegung erhält man, dass aus (A), dem Lemma 1 und der Bemerkung 1 die am Anfang genannte Formulierung von Banach folgt.

§ 2.

Wir wollen zunächst einige Sätze über ein Mass von einer ganz speziellen Struktur beweisen. Es handelt sich um eine Funktion, die nur der Werte 0 bzw 1 fähig ist — ein „zweiwertiges" Mass.

Wir beweisen den

Satz 1. *Lässt sich in einer Menge von der Mächtigkeit m kein zweiwertiges Mass definieren, so ist es auch unmöglich ein solches Mass für Mengen N mit der Mächtigkeit* 2^m *anzugeben* [1])

Beweis. Wir bemerken zunächst, dass es keine Zerlegung der Menge N in m disjunkte Teilmengen vom Masse 0 geben darf, soll ein zweiwertiges Mass in N existieren. Nach Anwendung der de Morganschen Regeln läuft diese Bemerkung dahin, dass der Durchschnitt von m Mengen vom Masse 1 wieder eine Menge vom Masse 1. ergibt. Diese Bemerkung lässt sich nämlich durch das Lemma 1 des ersten §. nur für das zweiwertige Mass transponiert, rechtfertigen.

[1]) dieser Satz wurde unabhängig auch von A. Tarski gefunden.

Die Menge N von der Mächtigkeit 2^m können wir uns als die Menge aller transfiniter Reihen von demselben Typus der Mächtigkeit m, deren Glieder 0 oder 1 sind, vorstellen.

Wir zerlegen die Menge N in zwei disjunkte Teilmengen, indem wir für irgendeine Stelle der transfiniten Reihe, alle Reihen die an dieser Stelle 0 bzw. 1 haben, gesondert zusammenfassen. Wenn wir diese Zerlegung für alle Stellen der Reihe vornehmen, so erhalten wir m Zerlegungen in zwei disjunkte Teilmengen. Da unser Mass zweiwertig sein soll, muss eine dieser Teilmengen das Mass 1, die andere das Mass 0 haben.

Wählen wir aus jeder Zerlegung die Menge vom Masse 1 und bilden den Durchschnitt der so erhaltenen m Mengen. Dieser Durchschnitt besteht, wie man leicht einsieht, aus einem einzigen Elemente — einer einzigen Reihe, muss also das Mass 0 haben. Nun sollte aber jeder Durchschnitt von m Mengen vom Masse 1, das Mass 1 haben. — Die Annahme der Existenz des zweiwertigen Masses in N führte zum Widerspruche

Satz 2. *Wenn sich in einer Menge Z eine Massfunktion f erklären lässt, dagegen kein „zweiwertiges“ Mass, so kann man für jedes natürliche n die Menge Z in endlichviele Teile zerlegen, die alle ein Mass $\leqq \frac{1}{n}$ haben.*

Beweis: Wir beweisen zunächst (a): Ist im Raume Z, wo eine Massfunktion f erklärt ist, die Bedingung erfüllt:

(δ) Es gibt keine Menge von positivem Masse δ so dass jede Teilmenge dieser Menge entweder das Mass δ oder das Mass 0 besitzt,

dann gilt: für jedes $\varepsilon > 0$ kann Z in eine endliche Anzahl von Mengen zerlegt werden:

$$Z = E_1 + E_2 + \dots + E_n$$

so dass

$$f(E_i) \leqq \varepsilon \qquad (i = 1 \dots k)\ ^{1)}.$$

Einfachkeitshalber legen wir $f(Z) = 1$.

Es genügt den Satz für $\varepsilon = \frac{1}{2}$ zu beweisen.

[1]) Den Satz kann mann auch so aussprechen: enthält Z keine Teilmenge mit „irreduziblem“ positivem Masse, so kann man Z mit endlichvielen Mengen „unbegrenzt fein überdecken“.

Denn wenn $Z = E_1 + E_2 + \ldots + E_k$ mit $f(E_i) \leqslant \frac{1}{2}$ ist, so kann, da in jedem E_i die Bedingung δ natürlich auch erfüllt ist, der Satz auf E_i wieder angewendet werden, so dass jedes E_i in eine endliche Anzahl von Mengen mit Massen $\leqslant \frac{1}{2^2}$ zerfällt; somit wird auch ganz Z in solche Mengen zerlegt, u. s. w.

Wir zeigen: Aus der Unrichtigkeit der Behauptung (a) folgt die Existenz einer transfiniten Folge: $P_1 \supset P_2 \supset \ldots P_\alpha \supset \ldots \; \alpha < \Omega$ so dass 1°: $f(P_\xi) < f(P_\eta)$, für $\xi > \eta$, 2°: $f(P_\alpha) > \frac{1}{2}$ für jedes $\alpha < \Omega$.

Das ist aber offenbar unmöglich: eine absteigende Folge von reellen Zahlen kann nicht die Mächtigkeit $\aleph_1$ besitzen.

Wir definieren die Folge P_α durch transfinite Induktion:

Es sei $P_1 = Z$ und die Folge $P_1, P_2, \ldots, P_\xi, \ldots \; \xi < \eta < \Omega$ bereits konstruiert.

Wir bilden den Durchschnitt $R = \prod_{\xi<\eta} P_\xi$. Es ist wegen der abzählbaren Additivität $f(R) \geqslant \frac{1}{2}$. Ist $f(R) = \frac{1}{2}$, dann steht $E = R + (E - R)$ $(f(E - R) = \frac{1}{2}!)$ im Widerspruch mit der Voraussetzung der Unrichtigkeit unserer Behauptung. Es ist also $f(R) > \frac{1}{2}$. Es sei $R = R_1 + R_2$, wo $f(R_1) \geqslant f(R_2) > 0$, eine Zerlegung von R (Eine solche Zerlegung gibt es immer, da (δ) erfüllt ist). Nun muss $f(R_1) > \frac{1}{2}$ sein: sonst würde die Zerlegung $E = (E - R) + R_1 + R_2$ uns wieder mit unserer Voraussetzung in Widerspruch bringen. Wir setzen $P_\eta = R_1$, Da $f(R_2)$ positiv ist, so ist

$$f(R_1) + f(P_\eta) < f(R) = f(\prod_{\xi<\eta} P_\xi) \leqslant f(P_\xi), \qquad (\xi < \eta).$$

Die Forderungen 1° und 2° sind so erfüllt. Die transfinite Folge ist konstruiert womit der Beweis der Behauptung (a) abgeschlossen ist.

Um jetzt den Beweis für den Satz 2. zu erbringen, genügt es offenbar aus der Nichtexistenz des zweiwertigen Masses das Zutreffen der Bedingung (δ) zu zeigen. Das ergibt sich aber so:

Existierte, entgegen der Bedingung (δ), eine Menge Y mit dem Masse δ, so dass jede Teilmenge dieser Menge entweder das Mass δ oder 0 hätte, so liesse sich ein zweiwertiges Mass in Z so definieren: Für eine Menge $Z' \subset Z$:

$$f(Z') = \frac{f(Z' Y)}{\delta}.$$

Wir werden jetzt einige Folgerungen aus der Hypothese (I) ziehen.

Wie gesagt, folgt aus der Richtigkeit von (I) das verschärfte Resultat von Vitali: Es lässt sich in dem Intervalle (0, 1) keine Massfunktion definieren.

Satz 3. *Wenn sich in einer Menge Z überhaupt ein Mass definieren lässt, so kann man (unter der Voraussetzung I) ein zweiwertiges Mass in Z definieren.*

Beweis: Wäre kein zweiwertiges Mass möglich, so würde sich nach dem Satze 2. dieses § für jedes n eine Zerlegung von Z in Mengen ergeben:

$$Z = A_1^1 + A_2^1 + \dots A_{m(1)}^1$$
$$Z = A_1^2 + A_2^2 + \dots A_{m(2)}^2$$
$$\dots\dots\dots$$
$$Z = A_1^n + A_2^n + \dots A_{m(n)}^n$$
$$\dots\dots\dots$$

so dass $f(A_k^n) \leqq \frac{1}{n}$ für $k = 1, 2 \dots m(n)$.

Bilden wir den Durchschnitt

$$Z = \prod_{n=1}^{\infty} \sum_{k=1}^{m(n)} A_k^n = \sum_{k_n \leq m(n)} A_{k_1}^1 \cdot A_{k_2}^2 \cdots A_{k_n}^n \cdots$$

Zur rechten Hand haben wir Kontinuum $\mathfrak{c}$ von Mengen, die alle das Mass 0 haben-als Durchschnitte von Mengen von beliebig kleinem Masse.

Wenn wir beachten, dass $\mathfrak{c}$ nach der ersten Folgerung aus der Hypothese (I) unmessbar ist, und das Lemma 1. anwenden, so sehen wir, dass das ganze Z auch unmessbar ist, entgegen der Voraussetzung.

Aus der Richtigkeit von (I) folgt weiter:

Satz 4. *Ist* $\mathfrak{m}$ *unmessbar, so ist auch* $2^{\mathfrak{m}}$ *unmessbar.*

Beispielweise $\mathfrak{f} = 2^{\mathfrak{c}}$, $2^{\mathfrak{f}}$, u. s. w.

Nach dem Satze 3. genügt es nur die Nichtexistenz des zweiwertigen Masses nachzuweisen. Dies ist aber der Satz 1. dieses §.

Zusammenfassend: Die „Klasse" der unmessbaren Alephs enthält 1°: $\aleph_0$, 2°: mit $\aleph_\xi$ auch $\aleph_{\xi+1}$, 3°: mit m alle Alephs $< m$ (Bemerkung 1), 4°: mit m auch jede Vereinigungsmenge von m Mengen, die zu der Klasse gehören (nach dem Lemma 1), endlich 5°: unter Zugrundelegung der Hypothese (*1*) nach dem Satze 4. mit m auch 2^m.

Was die Produkte von Kardinalzahlen anbetrifft, so sieht man, da sich jedes Produkt durch entsprechende Summe und Potenz majorisieren lässt, (das Produkt von m Kardinalzahlen:

$$\aleph_{\xi_1} \times \aleph_{\xi_2} \times \ldots \aleph_{\xi_\alpha} \times \ldots \quad \text{ist} \quad \leqslant 2^{\aleph_{\xi_1}+\aleph_{\xi_2}+\ldots\aleph_{\xi_\alpha}+\ldots})$$

und die Majorante nach unseren Bemerkungen unmessbar ist, dass zu der Klasse der unmessbaren Alephs auch solche hinzukommen, die sich als Produkte einer unmessbaren Anzahl von unmessbaren Kardinalzahlen darstellen lassen. Somit werden, freilich unter Benutzung der Hypothese (*1*), alle Kardinalzahlen, die keine im Sinne von Sierpiński u. Tarski unerreichbare[1]) Kardinalzahl majorisieren, unmessbar.

[1]) S. die Arbeit dieser Autoren, diese Fundamenta, Bd. XV, S. 292.

ON SYMMETRIC PRODUCTS OF TOPOLOGICAL SPACES*

BY KAROL BORSUK AND STANISLAW ULAM

1. *Introduction.* This paper is devoted to an operation that is defined for an arbitrary topological† space E and is analogous to the operation of constructing the combinatorial product spaces.‡ We shall be concerned with the topological properties of point sets defined by means of the above operation when executed on the segment $0 \leqq x \leqq 1$.

Let E be an arbitrary topological space. Let E^n denote the nth topological product of the space E, that is, the space whose elements are ordered systems $(x_1, x_2, \cdots, x_n)$ of points $x_i \epsilon E$. By a *neighborhood* of a point $(x_1, x_2, \cdots, x_n)$, we understand the set of all systems $(x_1', x_2', \cdots, x_n')$, where x_i' belongs to a neighborhood u_i of the point x_i in the space E.‡

The operation with which we are concerned in this paper consists in constructing a space which we shall call the nth *symmetric product* of the space E and denote by $E(n)$. Its elements are *non-ordered* systems of n points (which may be different or not) belonging to E. Two systems differing only by the order or multiplicity of elements are considered identical. A non-ordered system or simply a *set* consisting of n points $x_1, \cdots, x_n$ from the space E will be denoted by $\{x_1, x_2, \cdots, x_n\}$. If u_i is a neighborhood of the point x_i in the space E, then the set of all systems

* The definition of *symmetric products* is given below.

† In the sense of Hausdorff, *Grundzüge der Mengenlehre*, p. 228.

‡ See, for example, F. Hausdorff, *Grundzüge der Mengenlehre*, p. 102.

$\{x_1', x_2', \cdots, x_n'\}$ such that $\{x_1', x_2', \cdots, x_n'\} \subset \sum_{i=1}^{n} u_i$ and $u_i \cdot \{x_1', x_2', \cdots, x_n'\} \neq 0$ for $i=1, 2, \cdots, n$ will be, by definition, considered as the neighborhood of the point $\{x_1, x_2, \cdots, x_n\}$ in the space $E(n)$.

It may be important to observe that the sets $E(n)$ constitute a monotonic sequence of subsets of the space 2^E,* that is, the space whose elements are compact subsets of the space E.† The sets $E(n)$, with finite dimensionality,‡ if E is compact and of finite dimensions, approximate in a certain sense the space 2^E; we have, in fact, the following formula:

$$2^E = \overline{\sum_{n=1}^{\infty} E(n)}.$$

The study of $E(n)$ may, therefore, throw some light on the structure of 2^E.§

In the case where E is a metric space and $|x'-x|$ denotes the distance of two points x, $x' \epsilon E$, we may consider E^n, and $E(n)$ also, as metric and by means of the following formulas define the distance of two points of these spaces:

$$|(x_1, x_2, \cdots, x_n) - (x_1', x_2', \cdots, x_n')| = \left(\sum_{i=1}^{n} |x_i - x_i'|^2\right)^{1/2}$$

and

$$|\{x_1, x_2, \cdots, x_n\} - \{x_1', x_2', \cdots, x_n'\}|$$
$$= \text{Max}\left[\operatorname*{Sup}_{1 \leqq i \leqq n} \operatorname*{Inf}_{1 \leqq j \leqq n} |x_i - x_j'| ; \operatorname*{Sup}_{1 \leqq i \leqq n} \operatorname*{Inf}_{1 \leqq j \leqq n} |x_j - x_i|\right],$$

respectively.¶

We shall show that, generally speaking, the operations of constructing the combinatorial and the symmetric product of the space E lead to topologically different results. Thus in the

* This very convenient notation was introduced by C. Kuratowski, Fundamenta Mathematicae, vol. 17.

† F. Hausdorff, *Grundzüge der Mengenlehre*, p. 293.

‡ See (b), §2.

§ The space 2^E has been recently an object of some studies. See F. Hausdorff, *Grundzüge der Mengenlehre*, p. 145; S. Mazurkiewicz, Fundamenta Mathematicae, vol. 16, p. 151, and others.

¶ See Hausdorff, loc. cit.

case where E is a one-sphere, that is, the circumference of a circle, E^2 is the surface of an anchor ring (torus) and $E(2)$ is the well known one-sided Möbius strip. Neither of these sets is topologically contained in the other. This example may suffice to show that the symmetric products are apt to be used as means of simple definitions of interesting topological spaces.

In this paper, however, we shall be concerned with the space $I(n)$, I being the segment $0 \leqq x \leqq 1$. The first question to be treated is whether or not $I(n)$ is topologically equivalent with a subset of the n-dimensional euclidian space R^n (R is the set of all real numbers).

We formulate our problem in the following manner in order to emphasize certain algebraic analogies.

A real-valued function $\phi(x_1, x_2, \cdots, x_n)$ of n real variables will be called essentially symmetric if and only if its value depends upon the set $x_1, x_2, \cdots, x_n$ and not upon the order or multiplicity of the x's. Our problem may now be given the following formulation: Does there exist a system of functions $\phi_i(x_1, x_2, \cdots, x_n)$, $(i=1, 2, \cdots, n)$, which satisfies the conditions (a) the functions ϕ_i are essentially symmetric and continuous for $0 \leqq x \leqq 1$; (b) the system of equations $\phi_i(x_1, x_2, \cdots, x_n) = y_i$ has at most one solution for every system $y_1, y_2, \cdots, y_n$?

We show in §9 that the answer to our question is affirmative for $i=1, 2, 3$ and negative for $i \geqq 4$.

2. *Invariants*. Let us consider a function ϕ defined on the set E^n by the formula

$$\phi(x_1, x_2, \cdots, x_n) = \{x_1, x_2, \cdots, x_n\}.$$

This function transforms E^n on $E(n)$ continuously and every point belonging to $E(n)$ is an image of at most $n!$ points of the set E^n. If we recall well known properties* of the space E^n and continuous transformations, we obtain the propositions:

(a) *The properties local-connectedness, separability, compactness, arcwise connectedness, absolute G_δ, compactness and local connectedness at once, are invariants under the operation of constructing the symmetric product.*

(b) *If E is compact, then*

* F. G. van Dantzig, Fundamenta Mathematicae, vol. 15, p. 117.

$$\operatorname{Dim} E^n \leqq \operatorname{Dim}(E(n)) \leqq \operatorname{Dim} E^n + n! - 1 \leqq n^* \cdot \operatorname{Dim} E + n! - 1. \dagger$$

The question whether or not the following topological properties are invariants by the operation of symmetric product remains unsolved:

(α) *To be a locally connected unicoherent continuum.*
(β) *To possess a fixed point.*
(γ) *To be an n-dimensional Cantor manifold.*
(δ) *To be an absolute retract.*‡

It may also be interesting to know the exact relation between the dimensions of E and $E(n)$. Finally we wish to find the relation, if any, between the combinatorial characters, that is, the Betti and torsion numbers, of E and $E(n)$.

3. THEOREM 1. *Let A be an everywhere dense subset of the dense space E. Then $A(n) - E(n-1) = E(n)$ for $n = 2, 3, \cdots$.*

Let $\{x_1, x_2, \cdots, x_n\} \epsilon E(n)$ and U_i be, for $i = 1, 2, \cdots, n$, any arbitrary neighborhood of the point x_i in E.

We have to show that there exists a set $\{x_1', x_2', \cdots, x_n'\} \epsilon A(n) - E(n-1)$ such that $\{x_1', x_2', \cdots, x_n'\} \subset \sum_{i=1}^{n} U_i$ and

$$\{x_1', x_2', \cdots, x_n'\} \cdot U_i \neq 0 \text{ for } i = 1, 2, \cdots, n.$$

The set A constitutes an everywhere dense subset of E. Hence for each $i = 1, \cdots, n$, there exists a sequence $[x_n^{(i)}]$ consisting of different points of the set $A \cdot U_i$. Putting $x_1' = x_1^{(1)}$ and supposing that for a certain p such that $1 \leqq p \leqq n$ the points $x_2' = x_{k_p}^{2)}, \cdots, x_p' = x_{k_p}^{(p)}$ are already defined, we put $x_{p+1}^{1)}$ equal to the first term from the sequence $x_k^{(p+1)}$ different from all x_i'. Thus we obtain n different points $x_1', x_2', \cdots, x_n' \epsilon A$ or a point $\{x_1', x_2', \cdots, x_n'\}$ from the set $A(n) - E(n-1)$, such that $x_p' = x_{k_p}^{(p)} \epsilon U_p$. Hence the theorem is proved.

4. THEOREM 2. *The set $I(n) - I(n-1)$, for $n = 2, 3, \cdots$, is homeomorphic with a subset of R^n.*

Let T be a subset of R^n consisting of all points $(x_1, x_2, \cdots, x_n)$ which fulfill the conditions $0 \leqq x_1 < x_2 < \cdots < x_n \leqq 1$. For

* W. Hurewicz, Proceedings of the Amsterdam Academy, vol. 30, p. 164.

† K. Menger, *Dimensionstheorie*, p. 246.

‡ A subset B of A is a *retract* of A if there exists a continuous function f, defined on A, such that $f(A) = B$ and for every $x \epsilon B$, $f(x) = x$. An *absolute retract*, is, by definition, a homeomorph of a retract of the fundamental Hilbert cube. See K. Borsuk, Fundamenta Mathematicae, vol. 17, p. 153 and p. 159.

$(x_1, x_2, \cdots, x_n)\epsilon T$ let us put $\psi(x_1, x_2, \cdots, x_n) = \{x_1, x_2, \cdots, x_n\}$. It is evident that ψ is a homeomorphism which carries T on $I(n)-I(n-1)$, which was to be proved.

5. THEOREM 3. Dim $I(n)=n$.

In view of 1 (b) and the known fact that Dim $I^n=n$ it will be sufficient to prove that

$$\text{Dim } I(n) \leqq n. \tag{1}$$

The truth of inequality (1) is evident if $n=1$, that is, $I(1)$ is identical with the segment $0\leqq x\leqq 1$. Let us suppose now that for a certain k the inequality (1) is proved. The set $I(k+1)-I(k)$ is open ((a), §2); hence it is an F_σ of dimension at most $k+1$.

Regarding equality $I(k+1)=I(k)+[I(k+1)-I(k)]$ and applying Menger's "Summensatz,"* we obtain the desired inequality.

6. THEOREM 4. *Let A denote the segment I without its two ends: $A=I-(0)-(1)$. Then, for each pair of points, $\{x_1', x_2', \cdots, x_n'\}$ and $\{x_1'', x_2'', \cdots, x_n''\}$ of set $A(n)-I(n-1)$ for $n=2, 3, \cdots$, there exists a subset of $A(n)-I(n-1)$ homeomorphic with I^n and containing both points.*

PROOF. We do not diminish the generality if we put

$$\begin{cases} 0 < x_1' < x_2' < \cdots < x_n' < 1, \\ 0 < x_1'' < x_2'' < \cdots < x_n'' < 1, \end{cases} \tag{2}$$

and for a certain i_0, $(1\leqq i_0\leqq n)$,

$$x_{i_0}' < x_{i_0}''. \tag{3}$$

Let us put for each $(t_1, t_2, \cdots, t_n)\epsilon R^n$

$$\begin{cases} x_i(t_1, t_2, \cdots, t_n) = x_i' + t_{i_0}(x_i'' - x_i') + t_i, \text{ for } i \neq i_0, \\ x_{i_0}(t_1, t_2, \cdots, t_n) = x_{i_0} + t_{i_0}(x_{i_0}'' - x_{i_0}'). \end{cases} \tag{4}$$

From (2) and (4) it follows that there exists a positive number α such that the inequalities

$$0 \leqq t_{i_0} \leqq 1, \text{ and } |t_i| \leqq x \tag{5}$$

include the inequalities

* See K. Menger, *Dimensionstheorie*, p. 93.

$$(6)\ 0 < x_1(t_1, t_2, \cdots, t_n) < x_2(t_1, t_2, \cdots, t_n) < \cdots < x_n(t_1, t_2, \cdots, t_n) < 1.$$

The points of R^n with the coordinates $t_1, t_2, \cdots, t_n$ which satisfy the inequality (5) form in R^n a set P_n homeomorphic to I^n. For each point $(t_1, t_2, \cdots, t_n) \epsilon P_n$ let us put

$$f(t_1, t_2, \cdots, t_n) = \{x_1(t_1, t_2, \cdots, t_n), x_2(t_1, t_2, \cdots, t_n), \cdots, x_n(t_1, t_2, \cdots, t_n)\}.$$

This is a continuous function on the compact set P_n and, with regard to (3) and (4), a one-to-one correspondence (the second of the equations (4) can be solved for t_{i_0}, the rest for t_i, $i \neq i_0$). Since the function f carries P_n homeomorphically on a subset of the set $A(n) - I(n-1)$ (from (6)) and

$$f(0, 0, \cdots, 0) = \{x_1', x_2', \cdots, x_n'\},$$
$$f(0, 0, \cdots, 0, 1, 0, \cdots, 0) = \{x_1'', x_2'', \cdots, x_n''\},$$

our theorem is proved.

7. Theorem 5. *$I(n)$ is an n-dimensional, locally-connected Cantor manifold.**

In view of (a), §1, and Theorem 3, it remains to be shown that no compact $(n-2)$-dimensional subset C from $I(n)$ cuts $I(n)$. Since I^n is an n-dimensional Cantor-manifold,† it follows from (5) that $A(n) - I(n-1) - C$ is connected and everywhere dense in $A(n) - I(n-1)$. Hence, it follows that from §3

$$A(n) - I(n-1) - C \subset I(n) - C \subset I(n) = \overline{A(n) - I(n-1)} = \overline{A(n) - I(n-1) - C},$$

which includes the fact that $I(n) - C$ is connected.

8. Theorem 6. *For $n = 1, 2, 3$, $I(n)$ is a homeomorph of I^n.*

Proof. Let $\{x, y, z\} \epsilon I(3)$. Since order and multiplicity do not matter, we can suppose that $0 \leq x \leq y \leq z \leq 1$. Moreover, $x = y$ if and only if $x = y = z$. Let us put

$$(7) \qquad f(\{x, y, z\}) = (\xi(x, y, z), \eta(x, y, z), \zeta(x, y, z))$$

* See, for example, K. Menger, *Dimensionstheorie*, p. 217.
† K. Menger, *Dimensionstheorie*, p. 268.

where the coordinates ξ, η, ζ of the points $(\xi, \eta, \zeta) \epsilon R^3$ are defined by the formulas

$$(8) \qquad \xi(x, y, z) = \begin{cases} (z - x) \cdot \sin\left(2\pi \cdot \dfrac{y - z}{x - z}\right), & \text{if } z > x; \\ 0, & \text{if } z = x; \end{cases}$$

$$(9) \qquad \eta(x, y, z) = \begin{cases} (z - x) \cdot \cos\left(2\pi \cdot \dfrac{y - z}{x - z}\right), & \text{if } z > x; \\ 0, & \text{if } z > x; \end{cases}$$

$$(10) \qquad \zeta(x, y, z) = x.$$

It is easy to observe that f transforms $I(3)$ homeomorphically on a cone S, whose base is a circle in the plane $\zeta = 0$ with the center $(0, 0, 0)$ and radius 1 and whose vertex is the point $(0, 0, 1)$; $f(I(2))$ is a triangle with vertices $(0, 0, 0)$, $(0, 1, 0)$ and $(0, 0, 1)$. Finally $f(I(1))$ is a segment L with the ends $(0, 0, 0)$ and $(0, 0, 1)$. As a cone is homeomorphic with I^3 and a triangle with I^2, we obtain our theorem.

9. THEOREM 7. *For $n \geqq 4$, $I(n)$ is not homeomorphic with any subset of R^n.*

Let J denote the segment $-1 \leqq x \leqq 1$. Noting that $I(n)$ and $J(n)$ are homeomorphic, it will be sufficient to prove that there does not exist a homeomorphism h between $J(n)$ and a subset of R^n. Let

$$(11) \; Q = E_{\{x,y,z,x_1,\cdots,x_{n-3}\}} \left[0 \leqq y \leqq z \leqq 1, y = x \text{ if and only if } x = y = z, \text{ and } \left| x_i + \frac{i}{n} \right| \leqq \frac{1}{3n} \text{ for } 1 \leqq i \leqq n - 3\right].$$

Let us say, for each $\{x, y, z, x_1, x_2, \cdots, x_{n-3}\} \epsilon Q_3$, where $x, y, z, x_1, x_2, \cdots, x_{n-3}$ are chosen in such a way that the inequalities (11) are satisfied, that

$$\phi(\{x, y, z, x_1, x_2, \cdots, x_{n-3}\}) = (\xi, \eta, \zeta, \xi_1, \xi_2, \cdots, \xi_{n-3}),$$

where ξ, η, ζ are defined by (8), (9), (10), and

$$(12) \qquad \xi_i = x_i + \frac{i}{n}, \qquad (i = 1, 2, \cdots, n - 3).$$

From the definition of ϕ and the properties of f expressed by

(7) it follows immediately that $\phi(Q)$ is a combinatorial product* of S and an $(n-3)$-dimensional product of the segment $[-1/(3n), 1/(3n)]$, that is, topologically a homeomorph of I^n. The point $f(\{\frac{1}{2}, \frac{1}{2}, \frac{1}{2}\})$ is (by (7), (8), (9) and (10)) an inner point of the cone S; it follows that the point

$$\phi\left(\left\{\frac{1}{2}, \frac{1}{2}, \frac{1}{2}, -\frac{1}{n}, -\frac{2}{n}, \cdots, \frac{n-3}{n}\right\}\right) \tag{13}$$

is an inner point of $\phi(Q)$.

Applying Brouwer's theorem† of the invariance of region in R^n, we conclude that the point $h(\frac{1}{2}, \frac{1}{2}, \frac{1}{2}, -1/n, -2/n, \cdots, -(n-3)/n)$ is an inner point of $h(J(n))$. Now let us consider the sequence $[p_k]$ of points from the set $J(n)$:

$$p_k = \left\{\frac{1}{2}, \frac{1}{2}, -\frac{1}{n} + \frac{1}{3kn}, -\frac{1}{n}, -\frac{2}{n}, \cdots, -\frac{n-3}{n}\right\}.$$

We have p_k *non* ϵQ and

$$\lim_{k\to\infty} p_k = \left\{\frac{1}{2}, \frac{1}{2}, \frac{1}{2}, -\frac{1}{n}, -\frac{2}{n}, \cdots, -\frac{n-3}{n}\right\}.$$

Noting that h is a homeomorphism, we have further $h(p_k)\epsilon R^n - h(Q)$ and

$$\lim_{k\to\infty} h(p_k) = h\left(\left\{\frac{1}{2}, \frac{1}{2}, \frac{1}{2}, -\frac{1}{n}, -\frac{2}{n}, \cdots, -\frac{n-3}{n}\right\}\right),$$

which is impossible by (13). This proves our theorem.

It may be interesting to know whether or not $I(n)$ is homeomorphic with a subset of R^{n+1}.

LWOW, POLAND

* See Hausdorff, *Mengenlehre.*

† L. E. J. Brouwer, Mathematische Annalen, vol. 71, pp. 305–313.

THÉORIE DES FONCTIONS. — *Sur une propriété de la mesure de M. Lebesgue.* Note de MM. **J. Schreier** et **St. M. Ulam**, présentée par M. Émile Borel.

M. Lebesgue a prouvé le théorème suivant, d'unicité de la mesure :

« Une fonction F d'ensembles, définie pour les ensembles plans mesurables et remplissant les conditions :

1° $F(E_1 + E_2 + \ldots E_n + \ldots) = F(E_1) + F(E_2) + \ldots F(E_n) + \ldots$ pour chaque suite d'ensembles disjoints $\{E_n\}$;

2° $F(Q) = 1$, Q désignant le carré fondamental dans le plan;

3° $F(E) = F(E')$ pour deux ensembles E, E′ congruents (dans le sens de la géométrie élémentaire),

coïncide nécessairement avec la mesure qu'il a définie. »

Nous généraliserons ce théorème en introduisant d'une façon convenable la notion de congruence relative des ensembles.

Soient notamment φ et ψ deux fonctions continues de variable réelle, ne s'annulant que sur des ensembles dénombrables, finis ou vides.

Deux ensembles E et E' seront dits « congruents relativement aux fonctions φ et ψ », en symboles : $E \sim E'$, s'il existe deux constantes c et d, telles que la transformation $x' = x + c.\varphi(y)$, $y' = y + d.\psi(x)$ transforme E en E'.

Évidemment, si φ et ψ se réduisent à deux constantes $\neq 0$, l'opération considérée est une translation.

Voici la généralisation du théorème de M. Lebesgue :

Théorème. — *Toute fonction* F *d'ensembles plans mesurables* (L) *remplissant les conditions* 1°, 2° *et* 3' $F(E') = F(E)$ *si* $E' \sim E$, *est identique à la mesure de* E *au sens de M. Lebesgue.*

On déduit de ce théorème *l'unicité de la mesure des ensembles de droites du plan*, ce qui fournit une solution d'un problème proposé par M. H. Steinhaus [*Sur la pôrtée pratique et théorique de quelques théorèmes sur la mesure des ensembles de droites* (*C. R. du Ier Congrès de Math. des pays slaves*)] et qui se rattache à la Théorie des probabilités. Une droite du plan étant définie par sa distance $p = OP$ d'un point fixe O et par l'angle ϑ entre OP et l'axe fixe $O\infty$, à tout ensemble D de droites correspond un ensemble ponctuel E du plan cartésien (ϑ, p). MM. Cartan [cf. les renvois de la Note : *Sur les probabilités géométriques* de M. Hostinsky (*Publications de la Faculté des Sciences de l'Université Masaryk*, 50, 1925)] et Deltheil [*Probabilités géométriques. Traité du Calcul des Probabilités et de ses applications* de M. E. Borel, 2, II (Gauthier-Villars et Cie, Paris, 1926)] mesurent D en lui attribuant par définition la mesure de Lebesgue de l'ensemble E. Cette mesure ne change pas quand l'ensemble D se déplace en restant congruent (dans le sens de la géométrie élémentaire) à lui-même.

Réciproquement, chaque fonction des ensembles de droites absolument additive, normée d'une façon convenable et attribuant les mêmes valeurs à des ensembles congruents de droites, donne lieu, si l'on effectue la transformation indiquée, à une fonction des ensembles ponctuels du plan.

Cette fonction remplit les conditions 1°, 2° et, comme il est aisé à calculer, aussi 3', la fonction φ ayant dans le cas considéré la valeur constante 1 et ψ étant égale à $\cos\vartheta$.

Conformément à notre théorème, cette fonction doit donc être égale à la mesure de Lebesgue; la fonction des ensembles de droites est par suite identique à la mesure de Cartan et Deltheil, ce qui signifie l'unicité de cette mesure.

Démonstration du théorème. — D'après un résultat de M. Sierpinski [*Sur la définition axiomatique des ensembles mesurables* (L) (*Bull. de l'Acad. des Sc. Cracovie*, 1918)], pour qu'une fonction d'ensembles, remplissant les conditions 1° et 2°, soit identique à la mesure de M. Lebesgue, il faut et il suffit qu'elle prenne pour les carrés aux côtés parallèles aux axes des valeurs propres, c'est-à-dire égales à leur aire.

Il suffit donc de montrer qu'une fonction remplissant 3′ jouit de cette propriété. Comme il est aisé de le voir, il suffit dans ce but de montrer que la fonction F prend des valeurs égales pour deux carrés A et B, équivalents par une translation.

Les conditions étant symétriques par rapport à x et y, il est légitime d'admettre que le carré A a pour base l'intervalle (0, 1) de l'axe des x et que le carré B s'obtient de A en augmentant les ordonnées de ses points d'une valeur fixe a.

Il s'agit de prouver que $F(A) = F(B)$. Remarquons d'abord que s étant un segment parallèle à l'un des axes (à l'axe des y par exemple), on a $F(s) = 0$; car en effectuant les transformations

$$(x'_1, y') = \left(x + \frac{1}{n}\varphi(y), y\right) \qquad (n = 1, 2, 3, \ldots),$$

on obtient une suite d'ensembles congruents, qui deux à deux ont un ensemble au plus dénombrable de points communs. Ces ensembles ont donc tous la même mesure; l'ensemble somme étant mesurable, il vient $F(s) = 0$. Ceci établi, il résulte de l'additivité complète de la fonction F que, R étant un rectangle, F(R) tend vers zéro avec la largeur de la base de ce rectangle. Par hypothèse, les points x où $\psi(x)$ s'annule peuvent être enfermés dans un nombre fini d'intervalles de longueur totale très petite.

Le reste de l'intervalle (0, 1) peut être divisé en un nombre fini d'intervalles (α_i, β_i) $[i \leq k]$ si petits, que l'oscillation de la fonction ψ soit dans chacun d'eux très petite. Transformons chaque rectangle R_i de base (α_i, β_i) et de hauteur l, à l'aide de la transformation

$$x' = x, \qquad y' = y + \frac{a}{\psi(\alpha_i)}\psi(x).$$

Par conséquent

$$F(R_1) + \ldots + F(R_k) = F(R'_1) + \ldots + F(R'_k),$$

et comme, d'une part, l'ensemble $R_1 + \ldots + R_k$ diffère très peu du carré A et, d'autre part, $R'_1 + \ldots + R'_k$ diffère très peu de B (dans ce sens que la différence entre ces ensembles peut être renfermée dans un nombre fini de rectangles de base ou de hauteur très petite), la différence $|F(A) - F(B)|$ est aussi petite que l'on veut. Donc $F(A) = F(B)$.

Quelques propriétés topologiques du produit combinatoire.

Par

C. Kuratowski et St. Ulam (Lwów).

1. $\mathcal{X}$ et $\mathcal{Y}$ étant deux espaces métriques, on appelle produit combinatoire $\mathcal{X} \times \mathcal{Y}$ de $\mathcal{X}$ et $\mathcal{Y}$ l'ensemble des paires (x, y) où $x \in \mathcal{X}$ et $y \in \mathcal{Y}$ et où la distance de deux points $z_1 = (x_1, y_1)$ et $z_2 = (x_2, y_2)$ est définie par la formule

$$|z_1 - z_2| = \sqrt{|x_1 - x_2|^2 + |y_1 - y_2|^2},$$

$|x_1 - x_2|$ et $|y_1 - y_2|$ désignant la distance dans les espaces $\mathcal{X}$ et $\mathcal{Y}$ resp.

En vertu de cette définition l'espace $\mathcal{Z} = \mathcal{X} \times \mathcal{Y}$ est métrique; de plus, la condition $z = \lim z_n$ veut dire que $x = \lim x_n$ et $y = \lim y_n$. Par analogie à la géométrie analytique nous appellerons $\mathcal{X}$ et $\mathcal{Y}$ les „axes" de l'espace $\mathcal{X} \times \mathcal{Y}$, x et y les coordonnées (l'abscisse et l'ordonnée) du point (x, y).

La notion de projection d'un ensemble situé dans $\mathcal{X} \times \mathcal{Y}$ s'introduit alors d'elle même.

2. Dans beaucoup de cas, les propriétés topologiques du produit $A \times B$ se laissent déterminer par celles des facteurs A et B. Ainsi, par exemple, on prouve sans aucune difficulté, que pour que $A \times B$ soit respectivement: *fermé*, *ouvert* (en général *de classe borelienne* α), *dense*, *compact*, *complet*, *séparable*, *connexe* [1]), *localement connexe*, il faut et il suffit que A *et* B le soient également. Pour que $A \times B$ soit respectivement un *ensemble frontière*. *non-dense*, *dense-en-soi*, il faut et il suffit que A *ou* B le soit.

[1]) Voir, par ex., v. Dantzig, Fund. Math. XV (§ 5).

For commentary to this paper [6], see p. 676.

Complétons d'abord cette liste par l'énoncé suivant: pour que $A \times B$ soit *clairsemé* (c.-à-d. ne contienne aucun ensemble dense-en-soi), il faut et il suffit que A et B le soient [1]).

Pour démontrer cet énoncé, il suffit évidemment de prouver que, *Z étant un ensemble dense-en-soi (non-vide) situé dans l'espace $\mathcal{X} \times \mathcal{Y}$, une de ses projections, soit sur l'axe $\mathcal{X}$, soit sur l'axe $\mathcal{Y}$, contient un sous-ensemble dense-en-soi (non-vide).*

Or, désignons ces deux projections par A^* et B^* resp. Tout se réduit à démontrer que, a étant un point isolé de A^*, l'ensemble Z^a, composé des points de Z à abscisse a, est dense-en-soi.

Soit $(a, b) \in Z^a$. L'ensemble Z étant dense-en-soi, on a $(a, b) = \lim (a_n, b_n)$, où (a_n, b_n) est une suite de points extraits de Z et distincts de (a, b). Il vient $a = \lim a_n$ et, a étant un point isolé de A^*, on a, pour n suffisamment grand, $a_n = a$, donc $(a_n, b_n) \in Z^a$, ce qui prouve que (a, b) est un point d'accumulation de Z^a, c. q. f. d.

Dans la suite nous allons étudier, du point de vue du produit combinatoire, les notions d'ensemble *de première catégorie* (= somme d'une suite d'ensembles non-denses) et *de la propriété de Baire* (= somme d'un ensemble de I-re catégorie et d'un ensemble G_δ).

3. Nous allons supposer dorénavant que $\mathcal{X}$ et $\mathcal{Y}$ sont deux espaces métriques, dont le deuxième est *séparable.*

Théorème 1. *C étant un sous-ensemble non-dense du produit $\mathcal{X} \times \mathcal{Y}$, l'ensemble $C \cdot (x \times \mathcal{Y})$ est non-dense relativement à $(x \times \mathcal{Y})$, abstraction faite d'un ensemble des x de première catégorie* [2]) *(d'une façon moins précise, mais plus intuitive: un ensemble qui est superficiellement non-dense est linéairement non-dense sur „presque toute" droite parallèle à l'axe $\mathcal{Y}$).*

Démonstration. L'espace $\mathcal{Y}$ étant séparable, il existe une suite d'ensembles ouverts $R_1, \ldots, R_n, \ldots$, (non-vides), tels que tout sous-ensemble ouvert de $\mathcal{Y}$ s'obtient par la réunion de certains de

[1]) En cas de produit dénombrable (pour la définition, v. p. ex. Fund. Math. XVII, p. 265) cet énoncé est en défaut: le produit dénombrable des espaces tous identiques à la suite $(0, 1/2, \ldots, 1/n, \ldots)$ est dense-en-soi.

[2]) L'hypothèse que $\mathcal{Y}$ est séparable est essentielle. Supposons, en effet, que $\mathcal{X}$ désigne l'intervalle $(0, 1)$, $\mathcal{Y}$ soit un espace isolé de la puissance du continu, et que la fonction $y = f(x)$ établisse une correspondance biunivoque entre $\mathcal{X}$ et $\mathcal{Y}$. L'image de cette fonction est, pour tout x, linéairement ouverte et superficiellement non-dense.

termes de cette suite. Soit E_n l'ensemble des x tels que $(x \times R_n) \subset \subset \overline{C \cdot Y^x}$, Y^x désignant l'ensemble $(x \times \mathcal{Y})$. Il vient $E_n \times R_n \subset \subset \overline{C \cdot Y^x}$.

L'ensemble E_n est non-dense. Supposons, en effet, que l'ensemble ouvert G soit contenu dans $\overline{E}_n$. On aurait alors

$$G \times R_n \subset \overline{E}_n \times R_n \subset \overline{E_n \times R_n} \subset \overline{C \cdot Y^x} \subset \overline{C}.$$

Ainsi l'ensemble C, comme dense dans l'ensemble ouvert $G \times R_n$, ne serait pas un ensemble non-dense, contrairement à l'hypothèse.

Ceci établi, l'ensemble $P = \sum_{n=1}^{\infty} E_n$ est de première catégorie.

Reste à prouver que si x n'appartient pas à P l'ensemble $C \cdot Y^x$ est non-dense relativement à Y^x. Or, s'il n'en était pas ainsi, il existerait un ensemble R_n tel que $x \times R_n \subset \overline{C \cdot Y^x}$; mais alors x appartiendrait à E_n donc à P.

Corollaire 1. *C étant un ensemble de première catégorie situé dans $\mathcal{X} \times \mathcal{Y}$ l'ensemble $C \cdot Y^x$ est de première catégorie relativement à Y^x, abstraction faite d'un ensemble des x de première catégorie.*

Soit, en effet, $C = \Sigma C_n$, C_n non-dense. Soit (en vertu du théorème précédent) P_n un ensemble de première catégorie tel que pour x n'appartenant pas à P_n l'ensemble $C_n \cdot Y^x$ est non-dense dans Y^x. Or, si x n'appartient à aucun des P_n (dont la somme est évidemment de première catégorie), chacun des ensembles $C_n \cdot Y^x$ est non-dense dans Y^x; donc $C \cdot Y^x$, étant leur somme, est de première catégorie dans Y^x.

Corollaire 2. *Pour que le produit $C = A \times B$ soit de première catégorie, il faut et il suffit qu'un des ensembles A ou B le soit.*

En effet, si A n'est pas de première catégorie, A contient un x (en vertu du corollaire précédent) tel que $C \cdot Y^x$ est un ensemble de première catégorie relat. à Y^x. Or $C \cdot Y^x$ étant isométrique à B et Y^x à $\mathcal{Y}$ il s'ensuit que B est de première catégorie (relativement à $\mathcal{Y}$). Ainsi la condition est nécessaire. La suffisance résulte du fait, que le produit d'un ensemble non-dense par un ensemble arbitraire est non-dense.

Corollaire 3 [1]. *Soient $\mathfrak{X}$ et $\mathfrak{Y}$ deux espaces complets et soit $y = f(x)$ une fonction arbitraire transformant l'espace $\mathfrak{X}$ en un sous-ensemble de l'espace $\mathfrak{Y}$. C désignant l'ensemble des points (x, y) du produit $\mathfrak{X} \times \mathfrak{Y}$ qui n'appartiennent pas à l'image de cette fonction (c. à d. qui satisfont à l'inégalité $y \neq f(x)$), C n'est en aucun de ses points de première catégorie* [2].

En effet, dans le cas contraire il existeraient deux ensembles G et H ouverts dans $\mathfrak{X}$ et $\mathfrak{Y}$ resp. tels que $C \cdot (G \times H)$ soit de première catégorie. En remplaçant dans le corollaire 1: $\mathfrak{X}$ par G et $\mathfrak{Y}$ par H, on en conclut l'existence d'un point $x \in G$ tel que l'ensemble $C \cdot Y^x \cdot (G \times H)$ est de première catégorie dans $Y^x \cdot (G \times H)$.

Comme $Y^x \cdot (G \times H) = x \times H$, il en résulte que l'ensemble $C \cdot (x \times H)$ est de première catégorie dans $(x \times H)$. Or, l'ensemble $C \cdot (x \times H)$ est ouvert dans $(x \times H)$ puisqu'il ne diffère de celui-ci que par un seul point au plus (notamment par le point $(x, f(x))$. Nous arrivons ainsi à la conclusion qu'un ensemble ouvert dans $(x \times H)$, donc dans Y^x, y est de première catégorie. Or Y^x étant isométrique à $\mathfrak{Y}$ et $\mathfrak{Y}$, comme espace complet, n'étant en aucun de ses points de première catégorie, cela présente une contradiction.

Corollaire 4. *Dans les mêmes hypothèses sur $\mathfrak{X}$ et $\mathfrak{Y}$, si l'image d'une fonction jouit de la propriété de Baire (si par exemple la fonction jouit elle-même de la propriété de Baire* [3]), *cette image est de première catégorie.*

Cela résulte du corollaire précédent en vertu du fait que, si le complémentaire d'un ensemble à propriété de Baire n'est en aucun point de première catégorie, l'ensemble même est nécessairement de première catégorie [4].

4. Théorème 2. *C étant un sous-ensemble de $\mathfrak{X} \times \mathfrak{Y}$ jouissant de la propriété de Baire, l'ensemble $C \cdot Y^x$ jouit de la propriété de Baire relativement à Y^x, abstraction faite d'un ensemble des x de première catégorie (Y^x désigne l'ensemble $x \times \mathfrak{Y}$).*

[1]) Cf. C. Kuratowski, Fund. Math. V, p. 84 (théor. VI).

[2]) Un ensemble A est dit de première catégorie au point x, s'il existe un entourage E de x tel que $E \cdot A$ soit de première catégorie.

[3]) L'image d'une fonction jouissant de la propriété de Baire en jouit également; v. Fund. Math. XVII, p. 282.

[4]) Voir par ex Fund. Math. XVI, p. 390.

En effet C, comme ensemble à propriété de Baire, est de la forme $C = M + N$ où M est de première catégorie et N est un G_δ. En vertu du Corollaire 1, il existe un ensemble P de première catégorie tel que pour x n'appartenant pas à P l'ensemble $M \cdot Y^x$ est de première catégorie dans Y^x. Or $C \cdot Y^x = M \cdot Y^x + N \cdot Y^x$ et $N \cdot Y^x$ étant un G_δ dans Y^x, il en résulte que $C \cdot Y^x$ jouit de la propriété de Baire relativement à Y^x.

Corollaire 1. *Si $A \times B$ jouit de la propriété de Baire, un des ensembles A ou B en jouit également* [1]).

Corollaire 2. *Si $\mathcal{X}$ est un espace complet et $\mathcal{X} \times B$ jouit de la propriété de Baire, B en jouit également* [2]).

Car $\mathcal{X}$ n'étant pas de I-re catégorie, il existe un $x \in \mathcal{X}$ tel que $(x \times B)$ est à propriété de Baire rel. à Y^x. Comme l'ensemble $(x \times B)$ est isométrique avec B et Y^x est isométrique avec $\mathcal{Y}$, notre corollaire en résulte.

[1]) En cas de produit dénombrable ce corollaire (ainsi que le cor. 2 du th. 1) est en défaut: soit, dans l'intervalle (0, 1/2), A un ensemble qui ne jouit pas de la propriété de Baire; dans la „$\aleph_0$-potence" de l'intervalle 01 la $\aleph_0$-potence de A est non-dense.

[2]) Pour une application de ce corollaire, v. la note suivante de M. Kuratowski. Il est à remarquer que la plupart des énoncés de la note présente est applicable à la théorie de la *mesure*, lorsqu'on remplace la notion d'ensemble de I-re cat. par celle d'ensemble de mesure nulle, ainsi que la propriété de Baire par la mesurabilité. C'est un fait encore qui fait ressortir l'analogie entre les ensembles mesurables et ensembles à propriété de Baire, analogie à laquelle M. Szpilrajn a attiré l'attention (v. C. R. du Congr. math. pays Slaves, Varsovie 1929) Comme M. Szpilrajn nous a communiqué, quelques-uns parmi les énoncés de cette note lui ont été connus (mais non-publiés).

ANALYSE MATHÉMATIQUE. — *Sur les transformations isométriques d'espaces vectoriels, normés.* Note ([1]) de MM. **S. Mazur** et **S. Ulam**, présentée par M. Élie Cartan.

Définitions. — Un ensemble Z (d'éléments arbitraires) est dit un espace vectoriel ([2]) lorsque l'addition $z_1 + z_2$ des éléments de Z, ainsi que la multiplication tz par un nombre réel t, est définie de façon que les conditions suivantes soient remplies :

1° $$z_1 + z_2 = z_2 + z_1;$$

2° $$z_1 + (z_2 + z_3) = (z_1 + z_2) + z_3;$$

3° L'égalité $z + z_1 = z + z_2$ entraîne $z_1 = z_2$;

4° $$t(z_1 + z_2) = tz_1 + tz_2;$$

5° $$(t_1 + t_2)z = t_1 z + t_2 z;$$

6° $$t_1(t_2 z) = (t_1 t_2) z;$$

7° $$1.z = z.$$

L'espace vectoriel Z est normé s'il est un espace métrique tel que,

([1]) Séance du 7 mars 1932.

([2]) V. S. Banach, *Sur les opérations dans les ensembles abstraits* (*Fund. Math.*, 3, 1922, p. 133).

For commentary to this paper [7], see p. 676.

$\overline{z_1 z_2}$ désignant l'écart des points z_1, z_2, on ait

$$\overline{tz_1, tz_2} = |t|\,\overline{z_1 + (-1) z_2, \Theta},$$

Θ désignant le « zéro » de l'espace, c'est-à-dire que, pour tout z,

$$z + \Theta = z.$$

Plusieurs espaces à une infinité de dimensions, importants dans les recherches de l'analyse moderne, sont vectoriels et normés; tels sont aussi les espaces de Minkowski.

Une transformation $y = F(x)$ d'un espace métrique X en un espace métrique Y est dite isométrique, si l'on a

$$\overline{F(x_1), F(x_2)} = \overline{x_1, x_2},$$

quels que soient les éléments x_1 et x_2 de X.

Une transformation $y = F(x)$ d'un espace vectoriel X en un espace vectoriel Y est dite additive si, pour tout couple d'éléments x_1, x_2 de X, on a

$$F(x_1 + x_2) = F(x_1) + F(x_2).$$

Dans cette Note, nous démontrons le théorème suivant :

THÉORÈME [1]. — *Toute transformation isométrique $y = F(x)$ d'un espace vectoriel normé X en un espace vectoriel normé Y telle que $\Theta = F(\Theta)$ est additive* [2].

Démonstration. — Soit A un ensemble situé dans un espace métrique Z. Nous appelons le point z_0 centre du premier ordre de A si, pour tout $z \in A$, on a

$$\overline{z, z_0} \leqq \frac{1}{2} \delta(A),$$

le symbole $\delta(A)$ désignant le diamètre de A, c'est-à-dire la borne supérieure des nombres $\overline{z_1, z_2}$ pour z_1, $z_2 \in A$. Le point z_0 est centre du $n^{\text{ième}}$ ordre de A s'il est centre de l'ensemble C_{n-1} de tous les centres du $(n-1)^{\text{ième}}$ ordre de A qui appartiennent à A. Le point z_0 est centre métrique de l'ensemble A s'il est, pour tout n, centre du $n^{\text{ième}}$ ordre de A. Si l'ensemble A est borné, on a $\lim_{n=\infty} \delta(C_n) = 0$, et, par conséquent, l'ensemble A ne contient que tout au plus un seul centre métrique. Il est aisé de

---

[1] Ce théorème constitue une réponse positive à un problème posé par M. Banach.

[2] Cette transformation, étant additive et continue (comme isométrique), est *linéaire*.

vérifier que $y = F(x)$ étant une transformation isométrique définie sur un espace métrique X, si x_0 est centre métrique de $A \subset X$, $F(x_0)$ est centre métrique de $F(A)$.

A étant un ensemble situé dans un espace vectoriel Z, le point $z_0 \in A$ est dit centre de symétrie de A si pour tout $z \in A$, on a $(2z_0 - z) \in A$. On prouve que le centre de symétrie de A, s'il existe, est en même temps son centre métrique.

Soit à présent $y = F(x)$ une transformation isométrique de X en Y; $\theta = F(\Theta)$; x', x'' étant deux points de l'espace X, le point $1/2x' + 1/2x''$ est centre de symétrie de l'ensemble A composé de tous les points x pour lesquels $\overline{x, x'} = \overline{x, x''} = 1/2\,\overline{x', x''}$. Ce point est par suite centre métrique de A. Or, le centre métrique étant, évidemment, un invariant des transformations isométriques, il s'ensuit que le point $F(1/2x' + 1/2x'')$ est centre métrique de $F(A)$; comme d'autre part le point $1/2F(x') + 1/2F(x'')$ est centre de symétrie, donc centre métrique de $F(A)$, il vient

$$F\left(\frac{1}{2}x' + \frac{1}{2}x''\right) = \frac{1}{2}F(x') + \frac{1}{2}F(x''). \tag{1}$$

Posons dans (1) $x' = 2x_1$ et $x'' = 2x_2$; il en résulte

$$F(x_1 + x_2) = \frac{1}{2}F(2x_1) + \frac{1}{2}F(2x_2); \tag{2}$$

en posant dans (1) $x' = \Theta$, $x'' = 2x_1$, puis $x' = \Theta$, $x'' = 2x_2$, et en tenant compte de l'égalité $F(\Theta) = \theta$, on en conclut que

$$F(x_1) = \frac{1}{2}F(2x_1), \qquad F(x_2) = \frac{1}{2}F(2x_2). \tag{3}$$

Les formules (2) et (3) entraînent $F(x_1 + x_2) = F(x_1) + F(x_2)$. La transformation $y = F(x)$ est donc additive.

INTERNATIONALER MATHEMATIKERKONGRESS ZÜRICH 1932
CONGRÈS INTERNATIONAL DES MATHÉMATICIENS ZURICH 1932

ZUM MASSBEGRIFFE IN PRODUKTRÄUMEN

Von ST. ULAM, Lwów

Den Inhalt des Referates bildet die Darstellung einiger Resultate, die in Zusammenarbeit mit Herrn Z. Lomnicki erreicht wurden.

Es seien X, Y zwei abstrakte Räume, in denen für gewisse Mengen M (meßbare Mengen) ein Maß $m(M)$ definiert worden ist. Dabei werden keine topologische oder gruppentheoretische Voraussetzungen über die Natur der Räume gemacht. Es wird nun in dem Produktraume $X \times Y$ (d. h. dem Raume aller geordneten Paare (x, y), wo $x \varepsilon X$ und $y \varepsilon Y$) ein Maß eingeführt. Dies entspricht dem Probleme der Wahrscheinlichkeitsrechnung, das darauf beruht, aus gegebenen Wahrscheinlichkeiten (unabhängiger) Ereignisse auf die Wahrscheinlichkeiten zusammengesetzter Ereignisse zu schließen.

Ueber das Maß in X (und analog in Y) und die Klasse $\mathfrak{M}$ der dort meßbaren Mengen machen wir folgende Voraussetzungen:

I. Der ganze Raum X ist meßbar: $X \varepsilon \mathfrak{M}$, so auch einzelne Elemente (x) des Raumes $(x) \varepsilon \mathfrak{M}$.

II. Aus $M_i \varepsilon \mathfrak{M}$ für $i = 1, 2, \ldots n \ldots$ folgt $\sum\limits_{=1}^{\infty} M_i \varepsilon \mathfrak{M}$.

III. Aus M, $N \varepsilon \mathfrak{M}$ folgt $M - N \varepsilon \mathfrak{M}$.

IV. Ist $M \varepsilon \mathfrak{M}$ und $m(M) = 0$, $N \subset M$, so ist auch $N \varepsilon \mathfrak{M}$.

1. $m(X) = 1$; $m(M) \geq 0$.

2. $m\left(\sum\limits_{i=1}^{\infty} M_i\right) = \sum\limits_{i=1}^{\infty} m(M_i)$ bei $M_i \cdot M_j = 0$ wenn $i \neq j$.

3. Aus $m(M) = 0$ und $N \subset M$ folgt $m(N) = 0$.

Bei diesen Voraussetzungen gilt der

Satz. Man kann in dem Raume $X \times Y$ ein Maß einführen, so daß Mengen von der Gestalt $M \times N$ meßbar sind und zwar das Maß $m(M) \cdot m(N)$ haben (dabei bedeuten M und N meßbare Untermengen von X resp. Y), und alle unsere Postulate weiter erfüllt bleiben.

Ein entsprechender Satz gilt für die abzählbaren Produkte d. i. Mengen

$\mathfrak{P}(X_i) = X_1 \times X_2 \times \ldots X_K \times \ldots$ aller Folgen $\{x_i\}$, wo $x_i \varepsilon X_i$.

Im Zusammenhange mit Fragestellungen der Wahrscheinlichkeitsrechnung werden einige Eigenschaften dieses Maßes studiert. Eine ausführliche Darstellung erscheint demnächst in den *Fundamenta Mathematicae* (voraussichtlich Bd. 20).

THÉORIE DES GROUPES. — *Sur le groupe des permutations de la suite des nombres naturels.* Note [1] de MM. **J. Schreier** et **S. Ulam.**

Nous entendons par permutation de la suite infinie N des nombres naturels toute transformation biunivoque $f(n)$ de N en N tout entier. Ces transformations forment un groupe, désigné par S_∞, qui peut être considéré comme une généralisation naturelle, pour n infini, du groupe symétrique S_n composé de toutes les permutations de n lettres. L'importance du groupe S_∞ tient surtout au fait, qu'il contient, dans le sens d'isomorphie biunivoque, tout groupe dénombrable.

---

[1] Séance du 2 octobre 1933.

THÉORÈME I. — Il n'existe que deux sous-groupes invariants du groupe S_∞, outre le groupe S_∞ lui-même et le groupe composé de l'élément unité E.

1° Le groupe S de toutes les permutations f telles qu'on a $f(n) = n$ à partir d'un n suffisamment grand;

2° Le sous-groupe P de S composé des permutations f qui, considérées comme permutations du système, fini, des nombres n pour lesquels on a $f(n) \neq n$, sont *paires*, c'est-à-dire consistent d'un nombre pair d'inversions.

La série de composition de Jordan-Hölder pour le groupe S_∞ est la suivante :

$$S_\infty \supset S \supset P \supset E.$$

Remarque. — Il résulte du théorème I qu'il est impossible de généraliser le groupe « alternatif » (consistant de toutes les permutations paires) pour l'infini, c'est-à-dire qu'il n'existe aucun sous-groupe de S_∞ à l'indice 2 par rapport à S_∞. Il est impossible de diviser, comme dans le cas fini, toutes les permutations en deux classes A et B de façon que la composition de deux éléments d'une même classe donne un élément appartenant à A et que la composition de deux éléments extraits de classes différentes donne un élément de B. On ne peut donc orienter un simplex au nombre infini des sommets.

THÉORÈME II. — Il existe *trois* permutations f_1, f_2 et f_3, telles que le groupe Π engendré par elles permet d'approximer toute permutation donnée d'avance avec un degré arbitraire d'exactitude. Plus précisément, si f est une permutation donnée, k un entier positif et

$$f(i) = p_i \qquad (i = 1, 2, \ldots, k).$$

il existe une permutation g de Π, telle que $f(i) = g(i)$, quel que soit $i \leqq k$.

Remarque. — Le groupe S_∞ peut être considéré comme un espace topologique complet, si l'on définit comme distance entre deux éléments f et g le nombre (¹)

$$\rho(f, g) = \sum_{n=1}^{\infty} 2^{-n}\left(\frac{|f(n) - g(n)|}{1 + |f(n) - g(n)|} + \frac{|f^{-1}(n) - g^{-1}(n)|}{1 + |f^{-1}(n) - g^{-1}(n)|}\right)$$

(f^{-1} désigne la permutation inverse à f).

Notre théorème entraîne donc l'existence d'un sous-groupe (dénombrable) Π partout dense dans S_∞ et engendré par trois éléments.

Les démonstrations détaillées paraîtront dans un autre Recueil.

(¹) V. S. BANACH, *Théorie des opérations linéaires*, Varsovie, 1932, p. 229.

GÉOMÉTRIE. — *Sur les transformations continues des sphères euclidiennes.* Note (1) de MM. **J. Schreier** et **S. Ulam**, présentée par M. Élie Cartan.

Dans cette Note nous montrons l'existence d'une base finie de transformations continues des sphères euclidiennes, telle que leurs compositions

---

(1) Séance du 16 octobre 1933.

permettent d'obtenir toute transformation donnée d'avance avec une erreur arbitrairement petite. Pour des transformations biunivoques, la base peut être choisie aussi de telles transformations.

THÉORÈME I. — *Soit* K_n *la sphère euclidienne (l'intérieur y compris) à n dimensions. Il existe* quatre [1] *transformations continues de* K_n *en un sous-ensemble de* K_n : $\eta(p)$, $\varphi(p)$, $\chi(p)$, $\psi(p)$, *dont* η *est biunivoque et transforme* K_n *sur* K_n *tout entier, telles que* chaque *transformation continue* f *soit limite d'une suite uniformément convergente de fonctions de la forme*

$$\varphi^k \eta^l \chi \eta^{-l} \psi^k(p) \quad [2].$$

THÉORÈME II. — *Il existe trois* [1] *homéomorphies de* K_n *(c'est-à-dire transformations bicontinues de* K_n *sur* K_n *tout entier), telles que* Π *désignant le groupe engendré par leur composition (et inversion),* chaque *homéomorphie de* K_n *est, dans chaque sphère concentrique à* K_n*, limite d'une suite uniformément convergente de transformations appartenant à* Π.

Remarques. — Pour $n \leqq 3$ [3], on peut trouver un nombre fini de transformations pour la base, telles que la convergence devienne uniforme dans K_n tout entier. Il en est encore de même si l'on remplace dans notre théorème la sphère K_n par une circonférence de cercle ou par la surface de K_n [4], et aussi si l'on ne considère que des transformations de K_n (avec $n > 3$) qui laissent invariants les points de la surface de K_n. Si l'on définit comme distance de deux transformations f et g le nombre

$$\mathrm{Max}|f-g| + \mathrm{Max}|f^{-1}-g^{-1}| \quad [5],$$

l'ensemble G de toutes les homéomorphies devient un groupe métrique complet. Dans le cas $n \leqq 3$, il existe donc un sous-groupe π partout dense dans G et engendré par trois éléments.

Les démonstrations paraîtront dans un autre Recueil.

---

[1] On ne sait pas s'il n'existe pas un moindre nombre de fonctions jouissant d'une propriété analogue.

[2] φ^k désigne la $k^{\text{ième}}$ itérée de la transformation φ, φ^{-k} la $k^{\text{ième}}$ itérée de la transformation inverse de φ.

[3] La question si cette hypothèse est nécessaire reste ouverte.

[4] Il serait intéressant d'étudier ce problème pour d'autres variétés topologiques.

[5] V. S. BANACH, *Théorie des opérations linéaires*. Varsovie, 1932, p. 229.

Sur un coefficient lié aux transformations continues d'ensembles.

Par

C. Kuratowski et S. Ulam (Lwów).

1. Imaginons donnés deux ensembles compacts A et B tels qu'il existe une transformation continue de A en B, mais supposons, que A ne soit pas homéomorphe à B.

La deuxième hypothèse signifie que, pour chaque transformation continue $y=f(x)$ de A en B, il existe un $y \in B$ tel que l'ensemble $f^{-1}(y)$ (c. à d. l'ensemble des x tels que $y=f(x)$) ne se réduit pas à un seul point, donc que son diamètre [1] $\delta[f^{-1}(y)]$ soit >0; ou encore, en désignant par δ_f la borne supérieure des nombres $\delta[f^{-1}(y)]$:

$$\delta_f = \max_{y \in B} \delta f^{-1}(y), \quad \text{on a:} \quad \delta_f > 0.$$

Désignons par $\tau(A, B)$ la borne inférieure des nombres δ_f, f étant une fonction continue variable qui transforme A en B:

$$\tau(A, B) = \min_{f(A)=B} \delta_f = \min_{f(A)=B} \max_{f(x_1)=f(x_2)} |x_1 - x_2|.$$

L'étude de ce coefficient, attaché au couple A, B d'ensembles, s'impose d'une façon naturelle, lorsqu'il importe de transformer l'ensemble A en B de manière que les points „très" éloignés ne se transforment jamais en un seul point; $\tau(A, B)$ est bien le plus grand nombre tel qu'il existe nécessairement dans chaque transformation continue f de A en B un couple de points x_1, x_2 tels que $f(x_1)=f(x_2)$ et $|x_1 - x_2| \geqslant \tau(A, B)$.

[1]) $\delta(X) =$ borne supérieure des distances entre les éléments de X; c. à d. $= \max\limits_{x_1, x_2} |x_1 - x_2|$, le symbole $|x_1 - x_2|$ désignant la distance de x_1 à x_2.

For commentary to this paper [11], see p. 677.

Il est à remarquer que le nombre $\tau(A, B)$ ainsi défini ne dépend pas des propriétés métriques de B, tandis qu'il dépend bien des propriétés métriques de A. Mais on remarquera facilement que le fait que $\tau(A, B)$ est positif est de nature purement topologique.

On pourrait — peut être — étudier à la place du nombre $\tau(A, B)$ un coefficient qui n'est déterminé que par les propriétés topologiques des ensembles A et B: ce coefficient est le maximum des nombres $\tau(X, B)$ où X parcourt la famille de tous les ensembles homéomorphes à A et tels que $\delta(X) = 1$. Il mesure en certain sens la différence topologique entre A et B.

On est conduit à l'étude du coefficient $\tau(A, B)$ dans les problèmes des invariants des „petites" transformations; notamment comme l'a fait observer M. Alexandroff[1]) il y a des propriétés topologiques importantes des espaces compacts A, qui appartiennent non seulement à chaque ensemble homéomorphe à A mais aussi à chaque ensemble B qui — quel que soit $\varepsilon > 0$ — s'obtient de A à l'aide d'une transformation f, telle que $\delta_f < \varepsilon$, en d'autres termes: il s'agit d'une propriété telle que si A la possède, tandis que B ne la possède pas, on a: $\tau(A, B) > 0$.

Ainsi par exemple: 1) le fait que la dimension est $\geqslant n$, 2) le fait (pour les complexes, du moins) que le r-ième nombre de Betti est $\geqslant n$, 3) la non-unicohérence (v. N. 3), 4) l'existence d'une transformation „essentielle" (au sens de M. Hopf) en une surface sphérique [2]) sont des invariants des petites transformations.

Le problème plus précis — et de nature quantitative, dont nous nous occuperons dans quelques cas particuliers — est de calculer le nombre $\tau(A, B)$, au lieu de démontrer simplement que ce nombre est positif.

On n'a que très peu des renseignements jusqu'à présent sur ce nombre. On sait, par exemple que si A coupe l'espace euclidien n-dimensionnel tandis que B ne le coupe pas, on a $\tau(A, B) \geqslant \geqslant \varrho \sqrt{\frac{2(n+1)}{n}}$, ϱ désignant le rayon de la plus grande sphère inscrite dans une région bornée, complémentaire à A. Dans le cas particulier où A est une surface sphèrique à $n-1$ dimensions

[1]) V. surtout Ann. of Math. 30 (1928) „Gestalt und Lage..." (I Kap. „Kleine Transformationen").

[2]) K. Borsuk u. S. Ulam, Math. Ann. 1933.

de rayon ϱ, on a l'égalité Mais on ne connait pas de coefficient exact dans les cas général du r-ième nombre de Betti.

D'après un théorème publié dans ce volume [1]), si Q_n désigne la sphère n-dimensionnelle de rayon 1 et S_{n-1} sa surface, on a

$$\tau(Q_n, S_n) = \frac{2n+2-\sqrt{2n^2+2n}}{n+2}.$$

Un exemple entrant dans un ordre d'idées analogue où l'on connait le nombre τ est le suivant: d'après un théorème sur les „antipodes" [2]), si l'on transforme la surface S_n d'une sphère à $n+1$ dimensions en une partie de l'espace euclidien à n-dimensions, il y a toujours deux points „antipodiques" sur la sphère, qui se rencontrent dans cette transformation; cela veut dire qu'on a

$$\tau(S_n, X) = \delta(S_n),$$

quel que soit l'ensemble X situé dans l'espace n-dimensionnel; ou encore, en désignant d'une façon générale par $\tau(A, \mathfrak{B})$ la borne inférieure des nombres $\tau(A, B)$ où l'ensemble B parcourt la classe $\mathfrak{B}$:

$$\tau(S_n, \mathfrak{A}) = \delta(S_n),$$

$\mathfrak{A}=$ famille des ensembles situés dans l'espace n-dimensionnel.

2. *Dimension* [3]).

On appelle n-ième grade de dimension de l'espace compact A, $\sigma_n(A)$ (constante d'Urysohn), la borne inférieure des nombres ε tels que A se laisse décomposer en un système fini d'ensembles ouverts, de diamètres $\leqslant \varepsilon$ de façon qu'aucun point n'appartienne à $n+2$ ensembles de ce système.

D'après un théorème fondamental de la théorie de la dimension, l'égalité $\sigma_n(A)=0$ équivaut à la formule $\dim A \leqslant n$.

Théorème. A désignant un ensemble compact n-dimensionnel et $\mathfrak{B}$ désignant la famille des espaces compacts de dimensions inférieures, on a

$$\tau(A, \mathfrak{B}) = \sigma_{n-1}(A).$$

[1]) p. 206.

[2]) proposé par M. **Ulam** et démontré par M. K. **Borsuk**, *Drei Sätze über die n-dimensionale euklidische Sphäre*, ce vol.

[3]) Cf. **Alexandroff**, C. R. 183, p. 640.

Démonstration: 1) $\tau(A, \mathfrak{B}) \geqslant \sigma (= \sigma_{n-1}(A))$. Soit, en effet, f une fonction continue, définie sur A et telle que $\delta_f < \sigma$; il s'agit de prouver que $\dim f(A) \geqslant n$. L'espace A étant compact, il existe un nombre η tel que l'inégalité $|f(x) - f(x')| < \eta$ entraîne $|x - x'| < \sigma$. Soit $A = D_1 + \ldots + D_m$ une décomposition de A en ensembles fermés tels que $\delta[f(D_i)] < \eta/3$. Il vient $f(A) = f(D_1) + \ldots + f(D_m)$.

Supposons, par impossible, que $\dim f(A) < n$. Il existe alors un recouvrement de $f(A)$ par un système d'ensembles ouverts: $f(A) = Y_1 + \ldots + Y_m$ tels que 1^0 aucun point n'appartient à $n+1$ parmi les ensembles Y_i et 2^0 $Y_i \subset R_{\eta/3}[f(D_i)]$ [1]).

De là résulte que la formule $A = f^{-1}(Y_1) + \ldots + f^{-1}(Y_m)$ présente une décomposition de l'espace A en ensembles ouverts tels que 1^0: aucun point n'appartient à $n+1$ parmi les ensembles $f^{-1}(Y_i)$, car

$$f^{-1}(Y_{i_1} \cdot Y_{i_2} \cdot \ldots \cdot Y_{i_{n+1}}) = f^{-1}(Y_{i_1}) \cdot f^{-1}(Y_{i_2}) \cdot \ldots \cdot f^{-1}(Y_{i_{n+1}}),$$

2^0: $\delta[f^{-1}(Y_i)] < \sigma_{n-1}(A)$, car si x et x' appartiennent à $f^{-1}(Y_i)$, c. à d. si $f(x)$ et $f(x')$ appartiennent à Y_i, il existe deux points y et y' appartenant à $f(D_i)$ tels que $|f(x) - y| < \eta/3$, $|f(x') - y'| < \eta/3$, $|y - y'| < \eta/3$; d'où $|f(x) - f(x')| < \eta$ et par conséquent $|x - x'| < \sigma$. Une décomposition de ce genre est évidemment incompatible avec la définition de $\sigma_{n-1}(A)$.

2) $\tau(A, \mathfrak{B}) \leqslant \sigma$. Il s'agit de prouver qu'à chaque $\varepsilon > 0$ correspond une fonction continue f telle que $\delta_f < \sigma + \varepsilon$ et $\dim f(A) < n$.

Soit, en effet, $A = A_1 + \ldots + A_m$ une décomposition de A telle que 1^0: $\delta(A_i) < \sigma + \varepsilon$; 2^0: aucun point n'appartient à $n+1$ ensembles A_i

Cette dernière hypothèse équivaut à celle que la dimension du nerf (au sens de M. Alexandroff) du système $\{A_i\}$ est $< n$. Or, d'après un théorème fondamental concernant les nerfs [2]), on peut transformer A en une partie de ce nerf à l'aide d'une fonction continue f telle que $\delta_f \leqslant \max \delta(A_i) < \sigma + \varepsilon$. Notre théorème se trouve ainsi complètement démontré.

[1]) Le symbole $R_\eta(X)$ désigne l'ensemble des points dont la distance de X est $< \eta$.

[2]) Cf. le „Überführungssatz" de M. P. Alexandroff ainsi que C. Kuratowski ce vol. p. 196.

3. *Unicohérence.*

M étant un continu non-unicohérent [1]) et $\mathfrak{B}$ *désignant la famille des continus unicohérents, on a*: $\tau(M, \mathfrak{B}) > 0$. En d'autres termes: il existe une constante positive λ, ne dépendant que de l'ensemble M et telle que si f est une fonction continue, définie sur M et $\delta_f < \lambda$, alors $f(A)$ est un continu non-unicohérent.

Soit, en effet. $M = K + L$. $KL = A + B$. $AB = 0$, $A \neq 0$, $B \neq 0$ une décomposition de M en deux continus K et L. Soit η un nombre positif suffisamment petit pour que l'on ait $R_\eta(A) \cdot \cdot R_\eta(B) = 0$.

En tenant compte de l'égalité $\operatorname{Lim}_{n=0} [R_{\frac{1}{n}}(K) \cdot R_{\frac{1}{n}}(L)] = K \cdot L$, désignons par n un entier tel que

$$R_{\frac{1}{n}}(K) \cdot R_{\frac{1}{n}}(L) \subset R_\eta(K \cdot L) = R_\eta(A + B) = R_\eta(A) + R_\eta(B).$$

Nous allons prouver que f étant une fonction continue avec δ_f suffisamment petit $\left(\text{en tout cas } \delta_f < \frac{1}{n}\right)$, le continu $f(M)$ est non-unicohérent.

En effet: $f(M) = f(K) + f(L)$ et δ_f étant suffisamment petit, $f[R_\eta(A)] \cdot f[R_\eta(B)] = 0$. Il s'agit de prouver que l'ensemble $f(K) \cdot \cdot f(L)$ n'est pas un continu; il suffira à ce but de démontrer que 1⁰: $f(K) \cdot f(L) \subset f[R_\eta(A)] + f[R_\eta(B)]$, 2⁰: $f(K) \cdot f(L) \cdot f[R_\eta(A)] \neq 0$.

Or, l'inégalité $\delta_f < \frac{1}{n}$ implique $f^{-1}f(L) \subset R_{\frac{1}{n}}(L)$; donc l'identité: $f(K) \cdot f(L) = f[K \cdot f^{-1}f(L)]$ donne $f(K) \cdot f(L) \subset f[K \cdot R_{\frac{1}{n}}(L)] \subset \subset f[R_\eta(A) + R_\eta(B)]$ d'où l'inclusion 1⁰. D'autre part, l'inclusion $A \subset K$ implique $f(A) \subset f(K)$ et de même $f(A) \subset f(L)$; donc:

$$0 \neq f(A) \subset f[R_\eta(A)] \cdot f(K) \cdot f(L), \quad \text{c. q. f. d.}$$

Il résulte du théorème précédent que *le produit dénombrable* [2]) $X_1 \times \times X_2 \times \ldots X_n \ldots \times \ldots$ *(espace des suites infinies) d'une suite des continus unicohérents* X_i *est un continu unicohérent.*

[1]) Par définition, M est non-unicohérent, lorsque M est de la forme:

$$M = K + L, \quad KL = A + B, \quad AB = 0, \quad A \neq 0, \quad B \neq 0,$$

K et L étant des continus et A et B des ensembles fermés. D'après un théorème de MM. **Borsuk** et **Čech** (publié dans ce volume) le fait qu'un continu péanien est unicohérent veut dire que le premier nombre de Betti (au sens de **Vietoris**) s'annule.

[2]) Pour la définition v. p. ex., Fund. Math. XVII, p. 265.

En effet, cet énoncé est vrai — d'après un résultat de M. Borsuk [1]) dans le cas de produit *fini*. Or, un produit infini peut être transformé en un produit fini par une „projection" f ayant le nombre δ_f si petit que l'on veut. La non-unicohérence du produit infini serait donc incompatible avec notre résultat précédent.

4. *Superposition des fonctions.*

Soient A, B et C trois espaces compacts, soient h et g deux transformations continues de A en C et de C en B respectivement. Désignons par $f = g\,h$ la transformation superposée (de A en B). On voit aussitôt, que $\delta_h \leqslant \delta_f$.

Supposons à présent que *chaque* transformation continue de A en B soit une superposition de deux transformations g et h de ce genre: $f = g\,h$. Il vient alors:

$$\tau(A, C) \leqslant \tau(A, B).$$

On est dans ces conditions, lorsque A est un continu péanien (= image continue d'un intervalle) unicohérent et B la circonférence du cercle $|z| = 1$ (du plan complexe); car — d'après un théorème de M. Borsuk [2]) — chaque fonction continue $f(x)$ qui transforme A en B est de la forme $f(x) = e^{i\psi(x)}$, où $\psi(x)$ est une fonction continue de x, à valeurs réelles (appelée „rotation").

Par conséquent, *A étant un continu péanien unicohérent, on a*

$$\text{(I)} \qquad \tau(A, I) \leqslant \tau(A, S_1),$$

où I désigne un intervalle (de l'axe réel) et S_1 la circonférence d'un cercle.

Soit maintenant S_2 la surface de la sphère à 3 dimensions et T_2 la surface du tore. Toute transformation continue $g = f(p)$ de S_2 en T_2 peut être regardée comme superposition d'une transformation de S_2 en un carré C et d'une transformation de C en T_2.

En effet, toute transformation de S_2 en T_2 représente deux transformations f_1 et f_2 de S_2 en S_1 (la surface T_2 du tore étant le produit combinatoire $S_1 \times S_1$). S_2 étant unicohérent, il existe, d'après ce qui précède, deux fonctions $x = h_1(p)$ et $y = h_2(p)$ qui trans-

[1]) K. Borsuk, *Quelques théorèmes sur les ensembles unicohérents*, Fund. Math. XVII, p. 204.

[2]) K. Borsuk, l. c. p. 195. On rapprochera ce théorème du théorème classique sur la monodromie des fonctions analytiques

forment S_2 en l'intervalle (p. ex. en l'intervalle (0, 1)) ainsi que deux fonctions $g_1(x)$ et $g_2(y)$ qui transforment I en S_1 et telles que $f_1 = g_1 h_1$, $f_2 = g_2 h_2$; f est donc une superposition des transformations $g = [g_1(x), g_2(y)]$ et $(x, y) = h(p) = [h_1(p), h_2(p)]$; h transforme S_2 en le carré $C = I \times I$, et g transforme C en T_2.

Ceci établi, on en conclut, en vertu du théorème sur les „antipodes", cité dans l'introduction, que *dans chaque transformation continue de la surface sphérique en la surface du tore, il existe deux points antipodiques de la sphère qui se trouvent transformés en un seul point sur le tore*; ou encore, que

$$\tau(S_2, T_2) = \delta(S_2).$$

Tout cela se laisse généraliser à n dimensions, si l'on entend par le tore n-dimensionnel le produit combinatoire de n circonférences du cercle.

5. *Transformations en un arc ou en une courbe simple fermée.*

Soit T un „triode" composé de trois segments de droites A, B et C, chacun d'unité de longueur, chaque couple formant l'angle 120° et tous les trois n'ayant qu'un seul point en commun; $I = ab$ désignant un intervalle on a

$$(1) \qquad \tau(T, I) = 1.$$

Soit, en effet, $f(T) = ab$; pour des raisons de symétrie on peut admettre que $a \in f(A)$ et $b \in f(A) + f(B)$. Or, $A + B$ étant un continu, $f(A) + f(B)$ l'est également, est donc identique à l'intervalle ab. Il en résulte que, p désignant l'extrémité „libre" du segment C, on a $f(p) \in f(A) + f(B)$ et comme $\varrho(p, A + B) = 1$, il vient $\delta_f \geqslant 1$, donc $\tau(T, ab) \geqq 1$. Evidemment $\tau(T, ab) = 1$.

Ceci établi, nous allons prouver que *C étant un continu péanien tel que $\tau(C, I) = 0$, C est un arc simple.*

En effet, d'après le théorème du N 3, C, ainsi que chaque sous-continu de C, est unicohérent. Donc C, comme continu péanien qui ne contient aucune courbe simple fermée est une „dendrite". Si cette dendrite n'était pas un arc simple, elle contiendrait nécessairement un triode T (plus précisément: une courbe homéomorphe à un triode) mais ceci serait incompatible avec la formule (1).

Nous allons démontrer, à présent, que *C étant un continu péanien tel que $\tau(C, S_1) = 0$ (où S_1 est une circonférence du cercle), C est une courbe simple fermée.*

En outre:

$$\tau(T, S_1) = 1.$$

En vertu de la formule (I) du N 4, C ne peut pas être unicohérent, C contient donc une courbe simple fermée K. Le nombre δ_f de la transformation f de C en S_1 étant suffisamment petit, on a $f(K) = S_1$, car dans le cas contraire $f(K)$ serait un arc simple contrairement au théor. du N 3.

L'égalité $f(K) = S_1$ entraîne $K = C$, car autrement il existerait un point p appartenant à $C - K$ et la condition $f(p) \epsilon f(K)$ impliquerait $\delta_f > \varrho(p, K)$.

Donc C, comme identique à K, est une courbe simple fermée.

Remarque. Le théorème serait en défaut, si l'on omettait l'hypothèse que le continu C est péanien. Notamment, si $C =$ le continu indécomposable B_0 [1]) on a $\tau(B_0, S_1) = 0$. En effet, le continu B_0 est la partie commune d'une suite infinie de bandes, chacune contenue dans la précédente, les bandes étant de plus en plus étroites et les deux „extrémités" d'une même bande étant de plus en plus rapprochées. Ainsi à chaque $\varepsilon > 0$ correspond une bande qui se laisse ε-transformer en une courbe simple fermée, de sorte que le continu B_0 se trouve ε-transformé en cette courbe.

6. *Transformations de l'intervalle.*

C étant un espace péanien qui contient une sphère n-dimensionnelle ouverte K_n, $n \geqslant 2$ (ou un ensemble homéomorphe à K_n), qui constitue dans C un ensemble ouvert, on a

$$\tau(I, C) = 0$$

(I désignant un intervalle).

Divisons, en effet, l'intervalle I en un nombre fini d'intervalles $A_1, \ldots, A_k$ de longueur $< \varepsilon$. Soit $L_1, \ldots, L_k = K_n$ un système des sphères (ouvertes) concentriques. Soient: f_1 une transformation continue de A_1 en $\overline{L_1}$, f_2 de A_2 en $\overline{L_2 - L_1}$, f_3 de A_3 en $\overline{L_3 - L_2}$ et, ainsi de suite, f_k de A_k en $\overline{C - L_{k-1}}$. On peut évidemment s'arranger de façon que x_i étant l'extrémité commune des intervalles A_i et A_{i+1}, on ait $f_i(x_i) = f_{i+1}(x_i)$. On parvient ainsi à une transformation de l'intervalle I tout entier en l'espace C telle que $\delta_f < 2\varepsilon$.

[1]) Pour la définition de B_0 v. par ex., Fund. Math, III, p. 209, V, p. 40 ou XIX, p. 254.

Ainsi, en particulier, si C est une *multiplicité polyédrale* (telle que p. ex. la surface sphérique, la surface du tore etc.),

$$\tau(I, C) = 0.$$

7. *Quasi-homéomorphie.*

Nous appelons deux espaces compacts A et B *quasi-homéomorphes* lorsque $\tau(A, B) = \tau(B, A) = 0$.

La quasi-homéomorphie est une relation *transitive.* Plus précisément, si $\tau(A, B) = 0 = \tau(B, C)$, on a $\tau(A, C) = 0$.

Soit, en effet, f une transformation de A en B avec $\delta_f < \varepsilon$. Il existe un $\eta > 0$ tel que les conditions $Y \subset B$ et $\delta(Y) < \eta$ impliquent $\delta[f^{-1}(Y)] < \varepsilon$. Soit g une transformation de B en C avec $\delta_g < \eta$. Posons $h = gf$; il vient $h(A) = C$ et $\delta_h < \varepsilon$. Donc $\tau(A, C) = 0$.

La quasi-homéomorphie détermine donc une décomposition de la classe de tous les ensembles (compacts) en sous-classes disjointes. Deux ensembles quasi-homéomorphes possèdent toujours la même dimension, les mêmes groupes et nombres de Betti (lorsqu'il s'agit de complexes), ils ne peuvent être unicohérents que tous les deux. Mais ils ne sont pas nécessairement homéomorphes, (c. à d. que la classification en types topologiques est poussée plus loin que celle en ensemblesq uasi-homéomorphes), par exemple un cercle et deux cercles tangents (intérieur y compris) sont quasi-homéomorphes, sans être homéomorphes.

Il serait d'ailleurs intéressant de savoir si deux multiplicités (au sens combinatoire — donc espaces localement euclidiens) peuvent présenter la singularité en question.

Si A et C, B et D sont deux paires d'ensembles quasi-homéomorphes, les produits combinatoires $(A \times B)$ et $(C \times D)$ le sont aussi. Plus précisément on a la formule facile à vérifier:

$$\tau(A \times B, C \times D) \leqslant \sqrt{\tau(A, C)^2 + \tau(B, D)^2}.$$

Le problème suivant s'impose: l'existence d'une transformation continue sans point invariant est-elle invariante relativement à la quasi-homéomorphie?

Observons que cette propriété n'est pas invariante envers les „petites transformations". Nous allons notamment construire une courbe qui se laisse transformer d'une manière continue en une partie de soi-même sans point invariant, tandis que pour chaque $\epsilon > 0$, on peut la ϵ-transformer en un ensemble, qui possède un point invariant dans toute transformation continue en une partie de soi-même.

Soit C la courbe composée: 1° de la fermeture de la courbe $y = 1 + x \sin\left[\frac{\pi}{\sin\frac{\pi}{x}}\right]$, $0 < x \leqslant \frac{2}{3}$, 2° de la courbe précédente où l'on remplace 1 par -1 et x par $x - \frac{2}{3}$, 3° des deux segments verticaux: $-1 \leqslant y \leqslant +1$, $x = 0$ et $-1 \leqslant y \leqslant +1$, $x = \frac{2}{3}$.

La courbe C admet une transformation continue sans point invariant: notamment par symétrie radiale, effectuée du point $x = \frac{1}{3}$, $y = 0$.

Cependant, pour n fixe (impair), en remplaçant les y de la courbe supérieure, qui correspondent à $x < \frac{2}{n}$ par $y = 1$, on effectue sur C une transformation continue f avec $\delta_f < \frac{4}{n}$ et la courbe C ainsi transformée admet dans chaque transformation un point invariant; notamment en cas où l'on transforme $f(C)$ en $f(C)$ tout entier, le point $(+\frac{2}{3}, -1)$ est invariant; et dans le cas contraire, on n'aura qu'à tenir compte du fait qu'un continu qui ne coupe pas le plan admet un point invariant dans chaque transformation continue.

Über die Permutationsgruppe der natürlichen Zahlenfolge [1])

von

J. SCHREIER und S. ULAM (Lwów).

Wir betrachten die eineindeutigen Abbildungen $f(n)$, der Menge N aller natürlichen Zahlen auf sich selbst. Diese bilden in Bezug auf die Zusammensetzungsregel $fg = f\{g(n)\}$ eine Gruppe, die wir in Hinblick auf die Bedeutung der symmetrischen Gruppe S_n, die aus allen eineindeutigen Abbildungen einer Menge von n Elementen auf sich selbst besteht, mit S_∞ bezeichnen. In derselben Analogie nennen wir die Abbildungen auch Permutationen.

Wir werden in dieser Arbeit zwei Sätze über die Gruppe S_∞ beweisen. Der erste trägt einen rein gruppentheoretischen Charakter und führt zur Bestimmung aller Normalteiler von S_∞. Der zweite, der auch rein kombinatorisch formuliert werden könnte, erhält eine klarere Fassung, wenn man ihn in folgende topologische Form kleidet. Man bezeichnet als Abstand der Permutationen $f(n)$ und $g(n)$ die Zahl

$$\text{(B)} \qquad \sum_{n=1}^{\infty} 2^{-n}\left(\frac{|f(n)-g(n)|}{1+|f(n)-g(n)|}+\frac{|f^{-1}(n)-g^{-1}(n)|}{1+|f^{-1}(n)-g^{-1}(n)|}\right).$$

Dann wird S_∞ ein metrischer, vollständiger Raum [2]). Der Satz 2 besagt dann, daß dieser Raum eine überall dichte Untergruppe enthält, die von drei Elementen erzeugt wird.

§ 1. Die Normalteiler von S_∞.

1. Bezeichnungen. Es sei N die Menge der natürlichen Zahlen, $f(n)$ eine Permutation dieser Menge. $f^{-1}(n)$ bedeute die

[1]) S. unsere Note in C. R. 197 (1933) p. 54—55. Sur le groupe des permutations de la suite des nombres naturels.

[2]) S. Banach, Théorie des opérations linéaires, Warszawa 1932, p. 229.

For commentary to this paper [12], see p. 677.

zu $f(n)$ inverse Permutation, $f^k(n)$ die k-mal angewendete Iteration von $f(n)$, $f^{-k}(n)$ aber die k-mal angewendete Iteration von $f^{-1}(n)$, $f^0(n)$ die Identität.

Man kann für die unendlichen Permutationen eine der aus der endlichen Gruppentheorie wohlbekannten analoge Zerlegung in elementfremde Zyklen definieren. Am einfachsten geschieht dies auf folgende Weise. $f(n)$ sei die gegebene Permutation. Man teilt die natürlichen Zahlen in Klassen, indem man zwei Zahlen n_1 und n_2 zu derselben Klasse rechnet, wenn es zwei ganze Zahlen i_1 und i_2 gibt, so daß $f^{i_1}(n_1) = f^{i_2}(n_2)$ ist.

Nun sieht man leicht, daß wenn eine Klasse aus endlich vielen Elementen besteht, diese so angeordnet werden können

$$(n_1, n_2, \ldots n_s)$$

daß $n_2 = f(n_1)$, $n_3 = f(n_2)$ ... $n_1 = f(n_s)$ ist. Wir sagen dann: $n_1, n_2, \ldots n_s$ bilden einen s-Zyklus. Besteht dagegen eine Klasse aus unendlich vielen Elementen, dann können diese so in eine Folge vom Typus $\omega^* + \omega$ angeordnet werden

$$(\ldots n_{-3}, n_{-2}, n_{-1}, n_0, n_1, n_2, n_3, \ldots),$$

daß $f(n_j) = n_{j+1}$ für alle ganzen j ist. Wir sagen dann: $\ldots n_{-3}$, $n_{-2}, n_{-1}, n_0, n_1, n_2, n_3, \ldots$ bilden einen unendlichen Zyklus.

Bei gegebenen f und natürlichem s bezeichnen wir mit $k_f(s)$ bzw. $k_f(\infty)$ die evt. unendliche Anzahl von s-Zyklen bzw. unendlichen Zyklen, die bei der oben angegebenen Zerlegung von f in Zyklen entsteht. Nun kann man eine notwendige und hinreichende Bedingung dafür angeben, daß zwei Permutationen f und g miteinander konjugiert sind. (D. h. daß es eine Permutation h gibt, so daß $h^{-1} f h(n) = g(n)$ ist). Sie besteht darin, daß für jedes natürliche s, $k_f(s) = k_g(s)$ und $k_f(\infty) = k_g(\infty)$ ist. (Das Gleichheitszeichen ist im Sinne gleicher Mächtigkeit zu verstehen).

Eine Permutation nennen wir *endlich*, wenn $f(n) \neq n$ nur für endlich viele n gilt.

2. Wir formulieren jetzt den ersten Satz:

Satz 1. *Ein (von S_∞ verschiedener) Normalteiler D kann nur aus lauter endlichen Permutationen bestehen.*

Wir setzen also voraus, daß D wenigstens eine nicht endliche Permutation φ enthält und wollen beweisen, daß $D = S_\infty$ ist. Wie man leicht einsieht, ist dann entweder

$$(1) \qquad \sum_{s=2}^{\infty} k_{\varphi}(s) = \infty$$

oder

$$(2) \qquad k_{\varphi}(\infty) > 0.$$

3. Lemma 1. D enthält ein f mit $k_f(2) = \infty$.

Beweis. Es sei zunächst (1) erfüllt, und

$$(3) \qquad (n_1, n_2, \ldots n_{k_1})\,(n_{k_1+1}, \ldots n_{k_2}) \ldots (n_{k_i+1}), \ldots n_{k_{i+1}}) \ldots$$

die laut (1) in der Zerlegung von φ enthaltene unendliche Folge von endlichen Zyklen deren jeder mindestens zwei Zahlen enthält. Wir bestimmen eine Permutation $\psi(n)$ so, daß sie die Zyklen

$$(n_2, n_{k_1+1}, n_3, \ldots n_{k_1})\,(n_{k_1+2}, n_1, n_{k_1+3}, \ldots n_{k_2})$$
$$\ldots (n_{k_{2i}+2}, n_{k_{2i+1}+1}, n_{k_{2i}+3}, \ldots n_{k_{2i+1}})\,(n_{k_{2i+1}+2}, n_{k_{2i}+1},$$
$$(4) \qquad \ldots n_{k_{2i+1}+3}, \ldots n_{k_{2i+2}}) \ldots$$

bildet, für die Werte von n, die in (3) nicht auftreten, aber mit ψ übereinstimmt. Man sieht sofort, daß $k_{\varphi}(s) = k_{\psi}(s)$ für alle s und $k_{\varphi}(\infty) = k_{\psi}(\infty)$ ist, daß also ψ mit φ konjugiert ist. Da D Normalteiler ist und laut Voraussetzung $\varphi \in D$ ist, so ist auch $\psi \in D$. Wir setzen $f(n) = \psi\varphi(n)$; f gehört also zu D. Die Anwendung von (3) und (4) ergibt für jedes i: $\varphi(n_{k_{2i}+1}) = n_{k_{2i}+2}$, also $f(n_{k_{2i}+1}) = \psi\varphi(n_{k_{2i}+1}) = \psi(n_{k_{2i}+1+2}) = n_{k_{2i+1}+1}$, und $\varphi(n_{k_{2i+1}+1}) = n_{k_{2i+1}+2}$, also $f(_{k_{2i+1}+1}) = \psi\varphi(n_{k_{2i+1}+1}) = \psi(n_{k_{2i+1}+2}) = n_{k_{2i}+1}$. Bei jedem i bilden daher $n_{k_{2i+1}+1}$ und $n'_{k_{2i}+1}$ einen 2-Zyklus, der in der Entwicklung von f auftreten wird. Also ist $k_f(2) = \infty$.

Ist (1) nicht erfüllt, dann gilt (2). Es sei

$$(5) \qquad (\ldots n_0, n_1, n_2, n_3, \ldots)$$

einer der unendlichen Zyklen, die in der Entwicklung von φ, laut (2), auftreten. Wir bestimmen eine Permutation $\psi(n)$ so, daß sie den Zyklus

$$(6) \qquad (\ldots n_0, n_2, n_3, n_4, n_1, \ldots n_{4i+2}, n_{4i+3}, n_{4i+4}, n_{4i+1} \ldots)$$

bildet, für $n \neq n_0, n_1, n_2, n_3, \ldots$ dagegen mit φ übereinstimmt. Ebenso wie früher erkennt man, daß φ und ψ konjugiert sind, daher $\psi \in D$, und $f(n) = \psi\varphi(n) \in D$. Die Anwendung von (5) und (6) ergibt: $\varphi(n_{4i+1}) = n_{4i+2}$, also $f(n_{4i+1}) = \psi\varphi(n_{4i+1}) = \psi(n_{4i+2}) = n_{4i+3}$ und $f(n_{4i+3}) = \psi\varphi(n_{4i+3}) = \psi(n_{4i+4}) = n_{4i+1}$. Bei jedem natürlichen i bilden also n_{4i+1} und n_{4i+3} einen 2-Zyklus, der in der Entwicklung von f auftreten wird. Also ist $k_f(2) = \infty$.

4. Es bezeichne U die durch die Bedingungen $k_f(s) = \infty$, für $s = 1, 2, \ldots$ und $k_f(\infty) = \infty$ bestimmte Klasse konjugierter Permutationen.

Lemma 2. D enthält U.

Beweis. Laut Lemma 1 enthält D ein f_1 mit $k_{f_1}(2) = \infty$. Die Entwicklung von f_1 enthält also die 2-Zyklen $(n_1, n_2)(n_3, n_4)$ $(n_5, n_6), \ldots$ f_2 werde durch die Bedingungen bestimmt: $f_2(n) = f_1(n)$ für $n \neq n_1, n_2, n_3 \ldots$; für diese n aber bildet f_2 die Zyklen (n_1, n_2) (n_3, n_5) $(n_4, n_6) \ldots$ (n_{6k+1}, n_{6k+2}) (n_{6k+3}, n_{6k+5}) $(n_{6k+4}, n_{6k+6}) \ldots$. Man sieht leicht ein, daß $f_2(n)$ zu $f_1(n)$ konjugiert ist, daß also $f(n) = f_2 f_1(n)$ zu D gehört, daß $f(n_1) = n_1$, $f(n_2) = n_2$, $f(n_7) = n_7$, $f(n_8) = n_8 \ldots$, also

$$k_f(1) = \infty \tag{7}$$

und $f(n_3) = n_6$, $f(n_6) = n_3$, $f(n_4) = n_5$, $f(n_5) = n_4 \ldots$, also

$$k_f(2) = \infty \tag{8}$$

ist. Um mehrfache Indizes zu vermeiden, bezeichnen wir wieder die Zahlen, die in f laut (8) 2-Zyklen bilden, mit

$$(n_1, n_2)(n_3, n_4)(n_5, n_6) \ldots \tag{9}$$

Wir teilen die Zyklen (9) in unendlich viele unendliche Mengen $S_1, S_2, \ldots$. Die Menge S_{2q+1} bestehe aus den Zyklen

$$(\lambda_1^q, \lambda_2^q)(\lambda_3^q, \lambda_4^q) \ldots \qquad (q = 1, 2 \ldots). \tag{10}$$

Für die Zahlen n, die unter den λ in (10) auftreten, sei eine Permutation $g(n)$ so erklärt, daß sie die 2-Zyklen

$$\begin{gathered}(\lambda_2^q, \lambda_3^q)(\lambda_4^q, \lambda_5^q) \ldots (\lambda_{2q}^q, \lambda_1^q)(\lambda_{2q+2}^q, \lambda_{2q+3}^q) \ldots (\lambda_{4q}^q, \lambda_{2q+1}^q) \\ \ldots (\lambda_{2sq+2}^q, \lambda_{2sq+3}^q) \ldots (\lambda_{2(s+1)q}^q, \lambda_{2sq+1}^q)\end{gathered} \tag{11}$$

bildet. Die Menge S_{2q} bestehe aus den Zyklen

(12) $$(\mu_1^q, \mu_2^q)\ (\mu_3^q, \mu_4^q) \ldots .$$

Für die Zahlen n, die unter den μ in (12) auftreten, sei $g(n)$ so erklärt, daß sie die Zyklen

(13) $$(\mu_1^q)\ (\mu_2^q, \mu_3^q)\ (\mu_4^q, \mu_5^q) \ldots$$

bildet. Für alle übrigen n sei endlich $g(n) = f(n)$. Unter Beachtung von (7) folgt, daß f und g konjugiert sind. Also gehört $\varphi(n) = gf(n)$ zu D. Nun zeigen wir, daß φ zu U gehört. Wegen (7) ist zunächst $k_\varphi(1) = \infty$. Für $q > 1$ folgt aber aus (10) und (11), daß die Zahlen $\lambda_{2sq+1}^q, \lambda_{2sq+3}^q, \lambda_{2sq+5}^q, \ldots \lambda_{2(s+1)q-1}^q$ bei jedem natürlichen s in φ einen q-Zyklus bilden. Also ist $k_\varphi(q) = \infty$ für jedes q. Es ist aber auch $k_\varphi(\infty) = \infty$, denn wegen (12) und (13) bilden bei jedem natürlichen q die Zahlen $\mu_1^q, \mu_2^q, \ldots$ in φ den unendlichen Zyklus $(\ldots \mu_6^q, \mu_4^q, \mu_2^q, \mu_1^q, \mu_3^q, \mu_5^q, \ldots)$. Da D Normalteiler ist, enthält es mit $\varphi \in U$ ganz U.

5. Lemma 3. D enthält jedes h, für welches $k_h(1) = \infty$ ist.

Beweis. Es sei also ein beliebiges h mit

(14) $$k_h(1) = \infty,\ k_h(2) = r_2,\ k_h(3) = r_3, \ldots\ k_h(\infty) = r_\infty$$

gegeben. Die r sind ganz $\geqslant 0$, oder gleich ∞. Es bezeichne f eine beliebige Permutation, die zu U, also auch zu D gehört. Es ist $k_f(1) = \infty$. Es sei $f(n_1) = n_1$, $f(n_2) = n_2, \ldots$ diese unendliche Folge von 1-Zyklen in f. Die Permutation $h'(n)$ sei folgenderweise erklärt: Für $n \neq n_1, n_3, n_5, \ldots$ ist $h'(n) = n$, für diese n aber sei h' so erklärt, daß $k_{h'}(s) = k_h(s)$ ist, für jedes endliche und unendliche s. Daher ist h' mit h konjugiert. Wir setzen $g(n) = fh'(n)$. Für $n \neq n_1, n_3, n_5 \ldots$ ist $g(n) = f(n)$, und da, laut der Annahme $f \in U$, f unendlich viele Zyklen jeder Art aus den von $n_1, n_2, \ldots$ verschiedenen Zahlen bildet, gehört auch g zu U und daher zu D. Damit aber gehört auch $h' = f^{-1}g$ und h zu D, w. z. b. w.

6. Lemma 4. $D = S_\infty$.

Beweis. Die Permutation $h(n)$ erfülle zunächst die Bedingung

(15) $$k_h(\infty) + \sum_{s=1}^{\infty} k_h(s) = \infty,$$

d. h. die Menge der Zyklen, in die h zerfällt, sei unendlich. Wir teilen dann diese Menge in zwei unendliche Mengen M und N. Die Zahlen, die in den Zyklen der Menge M auftreten, bezeichnen wir der Reihe nach mit $n_2, n_4, n_6, \ldots$. Wir setzen $f(n)=n$ für $n=n_1, n_3, \ldots$ und $f(n)=h(n)$ für $n=n_2, n_4, \ldots$, ferner $g(n)=n$ für $n=n_2, n_4, \ldots$ und $g(n)=h(n)$ für $n=n_1, n_3, \ldots$ Es ist $k_f(1)=\infty$ und $k_g(1)=\infty$, also $f \in D$ und $g \in D$ laut Lemma 3.

Da aber $h(n)=fg(n)$ ist, gehört auch h zu D. Jede Permutation, die (15) erfüllt, gehört also zu D. Wenn man noch die Formel

$$[(1, 2)(3, 4)(5, 6)\ldots][(1)(2, 3)(4, 5)\ldots]=[(\ldots 6, 4, 2, 1, 3, 5, 7\ldots)]$$

beachtet, sieht man, daß jeder unendliche Zyklus zu D gehört, da er als Zusammensetzung zweier Permutationen, die (15) erfüllen, erhalten werden kann. Jeder endliche Zyklus gehört schon laut Lemma 3 zu D. Eine Permutation, die (15) nicht erfüllt, also in endlich viele Zyklen zerfällt, gehört als Zusammensetzung dieser Zyklen zu D. Damit ist der Beweis des Satzes 1 beendet.

7. Aus dem bewiesenen Satze folgt sofort, daß die Faktorgruppe nach dem von allen endlichen Permutationen gebildeten Normalteiler S einfach ist. Dieser Normalteiler besitzt die Gruppe aller geraden endlichen Permutationen A als Normalteiler mit einfacher Faktorgruppe. Man beweist die Einfachheit von A mit derselben Methode, die zum Nachweis der Einfachheit der alternierenden Gruppe A_n für $n>4$ dient. Wenn noch E die aus der Identität bestehende Gruppe bezeichnet, so bildet

$$S_\infty \supset S \supset A \supset E$$

die JORDAN-HÖLDERsche Kompositionsreihe von S_∞.

8. Da jede Untergruppe mit dem Index 2 notwendig Normalteiler sein muß, so folgt aus unserem Satze, daß S_∞ keine Untergruppe mit dem Index 2 enthält. Es ist also nicht möglich den Begriff „gerade Permutation“ auf unendliche Permutationen zu übertragen, d. h. alle Permutationen so in zwei Klassen A und B einzuteilen, daß die Zusammensetzung zweier Elemente aus derselben Klasse zu A, aus verschiedenen Klassen aber zu B gehöre. Dies hat zur Folge, daß man einen „Simplex“ mit unendlich vielen Eckpunkten nicht orientieren kann.

§ 2. Die Erzeugenden von S_∞.

1. Wir betrachten jetzt S_∞ als einen metrischen Raum, in dem der Abstand zweier Elemente mittels der Formel (B) der Einleitung erklärt ist. Wenn $\{f_i(n)\}$ eine Folge von Permutationen bezeichnet, so bestätigt man leicht, daß die notwendige und hinreichende Bedingung dafür, daß die Folge $\{f_i(n)\}$ gegen eine Permutation $f(n)$ konvergiert, sich so fassen läßt: zu jedem natürlichem N gibt es ein $I(N)$, so daß für $i>I(N)$, $n<N$, $f_i(n)=f(n)$ ist. Daher ist der Raum S_∞ separabel: die abzählbare Menge aller endlichen Permutationen liegt in ihm überall dicht. Wir werden jetzt drei Permutationen $\varphi(n)$, $\psi(n)$, $\chi(n)$ angeben, so daß die von ihnen erzeugte Gruppe, und zwar sogar die Permutationen der Gestalt

$$(1) \qquad \alpha(n)=\varphi^p\,\psi^{-q}\,\chi\,\psi^q\,\varphi^{-p}(n) \qquad (p=1,2,\ldots;\; q=1,2,\ldots)$$

in S_∞ überrall dicht liegen. Wir haben also folgenden Satz zu beweisen.

Satz 2. *Es gibt drei Permutationen $\varphi(n)$, $\psi(n)$, $\chi(n)$ derart, daß wenn $l_1, l_2, \ldots l_r$, r gegebene, untereinander verschiedene, natürliche Zahlen sind, es ein $\alpha(n)$ der Gestalt* (1) *gibt, so daß $\alpha(\nu)=l_\nu$ für $\nu=1,2,\ldots r$ ist.*

2. Definition von φ, ψ, χ. Wir teilen die Menge aller natürlichen Zahlen in unendlich viele, unendliche und elementfremde Mengen: $N=S_1+S_2+S_3+\ldots$. Als $\varphi(n)$ nehmen wir eine beliebige Permutation, die die Bedingungen

$$\varphi(S_1)=S_1+S_2,\ \varphi(S_2)=S_3,\ \varphi(S_3)=S_4,\ldots$$

erfüllt, als $\psi(n)$ dagegen eine Permutation, die die Bedingungen

$$\psi(S_1)=S_3,\ \psi(S_2)=S_1+S_2,\ \psi(S_3)=S_4,\ \psi(S_4)=S_5\ldots$$

erfüllt. Daraus folgt:

$$(2) \qquad \varphi^i(S_1)=S_1+S_2+\ldots+S_{i+1},\ \psi^i(S_1)=S_{i+2}.$$

Wir bezeichnen mit $\{T_k\}$ die Folge aller endlichen Systeme von $2r$ Zahlen (r beliebig) aus S_1: $A_1^{(k)}, A_2^{(k)}, \ldots A_r^{(k)}, B_1^{(k)}, B_2^{(k)}, \ldots B_r^{(k)}$, die die Bedingungen $A_\nu^{(k)} \neq A_\mu^{(k)}$, $B_\nu^{(k)} \neq B_\mu^{(k)}$ für $\nu \neq \mu$ erfüllen. Wir setzen

$$\psi^k(A_\nu^{(k)}) = C_\nu^{(k)}; \; \nu = 1, 2, \dots r \tag{3}$$

und

$$\psi^k(B_\nu^{(k)}) = D_\nu^{(k)}; \; \nu = 1, 2, \dots r. \tag{4}$$

Es ist nach (2) $C_\nu^{(k)} \subset S_{k+2}$ und $D_\nu^{(k)} \subset S_{k+2}$.

Als $\chi(n)$ nehmen wir eine beliebige Permutation, die jede der Mengen S_i in sich selbst überführt und dabei die Bedingung

$$\chi(C_\nu^{(k)}) = D_\nu^{(k)} \quad (\nu = 1, 2, \dots r). \tag{5}$$

erfüllt.

3. Beweis. Es seien die untereinander verschiedenen Zahlen $l_1, l_2, \dots l_r$ gegeben. Man wähle ϱ so groß, daß die Menge $S_1 + S_2 + \dots + S_\varrho$ die Zahlen $1, 2, \dots r$, $l_1, l_2, \dots l_r$ enthält. Die Zahlen $\varphi^{-\varrho+1}(1)$, $\varphi^{-\varrho+1}(2)$, $\dots \varphi^{-\varrho+1}(r)$, $\varphi^{-\varrho+1}(l_1), \dots \varphi^{-\varrho+1}(l_r)$, gehören laut (2) zu S_1, und bilden daher ein System T_j. Es ist also

$$\varphi^{-\varrho+1}(\nu) = A_\nu^{(j)} \quad (\nu = 1, 2, \dots r) \tag{6}$$

und

$$\varphi^{-\varrho+1}(l_\nu) = B_\nu^{(j)} \quad (\nu = 1, 2, \dots r.) \tag{7}$$

Wir setzen $p = \varrho - 1$, $q = j$ und behaupten, daß die mit diesen Werten laut (1) gebildete Permutation $\alpha(n)$ die Bedingung $\alpha(\nu) = l_\nu$ $(\nu = 1, 2, \dots r)$ erfüllt. Man hat aber, zu diesem Zwecke nur nacheinander die Formeln (6), (3), (5), (4), (7) anzuwenden. Damit ist aber Satz 2 bewiesen.

(Reçu par la Rédaction le 16. 11. 1933).

Über gewisse Invarianten der ε-Abbildungen.

Von

Karol Borsuk in Warschau und Stanisław Ulam in Lemberg.

Es sei f eine stetige Abbildung eines kompakten metrischen Raumes A auf eine Teilmenge irgendeines topologischen Raumes. Diese Abbildung wird nach Alexandroff[1]) eine ε-Abbildung genannt, wenn für jeden Punkt $y \in f(A)$ die Urbildmenge $f^{-1}(y)$ dieses Punktes einen Durchmesser $< 2\varepsilon$ hat. Es zeigt sich, daß gewisse topologische Eigenschaften, welche keine Invarianten von allgemeinsten stetigen Abbildungen sind, sich trotzdem gegenüber allen ε-Abbildungen bei hinreichend kleinem ε, invariant verhalten. Als solche Eigenschaften haben sich z. B. die Dimension $\geqq n$[2]) und die Nicht-Unikohärenz[3]) erwiesen.

In dieser Arbeit zeigen wir, daß auch die Existenz einer sogenannten wesentlichen[4]) Abbildung der Menge A auf die euklidische n-dimensionale Kugelfläche auch Invariante von ε-Abbildungen (bei hinreichend kleinem ε) ist, womit auch eine von Alexandroff aufgestellte Frage[5]) im positiven Sinne beantwortet wird.

Nebenbei ergibt sich daraus die Invarianz des Schnittes des euklidischen n-dimensionalen Raumes R_n gegenüber den ε-Abbildungen bei hinreichend kleinem ε. Obwohl diese Invarianz auch auf einem anderen Wege, und zwar mit Hilfe des Dualitätssatzes von Alexandroff[6]) erhalten werden könnte[7]), so ist unser Beweis vielleicht nicht ohne Interesse, weil wir mit ganz einfachen Mitteln und, insbesondere, ohne Gebrauch von Homologiebegriffen auskommen. Überdies erlaubt unsere

---

[1]) P. Alexandroff, *Untersuchungen über Gestalt und Lage abgeschlossener Mengen beliebiger Dimension*, Annals of Mathematics (2) **30** (1928), S. 103.

[2]) L. c., S. 120.

[3]) Vgl. auch eine in Fund. Math. demnächst erscheinende Arbeit von C. Kuratowski und S. Ulam.

[4]) Vgl. den Schluß von 5. Der Begriff der wesentlichen Abbildung stammt von H. Hopf. Siehe H. Hopf, Moskauer Math. Sammlung (1930), S. 53. Vgl. auch P. Alexandroff, *Dimensionstheorie*, Math. Ann. **106** (1932), S. 223.

[5]) P. Alexandroff, *Dimensionstheorie*, S. 226.

[6]) Derselbe, *Gestalt und Lage*, S. 156.

[7]) Nach einer brieflichen Bemerkung von P. Alexandroff.

For commentary to this paper [13], see p. 678.

Methode eine einfache und (in gewissem Sinne) scharfe Abschätzung der Zahl ε zu erhalten, bei welcher noch jede ε-Abbildung eines gegebenen Schnittes von R_n, wieder einen Schnitt dieses Raumes ergibt.

1. Im folgenden bedienen wir uns einer Bezeichnungsweise, die im allgemeinen mit der in der Arbeit von *K. Borsuk*, *Über Schnitte der n-dimensionalen euklidischen Räume*, Math. Ann. 106, S. 239 gebrauchten übereinstimmt. Insbesondere bezeichnet $\varrho(x, y)$ und $\varrho(X, Y)$ den Abstand zweier Punkte x, y bzw. der Punktmengen X, Y. $\delta(X)$ den Durchmesser der Menge X, $\underset{x \in X}{E}[\;]$, die Menge aller Punkte x von X, für welche die Eigenschaft [ ] gilt (das Lebesguesche Symbol); $\varphi(A)$ das durch die Abbildung φ erzeugte Bild von A, d. h. die Menge derjenigen Werte (Punkte) $y = \varphi(x)$, welche die Funktion $\varphi(x)$ für die Punkte x von A annimmt, $\overrightarrow{xy}$ bei $x \neq y$ die Halbgerade aus x durch y.

2. Es seien: K_n die in R_n liegende Kugelfläche vom Radius $r = 1$ und dem Mittelpunkte 0, ferner S ein Simplex[8]) mit den Eckpunkten $p_0, p_1, \dots p_k$ und P die Projektion von S aus dem Punkte 0 auf K_n.

Hilfssatz. *Liegen alle Eckpunkte von S auf K_n und ist $\delta(S) < \sqrt{2}$, so gilt:*

$$0 \in R_n - S, \tag{1}$$

$$\delta(P) < \sqrt{2}\ ^{9)}. \tag{2}$$

Beweis. Laut Voraussetzung ist $\varrho(p_i, p_j) < \sqrt{2}$ für $i, j = 0, 1, \dots k$, was die Ungleichung für die Winkel: $\measuredangle(p_i 0 p_j) < \frac{\pi}{2}$ ergibt. Infolgedessen befinden sich sämtliche Eckpunkte des Simplexes S nur an einer Seite der durch 0 durchgehenden $(n-1)$-dimensionalen zur Halbgeraden $\overrightarrow{p_i 0}$ orthogonalen Hyperebene in R_n, womit die Beziehung (1) bewiesen ist und woraus sich andererseits die Ungleichung $\measuredangle(p_i 0 x) < \frac{\pi}{2}$ für jeden Punkt $x \in S$ und für jeden Index $i = 0, 1, \dots k$ ergibt. Demzufolge befinden sich alle Eckpunkte von S auch an einer Seite der durch 0 durchgehenden $(n-1)$-dimensionalen und zu $\overrightarrow{x0}$ orthogonalen Hyperebene, so daß für jedes Punktepaar x, y von S die Ungleichung $\measuredangle(x 0 y) < \frac{\pi}{2}$ besteht. Aus dem Dreiecke $\triangle(0, x', y')$, wo $x' = \overrightarrow{0x} . K_n$ und $y' = \overrightarrow{0y} \cdot K_n$ ist, erhalten wir also $\varrho(x', y') < \sqrt{\varrho(0, x')^2 + \varrho(0, y')^2} = \sqrt{2}$, was wegen der Kompaktheit von S die Beziehung (2) ergibt.

---

[8]) Das heißt die kleinste konvexe Menge in R_n, welche die Punkte $p_0, p_1, \dots, p_k$ enthält. Ist S k-dimensional, so heißt es *eigentliches* Simplex. Ist $\varrho(p_i, p_j) = \text{const}$, für $i \neq j$, so wird S ein *regelmäßiges* Simplex genannt.

[9]) Man kann leicht zeigen, daß $\delta(P) = \delta(S)$ ist.

3. Von nun an wird mit A eine in sich kompakte (sonst aber beliebig gegebene) Punktmenge bezeichnet.

Hilfssatz. *Ist φ eine ε-Abbildung von A, so gibt es ein $\eta > 0$, derart, daß folgende Beziehung gilt:*

$$\text{(3)} \qquad aus\ \varrho[\varphi(x'), \varphi(x'')] \leqq \eta\ folgt\ \varrho(x', x'') < 2\varepsilon$$

für je zwei Punkte x' und x'' von A.

Beweis. Andernfalls würde es in A für jedes $n = 1, 2, \ldots$ ein Punktepaar x_n', x_n'' mit den Eigenschaften geben:

$$\text{(4)} \qquad \varrho[\varphi(x_n'), \varphi(x_n'')] \leqq \frac{1}{n},$$

$$\text{(5)} \qquad \varrho(x_n', x_n'') \geqq 2\varepsilon.$$

Da aber A kompakt ist, so gibt es eine Folge $\{n_k\}$ von natürlichen Zahlen, für welche die Limites

$$\text{(6)} \qquad x' = \lim_{k=\infty} x_{n_k}' \quad und \quad x'' = \lim_{k=\infty} x_{n_k}''$$

existieren. Aus (4) und (6) erhalten wir $\varphi(x') = \varphi(x'')$, dagegen aus (5) und (6), $\varrho(x', x'') \geqq 2\varepsilon$, was der Voraussetzung, daß φ eine ε-Abbildung ist, widerspricht.

4. Satz. *Ist f eine stetige Abbildung von A mit den Eigenschaften:*

$$\text{(7)} \qquad f(A) \subset K_n,$$

$$\text{(8)} \qquad aus\ \varrho(x', x'') < 2\varepsilon\ folgt\ \varrho[f(x'), f(x'')] < \sqrt{2}$$

für je zwei Punkte x' und x'' von A, und ist φ eine ε-Abbildung von A, so gibt es eine stetige Abbildung ψ der Menge $\varphi(A)$ derart, daß:

$$\text{(9)} \qquad \psi\varphi(A) \subset K_n,$$

$$\text{(10)} \qquad \varrho[f(x), \psi\varphi(x)] < 2\ für\ jedes\ x \in A$$

gilt.

Beweis. Wir beweisen die Existenz von ψ zunächst für den Fall, wo $\varphi(A) \subset R_k$ ist. Nach Hilfssatz **3.** gibt es ein $\eta > 0$, für welches die Bedingung (3) erfüllt ist. Sei Σ ein simpliziales Netz[10]) im R_k, dessen Simplexe vom Durchmesser $\leqq \frac{1}{3}\eta$ sind. Sei ferner T die Vereinigungsmenge aller zu $\varphi(A)$ nicht punktfremden Simplexe von Σ. Wählen wir nun[11]) für jeden Punkt $t \in T$ einen beliebigen Punkt $\alpha(t)$ in der Menge $\underset{y=\varphi(A)}{E}[\varrho(t, y) = \varrho(t, \varphi(A))]$ und für jeden Punkt $y \in \varphi(A)$ einen beliebigen Punkt $\beta(y)$ in der Menge $\varphi^{-1}(y)$, so gilt:

$$\text{(11)} \qquad \varrho[t, \alpha(t)] \leqq \tfrac{1}{3}\eta, \quad für\ jedes\ t \in T.$$

[10]) Das heißt eine Folge von eigentlichen Simplexen, die als Vereinigungsmenge R_k ergeben, wobei jeder Eckpunkt eines dieser Simplexe zugleich ein Eckpunkt aller anderen, ihn enthaltenden Simplexe ist.

[11]) Diese Auswahl kann, da diese Menge in sich kompakt ist, effektiv vorgenommen werden.

Ist also γ irgendein in T liegendes Simplex von Σ mit Eckpunkten $q_0, q_1, \dots q_k$, so folgt

$$\varrho[\alpha(q_i), \alpha(q_j)] \leqq \varrho[\alpha(q_i), q_i] + \varrho(q_i, q_j) + \varrho[q_j, \alpha(q_j)] \leqq 3 \cdot \tfrac{1}{3}\eta = \eta$$

und somit nach (3)

$$(12) \qquad \varrho[\beta\alpha(q_i), \beta\alpha(q_j)] < 2\varepsilon \quad \textit{für} \quad i, j = 0, 1, \dots k.$$

Betrachten wir nun das (eigentliche oder nicht eigentliche) Simplex S mit den Eckpunkten $p_i = f\beta\alpha(q_i) \in K_n$, so ist nach (12) und (8)

$$(13) \qquad \delta(S) < \sqrt{2}$$

und daher nach Hilfssatz 2 liegt der Mittelpunkt 0 von K_n außerhalb von S.

Sei τ die affine Abbildung von γ auf S, bei welcher die Punkte q_i in die Punkte p_i (mit den entsprechend gleichen Indizes) übergehen. Wir setzen

$$(14) \qquad \psi(s) = \overrightarrow{0\tau(s)} \cdot K_n \quad \textit{für jedes} \quad s \in \gamma.$$

Die in dieser Weise in allen in T liegenden Simplexen γ von Σ definierte Funktion ψ ist offenbar in T und somit in $\varphi(A) \subset T$ stetig. Da sich dabei aus (14) $\psi\varphi(A) \subset K_n$ ergibt, so ist die Bedingung (9) bewiesen. Um nun die Bedingung (10) zu beweisen, beachten wir, daß $\varphi(x) \in T$, daß also in T ein Simplex γ von Σ existiert, welches $\varphi(x)$ enthält. Wir haben nach (11): $\varrho[\alpha(q_0), \varphi(x)] \leqq \varrho[\alpha(q_0), q_0] + \varrho[q_0, \varphi(x)] \leqq \frac{1}{3}\eta + \frac{1}{3}\eta < \eta$ und nach (3) $\varrho[\beta\alpha(q_0), x] < 2\varepsilon$, woraus sich nach (8)

$$(15) \qquad \varrho[f\beta\alpha(q_0), f(x)] < \sqrt{2}$$

ergibt.

Andererseits folgt aber aus (14), daß $\psi\varphi(x)$ in der Projektion P (aus 0 auf K_n) des Simplexes S liegt.

Da der Punkt $p_0 = f\beta\alpha(q_0)$ ein Eckpunkt von S ist, so erhalten wir aus (13) und auf Grund des Hilfssatzes 2

$$(16) \qquad \varrho[f\beta\alpha(q_0), \psi\varphi(x)] < \sqrt{2}.$$

Die Formeln $\varrho(a, b) < \sqrt{2}$ und $\varrho(a, c) < \sqrt{2}$ haben aber auf der Kugelfläche K_n die Ungleichung $\varrho(b, c) < 2$ zur Folge, womit durch (15) und (16) die Bedingung (10) für den Fall $\varphi(A) \subset R_n$ bewiesen ist.

Um davon zum allgemeinen Fall überzugehen, dürfen wir annehmen, daß $\varphi(A)$ eine Teilmenge des Grundquaders Q_ω des Hilbertschen Raumes ist[12]). Wir wählen eine natürliche Zahl k so groß, daß wir für die

---

[12]) Nach dem bekannten Einbettungssatze von P. Urysohn, nach welchem jeder separable metrische Raum mit einer Teilmenge von Q_ω homöomorph ist. Vgl. z. B. K. Menger, *Dimensionstheorie*, S. 57.

positive Zahl η von der Eigenschaft (3) die Formel

$$\sum_{i=k}^{\infty} \frac{1}{i^2} < \frac{1}{4}\eta^2 \tag{17}$$

haben.

Setzen wir nun für jedes $(x_1, x_2, \ldots x_n, \ldots) \in Q_\omega$

$$\gamma((x_1, x_2, \ldots x_n, \ldots)) = (x_1, x_2, \ldots x_k) \in R_k,$$

so bildet die Funktion $\gamma\varphi$ die Punktmenge A auf eine Menge $\gamma\varphi(A) \subset R_k$ ab. Dabei hat $\gamma\varphi(x') = \gamma\varphi(x'')$ nach (17) $\varrho[\varphi(x'), \varphi(x'')] \leqq \eta$ und somit nach (3) $\varrho(x', x'') < 2\varepsilon$ zur Folge.

Die Voraussetzungen des schon erledigten Falles $\varphi(A) \subset R_k$ werden also erfüllt, wenn wir φ durch $\gamma\varphi$ ersetzen. Demnach existiert eine Funktion ψ', die $\gamma\varphi(A)$ auf eine Teilmenge von K_n abbildet und gemäß (10) der Bedingung

$$\varrho[f(x), \psi'\gamma\varphi(x)] < 2 \quad \textit{für jedes} \quad x \in A \tag{18}$$

genügt.

Setzen wir nun $\psi = \psi'\gamma$, so wird erstens die Beziehung (9) erfüllt, weil ψ die Menge $\varphi(A)$ auf eine Teilmenge von K_n abbildet, und zweitens auch die Beziehung (10), infolge von (18).

5. Sei K_n^A der metrische Raum[13]), den wir erhalten, indem wir die Menge sämtlicher stetigen Abbildungen von A auf eine Teilmenge von K_n durch die Formel $\varrho(f_1, f_2) = \operatorname{Sup}_{x \in A} \varrho[f_1(x), f_2(x)]$ metrisieren.

Im Falle, wo $A \subset R_n$, gilt der Satz[14]):

(B) *Der Funktionenraum K_n^A ist dann und nur dann zusammenhängend, wenn A den Raum R_n nicht zerschneidet.*

Der aus allen stetigen Abbildungen von A auf *einzelne Punkte* von K_n bestehende Teilraum von K_n^A wird mit $N_{n,A}$ bezeichnet. Da derselbe mit K_n isometrisch ist, so ist er für $n \geqq 2$ zusammenhängend. Infolgedessen liegt $N_{n,A}$ in *einer* Komponente von K_n^A, die mit $\mathfrak{N}_{n,A}$ bezeichnet wird. Da die Komponenten von K_n^A voneinander mindestens um 2 entfernt sind[15]), so gilt

$$\varrho(f_1, f_2) \geqq 2 \quad \textit{für jedes} \quad f_1 \in \mathfrak{N}_{n,A} \quad \text{und} \quad f_2 \in K_n^A - \mathfrak{N}_{n,A}. \tag{19}$$

[13]) Dieser Raum, der zahlreiche Anwendungen in der Topologie hat, steht in engster Beziehung zu den Untersuchungen von H. Hopf und P. Alexandroff. Explizit wurde er in topologischen Untersuchungen von K. Borsuk angewandt. Siehe Fund. Math. 18 (1932), S. 193—213 und die dort (in Zusatz 4) angegebene Literatur.

[14]) K. Borsuk, Math. Ann. 106 (1932), S. 247.

[15]) Derselbe, Fund. Math. 18 (1932), S. 202.

Da der Raum K_n^A lokal zusammenhängend ist[15]), so besteht $\mathfrak{N}_{n,A}$ aus sämtlichen unwesentlichen Abbildungen, d. h. aus solchen, die sich mit Elementen von $N_{n,A}$ durch einen in K_n^A liegenden Bogen (homöomorphes Streckenbild) verbinden lassen (mit anderen Worten, welche durch stetige Abänderungen in eine konstante Abbildung transformiert werden können). Der Unzusammenhang des Raumes K_n^A (wo $n \geqq 2$) ist somit mit der Existenz einer wesentlichen Abbildung von A auf K_n gleichbedeutend.

6. Satz. *Gibt es eine wesentliche Abbildung f von A auf K_n mit der Eigenschaft* (8), *so ist für jede ε-Abbildung φ von A, der Raum $K_n^{\varphi(A)}$ nicht zusammenhängend.*

Beweis. Auf Grund des Satzes in **4**, dessen Voraussetzungen die Funktionen φ und f genügen, gibt es (indem wir die Beziehungen (9) und (10) in der soeben eingeführten Bezeichnungsweise schreiben) eine Abbildung ψ derart, daß die Formeln

$$\psi \in K_n^{\varphi(A)}, \tag{20}$$

$$\varrho(f, \psi\varphi) < 2 \tag{21}$$

bestehen. Wäre nun der Raum $K_n^{\varphi(A)}$ zusammenhängend, so würde eine stetige einparametrige Schar von Abbildungen $\psi_t \in K_n^{\varphi(A)}$ existieren, wo $0 \leqq t \leqq 1$, $\psi_0 = \psi$ und $\psi_1 \in N_{n,\varphi(A)}$ ist. Dann aber würden die Abbildungen $\psi_t\varphi$ im Raume $K_n^{\varphi(A)}$ ein Kontinuum (sogar ein stetiges Streckenbild) bilden, welches $\psi_0\varphi = \psi\varphi$ mit $\psi_1\varphi \in N_{n,A} \subset \mathfrak{N}_{n,A}$ verbindet. Nach der Definition von $\mathfrak{N}_{n,A}$ folgt daraus $\psi\varphi \in \mathfrak{N}_{n,A}$, und somit nach (21) und (19) gilt $f \in \mathfrak{N}_{n,A}$, was der Voraussetzung, daß f eine wesentliche Abbildung ist, widerspricht.

7. Eine unmittelbare Folgerung aus dem letzten Satze ist das folgende

Korollar[16]). *Läßt sich A auf K_n wesentlich abbilden, so gilt dasselbe von jeder Menge A', die aus A bei hinreichend kleinem ε mittels einer ε-Abbildung hervorgeht*[17]).

Insbesondere ergibt sich daraus die Invarianz der Nicht-Unikohärenz von stetigen Streckenbildern A gegenüber solchen Abbildungen φ, da dieselbe mit dem Nicht-Zusammenhange von K_1^A, also mit der Existenz der wesentlichen Abbildungen A auf K_1 äquivalent ist[18]).

8. In gewissen Fällen läßt sich für die Menge A, für die der Raum K_n^A unzusammenhängend ist, eine nur von A abhängende Konstante ex-

---

[16]) Vgl. P. Alexandroff, *Dimensionstheorie*, Math. Ann. **106** (1932), S. 226, Korollar 1 und darunterstehende Bemerkung.

[17]) Diese Formulierung verdanken wir Herrn P. Alexandroff.

[18]) K. Borsuk, Fund. Math. **17** (1931), S. 195.

plizit angeben derart, daß der Raum für alle stetige Funktionen φ, bei welchen alle Durchmesser der Urbildmengen einzelner Bildpunkte unterhalb dieser Konstante bleiben, auch unzusammenhängend ist.

Sei r_A, wo $A \subset R_n$, die obere Schranke der Radien aller abgeschlossenen, in den beschränkten Komponenten von $R_n - A$ liegenden n-dimensionalen Kugeln. Wir beweisen den

Satz. *Zerschneidet A den Raum R_n und ist φ irgendeine ε_0-Abbildung von A auf A', wo $\varepsilon_0 = \frac{1}{2} r_A \sqrt{2}$ ist, so ist der Funktionenraum $K_n^{\varphi(A)}$ nicht zusammenhängend.*

Ist insbesondere $\varphi(A) \subset R_n$, so ist $\varphi(A)$ ebenfalls ein Schnitt von R_n.

Beweis. Da A kompakt ist, gibt es eine positive Zahl ε, daß φ eine ε-Abbildung und

$$\varepsilon < \tfrac{1}{2} r_A \sqrt{2} \tag{22}$$

ist. Es existiert also eine beschränkte Komponente von $R_n - A$, die eine Kugel Q vom Radius 2ε enthält. Man kann voraussetzen, daß der Mittelpunkt von Q der Punkt 0 ist.

Betrachten wir die Funktion $f \in K_n^A$, die folgendermaßen definiert ist:

$$f(x) = \overrightarrow{0x} . K_n \quad \text{für jedes} \quad x \in A.$$

Diese Funktion, im Gegensatz zu jeder Funktion aus $\mathfrak{N}_{n,A}$[19]), läßt sich nicht als eine Teilfunktion einer stetigen, den ganzen Raum R_n auf eine Teilmenge von K_n abbildenden Funktion auffassen[20]). Es ist also f eine wesentliche Abbildung von A auf K_n. Dabei folgt aus $\varrho(x' x'') < 2\varepsilon$, wo $x', x'' \in A$, die Ungleichung $\measuredangle(x' 0 x'') < \frac{1}{2}\pi$ und daher $\varrho[f(x'), f(x'')] < \sqrt{2}$. Somit ist die Bedingung (8) erfüllt. Nach Satz 6 ist folglich der Funktionenraum $K_n^{\varphi(A)}$ nicht zusammenhängend, also im Falle wo $\varphi(A) \subset R_n$ ist nach Satz (B), $\varphi(A)$ ein Schnitt von R_n.

9. Es entsteht die Frage, ob die im Satz **8.** angegebene Abschätzung sich nicht etwa verschärfen läßt. Wir zeigen allerdings, daß eine einheitliche, von der Dimension n unabhängige Verschärfung der Abschätzung von der Gestalt $r_A \cdot \lambda$ unmöglich ist.

Es seien: S ein regelmäßiges, n-dimensionales Simplex, dessen alle Eckpunkte $p_0, p_1, \dots p_n$ auf K_n liegen, S_i^{n-1} die gegenüber dem Eckpunkte p_i liegende $(n-1)$-dimensionale Seite von S und S_i das n-dimensionale Simplex mit Eckpunkten $0, p_0, p_1, \dots p_{i-1}, p_{i+1}, \dots p_n$. Setzen

[19]) Da die Menge aller Teilfunktionen stetiger, R_n auf eine Teilmenge von K_n abbildender Funktionen gleichzeitig offen und abgeschlossen im Raume K_n^A ist und die Menge $\mathfrak{N}_{n,A}$ enthält. Vgl. K. Borsuk, Monatsh. f. Math. u. Phys. 38 (1931), S. 382.

[20]) L. c., S. 384—385.

wir nun $T_i = \overline{B(S_i) - S_i^{n-1}}$ wo $B(S_i)$ die Oberfläche von S_i bezeichnet und betrachten die folgendermaßen in der Menge $B(S)$ definierte Funktion:

$$g(x) = \overrightarrow{p_i x} \cdot T_i \quad \textit{für } x \in S_i^{n-1} \quad \textit{und} \quad i = 1, 2, \dots n,$$

so ist die Funktion

$$\varphi(x) = g(\overrightarrow{0x} \cdot B(S)) \quad wo \quad x \in K_n$$

eine stetige Abbildung von K_n auf die Punktmenge $T = \sum_{i=0}^{n} T_i$. Da aber die Menge T offenbar den Raum R_n nicht zerschneidet, so ist nach Satz (B), K_n^T zusammenhängend. Andererseits aber, wie eine elementargeometrische Überlegung zu berechnen gestattet, gilt für jedes $y \in \varphi(T) = \varkappa_n$ die Beziehung

$$\delta[\varphi^{-1}(y)] \leqq \delta(S) = \sqrt{\frac{2(n+1)}{n}}.$$

Daraus und da $\lim_{n=\infty} \sqrt{\frac{2(n+1)}{n}} = \sqrt{2}$ ist, ergibt sich die Unmöglichkeit einer gleichmäßig für alle A geltenden Verbesserung der Abschätzung von der Form $r_A \cdot \lambda$ des Durchmesser der Urbilder; da schon für die Kugeln mit dem Radius 1 die Konstante nicht größer als $\frac{1}{2}\sqrt{2}$ sein kann. Die Frage, ob trotzdem eine solche Verbesserung für in sich kompakte Teilmengen eines einzelnen R_n möglich ist, bleibt noch offen.

(Eingegangen am 11. 7. 1932.)

Über gewisse Zerlegungen von Mengen [1].

Von

Stanisław Ulam (Lwów)

In einer früheren Arbeit [2]) habe ich den folgenden Satz bewiesen [3]).

Es sei Z eine Menge von der Mächtigkeit $\aleph_1$. Werden alle Teilmengen dieser Menge in zwei Klassen, M und N so eingeteilt, dass es in M nur höchstens abzählbar viele elementfremde Mengen gibt, so existieren in N abzählbar viele Mengen $\{A_n\}$, für welche die Menge $(Z - \sum_{n=1}^{\infty} A_n)$ höchstens abzählbar ist.

Als eine Folgerung aus diesem Satze ergab sich die Unmöglichkeit, ein abzählbar additives Maß [4]) für alle Teilmengen einer Menge zu definieren, deren Mächtigkeit kleiner ist, als die erste „unerreichbare" Kardinalzahl.

Nun hat unlängst Herr W. Sierpiński gezeigt [5]), *dass in dem am Anfang genannten Satze die Annahme, dass Z von der Mächtigkeit $\aleph_1$ ist, durch eine schwächere ersetzt werden kann, nämlich, dass die Mächtigkeit von Z kleiner ist als die erste unerreichbare Kardinalzahl.*

[1]) Die Sätze dieser Note wurden (bei der Voraussetzung dass die Mächtigkeit von $Z\aleph_1$ ist) am 7. III. 1931 an einer Sitzung der Poln. Math. Gesellschaft, Abteilung Lwów, dargestellt.

[2]) „Zur Masstheorie in der allgemeinen Mengenlehre" Fund. Math. XVI (1930). S. 140—150.

[3]) l. c., S. 145, Satz „B".

[4]) l. c., S. 140.

[5]) W. Sierpiński: „Sur un théorème de recouvrement dans la théorie générale des ensembles, dieser Band.

Ich will jetzt noch einige Folgerungen aus diesem verschärften Satze angeben, auf deren Bedeutung mich Herr W. Sierpiński aufmerksam machte.

Wir machen zunächst die Annahme, daß die Menge Z von einer kleineren Mächtigkeit ist als die erste unerreichbare Kardinalzahl. Es gelten:

Satz I. *Es sei Z (in einem beliebigen perfekten Raume gelegene) Menge von der II-ten Baireschen Kategorie* [1]. *Es gibt dann unabzählbar viele in Z enthaltene, elementfremde Menge, die alle auch von der II-en Baire'schen Kategorie sind.*

Zum Beweise teilen wir alle Untermengen von Z in zwei Klassen ein: M sei die Klasse aller solchen Untermengen, die von der II-ten Baire'schen Kategorie sind, N die Klasse aller anderen.

Wäre unser Satz nicht richtig, so würde diese Einteilung die Voraussetzung des am Anfang genannten Satzes erfüllen. Es würden sich also in N abzählbar viele Mengen finden, die definitionsgemäß von der ersten Baire'schen Kategorie sind, und die mit eventueller Zunahme einer abzählbaren Menge, zusammen Z, also eine Menge von der II-en Kategorie ergeben. Das ist aber nicht möglich: die Vereinigungsmenge von abzählbar vielen Mengen von der I-en Kategorie ergibt wieder eine solche; eine abzählbare Menge ist immer von der ersten Kategorie.

Satz II. *Es sei Z eine Menge, die die Baire'sche Bedingung* [2] *nicht erfüllt. Es gibt dann unabzählbar viele elementfremde Teilmengen von Z von deren keine die Bairesche Bedingung erfüllt.*

Hier sei die Einteilung der Telmengen von Z in zwei Klassen die folgende: M sei die Klasse aller Teilmengen von Z, die die Baire'sche Bedingung nicht erfüllen, N die Klasse aller anderen.

Wäre unser Satz nicht richtig, so würden wir genau wie bei Satz I schließen, das es abzählbar viele Teilmengen von Z gibt, die die Baire'sche Bedingung erfüllen, und die zusammen mit einer höchstens abzählbaren Mengen die Menge Z, die die Baire'sche Bedingung nicht erfüllt, ergeben. Dies ist aber nicht möglich: eine ab-

[1] D. h. Z läßt sich *nicht* als Vereinigungsmenge von abzählbar vielen, in Bezug auf den ganzen Raum nirgensdichten Mengen darstellen.

[2] D. h. Z läßt sich darstellen als Vereinigungsmenge von einer G_δ-Menge und einer Menge erster Kategorie von Baire.

zählbare Menge erfüllt die Baire'sche Bedingung immer, so auch die Vereinigungsmenge einer abzählbaren Folge von Mengen, die diese Bedingung erfüllen.

Satz III. *Es sei Z eine Menge von positivem äußerem (Lebesgueschem) Masse. Es gibt unabzählbar viele elementfremde Teilmengen von Z, die alle auch ein positives äußeres Maß haben.*

Wäre der Satz nicht richtig, so würden wir, wie bei Satz I und II, schließen dass es abzählbar viele Mengen vom äußeren Masse 0 gibt, die zusammen Z, also eine Menge von positivem äußerem Masse ergeben, was unmöglich ist.

Bemerkung I. Wenn wir uns im Satze I auf solche Mengen Z beschränken, die in einem separablen Raume (z. B. auf der Geraden) liegen, so ergibt sich die Möglichkeit einer Zerlegung von Z in unabzählbar viele elementfremde Mengen die in *jedem Punkte* einer Umgebung (z. B. eines Intervalles) von der II-ten Baireschen Kategorie sind.

Bemerkung II. Von den unabzählbar vielen Mengen, von denen im Satze III die Rede ist, müssen „fast alle", d. h. alle mit Ausnahme von höchstens abzählbar vielen *unmessbar* sein. Dies mag, wenn man beachtet, dass alle diese Mengen elementfremd sind, als paradox erscheinen.

Bemerkung III. Alle unsere Sätze behalten natürlich ihre Gültigkeit, wenn man von der Menge Z die Annahme macht, dass sie von der Mächtigkeit des Kontinuums ist, und dabei die Cantorsche Kontinuumshypothese, oder die schwächere Hypothese, dass das Kontinuum eine kleinere Mächtigkeit hat, als die erste unerreichbare Kardinalzahl, als richtig voraussetzt.

EXTRAIT DE „STUDIA MATHEMATICA“, T. V. (1935).

Eine Bemerkung über die Gruppe der topologischen Abbildungen der Kreislinie auf sich selbst

von

J. SCHREIER und S. ULAM (Lwów).

1. Es bezeichne G die Gruppe aller topologischen Abbildungen des Intervalls $(0, 1)$ auf sich selbst.

Wir wollen zunächst die verschiedenen Klassen äquivalenter Elemente, in die G zerfällt, angeben. (Dabei heißen, wie üblich, zwei Elemente f und g äquivalent, wenn es ein $h \in G$ gibt, so daß $f = hgh^{-1}$ gilt).

Wir können uns bei dieser Untersuchung auf die Einheitskomponente G^* von G, d. i. auf die Abbildungen, die durch monoton wachsende, stetige Funktionen $f(x)$ mit $f(0) = 0$, $f(1) = 1$ gegeben sind, beschränken.

Ein Fixpunkt x resp. ein Intervall von Fixpunkten X einer Abbildung f heiße *attraktiv*, wenn für jeden, in einer genügend kleinen Umgebung von x (resp. X) liegenden Punkt y, $\lim_{n \to \infty} f^n(y) = x$ resp. $\lim_{n \to \infty} f^n(y) \in X$ gilt.

Es bezeichne für eine Abbildung f, $Z(f)$ die Menge der attraktiven Fixpunkte und Fixintervalle, $X(f)$ die Menge der übrigen Fixpunkte.

Die notwendige und hinreichende Bedingung für die Äquivalenz zweier Abbildungen f und g aus G^* ist die gleichzeitige Äquivalenz der Mengen $Z(f)$ und $Z(g)$ und der Mengen $X(f)$ und $X(g)$ d. h. die Existenz eines $h \in G$, so daß $h(Z(f)) = Z()g$ und $h(X(f)) = X(g)$ ist.

For commentary to this paper [15], see p. 678.

Der einfache Beweis dieser Behauptung beruht auf folgender Bemerkung[1]). Zwei Abbildungen $f, g \in G^*$ sind immer in G^* äquivalent, wenn

$$f(x) > x \quad \text{und} \quad g(x) > x \quad \text{für} \quad 0 < x < 1$$

resp.

$$f(x) < x \quad \text{und} \quad g(x) < x \quad \text{für} \quad 0 < x < 1$$

gilt.

Man wähle nämlich ein x_0 $(0 < x_0 < 1)$, setze $h(x_0) > x_0$ beliebig, $x_n = g(x_{n-1})$, $x_{-n} = g^{-1}(x_{-(n-1)})$ und

$$h(x_n) = fh(x_{n-1}), \quad h(x_{-n}) = f^{-1}h(x_{-(n-1)}). \qquad (n = 1, 2, \ldots)$$

Dann verbinde man $h(x_0)$ mit $h(x_1)$ stetig und monoton wachsend, sonst aber beliebig und setze für jedes $x_0 < \xi_0 < x_1$: $\xi_n = g(\xi_{n-1})$, $\xi_{-n} = g^{-1}(\xi_{-(n-1)})$, $h(\xi_n) = fh(\xi_{n-1})$, $h(\xi_{-n}) = f^{-1}h(\xi_{-(n-1)})$.

Man bestätigt leicht, daß $h(x)$ eine topologische Abbildung des Intervalls $(0, 1)$ auf sich selbst ist und daß $hgh^{-1}(x) = f(x)$.

Es seien nun f und $g \in G^*$ mit $X(f) = X(g)$ und $Z(f) = Z(g)$ gegeben. Daraus folgt leicht, daß in jedem Intervalle (x_1, x_2) derart, daß $x_1 \in X + Z$, $x_2 \in X + Z$, $(x_1, x_2) \cdot (X + Z) = 0$ entweder $f(x) > x$ und $g(x) > x$, oder $f(x) < x$ und $g(x) < x$ gilt.

Laut der früheren Bemerkung läßt sich in jedem (x_1, x_2) ein $h(x)$ erklären, welches die Gleichung $hgh^{-1}(x) = f(x)$ für $x_1 \leqslant x \leqslant x_2$ erfüllt. Dabei ist $h(x_1) = x_1$ und $h(x_2) = x_2$. Damit ist aber in $(0, 1)$ ein $h(x) \in G^*$ erklärt, welches g in f transformiert, d. h. $hgh^{-1} = f$ erfüllt.

Wir betrachten jetzt einen Normalteiler D der Gruppe G^*; f bezeichne eine zu D gehörende Abbildung, für die 0 und 1 isolierte Fixpunkte sind.

Die Null sei z. B. für f nicht attraktiv, die Eins z. B. für f^{-1} attraktiv. (Da 0 und 1 isolierte Fixpunkte sind, muß jeder von ihnen entweder für f oder für f^{-1} attraktiv sein).

Man findet dann zwei Abbildungen $\varphi(x)$ und $\psi(x)$ mit folgenden Eigenschaften:

a) φ ist mit f, ψ mit f^{-1} äquivalent, daher gehören φ und ψ zu D.

b) $\varphi(x) > x$ in $0 < x < 1 - \varepsilon$

$\psi(x) > x$ in $\varepsilon < x < 1$

$|\varphi(x) - x| < \varepsilon(x)$ in $1 - \varepsilon < x < 1$

$|\psi(x) - x| < \varepsilon(x)$ in $0 < x < \varepsilon$

---

[1]) H. Kneser, Kurvenscharen auf Ringflächen, Math. Ann. 91 (1924) p. 135—154.

wobei ε und $\varepsilon(x)$ positiv und so klein sind, daß $\nu(x) = \varphi\psi(x)$ die Ungleichung

$$\nu(x) > x \quad \text{für} \quad 0 < x < 1$$

erfüllt.

Da $\nu = \varphi\psi$ ist, gehört laut a) ν zu D, also auch die Abbildung $\nu^{-1}(x)$, die die Ungleichung

$$\nu^{-1}(x) < x \quad \text{für} \quad 0 < x < 1$$

erfüllt. Daher gehören zu D alle mit ν oder ν^{-1} äquivalenten Abbildungen, also alle Abbildungen f, für die entweder $f(x) > x$ oder $f(x) < x$ in ganz $(0, 1)$ gilt. Es sei nun $p(x) \in G^*$ beliebig gegeben.

Wir nehmen ein $f(x)$, das die Bedingungen

$$(1) \qquad f(x) > x$$

$$\text{für } 0 < x < 1$$

$$(2) \qquad f(x) > p(x)$$

erfüllt. Wir setzen

$$(3) \qquad g(x) = f^{-1}p(x).$$

Da $f^{-1}(x)$ ebenfalls monoton wächst, ist nach (2) und (3)

$$(4) \qquad g(x) < f^{-1}f(x) = x.$$

Wegen (3) ist

$$(5) \qquad p(x) = fg(x).$$

Nach (1) und (4) gehören f und g zu D, also wegen (5) gehört auch p zu D. D ist also mit G^* identisch.

Alle Abbildungen f, für die $f(x) = x$ in $(0, \varepsilon(f))$ $(0 < \varepsilon(f) < 1)$ gilt, bilden, wie man leicht bestätigt, einen Normalteiler in G^*. Ebenso die Abbildungen $h(x)$, für die in beliebiger Nähe des Punktes 0 Punkte x und y liegen, so daß $h(x) > x$ und $h(y) < y$ gilt. Schließlich erhält man einen Normalteiler, wenn man Abbildungen $g(x)$ betrachtet, für die es ein $\varepsilon(g)$ gibt, so daß $g(x) = x$ für $0 \leqq x \leqq \varepsilon(g)$ gilt, und in beliebiger Nähe des Punktes $x = \varepsilon(g)$ Punkte y und z liegen, so daß $g(y) > y$ und $g(z) < z$ ist. Man kann dann noch dieselben Singularitäten bei $x = 1$ konstruieren.

Aus dem Bewiesenen folgt, daß G^* keinen abgeschlossenen nichttrivialen Normalteiler enthält.

2. Wir gehen nun zur Untersuchung der Gruppe K aller topologischen Abbildungen der Kreislinie auf sich selbst über. Wir können uns hier auch von vornherein auf die Einheitskomponente K^* von K, d. h. auf die die Indikatrix erhaltenden Abbildungen beschränken.

Wir beginnen mit folgender Bemerkung:

Die Untergruppe der Elemente von K, die ein gewisses x festlassen, ist stetig isomorph mit der Gruppe G[2]).

Für Abbildungen f, die mindestens einen Fixpunkt besitzen, sind daher die Bedingungen für Äquivalenz dieselben wie in der Gruppe G.

Wir betrachten zunächst zwei Abbildungen f, φ, deren n-te Iterationen Fixpunkte besitzen, während f, φ, f^2, $\varphi^2, \ldots f^{n-1}$, φ^{n-1} fixpunktfrei sind.

Wir behaupten: Zur Äquivalenz dieser Abbildungen ist die Äquivalenz von f^n und φ^n notwendig und hinreichend.

Die Bedingung ist offenbar notwendig. Um zu zeigen, daß sie hinreichend ist, können wir annehmen, daß φ^n und f^n miteinander identisch sind. Die bei f^n fixpunktfreien Intervalle zerfallen in Klassen von je n Intervallen: $I_1, I_2, \ldots I_n$, so daß $f(I_1) = I_2, \ldots f(I_{n-1}) = I_n$, $f(I_n) = I_1$ ist. Man setze

$$h(x) = f^{k-1}\varphi^{1-k}(x) \quad \text{für} \quad x \in I_k.$$

Man bestätigt, daß $h(x) \in K^*$ und

$$h^{-1}fh(x) = \varphi(x)$$

gilt.

Wir nehmen jetzt an, daß kein f^n Fixpunkte besitzt. Man kann dann mit Poincaré die sogenannte Rotationszahl α[3]) einführen, welche eine Klasseninvariante ist, d. h. für je zwei äquivalente Abbildungen übereinstimmt. Ist noch die Punktmenge $\{f^n(x)\}$ $(-\infty < n < \infty)$ für irgendein x überalldicht, so beweist man leicht, daß f einer Drehung um α äquivalent ist. Sonst ist die Ableitung der Menge $\{f^n(x)\}$ eine nirgendsdichte, perfekte Menge (sog. Cantorsche Menge). Ihre Restintervalle zerfallen in Klassen, indem zwei solche Intervalle I und J zu einer Klasse gerechnet werden, wenn $f^n(I) = J$ für ein gewisses n gilt. Da zwei Cantorsche Mengen immer äquivalent sind, so ist, wie man sich überzeugt, die Anzahl dieser Klassen nebst der Rotationszahl genügend, um die Klasse einer Abbildung zu charakterisieren.

Es bezeichne wieder D einen Normalteiler von K^*.

Enthält D ein f, für welches ein Punkt x_0 isolierter Fixpunkt ist, so enthält D laut obiger Bemerkung und dem Resultate von **1.** alle f, für die $f(x_0) = x_0$ ist, daher auch jede Abbildung, die überhaupt einen Fixpunkt besitzt. (Durch Drehung wird nämlich ihre Äquivalenz mit einer Abbildung, die x_0 zum Fixpunkt hat, bewiesen).

---

[2]) J. Schreier und S. Ulam, Über topologische Abbildungen euklidischer Sphären. Fund. Math. 23 (1934) p. 102—118.

[3]) S. die unter 1) zit. Arbeit.

Da jede fixpunktfreie Abbildung natürlich als Zusammensetzung zweier Abbildungen mit Fixpunkten erhalten werden kann, so ist D mit K^* identisch.

Nun enthalte D ein f, dessen Fixpunktmenge F perfekt ist. D enthält dann alle φ mit $X(\varphi) = X(f)$, $Z(\varphi) = Z(f)$. Man kann nun unter diesen φ ein solches wählen, welches in einem Restintervall I der Menge F so erklärt ist, daß die Zusammensetzung $\varphi^{-1}f$ in I einen isolierten Fixpunkt hat. Daher ist auch in diesem Falle D mit K^* identisch.

Enthält D eine fixpunktfreie Abbildung f, so kann unter den mit f^{-1} äquivalenten Abbildungen leicht eine Abbildung $g \not\equiv f^{-1}$ gefunden werden, so daß fg einen Fixpunkt besitzt. Nach dem Vorangehenden ist also D auch jetzt mit K^* identisch.

Unser Resultat können wir daher so aussprechen:

Die Gruppe der indikatrixerhaltenden topologischen Abbildungen der Kreislinie auf sich selbst ist einfach.

Die Frage, ob ein analoger Satz für die höheren Dimensionen gilt, bleibt offen.

(Reçu par la Rédaction le 12. 6. 1935).

Sur la théorie de la mesure dans les espaces combinatoires et son application au calcul des probabilités I. Variables indépendantes [1]).

Par

Z. Łomnicki et S. Ulam (Lwów).

L'analogie entre la mesure et la probabilité est connue depuis longtemps [2]).

La probabilité pour qu'un point appartienne à un ensemble A d'un espace donné remplit les postulats de la mesure d'ensemble. On admet notamment la règle des probabilités totales — c.-à-d. l'additivité finie ou dénombrable de la mesure. (Dans le cas où l'espace est dénombrable le postulat de l'additivité finie est souvent plus adéquat).

La théorie de la mesure pour *un* espace constitue cependant une théorie d'une seule variable éventuelle et ne semble pas donner un

[1]) Les résultats concernant la théorie de la mesure dans les produits ont été exposés par les auteurs dans un Séminaire de M. H. Steinhaus (Mai 1932). Les théorèmes relatifs ont été présentés à la séance de la Soc. Pol. Math. Section de Lwów du 2. VII. 1932 (v. aussi la note de l'un de nous insérée dans les „Verhandl. des Int. Math. Kongr. Zürich" 1932, Band II). Les applications au calcul des probabilités qui se trouvent dans la deuxième partie de ce travail ont été présentées à la séance de la Soc. Pol. Math. Section de Lwów le 18. III. 1933.

[2]) E. Borel, *Sur les probabilités dénombrables*... Rendiconti del Circolo Mat. di Palermo, 1909, p. 247—281. — A. Łomnicki, *Nouveaux fondements du calcul des probabilités.* Fund. Math. T. IV, p. 35—71. — H. Steinhaus, *Les probabilités dénombrables et leur rapport à la théorie de la mesure.* Fund. Math. T. IV, p. 287—310. — R. v. Mises, *Grundlagen der Wahrscheinlichkeitsrechnung.* Math. Zeit. Bd. 34, p. 568—619. — P. Lévy, *Calcul des probabilités.* Note (p. 325—345) Gauthier-Villars, Paris. — Cf. aussi une étude approfondie chez A. Kolmogoroff, *Grundbegriffe der Wahrscheinlichkeitsrechnung* dans les *Ergebnisse der Mathematik und ihrer Grenzgebiete,* Berlin 1933.

For commentary to this paper [16], see p. 679.

schéma mathématique assez général pour traiter la très grande variété des problèmes du calcul des probabilités. La règle des probabilités composées déjà ne se laisse pas interpréter à l'aide de la mesure dans un seul espace. En traitant des couples (x, y) d'événements on considère le produit [3]) combinatoire (ou cartésien) $X \times Y$ de deux espaces X, Y de ces événements [3*]) et la probabilité c.-à-d. la mesure dans ce produit.

Si l'on admet que le but d'une théorie des probabilités est de calculer à l'aide des probabilités connues des probabilités nouvelles, on doit pour le traitement méthodique des probabilités composées montrer comment on peut définir une mesure pour l'ensemble-produit en partant de la mesure dans les espaces composants. On doit déterminer, en particulier, dans le produit tous ces ensembles pour lesquels la mesure est définie d'une manière univoque par la règle des probabilités composées.

On doit montrer aussi, ce qui est important, que cette règle ne conduit pas à une contradiction, même dans les cas les plus généraux. L'existence de la mesure dans les produits (finis ou infinis) n'est, dans le cas général, nullement évidente *à priori*. Si l'on admet en particulier que la mesure donnée dans les espaces X, Y n'est pas complètement additive (l'additivité finie étant, bien entendu, admise) le problème de la détermination d'une mesure univoque pour le produit $X \times Y$ n'est pas encore resolu (Voir th. 4, Remarque).

Il est nécessaire de considérer des espaces très généraux. Dans les problèmes du calcul des probabilités on ne peut pas admettre *à priori* que tous les espaces d'événements possèdent quelques propriétés topologiques communes ou propriétés du groupe. On considère dans le calcul des probabilités des espaces très différents: p. ex. les ensembles finis dans le cas des problèmes élementaires, l'ensemble des nombres entiers dans les problèmes „dénombrables" (au sens de M. Borel) [4]), les espaces géométriques d'une nature souvent très compliquée dans la théorie des probabilités continues [5]). Enfin,

[3]) La définition des produits est donnée dans le N° 2.

[3*]) Cf. l'opération de la „Verbindung" de M. R. v. Mises, *Vorlesungen aus dem Gebiete der angewandten Mathematik*. Bd. I. *Wahrscheinlichkeitsrechnung*, Berlin 1931.

[4]) Cf. l. c. [2]).

[5]) *Traité du Calcul des probabilités et ses Applications* de M. Borel. T. II, fascicule 2: R. Deltheil, *Probabilités géométriques*.

on considère les espaces à une infinité de dimensions dans le cas des „lois limites“ [6]).

Les difficultés principales qui se présentent lorsqu'on veut définir un schéma mathematique général pour la théorie des probabilités proviennent du fait, qu'il est insuffisant de considérer un seul espace. Il faut plutôt établir une mesure pour une *classe d'espaces*. On obtient ces espaces des espaces à mesure donnée par quelques opérations combinatoires simples.

Nous avons insisté sur le rôle du produit de deux (ou d'un nombre fini quelconque) d'espaces. Or, il est nécessaire d'étudier aussi les produits infinis

$$X_1 \times X_2 \times \cdots \times X_n \times \cdots$$

d'une suite d'espaces $\{X_n\}$ [7]). Ces produits s'imposent d'une façon naturelle lorsqu'on étudie des probabilités des faits qui consistent en une réalisation simultanée d'une suite infinie d'événements (p. ex. la convergence d'une série à termes variables). L'étude de la mesure pour de tels produits permet d'opérer avec des probabilités des tels faits, au lieu de considérer des limites d'expressions qui, *à priori* au moins, ne semblent être choisies que d'une manière dans un certain dégré arbitraire.

On obtient aussi à l'aide d'une telle mesure des énoncés plus intuitifs et précis des différentes „lois des grands nombres“.

Pour le cas général des variables éventuelles *dépendantes* il faut introduire des nouvelles opérations combinatoires [8]).

Dans ce travail nous étudions la mesure dans le produit pour le cas des variables indépendantes. Cette mesure remplit un postu lat qui exprime la règle des probabilités composées. La mesure dans l'espace $X \times Y$ est assujettie à la condition suivante: Si un ensemble $C \subset X \times Y$ est de la forme $C = A \times B$ (où $A \subset X$ et $B \subset Y$) alors $m_{X \times Y}(C) = m_X(A) \cdot m_Y(B)$ [8a]) [8b]).

La mesure dans les produits correspondant au cas des chaînes de Markoff sera étudiée dans une note prochaine.

[6]) H. Steinhaus, *Sur la probabilité de la convergence des séries*, Studia Mat. T. II. 1930.

[7]) La définition du produit infini est donnée dans le texte p. 242.

[8]) P. ex. dans l'étude du problème de la „probabilité des causes“ on doit introduire, dans le cas général, l'espace des toutes les mesures dans un espace donné.

[8a]) Cf. A. Łomnicki, l. c. [3]), p. 43, cond. IV.

[8b]) Nous désignerons dans le texte la mesure toujours par le signe m même s'il s'agit des mesures dans des espaces différents.

Dans la première partie de ce travail nous demontrons les théorèmes sur l'existence de la mesure (multiplicative) dans les produits. Dans la deuxième nous montrons sur quelques exemples le rôle de la notion des produits dans le calcul des probabilités, ensuite, en nous servant des théorèmes sur la mesure nous donnons des énoncés uniformes des théorèmes fondamentaux de la théorie des variables éventuelles indépendantes [9]).

I

1. Les axiomes de la mesure pour une variable éventuelle [10]).

Soit E une ensemble abstrait. Nous supposons que dans l'ensemble E, pour une classe $\mathfrak{M}$ de sous-ensembles, est définie une fonction d'ensemble $m(X)$, appellée mesure de X admettant comme valeurs des nombres réels. (Cette fonction peut être interprétée comme probabilité pour qu'un élément x, pris au hasard, appartienne à l'ensemble donné X)

Supposons que les conditions suivantes sont remplies par la classe $\mathfrak{M}$ des ensembles mesurables:

(I) *L'espace total est mesurable, ainsi que les ensembles composés d'un seul point*: $E \epsilon \mathfrak{M}$; $\{x\} \epsilon \mathfrak{M}$

(II) *Si* $X \epsilon \mathfrak{M}$ *et* $Y \epsilon \mathfrak{M}$, *alors* $(X - Y) \epsilon \mathfrak{M}$

(III) $X_i \epsilon \mathfrak{M} (i = 1, 2, \ldots)$, *alors* $\sum_{\nu=1}^{\infty} X_\nu \epsilon \mathfrak{M}$

(IV) *Si* $Z \subset X$, $X \epsilon \mathfrak{M}$ *et* $m(X) = 0$, *alors* $Z \epsilon \mathfrak{M}$.

Nous supposons que la fonction $m(X)$ remplit les postulats:

$$(1) \qquad m(X) \geqslant 0, \quad m(E) = 1$$

$$(2) \qquad m\left(\sum_{\nu=1}^{\infty} X_\nu\right) = \sum_{\nu=1}^{\infty} m(X_\nu) \quad si \quad X_i X_j = 0 \quad pour \quad i \neq j$$

Remarques. La condition (2) exprime „l'additivité complète (dénombrable)" de la mesure. Ce postulat est indispensable dans beaucoup de problèmes dans le calcul des probabilités, même dans les problèmes élémentaires du calcul des pro-

[9]) Le lecteur qui ne s'intéresse qu'aux théorèmes du calcul des probabilités peut omettre les démonstrations des théorèmes de la partie première. La connaissance de la notion et des propriétés les plus élémentaires des produits et de ces théorèmes est suffisante.

[10]) Kolmogoroff, op. cit., Ch. I.

babilités il est nécessaire de l'utiliser p. ex. pour une solution rigoureuse du problème de la „ruine des joueurs" (v. les exemples dans la partie II-ième).

Observons que nos postulats dans le cas où E est l'espace euclidien, si l'on ajoute encore la condition que pour un cube euclidien C, $m(C)$ soit égal au volume de C dans le sens élémentaire, entraînent déjà, que $m(X)$ est identique avec la mesure de Lebesgue; la classe $\mathfrak{M}$ se confond avec la classe des ensembles mesurables (L).

Comme on le sait, dans le cas de la mesure de Lebesgue on postule outre les axiomes (I)—(IV) et (1), (2), l'axiome (C): Si X et Y sont congruents dans le sens de la géometrie élémentaire $m(X) = m(Y)$. Une telle mesure ne se laisse pas définir pour tous les sous-ensembles de l'espace (théorème de Vitali). On peut cependant montrer qu'une mesure qui remplit nos postulats seulement ne se laisse nonplus définir pour tous les sous-ensembles de E[11]). Il est donc nécessaire de parler de la classe des ensembles mesurables.

Parallèlement à la mesure qui remplit les postulats (I)—(IV) et (1), (2) il est nécessaire d'étudier dans le calcul des probabilités une mesure qui remplit au lieu des postulats (III), (2) les postulats suivants moins restrictifs:

(III'). *Si* $X_i \epsilon \mathfrak{M}$ $(i = 1, 2, \dots n)$, *on a* $\sum_{\nu=1}^{n} X_\nu \epsilon \mathfrak{M}$

$$(2') \qquad m\left(\sum_{\nu=1}^{n} X_\nu\right) = \sum_{\nu=1}^{n} m(X_\nu) \text{ si } X_i \cdot X_j = 0 \text{ pour } i \neq j.$$

Cela est nécessaire surtout dans les problèmes „dénombrables" du calcul des probabilités. Si E est l'ensemble des nombres entiers on ne peut pas évidemment définir pour des sous-ensembles de E une mesure qui remplirait les postulats (III), (2) d'une manière naturelle c.-à-d. de telle façon que la probabilité d'un ensemble composé d'un seul nombre soit égale à 0. Une telle mesure cependant peut être définie même pour tous les sous-ensembles de E, les axiomes (III') et (2') étant remplis.

Dans l'espace E à mesure donnée on peut définir d'une manière connue les intégrales des fonctions définies sur E et ayant pour valeur des nombres réels (ou plus généralement, des éléments d'un espace vectoriel), prises sur des sous-ensembles mesurables [12]). A l'aide de l'intégration de Stieltjes on peut définir, dans le cas où l'espace E se compose de nombres réels (ou de points d'un

[11]) V. S. Ulam, *Masstheorie in der allgemeinen Mengenlehre*, Fund. Math. T. XVI (1930).

[12]) Fréchet, *Sur l'intégrale d'une fonctionelle étendue à un ensemble abstrait*, Bull. Soc. Math. France, Bd. 43 (1915), S. 248.

espace vectoriel), l'espérance mathématique de E [13]) et les notions dérivées, qui jouent un grand rôle dans le calcul des probabilités, p. ex. les moments de n-ième ordre de la variable éventuelle.

L'étude de toutes ces notions concernant une seule variable éventuelle n'est pas encore, comme le remarque M. Kolmogoroff [14]), caractéristique pour le schéma mathématique du calcul des probabilités qui s'occupe des relations entre les mesures d'une classe d'espaces. Si l'on étudie l'ensemble de plusieurs variables éventuelles on a affaire au produit $\mathfrak{P} E_\nu = E_1 \times E_2 \ldots$ (v. les exemples dans la partie II-ième). Pour notre travail nous allons rappeler les définitions et quelques propriétés combinatoires de cette opération fondamentale.

2. L'opération du produit.

Nous entendons par produit (ou produit cartésien) de deux ensembles E_1 et E_2 l'ensemble de couples ordonnés (e_1, e_2) où $e_1 \epsilon E_1$, $e_2 \epsilon E_2$. D'une manière analogue, pour une suite (finie ou infinie) d'ensembles $\{E_i\}$, nous entendons par leur produit l'ensemble des suites d'éléments $\{e_i\}$ qui appartiennent aux ensembles respectifs E_i. Ce produit sera designé par

$$E_1 \times E_2 \times \ldots \times E_n = \mathop{\mathfrak{P}}_{\nu=1}^{n} E_\nu$$

ou

$$E_1 \times E_2 \times \ldots \times E_i \times \ldots = \mathop{\mathfrak{P}}_{\nu=1}^{\infty} E_\nu.$$

On prouve sans difficulté les formules suivantes concernant les propriétés algébro-logiques de l'opération du produit [15]):

1. $(A + B) \times (C + D) = A \times C + A \times D + B \times C + B \times D$

et plus généralement:

2. $(\sum_\nu A_\nu) \times (\sum_\mu B_\mu) = \sum_{\nu, \mu} A_\nu \times B_\mu$

3. $(AC) \times (BD) = (A \times B) \cdot (C \times D)$

[13]) Kolmogoroff, cf. l. c. [10]), ch. IV.

[14]) Kolmogoroff, cf. l. c. [10]), p. 8.

[15]) Kuratowski, *Topologie* I, Warszawa—Lwów 1933.

et plus généralement:

4. $(\underset{\nu}{\Pi} A_\nu) \times (\underset{\mu}{\Pi} B_\mu) = \underset{\nu,\mu}{\Pi}(A_\nu \times B_\mu)$

5. $\underset{i}{\Pi} \underset{\nu}{\mathfrak{P}} A_{i,\nu} = \underset{\nu}{\mathfrak{P}} \underset{i}{\Pi} A_{i\nu}$.

Si A est un sous-ensemble de l'espace X et B en est un de l'espace Y (ou plus généralement si $A_i \subset X_i$) on a les formules:

6. $C(A \times B) = CA \times Y + X \times CB$ [16])

7. $C(\underset{\nu}{\mathfrak{P}} A_\nu) = \underset{\nu}{\Sigma} D_\nu$ où $D_i = X_1 \times X_2 \ldots \times X_{i-1} \times (X_i - A_i) \times X_{i+1} \ldots$

Par $\boldsymbol{E}[\varphi(x)]$ nous désignons l'ensemble des x qui satisfont à la condition φ exprimée entre paranthèses.

Si l'on a affaire à une condition (propriété) $\varphi(x_1, \ldots, x_n, \ldots)$ concernant un système de variables on peut régarder φ comme une condition pour une variable $x = (x_1, \ldots x_n, \ldots)$ qui parcourt le produit $\underset{\nu}{\mathfrak{P}} X_\nu$, ou x_i sont les éléments de X_i.

Le symbole $\boldsymbol{E}[\ ]$ (fréquemment employé dans la deuxième partie de notre travail) désigne alors l'ensemble correspondant des x du produit. Les règles du calcul avec le symbole $\boldsymbol{E}$ et sa relation avec les opérateurs de la théorie des ensembles et avec l'opération du produit sont exposées d'une manière très claire et concise dans le livre de M. Kuratowski [17]).

Dans les applications, il est important de savoir que, f étant une fonction réelle mesurable, l'ensemble des x pour lesquels on a: $a \leqslant f(x) \leqslant b$ est mesurable.

3. Mesure complètement additive dans les produits finis.

Si les E_i désignent des espaces topologiques (c. à d. si dans E_i est définie la notion de la limite d'une suite de points), on peut définir une limite dans $\overset{n}{\underset{\nu=1}{\mathfrak{P}}} E_\nu$ de manière que cet ensemble devienne un espace topologique. Pareillement, si les E_i sont des groupes abstraits on peut définir une composition des systèmes d'éléments de telle façon que $\overset{n}{\underset{\nu=1}{\mathfrak{P}}} E_\nu$ devient un groupe. Ici, nous supposons

[16]) CA désigne le complémentaire (relativement à l'espace) de l'ensemble A.

[17]) Kuratowski, l. c. [15]). Introduction, § 1. Opération de la logique et de la théorie des ensembles, § 2. Produit cartésien, § 3. Fonctions.

qu'on a défini dans E_i une mesure (dans le sens du N⁰ 1.) et nous allons construire une mesure dans $\mathfrak{P}_{\nu=1}^{n} E_\nu$.

D'une manière plus précise, soient E_i $(i = 1, 2, \ldots, n)$ les espaces à mesure donnée, $\mathfrak{M}_i$ la classe des ensembles mesurables dans E_i, $E = \mathfrak{P}_{\nu=1}^{n} E_\nu$ leur produit. Notre but est de définir une mesure pour une classse $\mathfrak{M}$ de sous-ensembles de E de façon que les conditions suivantes soient remplies:

(I)—(IV); (1), (2) du N⁰ 1 et

(V) *Si* $X_i \epsilon \mathfrak{M}_i$ *pour* $i = 1, 2, \ldots, n$, *l'ensemble* $\mathfrak{P}_{\nu=1}^{n} X_\nu \epsilon \mathfrak{M}$.

La condition (V) exprime le postulat que l'existence des probabilités de faits A et B entraîne l'existence d'une probabilité pour la réalisation de A et B.

D'après (V) on obtient la mesurabilité dans E d'ensembles d'une nature assez spéciale: des produits d'ensembles mesurables. Mais à l'aide des postulats (I)—(IV) on obtient de (V) la mesurabilité d'une classe très étendue d'ensembles.

Par exemple si E_1 et E_2 sont identiques à l'intervalle $0 \leqslant x \leqslant 1$ et la mesure donnée est identique à celle de Lebesgue — les postulats (I)—(V) entraînent déjà que dans $E_1 \times E_2$ (c. à d. dans le carré) la classe $\mathfrak{M}$ contient tous les ensembles pour lesquels existe la mesure plane de Lebesgue.

La mesure dans E doit remplir le postulat: si $X_{i_j} \subset E_{i_j}$, (pour $j = 1, 2, \ldots k$), les autres composantes X_i du produit étant égales aux espaces entiers E_i, on a: $m(X_1 \times X_2 \ldots X_n) = m(X_{i_1}) \cdot m(X_{i_2}) \ldots m(X_{i_k})$.

Cette égalité exprime pour $n = 2$ le fait que la probabilité d'un système composé de l'événement A et d'un événement qui doit se produire nécessairement est la même que la probabilité de l'événement A seul. (Observons qu'il n'en résulte pas encore que le système composé d'un événement A et d'un événement de probabilité 1 a la même probabilité que A).

Dans ce travail nous nous occupons de variables indépendantes. L'indépendance s'exprime par le postulat

$$(3) \qquad m(X_1 \times X_2 \ldots \times X_n) = m(X_1) \cdot m(X_2) \cdot \ldots \cdot m(X_n).$$

Nous allons prouver le

Théorème 1. [18]) *Soient* E_i $(i = 1, 2, \ldots, n)$ *des espaces à mesure donnée remplissant les conditions* (I)—(IV) *et* (1), (2). *Il existe une mesure dans* $E = \mathfrak{P}_{\nu=1}^{n} E_\nu$ *remplissant les conditions* (I)—(V) *et* (1)—(3).

On peut évidemment sans diminuer la généralité supposer que $n = 2$, c. à d. qu'il s'agit d'un produit de deux espaces, le cas général se laissant réduire à celui-ci par induction.

Nous allons prouver deux lemmes qui jouent un rôle essentiel dans la démonstration.

Lemme 1. [19]) Soient E_1 et E_2 des espaces à mesure donnée. Pour les ensembles C dans le produit $E_1 \times E_2$ de la forme $A \times B$ où $A \subset E_1$, $B \subset E_2$ (A et B mesurables) posons $m(A \times B) = m(A) \cdot m(B)$.

Soit $C = A \times B = \sum_{\nu=1}^{m} A_\nu \times B_\nu$ une décomposition de l'ensemble C en un nombre fini d'ensembles-produits, et $(A_i \times B_i) \cdot (A_j \times B_j) = 0$ pour $i \neq j$, alors

$$m(C) = m(A) \cdot m(B) = \sum_{\nu=1}^{m} m(A_\nu) \cdot m(B_\nu).$$

Démonstration. Posons

$$\bar{A}_1 = A_1 - \sum_{\nu=1}^{m}{}' A_\nu$$

$$\cdots\cdots\cdots$$

$$\bar{A}_m = A_m - \sum_{\nu=1}^{m}{}' A_\nu$$

[18]) Une démonstration très élégante de ce théorème a été donnée par M. S. Saks dans son livre „*Theorie de l'intégrale*", Annexe, Warszawa 1934. La démonstration que nous réproduisons étant d'un caractère plus élémentaire et utilisant seulement les mesures des ensembles peut être traduite en langage du calcul élémentaire des probabilités.

[19]) Dans la théorie de la mesure n-dimensionnelle de Lebesgue on utilise pour la démonstration du théorème analogue au th. 1 le théorème connu de Heine-Borel, qui dans notre cas (des espaces abstraits) ne peut pas être formulé. Les lemmes 1. et 2. nous rendent les mêmes services.

$$\bar{A}_{m+1} = A_1 A_2 - \sum_{\nu,\mu}^{m,m}{}' A_\nu A_\mu$$

$$\cdots\cdots\cdots\cdots$$

$$\bar{A}_{m+\binom{m}{2}} = A_{m-1} A_m - \sum_{\nu,\mu}^{m,m}{}' A_\nu A_\mu$$

$$\cdots\cdots\cdots\cdots$$

$$\bar{A}_{2^m-1} = A_1 A_2 \ldots A_m,$$

où le signe Σ' exprime que dans la somme on doit omettre les termes écrits explicitement devant la somme.

Les ensembles $\bar{A}_i$ sont disjoints. Il est clair que les ensembles du même groupe (representés par le même nombre de facteurs) sont disjoints. On voit aussi que les ensembles des groupes différents sont disjoints, parce que au k-ième groupe appartiennent les ensembles, dont les points appartiennent à precisément k ensembles A_i.

Il est clair que $\sum_{\nu=1}^{2^m-1} \bar{A}_\nu \subset \sum_{\nu=1}^{m} A_\nu$. Mais on a réciproquement $\sum_{\nu=1}^{m} A_\nu \subset \sum_{\nu=1}^{2^m-1} \bar{A}_\nu$. En effet, si un point p appartient à l'ensemble $\sum_{\nu=1}^{m} A_\nu$, c'est-à-dire aux ensembles $A_{i_1}, A_{i_2}, \ldots, A_{i_k}$, et à aucun autre, alors p appartient à celui parmi les $\bar{A}$ qui a dans la représentation ci-dessus $A_{i_1} A_{i_2} \ldots A_{i_k}$ comme premier terme.

L'ensemble A peut donc être représenté comme une somme des ensembles disjoints $\bar{A}_i$ $(i = 1, 2, \ldots 2^m - 1)$. Tout l'ensemble A_ν possède une représentation analogue aussi. Il suffit dans ce but de sommer tous les ensembles $\bar{A}_i$ qui ont dans le premier terme de leur représentation l'indice ν.

L'ensemble $C = A_1 \times B_1 + A_2 \times B_2 + \ldots + A_m \times B_m$ peut donc être écrit

$$C = \left(\sum_{\mu=1}^{l_1} \bar{A}_\mu^1\right) \times B_1 + \left(\sum_{\mu=1}^{l_2} \bar{A}_\mu^2\right) \times B_2 + \ldots \left(\sum_{\mu=1}^{l_m} \bar{A}_\mu^m\right) \times B_m.$$

Si l'on renferme les termes appartenants aux mêmes $\bar{A}_i$, on obtient:

$$C = \bar{A}_1 \times \left(\sum_{\nu=1}^{m_1} B_\nu^1\right) + \bar{A}_2 \times \left(\sum_{\nu=1}^{m_2} B_\nu^2\right) + \ldots + \bar{A}_{2^m-1} \times \left(\sum_{\nu=1}^{m_{2^m-1}} B_\nu^{2^m-1}\right).$$

Observons que les ensembles B_i^s $(i=1, 2, \ldots m_s)$ sont disjoints et que $\sum_{\nu=1}^{m_s} B_\nu^s = B$. En effet, dans toute somme $\sum_{\nu=1}^{m_s} B_\nu^s$ figurent seulement ces B_ν, qui ont des indices qu'on trouve parmi les indices du premier terme (non vide) de l'ensemble $\bar{A}_s$. S'il existait donc des points communs pour B_λ et B_μ on aurait étant donné que $\bar{A}_s \neq 0$ et en particulier $A_\lambda \cdot A_\mu \neq 0$ une contradiction, car notre supposition $(A_\lambda \times B_\lambda) \cdot (A_\mu \times B_\mu) \neq 0$ implique pour $A_\lambda A_\mu \neq 0$ que $B_\lambda B_\mu = 0$.

Il est évident que $\sum_{\nu=1}^{m_s} B_\nu^s$ contient l'ensemble B tout entier.

Notamment si $b \in B$, alors, pour $a \in \bar{A}_s$, le couple $(a, b) \in A \times B$. Dans notre représentation de ce produit cet élément peut, en vertu du fait que les $\bar{A}_s$ sont disjoints, appartenir seulement au produit $\bar{A}_s \times \sum_{\nu=1}^{m_s} B_\nu^s$, d'où il résulte $b \in \sum_{\nu=1}^{m_s} B_\nu^s$. On a donc $m(\sum_{\nu=1}^{m_s} B_\nu^s) = m(B)$.

En s'appuyant sur l'additivité de la mesure dans les espaces E_1 et E_2, on obtient:

$$m(A_1) \cdot m(B_1) + m(A_2) \cdot m(B_2) + \ldots + m(A_m) \cdot m(B_m) =$$

$$= \sum_{\mu=1}^{l_1} m(\bar{A}_\mu^1) \cdot m(B_1) + \sum_{\mu=1}^{l_2} m(\bar{A}_\mu^2) \cdot m(B_2) + \ldots + \sum_{\mu=1}^{l_m} m(\bar{A}_\mu^m) \cdot m(B_m) =$$

$$= m(\bar{A}_1) \cdot \left(\sum_{\nu=1}^{m_1} m(B_\nu^1) \right) + m(\bar{A}_2) \cdot \left(\sum_{\nu=1}^{m_2} m(B_\nu^2) \right) + \ldots$$

$$+ \ldots m(\bar{A}_{2^m-1}) \cdot \left(\sum_{\nu=1}^{m_{2^m-1}} m(B_\nu^{2^m-1}) \right) = m(B) \cdot \sum_{\lambda=1}^{2^m-1} m(\bar{A}_\lambda) = m(A) \cdot m(B).$$

Lemme 2[20]. Soit $C = A \times B = \sum_{\nu=1}^{\infty} A_\nu \times B_\nu$, et $(A_i \times B_i) \cdot (A_j \times B_j) = 0$ pour $i \neq j$. Alors

$$m(C) = m(A) \cdot m(B) = \sum_{\nu=1}^{\infty} m(A_\nu) \cdot m(B_\nu).$$

D é m o n s t r a t i o n. Nous allons prouver d'abord que

$$\sum_{\nu=1}^{\infty} m(A_\nu) \cdot m(B_\nu) \leqslant m(A) \cdot m(B).$$

[20]) Ce lemme est très général. Sa démonstration n'utilise que l'additivé complète de la mesure. Il peut donc servir pour la construction dans le produit des mesures autres que celle de L e b e s g u e.

Il suffit de prouver que cette inégalité est remplie pour tout N:

$$\sum_{\nu=1}^{N} m(A_\nu) \cdot m(B_\nu) \leqslant m(A) \cdot m(B).$$

C'est une conséquence simple du lemme 1. On doit dans ce but compléter l'ensemble $\sum_{\nu=1}^{N} A_\nu \times B_\nu$ par une somme finie des ensembles-produits au produit $A \times B$. On est alors dans le cas du lemme 1 et en tenant compte du fait que la mesure est non négative, on obtient l'inégalité demandée.

Pour prouver que

$$\sum_{\nu=1}^{\infty} m(A_\nu) \cdot m(B_\nu) \geqslant m(A) \cdot m(B)$$

nous allons décomposer l'ensemble $A = \Sigma \bar{A}_\nu$ comme dans la dém. du lemme 1. Soit $\{k_i\}$ un système (fini ou infini) de nombres naturels. Soit $\bar{A}\{k_i\}$ l'ensemble de tous les points de A qui appartiennent aux ensembles $A_{k_1}, A_{k_2} \ldots, A_{k_i} \ldots$ et à aucun autre ensemble. En général on obtient de telle façon une décomposition de A en un c o n t i n u d'ensembles disjoints.

Soit $\{k_i\}$ une suite donnée. Les ensembles $B_{k_1}, B_{k_2}, \ldots, B_{k_l} \ldots$ sont disjoints. Si, en effet, il existait un point b tel que $b \in B_{k_i}$, $b \in B_{k_j}$ $(i \neq j)$, on déduirait du fait que l'ensemble $A_{k_1} A_{k_2} \ldots A_{k_i} \ldots A_{k_j} \ldots$ est non-vide l'existence d'un couple (a, b) qui appartiendrait aux ensembles $A_{k_i} \times B_{k_i}$ et $A_{k_j} \times B_{k_j}$, contrairement à notre hypothèse.

On a $\sum_{\nu=1}^{\infty} B_{k_\nu} = B$. En effet, si le point $b \in B$ n'appartenait à aucun B_{k_i}, on choisirait le point a de l'ensemble $A\{k_i\}$ et le couple (a, b) ne serait contenu dans aucun ensemble $A_i \times B_i$, contrairement à l'hypothèse.

La mesure dans les espaces E_1 et E_2 est complètement additive. On peut donc trouver pour tout $\varepsilon > 0$ et pour toute suite $\{k_i\}$ donnée un indice $N\{k_i\}$ tel qu'on ait

$$\sum_{\nu=1}^{N\{k_i\}} m(B_{k_\nu}) = m\left(\sum_{\nu=1}^{N\{k_i\}} B_{k_\nu}\right) > m(B) - \frac{\varepsilon}{m(A)}.$$

Soit $N\{k_i\}$ le plus petit nombre jouissant de cette propriété.

Au lieu de la classe des suites infinies $\{k_i\}=(k_1, k_2, \ldots, k_i, \ldots)$, qui est en général de la puissance du continu, considérons la classe au plus dénombrable des suites finies, obtenues de la précédente en remplaçant toute suite infinie par la suite correspondante finie $\{l_i\}=(k_1, k_2, \ldots, k_{N\{k_i\}})$.

A toute suite $\{l_i\}$ de cette classe correspond un ensemble $A^*_{\{l_i\}}$ qui est, par définition, la somme des ensembles $\bar{A}\{k_i\}$ dont les suites donnent par notre procédé la suite $\{l_i\}$. Observons que l'ensemble A est de la forme $A=\sum\limits_{\{l_i\}} A^*_{\{l_i\}}$, les ensembles $A^*_{\{l_i\}}$, comme il est facile de voir, étant disjoints.

Désignons par A^s_μ $(\mu=1, 2, \ldots)$ tous les ensembles $A^*_{\{l_i\}}$ dans la suite caractéristique desquels figure l'indice s. On voit que $\sum\limits_{\mu=1}^{\infty} A^s_\mu \subset A_s$, c. à d. que $m(A_s) \geqslant m(\sum\limits_{\mu=1}^{\infty} A^s_\mu)$.

Nous avons donc

$$\sum_{\nu=1}^{\infty} m(A_\nu)\cdot m(B_\nu) \geqslant \sum_{\nu=1}^{\infty} m\left(\sum_{\mu=1}^{\infty} A^\nu_\mu\right)\cdot m(B_\nu) = \sum_{\nu=1}^{\infty}\sum_{\mu=1}^{\infty} m(A^\nu_\mu)\cdot m(B_\nu),$$

la dernière égalité résultant de l'additivité complète de la mesure dans les espaces E_1 et E_2.

En rangeant la dernière somme suivant les ensembles A^ν_μ, c. à d. $A^*_{\{l_i\}}$, on obtient

$$\sum_{\{l_i\}} m(A^*_{\{l_i\}})\cdot m\left(\sum_{\nu=1}^{\infty} B^{\{l_i\}}_\nu\right).$$

où $\{B^{\{l_i\}}_\nu\}$ est la suite des ensembles dans la somme double qui sont multipliés par $A^*_{\{l_i\}}$

L'ensemble $\sum\limits_{\nu=1}^{\infty} B^{\{l_i\}}_\nu$ qui appartient à l'ensemble $A^*_{\{l_i\}}$ est égal à l'ensemble $\sum\limits_{i=1}^{N\{k_i\}} B_{k_i}$. On a donc

$$\sum_{\nu=1}^{\infty} m(A_\nu)\cdot m(B_\nu) \geqslant \sum_{\{l_i\}} m(A^*_{\{l_i\}})\cdot m\left(\sum_{i=1}^{N\{k_i\}} B_{k_i}\right) \geqslant$$

$$\geqslant \sum_{\{l_i\}} m(A^*_{\{l_i\}})\cdot\left(m(B)-\frac{\varepsilon}{m(A)}\right) = m(A)\cdot\left(m(B)-\frac{\varepsilon}{m(A)}\right) =$$

$$= m(A)\cdot m(B) - \varepsilon.$$

Démonstration du th. 1.

Soit $Z \subset E = E_1 \times E_2$ un ensemble-produit, c. à d. $Z = A \times B$ où $A \subset E_1$, $B \subset E_2$. Nous posons $m(Z) = m(A) \cdot m(B)$.

Si l'ensemble Z est de la forme $Z = A_1 \times B_1 + A_2 \times B_2 + \dots$ où $(A_i \times B_i) \cdot (A_j \times B_j) = 0$ pour $i \neq j$, nous posons $m(Z) = \sum_{\nu=1}^{\infty} m(A_\nu)\, m(B_\nu)$.

On conclut des lemmes 1 et 2 que pour les ensembles de cette forme la mesure est définie d'une manière univoque indépendamment de la manière suivant laquelle Z est décomposé en une somme de produits. Soient données, en effet, deux représentations de l'ensemble Z:

$$Z = A_1 \times B_1 + \dots + A_m \times B_m + \dots \qquad (A_i \subset E_1,\ B_i \subset E_2)$$

$$Z = C_1 \times D_1 + \dots + C_n \times D_n + \dots \qquad (C_i \subset E_1,\ D_i \subset E_2).$$

Nous allons prouver que

$$\sum_{\nu=1}^{\infty} m(A_\nu) \cdot m(B_\nu) = \sum_{\nu=1}^{\infty} m(C_\nu) \cdot m(D_\nu).$$

Formons dans ce but tous les ensembles des points communs aux ensembles $A_i \cdot C_j$ et $B_i \cdot D$. On voit que

$$\begin{aligned} Z &= A_1 C_1 \times B_1 D_1 + A_1 C_2 \times B_1 D_2 + \dots + A_1 C_n \times B_1 D_n + \dots \\ &= A_2 C_1 \times B_2 D_1 + A_2 C_2 \times B_2 D_2 + \dots + A_2 C_n \times B_2 D_n + \dots \\ &+ \dots \qquad\qquad\qquad\qquad \dots \qquad\qquad \dots \\ &+ A_m C_1 \times B_m D_1 + A_m C_2 \times B_m D_2 + \dots + A_m C_n \times B_m D_n + \dots \\ &+ \dots \end{aligned}$$

On obtient donc une décomposition de l'ensemble Z en une somme d'ensembles disjoints. Le lemme 2 appliqué aux décompositions des ensembles $A_i \times B_i$ et $C_i \times D_i$ nous donne, si l'on pose $\sigma_{\lambda\mu} = m(A_\lambda C_\mu \times B_\lambda D_\mu)$:

$$m(Z) = \sum_{\lambda,\mu} \sigma_{\lambda\mu} = \sum_{\lambda=1}^{\infty} \sum_{\mu=1}^{\infty} \sigma_{\lambda\mu} = \sum_{\nu=1}^{\infty} m(A_\nu \times B_\nu)$$

$$m(Z) = \sum_{\lambda,\mu} \sigma_{\lambda\mu} = \sum_{\mu=1}^{\infty} \sum_{\lambda=1}^{\infty} \sigma_{\lambda\mu} = \sum_{\mu=1}^{\infty} m(C_\nu \times D_\nu).$$

S'il s'agit de la définition de la mesure pour d'autres ensembles dans $E_1 \times E_2$ qui doivent être mesurables d'après nos axiomes, on peut procéder comme dans la théorie de la mesure n-dimensionelle de Lebesgue: Les sommes d'un nombre fini d'ensembles-produits jouent le rôle des „polygones" ou „figures élémentaires", les sommes d'un nombre dénombrable d'ensembles-produits jouent le rôle des ensembles „ouverts".

La mesure est bien définie pour les ensembles élémentaires et pour les ensembles ouverts. Si un ensemble F est „fermé" c.-à-d. est complément CZ d'un ensemble ouvert Z à mesure $m(Z)$ nous posons $M(F) = 1 - m(Z)$.

D'une manière analogue, si un ensemble P est somme dénombrable d'ensembles disjoints fermés, $P = \sum_{\nu=1}^{\infty} F_\nu$, nous rangeons l'ensemble P parmi les ensembles mesurables et nous posons $m(P) = \sum_{\nu=1}^{\infty} m(F_\nu)$. On définit pareillement une mesure pour les ensembles „G_δ" (qui sont partie commune d'un nombre dénombrable d'ensembles ouverts).

Si, enfin, l'ensemble P est de la forme $P = M + N$ où M est un ensemble G_δ et $N \subseteq \overline{M}$, $\overline{M}$ étant un ensemble G_δ à mesure 0, nous considérons P comme mesurable et posons $m(P) = m(M)$.

On voit facilement que les démonstrations de l'unicité d'une telle définition de la mesure et la vérification des postulats ne sont qu'une répetition verbale des démonstrations respectives dans la théorie de la mesure de Lebesgue.

Remarques. La classe $\mathfrak{B}$ des ensembles „boreliens" c.-à-d. la plus petite classe des ensembles qui contient les produits $X \times Y$ des ensembles mesurables et contient avec deux ensembles V et W leur différence $V - W$ ainsi qu'avec une suite des ensembles $\{V_i\}$ contient leur somme $\sum_{\nu=1}^{\infty} V_\nu$ — est contenue dans la classe des ensembles mesurables $\mathfrak{M}$.

On peut classifier les ensembles boreliens dans notre sens à l'aide des nombres ordinaux transfinis de la deuxième classe de Cantor, d'une manière analogue au procédé employé pour les ensembles boreliens dans le sens ordinaire. Mais il y a des différences essentielles entre la classe $\mathfrak{B}$ et la classe des ensembles boreliens dans le sens ordinaire dans le carré (si l'on admet p. ex. que

$E_1 = E_2 =$ l'intervalle $0 \leqslant x \leqslant 1$). La classe $\mathfrak{B}$ est de la puissance 2^c (la classe des ensembles-produits étant déjà de cette puissance!)

Il y a plusieurs problèmes non résolus relatifs à la notion de l'ensemble „borelien" par rapport aux ensembles-produits. P. ex. on ne sait pas s'il existe des ensembles qui soient d'une classe $\alpha < \Omega$ précise (c.-à-d. qui n'appartiennent pas à une classe β avec $\beta < \alpha$).

4. Mesure complètement additive dans les produits infinis [21]).

Dans l'étude des produits infinis on doit évidemment remplacer les postulats (V) et (3) par

(V') Si $X_i \in \mathfrak{M}_i$ pour $i = 1, 2 \ldots$ l'ensemble $\mathop{\mathfrak{P}}\limits_{\nu=1}^{\infty} X_\nu \in \mathfrak{M}$

$$(3') \qquad m\left(\mathop{\mathfrak{P}}\limits_{\nu=1}^{\infty} X_\nu\right) = \prod_{\nu=1}^{\infty} m(X_\nu).$$

Théorème 2. *Soit $\{E_i\}$ une suite d'espaces dans lesquels est définie une mesure remplissant les postulats* (I)—(IV) *et* (1), (2). *Il existe dans $E = \mathop{\mathfrak{P}}\limits_{\nu=1}^{\infty} E_\nu$ une mesure remplissant les postulats* (I)—(IV), (V') *et* (1), (2), (3').

D é m o n s t r a t i o n. Nous démontrerons deux lemmes (3 et 4) analogues aux lemmes 1, 2.

Les lemmes 3 et 4 établis, on peut pour achever la démonstration répéter verbalement les raisonnements de la démonstration du th 1.

Lemme 3. Soit $Z = \mathop{\mathfrak{P}}\limits_{\mu=1}^{\infty} A_\mu = \sum_{\nu=1}^{m} \mathop{\mathfrak{P}}\limits_{\mu=1}^{\infty} A_\mu^\nu$, où $(\mathop{\mathfrak{P}}\limits_{\mu=1}^{\infty} A_\mu^j) \cdot (\mathop{\mathfrak{P}}\limits_{\mu=1}^{\infty} A_\mu^i) = 0$ pour $i \neq j$. Pour les ensembles Z de la forme $\mathop{\mathfrak{P}}\limits_{\mu=1}^{\infty} A_\mu$ posons

$$m(Z) = \prod_{\mu=1}^{\infty} m(A_\mu).$$

[21]) Une étude approfondie de la mesure dans quelques produits infinis se trouve dans les travaux de MM. W. F e l l e r et E. T o r n i e r: Maß- und Inhaltstheorie des Baire'schen Nullraumes. Math. Ann. Bd. 107, S. 165—187, et de E. T o r n i e r, *Grundlagen der Wahrscheinlichkeitsrechnung*, Acta math. 1933, P. 239—380.

On a

$$m(Z) = \sum_{\nu=1}^{m} \prod_{\mu=1}^{\infty} m(A_\mu^\nu).$$

D é m o n s t r a t i o n. Il existe un indice $N(i,j)$ tel que les ensembles $(\mathfrak{P}_{\mu=1}^{N(i,j)} A_\mu^j) \cdot (\mathfrak{P}_{\mu=1}^{N(i,j)} A_\mu^i) = 0$ pour $i \neq j$. En effet, si pour tout n il existait des points communs des ensembles $\mathfrak{P}_{\mu=1}^{n} A_\mu^j$ et $\mathfrak{P}_{\mu=1}^{n} A_\mu^i$, on aurait des points communs pour les ensembles $\mathfrak{P}_{\mu=1}^{\infty} A_\mu^j$ et $\mathfrak{P}_{\mu=1}^{\infty} A_\mu^i$, contrairement à l'hypothèse.

Désignons par $\bar{N} = \max.\ N(i,j)$ (pour $i = 1 \ldots m$; $j = 1 \ldots m$). Soit $\varepsilon > 0$. Choisissons des entiers $n_0, n_1, \ldots n_m$ tels que pour $n > n_0$

$$\left| \prod_{\mu=1}^{n} m(A_\mu) - \prod_{\mu=1}^{\infty} m(A_\mu) \right| < \varepsilon,$$

et que pour $n > n_i\ (i = 1, \ldots m)$

$$\left| \prod_{\mu=1}^{n} m(A_\mu^i) - \prod_{\mu=1}^{\infty} m(A_\mu^i) \right| < \frac{\varepsilon}{m}.$$

Soit $N = \max.(n_0, n_1, \ldots n_m, \bar{N})$ Considérons l'espace $\mathfrak{P}_{\nu=1}^{N} E_\nu$ et la somme des ensembles disjoints $\sum_{\nu=1}^{m} \mathfrak{P}_{\mu=1}^{N} A_\mu^\nu$. On a d'après le lemme 1 (appliqué au produit de N espaces au lieu de deux)

$$\sum_{\nu=1}^{m} m(\mathfrak{P}_{\mu=1}^{N} A_\mu^\nu) = m(\mathfrak{P}_{\mu=1}^{N} A_\mu).$$

Le premier membre de cette égalité satisfait à la condition

$$\left| \sum_{\nu=1}^{m} m(\mathfrak{P}_{\mu=1}^{N} A_\mu^\nu) - \sum_{\nu=1}^{m} m(\mathfrak{P}_{\mu=1}^{\infty} A_\mu^\nu) \right| < \varepsilon$$

et le deuxième à la condition

$$\left| m(\mathfrak{P}_{\mu=1}^{N} A_\mu) - m(\mathfrak{P}_{\mu=1}^{\infty} A_\mu) \right| < \varepsilon,$$

d'où

$$\left| m(Z) - \sum_{\nu=1}^{m} \prod_{\mu=1}^{\infty} m(A_\mu) \right| < 2\varepsilon \qquad \text{c. q. f. d.}$$

Corollaire du lemme 2. *Si* $A \times B \subset \sum_{\nu=1}^{\infty} C_\nu \times D_\nu$ *(où* A, $C_i \subset E_1$ *et* $B, D_i \subset E_2$ *et où les ensembles dans la somme du deuxième membre ne sont nécessairement disjoints), on peut trouver pour tout* $\varepsilon > 0$ *un entier* M *tel que* $m(A \times B) < \sum_{\nu=1}^{M} m(C_\nu \times D_\nu) + \varepsilon$.

En effet, en posant $\overline{C}_i = A \cdot C_i$, $\overline{D}_i = B \cdot D_i$, on a $A \times B = \sum_{\nu=1}^{\infty} \overline{C}_\nu \times \overline{D}_\nu$. Au lieu des ensembles $\overline{C}_i \times \overline{D}_i$ (qui ne sont pas nécessairement disjoints) on peut introduire des ensembles M_i en posant p. ex. $M_i = \overline{C}_i \times \overline{D}_i - \sum_{\mu=1}^{i-1} M_\mu$, d'où $M_i \subset \overline{C}_i \times \overline{D}_i$ et en vertu du lemme 1 $m(M_i) \leqslant m(\overline{C}_i \times \overline{D}_i)$. En outre $A \times B = \sum_{\nu=1}^{\infty} M_\nu$

D'après le lemme 2

$$m(A \times B) = \sum_{\nu=1}^{\infty} m(M_\nu) < \sum_{\nu=1}^{M} m(M_\nu) + \varepsilon \leqslant \sum_{\nu=1}^{M} m(\overline{C}_\nu \times \overline{D}_\nu) + \varepsilon =$$

$$= \sum_{\nu=1}^{M} m(\overline{C}_\nu) \cdot m(\overline{D}_\nu) + \varepsilon \leqslant \sum_{\nu=1}^{M} m(C_\nu) \cdot m(D_\nu) + \varepsilon \quad \text{c. q. f. d.}$$

Lemme 4. Soit $Z = \mathfrak{P}_{\mu=1}^{\infty} A_\mu = \sum_{\nu=1}^{\infty} \mathfrak{P}_{\mu=1}^{\infty} A_\mu^\nu$ où $(\mathfrak{P}_{\mu=1}^{\infty} A_\mu^i) \cdot (\mathfrak{P}_{\mu=1}^{\infty} A_\mu^j) = 0$ pour $i \neq j$. On a

$$m(Z) = \prod_{\mu=1}^{\infty} m(A_\mu) = \sum_{\nu=1}^{\infty} \prod_{\mu=1}^{\infty} m(A_\mu^\nu).$$

D é m o n s t r a t i o n. Il résulte du lemme 3. que la série $\sum_{\nu=1}^{\infty} [\prod_{\mu=1}^{\infty} m(A_\mu^\nu)]$ converge vers une somme $\leqslant m(Z)$.

Pour prouver l'inégalité inverse observons que pour N suffisamment grand on a

$$\left| m(Z) - \prod_{\mu=1}^{N} m(A_\mu) \right| < \varepsilon.$$

L'ensemble $\mathop{\mathfrak{P}}\limits_{\mu=1}^{N} A_\mu$ est contenu dans l'ensemble $\sum\limits_{\nu=1}^{\infty} \mathop{\mathfrak{P}}\limits_{\mu=1}^{N} A_\mu^\nu$. A l'aide du corollaire du lemme 2 (appliqué au cas du produit de N au lieu de 2 espaces) on obtient

$$\prod_{\mu=1}^{N} m(A_\mu) - \sum_{\nu=1}^{M} m(\mathop{\mathfrak{P}}_{\mu=1}^{N} A_\mu^\nu) < \varepsilon$$

où

$$m(Z) < \sum_{\nu=1}^{M} \prod_{\mu=1}^{N} m(A_\mu^\nu) + 2\,\varepsilon.$$

Observons que, pour un $\varepsilon > 0$ donné, on peut trouver un entier N_i tel que, pour $n > N_i$, on a

$$|\, m(\mathop{\mathfrak{P}}_{\mu=1}^{\infty} A_\mu^i) - m(\mathop{\mathfrak{P}}_{\mu=1}^{n} A_\mu^i) \,| < \frac{\varepsilon}{M}.$$

Si l'on choisit N de telle façon que $N = \max.\ (N_1, \ldots, N_M)$, on a

$$m(Z) < \sum_{\nu=1}^{M} \prod_{\mu=1}^{\infty} m(A_\mu^\nu) + 3\,\varepsilon$$

ce qui achève la démonstration.

5. Mesure dans les produits à additivité finie.

Dans les numéros précédents nous avons construit, en partant de la mesure complètement additive dans les espaces composants, une mesure complètement additive dans les produits. Notre but est maintenant de construire une mesure dans les produits en partant d'une mesure additive au sens fini.

Dans le cas d'une telle mesure il est superflu de parler de la classe des ensembles mesurables. En effet, si une mesure est donnée pour une classe $\mathfrak{M}$ d'ensembles et remplit les postulats (I), (II), (III') et (1), (2') on peut la prolonger pour tous les sous-ensembles de l'espace. Ce fait est une conséquence du théorème de M. Banach sur le prolongement d'une fonctionelle additive[22]). On sait que la mesure d'un ensemble peut être considérée comme l'intégrale de la fonction caractéristique de cet ensemble. Dans l'en-

[22]) Banach, *Théorie des opérations linéaires,* Ch. II § 2, Warszawa 1932.

semble des fonctions réelles définies sur E, les combinaisons des fonctions caractéristiques constituent un ensemble linéaire, la mesure est, en vertu des proprietés (I), (II), (III′) et (1), (2′), une fonctionelle additive à norme 1, définie pour cet ensemble. Une telle fonctionelle peut être prolongée pour l'ensemble de toutes les fonctions définies sur E. En revenant à l'interprétation de la mesure, nous obtenons une mesure prolongée comme valeur de la fonctionelle pour les fonctions caractéristiques des ensembles respectifs.

Théorème 3. *Soient E_1 et E_2 des espaces à mesure remplissant les conditions* (1), (2′). *On peut définir dans $E = E_1 \times E_2$ une mesure pour tous les sous-ensembles qui remplit les postulats donnés ci-dessus et le postulat* (3).

Démonstration. Pour un ensemble dans E qui est un produit ou une somme d'un nombre fini de produits on définit une mesure de la même manière que dans le th 1. Il résulte du lemme 1, que cette mesure remplit les axiomes (1), (2′), (3). Somme et différence des figures élémentaires étant aussi des figures élémentaires on est dans les conditions du théorème sur le prolongement de la mesure, ce qui prouve notre théorème.

Remarque. D'après le théorème 3 seule la mesure des ensembles-produits et sommes finies de tels ensembles est définie d'une manière univoque par le postulat (3) de probabilité composée. C'est une classe d'ensembles très restreinte. Une définition de la mesure pour tous les autres sous-ensembles, bien que possible en vertu du théorème sur le prolongement, peut être effectuée de beaucoup de manières différentes. Or, on peut prouver qu'il existe dans $E_1 \times E_2$ des ensembles qui ne sont pas sommes finies d'ensembles-produits et qui prennent la même mesure pour tout prolongement. Il serait intéressant de trouver tous les ensembles dans $E_1 \times E_2$ qui jouissent de cette propriété. On peut dire que pour de tels ensembles la règle des probabilités composées entraîne une mesure univoque, les mesures de tous les autres ensembles n'étant pas encore determinées par cette règle.

Théorème 4. *Soit $\{E_i\}$ une suite d'espaces à mesure remplissant les postulats* (1), (2′). *On peut définir une mesure dans $\mathfrak{P}_{\nu=1}^{\infty} E_\nu$ remplissant les conditions* (1), (2′), (3′).

La démonstration est analogue à celle du théorème 3. La remarque sur le th. 3 conserve aussi sa validité.

II.

Nous allons montrer sur quelques exemples comment on emploie les définitions et théorèmes des paragraphes précédents dans les problèmes élémentaires du calcul des probabilités. Toutes les opérations combinatoires mentionnées dans la partie première seront illustrées [24]).

1. Les produits finis.

Exemple 1. *Produit des espaces à un nombre fini de points.*

Probabilité d'obtenir avec deux dés un nombre pair. Les espaces dans lesquels la probabilité (mesure) est donnée sont $E_1 = E_2$, composés chacun de six éléments. On admet habituellement que le jeu est „juste" c.-à-d. que chaque élément considéré comme un ensemble a la mesure $\frac{1}{6}$. La probabilité cherchée est évidemment la mesure de l'ensemble de ces points dans $E_1 \times E_2$, qui ont une somme paire de „coordonnées".

Le mesurabilité de cet ensemble est dans notre exemple élémentaire évidente *à priori*, l'espace $E_1 \times E_2$ étant composé d'un nombre fini de points. La nature combinatoire de l'ensemble, la mesure duquel est cherchée est souvent très compliquée, même dans les problèmes les plus simples, car leur interprétation mène à l'opération du produit prise un grand nombre de fois.

Exemple 2. *Produit d'espaces dénombrables.*

La probabilité pour que la somme de deux nombres naturels, pris au hasard, soit paire. Pour préciser notre problème on doit connaître les probablités „données" c.-à.-d. il faut admettre une mesure dans les espaces $E_1 = E_2 =$ ensemble de tous les nombres naturels. (Une telle mesure peut naturellement être introduite de différentes manières). On a à calculer dans $E_1 \times E_2$ la mesure de l'ensemble des points pour lesquels la somme des coordonnées est paire. Or, l'ensemble en question est mesurable dans le sens de nos théorèmes. Il se laisse représenter comme somme d'ensembles de la forme $X \times Y$. (On prendra pour ces ensembles: 1⁰: l'ensem-

[23]) V. les règles du calcul p. 243.

[24]) Cf. aussi la classification des problèmes de M. Borel, l. c. [2])

ble des (x, y) pour lesquels x et y sont pairs et 2°: l'ensemble des (x, y) où x et y sont impairs). La mesure adoptée dans E_1 et E_2 est habituellement telle que l'ensemble des nombres paires a pour mesure $\frac{1}{2}$. Il s'ensuit que la probabilité cherchée est égale à $\frac{1}{2}\cdot\frac{1}{2}+$ $+\frac{1}{2}\cdot\frac{1}{2}=\frac{1}{2}$.

Considérons cependant un autre problème, celui de Tchebycheff: la probabilité pour que deux nombres pris au hasard n'aient pas de diviseur commun. On cherche évidemment la mesure de l'ensemble des couples (x, y) dans $E_1 \times E_2$ pour lesquels x, y n'ont pas de diviseur commun. Or, cet ensemble n'appartient pas à la classe des ensembles mesurables dans le sens de notre théorie si la mesure donnée est d'additivité seulement finie, ce qui est le plus naturel dans les problèmes où l'espace est dénombrable [25]). Nous ne pouvons donc parler de ce problème comme d'un problème bien posé du point de vue de notre théorie. Les solutions habituelles de ce problème et des problèmes analogues ne sont, au fond, qu'un calcul de la limite des probabilités respectives correspondantes à l'approximation de l'espace total dénombrable par les ensembles finis.

Exemple 3. *Produit d'espaces continus* [26]).

La probabilité pour que deux points de la surface d'une sphère aient une distance plus grande qu'un nombre donné a. Il s'agit du produit de deux surfaces de la sphère. Le théorème 1 permet de définir une mesure pour cette variété à quatre dimensions située dans l'espace euclidien à 6 dimensions, en partant des mesures données sur la surface de la sphère (p. ex. étant égale à l'aire dans le sens ordinaire).

Exemple 4. *Probabilité pour que trois droites sur le plan forment un triangle aux angles aigus.*

L'espace à mesure donnée est ici l'ensemble de toutes les droites du plan. On peut définir une mesure dans cet ensemble (Crofton, Deltheil) [27]) et cela même dans un certain sens univoque [28]). Il

[25]) Cet ensemble n'est pas une somme finie d'ensembles-produits. V. aussi l'exemple 9.

[26]) Le livre de M. Deltheil (l. c. [5]) contient un grand nombre de problèmes de ce genre. Il serait instructif de les formuler dans le langage des produits.

[27]) Cf. Deltheil, l. c. [5]).

[28]) J. Schreier et S. Ulam, *Sur une propriété de la mesure de M. Lebesgue*, C. R. 1930.

s'agit de la mesure dans le produit de trois espaces de ce genre, notre problème concernant trois droites. On obtient dans ce produit une mesure d'après le théorème 1; notre problème est ainsi bien défini.

Un simple calcul montre que dans le produit l'ensemble correspondant aux droites formant des triangles de la propriété demandée a pour mesure le nombre $\frac{1}{6}$

Dans cet exemple, comme d'ailleurs dans tous le cas précedents, les théorèmes de la partie première assurent d'avance l'unicité de la solution du problème. S'il s'agit du calcul effectif des probabilités cherchées, ces théorèmes justifient les procédés ordinaires, consistant à calculer la mesure de l'ensemble correspondant par une décomposition en sous-ensembles convenablement choisis, indépendamment de la manière suivant laquelle cette décomposition était effectuée.

2. Les produits infinis.

Exemple 5. *Produit d'espaces finis.*

Dans le jeu de pile ou face la partie sera gagnée par un joueur, s'il a réussi de gagner k jeux consécutifs. On demande quelle est la probabilité pour que le jeu soit terminé?

Les espaces à mesure donnée sont les espaces E_i, chacun composé de deux éléments (pile — 0 et face — 1). La mesure donnée est constante: p pour l'élément 1 et $q = 1 - p$ pour l'élément 0. Il s'agit de la mesure de l'ensemble Z, situé dans l'espace $\mathfrak{P}_{\nu=1}^{\infty} E_\nu$ de toutes les suites des chiffres 0 et 1, composé des suites dans lesquelles il existe des séries de k zéros ou de k unités consécutives. Cet ensemble est

$$Z = \sum_{\nu=0}^{\infty} Z_\nu \quad \text{où} \quad Z_\nu = E\,[x_{\nu+1} = x_{\nu+2} = \ldots = x_{\nu+k}].$$

L'ensemble Z contient l'ensemble $\sum_{s=0}^{\infty} Z_{s\cdot k}$.

Si l'on désigne par

$$A_0 = Z_0, \quad A_s = Z_{s\cdot k} \cdot \prod_{j=0}^{s-1} C Z_{j\cdot k}, \qquad (s = 1, 2 \ldots),$$

les ensembles A_s sont disjoints et l'on a

$$Z \supset \sum_{s=0}^{\infty} A_s$$

et par conséquent $m(Z) \geqslant \sum_{s=0}^{\infty} m(A_s)$.

On voit aussitôt que

$$m(Z_{s \cdot k}) = p^k + q^k = \alpha < 1;$$

$$m(A_s) = (p^k + q^k) \cdot (1 - p^k - q^k)^s,$$

d'où

$$m(Z) \geqslant \sum_{s=0}^{\infty} \alpha (1 - \alpha)^s = 1.$$

Cela veut dire qu'il est „presque certain" que le jeu sera terminé en un temps fini.

Exemple 6. Le jeu entre deux personnes est le suivant: Le gain de deux parties consécutives dans le jeu de pile ou face décide sur le gain du „game" (de I-ier ordre). Si l'on gagne deux „games" consécutifs on gagne le „game de II-ième ordre" — et ainsi de suite. Avec le „game de n-ième ordre" on gagne le jeu. On cherche la probabilité pour que ce jeu se prolonge indéfiniment. Les espaces dans lesquels la mesure est donnée sont les mêmes que dans l'exemple 5. Ici, il s'agit de la mesure dans l'espace

$$E^{(n)} = \mathfrak{P}_{\nu=1}^{\infty} E_\nu^{(n-1)}, \quad \text{où} \quad E_i^j = \mathfrak{P}_{\nu=1}^{\infty} E_\nu^{(j-1)} \qquad (j = 2, 3, \ldots n), \ (i = 1, \ldots)$$

$$E^1 = \mathfrak{P}_{\nu=1}^{\infty} E_\nu \text{ où } E_i = E \text{ de l'exemple précédent } (i = 1, 2, \ldots).$$

On obtient de la mesure donnée dans les espaces $E_i^{(j-1)}$ les mesures pour les espaces nouveaux $E_i^{(j)}$, d'une manière analogue au procédé de l'exemple 5. Un simple calcul montre que l'ensemble qui correspond à tous les jeux gagnés est mesurable et sa mesure est $\dfrac{1}{1+\left(\dfrac{q}{p}\right)^{2n}}$; pour les jeux perdus on obtient $\dfrac{1}{1+\left(\dfrac{p}{q}\right)^{2n}}$; l'ensemble qui correspond aux parties jouées indéfiniment a la mesure 0. Par suite il est „presque certain" que le jeu sera décidé.

Exemple 7. *La ruine des joueurs.* Ce problème classique[29]) conduit à la considération des produits infinis. Dans le schéma mathématique le plus simple, les espaces E_i à mesure donnée se composent chacun de deux éléments $(-1, +1)$, et soit la mesure dans E_i $m(-1) = q$, $m(+1) = p$. Soit la fortune du joueur A de a, du joueur B de b unités.

Il s'agit de la mesure dans l'espace $\mathop{\mathfrak{P}}\limits_{\nu=1}^{\infty} E_\nu$ de l'ensemble

$$Z_A = \sum_{n=s}^{\infty} Z_n \text{ où } Z_n = \boldsymbol{E}[b > s_i > -a, \text{ pour } i < n;\ s_n \geqslant b]$$

s_i désignant la somme des i premiers termes de la suite.

En se servant de l'existence de la mesure dans $\mathop{\mathfrak{P}}\limits_{\nu=1}^{\infty} E_\nu$, on peut raisonner de la manière habituelle et en vertu du fait que le jeu est juste conclure que l'espérance mathématique du joueur A est 0 et par suite que

$$b \cdot m(Z_A) - a \cdot m(Z_B) = 0.$$

Sans l'emploi de la mesure dans $\mathop{\mathfrak{P}}\limits_{\nu=1}^{\infty} E_\nu$ le raisonnement ne serait pas rigoureux, car il n'est pas évident à priori que les probabilités $m(Z_A)$ et $m(Z_B)$ sont bien déterminées.

Exemple 8. *Produits infinis d'espaces continus.*

La probabilité pour qu'une suite de points choisis au hasard sur une surface (par exemple sur la surface d'une sphère) soit partout dense sur cette surface.

Les espaces aux mesures données sont ici $E_i = E$, où E est la surface donnée. Il s'agit de la mesure, dans $\mathop{\mathfrak{P}}\limits_{\nu=1}^{\infty} E_\nu$, de l'ensemble Z de toutes les suites dont les termes forment un ensemble partout dense dans E. L'ensemble Z est, comme il est aisé de voir, mesurable. Sa mesure est 1. Soit, pour la démonstration, $\{R_i\}$ la suite „des sphères

[29]) V. le remarque sur ce problème de M. S. Bernstein, *Théorie des probabilités* (en russe) Moscou 1927, p. 98.

rationnelles" partout denses dans E. En désignant par C_i l'ensemble de toutes les suites dont tous les termes appartiennent à CR_i, on a $CZ \subset \sum_{i=1}^{\infty} C_i$. La mesure de l'ensemble C_i est évidemment égale a $\lim_{n=\infty} (1-m(R_i))^n = 0$ et la mesure étant complètement additive on a $m(CZ) = 0$, d'où $m(Z)$, c. à d. la probabilité demandée est égale à 1, c. q. f. d.

Nous démontrerons dans la suite un théorème plus précis. On a la probabilité égale à 1 pour qu'une suite de points choisis au hasard soit uniformément dense. On peut de même abandonner l'hypothèse de la constance de la probabilité pour tout choix successif des points, c. à d. admettre différentes mesures dans E_i

Exemple 9. *Produit infini d'espaces dénombrables.*

Probabilité pour qu'il existe un point fixe dans une transformation de l'ensemble E des nombres entiers en lui-même.

Une telle transformation est donnée par une fonction $f(n)$ ayant pour valeur des nombres entiers et peut être considérée comme un point de l'espace $\mathfrak{P}_{\nu=1}^{\infty} E_\nu$, où $E_i = E$. Il s'agit de la mesure de l'ensemble Z de toutes les suites $\{f(n)\}$ pour lesquelles il existe un n tel que $f(n) = n$ [30].

Pour que notre problème soit bien posé, il faut qu'on ait défini une mesure dans E.

Or, si l'on a défini dans E une mesure complètement additive, l'ensemble Z est mesurable dans $\mathfrak{P}_{\nu=1}^{\infty} E_\nu$. Si, ce qui est plus naturel, la mesure dans E est seulement d'une additivité finie et même si tous les sous-ensembles de E sont mesurables, le théorème 4 ne nous apprend rien sur la mesurabilité de Z, cet ensemble n'étant pas une somme d'un nombre fini d'ensembles-produits.

La question si l'ensemble Z appartient à la classe des ensembles qui ont la même mesure pour tout passage de la mesure de la classe des ensembles mesurables (dans le sens du théorème 4)

[30]) Ce problème peut être regardé comme une extension pour l'infini du problème des rencontres de A. de Moivre, bien que dans notre cas il ne s'agit pas des permutations de la suite des nombres entiers, mais il est admissible que $f(n) = f(m)$ pour $n = m$; l'équation $f(x) = n$ peut ne pas être soluble.

à la classe de tous les sous-ensembles de $\mathfrak{P}_{\nu=1}^{\infty} E_\nu$, semble être difficile. Mais s'il en était ainsi, on pourrait régarder le problème comme bien posé même dans le cas où l'on part d'une mesure à additivité finie dans E.

3. Lois-limites.

La notion du produit infini nous permettra d'étudier d'une manière systématique les lois-limites du calcul des probabilités. Dans ce travail nous nous bornons, bien entendu, au cas de variables indépendantes. Ces théorèmes fondamentaux pour le calcul des probabilités remontent, comme on le sait, à Bernoulli, Poisson, Laplace; après Tchebycheff les démonstrations se sont simplifiées. Plus récemment, ces lois étaient étudiées par des savants nombreux (Borel, Cantelli, Mazurkiewicz, Khintchine et d'autres [31] [32]). Grâce à ces travaux on discerne aujourd'hui en particulier entre la loi forte et la loi faible des grands nombres. D'habitude ces théorèmes sont formulés de la manière suivante:

Théorème A. *Soit $x_1, x_2, \ldots, x_i \ldots$ une suite de variables éventuelles indépendantes, à espérance mathématique $E(x_i) = 0$ (ce qui ne diminue pas la généralité) et telles que la série $\sum_{\nu=1}^{n} b_\nu = o(n^2)$ [32a], b_i désignant le „deuxième moment", c.-à-d. $E(x_i^2)$. Pour tout $\varepsilon > 0$, il existe un $N(\varepsilon)$ tel que, pour tout $n > N(\varepsilon)$, on a*

$$\mathfrak{W}\left[\left|\frac{x_1 + x_2 + \ldots + x_n}{n}\right| < \varepsilon\right] > 1 - \varepsilon$$

$\mathfrak{W}[\ ]$ désignant la probabilité de l'inégalité exprimée entre parenthèses.

[31]) Borel, l. c. [7]); Cantelli, *Sulla legge dei grandi numeri*, Mem. Acad. Lincei, T. 11 (1916); Mazurkiewicz, *O pewnem uogólnieniu prawa wielkich liczb*, Wied. Mat. T. XXII, 1917; Khintchine, *Sur les lois fondamentales du calcul des probabilités*, (en russe), Moscou 1927.

[32]) Les hypothèses les plus générales et les énoncés les plus précis sont dus aux travaux récents des savants russes MM. Khintchine, Kolmogoroff et S. Bernstein dans le cas des variables éventuelles dépendantes. Khintchine, *Über das Gesetz der grossen Zahlen*, Math. Ann. (1926); Kolmogoroff, *Über die Summen durch den Zufall bestimmter unabhängiger Grössen*, Math. Ann. 99 (1928).

[32a]) $o(n)$ désigne le symbole de M. Landau: $\frac{o(n)}{n} \to 0$.

Théorème B. *Dans les hypothèses du théorème A,* [*à condition* $\sum_{\nu=1}^{\infty} \frac{b_\nu}{\nu^2} < +\infty$] *on a: Pour tout* $\varepsilon > 0$, *il existe un* $\overline{N}(\varepsilon)$ *tel que pour tout* $n > \overline{N}(\varepsilon)$ *et p naturel, on a*

$$\mathfrak{w}\left[\left|\frac{x_1+\ldots+x_n}{n}\right|<\varepsilon; \left|\frac{x_1+\ldots+x_{n+1}}{n+1}\right|<\varepsilon; \ldots \left|\frac{x_1+\ldots+x_{n+p}}{n+p}\right|<\varepsilon\right]>1-\varepsilon.$$

On voit que la loi faible des grands nombres peut s'exprimer à l'aide de la mesure dans les produits finis (d'ordre arbitrairement grand) de la forme $\mathfrak{P}_{\nu=1}^{n} E_\nu$; la loi forte au contraire conduit aux produits de la forme $\mathfrak{P}_{\nu=1}^{n} E_\nu \times \mathfrak{P}_{\nu=1}^{n+1} E_\nu \times \ldots \times \mathfrak{P}_{\nu=1}^{n+p} E_\nu$ [33].

Bien que les théorèmes en question concernent des probabilités dans les produits finis, on parle souvent de la convergence vers 0, (forte ou ordinaire) de la suite $\left\{\frac{x_1+x_2+\ldots+x_i}{i}\right\}$ dans le sens du calcul des probabilités. De plus, certains auteurs formulent le théorème *B* en disant qu'il existe une probablilité „arbitrairement voisine à 1", pour que tous les termes de la suite $\left\{\frac{x_1+x_2+\ldots+x_i}{i}\right\}$ soient petits à partir d'un *i* suffisamment grand. Mais — si l'on définit une probabilité d'un tel fait comme limite des probabilités correspondantes prises pour *i* fini, — on n'est pas sûr *à priori* si une telle expression possède les proprietés d'une probabilité. De plus, on peut exprimer le même fait mathématique de la convergence par une autre définition équivalente et il n'est pas aussi évident *à priori* que les limites correspondantes à ces nouvelles définitions donnerons le même nombre.

Cependant le théorème *B* possède pour l'intuition un sens invariant, indépendant de l'une ou l'autre définition du fait de la convergence de la suite $\left\{\frac{x_1+x_2+\ldots+x_i}{i}\right\}$.

En se servant de la notion du produit infini on peut exprimer

[33] Ces produits se laissent réduire aux produits plus simples de la forme $\mathfrak{P}_{\nu=1}^{n+p} E_\nu$.

les théorèmes A et B d'une manière plus intuitive et dans un certain sens invariante

On admet une suite de variables indépendantes, c.-à-d. dans notre interprétation une suite d'espaces $E_1, \ldots E_i, \ldots$ composés de nombres réels; pour tout E_i on définit une mesure (fonction de probabilité) telle que

$$\int x_i \, d(m(Z)) = 0; \int x_i^2 \, d(m(Z)) = b_i, \quad \sum_{\nu=1}^{n} b_\nu = o(n^2).$$

Dans l'espace $\mathfrak{P}_{\nu=1}^{\infty} E_\nu$, c.-à-d. dans l'espace des suites $x = (x_1, \ldots x_i, \ldots)$ des nombres choisis de E_i, on peut considérer la fonction $\frac{s_i(x)}{i}$, où $s_i(x) = x_1 + \ldots + x_i$. Sous lesdites conditions on a:

Théorème A*. *La suite des fonctions* $\frac{s_i(x)}{i}$ *converge, dans* $\mathfrak{P}_{\nu=1}^{\infty} E_\nu$, *„en mesure" vers* 0 [34]).

Cet énoncé n'est qu'une autre rédaction du th. A. Le théorème B au contraire admet un énoncé plus simple qu'auparavant.

Dans les hypothèses sur les espaces E_i, qui correspondent aux hypothèses du théorème B, on a:

Théorème B*. *La mesure de l'ensemble des suites (situé dans l'espace* $\mathfrak{P}_{\nu=1}^{\infty} E_\nu$*) dont les moyennes arithmétiques convergent vers* 0 *est* 1 [35]).

Démonstration du th. A^*.

Lemme a. (de Tchebycheff).

Désignons par $Z_n(\varepsilon)$ l'ensemble des points $(x_1, \ldots x_n)$ dans le produit $\mathfrak{P}_{\nu=1}^{n} E_\nu$, pour lesquels $\left|\frac{x_1 + \ldots + x_n}{n}\right| < \varepsilon$.

[34]) La convergence „en mesure" de la suite des fonctions $f_n(x)$ vers $f(x)$ est définie comme il suit: Pour tout $\varepsilon > 0$, il existe un $N(\varepsilon)$ tel que pour $n > N(\varepsilon)$ on a: $m(E[|f_n(x) - f(x)| < \varepsilon]) > 1 - \varepsilon$.

[35]) Ce théorème montre que la „convergence dans le sens du calcul des probabilités" (notion introduite par M. Cantelli) peut-être regardée comme une convergence presque partout dans l'espace $\mathfrak{P}_{\nu=1}^{\infty} E_\nu$.

On a $m(Z_n(\varepsilon)) > 1 - \frac{\sum_{\nu=1}^{n} b_\nu}{n^2 \varepsilon^2}$ [36]).

Dans le théorème il s'agit de la mesure de l'ensemble des points x dans $\mathfrak{P}_{\nu=1}^{\infty} E_\nu$, tels que $\left|\frac{s_n(x)}{n}\right| < \varepsilon$. Nous avons

$$E\left[\left|\frac{s_n(x)}{n}\right| < \varepsilon\right] = Z_n(\varepsilon) \times \mathfrak{P}_{\nu=n+1}^{\infty} E_\nu$$

où l'ensemble $Z_n(\varepsilon)$ se compose des points situés dans le produit fini $\mathfrak{P}_{\nu=1}^{n} E_\nu$.

Les variables x_i étant indépendantes, on a

$$m\left(E\left[\left|\frac{s_n(x)}{n}\right| < \varepsilon\right]\right) = m(Z_n(\varepsilon)) \cdot 1$$

comme $\sum_{\nu=1}^{n} b_\nu = o(n^2)$, on déduit du lemme a notre théorème.

Démonstration du th. B*.

Lemme b. (de M. Kolmogoroff).

Désignons par Z_{Nk}^p l'ensemble des points $(x_1, \ldots x_N, \ldots x_{N+p})$ (situés dans le produit $\mathfrak{P}_{\nu=1}^{N+p} E_\nu$) pour lesquels

$$\left|\frac{s_n(x)}{n}\right| < \frac{1}{k}, \quad \text{pour } n = N, N+1, \ldots N+p,$$

c. à d. l'ensemble

$$Z_{Nk}^p = \prod_{n=N}^{N+p} E\left[\left|\frac{s_n(x)}{n}\right| < \frac{1}{k}\right].$$

Si p croît, c'est une suite d'ensembles décroissants à partie commune Z_{Nk}.

La mesure de l'ensemble Z_{Nk}

$$m(Z_{Nk}) = \lim_{p=\infty} m(Z_{Nk}^p).$$

M. Kolmogoroff prouve que ce nombre (qui, d'après nos remarques générales, peut être interprété comme une probabilité) tend pour $N \to \infty$ vers 1 [37]).

[36]) V. p. ex. Castelnuovo, *Calcolo delle probabilita*, Bologna 1925, p. 60.

[37]) V. Kolmogoroff, l. c. [32]) C'est une conséquence simple du th. I.

Dans le théorème il s'agit de la mesure de l'ensemble Z des suites pour lesquelles l'expression $\lim_{n=\infty} \left| \frac{s_n(x)}{n} \right|$ existe et est égale a 0. Cet ensemble est de la forme:

$$Z = E\left[\lim_{n=\infty} \frac{s_n(x)}{n} = 0\right] = \prod_{k=1}^{\infty} \sum_{N=1}^{\infty} \prod_{n=N}^{\infty} E\left[\left|\frac{s_n(x)}{n}\right| < \frac{1}{k}\right].$$

(Cette forme s'obtient de la définition habituelle de la convergence d'une suite à l'aide p. ex. du procédé général des MM. Kuratowski et Tarski mentionné dans la première partie).

L'ensemble Z, s'exprimant par sommes et produits d'ensembles mesurables, est lui-même mesurable. Notre problème étant ainsi bien défini, il s'agit de calculer la mesure de Z.

Pour k croissant, les ensembles

$$Z_k = \sum_{N=1}^{\infty} \prod_{n=N}^{\infty} E\left[\left|\frac{s_n(x)}{n}\right| < \frac{1}{k}\right]$$

constituent une suite d'ensembles décroissants. Z étant leur partie commune, on a $m(Z) = \lim_{k=\infty} m(Z_k)$.

Les ensembles

$$Z_{Nk} = \prod_{n=N}^{\infty} E\left[\left|\frac{s_n(x)}{n}\right| < \frac{1}{k}\right]$$

constituent pour $N \to \infty$ une suite d'ensembles croissants. Z_k étant leur somme, on a $m(Z_k) = \lim_{N=\infty} m(Z_{Nk})$ et par conséquent

$$m(Z) = \lim_{k=\infty} \lim_{n=\infty} m(Z_{Nk})$$

Le lemme b nous assure que, pour tout $\eta > 0$ et N suffisamment grand, on a $m(Z_{Nk}) > 1 - \eta$.

Nous avons donc $m(Z) \geqslant 1 - \eta$, pour tout $\eta > 0$, ou $m(Z) = 1$, c. q. f. d.

Cette démonstration n'est, du reste, qu'une simple adaptation de celle de M. Kolmogoroff au langage des produits, mais on saisit de cette manière plus explicitement le contenu essentiel des lois-limites.

Remarques. 1. Si tous les espaces E_i se composent de deux éléments 0 et 1, et $m(0) = q$, $m(1) = 1 - q = p$, le théorème B^* nous donne le théorème de Cantelli dans le cas particulier de Bernoulli.

D'une manière analogue, si l'on fixe dans l'espace $E_i = (0, 1)$ la mesure: $m(0) = q_i$, $m(1) = 1 - q_i = p_i$, on obtient du théorème B^* l'énoncé suivant, connu comme théorème de Poisson:

$$\lim_{n \to \infty} \left| \frac{s_n}{n} - \frac{\sum_{\nu=1}^{n} p_\nu}{n} \right| = 0 \text{ avec une probabilité égale à 1.}$$

2. Dans le cas du théorème de Bernoulli notre interprétation identifie ce théorème au théorème de M. Borel, d'après lequel pour „presque tout" point x (dans le sens de la mesure de Lebesgue) de l'intervalle $0 \leqslant x \leqslant 1$, dans le développement dyadique de x le chiffre 0 et 1 ont la même fréquence.

En effet, l'intervalle des nombres réels peut être considéré (exception faite d'un ensemble dénombrable des x qui ont un développement fini) comme le produit des ensembles $E_i = (0, 1)$; si la mesure dans E_i est telle que $m(0) = m(1) = \frac{1}{2}$, on obtient la mesure de Lebesgue par notre procédé de la définition de la mesure dans $\mathfrak{P}_{\nu=1}^{\infty} E_\nu$, c. à d. dans l'intervalle $(0, 1)$, comme il est aisé de voir.

On peut obtenir une mesure différente de la mesure de Lebesgue en concevant, plus généralement, cet intervalle (abstraction faite d'un ensemble dénombrable) comme un produit des ensembles $E = (0, 1)$ avec la mesure $m(0) = q$, $m(1) = p$. Celle-ci correspond à la fonction de la probabilité totale $F(x) = x$. Dans le cas général on obtient pour la fonction de la probabilité totale la fonction suivante: Si $x = (0, \alpha_1 \alpha_2 \ldots \alpha_n \ldots)$, où α_i est le i-ième chiffre du développement dyadique de x,

$$F(x) = \sum_{\nu=1}^{\infty} \alpha_\nu \, p^{\sum_{\mu=1}^{\nu-1} \alpha_\mu} \, q^{\nu - \sum_{\mu=1}^{\nu-1} \alpha_\mu}$$

Cette fonction, qui se réduit évidemment à x pour $p = q = \frac{1}{2}$, est continue, mais sa dérivée (pour $p \neq q$) n'existe pas dans un ensemble partout dense. Nous avons tracé sur la figure quelques premières approximations de la fonction $F(x)$.

Dans le cas du théorème de Poisson, si l'on veut représenter sur l'intervalle toutes les parties possibles du jeu de pile ou face

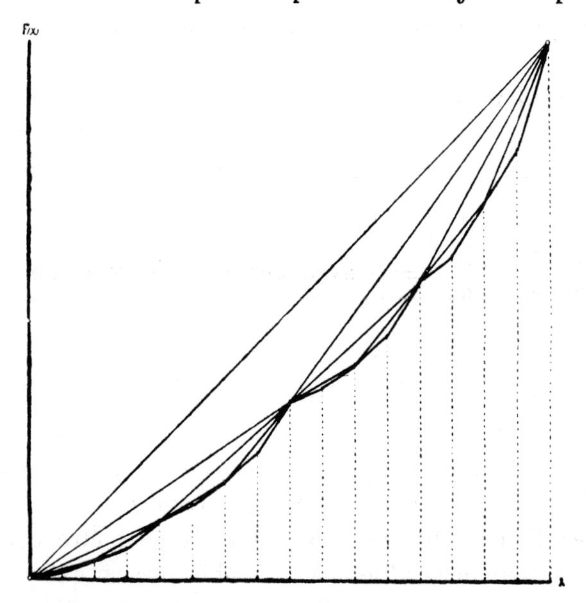

où la probabilité d'obtenir pile varie avec n, on aura

$$F(x) = \sum_{\nu=1}^{\infty} a_\nu \left(\prod_{\mu=1}^{\nu-1} p_\mu^{\alpha_\mu} q_\mu^{1-\alpha_\mu} \right) q_\nu^{\alpha_\nu}.$$

3. Les théorèmes A^* et B^* s'obtiennent à l'aide des lemmes qui servent à démontrer les théorèmes A et B et à l'aide des théorèmes sur la mesure dans les produits infinis.

On peut, d'une manière analogue, formuler pour les espaces des suites infinies d'autres théorèmes fondamentaux du calcul des probabilités.

Considérons comme exemple la „loi du logarithme itéré" de MM. Khintchine et Kolmogoroff.

Théorème C. *Conservons les notations du théorème A. A condition que*

$$\lim_{n=\infty} B_n = \sum_{\nu=1}^{\infty} b_\nu = +\infty$$

et

$$|x_n| \leqslant m_n = o\left(\sqrt{\frac{B_n}{\log\log B_n}}\right),$$

l'ensemble de ces suites pour lesquelles on a:

$$\limsup_{n=\infty} \frac{s_n}{\sqrt{2B_n \log\log B_n}} = 1$$

a la mesure 1 *dans l'espace* $\mathop{\mathfrak{P}}\limits_{\nu=1}^{\infty} E_\nu$.

MM. Khintchine et Kolmogoroff prouvent [38]) que, pour η et δ positifs, il existe un N tel que la probabilité pour qu'une au moins des inégalités

$$s_n > \sqrt{2B_n \log\log B_n}\ (1+\delta), \quad (n = N, N+1, \ldots N+p)$$

soit remplie est plus petite que η, et, pour η, δ, N arbitraires, il existe un nombre p naturel tel que la probabilité pour que toutes les inégalités

$$s_n > \sqrt{2B_n \log\log B_n}\ (1-\delta) \quad (n = N, N+1, \ldots N+p)$$

soient remplies est plus petite que η.

M. Kolmogoroff formule la „loi du logarithme itéré" de cette manière un peu compliquée parce qu'il ne voulait pas „employer des probabilités des relations qui ne peuvent être observées directement" [39]).

Il semble que l'existence de la mesure dans le produit infini permet d'éviter de telles objections en permettant de parler des probabilités des relations qui dépendent d'un nombre infini d'événements.

Il pourrait être intéressant d'étudier d'autres propriétés des séries infinies des variables éventuelles. Les lois des grands nombres expriment la sommabilité d'une suite de variables éventuelles par le procédé C_1 avec la probabilité 1. On peut demander de trouver tous les procédés de sommation pour lesquels il en est ainsi et cela déjà pour le cas des variables éventuelles les plus simples.

4. Suites de points uniformément denses.

Dans ce numéro nous formulons un théorème qui est intimement lié à la loi forte des grands nombres.

[38]) Khintchine l. c. [31]), Ch. III., Kolmogoroff, *Über das Gesetz des iterierten Logarithmus*, Math. Ann. 101, 1929.

[39]) Kolmogoroff, l. c. [38]), p. 127.

Soit E un espace métrique et séparable à mesure complètement additive. La mesure est définie en particulier pour des regions G dans E (Une région dans un espace séparable se compose toujours d'un nombre dénombrable de „sphères rationnelles denses").

Une suite de points $x = (x_1, x_2, \ldots, x_n, \ldots)$ est nommée uniformément dense si pour tout $\varepsilon > 0$ et pour tout G il existe un N tel que pour tout $n > N$ le nombre des points de la suite qui appartiennent à G divisé par n, diffère de la mesure de G par un nombre plus petit que ε.

Une suite de points pris au hasard dans E peut être considérée comme élément du produit infini $\mathfrak{P}_{\nu=1}^{\infty} E_\nu$, où $E_i = E$ $(i = 1, 2, \ldots)$. La mesure dans $\mathfrak{P}_{\nu=1}^{\infty} E_\nu$ est supposé déterminée conformément au théorème 2.

On peut alors se demander quelle est la probabilité pour qu'une suite de points pris au hasard soit uniformément dense.

Théorème D. *La mesure de l'ensemble des suites uniformément denses dans l'espace E est* 1.

Démonstration. L'espace E étant séparable et la mesure complètement additive, il en résulte aisément que pour qu'une suite soit uniformément dense dans le sens de notre définition, il suffit, qu'elle soit uniformément dense par rapport aux „sphères rationnelles denses" R_i.

Désignons par $l_{jn}(x)$ le nombre des points de l'intervalle $(x_1, x_2 \ldots x_n)$ de la suite $x = (x_1, x_2, \ldots)$ qui tombent dans R_j. En écrivant la définition d'une suite uniformément dense, on obtient l'ensemble de toutes ces suites sous la forme

$$Z = \prod_{j=1}^{\infty} \prod_{k=1}^{\infty} \sum_{N=1}^{\infty} \prod_{n=N}^{\infty} E\left[\left|\frac{l_{jn}(x)}{n} - m(R_j)\right| < \frac{1}{k}\right].$$

Les ensembles

$$Z_j = \prod_{k=1}^{\infty} \sum_{N=1}^{\infty} \prod_{n=N}^{\infty} E\left[\left|\frac{l_{jn}(x)}{n} - m(R_j)\right| < \frac{1}{k}\right].$$

ont la mesure 1. Pour s'en convaincre observons que pour un ensemble R_j donné $l_{jn}(x)$ est une somme des variables éventuelles indépendantes z définies de la manière suivante:

$$z_i^j = 1 \quad \text{avec la probabilité } m(R_j)$$
$$z_i^j = 0 \quad \text{avec la probabilité } 1 - m(R_j).$$

En utilisant le théorème B^* nous obtenons $m(Z_j) = 1$.

En tenant compte du fait que $CZ = \sum_{j=1}^{\infty} CZ_j$ et que la mesure donnée est complètement additive, on obtient $m(CZ) = 0$, c. q. f. d.

Remarques. Ce théorème d'un caractère géométrique qui est au fond une conséquence de la loi des grands nombres pour le cas de Bernoulli peut être regardé néanmoins comme une généralisation de cette loi.

En effet, dans le cas où l'ensemble E se compose des deux éléments 0 et 1 nous pouvons prendre pour des régions G les sous-ensembles (0) et (1). De la définition d'une suite $x = \{x_i\}$ uniformément dense il résulte que, si $s_n(x)$ désigne le nombre des points qui sont contenus dans la „région" (1), c. à d. des chiffres 1, on a $\lim_{n=\infty} \frac{s_n(x)}{n} = m(1) = p$ avec la probabilité 1.

Le théorème D se laisse généraliser pour le cas où la probabilité (la mesure dans E) varie pour les choix successifs, c. à d. avec n. Nous avons affaire à une suite $\{E_i\}$ d'espaces qui se composent des mêmes points mais où la mesure de la région G est $m_i(G)$. Une suite des points $x = \{x_i\}$ sera nommée uniformément dense si pour n suffisamment grand le nombre divisé par n des points qui tombent sur G approche la moyenne arithmétique des nombres $m_i(G)$. Le théorème D reste vrai et peut être regardé comme une généralisation de la loi des grands nombres dans le cas de Poisson.

5. Nombres normaux.

Dans un ordre d'idées analogues à celles du numéro précédent on peut envisager la notion d'un nombre normal, introduite par M. Borel [40]. Un nombre x est appelé normal (par rapport au développement dyadique) si $\lim_{n=\infty} \frac{s_n(x)}{n} = \frac{1}{2}$. Nous savons déjà que la

[40]) Borel, *Traité du calcul des probabilités et de ses applications*, T. II, fasc. I, Ch. I. Paris 1926.

mesure de l'ensemble des nombres normaux est 1. M. Borel a étudié les nombres totalement normaux c.-à-d., normaux par rapport au développement en puissance des bases 2, 4, 8 ... (La définition d'un nombre normal par rapport à un développement n-adique est évidente).

La définition d'un nombre normal peut être généralisée dans deux directions d'une manière évidente. Pour le cas de Poisson la mesure dans l'intervalle (0, 1) est établie par la fonction $F(x)$ de la page 269. Un nombre sera appelé totalement normal si pour tous les systèmes finis des zéros et des unités $(\beta_1, \beta_2 \ldots \beta_r)$ le nombre $s_n(x; \beta_1, \beta_2, \ldots, \beta_r)$ désignant combien de fois ce système se trouve dans la n-ième approximation dyadique du nombre x, jouit de la propriété suivante:

$$\lim_{n=\infty} \frac{s_n(x; \beta_1, \beta_2, \ldots \beta_r)}{n} = p^{\sum\limits_{\nu=1}^{r} \beta_\nu} q^{r - \sum\limits_{\nu=1}^{r} \beta_\nu}$$

Autrement dit, un nombre est totalement normal si la fréquence de tout système fini de chiffres est, à la limite, égale à la probabilité de ce système. On prouve facilement que la mesure des nombres totalement normaux est 1.

Or, on peut généraliser cette remarque de la même manière suivant laquelle la loi des grands nombres est généralisée par le théorème D.

Soit E un espace remplissant les conditions du théorème D. A tout elément $x = \{x_i\}$ de l'espace $\mathfrak{P}_{\nu=1}^{\infty} E_\nu$, où $E_i = E$, et à toute région G dans E correspond une suite des zéros et unités:

$$\varphi(x) = (\varphi(x_1), \ \varphi(x_2), \ldots \varphi(x_n), \ldots)$$

où $\varphi(y)$ est la fonction caractéristique de l'ensemble G (c. à d. $\varphi(y) = 0$ si $y \epsilon CG$, $\varphi(y) = 1$ si $y \epsilon G$). La suite x est appelée totalement uniformément dense si pour tout G et pour tout système $(\beta_1, \beta_2, \ldots \beta_r)$

$$\lim_{n=\infty} \frac{s_n(G; \beta_1 \beta_2 \ldots \beta_r)}{n} = [m(G)]^{\sum\limits_{\nu=1}^{r} \beta_\nu} [1 - m(G)]^{r - \sum\limits_{\nu=1}^{r} \beta_\nu}$$

où $s_n(G; \beta_1, \beta_2, \ldots \beta_r)$ exprime combien de fois le système $(\beta_1, \beta_2, \ldots \beta_r)$ se trouve dans la suite $(\varphi(x_1), \varphi(x_2) \ldots \varphi(x_n))$.

On prouve facilement le théorème suivant:

Dans l'espace $\mathfrak{P}_{\nu=1}^{\infty} E_\nu$ l'ensemble des suites totalement uniformément denses a la mesure 1.

On peut encore généraliser ce théorème pour le cas de Poisson, c.-à-d. pour le cas où la mesure dans l'espace E varie avec i. On doit modifier, bien entendu, la définition d'une suite totalement uniformément dense.

6. Convergence des séries.

Nous avons étudié la convergence des moyennes arithmétiques des suites à variables éventuelles. La notion du produit infini nous rend les mêmes services dans l'étude de la convergence des séries à variables éventuelles. Ce problème a été résolu dans quelques cas particuliers importants par M. Steinhaus[41]). Pour obtenir une mesure dans les espaces correspondants à une infinité de dimensions il transformait ces espaces d'une manière biunivoque en l'intervalle des nombres réels Une telle transformation étant convenablement choisie, on peut adopter comme mesure d'un ensemble la mesure de Lebesgue de l'ensemble linéaire correspondant. Ce procédé, un peu compliqué, conduit cependant à une solution rigoureuse du problème en question.

Le problème de la convergence des séries à variables éventuelles très générales a été résolu par MM. Khintchine, Kolmogoroff et P. Lévy[42]). La probabilité P de la convergence d'une série à variables éventuelles (à valeur réelle — cette supposition ne diminue pas la généralité) — y était définie de la manière suivante:

$$P = \lim_{i=\infty} \lim_{n=\infty} \lim_{N=\infty} \mathfrak{W}\left[\max_{n \leq p \leq r \leq N} \left| \sum_{\nu=p}^{r} x_\nu \right| < \frac{1}{i}\right].$$

[41]) Steinhaus, *Über die Wahrscheinlichkeit dafür, dass der Konvergenzkreis einer Potenzreihe ihre natürliche Grenze ist.* Math. Zeit. 31 (1929); *Sur la probabilité de la convergence des séries*, Studia Math. T. II (1930).

[42]) Khintchine et Kolmogoroff, *Über Konvergenz von Reihen*, Rec. math. Soc. Math. Moscou Bd. 32 (1925), Kolmogoroff l. c. [32]), P. Lévy, *Sur les séries dont les termes sont des variables éventuelles indépendantes.* Studia Math. T. III (1931).

Dans ce cas encore on pourrait soulever l'objection que l'adoption de différentes définitions équivalentes du fait de la convergence pourrait conduire à des valeurs différentes des probabilités et qu'il n'est pas admissible *à priori* de parler de cette expression limite comme d'une probabilité. Or, l'existence de la mesure dans le produit infini montre qu'il n'en est pas ainsi, la définition précédente étant de cette manière justifiée.

La probabilité de la convergence d'une série est égale à la mesure de l'ensemble Z des suites, dans l'espace $\mathfrak{P}_{\nu=1}^{\infty} E_\nu$, pour lesquelles les sommes partielles sont convergentes. Si l'on admet la définition habituelle (de Cauchy) de convergence, cet ensemble devient:

$$Z = \prod_{i=1}^{\infty} \sum_{N=1}^{\infty} \prod_{p=N}^{\infty} \prod_{r=p}^{\infty} \boldsymbol{E}\left[|x_p + x_{p+1} + \ldots + x_r| < \frac{1}{i}\right].$$

On peut écrire aussi: $Z = \prod_{i=1}^{\infty} \sum_{N=1}^{\infty} \prod_{n=N}^{\infty} Z(N, n, i)$, où

$$Z(N, n, i) = \boldsymbol{E}\left[\max_{N \leq p \leq r \leq n} |x_p + x_{p+1} + \ldots + x_r| < \frac{1}{i}\right].$$

Pour $n > n'$

$$Z(N, n, i) \subset Z(N, n', i)$$

donc

$$m\left(\prod_{n=N}^{\infty} Z(N, n, i)\right) = \lim_{n=\infty} m\left(Z(N, n, i)\right).$$

Les ensembles $\prod_{n=N}^{\infty} Z(N, n, i)$ sont tels que pour $N' > N$

$$\prod_{n=N'}^{\infty} Z(N', n, i) \supset \prod_{n=N}^{\infty} Z(N, n, i)$$

et

$$m\left(\sum_{N=1}^{\infty} \prod_{n=N}^{\infty} Z[N, n, i]\right) = \lim_{N=\infty} m\left(\prod_{n=N}^{\infty} Z[N, n, i]\right).$$

On a

$$\sum_{N=1}^{\infty} \prod_{n=N}^{\infty} Z[N, n, i] \supset \sum_{N=1}^{\infty} \prod_{n=N}^{\infty} Z[N, n, i+1]$$

d'où

$$m(Z) = \lim_{i=\infty} \lim_{N=\infty} \lim_{n=\infty} m\left(\mathop{E}\left[\max_{N \leq p \leq r \leq n} |x_p + \ldots + x_r| < \frac{1}{i}\right]\right).$$

De cette manière on parvient à la définition des MM. Khintchine et Kolmogoroff.

La méthode de M. Lévy nous permet de prévoir *à priori* que la probabilité en question peut être égale seulement à 1 ou à 0

L'ensemble Z peut être représenté de deux manières:

$$Z = \prod_{i=1}^{\infty} \sum_{N=1}^{\infty} \prod_{n=N}^{\infty} \mathop{E}\left[\max_{N \leq p \leq r \leq n} |s_r - s_p| < \frac{1}{i}\right]$$

$$Z = \prod_{i=1}^{\infty} \sum_{N=1}^{\infty} \prod_{n=N}^{\infty} \prod_{k=n}^{\infty} \mathop{E}\left[\max_{N \leq p \leq r \leq n} |s_r - s_p| < \frac{1}{i};\ \max_{n \leq p \leq r \leq k} |s_r - s_p| < \frac{1}{i}\right].$$

En comparant les mesures de Z obtenues de la première resp. deuxième expression, nous obtenons

$$m(Z) = m^2(Z), \quad \text{d'où} \quad m(Z) = 0 \quad \text{ou} \quad m(Z) = 1.$$

La marche de la démonstration peut être étudiée en la comparant avec la démonstration d'un théorème analogue où se présentent, du reste, quelques difficultés supplémentaires[43]).

Étant donnée une suite de variables éventuelles générales, on peut démontrer que l'alternative suivante a nécessairement lieu: la loi des grands nombres est remplie ou bien la suite $s_n(x):n$ ne converge pas vers 0 avec la probabilité 1. D'une manière plus précise nous avons le

Théorème E. *Soit, dans les notations du théorème A,* $|x_i| < a_i$ *et* $\int x_i\, d\,(m(Z)) = 0$. *L'ensemble Z des suites* $x = \{x_i\}$ *dans l'espace* $\mathfrak{P}_{\nu=1}^{\infty} E_\nu$, *pour lesquelles* $\lim_{n=\infty} \frac{s_n(x)}{n}$ *existe et est égale à* 0, *a la mesure* 0 *ou* 1[44]).

[43]) L'idée essentielle de la démonstration est celle de M. P. Lévy, l. c.[42]), p. 124.

[44]) Cf. aussi M. Kolmogoroff, *Grundbegriffe der Wahrscheinlichkeitsrechnung*, Anhang, Berlin 1933.

Démonstration. L'ensemble Z en question s'écrit:

$$Z=\prod_{k=1}^{\infty}\sum_{N=1}^{\infty}\prod_{n=N}^{\infty}E\left[\left|\frac{s_n(x)}{n}\right|<\frac{1}{k}\right].$$

En posant

$$Z(k,N,n,m)=E\left[\max_{N\leqslant p\leqslant n}\left|\frac{s_p(x)}{p}\right|<\frac{1}{3k};\ \max_{3k\bar{n}\leqslant r\leqslant m}\left|\frac{s_r(x)}{r}\right|<\frac{1}{3k}\right]$$

où $\bar{n}=\max(n,\sum_{\nu=1}^{n}a_\nu)$, on a

$$Z=\prod_{k=1}^{\infty}\sum_{N=1}^{\infty}\prod_{n=N}^{\infty}\prod_{m=3k\bar{n}}^{\infty}Z(k,N,n,m).$$

Mais

$$\frac{s_r(x)}{r}=\frac{s_n(x)}{r}+\frac{\sum_{\nu=n+1}^{r}x_\nu}{r}$$

et, comme pour $r>3k\bar{n}$

$$\left|\frac{s_n(x)}{r}\right|<\frac{\bar{n}}{3k\bar{n}}=\frac{1}{3k},$$

on voit que l'inégalité $\left|\frac{s_r(x)}{r}\right|<\frac{1}{3k}$ implique $\left|\frac{\sum_{\nu=n+1}^{r}x_\nu}{r}\right|<\frac{2}{3k}$.

En posant

$$\bar{Z}(k,N,n,m)=E\left[\max_{N\leqslant p\leqslant n}\left|\frac{s_p(x)}{p}\right|<\frac{1}{3k};\ \max_{3k\bar{n}\leqslant r\leqslant m}\left|\frac{\sum_{\nu=n+1}^{r}x_\nu}{r}\right|<\frac{2}{3k}\right],$$

nous avons donc

$$Z(k,N,n,m)\subset\bar{Z}(k,N,n,m).$$

Il en résulte que pour l'ensemble

$$\bar{Z}=\prod_{k=1}^{\infty}\sum_{N=1}^{\infty}\prod_{n=N}^{\infty}\prod_{m=3k\bar{n}}^{\infty}\bar{Z}(k,N,n,m)\quad\text{on a}\quad m(\bar{Z})\geqslant m(Z).$$

Posons

$$Z_{Nk}=\prod_{n=N}^{\infty} E\left[\max_{N\leqslant p\leqslant n}\left|\frac{s_p(x)}{p}\right|<\frac{1}{k}\right] \quad \text{et} \quad \bar{Z}_{Nk}=\prod_{n=N}^{\infty}\prod_{m=3\bar{n}k}^{\infty}\bar{Z}(k,N,n,m).$$

On voit facilement que

$$m(Z)=\lim_{k=\infty}\lim_{N=\infty} m(Z_{Nk}) \quad \text{et} \quad m(\bar{Z})=\lim_{k=\infty}\lim_{N=\infty} m(\bar{Z}_{Nk}).$$

Or, on a pour tout $M>N$, en posant $\bar{M}=\max(M,\sum_{\nu=1}^{M} a_\nu)$:

$$\bar{Z}_{Nk}=\prod_{n=N}^{\infty}\prod_{m=3k\bar{n}}^{\infty} E\left[\max_{N\leqslant p\leqq n}\left|\frac{s_p(x)}{p}\right|<\frac{1}{3k};\ \max_{3k\bar{n}\leqslant r\leqslant m}\left|\frac{\sum_{\nu=n+1}^{r} x_\nu}{r}\right|<\frac{2}{3k}\right]\subset$$

$$\subset E\left[\max_{N\leqslant p\leqslant M}\left|\frac{s_p(x)}{p}\right|<\frac{1}{3k}\right]\times\prod_{m=3k\bar{M}}^{\infty} E\left[\max_{3k\bar{M}\leqslant r\leqslant m}\left|\frac{\sum_{\nu=M+1}^{r} x_\nu}{r}\right|<\frac{2}{3k}\right]$$

et comme dans les deux facteurs du dernier produit on a affaire à des variables x différentes et indépendantes, il vient

$$m(\bar{Z}_{Nk})\leqslant m\left(E\left[\max_{N\leqslant p\leqslant M}\left|\frac{s_p(x)}{p}\right|<\frac{1}{3k}\right]\right)\cdot m\left(\prod_{m=3k\bar{M}}^{\infty} E\left[\max_{3k\bar{M}\leqslant r\leqslant m}\left|\frac{\sum_{\nu=M+1}^{r} x_\nu}{r}\right|<\frac{2}{3k}\right]\right).$$

Si M tend vers l'infini, le premier membre tend vers $m(Z_{N,3k})$. En majorant le deuxième membre par le nombre $m(Z_{3k\bar{M},k})$, on a pour $M=\infty$

$$m(\bar{Z}_{Nk})\leqslant m(Z_{N,3k})\cdot\lim_{M=\infty} m(Z_{Mk}).$$

En passant à la limite pour $N=\infty$, $k=\infty$, il vient

$$m(\bar{Z})\leqslant m(Z)\cdot m(Z)$$

d'où

$$m(Z)\leqslant m(\bar{Z})\leqslant m^2(Z),$$

ce qui entraîne $m(Z)=0$ ou $m(Z)=1$, c. q. f. d.

Über topologische Abbildungen der euklidischen Sphären [1]).

Von

J. Schreier und S. Ulam (Lwów).

Es seien $f(x)$, $\varphi(x)$, $\psi(x), \ldots$, stetige, im Intervall $\langle 0, 1 \rangle$ erklärte Funktionen deren Werte demselben Intervall angehören, gegeben.

Man kann dann durch Zusammensetzen, z. B. $f\varphi(x)$, $\varphi\psi f\varphi(x)$, $\varphi\varphi f(x), \ldots$ aus den gegebenen Funktionen unendlich viele neue bilden.

Wir fragen ob es eine *endliche* Anzahl von Funktionen gibt, so daß die aus ihnen mittels Zuusammensetzen gebildeten Funktionen jede andere mit beliebiger Genauigkeit zu approximieren erlauben.

Man könnte auch anders fragen, ob die stetigen Funktionen in Bezug auf die Zusammensetzung eine endliche *Basis* besitzen.

Wir werden u. a. in dieser Arbeit zu dem vielleicht unerwarteten Ergebnis gelangen, daß es schon *fünf* Funktionen gibt, die das Gewünschte leisten [2]).

Die eben gestellte Frage läßt sich in einen umfangreichen Problemkreis einreihen der, wie sich später herausstellen wird, einen gruppentheoretisch-topologischen Charakter trägt.

Wir werden auch den schwierigeren Fall der eineindeutigen Funktionen (man hat dann also nur alle eineindeutigen Funktionen zu approximieren, darf aber zu diesem Zwecke eben nur solche Funktionen anwenden!) und dies gleich im allgemeinen Falle, wo Argument- und Wertebereich der untersuchten Funktionen die n-dimensionale, euklidische Vollkugel ist, behandeln.

[1]) S. unsere Note in C. R., Nov. 1933.

[2]) Die Zahl 5 kann erniedrigt werden. Vgl. [11]).

For commentary to this paper [17], see p. 680.

Wir werden in verschiedenen Fällen, die Existenz einer endlichen Basis im obigen Sinne für die untersuchten Funktionen nachweisen.

Bei der Methode, die wir dazu anwenden werden, werden wir auf verschiedene topologische Schwierigkeiten stoßen, die sich in gewissen Fällen zurzeit nicht bewältigen ließen, so daß noch verschiedene Probleme offen geblieben sind. In dem am Anfang erwähnten Falle der stetigen Funktionen, die das Intervall (0, 1) in eine Teilmenge dieses Intervalls, abbilden, werden aber diese Schwierigkeiten gar nicht auftreten.

Ob unsere Sätze nicht nur Spezialfälle eines allgemeinen Satzes sind, steht dahin. Jedenfalls müßte der Beweis einen solchen Satzes ganz andere Methoden anwenden und dürfte sehr schwierig sein.

Wir kommen nun zur Präzisierung der angewendeten Begriffe und zur Aufzählung der in dieser Arbeit bewiesenen Sätze.

1. Es sei A ein metrischer, kompakter Raum. Unter einem topologischen *Automorphismus* [3]) $\varphi(p)$ von A werden wir eine eineindeutige und stetige Abbildung, von A auf sich selbst, verstehen. Die Menge aller topologischen Automorphismen von A bildet, wenn man als Verknüpfung zweier Elemente $f(p)$ und $\varphi(p)$ ihre Zusammensetzung $f\{\varphi(p)\}$, versteht, eine Gruppe, die wir mit $T(A)$ bezeichnen wollen. Im Anschluß an die in der Gruppentheorie üblichen Bezeichnungen versteht man unter $f\varphi$ die Abbildung $f\{\varphi(p)\}$, unter $\varphi^k(p)$ bei natürlichen k, die k-mal angewendete Iteration von φ, unter φ^{-k} die zu φ^k inverse Abbildung, unter $e(p)$ die identische Abbildung: $e(p) \equiv p$.

Es bezeichne weiter $K^{(n)}$ die n-dimensionale euklidische Vollkugel ($K^{(n)} = \underset{x_1,\dots,x_n}{E}(\sum_1^n x_i^2 \leqslant 1)$), $S^{(n-1)}$ ihre Oberfläche, $K_i^{(n)}$ die zu $K^{(n)}$ konzentrische Kugel mit dem Radius $1 - \frac{1}{i}$.

Wir bezeichnen mit $H_1^{(n)}$ den metrischen Raum, den wir erhalten, indem wir die Menge $T(K^{(n)})$ durch die Formel

$$\varrho_1(f, \varphi) = \operatorname*{Max}_{p \in K^{(n)}} \varrho[f(p), \varphi(p)]$$

metrisieren, mit $H_2^{(n)}$ den metrischen Raum, den wir aus derselben

[3]) S. C. Kuratowski, *Topologie I*, Warszawa-Lwów 1933, p. 70.

Menge mittels der Metrisationsformel

$$\varrho_2(f, \varphi) = \sum_{i=1}^{\infty} 2^{-i} \operatorname*{Max}_{p \in K_i^{(n)}} \varrho[f(p), \varphi(p)]$$

erhalten, endlich mit $G^{(n)}$ den metrischen Raum, den wir erhalten, indem wir die Menge $T(S^{(n)})$ durch die Formel

$$\varrho_1(f, \varphi) = \operatorname*{Max}_{p \in S^{(n)}} \varrho[f(p), \varphi(p)]$$

metrisieren. $F^{(n)}$ sei der durch die Bedingung

$$f(p) = p, \quad \text{für} \quad p \in S^{(n-1)}$$

bestimmte, abgeschlossene Teilraum von $H_1^{(n)}$. (Er ist zugleich Untergruppe von $T(K^{(n)})$!).

Man sieht sofort, daß in $H_1^{(n)}$ bzw. $G^{(n)}$ die Konvergenz einer Folge $\{f_\nu\}$, gegen f, mit der gleichmäßigen Konvergenz von $\{f_\nu\}$ gegen f, in K^n bzw. $S^{(n)}$, in $H_2^{(n)}$ mit der gleichmäßigen Konvergenz in jedem $K_i^{(n)}$, gleichbedeutend ist. Daraus folgt aber, daß $H_1^{(n)}$, $H_2^{(n)}$, $G^{(n)}$ und $F^{(n)}$ topologische Gruppen [4]) sind.

Man hat zu diesem Zwecke zu zeigen, daß aus der gleichmäßigen Konvergenz der Folge $\{f_\nu\}$ gegen f, und der Folge $\{g_\nu\}$ gegen g die gleichmäßige Konvergenz von $\{f_\nu g_\nu\}$ gegen fg und $\{f_\nu^{-1}\}$ gegen f^{-1} folgt. Es ist aber

$$\varrho(f_\nu g_\nu, fg) \leqq \varrho(fg, fg_\nu) + \varrho(fg_\nu, f_\nu g_\nu) \leqq \varrho(fg, fg_\nu) + \varrho(f, f_\nu).$$

Das erste Glied strebt wegen der Stetigkeit von f und wegen $\lim g_\nu = g$, das zweite wegen $\lim f_\nu = f$ gegen Null. Weiter ist

$$\varrho(f_\nu^{-1}, f^{-1}) = \varrho(f_\nu^{-1} f_\nu, f^{-1} f_\nu) = \varrho(e, f^{-1} f_\nu) = \varrho(f^{-1} f, f^{-1} f_\nu)$$

und dies strebt wegen $\lim f_\nu = f$ und wegen der Stetigkeit von f^{-1} gegen Null. Im Falle des Raumes $H_2^{(n)}$ hat man noch zu beachten, daß Bild und Urbild einer Kugel $K_i^{(n)}$ immer ganz in einer Kugel $K_j^{(n)}$ liegen.

Die betrachteten Räume sind nicht vollständig. Man kann aber nach S. Banach [5]) diese Räume mittels der Formel

$$\bar{\varrho}(f, g) = \varrho(f, g) + \varrho(f^{-1}, g^{-1})$$

ummetrisieren, so daß sie vollständig werden. Die gleichmäßige Konvergenz von $\{f_\nu\}$ gegen f, bleibt weiter notwendig und hinreichend für $\lim \varrho(f_\nu, f) = 0$, doch

[4]) Für den Begriff vgl. z. B. O. Schreier Abh. d. Sem. Hamb. IV *Abstrakte kontinuierliche Gruppen.*

[5]) S. Banach, *Théorie des opérations linéaires*, Warszawa 1932. p. 229.

erfüllt jetzt eine Folge von Automorphismen die gleichmäßig konvergent ist, deren Grenze aber kein Automorphismus ist, nicht mehr das Cauchysche Konvergenzkriterium. Die (Bogen-) Komponente einer topologischen Gruppe Φ in der das Element e liegt bildet selbst eine topologische Gruppe [6]) (u. zw. ist sie Normalteiler von Φ) die wir mit Φ^* bezeichnen wollen. $I(\Phi)$ bezeichne die Faktorgruppe des Normalteilers Φ^*, die auch die Abbildungstypengruppe genannt wird. Ihre Ordnung gibt die Anzahl der Komponenten des Raumes Φ an.

Zwei topologische Gruppen T und S, heißen *stetig isomorph*, wenn es eine eineindeutige Abbildung f, von T auf S gibt, die mit ihrer Umkehrung f^{-1} stetig ist und für $ab = c$ $(a, b, c \subset T)$ $f(a)f(b) = f(c)$ und $f(a^{-1}) = (f(a))^{-1}$ gilt.

Wenn A und B zwei homöomorphe, metrische Räume bezeichnen, so sind $T(A)$ un $T(B)$ stetig isomorph. In der Tat, bezeichnet h den Homöomorphismus der A auf B abbildet, h^{-1} seine Umkehrung, so ordne man dem Elemente $\varphi \in T(A)$ das Element $h\varphi h^{-1} \subset T(B)$ zu. Man bestätigt leicht, daß diese Zuordnung einen stetigen Isomorphismus zwischen $T(A)$ und $T(B)$ herstellt.

Sind A und B zwei topologische Gruppen, so nennt man die Menge aller geordneten Paare $a \in A$, $b \in B$ das *direkte Produkt* $A \times B$, wobei $A \times B$ selbst eine topologische Gruppe ist, mit dem Kompositionsgesetz $(a_1, b_1)(a_2, b_2) = (a_1 a_2, b_1 b_2)$ und der Limesrelation: $\lim (a_n, b_n) = (a, b) \equiv \lim a_n = a$ und $\lim b_n = b$.

Sei Φ eine topologische Gruppe, B eine Teilmenge von Φ. Mit $\Pi(B)$ bezeichnen wir die kleinste Untergruppe von Φ, die B enthält. Besteht B aus endlich vielen Elementen $a_1, a_2, \ldots, a_k$ so schreiben wir auch $\Pi(a_1, a_2, \ldots, a_k)$.

Wir sagen, daß Φ eine endliche Basis (kurz e. B.) besitzt, wenn es in Φ endlich viele Elemente $a_1, a_2, \ldots, a_k$ gibt, so daß $\Pi(a_1, a_2, \ldots, a_k)$ in Φ überall dicht liegt ($\overline{\Pi(a_1, a_2, \ldots, a_k)} = \Phi$).

Man sieht sofort, daß wenn Φ und Ψ stetig isomorph sind entweder beide eine e. B. besitzen, oder keines von ihnen. Dieselbe gilt also für $T(A)$ und $T(B)$, wenn A und B homöomorph sind.

Man sieht auch, daß wenn Φ^* eine e B. besitzt und $I(\Phi)$ endlich ist oder wenigstens endlich viele Erzeugende besitzt, auch Φ eine e. B. besitzt.

Besitzen A und B eine e. B., so gilt dies auch für $A \times B$. Ist nämlich $a_1, a_2, \ldots, a_k$ eine e. B. in A, $b_1, b_2, \ldots, b_l$ eine e. B. in B dann bilden die Elemente $(a_1, e), (a_2, e), \ldots, (a_k, e), (e, b_1), \ldots, (e, b_l)$ eine e. B. in $A \times B$.

Bemerkung. Ebenso zeigt man, daß wenn A und B zwei Untergruppen einer topologischen Gruppe C sind, und A und B beide

[6]) S. die unter [4]) zitierte Abhandlung. Unter Bogen-Komponente versteht man hier die Menge der Elemente, die sich mit e durch ein Bogen verbinden lassen.

eine e. B. besitzen, aus $\Pi(A+B)=C$ die Existenz einer e. B. für C folgt.

Nun können wir die folgenden Probleme formulieren:

Problem 1. Unter welchen Voraussetzungen kann man behaupten, daß eine topologische Gruppe Φ eine e. B. besitzt?

Problem 2. Ist dies insbesondere immer der Fall, wenn Φ die Einheitskomponente der Automorphismengruppe eines metrischen, kompakten Raumes ist: $\Phi = T(A)^*$.

Die Lösung dieser Probleme und in erster Linie die positive Beantwortung von Problem 2. scheint uns von großer Wichtigkeit für die Theorie der topologischen Gruppen Doch glauben wir, daß dies mit nicht geringen Schwierigkeiten verbunden sein wird.

Wir werden im folgenden eine Methode angeben, die in vielen interessanten Spezialfällen des Problems 2. zum Ziele führt. Der Nachteil dieser Methode besteht darin, daß sie direkt eigentlich nur für $H_2^{(n)}$ angewendet werden kann, in jedem anderen Falle noch verschiedene Hilfsbetrachtungen benötigt, die sich mit der topologischen Natur des untersuchten Raumes komplizieren, so daß wir schon z. B. die Existenz einer e. B. für $H_1^{(n)}$ bei $n > 3$ nicht nachweisen können. Die Vorteile unserer Methode bestehen darin, daß sie erstens jedesmal die e. B. effektiv zu konstruieren erlaubt, zweitens, daß diese e. B. aus einer verhältnismäßig kleinen Anzahl von Abbildungen besteht, daß sie, drittens, den Satz von der Existenz einer e. B. in einer verschärften Form aussprechen läßt (sie läßt nämlich behaupten, daß unter den Elementen von $\Pi(a_1, a_2, \ldots, a_k)$ wo $a_1, a_2, \ldots, a_k$ die e. B. bilden, nur gewisse von einer speziellen Gestalt zur Approximation aller anderen gebraucht werden) und viertens, daß sie sich auch im Falle der nicht notwendig eineindeutigen Abbildungen, die ja überhaupt keine Gruppe bilden, anwenden läßt und so die Lösung des zur Anfang der Arbeit gestellten Problems erlaubt. Es wäre interessant für die Funktionen der Basis solche zu finden, die aus dem Gesichtspunkte der Analysis regulär, vielleicht gar analytisch sind.

2. Wir werden folgende Sätze beweisen:

I. *In* $(H_2^{(n)})^*$ *gibt es eine aus drei Elementen bestehende e. B.*

II. *In* $F^{(n)}$ *gibt es eine aus drei Elementen bestehende e. B.*

III. *In*

$$\left.\begin{array}{ll} \alpha) & G^{(1)} \\ \beta) & G^{(2)} \\ \gamma) & H_1^{(1)} \\ \delta) & H_1^{(2)} \\ \varepsilon) & H_1^{(3)} \end{array}\right\} \textit{gibt es eine e. B.}$$

Dem Beweise der Satze I und II sind die Abschnitte 3. und 4. der Arbeit gewidnet. Jetzt schicken wir einige Betrachtungen, voraus, aus denen von Satz II, Satz III folgen wird.

Wir werden folgende Bezeichnungen gebrauchen. Sind p und q zwei Punkte im n-dimensionalen, euklidischen Raum R_n, so bezeichnen wir mit $|p-q|$ ihren Abstand. Den Durchmesser eine Menge A bezeichnen wir mit $\delta(A)$. Ist p ein von Null verschiedener Punkt im R_n, ϑ eine nicht negative Zahl so bezeichnet ϑp den auf der Halbgeraden $\overrightarrow{0p}$ liegenden, von 0 um $\vartheta \cdot |p-0|$ entfernten Punkt ($\vartheta \cdot 0 = 0$). Ist $f(p) \epsilon H_1^{(n)}$ ein Automorphismus der Vollkugel, so bildet er die Sphäre $S^{(n-1)}$ auf sich selbst ab und bestimmt auf diese Weise eindeutig einen Automorphismus $b_f(p) \subset G^{(n-1)}$. Es ist $b_{f'f''} = b_{f'} b_{f''}$, $b_{f-1} = (b_f)^{-1}$ und $\varrho(b_f, b_g) \leqslant \varrho(f, g)$.

Ist dagegen $\varphi(p) \subset G^{(n-1)}$ gegeben, so erklären wir ein $a^{(\varphi)}(p) \epsilon H^{(n)}$ durch die Festsetzung:

$$a^{(\varphi)}(\vartheta p) = \vartheta \varphi(p), \quad \text{für} \quad p \subset S^{(n-1)} \quad \text{und} \quad 0 \leqslant \vartheta \leqslant 1.$$

Es ist

$$a^{(\varphi' \varphi'')} = a^{(\varphi')} a^{(\varphi'')}, \quad a^{(\varphi^{-1})} = (a^{(\varphi)})^{-1},$$

$$\varrho(a^{(\varphi')}, a^{(\varphi'')}) = \varrho(\varphi', \varphi''), \quad b_{a^{(\varphi)}} \equiv \varphi.$$

Ein beliebiges $f(p) \epsilon H_1^{(n)}$ läßt sich nun so schreiben:

$$f(p) = a^{(b_f)} \cdot (a^{(b_f)})^{-1} f(p). \tag{1}$$

Dabei ist diese Zerlegung in Faktoren eindeutig d. h. zwei Abbildungen f und g sind dann und nur dann identisch, wenn gleichzeitig $a^{(b_f)} = a^{(b_g)}$ und $(a^{(b_f)})^{-1} f = (a^{(b_f)})^{-1} g$ ist. Das Element $a^{(b_f)}$ gehört einer topologischen Gruppe an, die offenbar mit $G^{(n-1)}$ stetig isomorph ist, die Elemente $(a^{(b_f)})^{-1} f$ durchlaufen aber die Gruppe $F^{(n)}$. Das heißt aber, daß $H_1^{(n)}$ das direkte Produkt von $F^{(n)}$ und $G^{(n-1)}$ ist:

$$H_1^{(n)} = G^{(n-1)} \times F^{(n)}. \tag{2}$$

Daraus folgt, daß aus II. und α), δ) folgt und ebenso aus II und β), ε). Da $S^{(0)}$ aus zwei Punkten besteht, folgt wegen (2) (für $n=0$) auch γ). aus II.

Nun werden wir beweisen, daß aus γ), α) und aus δ), β) folgt. Aus diesen Implikationen sieht man aber sofort daß III aus II folgt.

Wir kommen zum Beweise von $\gamma) \rightarrow \alpha)$.

Es bezeichne zu diesem Zwecke P einen festen Punkt auf $S^{(1)}$. $G(P)$ sei die Untergruppe der Abbildungen aus $G^{(1)}$, die P festlaßen, K dagegen die Gruppe der Drehungen des Kreises $S^{(1)}$. K besitzt eine e. B. da die Drehungen um einen mit π nicht kommensurablen Winkel bekanntlich eine in K dichte Untergruppe erzeugen. Wegen

$$G^{(1)} = G(P) \times K \tag{3}$$

genügt es nachzuweisen, daß $G(P)$ eine e. B. besitzt. Dies folgt aber wegen der vorausgesetzten Richtigkeit von α) daraus daß $G(P)$ mit $H_1^{(1)}$ stetig isomorph ist [7]).

Beweis von: $\delta) \rightarrow \beta)$.

Es sei P ein fester Punkt auf $S^{(2)}$, $G(P)$ bezeichne wieder die Untergruppe von $G^{(2)}$, der Abbildungen, die P festlaßen. R_i bezeichne eine Folge von Kugeln mit dem Mittelpunkt P und gegen Null konvergierenden Radien.

Lemma. Wenn $F(p) \subset G(P)$ und $\varepsilon > 0$ beliebig vorgegeben sind, dann gibt es ein $\Phi(p) \in G^{(2)}$ so, daß

$$|\Phi(p) - F(p)| < \varepsilon \tag{b}$$

und

$$\Phi(R_i \cdot S^{(2)}) = R \cdot S^{(2)} \tag{a}$$

für ein gewisses i gilt.

Beweis. Wir wählen zunächst ein j so groß, daß $\delta(F(R_j \cdot S^{(2)})) < \varepsilon$ wird, dann wird i so bestimmt, daß $R_i \subset R_j$, und $S^{(2)} \cdot R_i \subset F(S^{(2)} \cdot R_j)$ ist. Man setze in $S^{(2)} - S^{(2)} \cdot R_j$: $\Phi(p) \equiv F(p)$; in $S^{(2)} \cdot R_j - S^{(2)} \cdot R_i$ werde Φ so erklärt, daß es auf dem R_j begrenzendem Breitekreise noch mit F übereinstimmt und das Gebiet $S^{(2)} \cdot R_j - S^{(2)} \cdot R_i$ homöomorph auf das Gebiet $F(S^{(2)} \cdot R_j) - S^{(2)} \cdot R_i$ abbildet. (Dies ist sogar auf konforme Weise möglich). Φ erfüllt daher (a). In $S^{(2)} \cdot R_i$ kann man z. B. Φ so erklären, daß man die P mit dem R_i begrenzendem

[7]) Nach einer mündlichen Bemerkung von K. Borsuk.

Breitekreise verbindenden Halbmeridianbögen so aufeinander abbildet, wie Φ es mit ihren Endpunkten tut.

Für $p \in S^{(2)} - S^{(2)} R_j$ ist (b) trivial. Für $p \in R_j$ ist $F(p) \subset F(R_i)$ und auch $\Phi(p) \subset \Phi(R_i) - F(R_i)$. Wegen $\delta(F(S^{(2)} \cdot R_i)) \leqslant \delta(F(S^{(2)} \cdot R_j)) < \varepsilon$ ist auch hier (b) erfüllt.

Das Lemma ist daher bewiesen.

Wir führen auf jedem Halbmeridian, der P mit seiner Antipode P' verbindet, einen Parameter t ein, der von 0 in P', bis 1 in P läuft und erklären die Abbildung $N(p) \subset G(P)$ indem wir einem Punkte mit dem Parameter t, den auf demselben Halbmeridian gelegenen Punkt, mit dem Parameter $\sqrt{t}$ zuordnen. Es bezeichne R die Halbkugeloberfläche $t \leqslant \frac{1}{2}$. Die Mengen $N^i(R)$ werden dann durch Breitekreise begrenzt, die auf um P geschlagenen Kugeln R_i liegen, deren Radien gegen Null konvergieren. Da R mit $K^{(2)}$ homöomorph ist, so folgt aus δ) daß $T(R)$ eine e. B. besitzt. Ein $f \subset T(R)$ kann zu einem $f' \subset G(P)$ erweitert werden (z. B. so wie Φ im Lemma). Es sei $f_1, f_2, \ldots, f_k$ die e. B. in $T(R)$. Wir behaupten, daß $f'_1, f'_2, f'_3, \ldots, f'_k$, N eine e. B. in $G(P)$ bilden Zunächst sieht man sofort daß

$$N^{-i} f_1 N^i(p), \ldots, N^{-i} f_k N^i(p) \tag{3a}$$

eine e. B. in $T(N^i(R))$ bilden. Ist nun ein $F(p) \subset G(P)$ und die positive Zahl 2ε gegeben so nehme man ein Φ laut Lemma und außerdem i so groß, daß

$$\delta(R_i) < \varepsilon \tag{4}$$

ausfällt. Mittels der Abbildungen (3a) kann man wegen (a) und (4) Φ bis auf ε, also F bis auf 2ε approximieren.

Wir haben also bewiesen, daß $G(P)$ eine e. B. besitzt. Wir haben jetzt noch die Bemerkung auf S. 105 anzuwenden indem wir für C, die Gruppe $G^{(2)}$, $A = G(P)$ und für B die Gruppe aller Drehungen der Kugel [8]) nehmen.

Auf Grund der hier gemachten Bemerkungen kann man auch sofort beweisen, daß

$$I(G^{(1)}) = I(G^{(2)}) = I(H_1^{(1)}) = I(H_1^{(2)}) = I(H_1^{(3)}) = 2$$

ist (2 heißt die zyklische Gruppe aus zwei Elementen).

[8]) Diese Gruppe besitzt nämlich eine e. B. von zwei Elementen.

Beweis. Wir bemerken zunächst, daß $F^{(n)}$ zusammenhängend ist. In der Tat ist

$$f(\vartheta, p) = \vartheta f\left(\frac{1}{\vartheta} p\right) \qquad \text{wenn } |p-0| < \vartheta$$

$$f(\vartheta, p) = p \qquad \text{wenn } 1 \geqslant |p-0| \geqslant \vartheta$$

ein stetiges Streckenbild, welches das beliege Element $f(p) \subset F^{(n)}$, $f(p) = f(1, p)$ mit dem Element $f(0, p) = e(p)$ verbindet.

Aus der Formel (1) folgt, daß

$$(\times) \qquad I(H_1^{(n)}) = I(G^{(n-1)})$$

ist. Da $S^{(2)}$ aus zwei Punkten besteht, ist $I(G^{(0)}) = 2$, wegen $(\times)$ ist $I(H_1^{(1)}) = 2$.

Nun wenden wir (3) an. In dieser Formel ist $G(P)$ stetig isomorph mit $H_1^{(1)}$, die Drehungsgruppe K ist dagegen zusammenhängend, also $I(G^{(1)}) = 2$. Wegen $(\times)$ ist $I(H_1^{(2)}) = 2$.

Aus dem Lemma folgt, daß $I(G^{(2)}) = I(G(P)) = 2$ ist, denn jede Drehung der Kugel läßt sich stetig in die Identität überführen. Wegen $(\times)$ ist schließlich auch $I(H_1^{(3)}) = 2$.

Für höhere Dimensionen versagt diese Methode, wie auch alle anderen bekannten [9]).

3. Aus den Überlegungen des vorigen Abschnittes folgt, daß zum Beweise der Sätze I—III, der Beweis von I. und II. genügt. Dieser Abschnitt ist dem Beweise von I. gewidmet.

Wir wollen den Satz in der Fassung formulieren, die alles enthält was unsere Methode liefert.

Satz. *Es gibt drei Abbildungen $\varphi(p)$, $\psi(p)$ und $\chi(p) \subset H_2^{(n)*}$, derart, daß die Abbildungen*

$$(1) \qquad \Phi(p) = \varphi^k \psi^{-s} \chi \psi^s \varphi^{-k}(p) \quad k = 1, 2, \ldots; \; s = 1, 2, \ldots$$

im $H_2^{(n)*}$ überall dicht liegen d. h.:

wenn ein $F(p) \subset H_2^{(n)*}$, eine im Innern von $K^{(n)}$ enthaltene Kugel K und eine positive Zahl ε gegeben sind, dann gibt es ein $\Phi(p)$ der Gestalt (1), *so daß*

$$|F(p) - \Phi(p)| < \varepsilon \quad \text{für} \quad p \in K$$

gilt.

[9]) S. v. Kerékjártó, *Vorlesungen über Topologie,* J. Springer Berlin 1923, p. 186 ff.

Man sieht also, daß zur Approximation nur Abbildungen der speziellen Gestalt (1) verwendet werden. Die Abbildungen (1) sind aber alle *Transformierte* [10]) einer und derselben Abbildung $\chi(p)$ und dabei werden als *transformierende* Abbildungen nur Abbildungen einer Gruppe mit zwei Erzeugenden φ und ψ angewendet [11]).

Beweis. Wir betrachten ein Polarkoordinatensystem $(r, \xi_1, \xi_2, \dots, \xi_{n-1})$ mit dem Anfang in 0. $\varphi(p)$ ordne dem Punkte $r(\xi_1, \xi_2, \dots, \xi_{n-1})$ der Punkt $(\sqrt{r}, \xi_1, \xi_2, \dots, \xi_{n-1})$ zu. Es sei weiter $(\bar{r}, \bar{\xi}_1, \dots, \bar{\xi}_{n-1})$ ein zweites Polarkoordinatensystem dessen Anfang in $P = (1, 0, 0, \dots, 0)$, also auf $S^{(n-1)}$, liegt. Es sei $\bar{r} = \bar{r}(\bar{\xi}_1, \dots, \bar{\xi}_{n-1})$ die Gleichung von $S^{(n-1)}$ in diesen Koordinaten. $\bar{\psi}(p)$ ordne dem Punkte $(\varrho, \bar{\xi}_1, \dots, \bar{\xi}_{n-1})$ den Punkt $\left(\dfrac{\varrho^2}{\bar{r}(\bar{\xi}_1, \bar{\xi}_2, \dots, \bar{\xi}_{n+1})}, \bar{\xi}_1, \bar{\xi}_3, \dots, \bar{\xi}_{n-1}\right)$ zu. $(\bar{\psi}(P) = P)$.

Man sieht, das φ und $\bar{\psi}$ Automorphismen der Kugel $K^{(n)}$ sind.

Wir bezeichnen mit Q_0 die Kugel $K_{\frac{1}{2}}^{(n)}$ (vom Radius $\frac{1}{2}$) $Q_k = \varphi^k(Q_0)$ sind mit $K^{(n)}$ konzentrische Kugeln, deren Radien gegen Eins konvergieren.

Die Zahl l kann so groß gewählt werden, daß die Mengen

$$\bar{\psi}^l(Q_0), \quad \bar{\psi}^{2l}(Q_0), \quad \bar{\psi}^{3l}(Q_0), \dots$$

die wir mit $J_1, J_2, J_3, \dots$ bezeichnen wollen, paarweise elementfremd sind.

Man nehme $\psi(p) = \bar{\psi}^l(p)$.

Es ist:

$$\lim \delta(J_n) = 0 \tag{2}$$

und, wenn

$$p_n \subset J_n$$

(3) so

$$\lim p_n = P.$$

Der Raum aller topologischen Automorphismen der Kugel Q_0 ist Teilraum des Raumes $Q_0^{Q_0}$ aller stetigen Abbildungen von Q_0

[10]) f ist eine Transformierte von g, wenn es ein h gibt, so daß $f = h^{-1}gh$ gilt; h wird dann die transformierende Abbildung genannt.

[11]) Es kann gezeigt werden, daß als Erzeugende dieser Gruppe eine Abbildung φ und die Abbildung χ, die transformiert wird, gebraucht werden können. Daraus folgt, daß $H_2^{(n)*}$ eine Basis aus zwei Elementen besitzt, dies kann auch für $H_1^{(1)}$ bewiesen werden. Dieser Umstand kann zur in Fußnote [2]) erwähnten Erniedrigung der Funktionenzahl von 5 auf 4 benutzt werden. Dies geschieht jedoch direkt und einfacher in einer Note von W. Sierpiński, dieser Band, S. 119.

auf eine Teilmenge von Q_0, der bei der Metrik $\varrho(f,g) = \underset{p \subset Q_0}{\text{Max}} |f(p) - g(p)|$ separabel ist.

Also gibt es eine Folge von Automorphismen von Q_0:

$$f_1(p), f_2(p), f_3(p), \dots \tag{4}$$

derart, daß wenn $f(p)$ und $\varepsilon > 0$ gegeben sind, es ein $f_i(p)$ gibt, so daß

$$|f(p) - f_i(p)| < \varepsilon \quad p \subset Q_0$$

gilt. Dabei beschränken wir uns natürlich auf Abbildungen f und f_i die sich stetig in die Identität deformieren lassen.

Die Abbildung $\psi^s f_s \psi^{-s}(p)$ ist wegen $\psi^s(Q_0) = J_s$ ein topologischer Automorphismus von J_s.

Wir setzen für $p \in J_s$:

$$\chi(p) = \psi^s f_s \psi^{-s}(p). \tag{5}$$

Wir wollen diese Definition von χ, auf ganz $K^{(n)}$ erweitern. Wir betrachten zu diesem Zwecke die Bilder $J_s' = \psi_s(Q_0')$ einer mit Q_0 konzentrischen Kugel mit dem Radius $\frac{1}{2} + \eta$, wo $\eta > 0$ so klein ist, daß J_s' paarweise elementfremd sind und (2) und (3) (mit J' anstatt J) weiterhin erfüllt bleibt.

Es sei $f_s(\vartheta, p)$ (für $0 \leqslant \vartheta \leqslant 1$) eine stetige Schar von Automorphismen, so daß $f_s(1, p) = p$ und $f_s(0, p) = f_s(p)$ gilt.

Auf der Oberfläche der mit Q_0 konzentrischen Kugel mit dem Radius $\frac{1}{2} + \vartheta\eta$ erklären wir $f_s(p)$ durch die Formel

$$f_s(p) = \frac{\frac{1}{2} + \vartheta\eta}{\frac{1}{2}} f_s\left(\vartheta, \frac{\frac{1}{2}}{\frac{1}{2} + \vartheta\eta} p\right).$$

Diese $f_s(p)$ sind Automorphismen der Kugel Q_0', dabei ist auf der Oberfläche von Q_0', immer $f_s(p) = p$.

Formel (5) erklärt jetzt $\chi(p)$ für jedes $p \in J_s'$, dabei ist auf der Oberfläche von jedem J_s': $\chi(p) = p$. Setzt man noch $\chi(p) = p$ in $K^{(n)} - \Sigma J_s'$ und beachtet (2) und (3), so bietet der Nachweis, daß dieses $\chi(p)$ ein topologischer Automorphismus der Kugel $K^{(n)}$ ist, keinerlei Schwierigkeiten.

Wir werden jetzt zeigen, daß φ, ψ und χ die im Satz behauptete Eigenschaft besitzen.

Es sei also $F(p) \subset H_2^{(n)*}$, $\varepsilon > 0$, und $K' \subset K^{(n)}$ gegeben. Mit $K(\vartheta)$ bezeichnen wir, die mit $K^{(n)}$ konzentrische Kugel, mit dem

Radius ϑ. $F_\vartheta(p) = \vartheta F\left(\frac{1}{\vartheta} p\right)$ ist dann ein topologischer Automorphismus von $K(\vartheta)$. Dabei ist

$$|F(p) - F_\vartheta(p)| = \left|F(p) - \vartheta F\left(\frac{1}{\vartheta} p\right)\right| \leqslant$$

$$\leqslant \left|F(p) - F\left(\frac{1}{\vartheta} p\right)\right| + \left|F\left(\frac{1}{\vartheta} p\right) - \vartheta F\left(\frac{1}{\vartheta} p\right)\right| \quad \text{für } p \subset K(\vartheta).$$

Wenn also ϑ hinreichend nahe bei 1 liegt $(\vartheta > 1 - \alpha(\varepsilon))$, ist

$$(6) \qquad |F(p) - F_\vartheta(p)| < \tfrac{1}{2}\varepsilon \quad \text{für} \quad p \subset K(\vartheta).$$

Wir wählen k so groß, daß Q_k K' enthält und der Radius ϑ_k von Q_k größer als $1 - \alpha(\varepsilon)$ ausfällt.

Wegen der gleichmäßigen Stetigkeit von $\varphi^k(p)$, kann so ein $\delta(\varepsilon)$ gefunden werden, daß aus

$$|p - q| < \delta,$$

$$(7) \qquad |\varphi^k(p) - \varphi^k(q)| < \frac{\varepsilon}{2} \qquad p, q \in K^{(n)}$$

folgt.

Wir setzen

$$(8) \qquad f(p) = \varphi^{-k} F_{\vartheta_k} \varphi^k(p) \quad \text{für} \quad p \in Q_0$$

$f(p)$ ist ein topologischer Automorphismus von Q_0. Es gibt daher ein s, so daß

$$(9) \qquad |f(p) - f_s(p)| < \delta \qquad p \subset Q_0$$

ist. Wir setzen

$$\Phi(p) = \varphi^k \psi^{-s} \chi \psi^s \varphi^{-k}(p) \qquad p \subset K^{(n)}$$

und behaupten, daß $|\Phi(p) - F(p)| < \varepsilon$ für $p \subset K'$ ist.

Für $p \subset K' \subset Q_k$ ist nämlich $\varphi^{-k}(p) \subset Q_0$, also $\psi^s \varphi^{-k}(p) \subset J_s$, daher hat man in Φ für $\chi(p)$ laut Formel (5) $\psi^s f_s \psi^{-s}(p)$ einzusetzen. Dies gibt für $p \subset K'$

$$\Phi(p) = \varphi^k f_s \varphi^{-k}(p).$$

Für $p \subset K'$ haben wir

$$(10) \qquad |\Phi(p) - F_{\vartheta_k}(p)| = |F_{\vartheta_k}(p) - \varphi^k f_s \varphi^{-k}(p)| =$$

$$= |\varphi^k \varphi^{-k} F_{\vartheta_k}(p) - \varphi^k f_s \varphi^{-k}(p)| < \frac{\varepsilon}{2}$$

was aus (7) wegen

$$(11) \qquad |\varphi^{-k} F_{\vartheta_k}(p) - f_s \varphi^{-k}(p)| < \delta$$

folgt. Die Formel (11) aber, folgt aus (8) und (9) wenn man für p $\varphi^{-k}(p)$ einsetzt und so den Gültigkeitsbereich dieser Formeln von Q_0 auf $Q_k \supset K'$ erweitert.

Die Formeln (6) und (11), geben unter Beachtung von $\vartheta_k > 1 - \alpha(\varepsilon)$ die gesuchte Approximationsformel:

$$|\Phi(p) - F(p)| < \varepsilon \quad \text{für} \quad p \in K' \qquad \text{w. z. b. w.}$$

Wir schließen mit der folgenden Bemerkung. Die Abbildung $\chi(p)$ hat einen ziemlich komplizierten Charakter. Es wäre für verschiedene Anwendungen von Bedeutung unseren Satz mit solchen Abbildungen zu beweisen, die ähnlich wie φ und ψ, eine einfache anschaulich-geometrische oder analytische Definition erlauben würden.

4. Die im vorigen Abschnitt konstruierten drei Abbildungen φ, ψ, χ haben die Eigenschaft jeden Punkt auf $S^{(n-1)}$ festzulassen, gehören also zu $F^{(n)}$. Wenn man daher für ein $F(p) \subset F^{(n)}$, für eine Folge gegen Null konvergenter Zahlen $\varepsilon_i > 0$, und für die Folge der Kugeln $K_i^{(n)}$ den Satz aus 3. anwendet, so erhält man eine Folge von Abbildungen $\Phi_i(p) \subset \Pi(\varphi, \psi, \chi)$ für die $\lim \Phi_i(p) = F(p)$ in ganz $K^{(n)}$ gilt, doch braucht die Konvergenz nicht gleichmäßig zu sein. (Ausgenommen den Fall $n = 1$, da eine gegen einen Automorphismus konvergierende Folge von Automorphismen des Intervalls schon notwendig, wie sich leicht zeigen läßt, gleichmäßig konvergiert).

Wir wollen nun die Abbildungen φ, ψ, χ so abändern, daß wir die gleichmäßige Konvergenz in ganz $K^{(n)}$ behaupten werden können. Wir beschränken uns dabei, der Einfachheit halber, auf den Fall $n = 2$. Die höheren n können ebenso behandelt werden.

Der Kreis $K^{(2)}$ liege in der (x, y) Ebene mit den Polarkoordinaten (r, α). $r = \frac{1}{2}$ sei die Gleichung des Kreises Q_0. Wir betrachten auf Q_0 die Punkte $C\left(r = \frac{1}{2}, \alpha = \frac{\pi}{2} + \alpha_1\right)$, $D\left(r = \frac{1}{2}, \alpha = \frac{\pi}{2} - \alpha_1\right)$ und ihre symmetrischen Bilder $C'\left(r = \frac{1}{2}, \alpha = -\frac{\pi}{2} - \alpha_1\right)$, $D'\left(r = \frac{1}{2}, \alpha = -\frac{\pi}{2} + \alpha_1\right)$ $\left(\alpha_1 \text{ ein fester Winkel} < \frac{\pi}{4}\right)$. Den Punkt

$\left(r=\frac{1}{2}, \frac{\pi}{2}+\alpha\right)$ verbinde man mit dem Punkte $P(r=1, \alpha=0)$, wenn $\alpha \leqslant -\alpha_1$, mit dem Punkte $R(r=1, \alpha=\pi)$ für $\alpha \geqslant \alpha_1$ und mit dem Punkte $\left(r=1, \frac{\pi}{2}+\frac{\alpha\pi}{2\alpha_1}\right)$ wenn $|\alpha| \leqslant \alpha_1$ ist und ebenso symmetrisch unterhalb der x-Achse. Den in (0, 0) beginnenden, unter dem Winkel α bis zur Peripherie von Q_0 und von dort bis zur Peripherie von $K^{(2)}$ mit einer der oben definierten Verbindungsstrecken, laufenden Streckenzug nenner wir $L(\alpha)$.

Auf jedem $L(\alpha)$ führen wir einen Parameter $t (0 \leqslant t \leqslant 1)$, der von 0 in (0, 0) bis 1 auf $S^{(2)}$ läuft, ein. Dem Punkte mit dem Parameter t ordnen wir den auf demselben $L(\alpha)$ gelegen Punkt mit dem Parameter $\sqrt{t}$ zu. Auf diese Weise ist eine Abbildung $\varphi_1(p) \subset F^{(2)}$ erklärt worden. Wenn man nun, wie in Abschnitt 2. die Mengen $\varphi_1^k(Q_0)$ mit Q_k bezeichnet, so umfaßt Q_k einen beliebigen, ganz im Innern von $K^{(2)}$ gelegenen Kreis K', wenn nur k hinreichend groß ist.

Die Strecke RC, der Kreisbogen CD und die Strecke DP bestimmen zusammen das Bild einer, im Intervall $(-1, +1)$ erklärten, stetigen Funktion $y=f(x)$.

Ist nun $|\vartheta| \leqslant 1$ so bezeichne $M(\vartheta)$ das Bild der Funktion $y=\vartheta f(x)$. Auf jedem $M(\vartheta)$ führen wir einen Parameter t ein, der von 0 in R bis 1 in P läuft und ordnen dem Punkte mit dem Parameter t. den auf demselben $M(\vartheta)$ gelegenen Punkt, mit dem Parameter $\sqrt{t}$ zu. Diese Abbildung erweitern wir noch oberhalb von $M(1)$ und unterhalb von $M(-1)$ zu einem Automorphismus $\overline{\psi}_1(p)$, so daß $\overline{\psi}_1 \subset F^{(2)}$ ist

Die Mengen $\psi_1^s(Q_0)$ (bei positiven und negativen s) liegen zwischen $RCDP$ und seinem symmetrischem Bild unter der x-Achse.

Wir nennen T_1 das Gebiet RCC', T_2 das Gebiet PDD'. Das Gebiet $T_1+T_2+Q_0$ geht bei den Abbildungen $\overline{\psi}_1^s$ in sich selbst über.

Bei hinreichend großem j, hat $\psi_1=\overline{\psi}_1^j$ die Eigenschaft, daß die Mengen $J_s=\psi_1^s(Q_0)$ paarweise elementfremd sind und $\lim \delta(J_s)=0$ ist. Wieder ist auch, wenn $p_s \subset J_s$, $\lim p_s=P$.

Nun können die Überlegungen des vorigen Abschnittes wörtlich wiederholt werden (man braucht jetzt nur nicht mehr die Mengen J_s durch J_s' zu ersetzen weil die $f_i(p)$ der Folge (4), wegen der Beschränkung auf $F^{(2)}$, gleich so genommen werden können, daß

$f_i(p) = p$ ist, auf der Peripherie von Q_0) und wir erhalten, wenn $F(p) \subset F^{(2)}$, $\varepsilon > 0$, und k beliebig gegeben sind ein

$$\Phi(p) = \varphi_1^k \psi_1^{-s} \chi \psi_1^s \varphi_1^{-k}(p)$$

so daß

$$(1) \qquad |\Phi(p) - F(p)| < \varepsilon \quad \text{für} \quad p \in Q_k$$

gilt.

Wir wählen k so groß, daß erstens

$$(2) \qquad |F(p) - p| < \frac{\varepsilon}{2}, \quad p \in K^{(2)} - Q_k$$

gilt (Dies ist immer möglich, weil auf $S^{(2)}$ $F(p) = p$ ist, also (2) außerhalb einem mit $K^{(2)}$ konzentrischen Kreis, der von Q_k, bei hinreichend großem k umfaßt wird, erfüllt ist) und zweitens so, daß

$$(3) \qquad \delta(T_1(K^{(2)} - Q_k)) < \frac{\varepsilon}{2}$$

$$(4) \qquad \delta(T_2(K^{(2)} - Q_k)) < \frac{\varepsilon}{2}$$

wird.

Wir behaupten, daß dann (1) in ganz $K^{(2)}$ erfüllt ist. Es bleibt also (1) für $p \in K^{(2)} - Q_k$ zu beweisen. Wir unterscheiden zwei Fälle:

1) $p \bar{\in} T_1 + T_2$. Dann ist $S_1^{-k}(p) \bar{\in} Q_0 + T_1 + T_2$, also auch $\psi_1^s \varphi_1^{-k}(p) \bar{\in} Q_0 + T_1 + T_2$. ($\psi_1$ führt, wie bemerkt wurde $Q_0 + T_1 + T_2$ in sich über).

Außerhalb von $Q_0 + T_1 + T_2$ ist aber $\chi(p) = p$, daher ist in diesem Falle $\Phi(p) = p$, also wegen (2), gilt (1).

2) $p \in T_1 + T_2$. Dann ist $\varphi_1^{-k}(p) \in T_1 + T_2$, also $\psi_1^s \varphi_1^{-k}(p) \in Q_0 + T_1 + T_2$ und $\psi_1^s \varphi_1^{-k}(p) \bar{\in} J_s$, also $\chi \psi_1^s \varphi_1^{-k}(p) \in T_1 + T_2 + Q_0 - J_s$, daher $\psi_1^{-1} \chi \psi_1^s \varphi_1^{-k}(p) \in T_1 + T_2$ (und $\bar{\in} Q_0$!).

Es ist also

$$\Phi(p) \in (T_1 + T_2)(K^{(2)} - Q_0).$$

Dabei bestätigt man ebenso, daß aus $p \in T_1$, $\Phi(p) \in T_1$ und aus $p \in T_2$, $\Phi(p) \in T_2$ folgt. Die Anwendung von (3) und (4) ergibt

$$|\Phi(p) - p| < \frac{\varepsilon}{2}.$$

Daraus und aus (2) folgt (1) auch in diesem Falle, womit unsere Behauptung vollständig erwiesen ist.

5. Wir wollen in diesem Abschnitt den Fall der stetigen, nicht notwendig eineindeutigen, Abbildungen behandeln.

Wir werden folgenden Satz beweisen:

Es gibt drei stetige Abbildungen von $K^{(n)}$ auf seine Teilmenge. $\chi(p)$, $\mu(p)$, $\mu_1(p)$ und einem Automorphismus von $K^{(n)}$, $\psi(p)$ so, daß die Abbildungen.

$$(1^a) \qquad \Phi_s(p) = \mu\, \psi^{-s}\, \chi\, \psi^s\, \mu_1(p)$$

jede andere beliebig genau approximieren d. h. zu jeder stetigen Abbildung $F(p)$, von $K^{(n)}$ auf ein Teilmenge von $K^{(n)}$, und jeder positiven Zahl ε gibt es ein $s(F, \varepsilon)$ so, daß

$$|\Phi_s(p) - F(p)| < \varepsilon \qquad p \in K^{(n)}$$

gilt.

Beweis. Die Abbildungen μ, μ_1, ψ und χ werden, wie folgt, erklärt.

$\mu(p)$ ordnet dem Punkte mit den Polarkoordinaten $(r, \xi_1, \ldots, \xi_{n-1})$ den Punkt $(2r, \xi_1, \ldots, \xi_{n-1})$ wenn $r \leqslant \frac{1}{2}$ und den Punkt $(1, \xi_1, \ldots, \xi_{n-1})$ wenn $r \geqslant \frac{1}{2}$ ist, zu. $\mu(p)$ bildet also Q_0 auf ganz $K^{(n)}$ homöomorph ab.

$\mu_1(p)$ ordnet dem Punkte p de Punkt $\frac{1}{2}p$, zu, es bildet also $K^{(n)}$ auf Q_0 homöomorph ab.

$\psi(p)$ erklären wir genau so, wie in Abschnitt 3, und bezeichnen wieder mit J_s die Menge $\psi^s(Q_0)$.

Wenn noch

$$(1) \qquad f_1(p), f_2(p), \ldots$$

eine in Raume $K^{(n)}$ $K^{(n)}$ überall dichte Folge bezeichnet, so setzen wir

$$(2) \qquad \chi(p) = \psi^s f_s \psi^{-s}(p) \quad \text{für} \quad p \in J_s.$$

Nach allgemeinen Sätzen (Tietze) läßt sich $\chi(p)$ zu einer stetigen Abbildung von ganz $K^{(n)}$ erweitern.

Nun sei ein beliebiges $F(p) \subset K^{(n)} K^{(n)}$ und eine positive Zahl ε gegeben.

Man bemerkt, daß wenn $f(p)$ in Q_0 erklärt ist $\mu_1 F \mu(p)$ bzw. $\mu f \mu_1(p)$ in Q_0 bzw. $K^{(n)}$ erklärt sind. Dabei ist

$$(3) \qquad \varrho(f, \mu_1 F \mu) = \tfrac{1}{2}\varrho(\mu f \mu_1, F)$$

In der Folge (1) gibt es ein f_s so, daß

$$(4) \qquad |f_s(p) - \mu_1 F \mu(p)| < \frac{\varepsilon}{2} \qquad p \in Q_0$$

ist. Wir behaupten, daß

$$(5) \qquad |\Phi_s(p) - F(p)| < \varepsilon \qquad p \in K^{(n)}$$

ist. Es ist nämlich $\mu_1(p) \in Q_0$ also $\psi^s \mu_1(p) \in J_s$.

Laut (2) und (1^a) ist daher

$$\Phi_s(p) = \mu f_s \mu_1(p).$$

Dies gibt aber, mit Rücksicht anf (3) und (4) die gewünschte Ungleichung (5).

Dieser Satz, angewendet für $n = 1$, ergibt die Lösung des am Anfang dieser Arbeit gestellten Problems:

Es gibt fünf Funktionen ($\mu(x)$, $\mu_1(x)$, $\psi(x)$, $\psi^{-1}(x)$, $\chi(x)$) so daß jede andere im Intervall $\langle 0, 1 \rangle$ erklärte, stetige Funktion, deren Werte demselben Intervall angehören, durch Zusammensetzen dieser fünf Funktionen, beliebig genau approximiert werden kann.

THÉORIE DES GROUPES. — *Sur le nombre de générateurs d'un groupe semi-simple.* Note de MM. **H. Auerbach** et **S. Ulam.**

On sait que tout groupe topologique compact et connexe admet deux générateurs, c'est-à-dire deux éléments constituant un sous-groupe dénombrable partout dense ([1]). Dans cette Note nous nous proposons d'établir un théorème analogue, valable pour tout groupe semi-simple de Lie.

Nous commençons par le

Théorème I. — *Soit $\mathcal{G}$ un groupe infinitésimal semi-simple réel ou complexe d'ordre r. Dans la suite de crochets obtenue en partant de deux éléments déterminés du groupe $\mathcal{G}$ il existe en général r éléments indépendants.*

Ce théorème fut démontré dans le cas complexe par Mlle Junge ([2]), ensuite M. Cartan en a donné une démonstration très simple, non publiée.

Démonstration. — Désignons par a un élément général du groupe $\mathcal{G}$. Soient γ le groupe abélien formé par les éléments échangeables à a, e_α les éléments appartenant aux racines caractéristiques $\neq 0$ de la matrice infinitésimale du groupe adjoint correspondant à l'élément a et p le nombre de ces racines. Si le groupe est réel on peut supposer que les éléments e_α sont conjugués deux à deux en même temps que les racines α.

Posons

$$e = a_0 + \Sigma\, c_\alpha e_\alpha,$$

a_0 étant un élément du groupe γ et les c_α des nombres complexes différents de zéro, dans le cas réel conjugués deux à deux en même temps que les racines α.

Les éléments

$$(1)\quad e_1 = [a\,e] = \Sigma a c_\alpha e_\alpha,\ e_2 = [a e_1] = \Sigma \alpha^2 c_\alpha e_\alpha,\ \ldots,\ e_p = 1\,[a\, e_{p-1}] = \Sigma \alpha^p c_\alpha e_\alpha$$

([1]) J. Schreier et S. Ulam, *Fund. Math.*, **24**, 1935, p. 302-304.

([2]) M. E. Junge, *Leipz. Ber.*, **78**, 1926, p. 399-444.

For commentary to this paper [18], see p. 681.

sont indépendants (et réels dans le cas réel). Ils forment, avec ceux d'une base du groupe γ, une base du groupe $\mathcal{G}$. Si a' est un élément quelconque du groupe γ, les crochets $[a'e_k]$ s'expriment linéairement par les éléments (1). Par conséquent, pour former le groupe dérivé de $\mathcal{G}$, il suffit d'employer les éléments (1) ou, à plus forte raison, les éléments a, e. Le groupe $\mathcal{G}$ étant son propre groupe dérivé, il s'ensuit que ces deux éléments possèdent la propriété en question. Il résulte aisément de leur définition que les couples d'éléments pour lesquels cela n'est pas le cas sont exceptionnels.

Théorème II. — *Soit $\mathcal{G}$ un groupe de Lie, réel ou complexe, connexe et semi-simple. Dans tout voisinage de l'élément unité il existe quatre éléments constituant un sous-groupe d'inombrable partout dense.*

Démonstration. — Supposons d'abord que le groupe soit réel. Soient U, V deux transformations infinitésimales de $\mathcal{G}$ jouissant de la propriété formulée dans le théorème I. Elles engendrent deux sous-groupes à un paramètre e^{tU}, e^{tV} dont chacun admet deux générateurs dans le voisinage donné de l'élément unité ([1]). Soit Γ le plus petit groupe fermé dans $\mathcal{G}$ contenant ces quatre éléments. En vertu d'un théorème de M. Cartan ([2]), Γ est un groupe réel de Lie. Comme il admet les transformations infinitésimales U, V, son groupe infinitésimal coïncide avec celui de $\mathcal{G}$.

Supposons maintenant que le groupe soit complexe. D'après un théorème de M. Cartan ([3]) il contient un sous-groupe réel semi-simple unitaire $\mathcal{G}_1$ du même ordre. Désignons par $U_1, \ldots, U_r$ une base du groupe $\mathcal{G}_1$ et par U, V deux transformations infinitésimales choisies dans ce sous-groupe comme précédemment. Dans le voisinage donné de l'élément unité il existe trois générateurs ([1]) du sous-groupe complexe e^{zU} et un générateur ([4]) du sous-groupe réel e^{tV}. Soit Γ le plus petit sous-groupe fermé dans $\mathcal{G}$ contenant ces quatre éléments. On peut considérer le groupe $\mathcal{G}$ comme un groupe réel de

---

([1]) On peut choisir deux nombres réels α, β ou trois nombres complexes α, β, γ aussi petits qu'on le veut de manière que la suite $\{m\alpha + n\beta\}$ ou $\{m\alpha + n\beta + p\gamma\}$, m, n, p désignant des entiers arbitraires, soit partout dense sur la droite réelle ou complexe

([2]) E. Cartan, *Mém. Sc. Math.*, **42**, 1930, p. 24.

([3]) E. Cartan, *Ann. Éc. Norm.*, **31**, 1914, p. 263-355. Voir aussi H. Weyl, *Math. Zeitschr.*, **24**, 1926, p. 371-375.

([4]) En vertu du fait que le groupe $\mathcal{G}_1$ est clos (H. Weyl, *loc. cit.*, p. 380-381).

Lie d'ordre $2r$; par conséquent Γ est aussi un groupe réel de Lie. Il admet évidemment les transformations infinitésimales U, iU, V.

Dans la suite des crochets formés à l'aide de U, V, il existe par hypothèse r transformations infinitésimales indépendantes. Si l'on écrit, dans chacun de ces crochets, une fois iU au lieu de U, on obtiendra les mêmes éléments multipliés par i. Il en résulte que le groupe Γ admet les transformations infinitésimales $U_1, \ldots, U_r$; $iU_1, \ldots, iU_r$, qui forment une base réelle du groupe $\mathcal{G}$.

(Extrait des *Comptes rendus des séances de l'Académie des sciences*, t. 201, p. 117, séance du 8 juillet 1935.)

Sur le nombre des générateurs d'un groupe topologique compact et connexe.

Par

J. Schreier et S. Ulam (Lwów).

Dans un travail récent [1]) nous avons posé le problème de trouver, dans un groupe topologique donné G, un nombre fini d'éléments tels que le groupe (dénombrable) engendré par ces éléments soit partout dense dans G. Nous y avons déterminé de tels systèmes d'éléments pour quelques groupes importants à une infinité de dimensions.

M. Auerbach [2]) a démontré que dans un groupe linéaire [3]) compact (clos) et connexe, il existe toujours deux éléments jouissant de la propriété mentionnée. Il a démontré en plus que l'ensemble des couples de tels éléments, considéré dans l'espace de tous les couples, a la mesure 1.

Dans cette Note nous nous proposons de démontrer le théorème suivant:

Soient G un groupe topologique compact, connexe et satisfaisant à la deuxième condition de séparabilité de M. Hausdorff (un tel groupe est toujours métrisable [4]), on peut donc supposer qu'il s'agit d'un groupe métrique compact) *et G^2 l'ensemble des couples d'éléments de G.*

[1]) J. Schreier und S. Ulam, *Über topologische Abbildungen euklidischer Sphären.* Fund. Math. XXIII, p. 102—118.

[2]) H. Auerbach, *Sur les groupes linéaires bornés* (III). Studia Math. V. p. 43—49.

[3]) C. à d. un groupe de transformations linéaires d'un espace euclidien.

[4]) V. par ex. D. van Dantzig, *Zur topologischen Algebra.* Math. Ann. 107.

For commentary to this paper [19], see p. 681.

Abstraction faite d'un certain ensemble de première catégorie dans G^2 tout couple d'éléments de G engendre un sous-groupe partout dense dans G.

Dans la démonstration, le théorème de M. Auerbach et la représentation des groupes compacts de M. von Neumann joueront un rôle essentiel.

Soit R_i la suite des voisinages ouverts remplissant la condition de Hausdorff. Considérons dans l'ensemble G^2 l'ensemble Z_i des couples tels que le groupe engendré par chacun d'eux contient des éléments de R_i. Il est aisé de voir que Z_i est ouvert dans G^2.

L'ensemble $Z = \prod_i Z_i$, c. à d. la partie commune des ensembles Z_i, se compose de couples d'éléments (g_1, g_2) qui jouissent de la propriété suivante: le groupe engendré par g_1 et g_2 a des éléments communs avec tout R_i, ou, autrement dit, est partout dense dans G.

Il s'agit donc de prouver que l'ensemble $G^2 - Z$ est de première catégorie. Les ensembles Z_i étant ouverts, il suffit de prouver que tout ensemble Z_i est partout dense. En effet, il en résulte que les ensembles $G^2 - Z_i$ sont non denses et $G^2 - Z = \sum_i (G^2 - Z_i)$ de première catégorie.

Nous renvoyons, à présent, le lecteur au travail important de M. J. von Neumann [1]), où il est prouvé qu'un groupe compact peut être représenté par un groupe isomorphe de matrices infinies.

Soit $\mathfrak{G}^{(\infty)}$ une telle représentation du groupe G et désignons conformément aux notations de M. J. von Neumann par $\mathfrak{G}^{(\mu)}$ ses facteur-groupes linéaires, par $B^{(\infty)}$ resp. $B^{(\mu)}$ les matrices éléments de $\mathfrak{G}^{(\infty)}$ resp. $\mathfrak{G}^{(\mu)}$.

Il est aisé de voir que tout voisinage V dans $\mathfrak{G}^{(\infty)}$ contient un ensemble W constitué par tous les éléments de $\mathfrak{G}^{(\infty)}$ tels que les matrices $B^{(\mu)}$ correspondantes forment une sphère $W^{(\mu)}$ dans $\mathfrak{G}^{(\mu)}$ pour un μ suffisamment grand.

Soient V_1, V_2 deux voisinages dans $\mathfrak{G}^{(\infty)}$ et $W_1 \subset V_1$, $W_2 \subset V_2$, $P_i \subset R_i$ trois ensembles jouissant de la propriété mentionnée pour un certain μ. En vertu du théorème cité de M. Auerbach, il existe deux éléments $p^{(\mu)}$, $q^{(\mu)}$ de $\mathfrak{G}^{(\mu)}$ qui appartiennent resp. à $W_1^{(\mu)}$ et $W_2^{(\mu)}$ et qui engendrent un sous-groupe partout dense de $\mathfrak{G}^{(\mu)}$. Donc, p et q désignant deux éléments de $\mathfrak{G}^{(\infty)}$ tels que les matrices $B^{(\mu)}$

[1]) J. von Neumann, *Einführung analytischer Parameter in topologischen Gruppen.* Annals of Math. 34. Voir p. 177.

correspondantes sont $p^{(\mu)}$ et $q^{(\mu)}$, il vient $p \in V_1$, $q \in V_2$ et un certain produit de puissances de ces éléments appartient à P_i, donc aussi à R_i, c. q. f. d.

Par un raisonnement analogue, en utilisant le fait qu'un groupe linéaire compact, connexe et abélien a toujours un générateur, on peut démontrer que dans un groupe général topologique compact, connexe et abélien tout élément, abstraction faite d'un certain ensemble de première catégorie, engendre un sous-groupe partout dense, ou, autrement dit, qu'un tel groupe est monothétique dans le sens de M. van Dantzig [1]).

D'après un théorème de M. A. Markoff [2]) tout groupe topologique abélien localement compact et séparable est le produit direct d'un groupe compact et d'un groupe isomorphe au groupe des translations d'un espace euclidien. Par conséquent, un tel groupe admet toujours un nombre fini de générateurs.

[1]) V. D. van Dantzig, *Homogene Kontinua*. Fund. Math. XV.

[2]) A. Markoff, C. R. 197, p. 610—612.

Sonderabdruck aus den Monatsheften für Mathematik und Physik,
42. Band, 1. Heft, 1935.

Sur une propriété caractéristique de l'ellipsoïde.[1]

Par **H. Auerbach, S. Mazur** et **S. Ulam** à Lwów.

Dans cette Note, nous nous proposons d'établir le théorème suivant:

Soit S une surface convexe dans l'espace à trois dimensions jouissant de la propriété suivante:

Il existe à l'intérieur de S un point O tel que toutes les sections planes passant par O sont des courbes affines entre elles.

Alors la surface S est un ellipsoïde.

En particulier, si les sections planes par O sont congruentes ou semblables entre elles, la surface S est une sphère[2].

1. Nous prouverons d'abord *qu'une courbe convexe C admettant une infinité de transformations affines en elle-même est une ellipse.*

Prenons pour l'origine O le centre de gravité et pour points unités deux points de la courbe C. Les affinités transformant cette courbe en elle-même sont alors homogènes, leurs coefficients ne dépassant une limite fixe. Elles forment donc un groupe linéaire borné. Or, un tel groupe admet une forme quadratique positive définie invariante[3]. On peut donc, par une substitution linéaire homogène, transformer ce groupe en un groupe de rotations. Soit C' la courbe correspondant à C par cette transformation. C' est une courbe convexe invariante par le groupe transformé. Ce groupe étant infini, il s'ensuit que C' est invariante par des rotations arbitrairement petites et par conséquent par toute rotation autour O. On en conclut que la courbe C' est un cercle et C une ellipse.

---

[1] Communiqué à la Société Polonaise de Mathématique, section de Lwów, à la séance du 10. octobre 1931.

[2] Cf. W. Süß, Eine elementare Eigenschaft der Kugel. Tôhoku Math. Journ. **26** (1926), p. 125—127.

[3] H. Auerbach, Sur les groupes bornés de substitutions linéaires. Comptes Rendus 195 (1932), p. 1367. — On peut établir que la courbe C est une ellipse sans employer ce théorème, mais la démonstration serait plus longue.

2. *Si toutes les sections planes par O sont des ellipses, la surface S est un ellipsoïde.*

Soit α un plan passant par O. En faisant tourner dans ce plan une droite passant par O, on voit qu'à un certain moment les deux segments de la corde correspondante deviennent égaux. Désignons par a la corde ainsi obtenue. Nous déterminons ensuite dans un plan β passant par O et ne contenant la corde a une corde b jouissant de la même propriété. Soit γ le plan déterminé par les cordes a, b. D'après une propriété élémentaire de l'ellipse, le point O est le centre de l'ellipse suivant laquelle le plan γ coupe la surface S.

Considérons un plan d'appui de la surface S parallèle au plan γ. Il n'a avec S qu'un seul point en commun, car autrement une certaine section plane par O ne serait pas une ellipse. Désignons par d la corde de la surface S déterminée par ce point et le point O. Il est aisé de voir que toutes les sections planes passant par d ont cette corde pour diamètre.

Faisons une transformation affine telle que la section plane γ devienne un cercle et d une corde perpendiculaire au plan γ. Le diamètre d sera alors un axe pour toutes les sections passant par cette corde. Ces ellipses auront le même centre, le même axe d, et leurs cordes perpendiculaires à l'axe au point O seront égales. La surface transformée est donc un ellipsoïde de révolution.

3. Supposons, par impossible, qu'aucune section plane par O ne soit pas une ellipse. Soient C_0 une de ces sections et c_1^0 une corde quelconque de la courbe C_0. D'après 1. la courbe C_0 n'admet qu'un nombre fini de transformations affines en elle-même. Désignons par $c_1^0, \ldots, c_k^0$ les différentes cordes correspondantes à c_1^0 par ces transformations. On voit sans peine que ces cordes sont seulement permutées par toute affinité transformant la courbe C_0 en elle-même.

Soit C une section plane quelconque par O. Il existe par hypothèse une affinité T transformant la courbe C_0 en C. Soient $c_1, \ldots, c_k$ les cordes correspondantes aux cordes $c_1^0, \ldots, c_k^0$ par cette affinité. Ces cordes sont indépendantes de la transformation choisie. En effet, soit T' une autre affinité transformant la courbe C_0 en C. L'affinité $T' T^{-1}$ transforme la courbe C_0 en elle-même et, par conséquent, permute seulement les cordes $c_1^0, \ldots, c_k^0$, ce qui serait impossible si la transformation T' faisait correspondre à l'une d'elles une corde de la courbe C non contenue dans la série $c_1, \ldots, c_k$.

Nous avons ainsi défini dans toute section plane par O un système de k cordes passant par O. Elles seront désignées dans la suite comme *cordes distinguées.* Nous allons prouver qu'elles varient d'une façon continue avec le plan de section.

Soient C_0 une section déterminée et $c_{01}, \ldots, c_{0k}$ ses cordes distinguées. Traçons dans son plan γ_0 les courbes $(1-\varepsilon)\ C_0$, $(1+\varepsilon)\ C_0$, homothétiques par rapport au point O, ε désignant une petite constante positive. La première de ces courbes est située à l'intérieur, la seconde à l'extérieur de la surface S. Si l'on coupe les deux cylindres perpendiculaires au plan γ_0 dont les directrices sont ces courbes homothétiques par deux plans symétriques par rapport au plan γ_0, parallèles à ce plan et suffisamment rapprochés, on obtient deux nappes, la première intérieure, la seconde extérieure à la surface S. Si donc C est une section par un plan passant par O et faisant un angle suffisamment petit avec le plan γ_0, la projection orthogonale C' de la courbe C sur le plan γ_0 sera comprise entre les deux courbes homothétiques. Il existe par hypothèse une affinité plane A transformant la courbe C_0 en C' Comme les aires de ces deux courbes sont peu différentes, le module du déterminant de cette affinité est voisin de l'unité.

Soient $\{\gamma_n\}$ une suite de plans par O tendant vers le plan γ_0, $\{C_n\}$ la suite des sections correspondantes, $\{C'_n\}$ celle de leurs projections orthogonales sur le plan γ_0. Nous désignerons encore par $c_{n1}, \ldots, c_{nk}$ les cordes distinguées de la courbe C_n, rangées dans un ordre quelconque, et par $c'_{n1}, \ldots, c'_{nk}$ leurs projections orthogonales sur le plan γ_0. Enfin, soit A_n une affinité plane transformant la courbe C_0 en C'_n. Il est aisé de voir que les coefficients des affinités A_n sont uniformement bornés. En effet, si l'on choisit dans le plan γ_0 un système de coordonnées avec O comme origine et les points unités situés sur la courbe C_0, alors les coefficients de l'affinité A_n sont les coordonnées des points correspondant sur la courbe C'_n aux points unités. En outre, les modules des déterminants de ces affinités tendent vers 1.

Il s'ensuit qu'il existe une suite d'indices $\{n_i\}$ telle que

1°. Les affinités A_{n_i} tendent vers une affinité A à déterminant ± 1.

2°. On a $\lim\limits_{i \to \infty} c'_{n_i m} = \bar{c}_m$ $(m = 1, \ldots, k)$ $\bar{c}_1, \ldots, \bar{c}_k$ désignant certaines cordes de la courbe C_0.

L'affinité A transforme la courbe C_0 en elle-même. En effet, soit P un point quelconque de C_0. Le point $A_{n_i}(P)$ est situé sur la courbe C'_{n_i} et tend vers le point $A(P)$. Comme les courbes C'_{n_i} tendent vers C_0, le point $A(P)$ est situé sur C_0.

L'affinité $A^{-1} A_{n_i}$ transforme la courbe C_0 en C'_{n_i} et les cordes distinguées $c_{01}, \ldots, c_{0k}$ de C_0 en $c'_{n_i 1}, \ldots, c'_{n_i k}$. Comme elle tend vers l'identité, les cordes $\bar{c}_1, \ldots, \bar{c}_k$ se confondent avec $c_{01}, \ldots, c_{0k}$, à l'ordre près.

Les cordes $c_{n_i m}$ des courbes C_{n_i} différant très peu de leurs projections il s'ensuit qu'on peut les désigner de manière que $\lim_{i \to \infty} c_{n_i m} = c_{0m}$ $(m = 1, \ldots, k)$.

Nous avons ainsi démontré que toute suite des plans par O tendant vers γ_0 contient une suite partielle telle que les cordes distinguées des sections correspondantes convergent respectivement vers celles de C_0.

On en déduit aisément que les cordes distinguées varient d'une façon continue avec le plan de section. Cela veut dire qu'étant donnée une section déterminée C_0 et un nombre $\varepsilon > 0$, il existe un nombre $\eta > 0$ tel que pour toute section dont le plan fait avec celui de C_0 un angle $< \eta$ on peut faire correspondre les cordes distinguées à celles de C_0 de manière que chaque corde fasse avec la corde correspondante de C_0 un angle $< \varepsilon$.

4. Désignons par K la sphère de rayon 1 au centre O. Soit P un point quelconque de la sphère K et γ le plan par O parallèle au plan tangent en P. Les droites menées par P parallèlement aux cordes distinguées de la section par le plan γ seront des tangentes de la sphère en P.

D'après 3. les $2k$ directions tangentes ainsi obtenues varient d'une façon continue avec le point P. En vertu d'un théorème bien connu, on en peut choisir une qui soit continue sur toute la sphère. Or cela étant en contradiction avec un théorème célèbre de Brouwer, notre théorème se trouve démontré.

Zusatz während der Korrektur: Pendant la correction des épreuves nous nous sommes aperçu que le raisonnement du nro. 3. doit être modifié de la manière suivante. Il faut prendre pour c_1^0 une corde passant par le centre de gravité de la section C_0. On obtiendra un système de k cordes $c_1^0, \ldots, c_k^0$ passant par ce point et, de même, dans la section C un système de k cordes $c_1, \ldots, c_k$ passant par le centre de gravité de cette section. La démonstration s'achève comme précédemment, il faut seulement ajouter la remarque, facile à vérifier, que les centres de gravité des projections C'_n et, par suite, ceux des sections C_n, tendent vers le centre de gravité de C_0.

(Eingegangen: 7. II. 1934.)

Über die Automorphismen der Permutationsgruppe der natürlichen Zahlenfolge.

Von

J. Schreier und S. Ulam (Lwów).

In einer früheren Arbeit [1]) haben wir die Gruppe S_∞ aller eineindeutigen Abbildungen der Menge der natürlichen Zahlen auf sich selbst untersucht und u. a. die Normalteiler dieser Gruppe bestimmt [2]). Hier wollen wir die Automorphismen dieser Gruppe, d. h. alle eineindeutigen isomorphen Abbildungen der Gruppe auf sich selbst, behandeln.

Satz. *Ist $F(x)$ ein Automorphismus von S_∞, so gibt es ein Element $s \epsilon S_\infty$, derart daß*

$$F(x) = sxs^{-1}$$

identisch gilt.

In der üblichen Terminologie der Gruppentheorie kann man dies auch so aussprechen: *Es gibt keine äußeren Automorphismen von S_∞.*

Will man S_∞ als eine topologische Gruppe auffassen, so ergibt sich aus unserem Satze insbesondere, daß *jeder Automorphismus von S_∞ stetig ist.*

Beweis. Man bemerkt zunächst: wenn x und y zwei konjugierte Elemente sind $(x = tyt^{-1})$, so sind auch $F(x)$ und $F(y)$ konjugiert. Jede Menge von konjugierten Elementen geht daher wieder in eine solche über. Man ersieht auch sofort, daß die Elemente x und $F(x)$ immer von der gleichen *Ordnung* sind.

---

1) J. Schreier und S. Ulam, *Über die Permutationsgruppe der natürlichen Zahlenfolge,* Studia Math. 4 (1933), S. 134—141.

2) Herr R. Baer hat uns darauf aufmerksam gemacht, daß diese Gruppe und ihre Normalteiler von Herrn Onufri behandelt wurde.

For commentary to this paper [21], see p. 682.

Wir bezeichnen mit C die Menge aller mit der *Transposition* (1,2) konjugierten Elemente. Die Menge C ist die einzige abzählbare Menge konjugierter Elemente von der Ordnung 2, welche folgende Eigenschaft besitzt:

(α) *Die Produkte* $x_1 y_1$ *und* $x_2 y_2$, *wo* $x_1, x_2, y_1, y_2 \in C$, *wenn sie beide Elemente von der Ordnung* 2 *darstellen, sind miteinander konjugiert.*

Sie gehören nämlich zur Menge derjenigen Elemente, die mit der Permutation (1,2) (3,4) konjugiert sind.

Ist aber irgend eine andere abzählbare Menge von miteinander konjugierten Elemente der Ordnung 2 gegeben, so besitzt sie die Eigenschaft (α) nicht. Man kann dann nämlich immer zwei Paare von Elementen dieser Menge angeben, deren Produkte miteinander nicht konjugiert sind. Um dies einzusehen, betrachten wir eine Menge aller mit einer Permutation $(1,2), \ldots, (2n-1, 2n)$, wo $n > 1$, konjugierten Elemente [3]. Man nehme:

$$x_1 = (1, 2), \ldots, (2n-1, 2n), \quad y_1 = (2n+1, 2n+2), \ldots, \quad (4n-1, 4n),$$

$$x_2 = (1, 2), \ldots, (2n-1, 2n), \quad y_2 = (1,2)(2n+1, 2n+2), \ldots, (4n-3, 4n-2).$$

Man ersieht, daß die entsprechenden Produkte $x_1 y_1$ und $x_2 y_2$ beide von der Ordnung 2, doch nicht miteinander konjugiert sind.

Wir betrachten jetzt die Menge $F(C)$. Aus der einleitenden Bemerkung folgt, daß $F(C)$ mit einer gewissen abzählbaren Menge von miteinander konjugierten Elementen der Ordnung 2 identisch sein muß. Dann aber besitzt $F(C)$ auch die Eigenschaft (α), da F ein Isomorphismus ist, und bei einem solchen, wie man sofort ersieht, diese Eigenschaft erhalten bleibt. Da C die einzige abzählbare Menge von konjugierten Elementen von der Ordnung 2 mit der Eigenschaft (α) ist, so folgt $F(C) = C$.

Wir untersuchen jetzt genauer die Abbildung $C \to F(C)$.

Zunächst bemerken wir, daß das Bild einer Folge von Transpositionen (α, n), wo α eine fixe Zahl ist und n alle Zahlen durchläuft, wieder eine solche Folge (β, n) ergibt. Um dies zu beweisen, genügt es zu bemerken, daß ein Produkt zweier Elemente aus C dann und nur dann ein Element von der Ordnung 3 ist, wenn beide Elemente zu einer einzelnen Folge von obiger Gestalt gehören.

---

[3]) Dies sind nämlich alle möglichen abzählbaren Klassen konjugierter Elemente, die aus Elementen von der Ordnung 2 bestehen.

17*

Man ersieht ferner leicht, daß die Zuordnung von β zu α eine eineindeutige Abbildung der Menge der natürlichen Zahlen auf sich selbst ist. Wir bezeichnen diese Abbildung, d. h. dieses Element von S_∞, mit $s(n)$.

Betrachten wir nun den Automorphismus von S_∞:

$$\Phi(x) = s^{-1} F(x) s.$$

Es ist unser Zweck zu zeigen, daß

$$\Phi(x) \equiv x$$

ist. Man beachte, daß im Falle, wo x eine Transposition ist, dies direkt aus der Definition der Abbildung s folgt. Im Falle, wo x ein beliebiges Element von der Ordnung 2 ist, wird der Beweis indirekt geführt.

Es sei nämlich x_0 von der Ordnung 2 und $y_0 = \Phi(x_0) \neq x_0$. Ohne Einschränkung der Allgemeinheit kann man $x_0(n_1) = n_2$ und $y_0(n_1) \neq n_2$ voraussetzen. Es sei z die Transposition (n_1, n_2). Man sieht, daß während das Element zx_0 von der Ordnung 2, das Element zy_0 von einer höheren Ordnung ist. Dies ist ein Widerspruch, da Φ ein Isomorphismus ist, woraus

$$\Phi(zx_0) = \Phi(z)\Phi(y_0) = zy_0 \qquad (z = \Phi(z)\,!)$$

folgt.

Um jetzt zu zeigen, daß $\Phi(x)$ überall, auch für die Elemente x höherer Ordnung als 2. identisch gleich x ist, benutzen wir die Tatsache, daß jedes Element z von S_∞ Produkt zweier Elemente von der Ordnung 2 ist:

$$z = xy, \qquad x^2 = y^2 = 1.$$

Dies folgt aus der Zerlegung von z in Zyklen und aus der Formel

$$(\ldots, 6, 4, 2, 1, 3, 5, \ldots) = [(1,2,)\,(3,4)\ldots]\cdot[(1)\,(2,3)\,(4,5)\ldots]$$

sowie aus analogen Formeln für endliche Zyklen.

Es sei jetzt z ein beliebiges Element von S_∞. Wir können x und y von der Ordnung 2 finden, derart daß — wie wir schon wissen — $\Phi(x) = x$, $\Phi(y) = y$ gelte, und daß $z = xy$ sei. Es ist dann

$$\Phi(z) = \Phi(xy) = \Phi(x)\,\Phi(y) = xy = z, \qquad \text{w. z. b. w.}$$

Ein analoger Satz kann für höhere Mächtigkeiten bewiesen werden:

Ist N eine Menge von beliebiger unendlicher Mächtigkeit, $S(N)$ die Gruppe aller Permutationen der Menge N, so ist jeder Automorphismus von $S(N)$ in der Gestalt $F(x) = s\,x\,s^{-1}$, wo $s \in S(N)$, darstellbar.

Dagegen bleibt die Frage nach der allgemeinen Gestalt eines mehrstufigen Isomorphismus, der S_∞ auf eine Teilmenge abbildet, offen.

On the equivalence of any set of first category to a set of measure zero.

By

J. C. Oxtoby and S. M. Ulam (Cambridge, U. S. A.).

In n-dimensional euclidean space $R^{(n)}$ it is an easy matter to define sets of first category which have positive measure, and even, ones whose complements are of measure zero. For instance, the complement of the intersection of any sequence of dense open sets whose measures tend to zero defines such a set. Nevertheless one may ask whether such a set is equivalent to a set of measure zero under the group of homeomorphisms of the space onto itself. In other words, does there exist an automorphism (that is, a homeomorphism of the space onto itself) which carries the given set into one of measure zero. The object of the present note is to show that such a transformation always exists, indeed that the automorphisms which carry a given set of first category into one of measure zero form a residual set in the space of antomorphisms, provided the latter is suitably metrised.

To begin with, consider the unit cube $\mathfrak{I}^{(n)}$ in n-dimensional euclidean space, ($n \geqslant 1$). Let $[H]$ denote the space of all automorphisms of $\mathfrak{I}^{(n)}$ metrise by the formula [1])

$$\varrho(g, h) = \max_{x \in \mathfrak{I}^{(n)}} (|gx - hx|, |g^{-1}x - h^{-1}x|),$$

where $|x-y|$ denotes the euclidean distance between x and y. The space $[H]$ is complete and the group operations are continuous in the metric, so that it forms a metric group.

[1]) This metric is equivalent to the usual one. See S. Banach, *Théorie des opérations linéaires*, Monografie Matematyczne **1**, Warszawa 1932, p. 229.

For commentary to this paper [22], see p. 682.

Theorem 1. *Let A be any set of first category in $\mathfrak{I}^{(n)}$. The automorphisms of $\mathfrak{I}^{(n)}$ which carry A into a set of measure zero form a residual set in $[H]$.*

By hypothesis, $A=\sum_{i=1}^{\infty}F_i$, F_i nowhere dense. We may suppose all the F_i closed, for the union of their closures defines a first category set containing A, and if the theorem is proved for this set it will be all the more true for A. Let E_{ik} $(i,k=1,2,\ldots)$ denote the set of automorphisms h such that $m(hF_i)<1/k$, where $m(X)$ denotes the n-dimensional measure of X. Then the set E of automorphisms such that $m(hA)=0$ is represented by $E=\prod_{i,k=1}^{\infty}E_{ik}$. To prove the theorem it will be sufficient to show that the sets E_{ik} are all open and dense in $[H]$, for the complement of E will then be represented as a union of countably many nowhere dense sets. Let h be any element of E_{ik}. Then $m(hF_i)<1/k$. Since hF_i is closed, there is an $\varepsilon>0$ such that the ε-neighborhood $(hF_i)_\varepsilon$ of hF_i likewise has measure less than $1/k$. Consider any g in $[H]$ such that $\varrho(g,h)<\varepsilon$. We have $gF_i\subset(hF_i)_\varepsilon$, hence $m(gF_i)<1/k$, and so $g\epsilon E_{ik}$ and E_{ik} is open. To show that E_{ik} is dense, consider any $h\epsilon[H]$, $\varepsilon>0$. By the continuity of the group composition, there is a $\delta>0$ such that $\varrho(gh,h)<\varepsilon$ if $\varrho(g,I)<\delta$, I denoting the identical transformation. Divide $\mathfrak{I}^{(n)}$ into rectangular cells $\sigma_1,\sigma_2,\ldots,\sigma_l$ of diameter less than δ. In the interior of each σ_j select a closed sphere s_j which lies outside hF_i. This can be done since hF_i is nowhere dense. Now define g so as to expand each s_j within σ_j while leaving the boundary points of the cells σ_j fixed, the amount of expansion being such that $m(ghF_i)<1/k$. Such an automorphism g may conveniently be defined by a linear expansion and contraction along each radial line from the center of s_j to the boundary of σ_j. Since g moves no point by more than δ we have $\varrho(g,I)<\delta$. Hence $\varrho(gh,h)<\varepsilon$, $gh\epsilon E_{ik}$, which completes the proof.

Since $[H]$ is a complete space, it follows that there exists an automorphism of $\mathfrak{I}^{(n)}$ which carries A into a set of measure zero. If desired, the automorphism can be chosen so as to leave all boundary points of $\mathfrak{I}^{(n)}$ fixed, for the above proof applies without change to the complete sub-space of automorphisms which leave boundary points fixed.

One may ask whether a stronger theorem can be asserted, for instance, whether it is always possible to find a differentiable transformation, or an automorphism fulfilling a Lipschitz or Hölder condition, which will effect the desired transformation. That this is not always possible is shown by the following considerations.

Let $\delta_k(\varepsilon)$ be any sequence of positive monotone increasing functions defined for $\varepsilon>0$ and tending to zero with ε. There is a set A of first category in $\mathcal{I}^{(n)}$ such that no automorphism of $\mathcal{I}^{(n)}$ having a modulus of continuity $\leqslant$ than one of the functions $\delta_k(\varepsilon)$ carries A into a set of measure zero.

Let ε_i $(i=1,2,\ldots)$ be the radius of a sphere of n-dimensional measure $1/2^{i+1}$ and let E_i' be a positive number such that $\delta_k(E_i')\leqslant E_i$. Let $x_1, x_2, \ldots$ be a sequence of points dense in $\mathcal{I}^{(n)}$. Define F_k as the complement of the open set obtained by taking a sphere of radius $\delta_k(\varepsilon_i)$ about each point x_i. Then F_k is nowhere dense and any automorphism having modulus of continuity $\leqslant \delta_k(\varepsilon)$ will take F_k into a set of measure at least $1/2$. The set $A=\sum_{k=1}^{\infty} F_k$ is therefore carried into a set of measure at least $1/2$ by any automorphism which has for modulus of continuity a function $\leqslant$ than one of the functions $\delta_k(\varepsilon)$.

An automorphism is said to fulfill a Hölder condition if there exist constants $K>0$, $0<a\leqslant 1$ such that for all x, y in $\mathcal{I}^{(n)}$ we have $|hx-hy|\leqslant K|x-y|^{a}$. Such a transformation has a modulus of continuity $\leqslant$ than a function of the form $\delta_k(\varepsilon)=1/K\varepsilon^{1/k}$, whence it follows that not every first category set is equivalent to a set of measure zero under an automorphism fulfilling a Hölder condition.

Returning now to the consideration of unbounded sets we can deduce at once the following theorem.

Theorem 2. *Any set of first category in $R^{(n)}$ is equivalent to a set of n-dimensional measure zero under an automorphism of $R^{(n)}$.*

It is only necessary to divide the space into unit cubes and transform the part in each cube into a set of measure zero. If an automorphism of each is chosen which leaves boundary points fixed, the automorphisms of the separate cubes will join together to form an automorphism of the whole space which carries the given set into one of measure zero.

If one seeks to generalize the result that the transformations effecting the equivalence form a residual set in the space of automorphisms, one must first introduce a suitable metric. The metric previously used is not satisfactory because the domain is no longer compact. However, if one maps $R^{(n)}$ into the surface $S^{(n)}$ of the unit sphere in $R^{(n+1)}$ by projection from the north pole, one sees that automorphisms of $R^{(n)}$ are set in one-to-one correspondence with automorphisms of $S^{(n)}$ which leave the north pole fixed. Since $S^{(n)}$ is compact, the metric previously employed can be used, the absolute value signs now denoting distances in $R^{(n+1)}$. The proof given above for Theorem 1 carries through without change except that in the representation $A = \sum_{i=1}^{\infty} F_i$ the sets F_i should be taken to be bounded. Thus with this convention as to the metric in the space of automorphisms of $R^{(n)}$ one can again assert that the automorphisms which carry a given set of first category into a set of measure zero form a residual set in the space of automorphisms.

A set is said to be of *first category at a point* x if it intersects a neighborhood of x in a set of first category. It is said to have *the property of Baire* (in the weak sense) if it can be represented in the form $G - P_1 + P_2$, where G is open, P_1 and P_2 of first category [2]). Applying the above results to the set $P_1 + P_2$ we conclude:

Theorem 3. *Any set in $R^{(n)}$ (of $\mathfrak{I}^{(n)}$) having the property of Baire (in the weak sense) is equivalent under an automorphism of $R^{(n)}$ (or $\mathfrak{I}^{(n)}$) to a measurable set, indeed to a set differing from an open set by a set of measure zero. Points at which the set or its complement are of first category go into points of density* 0 *and* 1 *respectively. The automorphisms effecting such a transformation form a residual set in the space of automorphisms of $R^{(n)}$ (or $\mathfrak{I}^{(n)}$).*

Noting that a non-void open set has positive measure, on combining with previous results we obtain.

Theorem 4. *A set in $R^{(n)}$ (or $\mathfrak{I}^{(n)}$) having the property of Baire (in the weak sense) is of first or second category according as the automorphisms which carry it into a set of measure zero form a residual set or a first category set in the space of automorphisms of $R^{(n)}$ (or $\mathfrak{I}^{(n)}$).*

---

[2]) C. Kuratowski, *Topologie* I, Monografie matematyczne **3**, Warszawa-Lwów 1933, p. 49.

To set up an automorphism which will transform a given set of first category into a set of measure zero, one may proceed as follows. Represent the given set in the form $A = \sum_{i=1}^{\infty} N_i$ where the sets N_i are bounded, nowhere dense, and $N_i \subset N_{i+1}$ $(i = 1, 2, \ldots)$. First compress N_1 to a set of measure less than 1 by a transformation h_1 which expands spheres in the complement, as described in the proof of Theorem 1. Follow this by a smaller transformation h_2 compressing $h_1 N_2$ to measure less than $1/2$. At the i-th step let h_i compress $h_{i-1} \ldots h_2 h_1 N_i$ to measure less than $1/i$. Proceed in this manner making the transformations h_i grow small so rapidly that the successive products will converge uniformly together with their inverses. The limiting transformation is the desired automorphism. Now observe that at the i-th step the compression can be made continuously. The final automorphism can thus be joined to the identity by a continuous family of automorphisms, and we have in fact an isotopic deformation of the space. Hence we can assert the following more precise theorem.

Theorem 5. *Any set of first category in $R^{(n)}$ (or $\mathfrak{I}^{(n)}$) is equivalen to a set of measure zero under an isotopic deformation of the space. Likewise any set having the property of Baire (in the weak sense) can be deformed into a measurable set by an isotopic deformation of the space.*

The general question of the equivalence of an arbitrary set of first category to one of measure zero under automorphism has meaning in any complete metric space in which a measure is defined. One may ask how far the results here obtained can be generalized. Regarding more general measures in the spaces $R^{(n)}$ and $\mathfrak{I}^{(n)}$ it may be remarked at once that the proofs of Theorems 1, 2 and 3 apply to any Carathéodory measure, provided that each time a division into cells is made one chooses them so that their boundaries have measure zero. On the other hand, the methods of proof are restricted to euclidean or at least to locally euclidean spaces.

An example of a space requiring independent consideration is the Cantor set C with the usual probability measure. That is, C is assigned measure 1 and congruent subsets are assigned equal measure. Here the analog of Theorem 1 holds.

Theorem 6. *The automorphisms of C which carry a given set of first category (with respect to C) into a set of measure zero (in the probability measure) form a residual set in the space of automorphisms of C.*

The proof runs parallel to that of Theorem 1 until the proof that E_{ik} is dense. Here we divide C into disjoint segments $\sigma_1, \ldots, \sigma_l$ of length less than δ. In each σ_i select a prefect subset lying outside hF_i. Since any two perfect subsets of C are homeomorphic, g can be chosen so as to leave each σ_j invariant and yet map s_j into a subset of σ_j having an arbitrarily large fraction of the measure of σ_i. Thus ghF_i can be made to have measure less than $1/k$. The proof then proceeds similarly as before.

Isotopic deformation of C is of course impossible, so that no analog of Theorem 5 is to be sought.

The possible extension of the results here obtained to locally compact groups with Haar measure is a question which the authors hope to discuss subsequently.

Society of Fellows, Harvard University.

ANNALS OF MATHEMATICS
Vol. 40, No. 3, July, 1939

ON THE EXISTENCE OF A MEASURE INVARIANT UNDER A TRANSFORMATION

BY J. C. OXTOBY AND S. M. ULAM

(Received April 12, 1939)

1. Introduction

Transformations which leave a measure function invariant are of importance in many branches of mathematics, notably in dynamics and in the theory of probability. It is therefore of some interest to investigate under what circumstances an invariant measure is possible. This may be considered part of the general problem of determining conditions under which there exists in a space a measure which is invariant under a group of transformations acting on the space. This problem has been solved only in special cases. Thus in case the space is a locally compact separable topological group, considered as acted upon by the group of right (or left) translations by its own elements, Haar's theorem[1] asserts the existence of an invariant measure. We consider the case of a general complete separable metric space acted on either by a group consisting of an automorphism and its powers, or by a one-parameter group of automorphisms. We shall set up a method which yields a finite invariant measure whenever such a measure is possible.

One must, of course, specify what measure functions are to be admitted. In all cases they will be required to be completely additive, defined at least for all Borel sets, with real values $\geqq 0$, and not identically zero. For the most part, we shall be concerned with finite measures, that is, measure functions in which the whole space has finite measure. Only in §4 shall we consider infinite measures. A measure is said to be invariant under a transformation T if $m(TA)$ is defined and equal to $m(A)$ whenever $m(A)$ is defined. A measure is said to be invariant under a group of transformations if it is invariant under every transformation of the group.

In the case of a compact metric space, Kryloff and Bogoliouboff[2] have shown that every automorphism (that is, one-to-one bicontinuous transformation), and every one-parameter continuous group of automorphisms, possesses a finite

[1] A. Haar, *Der Massbegriff in der Theorie der kontinuerlichen Gruppen*, Annals of Math. (2), *34* (1933) pp. 147–169. See also the note by S. Banach in Saks, *Theory of the Integral*, Warsaw-Lwów 1937, pp. 314–319. Also J. von Neumann, *The uniqueness of Haar measure*, Recueil mathématique Moscou, (1), vol. 43 (1936).

[2] N. Kryloff and N. Bogoliouboff, *La théorie générale de la mesure dans son application à l'étude des systémes dynamiques de la mécanique non linéaire*, Annals of Math. (2), *38* (1937) pp. 65–113. Theorem 1, p. 92, states the result for a continuous group, and the same considerations apply to a single automorphism.

For commentary to this paper [23], see p. 683.

invariant measure. But in general, the existence of such a measure imposes a restriction on the transformation, and even in the compact case the only invariant measures may be quite trivial ones, confined to a finite set of points. By using a somewhat more general method than that of Kryloff and Bogolioubоff, we shall obtain a theorem which contains their result as a corollary, and shall also find necessary and sufficient conditions for the existence of non-trivial invariant measures.

2. Finite measures invariant under a transformation

THEOREM 1: *Let T be an automorphism of a complete separable metric space E. In order that there exist a finite invariant measure in E it is necessary and sufficient that for some point p and compact set $C \subset E$, we have*

$$\lim_{n\to\infty} \frac{1}{n} \sum_{k=1}^{n} f_C(T^k p) > 0,$$

where f_C is the characteristic function of C. It is sufficient that the limit superior be positive.

In words, the condition means that there exists a compact set some point of which returns with positive frequency (or superior frequency) under iteration of T.

Suppose such a measure exists. If $m(A)$ is any finite measure in a complete separable space, defined at least for Borel sets, there exists a compact set C with positive measure.[3] Since the measure is assumed invariant under T, it follows from the ergodic theorem of Birkhoff[4] that the limit

$$f^*(p) = \lim_{n\to\infty} \frac{1}{n} \sum_{k=1}^{n} f_C(T^k p)$$

exists for almost all p, is measurable (m), and that

$$\int_E f^*(p)\, dm(p) = m(C).$$

Hence the integrand must be defined and positive for at least one point p.

To prove sufficiency, it is convenient to make use of a notion of generalized limit due to Mazur and Banach. Namely, there exists a real-valued functional $\text{Lim}\, \xi_n$, defined for all bounded sequences $\{\xi_n\}$ of real numbers, such that

1) $\text{Lim}\,(a\xi_n + b\eta_n) = a\, \text{Lim}\, \xi_n + b\, \text{Lim}\, \eta_n \quad a, b$ real
2) $\text{Lim}\, \xi_n \geqq 0 \quad \text{if} \quad \xi_n \geqq 0 \quad \text{for} \quad n = 1, 2, \cdots$

[3] S. Ulam, *Sur la mesure dans un espace séparable et complet*, to appear in Comptes Rendus (Paris). The proof runs briefly as follows. Let $x_1, x_2, \cdots$ be a sequence of points dense in E. Let S_n^k denote the closed sphere with center x_n and radius $1/k$, and let $F_N^k = \sum_{n=1}^{n=N} S_n^k$. Suppose $0 < \epsilon < m(E)$ given. Define successively $N_1, N_2, \cdots$ where N_k is the least positive integer such that $m(\prod_1^k F_{N_i}^i) > m(E) - \epsilon$. The closed set $C = \prod_1^\infty F_{N_i}^i$ is totally bounded, therefore compact, and $m(C) \geqq m(E) - \epsilon$.

[4] See e. g. E. Hopf, *Ergodentheorie*, Ergebnisse der Mathematik und ihrer Grenzgebiete, Berlin 1937, §14.

3) $\text{Lim}\, \xi_{n+1} = \text{Lim}\, \xi_n$

4) $\liminf_{n \to \infty} \frac{1}{n} (\xi_1 + \cdots + \xi_n) \leqq \text{Lim}\, \xi_n \leqq \limsup_{n \to \infty} \frac{1}{n} (\xi_1 + \cdots + \xi_n)$.

The generalized limit is thus an extension of the limit by first means. In addition, it can be so defined that for any prescribed bounded sequence $\{\xi_n^0\}$ we have

5) $\text{Lim}\, \xi_n^0 = \limsup_{n \to \infty} \frac{1}{n} (\xi_1^0 + \cdots + \xi_n^0)$.

The existence of a generalized limit with these properties is an immediate consequence of a theorem of Banach on extension of linear functionals.[5] We shall take $\xi_n^0 = f_C(T^n p)$, so that under our hypotheses $\text{Lim}\, f_C(T^n p) > 0$. For every open set $G \subset E$ define $F(G) = \text{Lim}\, f_G(T^n p)$. Following the procedure of Carathéodory, for every set $A \subset E$ define $m^*(A) = \inf \sum_1^\infty F(G_n)$, taking all sequences of open sets G_n such that $A \subset \sum_1^\infty G_n$. The resulting set function is then an outer measure in the sense of Carathéodory, and hence defines a completely additive measure whose domain of definition includes all Borel sets. The measure is invariant under T in virtue of condition 3), it is always $\leqq 1$ since $F(E) = 1$. Furthermore $m(C) > 0$, because any covering of the compact set C contains a finite subfamily which covers C, and from the linearity of the functional $\text{Lim}\, \xi_n$ it follows that $m^*(C) \geqq \text{Lim}\, f_C(T^n p) > 0$. Hence $m(E) > 0$, and the measure can be normalized so that $m(E) = 1$. It may be remarked that the set function $m^*(A)$ could be defined starting with an arbitrary point p. The sufficient condition stated in the theorem merely plays the rôle of insuring that the resulting measure be not identically zero.

If the space is compact, the condition is trivially satisfied by taking $C = E$ and an arbitrary point p, so that in this case an invariant measure always exists, as was shown by Kryloff and Bogoliouboff. The condition is also satisfied if p is taken to be a periodic point and C the set of its images. But in this case the measure obtained is quite trivial, being confined to p and its images. Even in the compact case, however, the only finite invariant measures may be confined to a finite set of points, as is seen by considering the automorphism $Tx = x^2$ of the interval $0 \leqq x \leqq 1$. To obtain a condition for the existence of a non-trivial invariant measure, we distinguish two cases according as there are countably or uncountably many periodic points. We shall see that in the latter case a non-trivial invariant measure always exists, and in the former case a condition similar to that given in Theorem 1 is necessary and sufficient.

Suppose first that there are non-denumerably many periodic points. Let E_n, $n = 1, 2, \cdots$, be the set of points p such that $T^n p = p$. At least one of the sets E_n must be non-denumerable, since their union is. Let E_N be the first such set. Then the set $E_N - \sum_1^{N-1} E_n$ is a closed, non-denumerable subset of a complete separable space, and hence, by the theorem of Cantor-Bendixson,

---

[5] Banach, *Théorie des opérations linéaires*, Warsaw 1932, p. 27 Theorem 1. In the terminology there used one need merely take $p(\{\xi_n\}) = \limsup_{n \to \infty} \frac{1}{n} (\xi_1 + \cdots + \xi_n)$, and $f(\{t\xi_n^0\}) = t\, p(\{\xi_n^0\})$ in the linear manifold determined by the element $\{\xi_n^0\}$.

contains a non-void perfect subset. This, in turn, contains a subset C homeomorphic to the Cantor set.[6] Let $m_1(A)$ be a finite measure in C which is zero for points.[7] The measure $m(A) = \sum_1^N m_1(C \cdot T^k A)$, defined for all Borel sets, is a finite measure, zero for points, and invariant under T. The measure is, in fact, confined to C and its $N - 1$ disjoint images.

Suppose now that there are only countably many periodic points. The complementary invariant set E_1 of non-periodic points is G_δ, and therefore homeomorphic to a complete space E_1^*.[8] Let T_1 denote the transformation T considered as operating only within E_1, and let T_1^* be the corresponding automorphism of E_1^*. A non-trivial measure in E must be confined to E_1, since the rest of the space is countable. Thus T will admit a non-trivial invariant measure if and only if T_1^* does. But a measure invariant under T_1^* is necessarily zero for points, since a non-periodic point cannot have positive measure if the measure of the whole space is to be finite. Thus Theorem 1 applies to T_1^*, and since the condition stated there holds in E_1^* if and only if it holds in E_1, our conclusion can be stated thus,

THEOREM 2: *In order that an automorphism T of a complete separable metric space E admit a finite invariant measure which is zero for points, it is necessary and sufficient that one of the following conditions hold,*

1. *T has non-denumerably many periodic points.*
2. *There is a compact set C consisting of non-periodic points some one of which returns to C with positive frequency (or superior frequency) under iteration of T.*

3. Finite measures invariant under a one-parameter group of transformations

The above methods can be adapted to the treatment of steady flows as well as transformations. By a steady flow we shall mean a one-parameter group of automorphisms $T_\lambda x$, $-\infty < \lambda < +\infty$, of E such that $T_\lambda T_\mu x \equiv T_{\lambda+\mu} x$, $T_0 X \equiv x$. We shall assume further that the dependence of $T_\lambda x$ upon λ is Borel measurable for each fixed x. In other words, for each x and each closed set $F \subset E$ the set of numbers λ such that $T_\lambda x \in F$ is a Borel subset of the real number system. From these assumptions it follows that $T_\lambda x$ is a Borel measurable function of λ and x simultaneously.[9] It is therefore measurable in the sense of Hopf[10] with respect to any Borel measure, and the ergodic theorem can be applied in any such invariant measure.

THEOREM 3: *Let $T_\lambda x$ be a steady flow (in the sense just defined) of a complete separable metric space E into itself. There exists a finite invariant measure in E*

[6] See e.g. Kuratowski, *Topologie* I, Warsaw-Lwów 1933, p. 228.

[7] Such a measure may conveniently be defined by mapping C homeomorphically onto a Cantor set of Lebesgue measure one on the interval $0 \leqq x \leqq 2$ and taking the relative Lebesgue measure. It may be assumed further that the measure is positive for non-void relatively open subsets, a remark we shall use later.

[8] See e.g. Kuratowski, op. cit., p. 200.

[9] See e.g. Kuratowski, op. cit., p. 180.

[10] E. Hopf, op. cit., p. 9, Definition 3.4.

if and only if for some point p and compact set $C \subset E$ the limit (or limit superior) as $t \to \infty$ of the expression $\frac{1}{t}\int_0^t f_C(T_\lambda p)\,d\lambda$ *is positive.*

The necessity of the condition follows immediately from the ergodic theorem and the existence of a compact set of positive measure, just as in the proof of Theorem 1. To prove sufficiency, we define a generalized limit $\text{Lim}\, f(\lambda)$ for all bounded Lebesgue measurable functions $f(\lambda)$, $-\infty < \lambda < +\infty$, fulfilling the conditions,[11]

1) $\text{Lim}\,[af(\lambda) + bg(\lambda)] = a\,\text{Lim}\, f(\lambda) + b\,\text{Lim}\, g(\lambda) \quad a, b$ real
2) $\text{Lim}\, f(\lambda) \geqq 0$ if $f(\lambda) \geqq 0$
3) $\text{Lim}\, f(\lambda + \lambda_0) = \text{Lim}\, f(\lambda) \quad \lambda_0$ real
4) $\liminf_{t\to\infty} \frac{1}{t}\int_0^t f(\lambda)\,d\lambda \leqq \text{Lim}\, f(\lambda) \leqq \limsup_{t\to\infty} \frac{1}{t}\int_0^t f(\lambda)\,d\lambda$
5) $\text{Lim}\, f_0(\lambda) = \limsup_{t\to\infty} \frac{1}{t}\int_0^t f_0(\lambda)\,d\lambda$ where $f_0(\lambda) = f_C(T_\lambda p)$.

We define $F(G) = \text{Lim}\, f_G(T_\lambda p)$ for every open set $G \subset E$, and $m^*(A) = \inf \sum_1^\infty F(G_n)$ for all sequences of open sets G_n whose union contains A. By an argument exactly similar to that in the proof of Theorem 1, we conclude that $m^*(A)$ is a Carathéodory outer measure which defines a finite measure invariant under the group T_λ.

4. Infinite invariant measures

A simple example of a transformation which admits no finite invariant measure is the automorphism $Tx = x + 1$ of the real number system. A more interesting example is the following. Let T be a rotation of the circumference of the unit circle through an irrational angle. Let C be a nowhere dense perfect set on E with positive linear measure. The set $E_1 = E - \sum_{-\infty}^{+\infty} T^k C$ is then an invariant set of measure zero. Since it is G_δ, it may be supposed re-metrized so as to be complete. Let T_1 be the transformation T considered as an automorphism of E_1. T_1 admits no finite invariant measure, for such a measure would yield at once a measure in E invariant under T and positive for the set E_1. But the ordinary linear measure in E is the *only* finite measure invariant under T, and in this measure E_1 has measure zero.

The first of these examples, of course, admits an infinite invariant measure, and one may ask whether such a measure always exists. We proceed to show that this is the case, and indeed that the measure may be made to satisfy certain additional requirements,

THEOREM 4: *Let T be an automorphism of a metric space E, complete, separable, and dense in itself. There exists in E a finite or infinite measure $m(A)$, which is invariant under T, defined for Borel sets, zero for points, positive (possibly infinite) for non-void open sets, and such that there is a set A_0 of finite measure*

[11] See footnote 5. Here we take $p[x(\lambda)] = \limsup_{t\to\infty} \frac{1}{t}\int_0^t x(\lambda)\,d\lambda$ and $f[tx_0(\lambda)] = t\,p[x_0(\lambda)]$ in the linear manifold determined by $x_0(\lambda)$.

whose images $T^k A_0$, $k = -\infty, \cdots, +\infty$, *are all disjoint and together exhaust all but a finite amount of the measure in* E.[12]

Let us prove first of all that if S is any open sphere in which non-periodic points are dense, we can find a set $C \subset S$ which is homeomorphic to the Cantor set and whose images under powers of T and T^{-1} are all disjoint. Let us consider the open set $E_1 = \sum_{-\infty}^{+\infty} T^k S$ as our space. This will likewise be separable, dense in itself, and can be re-metrized so as to be complete. In it, non-periodic points form a dense G_δ set, therefore residual, and therefore non-denumerable in every neighborhood. We wish to define closed spheres $S_{i_1 \cdots i_n}$; $i_1, \cdots, i_n = 1, 2$; $n = 1, 2, \cdots$, such that 1) the sets $T^k S_{i_1 \cdots i_n}$, $k = 0, \pm 1, \cdots, \pm(n-1)$; $i_1, \cdots, i_n = 1, 2$ are disjoint, 2) $S_{i_1 \cdots i_n} \subset S_{i_1 \cdots i_{n-1}}$, and 3) $S_{i_1 \cdots i_n}$ has radius less than $1/n$. This is easily done step by step. S_1 and S_2 may be taken to be any two disjoint closed spheres of radius less than one. Suppose $S_{i_1 \cdots i_{n-1}}$, $i_1, \cdots, i_{n-1} = 1, 2$, already defined. In the interior of each of these spheres choose two points $x_{i_1 \cdots i_n}$, $i_n = 1, 2$, such that the 2^n points $x_{i_1 \cdots i_n}$ are distinct together with all of their images. This can be done, since, as we have seen, non-periodic points are non-denumerable in every sphere. Any sufficiently small spheres about the points $x_{i_1 \cdots i_n}$ will serve to define the spheres $S_{i_1 \cdots i_n}$. From the conditions 1), 2), and 3) it follows that the set

$$C = \prod_{n=1}^{\infty} \sum_{i_1, \cdots, i_n = 1,2} S_{i_1 \cdots i_n}$$

is homeomorphic to the Cantor set and that all its images are disjoint.[13]

We are now ready to prove the theorem. Let $S_1, S_2, \cdots$ be a sequence of open spheres which constitute a base for open sets in E. Let $S_{k_1}, S_{k_2}, \cdots$ be the subsequence of all spheres S_n in which non-periodic points are dense, and let $S_{l_1}, S_{l_2}, \cdots$ be the subsequence of spheres S_n which consist of periodic points only. These two subsequences, of which one or the other may be void, together constitute a base for open sets in E. For consider any sphere S_n. If the periodic points in it form a set of first category in E, it belongs to the subsequence S_{k_i}, and otherwise it contains one of the spheres S_{l_i}. For the present let us assume that neither subsequence is void. Let C_1 be a Cantor set contained in S_{k_1} whose images are all disjoint. Supposing $C_1, C_2, \cdots, C_{n-1}$ defined, let C_n be a Cantor set with disjoint images and contained in the first

[12] The condition that the space be dense in itself is clearly necessary, since an isolated point is itself an open set. This assumption can be dropped if it is required that the measure be positive only for non-denumerable open sets, but then it must be explicitly assumed that the space is not countable. It may be remarked that non-void open sets may necessarily have infinite measure. In the case of the automorphism T_1 of the space E_1 just considered, every non-void open set has a finite number of images which cover the space, and therefore must have infinite measure. In this space no open set is compact, but one cannot strengthen the theorem even to require that compact open sets have finite measure, as is seen again by the example $Tx = x^2$ on the interval $0 \leqq x \leqq 1$.

[13] From what has just been proved it follows that there exists a Cantor set on the circumference of the circle whose images under a given irrational rotation are all disjoint.

sphere S_{k_i} which overlaps no image of $C_1, \cdots, C_{n-1}$. We thus arrive at a finite or infinite sequence of sets C_i which are disjoint together with all their images. Let $m_i(A)$ be any normalized measure in C_i which is zero for points and positive for non-void relatively open subsets. The measure

$$m_0(A) = \sum_{k=-\infty}^{+\infty} \sum_i \frac{1}{2^i} m(T^k A \cdot C_i)$$

will be an infinite invariant measure, defined for Borel sets in E, zero for points and positive for all S_{k_i}. Now let $G_i = \sum_{-\infty}^{+\infty} T^k S_{l_i}$. This open invariant set may be considered a complete space, and since periodic points are non-denumerable in it, we can apply Theorem 2 and get a normalized invariant measure $\mu_i(A)$ which is zero for points. The measure $\mu(A) = \sum_1^{\infty} 2^{-i} \mu_i(A \cdot G_i)$, defined for Borel sets, will be finite, zero for points, positive for all S_{l_i}, and invariant under T. Finally, the measure $m(A) = m_0(A) + \mu(A)$ will satisfy all requirements of the theorem, with $A_0 = \sum C_i$. If the sequence S_{l_i} is void, the measure $m_0(A)$ alone suffices. If the sequence S_{k_i} is void, the measure $\mu(A)$ suffices, with A_0 taken to be the void set.

SOCIETY OF FELLOWS, HARVARD UNIVERSITY.

ANNALS OF MATHEMATICS
Vol. 42, No. 4, October, 1941

MEASURE-PRESERVING HOMEOMORPHISMS AND METRICAL TRANSITIVITY*

BY J. C. OXTOBY AND S. M. ULAM

(Received May 1, 1941)

INTRODUCTION

In the study of dynamical systems one is led naturally to the consideration of measure-preserving transformations. A Hamiltonian system of $2n$ differential equations induces in the phase space of the system a measure-preserving flow, that is, a one-parameter group of transformations that leave invariant the $2n$-dimensional measure. Making use of one or more integrals of the system, one obtains a reduced phase manifold of lower dimension which likewise undergoes a flow into itself, and in general admits an invariant measure related in a simple manner to that in the larger phase space. If the differential equations are sufficiently regular the flow will have corresponding properties of continuity and differentiability. Thus the study of one-parameter continuous groups of measure-preserving automorphisms[1] of finite dimensional spaces has an immediate bearing on dynamics and the theory of differential equations.

In statistical mechanics one is especially interested in time-average properties of a system. In the classical theory the assumption was made that the average time spent in any region of phase space is proportional to the volume of the region in terms of the invariant measure, more generally, that time-averages may be replaced by space-averages. To justify this interchange, a number of hypotheses were proposed, variously known as ergodic or quasi-ergodic hypotheses, but a rigorous discussion of the precise conditions under which the interchange is permissible was only made possible in 1931 by the ergodic theorem of Birkhoff.[2] This established the *existence* of the time-averages in question, for almost all initial conditions, and showed that if we neglect sets of measure zero, the interchange of time- and space-averages is permissible if and only if the flow in the phase space is *metrically transitive*. A transformation or a flow is said to be metrically transitive if there do not exist two disjoint invariant sets both having

* This paper includes results presented to the Amer. Math. Soc. in preliminary reports on April 15 and December 30, 1938. Most of the results were obtained while the authors were members of the Harvard Society of Fellows.

[1] An automorphism is a 1:1 bicontinuous transformation of a space onto itself. It is measure-preserving with respect to a measure μ if $\mu TA = \mu A = \mu T^{-1}A$ whenever A is measurable.

[2] G. D. Birkhoff, *Proof of the ergodic theorem*. Proc. Nat. Acad. USA. **17** (1931) 650–660. An interesting connection between this theorem and the fundamental theorem of the calculus has been shown by N. Wiener. See *The ergodic theorem*. Duke Jour. **5** (1939) 1–18.

For commentaries to this paper [24], see p. 683.

positive measure.[3] Thus the effect of the ergodic theorem was to replace the ergodic hypothesis by the hypothesis of metrical transitivity.[4]

Nevertheless, in spite of the simplification introduced by the ergodic theorem, the problem of deciding whether particular systems are metrically transitive or not has proved to be very difficult. Hedlund[5] showed that the flow defined by the system of geodesics on certain surfaces of constant negative curvature is metrically transitive, and this result was generalized by Hopf.[6] More recently, the result has been extended to surfaces of variable negative curvature, independently by both Hedlund[7] and Hopf.[8] This important class of systems is so far the only large class known to be metrically transitive. In fact, the only other known examples of metrically transitive continuous flows are the spiral motions on the n-dimensional torus, including the rotation of the circumference of a circle as the simplest example of all. The ergodic properties of these systems were established as long ago as 1916 by Weyl,[9] in a quite different connection.

Thus the known examples of metrically transitive continuous flows are all in manifolds, indeed in manifolds of restricted topological type, either toruses or manifolds of direction elements over surfaces of negative curvature.[10] An outstanding problem in ergodic theory has been the existence question—can a metrically transitive continuous flow exist in an arbitrary manifold, or in any space that is not a manifold? In the present paper we shall obtain a complete answer to this question, at least on the topological level, for polyhedra of dimension three or more. It will appear that the only condition that needs to be imposed is a trivially necessary kind of connectedness. In particular, there exists a metrically transitive continuous flow in the cube, in the solid torus, and in any pseudo-manifold of dimension at least three. Since the phase spaces of dynamical systems have the required kind of connectedness, it follows that the hypothesis of metrical transitivity in dynamics involves no *topological* contradiction. More precisely, in any phase space there can exist a continuous flow

[3] The notion was first introduced in a different connection by G. D. Birkhoff and P. A. Smith. See *Structure analysis of surface transformations*. Liouville's Jour. **7** (1928) 365.

[4] For a historical survey, see G. D. Birkhoff and B. O. Koopman, *Recent contributions to the ergodic theory*. Proc. Nat. Acad. USA. **18** (1932) 279.

[5] G. A. Hedlund, *On the metrical transitivity of the geodesics on closed surfaces of constant negative curvature*. Ann. of Math. **35** (1935) 787. See also his paper, *A new proof for a metrically transitive system*. Amer. J. Math. **62** (1940) 233–242.

[6] E. Hopf, *Fuchsian groups and ergodic theory*. Trans. Amer. Math. Soc. **39** (1936) 299. See also his booklet, *Ergodentheorie*. Ergebnisse der Mathematik und ihrer Grenzgebiete, vol. 5, Berlin 1937.

[7] G. A. Hedlund, Bull. Amer. Math. Soc. Abstract 46–3–173.

[8] E. Hopf, *Statistik der geodätischen Linien in Mannigfaltigkeiten negativer Krümmung*. Ber. Verh. Sächs. Akad. Wiss. Leipzig **91** (1939) 261–304.

[9] H. Weyl, *Über die Gleichverteilung von Zahlen mod. Eins*. Math. Ann. **77** (1916) 315.

[10] Some examples of discontinuous flows have been studied. See E. Hopf, *Ergodentheorie*, and J. von Neumann, *Zur Operatorenmethode in der klassischen Mechanik*. Annals of Math. **33** (1932) 587–642.

metrically transitive with respect to the invariant measure associated with the system.

It may be recalled that the original ergodic hypothesis of Boltzmann—that a single streamline passes through all points of phase space—had to be abandoned because it involved a topological impossibility.[10a] It was replaced by a quasi-ergodic hypothesis—that some streamline passes arbitrarily close to all points of phase space. But it is not obvious that even this weak hypothesis is topologically reasonable in general phase spaces, and in any case it is not sufficient to justify the interchange of time- and space-averages. It is therefore of some interest to know that the ergodic hypothesis in its modern form of metrical transitivity is at least free from any objection on topological grounds.

It must be emphasized, however, that our investigation is on the topological level. The flows we construct are continuous groups of measure-preserving automorphisms, but not necessarily differentiable or derivable from differential equations. Thus they correspond to dynamical systems only in a generalized sense. In this respect the flows studied by Hedlund and Hopf are closer to the dynamical problem. In any case, the problem is illuminated by considering it in a more general setting, and probably a certain amount of generalization is necessary in order to give meaning to some questions of the type with which we deal.

The conjecture has frequently been expressed, first by Birkhoff, then by Hopf and others, that metrical transitivity is probably the "general case." The conjecture was never given a precise formulation, but was based on the fact that the transitive case is the non-integrable case, in the sense that no uniform measurable integrals exist, and on the idea that a general transformation or flow should be expected to shuffle points in a more or less random fashion, and therefore leave invariant as few sets as possible. To see the precise connection between metrical transitivity and randomness, let us recall the statement of the ergodic theorem for transformations. It asserts that if μ is a completely additive finite measure in E, and T a measure-preserving transformation of E onto itself, then for any μ-integrable function $f(p)$ on E the average value $\lim_{n\to\infty} \frac{1}{n} \sum_{\nu=1}^{n} f(T^\nu p) = f^*(p)$ exists for almost all points p, is a measurable function of p, and that $\int f^*(p)\, d\mu(p) = \int f(p)\, d\mu(p)$. Metrical transitivity is precisely the condition that $f^*(p)$ should be constant for almost all p. Thus if T is metrically transitive and f_A is the characteristic function of any measurable set A, we have $\frac{\mu A}{\mu E} = \lim_{n\to\infty} \frac{1}{n} \sum_{\nu=1}^{n} f_A(T^\nu p)$ for almost all p. The left member of this equation represents the average number of images of p that fall in A, or the *frequency* with which the images of p fall in the set A under iteration of T, and the equation

[10a] A. Rosenthal, *Beweis der Unmöglichkeit ergodischer Gassysteme*, Ann. der Physik, (4) **42** (1913) 796–806.

asserts that this frequency is proportional to the measure of A. A sequence of points selected at random from E would be expected to have just such a distribution, in fact, the ratio $\mu A/\mu E$ may be interpreted as the probability that a point selected at random lies in A. Thus a metrically transitive transformation is one under which almost all points generate sequences that are distributed like random sequences in respect to the average number that fall in any measurable set. This perhaps makes it seem plausible that a measure-preserving transformation "selected at random" should be metrically transitive.

One might try to make precise the idea that metrical transitivity is the general case by introducing a *measure* in the space of all measure-preserving transformations of E, but it seems difficult to do this in any natural way. Nevertheless, there is a simple and natural *metric* in the space of automorphisms of E, in case E is compact, and one may ask whether in this space metrical transitivity is the general case in the *topological sense*, that is, whether such automorphisms constitute all but a set of first category.[11] This is what we are going to show, under suitable assumptions about E. Thereby we shall dispose of the existence problem at the same time.[12] The result is perhaps a little surprising in view of the fact that metrical transitivity is a purely measure-theoretic property, and it often happens that what is the general case in the sense of measure is exceptional in the sense of category, and vice versa.[13]

The fact that metrical transitivity turns out to be the general case in the topological sense raises the question how far residual sets of measure-preserving automorphisms may exhibit other random properties. For instance, a random sequence is not only distributed so that the average number of points in a set is proportional to its measure, but the limiting frequency is approached in the manner of "Laplace-Liapounoff." Presumably metrical transitivity is not sufficient to insure that almost all points should generate sequences exhibiting this more precise behavior. It would therefore be of interest to know whether transformations of this sort are likewise general, or indeed whether they can exist

[11] A set is said to be of first category if it can be represented as a sum of countably many nowhere dense sets. Any other set is said to be of second category. Complements of first category sets are called residual sets.

[12] Baire's Theorem asserts that in a complete metric space every residual set is of second category and is dense. It summarizes concisely a typical form of existence proof.

[13] For instance, the law of large numbers is false in the sense of category. That is, the set of numbers x in $0 < x < 1$ such that in their infinite dyadic development the number of ones in the first n places divided by n tends to one-half is of first category (although of measure one). If x_n denoted the n-th digit in the dyadic development of x, the set in question is represented by

$$\prod_k \sum_N \left[\prod_n \mathsf{E}_x \left\{ \left| \frac{x_1 + x_2 + \cdots + x_{N+n}}{N+n} - \frac{1}{2} \right| < \frac{1}{k} \right\} \right],$$

and for $N > 0$ and $k > 2$ the set enclosed in square brackets is nowhere dense, as may be seen by inserting a sufficiently long block of zeros far out in the development of a number.

at all. A result in this direction is given in §12. In the same order of ideas one may ask for automorphisms exhibiting the various types of *mixture* properties.[14]

In the course of our investigation we shall derive a number of results about transformations and measures which have an interest entirely apart from questions relating to transitivity. In particular, the results obtained in Part II concerning measures topologically equivalent to Lebesgue measure appear to be fundamental to the theory of measure-preserving homeomorphisms. In sections 13 to 15 a number of questions of group-theoretic interest are answered on the basis of these results. The paper involves an intimate combination of the methods of topology and measure theory. A suggestive example of the way in which such a combination of methods may lead to purely topological results not otherwise apparent is the sort of topological ergodic theorem obtained in §16. Another result of purely topological character is the Corollary to Lemma 16 concerning equivalence of Cantor sets under automorphism of the containing space. This result is of independent interest since it adds to the results obtained by Antoine[15] relating to this question.

A fundamental outstanding problem in topology is that of approximating to an arbitrary homeomorphism by a differentiable one.[16] In §18, as a by-product of our investigation, we obtain the result that any measure-preserving automorphism of the r-dimensional cube that leaves the boundary fixed can be approximated uniformly by one that is differentiable *almost everywhere,* in fact, it is locally linear about almost all points. This result in itself is not strong enough to have important topological implications, but it suggests that the special properties of measure-preserving automorphisms may enable one to obtain approximation theorems for them that are difficult or perhaps impossible to establish in general. In any case, the very precise properties of the approximating transformation obtained in Theorem 12 may well serve as a basis for answering other questions concerning approximation by automorphisms with special measure-theoretic properties.

I. Preliminary Results

1. Definitions and Principal Theorem

We assume the standard notions[17] of *polyhedron, euclidean polyhedron,* and *complex,* except that we shall always understand these to be finite. Thus a polyhedron is compact. Some of the results can be generalized to infinite polyhedra, but it seems best to leave such extensions out of the present paper.

[14] See G. A. Hedlund, *The dynamics of geodesic flows.* Bull. Amer. Math. Soc. **45** (1939) 241–260.

[15] L. Antoine, *Sur l'homéomorphie de deux figures et de leurs voisinage.* Liouville's Jour. (8) **4** (1921) 221–325 esp. p. 307 ff.

[16] J. W. Alexander, *Some problems in topology.* Verh. des Int. Math. Kong. Zurich 1932, vol. 1, p. 249–257.

[17] For all topological definitions see P. Alexandroff and H. Hopf, *Topologie I,* Berlin 1935, hereafter referred to as AH.

A *regular point* of an r-dimensional polyhedron is a point that has a neighborhood homeomorphic to an open sphere in r-space, any other point of the polyhedron is called *singular*. A polyhedron is called *regularly connected* if its regular points form a connected set dense in the polyhedron.[18]

A *finite outer measure* in a space E is a function μ^* defined for all subsets of E and satisfying the three conditions:

M1: $0 \leqq \mu^*A \leqq \mu^*E, \quad o < \mu^*E < +\infty, \quad \mu^*(\text{void set}) = 0$

M2: $\mu^*A \leqq \mu^*B \quad \text{if } A \subset B$

M3: $\mu^* \sum_{n=1}^{\infty} A_n \leqq \sum_{n=1}^{\infty} \mu^* A_n.$

A set A is *measurable* with respect to μ^* if $\mu^*W = \mu^*WA + \mu^*W(E - A)$ for every set W. The function μ^* is completely additive with respect to its class of measurable sets, and the measure function thus derived from μ^* by restricting its domain is denoted by μ.[19]

A *Carathéodory outer measure* is one that satisfies also the condition

M4: $\mu^*(A + B) = \mu^*A + \mu^*B$ if A and B are separated by a positive distance.

The measurable sets then include all Borel sets.

By a *Lebesgue-Stieltjes outer measure* in a polyhedron, we shall understand a finite Carathéodory outer measure that satisfies the further condition

M5: $\mu^*A = \inf_{G \supset A} \mu^*G \quad G$ open.

A measure derived from such an outer measure will be called a Lebesgue-Stieltjes measure (LS measure).[20]

We introduce the following special definition. A Lebesgue-Stieltjes measure in an r-dimensional polyhedron E, $r \geqq 1$, will be called *r-dimensional* if it is zero for points, zero for the set of singular points, and positive for neighborhoods of regular points. This definition is consistent with ordinary usage, since r-dimensional Lebesgue measure in any r-dimensional euclidean polyhedron evidently fulfills the requirements. The invariant measures associated with dynamical systems are usually defined by integrating a positive density function and are therefore r-dimensional in the present sense.

The set of all automorphisms of a polyhedron E (or of any compact space) is made into a metric space $H[E]$ by the definition

$$\rho(g, h) = \max\,[\max_{x \in E} \rho_E(gx, hx), \max_{x \in E} \rho_E(g^{-1}x, h^{-1}x)],$$

[18] This is equivalent to the ordinary definition. See AH pp. 400, 402.

[19] For the theory of measure, see Carathéodory, *Vorlesungen über reellen Funktionen*, Leipzig and Berlin 1918; or H. Hahn, *Theorie der reelen Funktionen I*, Berlin 1921.

[20] For polyhedra embedded in euclidean spaces, these measures are the same as those obtained by relativizing measures in the containing space that are derived from non-negative additive functions of an interval. The reasoning is essentially contained in S. Saks, *Theory of the integral*, Warsaw-Lwów 1937, Chap. 3. Thus our definition is consistent with the ordinary concept of a non-negative Lebesgue-Stieltjes measure, and has the advantage of being intrinsic. Also it is equally applicable to curved polyhedra, in which intervals have no meaning. An outer measure that satisfies conditions M1 to M5 is called by Hahn a (finite) Inhaltsfunktion. See Hahn, op. cit. p. 444.

where g, h are any two automorphisms of E, and ρ_E denotes distance in E. In this metric, convergence of a sequence of automorphisms means uniform convergence together with uniform convergence of the sequence of inverses. One can verify without difficulty that the space $H[E]$ is complete and that the group operations (composition and inverse of automorphisms) are continuous in this metric, so that $H[E]$ is a *metric group*, that is, a space of type (G) in the sense of (e.g.) Banach.[21] The distance of an automorphism from the identity will be called its *norm*, it is equal to the maximum distance through which any point is displaced by the automorphism. An *arbitrarily small* automorphism is one whose norm is arbitrarily small.

It may be remarked that whereas uniform convergence of a sequence of automorphisms h_n does not insure the existence of a limiting automorphism, nevertheless uniform convergence of h_n *to an automorphism* h implies uniform convergence of h_n^{-1} to h^{-1}, therefore convergence in $H[E]$.[22] Hence the metric in $H[E]$ is topologically equivalent to the metric $\max_{x \in E} \rho_E(gx, hx)$ usually used to metrize the space of continuous mappings of E into itself, but with respect to the latter metric the space of automorphisms is not complete.

The main object of our investigation, however, will not be the space $H[E]$ but rather the subspace consisting of all measure-preserving automorphisms with respect to a given LS measure μ.[23] This subspace, with the same metric as in $H[E]$, will be denoted by $M[E, \mu]$. To see that $M[E, \mu]$ is a closed subset of $H[E]$, and therefore complete, consider any sequence of measure-preserving automorphisms T_n converging to the limit T in $H[E]$. Let F be any closed subset of E, and G any open set containing TF. From the uniform convergence of the sequence T_n it follows that $T_nF \subset G$, for all sufficiently large n, so that $\mu^*G \geqq \mu^*T_nF = \mu^*F$. Therefore $\mu^*F \leqq \mu^*TF$, by condition M5. Similar reasoning applied to the sequence T_n^{-1} yields the inverse inequality, so that $\mu^*TF = \mu^*F$ for all closed sets F. Hence equality holds also for all open sets, therefore for all sets, and the limiting automorphism T is therefore measure-preserving. Since the measure-preserving automorphisms of E form a group, it follows that $M[E, \mu]$ is a metric group, in fact, a closed subgroup of $H[E]$.

In case E is a rectangular r-cell R, and μ is the ordinary r-dimensional Lebesgue measure m in R, we shall write simply $M[R]$ instead of $M[R, m]$. The closed subgroups of $M[R]$ and $H[R]$ consisting of automorphisms that leave all boundary points fixed, we shall denote by $M_0[R]$ and $H_0[R]$ respectively.

THEOREM 1: *Let E be any regularly connected polyhedron of dimension $r \geqq 2$,*

---

[21] S. Banach, *Théorie des opérations linéaires*, Warsaw 1932, Chap. 1 and p. 229. Our metric in $H[E]$ is topologically equivalent to the one there assigned to this group.

[22] J. Schreier and S. Ulam, *Über topologische Abbildungen der euklidischen Sphären.* Fund. Math. **23** (1934) 102–118, esp. p. 104.

[23] We shall denote general measures by μ or ν. The letter m will be reserved for Lebesgue measure. General automorphisms will be denoted by g or h, measure-preserving automorphisms by S or T. For LS measures, the definition of a measure-preserving automorphism may be taken to be the condition $\mu^*TA = \mu^*A$ for every A.

and let μ be any r-dimensional Lebesgue-Stieltjes measure in E. In the space $M[E, \mu]$ of measure-preserving automorphisms of E the metrically transitive automorphisms form a residual G_δ set.[24]

The proof of this theorem will only be completed in Part III. In Part IV we shall discuss the generality of the result and its significance for metrically transitive flows.

In a previous note[25] one of us has obtained a similar result for the set of *topologically transitive* automorphisms, for slightly different assumptions on E. A transformation is topologically transitive if there do not exist two disjoint invariant open sets, both non-void, (or, equivalently, if the sequence of images of some point is dense in E). Metrical transitivity with respect to an r-dimensional LS measure obviously implies topological transitivity. Hence Theorem 1 may be considered as a generalization of the earlier result, but whereas the latter required only the most elementary constructions and properties of measure-preserving automorphisms, the present theorem will require a much more extensive investigation.

As regards the interpretation of Theorem 1 as a proof of the conjecture that metrical transitivity is the general case, it is interesting to note that the topological notion of probability in the sense of category can be subsumed under the general theory of measure and probability. In any complete metric space, if we define $\mu^*A = 0$ or 1 according as A is of first or second category, it is easily verified that conditions M1, M2, M3 are satisfied,and that the class of measurable sets consists of the first category sets and the residual sets, which have measures zero and one respectively. The notion of probability in the sense of category is therefore a special case of the general notion of (non-Borel) measure.

2. Preliminary Lemmas

The object of this first sequence of lemmas is to show that the general situation can be reduced, in all essential respects, to the consideration of ordinary Lebesgue measure in a rectangular r-cell. In making this reduction, a central role is played by the notion of a continuous map f of a convex cell Z onto a polyhedron E which is a *homeomorphism up to the boundary*. By this we shall mean that f maps the interior Z_0 of Z homeomorphically onto fZ_0. For such continuous maps $fBdZ = E - fZ_0$, so that $f^{-1}f$ is single-valued on Z_0, though possibly multiple-valued on BdZ.

LEMMA 1: *Let E be the polyhedron of a regularly connected euclidean complex K of dimension $r \geqq 1$. It is possible to represent E as the continuous image of a convex r-cell Z under a map f which is a homeomorphism up to the boundary and which is a simplicial map of a certain subdivision of Z onto K.*

This is not a new result, but rather a corollary of the known result that any

[24] A G_δ set is one that can be represented as a countable intersection of open sets.

[25] J. C. Oxtoby, *Note on transitive transformations*. Proc. Nat. Acad. USA. **23** (1937) 443–446. See also E. Hopf, *Statistische Probleme und Ergebnisse in der klassischen Mechanik*. Actualités Scientifiques et Industrielles, No. 737 (1938) 5–16.

regularly connected complex can be obtained from a convex cell by suitable boundary identifications.[26] To prove it, let $\sigma_1, \cdots, \sigma_n$ be the base simplexes of K, so numbered that each, after the first, has at least one regular face $\sigma_{i-1,i}$ in common with one of its predecessors. The possibility of such a numbering follows at once from the definition of a regularly connected complex.[27] Let τ_1 be any simplex in r-space, and set its vertices in correspondence with those of σ_1. Let $\tau_{1,2}$ be the face of τ_1 that corresponds to $\sigma_{1,2}$. Take a new vertex outside τ_1 which forms with $\tau_{1,2}$ a simplex τ_2 such that $\tau_1 + \tau_2$ is still convex. Any point sufficiently near the center of $\tau_{1,2}$ will suffice. Let this vertex correspond to the remaining vertex of σ_2. Evidently we have a simplicial map of the complex $\tau_1 + \tau_2$ onto the complex $\sigma_1 + \sigma_2$, and the only $(r - 1)$-simplex in $\tau_1 + \tau_2$ which does not lie on the boundary is $\tau_{1,2}$. Proceeding by induction, suppose that we have a complex K_i with r-dimensional base simplexes $\tau_1, \cdots, \tau_i$ whose union is a convex cell, and a simplicial map onto the complex $\sigma_1 + \sigma_2 + \cdots + \sigma_i$ under which τ_j corresponds to σ_j, $1 \leqq j \leqq i$. Also suppose that the only $(r$-$1)$-simplexes of K_i that do not lie on the boundary are $\tau_{1,2}, \cdots, \tau_{i-1,i}$ which correspond to $\sigma_{1,2}, \cdots, \sigma_{i-1,i}$. The simplex $\sigma_{i,i+1}$ is a face of some one of $\sigma_1, \cdots, \sigma_i$. Let $\tau_{i,i+1}$ be the corresponding face of the corresponding one of the simplexes $\tau_1, \cdots, \tau_i$. Since $\sigma_{i,i+1}$ is regular, it is distinct from $\sigma_{1,2}, \cdots, \sigma_{i-1,i}$, and it follows that $\tau_{i,i+1}$ lies on the boundary of the convex cell. We can therefore adjoin a new simplex τ_{i+1} having $\tau_{i,i+1}$ for a face, and such that $\tau_1 + \cdots + \tau_{i+1}$ is again convex. If we map the new vertex of τ_{i+1} onto the remaining vertex of σ_{i+1} the simplicial map will be extended to map $K_{i+1} = \tau_1 + \cdots + \tau_{i+1}$ onto $\sigma_1 + \cdots + \sigma_{i+1}$, and the only $(r - 1)$-dimensional simplexes of K_{i+1} not on its boundary will be $\tau_{1,2}, \cdots, \tau_{i,i+1}$. At the n-th stage we obtain a simplicial map of a complex K_n onto K. If we map the simplexes $\tau_1, \cdots, \tau_n$ affinely onto $\sigma_1, \cdots, \sigma_n$, we obtain a continuous map of a convex cell Z onto E. The interior of Z is the union of the interiors of $\tau_1, \cdots, \tau_n$ and of $\tau_{1,2}, \cdots, \tau_{n-1,n}$. These correspond to the interiors of $\sigma_1, \cdots, \sigma_n$ and of $\sigma_{1,2}, \cdots, \sigma_{n-1,n}$ respectively, which are disjoint since these simplexes are all distinct. Hence distinct interior points of Z go into distinct points, and the map is therefore a homeomorphism up to the boundary.

LEMMA 2: *Let f be a continuous map of a rectangular r-cell R, $r \geqq 1$, onto a polyhedron E which is a homeomorphism up to the boundary. Let μ be an r-dimensional Lebesgue-Stieltjes measure in E, and suppose that $\mu fBdR = 0$. Then the function $\nu^*A = \mu^*fA$ defines an r-dimensional Lebesgue-Stieltjes measure in R.*

That ν^* satisfies conditions M1 to M3 may be verified by inspection. Con-

---

[26] See AH p. 264.

[27] A complex is *homogeneous r-dimensional* if every simplex lies on an r-simplex. A *regular face* is an $(r - 1)$-simplex that lies on two and only two r-simplexes of the complex. A homogeneous r-dimensional complex K is *regularly connected* if in any division of K into two r-dimensional subcomplexes, these have in common at least one regular face. It can be shown that a polyhedron is regularly connected if and only if it is the polyhedron of a regularly connected complex. See AH p. 402.

dition M4 is less evident in view of the fact that f need not be 1:1. To verify it, consider any two sets A, B contained in R and separated by a positive distance. Their closures $\bar{A}$, $\bar{B}$ are disjoint, and therefore $f\bar{A}$ and $f\bar{B}$ intersect at most in a subset of $fBdR$, since all other points have unique antecedents. Hence $\mu f\bar{A}\cdot f\bar{B} = 0$, by hypothesis. It is possible to enclose $f(A + B)$ in a G_δ set C such that $\mu^*f(A + B) = \mu C$. Let $A_1 = C\cdot f\bar{A}$ and $B_1 = C\cdot f\bar{B}$. Then we have $fA \subset A_1$ and $fB \subset B_1$, and so $\mu^*f(A + B) = \mu^*(fA + fB) \leqq \mu^*(A_1 + B_1) \leqq \mu^*C = \mu^*f(A + B)$, that is, $\mu^*f(A + B) = \mu(A_1 + B_1)$. But A_1 and B_1 are Borel sets that intersect in a set of measure zero, therefore $\mu(A_1 + B_1) = \mu A_1 + \mu B_1 \geqq \mu^*fA + \mu^*fB$, and so $\nu^*(A + B) \geqq \nu^*A + \nu^*B$. The inverse inequality follows from M2, and so M4 is satisfied. To verify condition M5, we have $\nu^*A = \mu^*fA = \inf_{G\supset fA} \mu G$. But whenever G is an open set containing fA, $f^{-1}G$ is an open set in R containing A. Hence $\nu^*A \geqq \inf_{G\supset A} \nu G$. Again the inverse inequality follows from M2, and the verification that ν is a LS measure is complete. Furthermore, ν is positive for non-void open sets in R, because the image of such a set contains a neighborhood of a regular point of E. Finally, it is evident that single points have measure zero, and by hypothesis $\nu BdR = \mu fBdR = 0$. Hence ν is an r-dimensional LS measure in R.

Lemma 3: *Let μ be an r-dimensional LS measure in a regularly connected polyhedron E, $r \geqq 1$. It is possible to represent E as the continuous image of a rectangular r-cell R under a map f which is a homeomorphism up to the boundary, and which is such that $\mu^*fA = m^*A$ for all $A \subset R$, where m^* denotes ordinary Lebesgue outer measure in R.*

E is homeomorphic to the polyhedron E_1 of a euclidean complex K_1. It is possible to choose the correspondence in such a way that all $(r - 1)$-dimensional simplexes of K_1 correspond to sets of measure zero. Because if μ_1 is the measure in E_1 corresponding to μ under any homeomorphism, the singular $(r - 1)$-simplexes have measure zero by hypothesis, and a suitably chosen automorphism of E_1 will displace all regular $(r - 1)$-simplexes into sets of μ_1-measure zero. (First displace the interiors of all 1-simplexes not contained in the set of singular points to nearby positions having measure zero. Then do the same for 2-simplexes, and so on until all regular $(r - 1)$-simplexes have been displaced to sets having μ_1-measure zero. The automorphism need displace only regular points of E_1 and may leave all vertices of K_1 fixed.[28]) We may therefore suppose that E is the polyhedron of a euclidean complex K and that all lower dimensional simplexes of K have μ-measure zero. By Lemma 1 there exists a continuous map of a convex cell Z onto E which is a homeomorphism up to the boundary and is a simplicial map of a certain subdivision of Z onto K. Let us combine this map with a homeomorphism of Z onto a rectangular r-cell R, of volume $mR = \mu E$,

[28] More generally, any set of first category can be displaced to a set of measure zero by an automorphism. See our joint paper, *On the equivalence of any set of first category to a set of measure zero*. Fund. Math. **31** (1938) 201–206.

and let f denote the resulting map of R onto E. Then f is a homeomorphism up to the boundary, and since BdR corresponds to certain $(r - 1)$-dimensional simplexes of K, we see that $\mu f BdR = 0$. By Lemma 2, the function $\nu^*A = \mu^*fA$ defines an r-dimensional LS measure in R, and also we have $\nu R = mR$. In Part II (Theorem 2) it will be shown that for any such measure there exists an automorphism h of R such that $\nu^*A = m^*hA$ for all $A \subset R$. Assuming this result for the moment, we conclude that fh^{-1} is a continuous map of R onto E which is a homeomorphism up to the boundary, and that $m^*A = \mu^*fh^{-1}A$ for all $A \subset R$, as required.

LEMMA 4: *Let f be a map of R onto E with the properties stated in Lemma 3. Any Lebesgue-measure-preserving automorphism T of R that leaves the boundary fixed corresponds to a μ-measure-preserving automorphism fTf^{-1} of E. On the other hand, any μ-measure-preserving automorphism T of E defines a transformation $f^{-1}Tf$ in R which is a Lebesgue-measure-preserving homeomorphism of the open set $R_0 \cdot f^{-1}T^{-1}fR_0$ onto the open set $R_0 \cdot f^{-1}TfR_0$, both of which are contained in the interior R_0 of R and have measure equal to mR.*

To prove the first assertion we first show that fTf^{-1} is a 1:1 map of E onto itself. Consider any point p in fR_0. Then $f^{-1}p$ is single-valued, and therefore also $fTf^{-1}p$. If $p \notin fR_0$, then $p \epsilon fBdR$, and so $fTf^{-1}p = p$. Any point of $fBdR$ is its own image, and any point $p \epsilon fR_0$ is the image of $fT^{-1}f^{-1}p$. Hence fTf^{-1} is a 1:1 map of E onto itself, with inverse $fT^{-1}f^{-1}$. Both these transformations carry closed sets into closed sets, and so they are automorphisms. That fTf^{-1} preserves μ-measure is immediate, since $\mu^*A = m^*f^{-1}A = m^*Tf^{-1}A = \mu^*fTf^{-1}A$ for every set $A \subset E$.

To prove the second assertion of the lemma, consider any $p \in R_0 \cdot f^{-1}T^{-1}fR_0$. Then $Tfp \in fR_0$ and so $f^{-1}Tfp$ is single-valued on this domain. Its inverse is $f^{-1}T^{-1}f$ with domain $R_0 \cdot f^{-1}TfR_0$. Both of these are continuous throughout these domains. Thus $f^{-1}Tf$ maps the open set $R_0 \cdot f^{-1}T^{-1}fR_0$ homeomorphically onto $R_0 \cdot f^{-1}TfR_0$, and does so in a measure-preserving manner, as follows from the measure-preserving properties of f and T.

The partial correspondence between measure-preserving automorphisms of E and R described in Lemma 4 makes it possible to reduce the proof of Theorem 1 to the following lemma concerning transformations in R and ordinary Lebesgue measure.

LEMMA 5: *Let R_0 be the interior of a rectangular r-cell R, $r \geqq 2$, and let T be a measure-preserving homeomorphism of an open set $G \subset R_0$ onto an open set $TG \subset R_0$, where $mG = mR$, and let $\sigma_1, \cdots, \sigma_N$ be the cells of any given dyadic subdivision. There exist arbitrarily small automorphisms h_1 and h_2 of R that leave the boundary fixed such that h_1Th_2 is a measure-preserving homeomorphism of $h_2^{-1}G$, which has measure mR, onto h_1TG, and such that under this transformation a certain closed set F is transformed in the following manner: There exists a positive integer K such that the first KN images of F under iteration of h_1Th_2 are disjoint and exactly K are contained in the interior of each cell σ_i and contain exactly half the measure of σ_i.*

We shall prove this lemma in §7. We proceed to show that from it we can deduce

Theorem 1, as regards category. Let f be the map of R onto E defined in Lemma 3, and let $\sigma_1, \sigma_2, \cdots$, be an enumeration of all dyadic cells in R. If σ_i and σ_j belong to the same dyadic subdivision, let $E_{i,j}$ be the set of all automorphisms $T \epsilon M[E, \mu]$ such that for some Borel set A we have $TA = A$, $\mu(A \cdot f\sigma_i) > \frac{3}{4}\mu f\sigma_i$, and $\mu(A \cdot f\sigma_j) < \frac{1}{4}\mu f\sigma_j$, otherwise undefined. Every metrically intransitive automorphism in $M[E, \mu]$ belongs to one of the sets $E_{i,j}$. Because if T is metrically intransitive, there exists a measurable set, and therefore a Borel set A, such that $TA = A$ and $0 < \mu A < \mu E$. Therefore $0 < mf^{-1}A < mR$. Let p, q be points of R at which the Borel set $f^{-1}A$ has metric density 1 and 0 respectively. It follows from Lebesgue's density theorem[29] that in any sufficiently fine dyadic subdivision the cells σ_i and σ_j that contain p and q respectively are such that $m(f^{-1}A \cdot \sigma_i) > \frac{3}{4}m\sigma_i$ and $m(f^{-1}A \cdot \sigma_j) < \frac{1}{4}m\sigma_j$. Therefore $\mu(A \cdot f\sigma_i) > \frac{3}{4}\mu f\sigma_i$ and $\mu(A \cdot f\sigma_j) < \frac{1}{4}\mu f\sigma_j$, so that T belongs to $E_{i,j}$. Thus, to prove that the metrically transitive automorphisms form a residual set in $M[E, \mu]$ it suffices to show that each of the sets $E_{i,j}$ is nowhere dense, because there are only countably many of them and their union contains all metrically intransitive automorphisms in $M[E, \mu]$.

Consider any $T \epsilon M[E, \mu]$ and any set $E_{i,j}$. We shall show that arbitrarily near T there exists an automorphism S which together with a neighborhood lies outside $E_{i,j}$. According to Lemma 4, the transformation $f^{-1}Tf$ is a measure-preserving homeomorphism of the open set $G = R_0 \cdot f^{-1}T^{-1}fR_0$ onto another open set contained in R_0 and likewise having measure mR. Applying Lemma 5 to this transformation, there exist arbitrarily small automorphisms h_1, h_2 of R which leave the boundary fixed, such that $h_1f^{-1}Tfh_2$ is a measure-preserving homeomorphism of $h_2^{-1}G$ onto another open set with measure mR. Also there is a closed set F whose first KN images under this transformation are disjoint and equally distributed among the cells $\tau_1, \cdots, \tau_N$ of the dyadic subdivision to which σ_i and σ_j belong, and contain half their measure. By Lemma 4, the transformations fh_1f^{-1} and fh_2f^{-1} are automorphisms of E. Hence $S = (fh_1f^{-1})T(fh_2f^{-1})$ is an automorphism of E, and from the properties of $h_1f^{-1}Tfh_2$ it follows that S transforms the closed set fF in such a way that its first KN images are disjoint and equally distributed among $f\tau_1, \cdots, f\tau_N$, and contain half their measure. Now consider any Borel set A invariant under S. Let α be the fraction of the measure of fF contained in A. Suppose $\alpha \leqq \frac{1}{2}$, then the K disjoint images of fF in σ_i contain the same fraction of the measure of A, and therefore $mA\sigma_i \leqq \left(\frac{1}{2} + \frac{\alpha}{2}\right)m\sigma_i \leqq \frac{3}{4}m\sigma_i$, so that S does not belong to $E_{i,j}$. On the other hand, if $\alpha \geqq \frac{1}{2}$ the K disjoint images of fF in σ_j contain measure $\frac{\alpha}{2}m\sigma_j$, and therefore $mA\sigma_j \geqq \frac{\alpha}{2}m\sigma_j \geqq \frac{1}{4}m\sigma_j$, so that in this case also S does not belong to $E_{i,j}$. Furthermore, the first KN images of fF under any automorphism sufficiently close to S will be distributed among $f\tau_1, \cdots, f\tau_N$ in the same

[29] See e.g. Hobson, *The theory of functions of a real variable*, 2nd edition, Cambridge 1921, vol. 1, p. 181.

way as under S, so that the same reasoning shows that a whole neighborhood of S lies outside $E_{i,j}$. But h_1 and h_2 were arbitrarily small, hence also the automorphisms fh_1f^{-1} and fh_2f^{-1}, so that S is arbitrarily close to T. Hence $E_{i,j}$ is nowhere dense in $M[E, \mu]$.

It will be observed that S may be written in the form S_1T, where S_1 is the measure-preserving transformation ST^{-1}. But S_1 may also be written $(fh_1f^{-1})T(fh_2f^{-1})T^{-1}$, from which it follows that S_1 leaves all singular points fixed. We have therefore proved the more precise result that $E_{i,j}$ is nowhere dense with respect to every coset of the closed normal subgroup of automorphisms that leave all singular points fixed. Since these cosets are complete spaces, we may state Theorem 1 in the following slightly stronger form which we shall need later.

COROLLARY: *For every $T \in M[E, \mu]$ there exists an arbitrarily small automorphism $S \in M[E, \mu]$, that leaves all singular points fixed, such that ST is metrically transitive with respect to μ.*

II. MEASURES TOPOLOGICALLY EQUIVALENT TO LEBESGUE MEASURE

3. The Fundamental Theorem

Let μ be any measure function in a space, and h an automorphism of the space. The function μhA, considered as defined for all sets such that hA is measurable, is easily seen to be a measure function. It would naturally be described as a measure automorphic to μ, (or equivalent to μ under automorphism). It is easily seen that any measure automorphic to a LS measure is again a LS measure, and that the same transformation also effects a correspondence between their outer measures. Likewise, any measure automorphic to an r-dimensional measure is r-dimensional. The object of the present Part will be to obtain a simple characterization of measures automorphic to Lebesgue measure. The basic result, stated as Theorem 2, was originally proposed by one of us in 1936, in connection with some other group-theoretic investigations, and a proof was obtained at that time by J. von Neumann, but was not published. The present proof, based on somewhat different considerations, was worked out subsequently. The result is here published for the first time.

THEOREM 2: *In order that a measure μ in a rectangular r-cell R, $r \geqq 1$, be automorphic to Lebesgue measure it is necessary and sufficient that it be an r-dimensional Lebesgue-Stieltjes measure and that $\mu R = mR$. The correspondence can always be effected by an automorphism that leaves the boundary of R fixed.*

An equivalent formulation, from the standpoint of the Carathéodory theory, runs as follows.

THEOREM 2_1: *In order that a function μ^* defined for all subsets of R be automorphic to the Lebesgue outer measure m^* it is necessary and sufficient that it satisfy conditions $M1$ to $M5$, and also*

M6: $\mu^*G > 0$ *if G is non-void, open*

M7: $\mu^*p = 0$ *for every point p*

M8: $\mu^*BdR = 0$

and finally that $\mu^*R = mR$. *If* μ^* *satisfies these conditions there exists an automorphism* h *of* R *such that* $\mu^*A = m^*hA$ *for every* $A \subset R$, *and such that* h *leaves the boundary fixed.*

The equivalence of the two formulations is evident. Before proving the theorem we shall derive from it a number of corollaries and generalizations. In Part V, §13 to §15 and §17, we shall give some further applications.

COROLLARY 1: *Let* E *be any regularly connected polyhedron of dimension* $r \geqq 1$, *and let* μ_1 *and* μ_2 *be two r-dimensional LS measures in* E *such that* $\mu_1 E = \mu_2 E$. *There exists an automorphism* h *such that* $\overset{*}{\mu_1} A = \overset{*}{\mu_2} hA$ *for all* $A \subset E$, *and such that* h *leaves all singular points of* E *fixed. In particular, any r-dimensional LS measure in a regularly connected euclidean polyhedron* E *is automorphic to the Lebesgue measure in* E, *provided only that* $\mu E = mE$.

Let ν be the r-dimensional LS measure in E defined by $\nu^*A = \frac{1}{2}(\overset{*}{\mu_1} A + \overset{*}{\mu_2} A)$, and let f be the map of a rectangular r-cell R onto E given by Lemma 3. Then $m^*A = \nu^*fA$, and in particular, $\nu fBdR = 0$, so that we have $\mu_1 fBdR = \mu_2 fBdR = 0$. Let $\overset{*}{\nu_1}$ and $\overset{*}{\nu_2}$ be the outer measures in R defined by $\overset{*}{\nu_1} A = \overset{*}{\mu_1} fA$ and $\overset{*}{\nu_2} A = \overset{*}{\mu_2} fA$. By Lemma 2, these are both r-dimensional LS measures with $\nu_1 R = \nu_2 R = mR$. Hence, by Theorem 2_1, there exist automorphisms h_1, h_2 of R that leave the boundary fixed, such that $\overset{*}{\nu_1} A = m^*h_1A$ and $\overset{*}{\nu_2} A = m^*h_2A$. Then $h = fh_2^{-1}h_1f^{-1}$ is an automorphism of E that leaves singular points fixed and carries $\overset{*}{\mu_1}$ into $\overset{*}{\mu_2}$ as required.

COROLLARY 2: *Let* E *be a regularly connected polyhedron of dimension* $r \geqq 1$, *and* E_1 *any homeomorphic euclidean polyhedron. The r-dimensional Lebesgue-Stieltjes measures in* E *are the same as the measures of the form* $C \cdot mhA$, *where* h *is an arbitrary homeomorphism of* E *onto* E_1, C *is an arbitrary positive constant, and* m *denotes Lebesgue measure in* E.

This follows at once from Corollary 1 and the fact that any multiple of a measure homeomorphic to an r-dimensional measure is again an r-dimensional measure.

COROLLARY 3: *Let* E_1 *and* E_2 *be homeomorphic regularly connected polyhedra of dimension* $r \geqq 1$, *and let* μ_1 *and* μ_2 *be r-dimensional Lebesgue-Stieltjes measures in* E_1 *and* E_2 *respectively, such that* $\mu_1 E = \mu_2 E$. *Let* h *be any homeomorphism of* E_1 *onto* E_2. *There exists a homeomorphism* g *of* E_1 *onto* E_2 *which carries* μ_1 *into* μ_2 *and which is equal to* h *for all singular points. In particular, there exists a measure-preserving automorphism of a rectangular r-cell which is equal to any given automorphism on the boundary.*

The measure $\nu_1 A = \mu_2 hA$ is an r-dimensional LS measure in E_1, and $\nu_1 E_1 = \mu_1 E_1$. By Corollary 1, there exists an automorphism h_1 of E_1 such that $\mu_1 A = \nu_1 h_1 A$, and such that it leaves singular points fixed. Hence $g = hh_1$ is a homeomorphism of E_1 onto E_2 that takes μ_1 into μ_2 and is equal to h for all singular points.

For later use the following corollary is convenient.

COROLLARY 4: *Let* μ *be a Lebesgue-Stieltjes measure in a rectangular r-cell* R, $r \geqq 2$, $\mu R = mR$, *and let* L *be any straight line segment contained in the interior* *of R such that $\mu L = 0$. Then the automorphism that carries μ into m may be so chosen that it leaves L fixed, as well as the boundary.*

To see this, let R_1 and R_2 be rectangular r-cells with R_1 contained in the interior of R_2, and let E be the polyhedron obtained from R_2 by removing the interior of R_1. In the case $r \geqq 2$ it is evident that E is regularly connected. From E we can obtain R by identifying points of the inner boundary that have the same x_1 coordinate, say. That is, we can define a continuous map f of E onto R which carries the inner boundary into L and elsewhere is 1:1. The two measures $\mu_1 A = mfA$ and $\mu_2 A = \mu f A$ are easily verified to be r-dimensional LS measures in E, and $\mu_1 E = \mu_2 E$. By Corollary 1, there exists an automorphism h which leaves the inner and outer boundary of E fixed and carries μ_1 into μ_2. Hence fhf^{-1} is an automorphism of R that carries μ into m and leaves both L and BdR fixed.

The next two corollaries characterize measures automorphic to Lebesgue measure in the whole of euclidean space $E^{(r)}$.

COROLLARY 5: *A function μ^* defined for all subsets of $E^{(r)}$, $r \geqq 2$, is automorphic to Lebesgue outer measure m^* if and only if it satisfies the conditions stated in Theorem 2_1, except that conditions* M1 *and* M8 *are to be replaced by*

M1′: $0 \leqq \mu^* A \leqq \mu^* E^{(r)} = +\infty$

M8′: $\mu^* A < +\infty$ *for every bounded set A.*

The necessity of each of the eight conditions is evident. To prove that they are sufficient, let μ be the measure defined by any such outer measure. Let R_α be the closed cube in $E^{(r)}$ with edges parallel to the axes, center at the origin, and edge length α. Under our hypotheses, the function $f(\alpha) = \mu R_\alpha$ is a finite, strictly increasing function in $0 < \alpha < +\infty$. Such a function can have at most countably many discontinuities, hence we can select a sequence of values $0 < \alpha_1 < \alpha_2 < \cdots$, tending to infinity, at each of which $f(\alpha)$ is continuous. This means that $\mu BdR_{\alpha_i} = 0$. By hypothesis we have also that $\mu R_{\alpha_i} \to +\infty$ as $i \to \infty$. Let β_i be the edge length such that $mR_{\beta_i} = \mu R_{\alpha_i}$. Then $0 < \beta_1 < \beta_2 < \cdots$, and $\beta_i \to +\infty$ as $i \to \infty$. We can therefore define a radial transformation h_1 of $E^{(r)}$ to carry R_{α_i} into R_{β_i}, $i = 1, 2, \cdots$. Consider the new outer measure $\overset{*}{\mu_1} A = \mu^* h_1^{-1} A$. Then $\mu_1 R_{\beta_i} = mR_{\beta_i}$ and $\mu_1 BdR_{\beta_i} = 0$, $i = 1, 2, \cdots$. Let $E_1 = R_{\beta_1}$ and for $n > 1$ let E_n be R_{β_n} minus the interior of $R_{\beta_{n-1}}$. Then each E_n is a regularly connected polyhedron, since we have assumed $r \geqq 2$, and μ_1 is an r-dimensional measure in it such that $\mu_1 E_n = mE_n$. By Corollary 1, there exists an automorphism h_2 of E_n such that $\overset{*}{\mu_1} A = m^* h_2 A$, for all $A \subset E_n$. Since h_2 leaves the inner and outer boundaries of E_n fixed, these automorphisms join up to form an automorphism of $E^{(r)}$. In view of the measurability of the sets E_n and the fact that their boundaries have measure zero, we have, for any set $A \subset E^{(r)}$,

$$\overset{*}{\mu_1} A = \sum_{n=1}^{\infty} \overset{*}{\mu_1} AE_n = \sum_{n=1}^{\infty} m^* h_2(AE_n) = \sum_{n=1}^{\infty} m^*(h_2 A)E_n = m^* h_2 A.$$

Therefore $\mu^* A = m^* h_2 h_1 A$ for all $A \subset E^{(r)}$.

It should be added that in the case $r = 1$ the conditions stated in Corollary 5 are not sufficient, but a complete characterization is obtained by merely adding the requirement that the positive and negative halves of the line both have infinite measure. The function $f(x) = \mu(0 \leqq t \leqq x)$ for $x \geqq 0$, $f(x) = -\mu(x \leqq t \leqq 0)$ for $x < 0$ then defines an automorphism of the line that carries μ into m.

COROLLARY 6: *A measure in* $E^{(r)}$, $r \geqq 2$, *is automorphic to Lebesgue measure if and only if it is an infinite Lebesgue-Stieltjes measure that is zero for points and positive for non-void open sets.*

Here we may understand by a *LS* measure one that is derived from a non-negative additive function of an interval. Such a measure is necessarily finite for bounded sets, and so condition M8′ is implicit. The rest then follows from Corollary 5.

4. Proof of Theorem 2

It may be remarked first of all that in the case $r = 1$ Theorem 2 is almost trivial. For suppose R is the unit interval $0 \leqq x \leqq 1$, and let $hx = \mu(0 \leqq t \leqq x)$. Then hx is strictly increasing, in virtue of the additivity of μ and the fact that all subintervals have positive measure. It is also continuous, since points have measure zero. As x describes the interval $(0, 1)$, hx describes the same interval. Hence h is an automorphism of R. By definition, we have $\mu A = mhA$ for every interval A of the form $(0, x)$. From additivity it follows that equality holds for all intervals, and therefore for all open sets. But this implies that $\mu^*A = m^*hA$ for all sets A, by condition M5. Hence h effects the desired transformation, and also leaves the end points fixed.

However, even in the case $r = 2$ the difficulties of the general case already appear. Our proof will be based on a sequence of lemmas whose motivation lies in the idea of securing first that $\mu A = mhA$ for all sets of a division automorphic to a dyadic subdivision. Then h is modified within each of these sets so as to secure equality for the sets of a finer subdivision. Finally, a convergent sequence of such modifications is obtained and the limiting automorphism effects the desired transformation for all sets.

LEMMA 6: *Let μ be any Lebesgue-Stieltjes measure in R that is zero for points and for the boundary, and let α be any number in the interval $0 < \alpha < \mu R$. There exists an open set G contained in the interior of R such that $\mu G = \alpha$.*

Since μR is finite, there can be at most countably many planes parallel to the faces of R that intersect R in sets of positive μ-measure. Hence we can divide R into a finite number of rectangular r-cells $\sigma_1, \cdots, \sigma_N$ of diameter less than $\frac{1}{2}$, whose boundaries all have μ-measure zero. Let i be the least integer such that $\mu(\sigma_1 + \sigma_2 + \cdots + \sigma_i) \geqq \alpha$, and let G_1 be the union of the interiors of $\sigma_1, \sigma_2, \cdots, \sigma_{i-1}$. (Take G_1 equal to the void set in case $i = 1$). Then $\mu G_1 < \alpha$, but $\mu G_1 + \mu\sigma_i \geqq \alpha$. Now consider the cell σ_i, calling it R_1, and divide it into rectangular r-cells $\sigma_1^{(1)}, \sigma_2^{(1)}, \cdots, \sigma_{N_1}^{(1)}$ of diameter less than $\frac{1}{4}$, whose boundaries all have μ-measure zero. Again we find an open set $G_2 \subset R_1$, either void or consisting of the interiors of some of the cells $\sigma_i^{(1)}$, such that $\mu(G_1 + G_2) < \alpha$, while $\mu(G_1 + G_2) + \mu R_2 \geqq \alpha$, where R_2 is one of the cells $\sigma_i^{(1)}$ and is disjoint to $G_1 + G_2$. Proceeding in this manner, we find disjoint open sets $G_1, G_2, \cdots$ interior to R, and a nested sequence of rectangular r-cells $R_1, R_2, \cdots$ whose diameters tend to zero, such that $\alpha - \mu R_n \leqq \mu(G_1 + G_2 + \cdots + G_n) < \alpha$, $n \geqq 1$. The cells R_n intersect in a point p, and we have $\lim_{n \to \infty} \mu R_n = \mu p = 0$, by hypothesis. Hence $\mu G = \alpha$, where G is the union of the sets G_n.

LEMMA 7: *Let μ be any Lebesgue-Stieltjes measure in R that is zero for points and for the boundary. Let R_1 and R_2 be the two cells obtained by bisecting R perpendicularly to one of its edges. Let α_1 and α_2 be any two positive numbers such that $\alpha_1 + \alpha_2 = \mu R$. There exists an automorphism $h \in H_0[R]$ such that $\mu h R_1 = \alpha_1$ and $\mu h R_2 = \alpha_2$.*

We shall establish the existence of h by a category argument. Let H_1 be the set of automorphisms h in $H_0[R]$ such that $\mu h R_1 \geqq \alpha_1$ and $\mu h R_2 \geqq \alpha_2$. This is a closed set, for if h_n is any sequence of automorphisms in H_1 tending uniformly to h, then any ϵ-neighborhood of hR_i, $i = 1, 2$, contains $h_n R_i$ for all sufficiently large n, and therefore has μ-measure at least α_i. This being true for every $\epsilon > 0$, it follows that $\mu h R_1 \geqq \alpha_1$ and $\mu h R_2 \geqq \alpha_2$. Furthermore, the set H_1 is non-void. For, unless it contains the identity, we have either $\mu R_1 < \alpha_1$ or $\mu R_2 < \alpha_2$. Suppose $\mu R_1 < \alpha_1$. We can deform R by a continuous family of automorphisms $h_\lambda \in H_0[R]$ in such a way that as λ increases, $h_\lambda R_1$ includes more and more of the interior of R_2. At the limiting value λ_0 where $\mu h_\lambda R_1$ first becomes greater than or equal to α_1 we have also $\mu h_\lambda R_2 \geqq \alpha_2$, because $\mu h_{\lambda_0} R_2 = \lim_{\lambda \to \lambda_0^-} \mu h_\lambda R_2$, and for $\lambda < \lambda_0$ we have $\mu h_\lambda R_2 \geqq \mu R - \mu h_\lambda R_1 > \mu R - \alpha_1 = \alpha_2$. Hence the set H_1, as a non-void closed subset of a complete space, is itself a complete space. We shall show that in it the set of automorphisms that fulfill the requirements of the lemma is residual.

Let E_n be the set of all h in H_1 such that $\mu h R_1 \geqq \alpha_1 + \frac{1}{n}$. This set is closed in H_1, since we have already seen that it is closed in $H_0[R]$. To show that it is nowhere dense in H_1 it suffices to show that if $h \in E_n$ there exists an arbitrarily small automorphism $g \in H_0[R]$ such that $hg \in H_1 - E_n$. Suppose that $h \in E_n$ and consider the new measure $\nu A = \mu h A$. Then $\nu R_1 \geqq \alpha_1 + \frac{1}{n}$ and $\nu R_2 \geqq \alpha_2$. Let R_3 denote the $(r - 1)$-dimensional face common to R_1 and R_2. Then $\nu R_1 + \nu R_2 - \nu R_3 = \nu R = \alpha_1 + \alpha_2$. Hence $\nu R_1 - \alpha_1 \leqq \nu R_1 - \alpha_1 + \nu R_2 - \alpha_2 = \nu R_3$, and this with the inequality $\nu R_1 \geqq \alpha_1 + \frac{1}{n}$ gives $0 < \nu R_1 - \alpha_1 - \frac{1}{2n} < \nu R_3$. Considered with respect to R_3, ν is a LS measure that is zero for points and for the boundary. Hence, by Lemma 6, there exists a set G, open with respect to R_3 and contained in its interior, such that $\nu G = \nu R_1 - \alpha_1 - \frac{1}{2n}$. Now let g be an automorphism of R that leaves the boundary fixed and also the points of $R_3 - G$, but let it displace all points of G slightly into the interior of R_1. Such an automorphism can be found arbitrarily close to the identity. Since $gR_2 \supset R_2$, we have $\nu gR_2 \geqq \alpha_2$; and νgR_1 differs only slightly from $\nu R_1 - \nu G$, which is equal to $\alpha_1 + \frac{1}{2n}$. If g is sufficiently near the identity we shall have $\alpha_1 < \nu gR_1 < \alpha_1 + \frac{1}{n}$. Going back to the measure μ, we have $\alpha_1 < \mu hgR_1 < \alpha_1 + \frac{1}{n}$ and $\mu hgR_2 \geqq \alpha_2$. Hence $hg \in H_1 - E_n$, and so E_n is nowhere dense in H_1. Therefore the set $H_1 - \sum_1^\infty E_n$ is residual, that is, the equation $\mu hR_1 = \alpha_1$ holds for a residual set of automorphisms in H_1. The rôle of R_2 being similar, we see that the automorphisms h such that $\mu hR_2 = \alpha_2$ is also residual in H_1. The intersection of these two residual sets is the set of automorphisms having the properties required by the lemma, and therefore such automorphisms exist.

It may be remarked that in the proof of this lemma it is possible to confine attention to automorphisms that involve only a single coordinate, namely that in the direction along which R is bisected. Thus, if desired, the automorphism h can be chosen so as to leave all coordinates unchanged except this one.

LEMMA 8: *Let μ be any Lebesgue-Stieltjes measure in R that is zero for points and for the boundary. Let $\sigma_1, \cdots, \sigma_N$ be the cells of any dyadic subdivision of R, and let $\alpha_1, \cdots, \alpha_N$ be associated positive numbers whose sum is equal to μR. There exists an automorphism $h \in H_0[R]$ such that $\mu h\sigma_i = \alpha_i$, $i = 1, \cdots, N$.*

The proof consists in applying Lemma 7 a finite number of times. The cells of any dyadic subdivision are obtained from R by making a finite number of bisections. Let R_1, R_2 be the cells obtained from the first bisection. Let R_{11}, R_{12}; R_{21}, R_{22} be the cells obtained by bisecting R_1 and R_2. After n bisections we get 2^n cells $R_{i_1 \cdots i_n}$, $i_1, \cdots, i_n = 1, 2$, and for $n = K \equiv \log_2 N$ the cells $R_{i_1 \cdots i_K}$ are the cells $\sigma_1, \cdots, \sigma_N$ renamed. Let $\beta_{i_1 \cdots i_K}$ be the number α_i associated with $R_{i_1 \cdots i_K}$. For $1 \leqq n < K$ let $\beta_{i_1 \cdots i_n} = \sum \beta_{i_1 \cdots i_K}$ summed over $i_{n+1}, \cdots, i_K = 1, 2$. For $n = 1$ we have $\beta_1 + \beta_2 = \mu R$, and by Lemma 7 we can find h_1 such that $\mu h_1 R_i = \beta_i$, $i = 1, 2$. Suppose we have defined h_n such that $\mu h_n R_{i_1 \cdots i_n} = \beta_{i_1 \cdots i_n}$, $i_1, \cdots, i_n = 1, 2$. Form the new measure $\mu_n A = \mu h_n A$. Then μ_n satisfies the conditions of Lemma 7 with respect to each of the cells $R_{i_1 \cdots i_n}$, and since $\beta_{i_1 \cdots i_n 1} + \beta_{i_1 \cdots i_n 2} = \mu_n R_{i_1 \cdots i_n}$, we can find an automorphism $g_{i_1 \cdots i_n}$ in $H_0[R_{i_1 \cdots i_n}]$ such that $\mu_n g_{i_1 \cdots i_n} R_{i_1 \cdots i_{n+1}} = \beta_{i_1 \cdots i_{n+1}}$, $i_{n+1} = 1, 2$. Since $g_{i_1 \cdots i_n}$ leaves the boundary of $R_{i_1 \cdots i_n}$ fixed, these automorphisms join up to form an automorphism $g \in H_0[R]$, and we have $\mu h_n g R_{i_1 \cdots i_{n+1}} = \beta_{i_1 \cdots i_{n+1}}$. Setting $h_{n+1} = h_n g$, the induction hypotheses are again satisfied, and for $n = K$ we get an automorphism $h = h_K$ that fulfills the requirements of the lemma.

LEMMA 9: *Let μ be any Lebesgue-Stieltjes measure in R that is zero for points and for the boundary. There exists an automorphism $h \in H_0[R]$ such that for every dyadic cell σ we have $\mu hBd\sigma = 0$.*

We shall show that the automorphisms having the desired property form a residual set in $H_0[R]$.[30] Let R_1 be the $(r - 1)$-dimensional cell in which R is intersected by any one of the planes used in forming dyadic subdivisions. Let E_n be the set of automorphisms $h \epsilon H_0[R]$ such that $\mu h R_1 \geqq \frac{1}{n}$. This is a closed set. If h is any element of E_n, let $\nu A = \mu h A$. Choose $\epsilon > 0$ so that the ϵ-neighborhood of R_1 has ν-measure less than $\nu R_1 + \frac{1}{n}$. Let $g \epsilon H_0[R]$ be an automorphism that displaces the interior of R_1 into a set disjoint to R_1 but contained in its ϵ-neighborhood. Then, since the intersection of R_1 and gR_1 has ν-measure zero, we have $\nu R_1 + \nu g R_1 = \nu(R_1 + gR_1) \leqq \nu(\epsilon\text{-neighborhood of } R_1) < \nu R_1 + \frac{1}{n}$. Hence $\mu h g R_1 < \frac{1}{n}$, and $hg \epsilon H_0 - E_n$. Since hg is arbitrarily near h, the set E_n is nowhere dense, and $H_0 - \sum_1^\infty E_n$ is residual. The intersection of the residual sets corresponding to each of the countably many planes used in forming dyadic subdivisions is therefore also residual, and the automorphisms belonging to this set have the required property.

In the following lemma we require for the first time that the measures under consideration be positive for non-void open sets.

LEMMA 10: *Let μ, ν be two r-dimensional Lebesgue-Stieltjes measures in R such that $\mu R = \nu R$, and let $\epsilon > 0$ be given. There exist automorphisms g, h in $H_0[R]$ such that for each cell σ of a certain dyadic subdivision of R we have $\mu g \sigma = \nu h \sigma$; $\mu g B d \sigma = \nu h B d \sigma = 0$; diam $g\sigma < \epsilon$, diam $h\sigma < \epsilon$, diam $\sigma < \epsilon$.*

By Lemma 9, there exists an automorphism g_1 such that for every dyadic cell σ we have $\mu g_1 B d \sigma = 0$. Let μ_1 be the measure $\mu_1 A = \mu g_1 A$. Let $\sigma_1, \cdots, \sigma_N$ be the cells of a dyadic subdivision such that diam $\sigma_i < \epsilon$ and diam $g_1 \sigma_i < \epsilon$, $i = 1, \cdots, N$. The numbers $\mu_1 \sigma_i$ are all positive and their sum is equal to νR. Hence, by Lemma 8, there exists an automorphism h_1 such that $\nu h_1 \sigma_i = \mu_1 \sigma_i$, $i = 1, \cdots, N$. Put $\nu_1 A = \nu h_1 A$. Then since $\nu_1 B d \sigma_i = 0$, we can apply Lemma 9 to each cell σ_i and get an automorphism h_2 that transforms each cell σ_i into itself, such that $\nu_1 h_2 B d \sigma = 0$ for every dyadic cell σ. Let σ_{ij}, $i = 1, \cdots, N$, $j = 1, \cdots, M$, $\sigma_{ij} \subset \sigma_i$, be the cells of a finer dyadic subdivision of R, such that diam $h_1 h_2 \sigma_{ij} < \epsilon$. Since the numbers $\nu_1 h_2 \sigma_{ij}$ are all positive, and $\sum_j \nu_1 h_2 \sigma_{ij} = \mu_1 \sigma_i$, we can apply Lemma 8 to each cell σ_i using the measure μ_1 and get an automorphism g_2 that transforms each cell σ_i into itself, such that $\mu_1 g_2 \sigma_{ij} = \nu_1 h_2 \sigma_{ij}$. That is, $\mu g_1 g_2 \sigma_{ij} = \nu h_1 h_2 \sigma_{ij}$. Since the sum of these MN numbers is equal to μR (or νR), we have also $\mu g_1 g_2 B d \sigma_{ij} = \nu h_1 h_2 B d \sigma_{ij} = 0$. Finally, since $h_1 h_2 \sigma_{ij} \subset h_1 \sigma_{ij}$, $g_1 g_2 \sigma_{ij} \subset g_1 \sigma_{ij}$, and $\sigma_{ij} \subset \sigma_i$, we have also diam $h_1 h_2 \sigma_{ij} < \epsilon$, diam $g_1 g_2 \sigma_{ij} < \epsilon$, and diam $\sigma_{ij} < \epsilon$, and so the automorphisms $g = g_1 g_2$, $h = h_1 h_2$ fulfill all the requirements of the lemma.

LEMMA 11: *Any two r-dimensional Lebesgue-Stieltjes measures μ, ν in R such*

[30] Cf. footnote 28.

that $\mu R = \nu R$ are automorphic to each other under an automorphism that leaves the boundary fixed.

The proof consists in finding a sequence of partitions of R into cells[31] $\sigma_1^{(n)}, \cdots, \sigma_{N_n}^{(n)}$ and two sequences of automorphisms $g_n, h_n \in H_0[R]$, with the following properties, where $\epsilon_n = \frac{1}{2^n}$ diam R.

1° diam $\sigma_i^{(n)} \leqq \epsilon_n$, diam $g_n\sigma_i^{(n)} \leqq \epsilon_n$, diam $h_n\sigma_i^{(n)} \leqq \epsilon_n$

2° $g_n\sigma_i^{(n-1)} = g_{n-1}\sigma_i^{(n-1)}$, $h_n\sigma_i^{(n-1)} = h_{n-1}\sigma_i^{(n-1)}$, $i = 1, \cdots, N_{n-1}$

3° $\mu g_n Bd\sigma_i^{(n)} = \nu h_n Bd\sigma_i^{(n)} = 0$, $i = 1, \cdots, N_n$

4° $\mu g_n\sigma_i^{(n)} = \nu h_n\sigma_i^{(n)}$, $i = 1, \cdots, N_n$

These conditions are satisfied for $n = 0$ if we take g_0, h_0 equal to the identity, and $\sigma^{(0)} = R$. Suppose $g_n, h_n; \sigma_1^{(n)}, \cdots, \sigma_{N_n}^{(n)}$ have been defined so that conditions 1° to 4° are satisfied. Let $\mu_n A = \mu g_n A$ and $\nu_n A = \nu h_n A$. Then for each cell $\sigma_i^{(n)}$ we have $\mu_n\sigma_i^{(n)} = \nu_n\sigma_i^{(n)}$ and $\mu_n Bd\sigma_i^{(n)} = \nu_n Bd\sigma_i^{(n)} = 0$. Hence μ_n and ν_n satisfy the conditions of Lemma 10 relative to $\sigma_i^{(n)}$, and there exist automorphisms $g', h' \in H_0[\sigma_i^{(n)}]$ such that for each cell $\sigma_{ij}^{(n)}$ of a certain dyadic subdivision of $\sigma_i^{(n)}$ we have $\mu_n g'\sigma_{ij}^{(n)} = \nu_n h'\sigma_{ij}^{(n)}$; $\mu_n g' Bd\sigma_{ij}^{(n)} = \nu_n h' Bd\sigma_{ij}^{(n)} = 0$; and diam $g'\sigma_{ij}^{(n)} < \epsilon$, diam $h'\sigma_{ij}^{(n)} < \epsilon$, diam $\sigma_{ij}^{(n)} < \epsilon$, where we suppose ϵ chosen less than ϵ_{n+1}, and so small that both g_n and h_n take sets of diameter less than ϵ into sets of diameter less than ϵ_{n+1}. Then we have $\mu g_n g'\sigma_{ij}^{(n)} = \nu h_n h'\sigma_{ij}^{(n)}$; $\mu g_n g' Bd\sigma_{ij}^{(n)} = \nu h_n h' Bd\sigma_{ij}^{(n)} = 0$; and diam $g_n g'\sigma_{ij}^{(n)} < \epsilon_{n+1}$, diam $h_n h'\sigma_{ij}^{(n)} < \epsilon_{n+1}$, diam $\sigma_{ij}^{(n)} < \epsilon_{n+1}$. The automorphisms g', h' thus defined in each of the cells $\sigma_i^{(n)}$ join up to form automorphisms of R, which we shall also denote by g', h'. Putting $g_{n+1} = g_n g'$, $h_{n+1} = h_n h'$ and taking the totality of cells $\sigma_{ij}^{(n)}$ for $\sigma_1^{(n+1)}, \cdots, \sigma_{N_{n+1}}^{(n+1)}$, we see that conditions 1° to 4° are again satisfied, and the inductive definition is complete.

From conditions 1° to 4° it follows that the sequences g_n, h_n converge in $H_0[R]$ to automorphisms g, h. Because, if $x \in \sigma_i^{(n)}$ then $g_{n+1}x$ and $g_n x$ both belong to $g_n\sigma_i^{(n)}$, by 2°. Hence $| g_{n+1}x - g_n x | < \epsilon_n$,[32] by 1°. Similarly, if $x \in g_{n+1}\sigma_i^{(n)}$ then $\overset{-1}{g_n}x$ and $\overset{-1}{g_{n+1}}x$ both belong to $\sigma_i^{(n)}$. Hence $| \overset{-1}{g_{n+1}}x - \overset{-1}{g_n}x | < \epsilon_n$. Therefore $\rho(g_{n+1}, g_n) < \epsilon_n$, and because of our choice of ϵ_n, $\rho(g_{n+k}, g_n) < \epsilon_{n-1}$. Hence the sequence g_n converges in $H_0[R]$, and the sequence h_n for similar reasons. From 2° it follows that $g\sigma_i^{(n)} = g_n\sigma_i^{(n)}$ and $h\sigma_i^{(n)} = h_n\sigma_i^{(n)}$, and therefore that $gBd\sigma_i^{(n)} = g_n Bd\sigma_i^{(n)}$ and $hBd\sigma_i^{(n)} = h_n Bd\sigma_i^{(n)}$. From 3° and 4° we then find that $\mu g\sigma_i^{(n)} = \nu h\sigma_i^{(n)}$ and $\mu gBd\sigma_i^{(n)} = \nu hBd\sigma_i^{(n)} = 0$. Hence the measures $\mu' A = \mu gA$ and $\nu' A = \nu hA$ are equal for all cells $\sigma_i^{(n)}$ and are zero for their boundaries. Now from 1° it follows that any open set G can be represented as the union of a sequence of non-overlapping cells $\sigma_i^{(n)}$, say $G = \sum_{s=1}^{\infty} \sigma_s$. Then, since the intersection of any two cells σ_s has measure zero with respect to μ'

[31] The cells will all be dyadic cells, but not necessarily of the same subdivision.

[32] We use the vector notation $| x - y |$ to denote the euclidean distance between points x and y.

and ν', we have $\mu'G = \sum_{s=1}^{\infty} \mu'\sigma_s = \sum_{s=1}^{\infty} \nu'\sigma_s = \nu'G$. Hence μ'^* and ν'^* agree for all open sets, and therefore for all sets. Thus $\mu^*A = \nu^*hg^{-1}A$ for every set, and the lemma is proved.

Theorem 2 follows at once on taking $\nu = m$.

In the proof of Lemma 7 it was remarked that in that case the automorphism could be chosen so as to involve only a single coordinate. A simple refinement of the proof of Lemma 9 shows that the automorphism there obtained may be taken to be a finite product of automorphisms each of which involves only a single coordinate. The final automorphism in Lemma 11 is obtained by composing such automorphisms and then passing to the limit. Thus it is possible to assert that the correspondence of the two measure functions can be effected by an automorphism that can be approximated uniformly by finite products of automorphisms each of which involves only a single coordinate. This remark is of interest in view of the fact that it is still an open question whether an arbitrary automorphism of R can be so approximated.[33] If true, this would furnish a strong inductive method for proving topological theorems.

Since the transformation can always be effected by an automorphism that is special at least to the extent of leaving the boundary fixed, and perhaps in other respects, as indicated in the last paragraph, it is natural to inquire whether it would suffice to consider only differentiable automorphisms, or ones whose modulus of continuity is otherwise restricted. That this is not the case may be seen as follows.

Consider the r-dimensional unit cube and let h be a radial contraction that leaves the boundary fixed and takes each concentric cube with edge length $2d < \frac{1}{2}$ into the one whose circumscribed sphere has radius $\delta = e^{-\frac{1}{d}}$, and let $\mu A = mh^{-1}A$. Then the μ-measure of any sphere about the center with radius $\delta < e^{-4}$ is greater than the Lebesgue measure of a sphere with radius $d = \left(\log \frac{1}{\delta}\right)^{-1}$. Hence any automorphism that takes μ into m must have a modulus of continuity at the center greater than $\left(\log \frac{1}{\delta}\right)^{-1}$. But this function dominates any of the form $C\delta^{\alpha}$, $\alpha > 0$, and so the automorphism cannot be differentiable at the center, nor can it satisfy even a Lipschitz or Hölder condition there.

III. Proof of Theorem 1

In Part I, the proof of Theorem 1, as regards category, was reduced to Lemma 5, except for an assumption in the proof of Lemma 3 which has now been removed. Lemma 5 is an approximation theorem. To prove it, it will be necessary to devise methods for modifying a given transformation so as to secure control over the distribution of the images of a certain closed set. We shall do

[33] Problem proposed by one of us in Fund. Math. **24** (1935) 324.

this in three main steps. In §5 it is shown how a transformation can be modified so that it will have a periodic point with images equally distributed among the cells of a given dyadic subdivision. In §6 a lemma is derived which, in effect, enables one to obtain a periodic Cantor set, starting with only a periodic point. These two constructions are combined in §7, and the final step consists in an equivalence transformation which expands the Cantor set to one having the required amount of measure. The result is that an arbitrary measure-preserving automorphism of a rectangular r-cell (or, more generally, a transformation in the r-cell that corresponds to a measure-preserving automorphism of another polyhedron as described in Lemma 4) can be modified so that the resulting transformation is measure-preserving and has a periodic Cantor set whose images carry a prescribed fraction of the total measure and are equally distributed among the cells of a given dyadic subdivision. All modifications are effected by left and right multiplication with automorphisms that leave the boundary fixed. Thus Lemma 5 is established, and therefore Theorem 1. In §18 it will be shown that for automorphisms in $M_0[R]$ one can obtain still more precise approximation theorems.

5. Lemmas Concerning Distribution of Finite Sets

LEMMA 12: *Let p, q be any two interior points of a rectangular r-cell R, $r \geqq 2$. There exists an automorphism $T \epsilon M_0[R]$ such that $Tp = q$.*

Observe first that there is no difficulty in defining an automorphism g that carries p into q and leaves BdR fixed. Most simply, join p to the vertices of R and map the resulting cell complex affinely onto the similar complex obtained from q. Thus the only point to the lemma is to show that the transformation can be effected by a measure-preserving automorphism. It is possible to do this directly by a construction involving displacements around closed tubes, but a more elegant proof is based on Theorem 2. Let g be the automorphism just defined, and consider the outer measure $\mu^*A = m^*g^{-1}A$. By Theorem 2_1 there exists an automorphism $h \epsilon H_0[R]$ such that $\mu^*A = m^*hA$. By Corollary 4, h can be so chosen as to leave q fixed, in fact it may leave a line segment through q fixed. The composed automorphism hg then carries p into q, leaves BdR fixed, and since $m^*hgA = \mu^*gA = m^*A$ we see that $hg \epsilon M_0[R]$.

LEMMA 13: *Given two sets of points $p_1, \cdots, p_N$; $q_1, \cdots, q_N$, each consisting of N distinct interior points of a rectangular r-cell R, $r \geqq 2$, and $| p_i - q_i | < \epsilon$, there exists an automorphism $T \epsilon M_0[R]$ such that $Tp_i = q_i$, $i = 1, \cdots, N$, and $\rho(T, I) < \epsilon$.*

In effect, the lemma asserts that any two finite sets of interior points that could possibly be equivalent under automorphism are equivalent under a measure-preserving automorphism whose norm is no larger than it need be. Let L_i denote the straight line segment joining p_i to q_i. Then L_i is contained in the interior of R. Suppose first that these segments are all disjoint. Then we can enclose each in the interior of a rectangular r-cell σ_i so chosen that the cells $\sigma_1, \cdots, \sigma_N$ are disjoint and each has diameter less than ϵ. By the previous lemma, we can find an automorphism T_i of each cell σ_i which preserves measure and carries p_i into q_i. Since these automorphisms leave the boundary of each cell σ_i fixed, they can be extended to an automorphism T of R by defining T equal to T_i in each cell σ_i and equal to the identity elsewhere. Since T only permutes points within each cell, it moves no point by more than ϵ, and so $\rho(T, I) < \epsilon$. In case any of the segments L_i intersect, we shall join each p_i to q_i by a chain of $k + 1$ points $p_i = p_{i0}, p_{i1}, \cdots, p_{ik} = q_i$ in such a way that $| p_{i,j} - p_{i,j+1} | < \epsilon/k$ and such that for each $1 \leqq j \leqq k$ the segment L_{ij} joining $p_{i,j-1}$ to $p_{i,j}$ are disjoint. This can always be done; in fact, unless two of the segments L_i intersect in a point that divides each in the same ratio, it will suffice to take the points p_{ij} as the points that divide L_i into k equal segments. In the exceptional case a slight displacement of some of the intermediate points will make the segments $L_{1j}, \cdots, L_{Nj}$ disjoint and will not disturb the inequalities $| L_{ij} | < \epsilon/k$. By the argument just given above, we can find an automorphism T_j of norm less than ϵ/k such that $T_j p_{i,j-1} = p_{i,j}$, $i = 1, \cdots, N$. Hence the composed transformation $T = T_k T_{k-1} \cdots T_1$ will carry each p_i into q_i. Furthermore, it will have norm less than ϵ, because the product of k transformations whose norms are less than ϵ/k always has norm less than ϵ, as may be deduced at once from the triangle inequality.

LEMMA 14: *In a rectangular r-cell R, $r \geqq 2$, let D be a set with measure equal to mR. Let U_δ denote the operation of "taking the δ-neighborhood with respect to D," so that $U_\delta A$ means the set of all points in D at distance less than δ from A. To each $\delta > 0$ corresponds a positive integer λ with the following property: If T is any* 1:1 *measure-preserving transformation of D onto itself and p is any point of D then $(TU_\delta)^\lambda p = D$.*

It is to be emphasized that λ depends only upon δ and the dimension number r, is independent of the transformation T and of the set D. The transformation need not be assumed to be an automorphism, although it will be in the only application we shall make.

Essentially, the proof consists in the observation that the composite operation TU_δ operating on any non-void set either increases its measure by at least a certain minimum amount or else yields the whole domain D. The important thing is to obtain a uniform estimate of this minimum amount.

The integer λ is determined as follows. Let η be the minimum volume of the intersection of the δ-neighborhoods of any two points of R at distance δ apart. This number η is not simply the common volume of two such spheres, because when the two points are near the boundary of R part of the intersection of their δ-spheres may fall outside R. However, we are concerned only with the fact that $\eta > 0$, and a lower bound to η is furnished by the number $V/2^r$, where V is the r-dimensional volume of a sphere of radius $\delta/2$. This is because the intersection of the δ-neighborhoods of any two points at distance δ certainly contains the $\delta/2$-neighborhood of the point of R midway between them, and the minimum volume of the $\delta/2$-neighborhood of any point of R is $V/2^r$, that being the volume of the $\delta/2$-neighborhood of a vertex of R. We shall take for λ the least integer greater than mR/η.

Now consider the composite operation TU_δ. We shall show that λ iterations of this operation, starting from any point of D, yields the whole of D. For let us suppose that $(TU_\delta)^\lambda p$ is not equal to D, and consider any integer n, $0 \leqq n < \lambda$. The set $(TU_\delta)^n p$ is then not δ-dense in D, and so there is a point q in R whose distance from it is exactly δ. Let q' be a limit point of the set $(TU_\delta)^n p$ at distance δ from q. The operation U_δ applied to the set $(TU_\delta)^n p$ therefore adjoins at least the common part in D of the δ-neighborhoods of q and q', and therefore increases the measure of the set by at least η. That is, $mU_\delta(TU_\delta)^n p \geqq m(TU_\delta)^n p + \eta$. Hence, recalling that T preserves measure, $m(TU_\delta)^{n+1} p \geqq m(TU_\delta)^n p + \eta$. This being true for $n = 0, 1, \cdots, \lambda - 1$, it follows that $m(TU_\delta)^\lambda p \geqq \lambda\eta$, which is impossible, since by definition $\lambda\eta > mR$.

LEMMA 15: *Let T be any measure-preserving automorphism of a set D contained in the interior of an r-cell R, $r \geqq 2$, with $mD = mR$. There exist arbitrarily small measure-preserving automorphisms S_1, $S_2 \in M_0[R]$ such that under the transformation $S_2S_1TS_2^{-1}$ the centers of the cells of a certain arbitrarily fine dyadic subdivision of R undergo a cyclic permutation.*

This lemma is in a sense the key to the present problem. It enables one to gain control over the successive images of a certain point, in fact, it secures their equal distribution among the cells of a certain dyadic subdivision. Since the proof is somewhat involved, it may be well to sketch the idea in advance. It is first shown that one can find points $p_1, \cdots, p_K$ whose first L images under T constitute a set which is "nearly" uniformly distributed among the cells of a given dyadic subdivision. This much is deduced from the ergodic theorem. (This is the only place in the entire proof of Theorem 1 where the ergodic theorem of Birkhoff is used.) But some of the points of this set may coincide. The next part of the proof consists in modifying T by composition with a small transformation S_1 to secure that a similarly distributed set of points are all distinct and *also* are linked together to form a single cycle. In joining the chains together, it is necessary to add more points, but it is secured that only a "few" points are added. The result is that under S_1T there is a cycle consisting of points $q_1, \cdots, q_s$ of which KL are distributed exactly like the first L images of $p_1, \cdots, p_K$. It is further secured that the number s is equal to the number of cells in a finer subdivision, and that the points $q_1, \cdots, q_s$ are sufficiently nearly uniformly distributed that they can be set in correspondence with the centers of the cells of this finer subdivision in such a way that corresponding points are close together. By a final transformation S_2 the points $q_1, \cdots, q_s$ are carried into these centers, so that the latter constitute a cycle under $S_2S_1TS_2^{-1}$.

Proceeding now to a more precise formulation of this proof, let δ be an arbitrary positive number, and let $\sigma_1, \cdots, \sigma_N$ be the cells of a dyadic subdivision with diameter less than δ, say $N = 2^{\alpha r}$. Let them be numbered in such a way that σ_i has at least one vertex in common with σ_{i+1}, $1 \leqq i < N$. This is easily seen to be possible. Let $f_i(p)$ be the characteristic function of the interior of σ_i. The ergodic theorem asserts that the limit $f_i^*(p) = \lim_{n\to\infty} \frac{1}{n} \sum_{\nu=1}^{n} f_i(T^\nu p)$

exists for almost all p, and that $\int_D f_i^*(p)\,dm(p) = \int_D f_i(p)\,dm(p) = 1/N$, provided we suppose that $mR = 1$. We shall show first that we can select a finite set of points $p_1, \cdots, p_K$ such that the average of each of the functions $f_i^*(p)$ over this set is nearly equal to its integral, that is, $\left| \frac{1}{K} \sum_{k=1}^{K} f_i^*(p_k) - \frac{1}{N} \right| < \eta$. Here η may be any positive number, but for our purpose it will suffice to take $\eta = 1/N^2$. Let J be an integer greater than $2/\eta$, and for each set of positive integers $i_1, \cdots, i_N$ not exceeding J let $A_{i_1 \cdots i_N}$ denote the set of points p in D such that $\frac{i_j - 1}{J} < f_j^*(p) \leqq \frac{i_j}{J}$ for all integers $1 \leqq j \leqq N$. Let $p_{i_1 \cdots i_N}$ be any point of $A_{i_1 \cdots i_N}$, provided this set has positive measure, otherwise undefined. Then, since on $A_{i_1 \cdots i_N}$ the functions $f_j^*(p)$ differ from $f_j^*(p_{i_1 \cdots i_N})$ by not more than $1/J$, we have $| 1/N - \sum' f_j^*(p_{i_1 \cdots i_N}) \cdot mA_{i_1 \cdots i_N} | \leqq 1/J < \eta/2$, where $\sum'$ denotes summation over all sets of indices $i_1, \cdots, i_N$ such that $mA_{i_1 \cdots i_N} > 0$. Now let $r_{i_1 \cdots i_N}$ be rational numbers such that $| 1/N - \sum' f_j^*(p_{i_1 \cdots i_N}) r_{i_1 \cdots i_N} | < \eta/2$ and $\sum' r_{i_1 \cdots i_N} = 1$. Such numbers exist, since $\sum' mA_{i_1 \cdots i_N} = mR = 1$. Let K be a common denominator of all the numbers $r_{i_1 \cdots i_N}$, and select $Kr_{i_1 \cdots i_N}$ points from each set $A_{i_1 \cdots i_N}$ that has positive measure. Call these points $p_1, \cdots, p_K$. Then we have

$$\left| \sum' f_j^*(p_{i_1 \cdots i_N}) r_{i_1 \cdots i_N} - \frac{1}{K} \sum_{k=1}^{K} f_j^*(p_k) \right| \leqq \frac{1}{J} < \eta/2$$

and therefore $| 1/N - 1/K \sum_{k=1}^{K} f_j^*(p_k) | < \eta$, $j = 1, \cdots, N$. We may suppose the points $p_1, \cdots, p_K$ so chosen that no image falls on the boundary of a cell σ_j, since this means avoiding only a set of measure zero in making the selection. Recalling that $f_j^*(p_k) = \lim_{n \to \infty} \frac{1}{n} \sum_{\nu=1}^{n} f_j(T^\nu p_k)$, we see that for all sufficiently large n we have $\left| 1/N - 1/K \sum_{k=1}^{K} \frac{1}{n} \sum_{\nu=1}^{n} f_j(T^\nu p_k) \right| < \eta$, $j = 1, \cdots, N$. Hence for all sufficiently large n we have also

$$\left| \frac{1}{N} - \frac{1}{Kn} \sum_{k=1}^{K} \sum_{\nu=1}^{n} f_j(T^\nu p_k) \right| + \frac{2(\lambda + 1)}{n} < \eta \tag{1}$$

where λ is the integer corresponding to δ determined in Lemma 14. Let L be such a value of n, and in addition let it be so chosen that for some integer $\beta > \alpha$ we have $2^{\beta r} - K \leqq K(L + \lambda) < 2^{\beta r}$. Such a number L can be found since the numbers $K(n + \lambda)$ include all multiples of K greater than some number. If we let $s = 2^{\beta r}$ and $K_1 = s - K(L + \lambda)$, then $0 \leqq K_1 \leqq K$.

We now proceed to select s distinct points $q_1, \cdots, q_s$ from D in such a way that Tq_i is within distance δ of q_{i+1}, $1 \leqq i \leqq s$, (letting $q_{s+1} = q_1$), and such that KL of these points are interior to the same cells σ_j as corresponding points

$T^{\nu}p_k$, $1 \leqq \nu \leqq L$, $1 \leqq k \leqq K$. To secure this, we proceed step by step as follows. Take $q_1 = Tp_1$. For $1 < i \leqq L$, take q_i in the same cell as $T^i p_1$ and such that Tq_i is in the same cell as $T^{i+1}p_1$. By Lemma 14 applied to T^{-1}, $q_L \in (T^{-1}U_\delta)^\lambda p_2$. Hence $Tq_L \in U_\delta(T^{-1}U_\delta)^{\lambda-1}p_2$. Choose q_{L+1} in $(T^{-1}U_\delta)^{\lambda-1}p_2$ within distance δ of Tq_L. For $1 < i \leqq \lambda - 1$, take $q_{L+i} \in (T^{-1}U_\delta)^{\lambda-i}p_2$ within distance δ of Tq_{L+i-1}. Then $Tq_{L+\lambda-1}$ is within distance δ of p_2. Take $q_{L+\lambda}$ within distance δ of $Tq_{L+\lambda-1}$ and such that $Tq_{L+\lambda}$ is in the same cell as Tp_2. For $1 \leqq i \leqq L$, take $q_{L+\lambda+i}$ in the same cell as $T^i p_2$ and such that $Tq_{L+\lambda+i}$ is in the same cell as $T^{i+1}p_2$. The next λ points are then chosen to lead up to p_3, similarly to the way in which q_{L+i}, $1 \leqq i \leqq \lambda$, were chosen leading up to p_2. Eventually, the point $q_{(K-1)(L+\lambda)+L}$ is chosen in the same cell as $T^L p_K$. We then close the cycle by a chain of $\lambda + K_1$ points (instead of λ points as in the preceding cases) in order to get a cycle with exactly s points. This can be done since $q_{(K-1)(L+\lambda)+L} \in (T^{-1}U_\delta)^{\lambda+K_1}p_1$. At the last step we have $q_{s-1} \in T^{-1}U_\delta p_1$ and choose q_s near enough to p_1 so that it is within distance δ of Tq_{s-1} and so that Tq_s is within distance δ of $q_1 = Tp_1$. At every step the point to be selected can be chosen arbitrarily from an open sphere with respect to D. Hence there is no difficulty in avoiding points previously chosen and thus securing that the points $q_1, \cdots, q_s$ are distinct, and therefore also the points $Tq_1, \cdots, Tq_s$. We see that the KL points $q_1, q_2, \cdots, q_L, q_{L+\lambda+1}, \cdots, q_{L+\lambda+L}, \cdots, q_{(K-1)(L+\lambda)+1}, \cdots, q_{(K-1)(L+\lambda)+L}$ are respectively interior to the same cells σ_j as the points Tp_1, $T^2p_1, \cdots, T^Lp_1, Tp_2, \cdots, T^Lp_2, \cdots, Tp_K, \cdots, T^Lp_K$. Finally, it is seen that q_{i+1} is always within distance δ of Tq_i, since they either lie in the same cell, or were specifically chosen so as to be within this distance.

In virtue of Lemma 13, we can find an automorphism $S_1 \in M_0[R]$, with norm less than δ, which carries Tq_i into q_{i+1}, $1 \leqq i \leqq s$. Under the composed transformation S_1T the points $q_1, \cdots, q_s$ thus form a single cycle. Furthermore, since KL of these lie in the same cells as corresponding points $T^\nu p_k$, it follows that

$$\left|\frac{1}{N} - \frac{1}{s}\sum_{i=1}^{s} f_j(q_i)\right| \leqq \left|\frac{1}{N} - \frac{1}{s}\sum_{k=1}^{K}\sum_{\nu=1}^{L} f_j(T^\nu p_k)\right| + \frac{K\lambda + K_1}{s}$$

$$\leqq \left|\frac{1}{N} - \frac{1}{KL}\sum_{k=1}^{K}\sum_{\nu=1}^{L} f_j(T^\nu p_k)\right| + \frac{s - KL}{s} + \frac{K\lambda + K_1}{s}.$$

From equation (1) it follows that this is less than η, because

$$\frac{s - KL}{s} + \frac{K\lambda + K_1}{s} = \frac{2(K\lambda + K_1)}{s} \leqq \frac{2K(\lambda + 1)}{K(L + \lambda) + K_1} < \frac{2(\lambda + 1)}{L}.$$

Hence the relative number of points $q_1, \cdots, q_s$ in any cell σ_j differs from $1/N$ by less than $\eta = 1/N^2$. Let us now re-number the points $q_1, \cdots, q_s$ counting first all those in σ_1, then those in σ_2, and so on. Call the resulting sequence $q_1', q_2', \cdots, q_s'$. Let $q_1'', q_2'', \cdots, q_s''$ be the centers of the $s = 2^{\beta r}$ cells of the β-th dyadic subdivision, again counting first those in σ_1, then those in σ_2, and so on. We shall show that corresponding points q_i' and q_i'' always lie either in the same cell σ_j or in adjacent cells. The number of points q_i in any cell σ_j differs from s/N by less than s/N^2. Hence the indices of the points q_i' that lie in σ_j fall between the limits $\frac{(j-1)s}{N} - \frac{js}{N^2}$ and $\frac{js}{N} + \frac{js}{N^2}$, which are contained between $\frac{(j-2)s}{N}$ and $\frac{(j+1)s}{N}$, the limits of the indices of points q_i'' lying in σ_{j-1}, σ_j, and σ_{j+1}. Since these are adjacent cells, the distance from q_i' to q_i'' is less than 2δ, and by Lemma 13 we can find an automorphism $S_2 \in M_0[R]$ with norm less than 2δ which carries q_i' into q_i'', $1 \leqq i \leqq s$. The points q_i'' then constitute a single cycle under the transformation $S_2S_1TS_2^{-1}$. Since S_1 and S_2 are both arbitrarily near the identity, the lemma is established.

6. Equivalence of Cantor sets under automorphism of the containing space

LEMMA 16: *Let C be a linear Cantor set*[34] *contained in the interior of an r-cell σ_0, which in turn is contained in an r-cell R, $r \geqq 2$, and let f be a homeomorphism of σ_0 onto a subset of R. There exists an automorphism h of R that leaves the boundary fixed and carries fC back into coincidence with C in such a way that $hfp = p$ for every p in C.*

The lemma implies that any topological Cantor set C' contained in the interior of R is equivalent under automorphism of R to a linear Cantor set C provided some homeomorphism of C onto C' can be extended to an r-cell containing C. It should be remarked that such an extension is not always possible even when one of the sets is linear, as is shown by Antoine's example in the 3-dimensional cube (See footnote 15).

The first step in the proof consists in showing that if σ_1 and σ_2 are disjoint r-cells contained in the interior of σ_0, there exists an automorphism $h \in H_0[R]$ such that $hf\sigma_1 \subset \sigma_1$ and $hf\sigma_2 \subset \sigma_2$. This may be seen as follows. Let p_1, p_2 be the centers of σ_1, σ_2 respectively. Then fp_1, fp_2 are distinct interior points of R, and we can find an automorphism $h_1 \in H_0[R]$ which carries them back into p_1, p_2 respectively. Then h_1f is a homeomorphism of σ_0 onto a subset of R which leaves p_1, p_2 fixed. By continuity, there exist r-cells σ_1', σ_2' about p_1, p_2 such that $h_1f\sigma_1' \subset \sigma_1$ and $h_1f\sigma_2' \subset \sigma_2$. Let $g \in H_0[\sigma_0]$ be such that $g\sigma_1 \subset \sigma_1'$ and $g\sigma_2 \subset \sigma_2'$. Such an automorphism is easily defined by dividing σ_0 into two cells containing σ_1, σ_2 in their interiors and then drawing the interior points of each part toward p_1, p_2 respectively. The homeomorphism h_1fg then carries σ_1 and σ_2 into subsets of themselves. Now consider the transformation h_2 of R equal to fgf^{-1} in $f\sigma_0$ and equal to the identity in the rest of R. That h_2 is an automorphism of R follows from Brouwer's theorem on invariance of region.[35] According to this theorem, $f\sigma_0$ is a closed region, that is, the closure of a connected open set, and interior and boundary points of σ_0 correspond respectively to

[34] That is, a topological Cantor set contained in a straight line segment.
[35] See e.g. AH p. 396.

interior and boundary points of $f\sigma_0$. Hence fgf^{-1} is an automorphism of $f\sigma_0$ that leaves all boundary points fixed, and can therefore be extended to the rest of R by defining it equal to the identity outside. The automorphism $h = h_1 h_2$ belongs to $H_0[R]$ and hf carries σ_1 and σ_2 into subsets of themselves, as required.

Now let us choose a family of r-cells $\sigma_{i_1 \cdots i_n}$; $i_1, \cdots, i_n = 1, 2; n = 1, 2, \cdots$; contained in the interior of σ_0, so as to fulfill the following conditions:

1° For each n, the cells $\sigma_{i_1 \cdots i_n}$ are disjoint.

2° $\sigma_{i_1 \cdots i_n}$ is contained in the interior of $\sigma_{i_1 \cdots i_{n-1}}$.

3° The diameter of $\sigma_{i_1 \cdots i_n}$ tends to zero with $1/n$.

4° $C = \prod_{n=1}^{\infty} \sum_{i_1 \cdots i_n} \sigma_{i_1 \cdots i_n}$.

Such a representation of any linear Cantor set is immediate from its definition. We have shown that there exists an automorphism $h_1 \in H_0[R]$ such that $h_1 f\sigma_1 \subset \sigma_1$ and $h_1 f\sigma_2 \subset \sigma_2$. Similarly, let h_2 be an automorphism of R that is equal to the identity outside σ_1 and σ_2 and such that $h_2 h_1 f\sigma_{i_1 i_2} \subset \sigma_{i_1 i_2}$; $i_1, i_2 = 1, 2$. This involves only an application of the previous result to the two cells σ_1, σ_2 separately. At the n-th stage, we find an automorphism $h_n \in H_0[R]$ which is equal to the identity outside the cells $\sigma_{i_1 \cdots i_{n-1}}$ and is such that $h_n h_{n-1} \cdots h_1 f\sigma_{i_1 \cdots i_n} \subset \sigma_{i_1 \cdots i_n}$. The successive products $h_n h_{n-1} \cdots h_1$ converge uniformly, in virtue of condition 3°, and therefore have for limit a continuous mapping h of R onto itself. That h is likewise 1:1 may be seen as follows. Consider two distinct points p, q outside fC. They are outside the sets $f\sigma_{i_1 \cdots i_n}$ for some n. Hence their images under h are the same as under the finite product $h_n \cdots h_1$, and therefore distinct and outside C. On the other hand, h carries distinct points of fC into distinct points of C. In fact, h is equal to f^{-1} on fC, because $hf\sigma_{i_1 \cdots i_n} \subset \sigma_{i_1 \cdots i_n}$ and therefore hf leaves every point of C fixed. Thus h is a 1:1 continuous map of R onto itself, therefore an automorphism, and it leaves boundary points fixed since it is equal to h_1 there. Furthermore, we have just seen that $hfp = p$ for all p in C.

Before proceeding with our main investigation, which will be resumed beginning with Lemma 17, we shall digress to consider the bearing of the present lemma on the work of Antoine.[36] Given two Cantor sets C and C' situated in a euclidean space $E^{(r)}$ three possibilities are conceivable: (i) there is a homeomorphism of C onto C' that can be extended to the whole space; (ii) there is a homeomorphism that can be extended to a neighborhood of C but none can be extended to the whole space; (iii) no homeomorphism can be extended even to a neighborhood of C. Antoine showed that in the plane only case (i) can arise, indeed that the homeomorphism can be taken equal to the identity outside any given rectangle to which C and C' are interior. But he showed by examples that in 3-space (and therefore in r-space, $r > 3$) all three possibilities can arise, and that case (iii) can be realized even when one of the sets is linear. In his example illustrating case (ii), however, both sets are skew. We shall show

[36] See footnote 15.

that this is necessarily the case, that is, that if either of the sets is plane or linear case (ii) cannot arise in any space $E^{(r)}$. More precisely, the result is as follows.

COROLLARY: *If C and C' are Cantor sets contained in $E^{(r)}$, $r \geqq 2$, and C is plane or linear then C and C' are equivalent under automorphism of $E^{(r)}$ if and only if some homeomorphism of C onto C' can be extended to a neighborhood of C. If f is any homeomorphism of C onto C' that can be extended to a neighborhood of C then f can be extended to $E^{(r)}$ in such a way that it is equal to the identity outside any given rectangular r-cell to which C and C' are interior.*

Only the second assertion need be proved since it obviously implies the first. We shall consider first the case in which C is linear. Let R be the given rectangular r-cell containing C and C' in its interior and let f_1 be an extension of f to an open set G containing C. We may suppose that both G and f_1G are contained in R. Using the Heine-Borel theorem we can find a finite number of disjoint rectangular r-cells $R_1, \cdots, R_N$ contained in G whose interiors cover C. Let p_i be the center of R_i. Then $f_1p_1, \cdots, f_1p_N$ are distinct interior points of R and by Lemma 13 there exists $h_1 \in H_0[R]$ such that $h_1f_1p_i = p_i$. Let σ_i' be an r-cell about p_i so small that $h_1f_1\sigma_i' \subset R_i$, and let σ_i be an r-cell interior to R_i and containing CR_i in its interior. Let g be a shrinking transformation of the interior of each cell R_i such that $g\sigma_i \subset \sigma_i'$ and such that g leaves boundary points of $R_1, \cdots, R_N$ fixed. Then h_1f_1g is a homeomorphism of each σ_1 onto a subset of R_i. By Lemma 16 applied to each cell R_i there exists $h_2 \in H_0[R]$ such that $h_2h_1f_1g$ is equal to the identity on C. From the theorem on invariance of region it follows as before that the transformation h_0 equal to $f_1gf_1^{-1}$ in $f_1R_1 + \cdots + f_1R_N$ and equal to the identity elsewhere is an automorphism, and it evidently belongs to $H_0[R]$ since $f_1G \subset R$. Hence $h = h_2h_1h_0$ is in $H_0[R]$ and $hfp = h_2h_1f_1gp = p$ for every p in C. Therefore h^{-1} is an extension of f to R, and it may be defined equal to the identity outside R.

It remains to consider the case in which C is plane but not linear. Let R be a rectangular r-cell containing both C and C' in its interior and let C'' be a linear Cantor set in the same plane as C and likewise interior to R. Antoine has shown that there is an automorphism of this plane that carries C into C'' and is equal to the identity outside R. This automorphism can be extended to $E^{(r)}$ in such a way that it is still equal to the identity outside R. This can be done conveniently by projecting the transformation from two points in R on opposite sides of the plane in which C lies, then projecting again from two points in R on opposite sides of the 3-space in which the transformation is already defined. After $r - 2$ steps the desired extension is obtained. Thus we have an automorphism $h \in H_0[R]$ that carries C into C''. The transformation fh^{-1} is a homeomorphism of C'' onto C' that can be extended to a neighborhood of C''. Since C'' is linear there exists an automorphism $g \in H_0[R]$ equal to fh^{-1} on C'', as we have already shown. The automorphism gh is therefore equal to f on C and it can be defined equal to the identity outside R.

LEMMA 17: *If the given homeomorphism f in Lemma 16 is measure-preserving then the automorphism h can also be taken to be measure-preserving.*

Let h be any automorphism fulfilling the requirements of Lemma 16. In R consider the measure function $\mu A = mh^{-1}A$. Then $\mu C = mh^{-1}C = mfC = 0$, since f is now assumed to be measure-preserving. Let L denote a line segment containing C and contained in the interior of R. It is possible that μL may be positive, but in that case we can displace the points of L not in C by an automorphism $g_1 \in H_0[R]$ so chosen that $\mu g_1 L = 0$ and $g_1 p = p$ for $p \in C$. Then consider the measure $\mu_1 A = \mu g_1 A = mh^{-1}g_1 A$. This is an r-dimensional LS measure and in addition $\mu_1 L = 0$. Hence, by Theorem 2 Corollary 4, there exists an automorphism $g_2 \in H_0[R]$ such that $\mu_1 A = mg_2 A$ and $g_2 p = p$ for $p \in L$. Thus we have $mh^{-1}g_1 A = mg_2 A$ for all $A \subset R$, that is, $mg_2 g_1^{-1}hA = mA$. Hence $g_2 g_1^{-1}h$ is measure-preserving, and since both g_1 and g_2 leave the points of C fixed the measure-preserving automorphism $T = g_2 g_1^{-1}h$ fulfills the requirements of the lemma.

7. Proof of Lemma 5

LEMMA 18: *Let R_0 be the interior of a rectangular r-cell R, $r \geqq 2$, and let T be a measure-preserving homeomorphism of an open set $G \subset R_0$ onto an open set $TG \subset R_0$, where $mG = mR$, and let $\sigma_1, \cdots, \sigma_N$ be the cells of any given dyadic subdivision. There exist arbitrarily small automorphisms h_1, $h_2 \in H_0[R]$ such that $h_1 T h_2$ is a measure-preserving homeomorphism of $h_2^{-1}G$ onto $h_1 TG$, where $mh_2^{-1}G = mR$, and such that $h_1 T h_2$ transforms a certain Cantor set C_1 in the following manner: The points of C_1 are all periodic with the same period; the distinct images of C_1 are disjoint and equally distributed among the cells $\sigma_1, \cdots, \sigma_N$; the images of C_1 contain a prescribed fraction $\alpha < 1$ of the measure of each of the cells σ_i.*

In the proof we shall assume $mR = 1$, which involves no loss of generality. Observe first that T may be considered as a measure-preserving automorphism of the set $D = \prod T^n G$, $-\infty < n < +\infty$, which has measure one, since it is the intersection of countably many sets of measure one. By Lemma 15, there exist arbitrarily small automorphisms S_1, $S_2 \in M_0[R]$ such that $T_1 = S_2 S_1 T S_2^{-1}$ permutes cyclically the centers of the cells $\tau_1, \cdots, \tau_{KN}$ of an arbitrarily fine dyadic subdivision, in particular, finer than the given subdivision $\sigma_1, \cdots, \sigma_N$. Now consider T_1 to have domain $S_2 G$, and let $\tau \subset \tau_1$ be an r-cell about the center of τ_1 so small that its first KN images under T_1 are defined and respectively interior to $\tau_2, \tau_3, \cdots, \tau_{KN}, \tau_1$. Let C be a linear Cantor set interior to τ. Since T_1^{KN} is a measure-preserving homeomorphism of τ onto a subset of τ_1, by Lemma 17 there exists an automorphism $S_3 \in M_0[\tau_1]$, which we define equal to the identity outside τ_1, such that under $T_2 = S_3 T_1$ the set C is brought back to coincidence with itself. Under T_2 the points of C are all periodic with period KN, and there is one image of C in each of the cells $\tau_1, \cdots, \tau_{KN}$. Now introduce a measure μ_1 in C by mapping it onto a linear Cantor set with measure $1/KN$, and extend this measure to the images of C by the transformation T_2. Then μ_1 will be a normalized measure invariant under T_2 such that $\mu_1 \tau_i = 1/KN$, $i = 1, \cdots, KN$. Since T_2 preserves both measures μ_1 and m, it will also preserve the measure $\mu A = (1 - \alpha)mA + \alpha\mu_1 A$. It is easily verified that μ^* fulfills the conditions of Theorem 2_1 with respect to each of the cells τ_i. Hence there exists an automorphism $h \in H_0[R]$, which transforms each cell τ_i into itself, such that $\mu^*A = m^*hA$ for all $A \subset R$. Hence $T_3 = hT_2h^{-1}$ will preserve Lebesgue measure, and the Cantor set $C_1 = hC$ will be periodic under T_3, will have one image in each cell τ_i, and this image will contain the fraction α of the measure of τ_i. Hence the images of C_1 under T_3 are equally distributed among $\sigma_1, \cdots, \sigma_N$, and contain the fraction α of their measure. Expressing T_3 in terms of T, we have $T_3 = hS_3S_2S_1TS_2^{-1}h^{-1}$. The automorphisms S_1 and S_2 could be taken arbitrarily small, and S_3 and h have norms no greater than the diameter of the cells τ_i. Hence T_3 is of the required form h_1Th_2, where h_1 and h_2 belong to $H_0[R]$ and are arbitrarily small. It should be noted that T_3 is measure-preserving even though h_1 and h_2 are not.

Taking $\alpha = \frac{1}{2}$, it is evident that Lemma 18 implies Lemma 5, in fact the periodicity of the set C_1 is for this purpose superfluous.

8. The Borel class of the set of metrically transitive automorphisms

To complete the proof of Theorem 1, it only remains to show that the set of metrically transitive automorphisms is G_δ in $M[E, \mu]$. This is most easily done by utilizing what amounts to a necessary and sufficient condition for metrical transitivity, rather than the definition itself. Consider the space L_2 of functions quadratically integrable over E. Any automorphism $T \in M[E, \mu]$ induces a unitary transformation $Uf(x) = f(Tx)$ of L_2, as is well-known.[37] Let $f_1, f_2, \cdots$ be a sequence of continuous functions dense in L_2. Let $E(i, j, n)$ be the set of all T in $M[E, \mu]$ such that

$$\int_E \left[\frac{1}{n}\sum_{\nu=0}^{n-1} f_i(T^\nu x) - (f_i, 1)\right]^2 dx < \frac{1}{j},$$

where $(f_i, 1)$ denotes the number $\int_E f_i(x)\,dx$. This set is open in $M[E, \mu]$. For suppose T_k, $k = 1, 2, \cdots$, belongs to the complement of $E(i, j, n)$ and that $\rho(T_k, T) \to 0$ as $k \to \infty$. The integrand converges boundedly to the limit and so T also belongs to the complement. Now, the set M_T of metrically transitive automorphisms is represented by

$$M_T = \prod_{i=1}^{\infty}\prod_{j=1}^{\infty}\sum_{n=1}^{\infty} E(i, j, n).$$

For if T is metrically transitive it will belong to $E(i, j, n)$ for all sufficiently large n, in virtue of the mean ergodic theorem of von Neumann.[38] On the other hand, if T is metrically intransitive there exists an invariant function φ not constant, namely, the characteristic function of an invariant set with measure inter-

---

[37] B. O. Koopman, *Hamiltonian systems and transformations in Hilbert space.* Proc. Nat. Acad. USA. **17** (1931) 315.

[38] J. von Neumann, *Proof of the quasiergodic hypothesis.* Proc. Nat. Acad. USA. **18** (1932) 70.

mediate between zero and μE. Let d denote the distance in L_2 from φ to the axis of constant functions, and choose $1/j < \frac{1}{4} d^2$. Choose f_i from the sphere of radius $d/2$ about φ. Since φ is a fixed point under U, and U is unitary, all images of f_i under U also lie in the sphere, and therefore also all averages $\frac{1}{n} \sum_{\nu=0}^{n-1} f_i(T^\nu x)$. For no n does T belong to $E(i, j, n)$, because the distance from $\frac{1}{n} \sum_{\nu=0}^{n-1} f_i(T^\nu x)$ to the constant function $(f_i, 1)$ is always at least $d/2$, and therefore the squared distance is greater than $1/j$.

This representation of M_T evidently exhibits it as a G_δ set in $M[E, \mu]$.

IV. Metrically Transitive Flows

9. Existence Theorem

The results concerning metrically transitive *automorphisms* contained in the preceding sections make it possible to set up a general procedure for defining metrically transitive *continuous flows*.[39] By a continuous flow we shall mean a one-parameter group of automorphisms T_λ, $-\infty < \lambda < +\infty$, of a space E such that $T_\lambda x$ is continuous in x and λ, and such that $T_{\lambda+\mu} = T_\lambda T_\mu$ and $T_0 = I$.

THEOREM 3: *Let E be a regularly connected polyhedron of dimension $r \geqq 3$, and let μ be any r-dimensional Lebesgue-Stieltjes measure in E. There exists a continuous flow in E which is metrically transitive with respect to μ, and which leaves all singular points of E fixed.*

It is sufficient to define a continuous flow in a rectangular r-cell which leaves the boundary fixed and is metrically transitive with respect to ordinary Lebesgue measure. Because the image of such a flow under the map defined in Lemma 3 will be a continuous flow in E, in virtue of Lemma 4, and metrically transitive with respect to μ.

We first define a flow in a space Q_1 defined as follows. Let B denote the $(r-1)$-dimensional unit cube and introduce in it a measure $\nu A = \int_A f(p)\, dp$, where the integral is an ordinary $(r-1)$-dimensional Lebesgue integral and $f(p)$ is a continuous function positive over the interior of B and tending to infinity at the boundary in such a way that $\int_B f(p)\, dp = 1$. Then νA is an $(r-1)$-dimensional LS measure in B, and by Theorem 1 there exists an automorphism T of B which leaves the boundary fixed and is metrically transitive with respect to ν. Consider the product space of B with the unit interval $0 \leqq \lambda \leqq 1$, and identify points $(p, 1)$ and $(Tp, 0)$. In this space Q_1 define a flow upward along streamlines perpendicular to B, taking the velocity at any point to be $1/f(p)$, where p is the last intersection of the streamline with B. The velocity along a streamline undergoes a discontinuous change when it crosses B. Nevertheless,

---

[39] This is an adaptation of a standard method. See E. Hopf, *Ergodentheorie*, p. 41.

the flow defined by this velocity field is continuous, and since $1/f(p)$ tends to zero at the boundary, the boundary remains fixed. This flow preserves r-dimensional Lebesgue measure in Q_1. To see this, consider any small cube interior to Q_1 with height h and base σ. Until it crosses B, the segments along the streamlines are rigidly translated, so that its volume is not changed. After it crosses B it has a new cross-section $T\sigma$, and the length along any streamline is now $h \cdot f(p)/f(Tp)$. The volume is therefore

$$\int_{T\sigma} h \frac{f(p)}{f(Tp)}\, dp = \int_{T\sigma} \frac{h}{f(Tp)}\, d\nu(p) = \int_{\sigma} \frac{h}{f(p)}\, d\nu(p) = \int_{\sigma} h\, dp.$$

Thus, the change in velocity exactly compensates for the change in cross-section, and therefore the flow is measure-preserving, since a flow that preserves the measure of small cubes preserves the measure of all sets. To see that it is metrically transitive, consider any invariant measurable set. The set must be cylindrical, and so its intersection A with B is Lebesgue measurable and therefore ν-measurable. Since A is invariant under T its ν-measure is either zero or one, and consequently also its Lebesgue measure. Hence the cylindrical set over it has Lebesgue measure zero or one.

The space Q_1 in which we have just defined a flow is homeomorphic to the r-dimensional tube Q, that is, the product space of B with $(0, 1)$ where points $(p, 0)$ and $(p, 1)$ are identified. To prove this, observe that since T leaves the boundary fixed, it can be joined isotopically to the identity. That is, there exists a continuous family of automorphisms T_λ of B, $0 \leqq \lambda \leqq 1$, such that $T_1 = T$ and T_0 is the identity.[40] The correspondence $(p, \lambda) \rightarrow (T_\lambda p, \lambda)$ is a homeomorphism of Q onto Q_1.

The r-dimensional unit cube R can be represented as the continuous image of Q under a map which is a homeomorphism up to the boundary of Q. This may be done as follows. The region of r-space defined by the inequalities $1 \leqq (x_1^2 + x_2^2)^{\frac{1}{2}} \leqq 2$, $0 \leqq x_i \leqq 1$, $i > 2$, is evidently homeomorphic to Q. The map defined by $x_1' = x_1[(x_1^2 + x_2^2)^{\frac{1}{2}} - 1]$, $x_2' = x_2[(x_1^2 + x_2^2)^{\frac{1}{2}} - 1]$, $x_i' = x_i$, $i > 2$, is continuous and has for range the set $0 \leqq (x_1'^2 + x_2'^2)^{\frac{1}{2}} \leqq 2$, $0 \leqq x_i' \leqq 1$, $i > 2$, which is homeomorphic to R. This map is 1:1 except at points where $(x_1^2 + x_2^2)^{\frac{1}{2}} = 1$, which belong to the boundary of Q. Combining this map with an automorphism of R based on Theorem 2, the map from Q_1 to R can be made measure-preserving. The image in R of the flow already defined in Q_1 is therefore metrically transitive and leaves the boundary of R fixed. From this we can derive a metrically transitive continuous flow in E, as already explained.

10. Most general polyhedra that can support metrically transitive automorphisms or flows

In the present section we consider arbitrary finite polyhedra, not necessarily connected or even homogeneous-dimensional, and seek to characterize those in

[40] J. W. Alexander, *On the deformation of an n-cell.* Proc. Nat. Acad. USA. **9** (1923) 406–407.

which metrically transitive continuous flows or automorphisms are possible. It will be seen that a complete answer is obtained except in the case $r = 2$ for flows.

THEOREM 4: *A finite polyhedron of dimension $r \geqq 3$ can support a metrically transitive continuous flow with respect to any given r-dimensional Lebesgue-Stieltjes measure if and only if its regular points form a connected set.*

If the regular points form a connected set, the closure of this set is a regularly connected polyhedron, and we have already seen how to construct a metrically transitive flow in this part. Since the flow given by Theorem 3 leaves singular points fixed, it can be extended to the rest of the polyhedron, which is a set of measure zero, by defining it equal to the identity there. On the other hand, if the regular points fall into two or more disjoint open sets, these will be invariant under any continuous flow, since if one point of a streamline is regular, all must be. There will therefore necessarily exist disjoint invariant sets with positive measure, so that no metrically transitive, or even topologically transitive, continuous flow is possible.

The case $r = 2$ is not covered by our present method of construction, and it appears likely that further conditions on the polyhedron are necessary in this case. The case $r = 1$ is trivial, the only possibility is for the polyhedron to be homeomorphic to the circumference of a circle, plus possibly some isolated points, the flow being topologically equivalent to a steady rotation of the circle. The details are left to the reader.

THEOREM 5: *A finite polyhedron of dimension $r \geqq 2$ can support a metrically transitive automorphism with respect to any given r-dimensional Lebesgue-Stieltjes measure if and only if its regular components have equal measure and can be permuted cyclically by an automorphism.*

Suppose the conditions satisfied. Let h be an automorphism that permutes the regular components cyclically. By Theorem 2 Corollary 3, we can modify h within each regular component so as to make it measure-preserving. Call the resulting automorphism T. Suppose there are n regular components, and let E_1 be one of them. Then T^n is a measure-preserving automorphism of E_1. By the Corollary to Theorem 1, there exists an arbitrarily small automorphism S of E_1 such that ST^n is a metrically transitive automorphism of E_1, and since S leaves all singular points fixed, it can be extended to the rest of E by defining it equal to the identity outside E_1. Then ST is a metrically transitive automorphism of E, because any invariant set of positive measure must contain almost all points of E_1 and therefore almost all points of every regular component. Conversely, if there exists a metrically transitive automorphism it must permute the regular components cyclically, and consequently they must have equal measure. In fact, these conditions are necessary if there is to exist even a topologically transitive measure-preserving automorphism.

It may be added that in the case $r = 1$ the only polyhedron that can support a metrically transitive automorphism is one that is homeomorphic to a finite number of disjoint circumferences of circles of equal measure, plus possibly some isolated points. In such a polyhedron a metrically transitive automorphism is easily defined by composing a non-periodic rotation of one of the circles with a cyclic permutation of the circles. The isolated points, if any, may be left fixed. Any other 1-dimensional polyhedron will have a regular component with at least one singular point. This component will be either a line segment or a circle with one singular point. Some power of any given automorphism will leave all singular points fixed and each regular component invariant, since there are only a finite number of each. But a measure-preserving automorphism of a line segment that leaves the ends fixed must be the identity, and a measure-preserving automorphism of a circle that leaves one point fixed must be either the identity or a reflection in a diameter. In none of these cases can the automorphism be metrically transitive, since a power of it will be equal to the identity on some regular component.

Thus we have obtained a characterization of all finite polyhedra that can support metrically transitive automorphisms.

It may be remarked that although metrically transitive automorphisms may exist in polyhedra having more than one regular component, nevertheless Theorem 1 is not true for such polyhedra, because we have just seen that a transitive automorphism must permute the regular components cyclically, and therefore cannot be near the identity. However, the following generalization of Theorem 1 holds. *In the space* $M[E, \mu]$ *of measure-preserving automorphisms of any polyhedron that can support at least one metrically transitive automorphism, the metrically transitive automorphisms form a residual set with respect to the subspace of automorphisms that permute the regular components cyclically.* One need only observe that in the existence proof given above it was really shown that the metrically transitive automorphisms are dense in this subspace. Since they form a G_δ set with respect to the whole space, they are G_δ with respect to the subspace also, and therefore residual.

11. Transitive automorphisms in transitive flows

In the preceding sections we have made use of a general method whereby a transformation can be used to define a flow in a product space. By this construction a transitive transformation gives rise to a transitive flow. But there is an even simpler relation connecting transformations and flows, namely, the individual transformations that make up a continuous flow are themselves automorphisms of the space. If even one automorphism in the flow is transitive, either metrically or topologically, then the flow is, because by definition a set that is invariant under a flow is invariant under each of the transformations that make up the flow. This might seem to suggest another way of deriving a transitive flow from a transformation, but the method is of little use because in general an automorphism cannot be embedded in a flow.[41] Nevertheless, it is natural

[41] Any transformation embedded in a continuous flow must, for example, have roots of all orders. A simple example of an automorphism which hasn't even a square root is the following. Let T_1 be a rotation of the circumference of a circle through an angle π/k,

to inquire whether any of the automorphisms that make up a transitive continuous flow are necessarily transitive. In the case of metrical transitivity this question appears to be open.[42] It is therefore of interest to note that in the topologically transitive case the answer is in the affirmative, as we proceed to show.

THEOREM 6: *Let T_λ, $-\infty < \lambda < +\infty$, be a topologically transitive continuous flow in a separable metric space E, and suppose there is no isolated streamline. For all values of λ, except a set of first category on the line $-\infty < \lambda < +\infty$, the automorphisms T_λ are topologically transitive.*

Let $G_1, G_2, \cdots$ be a countable base for all open sets in E. Let $A_{i,j}$ be the set of values of λ such that for some positive or negative integer k the set $T_{k\lambda}G_i$ overlaps G_j, that is, such that G_i overlaps G_j under some power of T_λ. The set $A_{i,j}$ is evidently open, we proceed to show that it is dense on the line $-\infty < \lambda < +\infty$. Consider any interval I and form the set of all numbers $k\lambda$, where $\lambda \in I$ and $k = 0, \pm 1, \pm 2, \cdots$. This set includes all numbers greater in absolute value than some number Λ. Form the set $H = \sum_{|\lambda|>\Lambda} T_\lambda G_i$ and consider a sphere $\sigma \subset G$ so small that the set swept out by it as λ describes the interval $-\Lambda \leqq \lambda \leqq \Lambda$ is not dense in G_j. Any sufficiently small sphere about a point of G_i will do since by hypothesis there is no isolated streamline. The set H must overlap G_j, because otherwise the invariant open set swept out by σ would not be dense in G_j, contrary to the hypothesis that the flow is topologically transitive. Hence there exist values of λ greater in absolute value than Λ such that $T_\lambda G_i$ overlaps G_j, and so the set $A_{i,j}$ contains points of the interval I. Since the sets $A_{i,j}$ are open and everywhere dense, their intersection as i and j run independently over all positive integers is a residual set. For any value of λ in this set, the automorphism T_λ is topologically transitive, because if it had an invariant open set not everywhere dense in E there would exist a pair G_i, G_j such that no image of G_i overlaps G_j, and this would contradict the fact that λ belongs to $A_{i,j}$.

V. SOME RELATED QUESTIONS

12. Generation of Random Sequences by Automorphisms

In the Introduction it was explained that metrical transitivity implies that the images of almost all points are distributed like random sequences in respect

---

where k is a positive integer. Let T_2 be the automorphism of the segment $0 \leqq \theta \leqq \frac{\pi}{k}$ defined by $T_2\theta = \left(\frac{k\theta}{\pi}\right)^2$, and let it leave all other points fixed. Under the automorphism $T = T_2T_1$, the points $0, \frac{\pi}{k}, \frac{2\pi}{k}, \cdots, \frac{(2k-1)\pi}{k}$ form a single cycle of period $2k$, and no other points are periodic. It is clear that T cannot be the square of a transformation S, because a cycle of period $2k$ in T could arise only by the splitting of a cycle of period $4k$ in S, and then T would have two cycles of period $2k$. Hence T has no square root.

[42] The question was raised by H. E. Robbins.

to the frequency with which they fall in any given measurable set. A random sequence has also the property that of its first n points the number that fall in A differs from $n \cdot mA$ by something of the order of $\sqrt{n}$. Furthermore, this difference oscillates in sign. Let us consider the r-dimensional unit cube R and take for A any dyadic cell σ. Let the characteristic function of σ be f. The difference in question is then $\sum_{\nu=1}^{n} f(T^{\nu}p) - nm\sigma$. Consider instead the difference $K \cdot \sum_{\nu=1}^{n} f(T^{\nu}p) - n$, where $K = 1/m\sigma$ is the number of cells in the dyadic subdivision to which σ belongs. This latter difference is an integer, and as n increases by unity the difference either increases by $K - 1$ or decreases by 1. To pass from a positive to a negative value it must therefore pass through the value zero. For a random sequence this must happen infinitely often. This means that a random sequence from R has arbitrarily long segments distributed between σ and its complement in *exact* proportion to their measures. We proceed to show that there exist automorphisms of R under which almost all points generate sequences that share this property of random sequences with respect to every dyadic cell, and indeed that such automorphisms are the "general case."

THEOREM 7: *Let R denote the r-dimensional unit cube, $r \geqq 2$. There is a residual set of automorphisms in $M[R]$ under which almost all points p generate sequences distributed in the following manner: Given any dyadic subdivision, there exist arbitrarily large values n such that the first n images of p are equally distributed among the cells of this subdivision.*

Let $\sigma_1^{(j)}, \cdots, \sigma_{N_j}^{(j)}$ be the cells of the j-th dyadic subdivision, and let $f_i^{(j)}$ be the characteristic function of $\sigma_i^{(j)}$. Let $E(i, j, k, n)$ be the set of all T in $M[R]$ such that the measure of the set $A(i, j, n, T)$ of points p for which $\frac{1}{n} \sum_{\nu=1}^{n} f_i^{(j)}(T^{\nu}p) = m\sigma_i^{(j)}$ is greater than $1 - \frac{1}{k}$. The set $E(i, j, k, n)$ is open, because if T belongs to it and G_ϵ denotes the ϵ-neighborhood of the boundary of $\sigma_i^{(j)}$, where ϵ is so chosen that $mG_\epsilon < \frac{1}{n}\left[mA(i, j, n, T) - \left(1 - \frac{1}{k}\right)\right]$, we have $m\left[A(i, j, n, T) - \sum_{\nu=1}^{n} T^{-\nu}G_\epsilon\right] > 1 - \frac{1}{k}$. If p belongs to $A(i, j, n, T) - \sum_{\nu=1}^{n} T^{-\nu}G_\epsilon$, then $Tp, T^2p, \cdots, T^np$ all lie outside G_ϵ, and so $\sum_{\nu=1}^{n} f_i^{(j)}(T_1^{\nu}p) = \sum_{=1}^{n} f_i^{(j)}(T^{\nu}p)$ for any automorphism T_1 so near to T that $\rho(T_1^{\nu}, T^{\nu})$ is less than ϵ for $\nu = 1, \cdots, n$. Consequently, the set $A(i, j, n, T_1)$ contains $A(i, j, n, T) - \sum_{=1}^{n} T^{-\nu}G_\epsilon$, and therefore has measure exceeding $1 - \frac{1}{k}$, so that T_1 belongs to $E(i, j, k, n)$. Hence $\sum_{n=N}^{\infty} \prod_{i=1}^{N_j} E(i, j, k, n)$ is likewise open. But this is the set of automorphisms such that for some $n \geqq N$ the set of points whose images are distributed equally among $\sigma_1^{(j)}, \cdots, \sigma_{N_j}^{(j)}$ has measure greater than $1 - \frac{1}{k}$. Lemma 18 implies that such automorphisms are everywhere dense in $M[R]$, as may be seen by taking $\alpha > 1 - \frac{1}{k}$, because an automorphism that has a periodic set whose images are equally distributed among $\sigma_1^{(j)}, \cdots, \sigma_{N_j}^{(j)}$ and have combined measure α will belong to $E(i, j, k, n)$ for any value of n divisible by the period of the set. Therefore the set $\prod_{j=1}^{\infty} \prod_{N=1}^{\infty} \prod_{k=1}^{\infty} \sum_{n=N}^{\infty} \prod_{i=1}^{N_j} E(i, j, k, n)$ is residual, since it is an intersection of countably many dense open sets. If T belongs to this set, then, for every pair j, N, almost all points will generate sequences which have segments of length greater than N that are equally distributed among the cells of the j-th dyadic subdivision. The intersection of these sets as j and N run over all positive integers will still have measure one, and the points of this set will have images distributed in the required manner.

13. Automorphisms that preserve sets of measure zero

We shall say that an automorphism *preserves zero sets* if it has the property that $mhA = 0$ if and only if $mA = 0$. This is equivalent to requiring that hA be measurable if and only if A is measurable, because any Lebesgue measurable set is the sum of a Borel set and a set of measure zero, and therefore its image is the sum of a Borel set and the image of a zero set. Consequently, if h preserves zero sets it will also preserve measurable sets. On the other hand, if it does not preserve zero sets either h or its inverse must take some zero set into a set with positive outer measure. Therefore it takes some zero set into a non-measurable set, because every set with positive outer measure contains a non-measurable subset.[43] Hence automorphisms that preserve zero sets might equally well be called *measurability preserving*. It is well-known that an automorphism need not preserve zero sets, but we proceed to show that any automorphism is topologically equivalent to one that does.

THEOREM 8: *Let h be any automorphism of a rectangular r-cell R (or of $E^{(r)}$), $r \geqq 1$. There exists an automorphism g of R (or $E^{(r)}$) such that ghg^{-1} preserves zero sets.*

Consider first the case of an automorphism of R. Define $\mu^*A = \frac{1}{2}m^*A + \sum_{n=1}^{\infty} \frac{1}{2^{n+2}} (m^*h^nA + m^*h^{-n}A)$. Then μ is an r-dimensional LS measure in R, and $\mu R = mR$. By Theorem 2 there exists an automorphism g such that $\mu A = mgA$. Evidently $\mu A = 0$ if and only if $mh^nA = 0$ for every positive or negative integer n. Therefore $\mu hA = 0$ if and only if $\mu A = 0$; that is, $mghg^{-1}A = 0$ if and only if $mA = 0$.

Now consider any automorphism h of $E^{(r)}$. Divide the space into a lattice

[43] Carathéodory, *Vorlesungen über reelle Funktionen*, Leipzig and Berlin 1918, p. 354.

of cubes $R_1, R_2, \cdots$ and let G_i be the interior of R_i. Form the outer measure

$$\mu^*A = \sum_{i=1}^{\infty}\left\{\frac{1}{2}\cdot\frac{m^*AG_i}{mG_i} + \sum_{n=1}^{\infty}\frac{1}{2^{n+2}}\left[\frac{m^*h^n(AG_i)}{mh^nG_i} + \frac{m^*h^{-n}(AG_i)}{mh^{-n}G_i}\right]\right\}.$$

This measure is evidently zero for points and positive for non-void open sets. Furthermore, $\mu R_i = 1$, and therefore bounded sets have finite measure, while $\mu E^{(r)} = +\infty$. Hence, by Theorem 2 Corollary 5, there exists an automorphism g such that $\mu^*A = m^*gA$, at least in case $r \geqq 2$. The same argument as above then shows that ghg^{-1} preserves zero sets. In the case $r = 1$, the positive and negative half lines both have infinite μ-measure, so that the proof is valid in this case also.

14. Sets automorphic to zero sets

We shall say that two subsets A and B of $E^{(r)}$ are automorphic if there exists an automorphism h of $E^{(r)}$ such that $hA = B$. This is evidently a stronger notion than that of homeomorphism of the two sets. We shall obtain a simple characterization of sets that are automorphic to sets of measure zero.

THEOREM 9: *Let B be any subset of $E^{(r)}$, $r \geqq 1$. In order that there exist an automorphism h of $E^{(r)}$ such that hB has Lebesgue measure zero it is necessary and sufficient that the complement of B contain a sequence of perfect sets whose union is dense in $E^{(r)}$. If a bounded set B satisfies this condition, the automorphism h can be taken equal to the identity outside of any cube that contains B.*

The condition is evidently necessary, since if $mhB = 0$ the complement of hB contains a sequence of perfect sets P_n whose union contains all the measure in $E^{(r)}$, and is therefore dense. The perfect sets $h^{-1}P_n$ are therefore outside B and their union is also dense in $E^{(r)}$. To prove that the condition is also sufficient, suppose that B satisfies it and consider any cube R with interior R_0. By hypothesis $R_0 - R_0B$ contains a dense sequence of perfect sets. In $R_0 - R_0B$ it is therefore possible to find a sequence of perfect sets P_n such that every neighborhood in R contains at least one member of the sequence. In each set P_n there exists a Cantor set C_n. Introduce a normalized LS measure μ_n in C_n by mapping it homeomorphically onto a linear Cantor set with linear measure one. Form the outer measure $\mu^*A = mR\cdot\sum_{n=1}^{\infty}\frac{1}{2^n}\mu_n^*AC_n$. This is evidently a LS outer measure in R, zero for points and positive for non-void open sets in R. Also $\mu R = mR$ and $\mu BdR = 0$. By Theorem 2_1 there exists an automorphism $h \in H_0[R]$ such that $\mu^*A = m^*hA$. Since $\mu BR = 0$, we have $mh(BR) = 0$. If B is contained in R we can take h equal to the identity outside R. If B is an unbounded set we can divide $E^{(r)}$ into a lattice of cubes and transform each into itself in such a way as to compress the part of B contained in it to a set of measure zero. The resulting automorphism of $E^{(r)}$ will then carry B into a zero set.

Since any residual set in $E^{(r)}$ contains a dense sequence of perfect sets, it follows as a corollary that any set of first category is automorphic to a zero set.

In a previous paper[44] we have obtained this result directly by a simple category argument. In the case $B \subset R$, we showed in fact that the automorphisms h such that $mhB = 0$ form a residual set in $H[R]$. It is interesting to note that it is *only* for first category sets that this method of proof is available, at least for sets having the property of Baire, because in the same paper it was shown (Theorem 4) that a set having the property of Baire is of first or second category according as the automorphisms that carry it into a set of measure zero form a residual set or a set of first category in $H[R]$. The present theorem may be regarded as completing the earlier result by showing precisely which second category sets are automorphic to zero sets.

15. Transformations topologically equivalent to measure-preserving automorphisms

G. D. Birkhoff has formulated the following problem: Given an automorphism h, when can one assert that there exists an automorphism g such that ghg^{-1} is Lebesgue measure-preserving? In other words, the problem is to characterize *in topological terms* automorphisms that are topologically equivalent to measure-preserving ones. It is well-known that measure-preserving automorphisms have special topological properties, such as the recurrence property discovered by Poincaré and similar topological properties that can be deduced from the ergodic theorem. Again, it is obvious that a measure-preserving automorphism cannot transform a closed sphere into a subset of its interior. Thus it is clear that h must be restricted, in contrast to what we found in the case of the related problem of characterizing automorphisms equivalent to ones that preserve zero sets (§13). In the present section we shall obtain a result that constitutes a solution of this problem, though it must be admitted that the characterization obtained is neither particularly simple nor easy to apply. It is to be hoped that a more elegant solution may be found.

Let us restrict attention to an r-dimensional cube R. If there exists an automorphism g such that ghg^{-1} preserves Lebesgue measure, then h preserves the measure $\mu A = mgA$; and conversely, if h preserves a measure μ automorphic to Lebesgue measure, there will exist a g such that ghg^{-1} preserves Lebesgue measure, namely, any automorphism such that $\mu A = mgA$. Thus Theorem 2 provides the following equivalent formulation of the problem: Given an automorphism h, when can one assert that it preserves some r-dimensional LS measure? It is from this point of view that we shall approach the problem. In a previous paper[45] we have discussed the question of the existence of somewhat more general invariant measures. We obtained the result (loc. cit. Th. 2) that an automorphism T of a complete separable metric space admits a finite invariant Borel measure that is zero for points if (i) T has non-denumerably many

[44] See footnote 28.

[45] J. C. Oxtoby and S. M. Ulam, *On the existence of a measure invariant under a transformation.* Ann. of Math. **40** (1939) 560–566.

periodic points, or (ii) there exists a compact set C consisting of non-periodic points of which at least one returns to C with positive frequency under iteration of T. (A point p of C is said to return with positive frequency if $\lim_{n\to\infty} \frac{1}{n} \sum_{\nu=1}^{n} f_C(T^\nu p) > 0$, where f_C is the characteristic function of C.) It should be added that in case (i) the invariant measure was defined in such a way that any prescribed sphere containing non-denumerably many periodic points could be made to have positive measure, and in case (ii) positive measure was ascribed to the set C. Assuming this result, we can obtain the following theorem.

THEOREM 10: *Let h be an automorphism of the r-dimensional unit cube R, $r \geqq 2$. In order that there exist an automorphism g such that ghg^{-1} preserves Lebesgue measure it is necessary and sufficient that in every sphere there exist a perfect set to which some point returns with positive frequency under iteration of h, and that these perfect sets can all be chosen outside of an arbitrarily prescribed countable set.*

The necessity of the condition may be deduced from the ergodic theorem as follows. Suppose h is measure-preserving and let A_0 be any given countable set. Then h is a measure-preserving automorphism of the set $E = R - \sum_{n=-\infty}^{+\infty} h^n A_0$, which has measure one. In any sphere there exists a perfect set $P \subset E$ such that $mP > 0$. Denote its characteristic function by $f(p)$. According to the ergodic theorem $f^*(p) = \lim_{n\to\infty} \frac{1}{n} \sum_{\nu=1}^{n} f(T^\nu p)$ exists for almost all p, and $\int f^*(p)\, dp = mP$. Hence $f^*(p)$ is positive for some point p, that is, p returns to P with positive frequency. Thus h has the required property, and therefore also any equivalent automorphism ghg^{-1}, since the condition is topologically invariant.

Conversely, suppose h fulfills the condition. Let $\sigma_1, \sigma_2, \cdots$ be an enumeration of all dyadic cells in R. Let A_n denote the set of periodic points contained in σ_n provided there are only countably many, otherwise let A_n be void. Then the union A_0 of the sets A_n is countable. Consider any cell σ_n. If it contains non-denumerably many periodic points there exists an invariant measure μ_n which is zero for points and positive for σ_n, in virtue of the theorem quoted above. If σ_n contains only countably many periodic points, then by hypothesis there exists a perfect set $P \subset \sigma_n - A_0$ to which some point returns with positive frequency. Since P consists of non-periodic points, the theorem quoted asserts again the existence of an invariant measure μ_n. Thus we obtain a sequence of invariant Borel measures μ_n, each zero for points and such that $\mu_n\sigma_n > 0$. We suppose them normalized. Then $\mu A = \sum_{n=1}^{\infty} \frac{1}{2^n} \mu_n A$ is a normalized invariant Borel measure in R that is zero for points and positive for non-void open sets. To obtain a measure which in addition is zero for the boundary it is sufficient to form $\nu A = \mu A R_0 / \mu R_0$. The Borel measure ν admits an r-dimensional LS extension defined by $\nu^* A = \inf \nu G$, G open, $G \supset A$, which is invariant under h.

By Theorem 2_1 there exists an automorphism g such that $\nu^*A = m^*gA$, and therefore ghg^{-1} preserves Lebesgue measure, as already remarked.

16. A topological ergodic theorem

The paper[45] referred to in the last section also contains the result (Th. 1) that an automorphism T of a complete separable metric space admits a finite invariant Borel measure if there exists a compact set C to which some point returns with positive *superior* frequency $\left(\text{that is, } \limsup_{n\to\infty} \frac{1}{n} \sum_{\nu=1}^{n} f_C(T^\nu p) > 0\right)$ Again it should be added that the measure is positive for C. As a corollary of this, we may assert the following theorem, which properly belongs to the earlier paper since it involves no other results.

THEOREM 11: *Let T be an automorphism of a separable metric space E, and let C be any compact set contained in E. There exists at least one point p in C such that $\lim_{n\to\infty} \frac{1}{n} \sum_{\nu=1}^{n} f_C(T^\nu p)$ exists, where f_C is the characteristic function of C.*

The interest of this theorem is that, like the ergodic theorem, it asserts the existence of a limiting frequency of return for certain points, but *without any measure-theoretic assumptions.* Nevertheless, the proof requires a very considerable excursion into measure theory. It would be interesting to know whether the result can be obtained directly, even in the special case in which C is a square and T an automorphism of the plane.

To prove the theorem, observe that either $\lim_{n\to\infty} \frac{1}{n} \sum_{\nu=1}^{n} f_C(T^\nu p) = 0$ for every point of C, in which case the theorem is true, or there exists a point p in C such that $\limsup_{n\to\infty} \frac{1}{n} \sum_{\nu=1}^{n} f_C(T^\nu p) > 0$. In the latter case there exists a finite invariant Borel measure μ such that $\mu C > 0$, by the theorem quoted above. The ergodic theorem is applicable to this measure and asserts that the limit in question exists for almost all points (in this measure) and is positive for at least one point, since its integral is equal to μC. Therefore there exists a point of C, not necessarily equal to p, which returns with positive frequency.

17. Connections with Haar measure

Haar's Theorem[46] asserts that in any locally compact separable topological group there exists a left (or right) invariant measure which is finite and positive for every compact neighborhood. It is evidently zero for points, and it is unique up to a constant multiplier.[47] In the compact case it is finite, and then left and

[46] A. Haar, *Der Massbegriff in der Theorie der kontinuierlichen Gruppen.* Ann. of Math. **34** (1933) 147–169.

[47] J. von Neumann, *The uniqueness of Haar's measure.* Recueil Mathématique, Moscou, N.S. **1** (1936) 721–734.

right invariant measures coincide and are also inverse invariant.[48] The measure is derived from a Carathéodory outer measure that satisfies condition M5.[49] In case the group manifold is a polyhedron, the Haar measure is therefore r-dimensional, and Theorem 2 Corollary 1 establishes the following connection with Lebesgue measure. *If the group manifold of a topological group is a finite euclidean polyhedron, the Haar measure, suitably normalized, is automorphic to the Lebesgue measure in the polyhedron.* Again, we may say that *if two groups have for group manifold the same polyhedron, their Haar measures are automorphic to each other, provided only that they are similarly normalized.*

18. Approximation by locally linear automorphisms

In the proof of Theorem 1, methods were developed that made it possible to approximate any measure-preserving automorphism by one that has a periodic Cantor set whose images are distributed in a regular manner. It is natural to ask whether one can find similarly an approximating automorphism that has a periodic cube. In the present section we shall show that this is possible for automorphisms of a cube that leave the boundary fixed, indeed that an approximating automorphism can be found that has a sequence of periodic cubes which together include almost all points and which the transformation permutes among themselves as if by translation.

It may be remarked that the same methods can be extended to apply to r-dimensional polyhedra embedded in r-space, and to automorphisms that are merely isotopic to the identity. An attempt to extend them to general automorphisms and polyhedra seems to encounter more serious difficulties, and for this reason the proof of Theorem 1 was based on periodic Cantor sets rather than on periodic cubes, which might have seemed more natural.

LEMMA 19: *Let R be the r-dimensional unit cube, $r \geqq 2$, and let T be a measure-preserving automorphism of R that has an invariant set consisting of a finite number of disjoint cubes $R_1, \cdots, R_M$ interior to R and each a sum of cubes of some dyadic subdivision. There exists an arbitrarily small automorphism $S \in M_0[R]$, equal to the identity on $R_1, \cdots, R_M$, such that ST permutes cyclically the centers of the cubes of a certain arbitrarily fine dyadic subdivision of R that are not contained in $R_1, \cdots, R_M$.*

Let E be the polyhedron obtained from R by removing the interiors of $R_1, \cdots, R_M$. Then E is regularly connected, and T may be considered as an automorphism of E. By the Corollary to Theorem 1, T can be made metrically transitive by composing it with an arbitrarily small automorphism that leaves the boundary of E fixed. We may therefore assume without loss of generality that T itself is metrically transitive with respect to E. By Lemma 3, we can represent E as the continuous image of a rectangular r-cell R^* under a measure-

[48] J. von Neumann, *Zum Haarschen Mass in topologischen Gruppen.* Compositio Mathematica **1** (1934) 106–114.

[49] See the note of S. Banach in Saks, *Theory of the Integral*, Appendix 1.

preserving map f which is a homeomorphism up to the boundary. Let ϵ be a given positive number and let $\delta > 0$ be such that any automorphism in $H_0[R^*]$ with norm less than δ corresponds to an automorphism of E with norm less than ϵ. Let $\sigma_1, \cdots, \sigma_N$ be the cubes in E belonging to a dyadic subdivision of diameter less than ϵ. Order these in a sequence $\sigma_{i_1}, \cdots, \sigma_{i_K}$ such that each cube σ_i appears at least once and such that σ_{i_k} and $\sigma_{i_{k+1}}$ have a regular face in common, $1 \leqq k \leqq K - 1$. (It may be possible to find such an ordering without repetitions, but this is not necessary.) Since T is metrically transitive on E there exists a non-periodic point whose images under iteration of T fall in the interiors of each of the cubes σ_i with frequency $1/N$, in fact, such points form a set of measure mE. Let p be such a point, then for any sufficiently large n the number of points $Tp, \cdots, T^n p$ in any cube σ_i will differ from n/N by an arbitrarily small fraction of n, in particular by less than n/NK^2. In addition, let us choose n so that $n > NK^2\lambda$ and so that $n + \lambda$ is a number of the form $N \cdot 2^{\alpha r}$, where λ is the integer corresponding to δ defined in Lemma 14. As in the proof of Lemma 15, we can modify the transformation $f^{-1}Tf$ by composition with an automorphism $S_0 \epsilon M_0[R^*]$ such that under $S_0 f^{-1}Tf$ the point $f^{-1}p$ has period $n' = n + \lambda$ and its first n images are the same as under $f^{-1}Tf$. (Let D be the set on which $T_0 = f^{-1}Tf$ is 1:1, and let $p_i^* = f^{-1}T^i p$, $0 \leqq i \leqq n$. Then $p_n^* \epsilon (T_0^{-1}U_\delta)p_0^*$ and step by step we can choose points $p_{n+1}^*, \cdots, p_{n+\lambda-1}^*$ such that $p_0^*, \cdots, p_{n+\lambda-1}^*$ are distinct, and therefore also $T_0 p_0^*, \cdots, T_0 p_{n+\lambda-1}^*$, and such that $T_0 p_i^*$ is always within distance δ of p_{i+1}^*, $p_{n+\lambda}^*$ being taken equal to p_0^*. Then define S_0 with norm less than δ so that $S_0 T_0 p_i^* = p_{i+1}^*$, $0 \leqq i \leqq n + \lambda - 1$.) Then $S_1 = fS_0 f^{-1}$ will be an automorphism of E with norm less than ϵ that leaves the boundary fixed, and under $S_1 T$ the point p will have period n' and its first n images will still be $Tp, \cdots, T^n p$. It follows that the number of distinct images of p under $S_1 T$ in any cube σ_i differs from n'/N by less than $\eta = 2n'/NK^2$, because if σ_i contains n_i of the points $Tp, \cdots, T^n p$, and λ_i of the points $T_1^{n+1}p, \cdots, T_1^{n+\lambda}p$, we have

$$\left| n_i + \lambda_i - \frac{n'}{N} \right| \leqq \left| n_i - \frac{n}{N} \right| + \left| \lambda_i - \frac{\lambda}{N} \right| \leqq \frac{n}{NK^2} + \lambda < \frac{2n}{NK^2} < \eta.$$

Let $p_1, \cdots, p_{n'}$ denote the n' distinct images of p under $S_1 T$. We proceed to assign each of these to one of the cubes σ_i in such a way that exactly n'/N are assigned to each cube σ_i, and each point is assigned either to the cube in which it lies or to an adjacent one. To do this we proceed as follows. First assign each point tentatively to the cube in which it lies. If the number of points in σ_{i_1} is less than n'/N, make up the deficiency by assigning to it points from σ_{i_2}. If the number in σ_{i_1} is greater than n'/N, assign enough points from it to σ_{i_2} to remove the excess. Next consider σ_{i_2} and eliminate the excess or deficiency of points now assigned to it by assigning points of it to σ_{i_3}, or to it from σ_{i_3}. After the k-th step there will be exactly n'/N points assigned to each of the cubes $\sigma_{i_1}, \sigma_{i_2}, \cdots, \sigma_{i_k}$ except those (if any) which are equal to $\sigma_{i_{k+1}}$. After the $(K - 1)$-st step there will be exactly n'/N points assigned to each cell

σ_i. At the first step the number of points assigned to other cubes is at most η, at the second step at most 2η are assigned, and at the last step at most $(K - 1)\eta$. The total number of points assigned to other cubes is therefore at most $\frac{1}{2}K(K - 1)\eta$, which is less than the number of points p_j in any one of the cubes σ_i. Hence the assignments can be made in such a way that no point is assigned to another cube more than once, and therefore every point is assigned finally either to the cube in which it lies or to an adjacent one. We can therefore number the centers $q_1, \cdots, q_{n'}$ of the cubes of the α-th dyadic subdivision of $\sigma_1, \cdots, \sigma_N$ in such a way that the points q_j and p_j always lie in the same cube σ_i or in cubes having an $(r - 1)$-face in common. The distance between them is therefore less than 2ϵ, and we can define a measure-preserving automorphism S_2 of E that leaves the boundary fixed, has norm less than 2ϵ, and carries each p_j into q_j. (This does not follow directly from Lemma 13 as stated, but the same method of proof used there evidently applies in the present case since the line segment joining p_j to q_j lies in the interior of E.) It follows that the points $q_1, \cdots, q_{n'}$ are permuted cyclically by $S_2S_1TS_2^{-1}$, which may be written in the form ST by letting $S = S_2S_1TS_2^{-1}T^{-1}$. This completes the proof, since S leaves the entire boundary of E fixed and can be made arbitrarily small by suitable choice of ϵ, since S_1 and S_2 have norms less than 2ϵ.

LEMMA 20: *Let R be the r-dimensional unit cube, $r \geqq 2$, and let $T \epsilon M_0[R]$ be such that it permutes cyclically the centers of some of the cubes of a certain dyadic subdivision of R. Let $\sigma_1, \cdots, \sigma_N$ be the cubes whose centers are permuted. There exist automorphisms $S \epsilon M_0[R]$ and $h \epsilon H_0[R]$, equal to the identity outside the cubes σ_i and on their boundaries, such that $hSTh^{-1}$ is measure-preserving and under it the cubes concentric[50] to $\sigma_1, \cdots, \sigma_N$ with half their diameter are rigidly permuted in a single cycle.*

Let p_i be the center of σ_i. We may suppose the cubes so numbered that $Tp_i = p_{i+1}$, $i < N$, and $Tp_N = p_1$. Let $\tau_1, \cdots, \tau_N$ be cubes concentric to $\sigma_1, \cdots, \sigma_N$, all with the same diameter, and small enough so that $T\tau_i$ is interior to σ_{i+1}, $i < N$, and $T\tau_N$ to σ_1. Let g_1 be an automorphism of R that leaves the boundary fixed and translates τ_1 into τ_2. Extend both g_1 and T to the whole space $E^{(r)}$ by defining them equal to the identity outside R. Let g_2 be a radial shrinking transformation of $E^{(r)}$ such that $g_2R \subset \tau_2$ and such that g_2 leaves both τ_2 and $T\tau_1$ fixed. It is easily verified that $(g_2g_1T^{-1}g_2^{-1})T$ translates τ_1 into τ_2, and $g_2g_1T^{-1}g_2^{-1}$ is an automorphism equal to the identity outside τ_2 that carries $T\tau_1$ into τ_2 in a measure-preserving manner. By Theorem 2 Corollary 3, we can modify $g_2g_1T^{-1}g_2^{-1}$ in $\tau_2 - T\tau_1$ in such a way that it becomes a measure-preserving automorphism of τ_2 that leaves the boundary fixed. Call this modified automorphism S_2 and define it equal to the identity outside τ_2. Then S_2T carries τ_1 rigidly into τ_2. In exactly the same way we can define $S_3 \epsilon M_0[\tau_3]$ such that S_3T carries τ_2 rigidly into τ_3. Finally we define $S_1 \epsilon M_0[\tau_1]$

[50] We shall use the term "concentric" to mean that the cubes have the same center and also have corresponding faces parallel.

such that S_1T carries τ_N rigidly into τ_1. The product of these N transformations S_i defines an automorphism S of R that carries each cube σ_i into itself and is equal to the identity on their boundaries and outside them. Under ST each cube τ_i is translated into τ_{i+1}, $i < N$, and τ_N into τ_1. Finally, let h be a radial transformation of each cube σ_i into itself which carries τ_i into the concentric cube σ_i^* with diameter half that of σ_i. Then $hSTh^{-1}$ is a measure-preserving automorphism that permutes the cubes σ_i^* rigidly and cyclically.

LEMMA 21: *Let T be a measure-preserving automorphism of the r-dimensional unit cube R, $r \geqq 2$, and suppose that T permutes rigidly certain disjoint cubes $R_1, \cdots, R_K$ contained in the interior of R which are sums of cubes of a dyadic subdivision. There exists an arbitrarily small automorphism $S \in M_0[R]$ equal to the identity on $R_1, \cdots, R_K$ such that ST permutes rigidly in a single cycle the cubes concentric to and with half the diameter of those of a certain dyadic subdivision of the complementary set $E = R - (R_1 + \cdots + R_K)$.*

We shall consider the degenerate case $K = 0$ to be included in this statement.

Let $\epsilon > 0$ be given. By the continuity of the group product, there exists a positive number δ such that if h, S_1, S_2 are any automorphisms with norm less than δ then $hS_2S_1Th^{-1}T^{-1}$ will have norm less than ϵ. By Lemma 19 (or by Lemma 15 in the case $K = 0$), there exists $S_1 \in M_0[R]$ equal to the identity on $R_1, \cdots, R_K$ with norm less than δ such that S_1T permutes cyclically the centers of the cubes $\sigma_1, \cdots, \sigma_N$ of a dyadic subdivision of E, and we may suppose the cubes σ_i to have diameter less than δ. By Lemma 20, there exist automorphisms h and S_2 that carry these cubes into themselves and leave their boundaries fixed such that $hS_2(S_1T)h^{-1}$ is measure-preserving and permutes rigidly in a single cycle the cubes σ_i^* concentric to and with half the diameter of σ_i. But this transformation may be written in the form ST, where $S = hS_2S_1Th^{-1}T^{-1}$. The automorphism S is measure-preserving, since it is a product of two such transformations; it is equal to the identity on $R_1, \cdots, R_K$ and on the boundary of R, since h, S_1, S_2 are; and it has norm less than ϵ, since h, S_1, S_2 have norms less than δ.

THEOREM 12: *Any measure-preserving automorphism T of the r-dimensional unit cube R, $r \geqq 2$, that leaves the boundary fixed can be approximated uniformly by another such automorphism T^* which is locally linear almost everywhere. More precisely, there is a sequence of disjoint cubes interior to R, having total measure one, which under T^* are rigidly permuted among themselves.*

The proof is based on a sequence of modifications of T defined inductively. Let $\delta_1, \delta_2, \cdots$ be a sequence of positive numbers with sum less than a given number $\epsilon > 0$. We shall show that it is possible to find a sequence of automorphisms $S_1, S_2, \cdots$ with the following properties:

1° $S_n \in M_0[R]$, $\rho(S_n, I) < \delta_n$

2° $S_nS_{n-1} \cdots S_1T$ permutes rigidly among themselves a finite number of disjoint cubes $R_1, \cdots, R_{N_n}$ in n cycles, each of these cubes being a sum of interior cubes of a certain dyadic subdivision of R.

3° S_n is equal to the identity on $R_1 + R_2 + \cdots + R_{N_{n-1}}$, $n > 1$.

$4°$ $m(R_1 + \cdots + R_{N_n}) = 1 - \left(1 - \frac{1}{2^r}\right)^n.$

By Lemma 21, there exists $S_1 \epsilon M_0[R]$ with norm less than δ_1 such that S_1T permutes rigidly in a single cycle the cubes $R_1, \cdots, R_{N_1}$ concentric to and with half the diameter of those of a certain dyadic subdivision. These cubes therefore have total measure $1/2^r$. Thus conditions 1° to 4° are satisfied in the case $n = 1$, condition 3° being vacuous. Now suppose that $S_1, \cdots, S_n$ and $R_1, \cdots, R_{N_n}$ have been defined so that conditions 1° to 4° are satisfied for values of $n \leqq k$. Then $T_k = S_kS_{k-1} \cdots S_1T$ fulfills the hypotheses of Lemma 21 and there exists an automorphism $S_{k+1} \epsilon M_0[R]$, with norm less than δ_{k+1}, equal to the identity on $R_1 + \cdots + R_{N_k}$, such that $S_{k+1}T_k$ permutes rigidly in a single cycle the cubes $R_{N_k+1}, \cdots, R_{N_{k+1}}$ concentric to and with half the diameter of those of a certain dyadic subdivision of R that are not contained in $R_1 + \cdots + R_{N_k}$. The total measure of these additional cubes is therefore $\frac{1}{2^r}\left(1 - \frac{1}{2^r}\right)^k$, so that conditions 1° to 4° are now satisfied for values of $n \leqq k + 1$, and the inductive definition of the successive modifications S_n is complete.

From 1° it follows that the limit $S = \lim_{n \to \infty} S_nS_{n-1} \cdots S_1$ exists and has norm less than ϵ. Furthermore, ST is equal to $S_n \cdots S_1T$ on $R_1 + R_2 + \cdots + R_{N_n}$, in virtue of 3°, and so it permutes these cubes rigidly among themselves. Thus under ST the cubes $R_1, R_2, \cdots$ are all rigidly permuted among themselves in finite cycles. Finally, by 4°, these cubes contain almost all points of R. Hence $T^* = ST$ has all the properties described in the theorem.

Bryn Mawr College
The University of Wisconsin

WHAT IS MEASURE?

S. M. ULAM, University of Wisconsin

The concept of measure includes the notions generalizing the old ideas of length, area and volume of figures; all of which are among the oldest in mathematics and, in fact, as basic as the idea of number itself.

1. The elementary approach. The first mathematical approach to the idea of length of a curve, area of a surface, or volume of a solid consisted in assuming them as "evident" or given for some elementary figures, for instance line segments, rectangles, parallelepipeds or prisms. One calculated these numbers for polygons or polyhedra by decomposing them into elementary figures; and for general curved figures, passages to the limit were used. Archimedes followed this procedure for the computation of areas and volumes of a few of the simplest geometrical objects. The question of the consistency of this method was, of course, raised much later. Not until the development of the infinitesimal calculus, which gave the necessary tools for working systematically with the limit processes, was the class of figures whose areas or volumes could be defined and calculated significantly enlarged.

Cauchy, Riemann and Jordan gave rigorous definitions and statements of properties of the definite integral, and thus gave solid foundations for a treatment of the areas and volumes of the objects that formed the domain of study of most of the 19th century mathematics—sets defined by inequalities satisfied by continuous functions on the n-dimensional Euclidean space. [Let us point out here that in general the problem of the integral and the problem of measure are intimately related. A general notion of integral permits one to define a general measure, and vice-versa.] In the second half of the 19th century, in many parts of mathematics, it became necessary to investigate more general sets of points in Euclidean space. In the study of trigonometric series, in function theory, and especially in the investigations of Poincaré in the theory of probability and the general theory of dynamical systems, there appeared sets of points defined by discontinuous functions (obtained through passages to the limit effected on continuous functions).

Cantor's creation of set theory, where the notion of a geometrical figure was generalized into that of an arbitrary subset of points of a given space, introduced also a need for an axiomatic investigation of the problem of measure of sets and, at the same time, made possible a logical analysis of the notion of measure in general.

2. Lebesgue's measure. Borel, and above all Lebesgue [0], applied the ideas of set theory to the problem of measure. Lebesgue's procedure for introducing a measure for sets situated in the Euclidean space was essentially this:

There is given a collection (class) of sets situated in the Euclidean space. This class contains the elementary figures and is large enough to include all sets that can be obtained by the processes usually employed in analysis; in particular the complement of a set that is in the class also belongs to the class, and

For commentary to this paper [25], see p. 684.

the union of any denumerable number of sets in the class yields sets belonging to this class (the denumerable union of sets corresponds to the process of summing infinite series). One has to attach to every set in the class a non-negative, real number, called its measure, so that the following postulates will be satisfied:

I. All sets of a specific subclass should have measure, for example, all sets consisting of a single point.

II. The measure of a set should coincide with its ordinary value in the case when the set is an elementary figure.

III. The postulate of additivity: two forms, a weaker and a stronger form are possible. This requires that the measure of a finite (or in a stronger form, denumerably infinite) sum of mutually disjoint sets should be equal to the numerical sum of the measures of the individual sets.

IV. The invariance or congruence postulate: This requires that sets congruent in the sense of elementary geometry, should have equal measure.

The smallest class of sets for which it is meaningful to discuss this general problem of measure, consists of the so-called Borel sets. These are all sets that one can obtain from intervals (or parallelepipeds in the n-dimensional case) by the two operations of taking a complement of a set and taking infinitely denumerable sums of sets—repeated any number of times. Lebesgue recognized that one can enlarge this class by considering all sets that differ from a Borel set by a subset of any Borel set of measure. He succeeded in giving a constructive definition of a measure of this kind. His measure meets many but, as we shall see, not all the needs of analysis.

We give these well known historical facts because, even in this abbreviated and schematic form, they throw light on the developments that followed Lebesgue.

3. The abstract point of view. If we examine the problem critically, the following questions arise at once:

How large can the class of sets be for which a measure, in the sense given above, can be defined? Lebesgue's class is closed under the operations of denumerable addition and intersection of sets, and so meets many but not all of the situations arising in analysis. Using simply the operation of projection of a Borel set from the n-dimensional space into the $n-1$ dimensional space (for example from the plane into the straight line) one can define sets that do not belong to Borel's class. By operating again with complements of such sets and projections, one obtains a wide class of sets, the so-called projective sets. Whether these belong to Lebesgue's class is still an open question. Such sets arise very naturally, especially in function spaces. It seems very natural to include the operation of projection of sets (or what would be the same, taking the images through continuous transformations) in the consideration of the class of measurable sets. Lebesgue's class certainly does not contain all sets. This was shown first by Vitali [1].

The n-dimensional Euclidean space still remains the most important space of mathematics, but it is only the most important special case among the many spaces studied in geometry and analysis. It may be variously looked upon as an example of a general topological space, of a group manifold, of a Riemannian space, or of a finite dimensional vector space. Is a theory of measure for subsets of these spaces possible? The great development of the theory of probability creates the need of a theory of measure for sets in general "phase spaces" some of which are infinitely dimensional function spaces.

In analogy with the property of invariance of Lebesgue's measure for congruent sets, we might have to examine, in spaces more general than the Euclidean, the various meanings of congruence, and the corresponding properties of a measure.

Let us study all these points in greater detail.

4. The problem of the class of measurable sets. Let us make clear from the beginning that this discussion is necessary only because it is impossible to have measure defined for all subsets of a set [2]. To make things precise, we shall examine the situation where all sets for which a measure has to be defined, are subsets of the interval $0 \leqq x \leqq 1$. The class of sets should be such that the sums of denumerably many sets belonging to it also belong to the class, and the complement of a set in the class is also in the class. We seek a real valued function $m(A)$, the measure of the set, having the following properties:

I. $m(E)=1$, where E denotes the entire interval $[0-1]$.

II. $m(p)=0$, when p denotes a set composed of any single point p.

III. The additivity property, that is, the measure of the sum of two disjoint sets, is equal to the numerical sum of the measures: $m(A+B)=m(A)+m(B)$ whenever $A \cdot B=0$.

III'. The additivity for a denumerably infinite number of disjoint sets: $m(\sum_{i=1}^{\infty} A_i)=\sum_{i=1}^{\infty} m(A_i)$ if $A_i \cdot A_j=0$ for $i \neq j$.

It is impossible to have a measure function defined for all sets on the interval with the properties I, II, III' [2]. But if one requires only I, II, III, one can define a measure with these properties for *all* subsets, and as a matter of fact, one can have an additional property of the measure, namely, that any two sets congruent in the sense of elementary geometry, will have equal measure [3].

Quite generally it is possible to have a finitely additive measure function, in any additive class of sets, the measure assuming only the two values 0 and 1, and being equal to 0 for all sets of a *prescribed* additive subclass of the given class [4].

The fact that it is not possible to have an infinitely additive measure for all sets on the interval, makes it necessary, *even in the case where one does not require invariance or congruence properties* for the measure to consider *subclasses* of measurable sets; we see that the problem is one of algebra and set theory at this stage. The problem of determining for what additive classes of sets there will exist a real valued infinitely additive set function is not yet solved. An answer can be given in the case when the given class is generated by denumerably many sets (generated by the operations of taking the complement of a set and adding denumerably many sets in the class) [5].

5. Invariant measures. The problem of measure becomes more interesting when one requires that sets that are congruent or equivalent should have the same measure.

We shall now explain the meaning of congruence or equivalence by means of a few examples.

If the space E is Euclidean or more generally metric, the congruence of two sets A and B means that one can map A into B by a one-to-one transformation, so as to preserve distances between pairs of points. (The transformation needs to be defined only on A, and not necessarily on the whole space E.) It is natural to require that any two congruent sets should have equal measure.

Another case: If the space E is Euclidean, we have a group of point transformations of E into itself, for example, the rigid motions. In general, when E is a manifold on which a given group $\mathfrak{G}$ of point transformations is specified, one might call two sets A and B equivalent if there is a transformation T in the group $\mathfrak{G}$ such that $T(A)=B$. In any space E that is also a group manifold a group of transformations is obtained in a natural way, for example, by multiplying all elements x of the space by a fixed element a on the left: $T(x)\equiv a\cdot x$.

Finally, we could conceive in the most general way of equivalence among subsets of a given space E as a given (but otherwise arbitrary) reflexive and transitive relation between sets, thus dividing the sets into classes of mutually "equivalent" ones.

In all these cases one can ask for a measure function which, in addition to the additivity properties, would assign equal values to equivalent sets.

Let us indicate briefly the present state of knowledge concerning the existence of measures with specified invariance properties. Lebesgue's measure is defined only for certain sets, but it has the properties I, II, III', IV [6]. Banach has constructed a measure function for all sets on the real line, or all sets in the Euclidean plane with properties I, II, III, IV. The fact that a measure of this sort is not possible for all subsets of the Euclidean space of *three dimensions* was first proved by Hausdorff [7]. A later proof was given by Banach and Tarski [8] in their famous paradoxical decomposition of two solid spheres, of different radii. Each sphere can be decomposed into the same finite number of mutually disjoint sets, the sets forming the bigger sphere being respectively congruent to those used in the decomposition of the smaller sphere!

In a very interesting paper [9] Von Neumann has discussed the problem of measure for all subsets of any given group (thus generalizing the problem from the Euclidean vector groups to an arbitrary group) and characterized those groups for which a measure function satisfying I, II, III, IV can be defined for all subsets of the group. Let us note that IV refers to congruence as defined above for a general group.

Since in general we cannot define a measure function for all subsets, we have to consider special classes of subsets. In an important paper [10] A. Haar has established the existence of a denumerably additive and invariant measure function for subsets of any locally compact topological group. The class of sets for which he succeeded in defining his measure includes all Borel sets (we recall that these are sets obtainable from open sets by the operations of complementation and denumerable summation).

The problem of the existence of a measure function was treated for a general notion of equivalence of sets by Tarski [11]. He obtained necessary and sufficient conditions under which, if *one requires finite additivity only*, a measure function satisfying the congruence postulate is possible. But the solution of this general problem for denumerably infinite additivity is yet to be obtained. Special notions of equivalence for sets, and the existence of a measure having equal values for equivalent sets, have been considered by several others [12].

We cannot here enter into the discussion of properties and applications of measure functions. Let us remark, however, that in many cases the properties which we have postulated for measure determine it uniquely. Lebesgue himself showed that a measure function for sets in the Euclidean space satisfying I, II, III, IV, must necessarily coincide with the one he has constructively defined. This also holds for measures in Haar's groups [13].

Limitations of space forbid the discussion of applications of measure theory to the general theory of probability where one deals with some of the most abstract aspects of the theory and constructs measures in composite spaces, using given measures in given spaces [14], or to ergodic theory [15]. For connections between measures and topologies in algebraical structures the reader is referred to the book of A. Weil [10]. Likewise, we cannot discuss here the general point of view under which one would define measures, not necessarily for sets that are subsets of a given space, but rather for elements of a general Boolean algebra [16]. Also one could study measures with other than real values; p-adic values for example seem indicated in certain general situations arising in topology.

6. The general problem. To summarize, we note that measure is a set function associating a non-negative, real number with every set in a certain class of sets, all forming subsets of a given space. Characteristic properties of measure are its additivity for disjoint sets (finite additivity or, if one can obtain it, denumerably infinite additivity). The other characteristic property is equality of measure for sets that are congruent or otherwise considered equivalent, the notion of congruence or equivalence being provided by the geometry or algebra of the given situation. The problem of existence of such set functions has been solved in many special cases, including some most important ones in Euclidean space. However the general problem, that is, to determine whether or not, in a given class of sets with a given equivalence relation, a measure is possible, seems very difficult.

Bibliography

For an account of Lebesgue's measure see:

0. H. Lebesgue, Leçons sur l'intégration et la recherche des fonctions primitives, deuxième édition, Paris, 1928.

For notions of set theory and an excellent treatment of measure functions see also:

F. Hausdorff, Grundzüge der Mengenlehre, Leipzig, 1914.

Charles J. de la Vallée Poussin, Intégrales de Lebesgue, Fonctions d'ensembles, classes de Baire, Paris, 1916.

1. G. Vitali, Sulle funzione integrali, Atti Torino, 40, 1905; also Hausdorff, *loc. cit.*
2. S. Ulam, Zum Massbegriffe in der allgemeinen Mengenlehre, Fund. Math., vol. 16, 1930.
3. S. Banach, Sur le problème de mesure, Fund. Math., vol. 4, 1923.
4. S. Ulam, On functions of sets, Fund. Math., vol. 14, 1929.
5. D. Bernstein and S. Ulam, Abstract of the B. A. M. S., vol. 48, p. 351.
6. See Hausdorff, *loc. cit.*
7. Hausdorff, pp. 401, 469.
8. S. Banach and A. Tarski, Sur la décomposition des ensembles de points en parties repectivement congruentes, Fund. Math., vol. 6, 1924.
9. J. von Neumann, Fund. Math., vol. 13, 1929.
10. A. Haar, Der Massbegriff in der Theorie kontinuierlichen Gruppen, Ann. of Math., vol. 34 1933.

There is an excellent account of integration and measure theory in topological groups due to Haar and A. Weil in

A. Weil, L'Intégration dans les Groupes Topologiques et ses Applications, Paris, 1938.

11. A. Tarski, Algebraische Fassung des Maßproblems, Fund. Math., vol. 31.
12. Bogoliouboff and Kryloff, La théorie générale de la mesure dans son application à l'étude des systémes dynamiques. Ann. of Math., vol. 38, 1937.

J. Oxtoby and S. Ulam, On the existence of a measure invariant under a transformation. Ann. of Math., vol. 40, 1939.

13. J. von Neumann, The uniqueness of Haar's measure, Mat. Sbornik, vol. 1/43, 1936. See also, A. Weil, *loc. cit.*
14. Z. Lomnicki and S. Ulam, Sur la théorie de la mesure . . . et son application au calcul des probabilités, Fund. Math., vol. 22, 1934.
15. G. D. Birkhoff, Proof of the ergodic theorem, Proc. Nat. Ac. Sc., vol. 1931; G. Birkhoff, Lattice theory, A.M.S. Colloquium Publications, 1940, Chapter IX.
16. M. H. Stone, Trans. Am. Math. Soc., vol. 40, 1936; M. H. Stone and J. von Neumann, Fund. Math., vol. 25, 1935.

ON APPROXIMATE ISOMETRIES

D. H. HYERS AND S. M. ULAM

In a previous paper, a problem of mathematical "stability" for the case of the linear functional equation was studied.[1] It was shown that if a transformation $f(x)$ of a vector space E_1 into a Banach space E_2 satisfies the inequality $\|f(x+y)-f(x)-f(y)\|<\epsilon$ for some $\epsilon>0$ and all x and y in E_1, then there exists an additive transformation $\phi(x)$ of E_1 into E_2 such that $\|f(x)-\phi(x)\|<\epsilon$.

In the present paper we consider a stability problem for isometries. By an ϵ-isometry of one metric space E into another E' is meant a transformation $T(x)$ which changes distances by at most ϵ, where ϵ is some positive number; that is, $|\rho(x, y)-\rho(T(x), T(y))|<\epsilon$ for all x and y in E. Given an ϵ-isometry $T(x)$, our object is to establish the existence of a true isometry $U(x)$ which approximates $T(x)$; more precisely, to establish the existence of a constant $k>0$ depending only on the metric spaces E and E' such that $\rho(T(x), U(x))<k\epsilon$ for all x in E. In this paper this result will be proved for the case in which $E=E'$, where E is n-dimensional Euclidean space or Hilbert space (not necessarily separable). The case in which E is the space C of continuous functions will be treated in another paper.

The above problem of ϵ-isometries is related to the problem of constructing space models for sets in which distances between points are given only with a certain degree of exactness (measurements are possible only with a certain degree of precision). The question of the uniqueness of the idealized model corresponding to the given measurements and the extrapolation from the measurements to the model could be looked upon as a problem in determining a strict isometry from an approximate isometry.

In the case of certain simple metric spaces, for example the surface of the Euclidean sphere, this question can be answered in the affirmative, but it may be more difficult for other bounded manifolds. A simple but interesting example showing a case where the answer is *negative* has been worked out by R. Swain.

THEOREM 1. *Let E be a complete abstract Euclidean vector space.*[2]

Presented to the Society, September 5, 1941; received by the editors October 10, 1944.

[1] D. H. Hyers, *On the stability of the linear functional equation*, Proc. Nat. Acad. Sci. U. S. A. vol. 27 (1941) pp. 222–224.

[2] A complete Euclidean vector space is a Banach space whose norm is generated by an inner product, (x, y). It includes real Hilbert space and n-dimensional Euclidean spaces as special cases.

For commentary to this paper [27], see p. 685.

Let $T(x)$ be an ϵ-isometry of E into itself such that $T(0)=0$. The limit $U(x)=\lim_{n\to\infty}(T(2^n x)/2^n)$ exists for every x in E and $U(x)$ is an isometric transformation.

PROOF. Put $r=\|x\|$. Then $|\|T(x)\|-r|<\epsilon$ and $|\|T(x)-T(2x)\|-r|<\epsilon$. Put also $y_0=T(2x)/2$, so that $|r-\|y_0\||<\epsilon/2$. Consider the intersection of the two spheres: $S_1=[y;\ \|y\|<r+\epsilon]$, $S_2=[y;\ \|y-2y_0\|<r+\epsilon]$. Now $T(x)$ belongs to this intersection, and for any point y of $S_1\cap S_2$ we have

$$2\|y-y_0\|^2=2\|y\|^2+2\|y_0\|^2-4(y,y_0);$$
$$\|y-2y_0\|^2=\|y\|^2+4\|y_0\|^2-4(y,y_0)<(r+\epsilon)^2$$

and $\|y\|^2<(r+\epsilon)^2$. It follows that

$$2\|y-y_0\|^2<(r+\epsilon)^2+\|y\|^2-2\|y_0\|^2<2(r+\epsilon)^2-2\|y_0\|^2$$
$$<2(r+\epsilon)^2-2(r-\epsilon/2)^2=6\epsilon r+3\epsilon^2/2.$$

Hence, $\|T(x)-T(2x)/2\|<2(\epsilon\|x\|)^{1/2}$ if $\|x\|\geqq\epsilon$, and $\|T(x)-T(2x)/2\|<2\epsilon$ in the contrary case.

Therefore, for all x in E the inequality

$$\|T(x/2)-T(x)/2\|<2^{-1/2}k(\|x\|)^{1/2}+2\epsilon \tag{1}$$

is satisfied, where $k=2\epsilon^{1/2}$. Now let us make the inductive assumption

$$\|T(2^{-n}x)-2^{-n}T(x)\|<2^{-n/2}k(\|x\|)^{1/2}\left(\sum_{i=0}^{n-1}2^{-i/2}\right)+(1-2^{-n})\cdot 4\epsilon. \tag{2}$$

The inequality (2) is true for $n=1$. Assuming it true for any particular value of n we shall prove it for $n+1$.

Dividing the inequality (2) by 2 we have

$$\|T(2^{-n}x)/2-2^{-n-1}T(x)\|$$
$$<2^{-(n+1)/2}k(\|x\|)^{1/2}\left(\sum_{i=1}^{n}2^{-i/2}\right)+(1/2-2^{-n-1})\cdot 4\epsilon.$$

Replacing x by $2^{-n}x$ in the inequality (1) we get

$$\|T(2^{-n-1}x)-T(2^{-n}x)/2\|<2^{-(n+1)/2}k(\|x\|)^{1/2}+2\epsilon.$$

On adding the last two inequalities we obtain

$$\|T(2^{-n-1}x)-2^{-n-1}T(x)\|$$
$$<2^{-(n+1)/2}k(\|x\|)^{1/2}\left(\sum_{i=0}^{n}2^{-i/2}\right)+(1-2^{-n-1})\cdot 4\epsilon.$$

This proves the induction. Therefore inequality (2) is true for all x

in E and for $n=1, 2, 3, \cdots$. If we put $a=k\sum_{i=0}^{\infty}2^{-i/2}$, we have

$$\|T(2^{-n}x) - 2^{-n}T(x)\| < 2^{-n/2}a(\|x\|)^{1/2} + 4\epsilon.$$

Hence, if m and p are any positive integers,

$$\|2^{-m}T(2^m x) - 2^{-m-p}T(2^{m+p}x)\|$$
$$= 2^{-m}\|T(2^{m+p}x/2p) - 2^{-p}T(2^{m+p}x)\| < 2^{-m/2}a(\|x\|)^{1/2} + 2^{2-m}\epsilon,$$

for all x in E.

Therefore since E is a complete space, the limit $U(x) = \lim_{n\to\infty} (T(2^n x)/2^n)$ exists for all x in E.

To prove that $U(x)$ is an isometry, let x and y be any two points of E. Divide the inequality

$$|\,\|T(2^n x) - T(2^n y)\| - 2^n\|x - y\|\,| < \epsilon$$

by 2^n and take the limit as $n\to\infty$. The result is $\|U(x)-U(y)\| = \|x-y\|$. This completes the proof of Theorem 1.

THEOREM 2. *Let T satisfy the hypotheses of Theorem 1 and let u and x be any points of E such that $\|u\|=1$ and $(x, u)=0$. Then $|(T(x), U(u))| \leqq 3\epsilon$, where $U(x)$ is defined as in the statement of Theorem 1.*

PROOF. For an arbitrary integer n put $z=2^n u$. Let y denote an arbitrary point of the sphere S_n of radius 2^n and center at z. Then $\|y-z\|=\|z\|$ and it follows that $(y, u)=2^{-n-1}(y, y)$. Since T is an ϵ-isometry, $\|T(y)-T(z)\|=\eta(y, z)+\|T(z)\|$ where $|\eta(y, z)|<2\epsilon$.

The last equality may be written

$$2(T(y), T(z)) = (T(y), T(y)) - 2\eta\|T(z)\| - \eta^2.$$

Dividing by 2^{n+1} and remembering that $z=2^n u$, we obtain the equality

$$(3)\quad (T(y), 2^{-n}T(2^n u)) = \frac{1}{2^{n+1}}[(T(y), T(y)) - \eta^2] - \eta\left\|\frac{T(2^n u)}{2^n}\right\|.$$

Now let x be any point of the hyperplane $(x, u)=0$.

Then $y=x+ru$, where $r=2^n-(2^{2n}-\|x\|^2)^{1/2}$, is a point of the sphere S_n. For,

$$\|y - z\|^2 = (y, y) - 2(y, z) + (z, z)$$
$$= (x, x) + r^2 - 2(x, z) - 2(u, z)\cdot r + (z, z)$$
$$= r^2 - 2^{n+1}r + \|x\|^2 + \|z\|^2 = \|z\|^2.$$

Moreover, $\|y-x\|=r\to 0$ as $n\to\infty$. By Theorem 1, $t=\lim_{n\to\infty}2^{-n}T(2^n u)$ exists and is a unit vector. Finally, for an arbitrary positive δ and n sufficiently large, one can easily establish the following inequalities by means of equality (3) and the above remarks:

$$\begin{aligned} |(T(x),t)| &\leqq \left|\left(T(x), t-\frac{T(2^n u)}{2^n}\right)\right| + \left|\left(T(y), \frac{T(2^n u)}{2^n}\right)\right| \\ &\quad + \left|\left(T(x)-T(y), \frac{T(2^n u)}{2^n}\right)\right| \\ &< \|T(x)\|\cdot\left\|t-\frac{T(2^n u)}{2^n}\right\| + \frac{\delta}{2} + 2\epsilon\left\|\frac{T(2^n u)}{2^n}\right\| \\ &\quad + \|T(x)-T(y)\|\cdot\left\|\frac{T(2^n u)}{2^n}\right\| < \delta + 3\epsilon(1+\delta). \end{aligned}$$

It follows that

$$|(T(x), U(u))| = |(T(x), t)| \leqq 3\epsilon.$$

THEOREM 3. *Let $T(x)$ satisfy again the hypotheses of Theorem 1, and let it take E into the whole of E. Then the transformation $U(x)$ also takes E into the whole of E.*

PROOF. For each point z of E, let $T^{-1}(z)$ denote any point whose T-image is z. Then $T^{-1}(z)$ is an ϵ-isometry of E. By Theorem 1, the limit $U^*(z)=\lim_{n\to\infty}(T^{-1}(2^n z)/2^n)$ exists, and U^* is an isometry of E. Now clearly

$$\begin{aligned} \|2^n z - T(2^n U^*(z))\| &= \left\| T\left(2^n\frac{T^{-1}(2^n z)}{2^n}\right) - T(2^n U^*(z))\right\| \\ &< 2^n\left\|\frac{T^{-1}(2^n z)}{2^n} - U^*(z)\right\| + \epsilon. \end{aligned}$$

On dividing by 2^n and letting $n\to\infty$, we see, for each point z of E, that $z=UU^*(z)$. Therefore $U(E)=E$.

THEOREM 4. *Let E be a complete abstract Euclidean vector space. If $T(x)$ is an ϵ-isometry which takes E into the whole of E such that $T(0)=0$, then the transformation $U(x)=\lim_{n\to\infty}(T(2^n x)/2^n)$ is an isometry of E into the whole of itself, and the inequality $\|T(x)-U(x)\|<10\epsilon$ is satisfied for all x in E.*

PROOF. For a given point $x\neq 0$ let M denote the linear manifold orthogonal to x. By Theorem 3, U is an isometric transformation which takes E into the whole of E. Hence $U(M)$ is the linear manifold orthogonal to $U(x)$. Let w be the projection of $T(x)$ on $U(M)$. If $w=0$ put $t=0$. Otherwise put $t=w/\|w\|$. In either case (cf. Theorem 2), the inequality $|(T(x), t)| \leqq 3\epsilon$ is satisfied. Put $v=(1/\|x\|)U(x)$. Then v is a unit vector orthogonal to t and is coplanar with $T(x)$ and t.

Hence, by the pythagorean theorem we have the identity:

$$\|T(x) - U(x)\|^2 = (T(x), t)^2 + [\|x\| - (T(x), v)]^2. \tag{4}$$

Let $z_n=2^n x$ and if the projection w_n of $T(z_n)$ on $U(M)$ is not zero, put $t_n=w_n/\|w_n\|$. Otherwise we shall put $t_n=0$. In either case $(t_n, v)=0$, and $|(T(z_n), t_n)| \leqq 3\epsilon$. If $\|T(z_n)\|<3\epsilon$, it is obvious that $\|T(z_n)\|-|(T(z_n), v)| \leqq 3\epsilon$. If $\|T(z_n)\| \geqq 3\epsilon$, we have $0 \leqq \|T(z_n)\| - |(T(z_n), v)| = \|T(z_n)\| - (\|T(z_n)\|^2-(T(z_n), t_n)^2)^{1/2} \leqq 3\epsilon$.

Hence the inequality:

$$\big|\,\|z_n\| - |(T(z_n), v)|\,\big| < 4\epsilon \tag{5}$$

is satisfied, since $\|z_n\| < \|T(z_n)\|+\epsilon$.

Two cases arise. If $(T(x), v) \geqq 0$, we put $n=0$ in the inequality (5) and use the identity (4) to obtain the inequality $\|T(x)-U(x)\|<5\epsilon$. If $(T(x), v)<0$, then for some integer $m \geqq 0$ we must have $(T(z_m), v)<0$ and $((T(2z_m), v) \geqq 0$, since $(U(x), v)$ is positive and $U(x)=\lim_{n\to\infty} (T(z_n)/2^n)$. Hence, by inequality (2),

$$\|T(2z_m) - T(z_m)\| \geqq (T(2z_m), v) - (T(z_m), v) > 3\|z_m\| - 8\epsilon.$$

But we know that $\|T(2z_m)-T(z_m)\| < \|z_m\|+\epsilon$. Therefore,

$$\|x\| \leqq \|z_m\| < (9/2)\epsilon, \quad \text{and} \quad \|T(x) - U(x)\| < 2\|x\| + \epsilon \leqq 10\epsilon.$$

In order to prove the above theorem we had to assume that $T(x)$ takes E into itself. We now show that the theorem is not always true for ϵ-isometric transformations of one Euclidean space into *part* of another. Consider the transformation $T(x)$ of the real axis into a subset of the plane defined as follows: the coordinates x, y of $T(x)$ are $(x, 0)$ for $x \leqq 1$, and $(x, c\cdot\log x)$ for $x>1$. It is easy to verify that T will be an ϵ-isometry if we choose c in such a way that $\epsilon>c^2 \cdot\max_{x>1}((\log x)^2/(2x-2))$.

On the other hand, $T(x)$ obviously cannot approximate an isometry in the sense of our theorem.

THE UNIVERSITY OF SOUTHERN CALIFORNIA AND
THE UNIVERSITY OF WISCONSIN

ON ORDERED GROUPS

BY

C. J. EVERETT AND S. ULAM

1. **Introduction.** We base our discussion upon the concept of an ordered group, that is, the generalization of the l-group studied by G. Birkhoff [1](1) in which the lattice property is replaced by the weaker "Moore-Smith" or directed set axiom. An ordered group is embeddable in a complete ordered group if and only if it is integrally closed. We prove that if the commutator group of an ordered group is in its center, then integral closure of the group implies commutativity. Thus the conjecture of Birkhoff (see Problem 2, loc. cit.) is proved, although negative examples are given showing the falsity of Problems 1 and 2. Problem 3 is left open; the authors hope to settle it in a later paper. The linear group of functions $ax+A$, a, A real, $a>0$, admits no integrally closed order under composition.

Every ordered group is embeddable in a group which is sequence-complete in the sense of o-convergence. The results of [5] are thus extended to the non-commutative case.

Properties of the group of monotone continuous functions on $(0, 1)$ to itself under composition are studied, and various orders in the free group with two generators are used to establish a curious property of the function group.

2. **Completion of ordered groups.** A group G is called an *ordered* or *o-group* in case G is (1) *partially ordered* by a relation $\geqq$ ($a\geqq a$; $a\geqq b$ and $b\geqq a$ imply $a=b$; $a\geqq b$ and $b\geqq c$ imply $a\geqq c$), (2) a *directed set*: for every a, b there is a $c\geqq a$, b, (3) *homogeneous*: $a\geqq b$ implies $c+a+d\geqq c+b+d$.

G is called an *l-group* in case the order is a lattice order, that is, every two elements a, b possess a l.u.b. $a\vee b$ and a g.l.b. $a\wedge b$ [1].

G is said to be *conditionally complete* in case every set of elements a_α bounded above has a l.u.b. $\vee a_\alpha$ (and hence also a g.l.b. $\wedge a_\alpha$ when bounded below).

G is *integrally closed* in case $na=a+\cdots+a\leqq b$, $n=1, 2, \cdots$, implies $a\leqq 0$ [4], and *archimedean* if $na\leqq b$, $n=0, \pm 1, \pm 2, \cdots$, implies $a=0$ [1].

LEMMA 1. *If G is an o-group, integral closure implies archimedean order.*

For if $na\leqq b$, $n=0, \pm 1, \pm 2, \cdots$, then $na\leqq b$ and $n(-a)\leqq b$, $n=1, 2, \cdots$. By integral closure, $a\leqq 0$ and $-a\leqq 0$, $a=0$.

We shall see later (Theorem 12) that the converse is false.

Presented to the Society, August 14, 1944; received by the editors April 25, 1944.

1 Numbers in brackets refer to the Bibliography at the end of the paper.

For commentary to this paper [28], see p. 685.

LEMMA 2 [1, p. 313]. *If G is an l-group, integral closure and archimedean order are equivalent.*

Let G be archimedean, and $na \leqq b$, $n=1, 2, \cdots$. Then $n(a\vee 0) = na\vee(n-1)a\vee \cdots \vee a\vee 0 \leqq b\vee 0$, and $-n(a\vee 0) = n(-(a\vee 0)) \leqq 0 \leqq b\vee 0$. Hence $n(a\vee 0) \leqq b\vee 0$, $n=0, \pm 1, \cdots$, and $a\vee 0 = 0 \geqq a$.

A. H. Clifford [4] has proved that a commutative o-group is embeddable in a conditionally complete group if and only if it is integrally closed. We here extend this result to noncommutative groups.

THEOREM 1. *An o-group G is embeddable in a conditionally complete o-group with preservation of order,* g.l.b., *and* l.u.b. *if and only if G is integrally closed.*

If G is so embeddable, and $na \leqq b$, $n=1, 2, \cdots$, let $u = \bigvee(na; \text{all } n)$; then $u-a \geqq na$, $n=1, 2, \cdots$, $u-a \geqq u$, $a \leqq 0$ (cf. [1, p. 322]).

It is well known that if $L(X)$, $U(X)$ denote the sets of all lower and of all upper bounds, respectively, for all the elements of a subset $X \subset G$, then the operation $X^* = L(U(X))$ has the closure properties: $X^* \supset X$; $X^{**} = X^*$; $X \supset Y$ implies $X^* \supset Y^*$; and that the class $\mathfrak{C}$ of all "closed" sets $C = C^*$ is a conditionally complete lattice under set inclusion, with set intersection effective as g.l.b., and closure of set union as l.u.b. [2, p. 25; 7]. The correspondence $a \rightarrow (a)^*$ embeds G in $\mathfrak{C}$ with preservation of order, g.l.b. and l.u.b. Moreover one has the important property: $\bigvee(x_\alpha) = x$ in G if and only if $(x_\alpha; \alpha)^* = (x)^*$.

Defining $X+Y = (\text{all } x+y;\ x \in X,\ y \in Y)^*$, $X, Y \in \mathfrak{C}$, one readily verifies $X+(Y+Z) = (X+Y)+Z$, $0+X = 0 = X+0$, where $0 = (0)^*$, and $A \supset B$ implies $X+A+Y \supset X+B+Y$. Since $(a+b)^* = (a)^* + (b)^*$, G is an o-subgroup of the lattice semigroup $\mathfrak{C}$. In general, $\mathfrak{C}$ is not a group.

LEMMA 3. *For $X \in \mathfrak{C}$, there exists $Y \in \mathfrak{C}$ such that $Y+X=0$ if and only if $0 = \bigwedge(-x+u;\ x \in X,\ u \in U(X))$; similarly $X+Y=0$ if and only if $\bigwedge(u-x) = 0$. There is a $Y \in \mathfrak{C}$ for which $Y+X = 0 = X+Y$ if and only if $\bigwedge(-x+u) = 0 = \bigwedge(u-x)$ and when such Y exists $Y = L(\text{all } -x;\ x \in X)$.*

If $\bigwedge(-x+u) = 0 = \bigvee(-u+x) = \bigvee(l+x;\ l \in L(-x),\ x \in X)$, then $(l+x)^* = (0)^*$ and $L(-x)+X = 0$. Conversely, if $Y+X=0$, $(y+x)^* = (0)^*$, $\bigvee(y+x) = 0$, $y+x \leqq 0$, $y \leqq -x$, $Y \subset L(-x)$. Hence $(y+x) \subset (l+x;\ l \in L(-x))$, $0 = Y+X \subset L(-x)+X$. But $L(-x)+X \subset 0 = Y+X$, thus $0 = L(-x)+X$. Hence $0 = \bigvee(l+x) = \bigvee(-u+x) = \bigwedge(-x+u)$. Similarly for the second implication.

If both intersections are zero, $L(-x)+X = 0 = X+L(-x)$, and if $Y+X = 0 = X+Y$, both intersections are zero, $L(-x)+X = 0$, $L(-x)+X+Y = Y = L(-x)$.

LEMMA 4. *If G is integrally closed, every $X \in \mathfrak{C}$ has an inverse.*

Let $X \in \mathfrak{C}$, $y \leqq u-x$, all $u \in U(X)$, $x \in X$. Then $x \leqq -y+u$, $-y+u \in U(X)$ for all $u \in U(X)$. Let u_0 be any element of $U(X)$. Then $-y+u_0 = u_1$, $-y+u_1$

$=u_2, \cdots$ where $u_i \in U(X)$, and $u_0 = ny + u_n$. Thus $ny = u_0 - u_n \leqq u_0 - x_0$, x_0 any fixed element of X. By integral closure, $y \leqq 0$. Hence $\wedge(u-x)=0$. Similarly $\wedge(-x+u)=0$.

3. **Commutativity in integrally closed groups.** G. Birkhoff has raised the question whether archimedean order in l-groups does not imply commutativity [1, p. 329]. In this connection he quotes the following theorem [3].

THEOREM 2 (H. CARTAN). *If G is an archimedean ordered o-group, then linear order implies commutativity.*

He also conjectures the truth of the following theorem.

THEOREM 3. *If G is an integrally closed o-group (or archimedean l-group), and if the commutator subgroup of G is in the center of G, then G is commutative.*

Birkhoff bases a tentative proof on previous conjectures which we shall later show false. The following proof however establishes Theorem 3.

Let $c=b+a-b-a$ and note $b+a=b+(a+b)-b$. Then $c+(a+b)=(b+a)$ $= b + (a + b) - b$; $2c+(a+b)=c+b+(a+b)-b=(c+b-c)+c+(a+b)-b$ $=(c+b-c)+b+(a+b)-2b$; $3c+(a+b)=(2c+b-2c)+(c+b-c)+b+(a+b)$ $-3b$; by induction, $nc+(a+b)=((n-1)c+b-(n-1)c)+ \cdots +b+(a+b)$ $-nb$. For a, $b \geqq 0$, and c in the center, we have $nc+(a+b)=nb+(a+b)-nb$ $\geqq 0$. By integral closure, $b+a-b-a \geqq 0$, $b+a \geqq a+b$. Similarly $a+b \geqq b+a$, and $a+b=b+a$, all a, $b \geqq 0$. But then $-b+a=a-b$, and every positive element commutes with every negative element. Since every $g=g\vee 0+g\wedge 0$ [1, p. 306] and $h=h\vee 0+h\wedge 0$ we have $g+h=h+g$, all h, g of G.

THEOREM 4. *The linear group of elements $ax+A$, a, A real, $a>0$, under composition, admits no integrally closed o-group order.*

One computes $(gx+G)^{-1}=(x-G)/g$ and $(gx+G)(fx+F)(gx+G)^{-1}=fx$ $+gF+G(1-f)$. The conjugates of $fx+F$ $(f\neq 1)$ consist of all $fx+R$, R real; for example, use $g=1$, $G=(R-F)/(1-f)$.

There exists $f_0x+F_0 \geqq x$ $(f_0 \neq 1)$, for since G is a directed set, let $fx+F \geqq x$, $x/2$. If $f=1$, $x+F \geqq x$, $x/2$. Substitution into $2x$ yields $2(x+F)=2x+2F$ $\geqq 2(x/2)=x$.

Now all conjugates of f_0x+F_0, namely all f_0x+R (R real), are greater than or equal to x. Thus $f_0x-nk \geqq x$, k real, $n=1, 2, \cdots$, and substitution into $(x+k)^n=x+nk$ yields $f_0x \geqq x+nk=(x+k)^n$. By integral closure, $x+k \leqq x$, all k. Thus $(x+1) \leqq x$ and $(x+1)^{-1}=x-1 \leqq x$; $x+1=x$, a contradiction.

COROLLARY 1. *The group of functions αx^A, α, A real and positive, under composition and the group* [1, p. 303] *of real pairs (x, y) under $(x, y)+(x', y')$ $=(x+x', e^{x'}y+y')$ are isomorphic to the group of Theorem 4, hence cannot be integrally closed o-groups under any order.*

For consider the correspondences $ax+A \rightarrow e^A x^a$, $ax+A \rightarrow (-\log a, -A/a)$

4. **On o-convergence and sequence completion in l-groups.** The absolute $|a| = a \vee -a$ has proved of importance in the study of commutative l-groups, since its fundamental properties: $|a| = |-a| \geqq 0$; $|a| = 0$ if and only if $a = 0$; $|a \vee b - a' \vee b| \leqq |a - a'|$ and dually; and $|a+b| \leqq |a| + |b|$ serve to establish an intrinsic topology in G via o-convergent sequences [2; 5; 6]. All these properties except the last are valid in l-groups. We suggest the following as a generalization which seems adequate:

LEMMA 5. *In an l-group,* $|a+b| \leqq (|a| + |b|) \vee (|b| + |a|) \leqq |a| + |b| + |a|$.

$|a| \geqq a, -a$, $|b| \geqq b, -b$ yields $|a| + |b| \geqq a+b$ and $|b| + |a| \geqq -b-a$. Hence the first inequality. But $|a| + |b| \leqq |a| + |b| + |a|$ and $|b| + |a| \leqq |a| + |b| + |a|$.

We write $x_n \uparrow x$ for $x_1 \leqq x_2 \leqq \cdots$ with $x = \bigvee x_n$ and $x_n \downarrow x$ dually. Define $x_n \rightarrow x$ (o-convergence) as usual [2, pp. 28, 112] to mean there exist sequences l_n, u_n such that $l_n \leqq x_n \leqq u_n$ where $l_n \uparrow x$ and $u_n \downarrow x$. One proves $x_n \rightarrow x$ if and only if for some $w_n \downarrow 0$, $|x_n - x| \leqq w_n$; also if and only if $|-x+x_n| \leqq w_n' \downarrow 0$.

THEOREM 5. *o-convergence is a Fréchet convergence for which $a_n \rightarrow a$, $b_n \rightarrow b$ implies $a_n + b_n \rightarrow a+b$, $-a_n \rightarrow -a$, $a_n \vee b_n \rightarrow a \vee b$, and dually.*

One easily verifies $a, a, a, \cdots \rightarrow a$; $a_{n_i} \rightarrow a$; and $a_n \rightarrow a$, $a_n \rightarrow b$ implies $a = b$. Moreover $|(a_n + b_n) - (a+b)| = |a_n + b_n - b - a| = |(a_n - a) + (a + b_n - b - a)| \leqq |a_n - a| + |a + (b_n - b) - a| + |a_n - a| \leqq w_n + (a + w_n' - a) + w_n \downarrow 0$.

Finally, $|a_n \vee b_n - a \vee b| = |a_n \vee b_n - a_n \vee b + a_n \vee b - a \vee b| \leqq |a_n \vee b_n - a_n \vee b| + |a_n \vee b - a \vee b| + |a_n \vee b_n - a_n \vee b| \leqq |b_n - b| + |a_n - a| + |b_n - b| \leqq w_n' + w_n + w_n' \downarrow 0$.

We say a sequence is *o-regular* in case, for some $w_n \downarrow 0$, $|-a_{n+p} + a_n| \vee |a_n - a_{n+p}| \leqq w_n$, all $n, p = 1, 2, \cdots$.

THEOREM 6. *Every o-convergent sequence is o-regular.*

For $|a_n - a_{n+p}| = |(a_n - a) + (a - a_{n+p})| \leqq |a_n - a| + |a - a_{n+p}| + |a_n - a| \leqq 3w_n$. Similarly, $|-a_{n+p} + a_n| \leqq 3w_n'$, by the remark preceding Theorem 5.

G is called *o-complete* in case every o-regular sequence o-converges. In a previous paper conditions were given for a commutative l-group to be o-complete. The conditions and proofs given there [5] extend readily to the noncommutative case. We merely state the following theorem.

THEOREM 7. *In an l-group G the following conditions are equivalent:*

(i) *G is o-complete.*

(ii) *Every o-regular monotone sequence $y_1 \geqq y_2 \geqq \cdots$ o-converges.*

(iii) *For every o-regular monotone sequence $y_1 \geqq y_2 \geqq \cdots$, $\bigwedge y_n$ exists.*

(iv) *For every o-regular sequence x_n, $\bigwedge x_n$ exists.*

COROLLARY 2. *Every conditionally complete l-group is o-complete.*

(Cf. Kantorovitch [6] for the commutative case.) For if a_n is o-regular, $|a_n - a| \leqq w_n$, $-w_n \leqq a_n - a$, $-w_1 + a \leqq -w_n + a \leqq a_n$ (all n). Hence by completeness, $\wedge a_n$ exists and (iv) above is satisfied.

One naturally asks whether an arbitrary l-group may be embedded in an o-complete l-group. Let $\mathfrak{G}$ be the set of all elements X of $\mathfrak{C}$ (cf. §2) for which an inverse exists: $X + Y = 0 = Y + X$.

THEOREM 8. *Every l-group G is embeddable with preservation of order*, g.l.b., l.u.b., *and o-convergence in the l-group $\mathfrak{G}$, which is o-complete.*

The correspondence $a \rightarrow (a)^*$ maps G into $\mathfrak{G}$ with preservation of order, g.l.b., l.u.b. (and hence o-convergence), by the discussion in §2. That $\mathfrak{G}$ is an o-group is obvious. We must show that the order is a lattice order. We need only the following lemma.

LEMMA 6. *If X, Y, are in $\mathfrak{G}$ then the $X \wedge Y$ (of $\mathfrak{C}$) is in $\mathfrak{G}$.*

The proof is given in [5].

Finally we verify Theorem 7 (iii) using the following lemma.

LEMMA 7. *If Y_p, $W_p \in \mathfrak{G}$, $W_p \downarrow 0$ in $\mathfrak{G}$, and $Y \in \mathfrak{C}$, where $Y_p - W_p \subset Y \subset Y_p$* and *$-W_p + Y_p \subset Y \subset Y_p$, all p, then $Y \in \mathfrak{G}$.*

The proof is given in [5].

Hence condition (iii) holds; for let Y_n be o-regular in $\mathfrak{G}$, $Y_1 \supset Y_2 \supset \cdots$, $|Y_n - Y_{n+p}| = Y_n - Y_{n+p} \subset W_n \downarrow 0$ in $\mathfrak{G}$. Then $-W_1 + Y_1 \subset Y_n$ (all n), and $Y = \wedge(Y_n)$ exists in $\mathfrak{C}$ with $-W_n + Y_n \subset Y \subset Y_n$ for $-W_n + Y_n \subset Y_{n+p}$. Similarly we fulfill the other condition of Lemma 7.

5. **The group of topological transformations of the line into itself.** Let T be the class of all functions $f(x)$ on $0 \leqq x \leqq 1$, having the properties:

(1) $f(x)$ is continuous monotone increasing on $0 \leqq x \leqq 1$,

(2) $f(0) = 0$, $f(1) = 1$.

For f, g in T, define $fg = f(g(x))$, and $f \geqq g$ to mean $f(x) \geqq g(x)$, $0 \leqq x \leqq 1$.

THEOREM 9. *T is an l-group, non-integrally closed.*

Under composition, T is the well known [8] group of topological transformations of (0, 1) into itself, with identity $e(x) = x$. Verification of the order postulates is trivial. The function $u(x) = \max(f(x), e(x))$, $0 \leqq x \leqq 1$, is in the class T and has the properties of a l.u.b. for $f(x)$ and $e(x)$. This is sufficient for lattice order. The function $b(x)$ in T defined by a broken line of four segments: $b(x) \equiv x$ on (0, 1/4) and on (3/4, 1) with $b(1/2) = 5/8$ is obviously bounded: $b^n \leqq f$ for some f, all n, but b not less than or equal to e.

THEOREM 10. *In an l-group, the relation $(aba^{-1}b^{-1})^n < ab$, a, $b > e$, $n = 1, 2, \cdots$, need not be true. Indeed in the l-group T, there are elements f, $g > e$ for which the elements $(fgf^{-1}g^{-1})^n$, $n = 0, \pm 1, \pm 2, \cdots$, are unbounded by any element.*

For, any two polynomials f, $g>e$ have the property that either $fgf^{-1}g^{-1}$ or $(fgf^{-1}g^{-1})^{-1}=gfg^{-1}f^{-1}$ is greater than x on an open interval $(0, d)$, where d is the first fixed point of the commutator. The powers of this commutator $c(x)$ are then unbounded, since, if $\lim_n c^n(x)=L(x)$, $0\leqq x\leqq d$, then $\lim_n c(c^n(x))=c(L(x))=L(x)$. But since $x\leqq d$, $c^n(x)\leqq c(d)=d$. Hence $L(x)\equiv d$ and no *continuous* function can serve as upper bound.

This settles Problem 2 [1, p. 329].

THEOREM 11. *In an l-group, $a^m>b^m$ for some m, a, $b>e$ does not imply $a\geqq b$.*

In T, we exhibit the broken line $f(x)$ of two segments with vertex at $f(3/8)=3/4$, and the broken line of four segments $g(x)\equiv x$ on $(0, 1/4)$ and on $(3/4, 1)$ with vertex at $g(5/16)=11/16$. One verifies $f^2(x)>g^2(x)$ for all x on open $(0, 1)$, but $f(x)$ not greater than or equal to $g(x)$.

This settles Problem 1 [1, p. 329].

THEOREM 12. *The subgroup of algebraic functions of T is an archimedean ordered o-group, but is not integrally closed.*

If $f(x)\neq x$ and $g(x)$ are algebraic and $f^n(x)<g(x)$, $n=0, \pm 1, \pm 2, \cdots$, then either $f(x)$ or $f^{-1}(x)$ must be greater than or equal to x on an interval $(0, d)$, hence its powers cannot be bounded by any continuous function (see the argument of Theorem 10). It is clear however that an algebraic function similar to the broken line of Theorem 9 can be defined, hence the group is not integrally closed.

6. **The free group, combinatorial order.** Let F be the free group with two generators a, b. We recall that an order may be established in a group G by defining a subsemi-group of "positive" elements K which (a) is closed under multiplication, (b) is closed under conjugation: $gKg^{-1}\subset K$, all $g\in G$, (c) contains e, and no other element along with its inverse. Then $a\geqq b$ is defined to mean $ab^{-1}\in K$ [1]. These conditions are equivalent to (1, 3) of §2. A first attempt to define an o-group on F consists in letting K_0 be the set of all elements expressible as products of conjugates of a and of b, together with the identity e.

THEOREM 13. *The order $x\geqq y$ meaning $xy^{-1}\in K_0$ defines a non-integrally closed o-group on F.*

Properties (a, b, c) are trivial, the latter because the sum of exponents of any K_0 element not e is positive. Moreover, F is a directed set, since $a^n\leqq a^{|n|}$, hence $a^m b^n \cdots \leqq a^{|m|}b^{|n|}\cdots$. Thus both $a^m b^n \cdots$ and $a^{m'}b^{n'}\cdots \leqq (a^{|m|}b^{|n|}\cdots)(a^{|m'|}b^{|n'|}\cdots)$. The remainder of the proof requires the following lemma.

LEMMA 8. *In any l-group, $a^n\geqq e$ for some n implies $a\geqq e$.*

Proof. $(a\wedge e)^n=a^n\wedge a^{n-1}\wedge\cdots\wedge a\wedge e=(a\wedge e)^{n-1}$, $a\wedge e=e\leqq a$.

However, in the group F one has $A = ab^{-2}a^2b^2a^{-1}ba^{-1}b$ not greater than or equal to e and $A^2 = (ab^{-2}a^2b^2a^{-1})(ba^{-1}bab^{-1})(b^{-1}a(ab^2a^{-1}b)a^{-1}b) > e$.

Now if F were integrally closed, by Theorem 1, F would be embeddable in an l-group. Or one may argue directly that $A^{2n} > e$, $A^{2n+1} > A$, hence $A^n > B$ for any $B < e$, A (directed set property), and integral closure would imply $A \geqq e$.

7. **The free group, function order.** Again let F be the free group with generators a, b, and let T be the group of continuous monotone functions of §5. Denote by T^+ the functions $f(x) \geqq x$, $0 \leqq x \leqq 1$, of T. We now introduce an order into F by defining a positive class F^+ consisting of e, and of all formal products $a^m b^n \cdots$ of F for which $f^m g^n \cdots (x) \in T^+$ for all f, g of T^+.

It is clear that F^+ is closed under multiplication and conjugation, inasmuch as T^+ is. Moreover, if a formal product $a^m b^n \cdots \neq e$ were in F^+ along with its inverse, we should have $f^m g^n \cdots (x) \equiv x$, $0 \leqq x \leqq 1$, for all f, g of T^+. We show that this is impossible.

LEMMA 9. *If $a^m b^n \cdots \neq e$, there exist functions f, g of T^+ for which $f^m g^n \cdots (x) \not\equiv x$.*

Assume $P = a^m b^n \cdots$ completely reduced, that is, with no adjacent a, a^{-1} or b, b^{-1}. Define the first function (f, f^{-1}, g, or g^{-1}) from the right at $x = 1/2$ by $f(1/2) = 3/4$, or $g(1/2) = 3/4$ if $P = a^m \cdots a$ or $P = a^m \cdots b$ respectively, and $f(1/4) = 1/2$ or $g(1/4) = 1/2$, that is, $f^{-1}(1/2) = 1/4$, or $g^{-1}(1/2) = 1/4$ in case $P = a^m \cdots a^{-1}$ or $P = a^m \cdots b^{-1}$. Now suppose the functions f, g have been defined on a finite set of points of $(0, 1)$ so that

(1) f, g are monotone increasing and greater than or equal to x,

(2) the values $1/2$, $F_1(1/2), \cdots, F_m(1/2)$ are all distinct, where $F_i(x) = (\cdots f \cdots g \cdots)$ is the product of the i right-most factors of P with f, g substituted for a, b.

Suppose $F_{m+1}(x) = f(F_m(x))$. We may define $F_{m+1}(1/2)$ distinct from $F_m(1/2), \cdots, 1/2$, and f monotone with the previous values, and $f(F_m(1/2)) > F_m(1/2)$ provided only that $F_m(x)$ was not previously a point of definition for f. But previous definition of f occurred in only two ways. Either $F_k = f(F_{k-1})$, $k \leqq m$ (and this is impossible by (2)), or $F_k = f^{-1}(F_{k-1})$, $f(F_k) = F_{k-1}$. Again by (2), $k < m$ is impossible, and if $k = m$, we have $F_{m+1} = fF_m = ff^{-1}F_{m-1}$ contradicting the hypothesis on irreducibility of P.

Finally suppose $F_{m+1}(x) = f^{-1}(F_m(x))$. We can define f^{-1} (and hence f) and $F_{m+1}(1/2)$ as we wish provided only that $F(1/2)$ has not previously occurred as a point of definition of f^{-1}. If previously $F_k = f^{-1}F_{k-1}$, $k \leqq m$, we would contradict (2). If $F_k = fF_{k-1}$ and thus $f^{-1}(F_k) = F_{k-1}$ either $k < m$ is impossible by (2) or $k = m$ means $F_{m+1} = f^{-1}F_m = f^{-1}fF_{m-1}$ contradicting irreducibility of P.

It follows that we have the theorem:

THEOREM 14. *The order $x \geqq y$ meaning $xy^{-1} \in F^+$ defines a non-integrally* *closed o-group on F, for which, however, $x^n \geqq e$ for some n implies $x \geqq e$. The functionally positive elements F^+ properly include the elements of K_0.*

Conditions (1, 3) of §2 are immediate. Since every element of K_0 is a product of conjugates of a and b, clearly $K_0 \subset F^+$. Hence function order defines a directed set since K_0 did. Hence also (2).

Suppose now $P = a^m b^n \cdots \in F$ and $P^\nu \in F^+$ for some ν. Then for every f, g in T^+, $(f^m g^n \cdots)^\nu \in T^+$, and since T is an l-group (Theorem 9), $(f^m g^n \cdots) \in T^+$, hence $P \in F^+$ (Lemma 8). Note that the element A of Theorem 12 (proof) is in F^+, not in K_0.

The fact that F^+-order is non-integrally closed seems deeper, and has curious consequences for function theory which we shall point out later.

COROLLARY 3. *The composite function $fg^{-2}f^2g^2f^{-1}gf^{-1}g(x)$ is in T^+ for all f, g of T^+.*

For $(ab^{-2}a^2b^2a^{-1}ba^{-1}b)^2 \in K_0 \subset F^+$.

LEMMA 10. *In F, let $A = ab^{-2}a^2b^2a^{-1}ba^{-1}b$, and $B = ab^{-2}ab^2a^{-1}$. Under function order, $A^2 > B^2$, but A, B are incomparable.*

For $A^2 \geqq B^2$, see the computation of §6. Indeed, $B^{-2}A^2 \in K_0 \subset F^+$. One easily constructs broken lines f, g of T^+ for which $P = B^{-1}A = ab^{-2}ab^2a^{-1}ba^{-1}b$ under substitution yields a function $P(f, g) = fg^{-2}fg^2f^{-1}gf^{-1}g$ which has $P(f, g)(1/2) > 1/2$, and other broken lines for which $P(f, g)(1/2) < 1/2$. (This is most easily accomplished graphically by a point by point construction of the functions and their inverses from the right end of P, in the first case always assigning f, g values as great as is possible, consistent with monotonicity, in the second case, as small.)

LEMMA 11. *If for some f_0, g_0 of T^+ and x_0 on (0, 1) one has $P(f_0, g_0)(x_0) = f_0g_0^{-2}f_0g_0^2f_0^{-1}g_0f_0^{-1}g_0(x_0) < x_0$, then $P(f_0, g_0)(x_1) > x_1$, and $P(f_0, g_0)(x_2) > x_2$, where $x_1 = B^{-1}(f_0, g_0)(x_0) < x_0 < x_2 = A(f_0, g_0)(x_0)$.*

Let $P(x_0) = B^{-1}A(x_0) < x_0$ (where throughout we understand that f_0, g_0 are substituted for a, b respectively). Then $B^{-1}A(A(x_0)) \geqq B(x_0) > A(x_0) > x_0$. Similarly, $B^{-1}A(B^{-1}(x_0)) \geqq B^{-1}A^{-1}B(x_0) > B^{-1}(x_0)$.

LEMMA 12. *In the group F with function order F^+, one has $P^\nu B > e$ for all ν, with P not greater than or equal to e.*

For all $\nu = 1, 2, \cdots$, and all x on (0, 1), all f, g of T^+, one has $P^\nu B(f, g)(x) \geqq x$. For if not, then there are functions f_0, g_0 of T^+, such that $P^\nu B(f_0, g_0)(x) \geqq x$, all x on (0, 1), $\nu = 1, 2, \cdots, N-1$, but for $\nu = N$ there exists an x_0 on (0, 1) for which $P^N B(f_0, g_0)(x_0) < x_0 \leqq P^{N-1}B(f_0, g_0)(x_0)$. Hence $PB(f_0, g_0)(x_0) < B(f_0, g_0)(x_0)$. By Lemma 11, $PB^{-1}B(f_0, g_0)(x_0) = P(f_0, g_0)(x_0) > B^{-1}B(f_0, g_0)(x_0) = x_0 > P^N B(f_0, g_0)(x_0)$. Hence $x_0 > P^{N-1}B(f_0, g_0)(x_0)$.

This concludes the proof of Theorem 13.

COROLLARY 4. *Given functions f, g of T^+ for which some $f^m g^n \cdots (x) \leqq x$ on certain subintervals of* $(0, 1)$, *there do not in general exist functions f', g' of T^+ arbitrarily close to $e(x) = x$ for which $f'^m g'^n \cdots (x) \leqq x$ on these subintervals.*

For if this were true, we could prove F integrally closed under function order. Either $S \geqq e$, or if not, for some f_0, g_0 of T^+ one has $S(f_0, g_0)(x_0) < x_0$. One would then have $S(f_0', g_0')(x_0) < x_0$ with f_0', g_0' arbitrarily close to $e(x) = x$. But then S could not be bounded; indeed if $S^\nu T(f_0', g_0')(x) \geqq x$, we should have $S^\nu(f_0', g_0')(x_0) > T^{-1}(f_0', g_0')(x_0)$. Since T is fixed, this would be a contradiction.

BIBLIOGRAPHY

1. G. Birkhoff, *Lattice ordered groups*, Ann. of Math. vol. 43 (1942) pp. 298–331.
2. ———, *Lattice theory*, Amer. Math. Soc. Colloquium Publications, vol. 25, 1940.
3. H. Cartan, *Un théorème sur les groupes ordonnes*, Bull. Sci. Math. vol. 63 (1939) pp. 201–205.
4. A. H. Clifford, *Partially ordered abelian groups*, Ann. of Math. vol. 41 (1940) pp. 465–473.
5. C. J. Everett, *Sequence completion of lattice moduls*, Duke Math. J. vol. 11 (1944) pp. 109–119.
6. L. V. Kantorovitch, *Lineare halbgeordnete Räume*, Rec. Math. (Mat. Sbornik) N.S. vol. 44 (1937) pp. 121–165.
7. H. M. MacNeille, *Partially ordered sets*, Trans. Amer. Math. Soc. vol. 42 (1937) pp. 416–460.
8. J. Schreier and S. Ulam, *Eine Bemerkung über die Gruppe der topologischen Abbildungen der Kreislinie auf sich selbst*, Studia Mathematica vol. 5 (1935) pp. 155–159.

UNIVERSITY OF WISCONSIN,
MADISON, WIS.

PROJECTIVE ALGEBRA I.*

C. J. EVERETT and S. ULAM.

1. Projective algebra of subsets of a direct product. Let X and Y be any two sets of points and $[X;Y]$ the class of all pairs $[x;y]$, $x \in X$, $y \in Y$. Fix $[x_0;y_0]$ arbitrarily in $[X;Y]$. Suppose now that $\mathfrak{B}$ is a boolean algebra of subsets A of $[X;Y]$, including as the identity I the entire set $[X;Y]$, and the null set 0. Define for every $A \in \mathfrak{B}$, the set A_x as the class of all pairs $[x;y_0]$ for which there exists a $y \in Y$ such that $[x;y] \in A$. Define A_y similarly. For every $A = [X_1;y_0]$ in $\mathfrak{B}$, $X_1 \subset X$, and $B = [x_0;Y_1] \in \mathfrak{B}$, $Y_1 \subset Y$, define the "product" set $A \square B$ as the class of all pairs $[x_1;y_1]$, $x_1 \in X_1$, $y_1 \in Y_1$. Note that if A or B is 0, $A \square B = 0$. We say that $\mathfrak{B}$ is a *projective algebra of subsets of the product* $[X;Y]$ in case for all A of $\mathfrak{B}$, A_x and A_y also are in $\mathfrak{B}$, and $A \square B$ is in $\mathfrak{B}$ for A, B of the above type. Thus we demand that the boolean algebra $\mathfrak{B}$ be closed under projection and $\square$-product formation. In particular, the boolean algebra of *all* subsets of $[X;Y]$ is such a projective algebra.

It may be remarked that the following is a modest beginning for a study of logic with quantifiers from a boolean point of view, since, for example, the set A_x is essentially (the y_0 being a dummy) the class of all x for which there exists a y such that the proposition $A(x,y)$ is true.

Our object is to discover a set of properties of the above model which, adopted as postulates for an abstract "projective algebra," will permit a representation theorem to the effect that every such abstract algebra is isomorphic, with preservation of union, intersection, complement, x- and y-projection, and $\square$-product, to a projective algebra of subsets of some direct product $[X;Y]$. This is accomplished only for the atomic case in the present paper. The authors hope to study the general problem and related matters subsequently.

We have derived many of the more immediate properties of the abstractly defined projective algebra, and have shown, in particular, that every such algebra is embeddable in a complete ordered projective algebra.

The essential properties of the above model which we use later as axioms are the following.

PROPERTY 1. $(A \cup B)_x = A_x \cup B_x$; $(A \cup B)_y = A_y \cup B_y$.

* Received February 16, 1945.

Reprinted from American Journal of Mathematics, vol. 68 (1946), pp. 77–88. Copyright 1946, The Johns Hopkins Press. For commentary to this paper [30], see p. 686.

For if $[x;y_0] \epsilon (A \cup B)_x$, then for some y, $[x;y]$ is in A or B; hence $[x;y_0]$ is in A_x or B_x. Similarly, $A_x \cup B_x \subset (A \cup B)_x$. If either A or B is 0, Property 1 is trivial. (See Property 3.)

PROPERTY 2. *$I_{xy} = [x_0;y_0] = I_{yx}$, where $[x_0;y_0]$ is an atom in $\mathfrak{B}$, i. e., an element minimal over 0 in $\mathfrak{B}$.*

For $I_x = [X;y_0]$, and I_{xy} is the class of all $[x_0;y]$ such that for some x, $[x;y] \epsilon I_x$; hence $I_{xy} = [x_0;y_0]$.

PROPERTY 3. *$A_x = 0$ if and only if $A = 0$, and similarly for y-projection.*

If $A = 0$, then the set of all $[x;y_0]$ for which there exists an $[x;y]$ in A is empty. If $A \neq 0$, there is at least one point $[x;y] \epsilon A$, and, hence, at least one $[x;y_0]$ in A_x.

PROPERTY 4. *$A_{xx} = A_x$; $A_{yy} = A_y$.*

If $[x_1;y_0] \epsilon A_x$, then $[x_1,y_0] \epsilon A_{xx}$. If $[x_1;y_0] \epsilon A_{xx}$ then for some y, $[x_1;y] \epsilon A_x$; hence $y = y_0$, and $[x_1;y_0] \epsilon A_x$. If $A = 0$, Property 4 is trivial.

PROPERTY 5. *For $0 \neq A = [X_1;y_0] \subset I_x$, $0 \neq B = [x_0;Y_1] \subset I_y$, the set $A \square B = \{[x_1;y_1];\ x_1 \epsilon X_1,\ y_1 \epsilon Y_1\}$ has the property*

$$(A \square B)_x = A, \qquad (A \square B)_y = B,$$

and if $S \epsilon \mathfrak{B}$, $S_x = A$, $S_y = B$, then $S \subset A \square B$.

For $(A \square B)_x$ is the set of all $[x;y_0]$ such that $[x;y] \epsilon [X_1;Y_1]$ for some y; hence $(A \square B)_x = [X_1;y_0] = A$. Suppose now $S_x = [X_1;y_0]$, $S_y = [x_0;Y_1]$ and $[x;y] \epsilon S$. Then $[x;y_0] \epsilon S_x = [X_1;y_0]$ and $[x_0;y] \epsilon S_y = [x_0;Y_1]$. Hence $x \epsilon X_1$, $y \epsilon Y_1$, and $[x;y] \epsilon [X_1;Y_1] = A \square B$. Thus $S \subset A \square B$.

PROPERTY 6. *$I_x \square p_0 = I_x$, and $p_0 \square I_y = I_y$, where $p_0 = [x_0;y_0]$.*

This is immediate from the $\square$-definition.

PROPERTY 7. *For $A_1 = [X_1;y_0]$, $A_2 = [X_2;y_0]$ in $\mathfrak{B}$, one has $(A_1 \cup A_2) \square I_y = (A_1 \square I_y) \cup (A_2 \square I_y)$, and similarly for $B_i = [x_0;Y_i]$ and I_x.*

For $(A_1 \cup A_2) \square I_y = [X_1 \cup X_2;y_0] \square [x_0;Y] = [X_1 \cup X_2;Y]$ and $(A_1 \square I_y) \cup (A_2 \square I_y) = [X_1;Y] \cup [X_2;Y] = [X_1 \cup X_2;Y]$.

If A_1 or A_2 is 0, Property 7 is trivial.

2. Projective algebra. Let $\mathfrak{B}$ be a boolean algebra with unit i, zero 0, $i > 0$, so that for all $a \epsilon \mathfrak{B}$, $0 \leqq a \leqq i$. $\mathfrak{B}$ is said to be a *projective algebra* if two mappings $a \to a_x$ and $a \to a_y$ of $\mathfrak{B}$ into $\mathfrak{B}$ are defined, satisfying the following postulates.

P1. $(a \vee b)_x = a_x \vee b_x$; $(a \vee b)_y = a_y \vee b_y$.

P2. $i_{xy} = p_0 = i_{yx}$ *where p_0 is an atom of $\mathfrak{B}$, that is, an element minimal over 0 in $\mathfrak{B}$.*

P3. $a_x = 0$ *if and only if* $a = 0$, *and* $a_y = 0$ *if and only if* $a = 0$.

P4. $a_{xx} = a_x$; $a_{yy} = a_y$.

P5. For $0 < a \leqq i_x$, $0 < b \leqq i_y$, *there exists an element* $a \square b$ *such that* $(a \square b)_x = a$, $(a \square b)_y = b$, *with the property that* $t \in \mathfrak{B}$, $t_x = a$, $t_y = b$ *implies* $t \leqq a \square b$.

P6. $i_x \square p_0 = i_x$; $p_0 \square i_y = i_y$.

P7. $0 < a_1, a_2 \leqq i_x$ *implies* $(a_1 \vee a_2) \square i_y = (a_1 \square i_y) \vee (a_2 \square i_y)$; *and* $0 < b_1, b_2 \leqq i_y$ *implies* $i_x \square (b_1 \vee b_2) = (i_x \square b_1) \vee (i_x \square b_2)$.

We prove a number of immediate consequences of these postulates.

C1. $a \leqq b$ *implies* $a_x \leqq b_x$ and $a_y \leqq b_y$.

From $a \vee b = b$ one has $(P1)$ $(a \vee b)_x = b_x = a_x \vee b_x$.

C2. For all $a \in \mathfrak{B}$, $a_x \leqq i_x$, $a_y \leqq i_y$.

By $C1$, $a \leqq i$; hence $a_x \leqq i_x$.

C3. $i_x \wedge (a_x)' \leqq (a')_x$ *and similarly for y-projection.*

For $i = a \vee a'$, $i_x = a_x \vee (a')_x$ $(P1)$, $i_x \wedge (a_x)' = (a_x \vee (a')_x) \wedge (a_x)'$ $= 0 \vee ((a')_x \wedge (a_x)') \leqq (a')_x$.

C4. $(a \wedge b)_x \leqq a_x \wedge b_x$; $(a \wedge b)_y \leqq a_y \wedge b_y$.

For $(a \wedge b) \leqq a, b$, hence $(C1)$ $(a \wedge b)_x \leqq a_x, b_x$.

C5. $a > 0$ *implies* $a_{xy} = p_0 = a_{yx}$.

Since $a \leqq i$, $a_{xy} \leqq i_{xy} = p_0$ $(C1, P2)$. Since p_0 is an atom, $a_{xy} = p_0$ for $a_{xy} = 0$ is impossible by $P3$.

C6. $(p_0)_x = p_0 = (p_0)_y \leqq i_x \wedge i_y$.

For $p_0 = i_{xy}$ and hence $(p_0)_y = i_{xyy} = i_{xy} = p_0$, $(P2, P4)$. The final inequality follows from $C2$.

In connection with the $\square$-product, it is convenient to make two definitions.

D1. For $a \leqq i_x$, $b \leqq i_y$, define $a \square 0$, $0 \square b$, and $0 \square 0$ all to be 0.

It is seen that for all of these the x- and y-projections are 0.

D2. For all $a \in \mathfrak{B}$, define the "closure" a^* of a to be $a_x \square a_y$.

C7. $a \square b = 0$ *if and only if* $a = 0$ *or* $b = 0$.

For if $a > 0$, $b > 0$, by $P5$, $(a \square b)_x = a > 0$, hence $a \square b > 0$ by $P3$.

$C8$. *The following are equivalent*: (a) $a \leqq i_x$; (b) *there is a* $t \in \mathfrak{B}$ *such that* $t_x = a$; (c) $a_x = a$; *and similarly for y-projections.*

(a) implies (b), for if $a = 0$, $a_x = a$ by $P3$, whereas for $a > 0$, $(a \square i_y)_x = a$ by $P5$. Next, (b) implies (c), since $t_x = a$ yields $t_{xx} = a_x = t_x = a$ ($P4$). Finally, (c) implies (a) by $C2$.

$C9$. $0 < a_1 \leqq a_2 \leqq i_x$, $0 < b_1 \leqq b_2 \leqq i_y$ *if and only if* $0 < a_1 \square b_1 \leqq a_2 \square b_2$.

If $0 < a_1 \square b_1 \leqq a_2 \square b_2$, then ($C1$, $P5$) $0 < a_1 \leqq a_2$, $0 < b_1 \leqq b_2$. Conversely, assume the first inequalities and let $u = (a_1 \square b_1) \vee (a_2 \square b_2)$. Then $u_x = a_1 \vee a_2 = a_2$, $u_y = b_1 \vee b_2 = b_2$ ($P1$, $P5$); hence $u \leqq a_2 \square b_2$, by $P5$, and $a_1 \square b_1 \leqq a_2 \square b_2$.

$C10$. *The* (*) *operation has the following closure properties*:

$$0^* = 0;\quad a^* \geqq a;\quad a^{**} = a^*;\quad a \geqq b \text{ implies } a^* \geqq b^*;\quad i^* = i.$$

First, $0^* = 0_x \square 0_y = 0 \square 0 = 0$. Second, $a^* = a_x \square a_y \geqq a$ by the maximality property of $\square$-product in $P5$, for $a > 0$. For $a = 0$, this becomes trivial. Third, $a^{**} = (a_x \square a_y)_x \square (a_x \square a_y)_y = a_x \square a_y = a^*$, for $a \geqq 0$. Fourth, $a \leqq b$ implies $a^* \leqq b^*$ by $C1$, $C9$. Fifth, $i^* \geqq i$, hence $i^* = i$.

$C11$. $a_1, a_2 \leqq i_x$, $b_1, b_2 \leqq i_y$ *imply* $(a_1 \wedge a_2) \square (b_1 \wedge b_2) = (a_1 \square b_1) \wedge (a_2 \square b_2)$.

If any one of the a_i, b_i is 0, $C11$ is trivial. Otherwise, $a_1 \wedge a_2 \leqq a_1, a_2$, $b_1 \wedge b_2 \leqq b_1, b_2$, hence $(a_1 \wedge a_2) \square (b_1 \wedge b_2) \leqq a_1 \square b_1$, $a_2 \square b_2$ by $C9$. Hence the left member is contained in the right. Let t denote the right member. We have $t_x \leqq a_1 \wedge a_2$, $t_y \leqq b_1 \wedge b_2$ ($C4$, $P5$); hence $t \leqq t^* = t_x \square t_y \leqq (a_1 \wedge a_2) \square (b_1 \wedge b_2)$ by $C10$, $C9$.

$C12$. $a \leqq i_x$, $b \leqq i_y$ *implies* $a \square b = (a \square i_y) \wedge (i_x \square b)$.

In $C11$, let $a_1 = a$, $a_2 = i_x$, $b_1 = i_y$, $b_2 = b$.

$C13$. $t \leqq a \square b$ *implies* $t_x \leqq a$, $t_y \leqq b$.

This requires only $C1$ and $P5$ for $t > 0$. It is trivial for $t = 0$.

$C14$. $((a_1 \square b_1) \vee (a_2 \square b_2))^* = (a_1 \vee a_2) \square (b_1 \vee b_2)$ *for* a_i, b_i *not* 0. We require $P1$, $P5$.

$C15$. $a \leqq i_x$ *implies* $(a \square i_y) \wedge i_x = a$.

For $a \leqq i_x$, $a \leqq a^* = a \square p_0 < a \square i_y$. Hence $a \leqq (a \square i_y) \wedge i_x$. Now let d be this intersection. Then $d \leqq i_x$, $d_x = d \leqq a \wedge i_x = a$.

$C16$. *All* a, b *satisfying* $0 < a \leqq i_x$, $0 < b \leqq i_y$ *are closed, i. e.*, $a \square p_0 = a$, $p_0 \square b = b$.

$a^* = a \square p_0$ since $a_x = a$ and $a_y = a_{xy} = p_0$. By $C12$, $a \square p_0 = (a \square i_y) \wedge (i_x \square p_0) = (a \square i_y) \wedge i_x$ by $P6$. Then $a \square p_0 = a$ by $C15$.

$C17$. $(p_0)^* = p_0 = p_0 \square p_0 = (p_0 \square i_y) \wedge (i_x \square p_0) = i_x \wedge i_y$.

This follows from $C6$, $C16$, $C12$, $P6$.

$C18$. *If* $t_y = p_0$ *then* $t \leqq i_x$. *If* $t_x = p_0$ *then* $t \leqq i_y$.

For $t \leqq t^* \leqq i_x \square p_0 = i_x$ by $C10$, $C9$, $P6$.

$C19$. For $a_1, a_2 \leqq i_x$, $b \leqq i_y$, *one has* $(a_1 \vee a_2) \square b = (a_1 \square b) \vee (a_2 \square b)$. *Similarly for* $b_1, b_2 \leqq i_y$, $a \leqq i_x$.

By $C12$, $P7$, $(a_1 \vee a_2) \square b = ((a_1 \vee a_2) \square i_y) \wedge (i_x \square b) = ((a_1 \square i_y) \vee (a_2 \square i_y)) \wedge (i_x \square b) = ((a_1 \square i_y) \wedge (i_x \square b)) \vee ((a_2 \square i_y) \wedge (i_x \square b)) = (a_1 \square b) \vee (a_2 \square b)$.

$C20$. *For* $a_1, a_2 \leqq i_x$, $b_1, b_2 \leqq i_y$ *one has* $(a_1 \vee a_2) \square (b_1 \vee b_2) = (a_1 \square b_1) \vee (a_1 \square b_2) \vee (a_2 \square b_1) \vee (a_2 \square b_2)$.

This requires two applications of $C19$.

$C21$. $(a \square i_y)' = (a' \wedge i_x) \square i_y$; $(i_x \square b)' = i_x \square (b' \wedge i_y)$.

For $(a \square i_y) \vee ((a' \wedge i_x) \square i_y) = (a \vee (a' \wedge i_x)) \square i_y = i_x \square i_y = i$, by $P7$. and $(a \square i_y) \wedge ((a' \wedge i_x) \square i_y) = (a \wedge (a' \wedge i_x)) \square i_y = 0 \square i_y = 0$, by $C11$. Since complementation is unique in a boolean algebra, $C21$ follows.

$C22$. $(a \square b)' = ((a' \wedge i_x) \square i_y) \vee (i_x \square (b' \wedge i_y))$.

This follows from $C12$ and $C21$.

$C23$. *For* $c = c^*$, $c_y < i_y$, *one has* $(c')_x = i_x$.

If $c = 0$, this is trivial. If $0 < c = c_x \square c_y$, then $c' = (((c_x)' \wedge i_x) \square i_y) \vee (i_x \square ((c_y)' \wedge i_y))$ by $C22$, and $(c')_x = (((c_x)' \wedge i_x) \square i_y)_x \vee i_x = i_x$ by $P1$, $P5$. We need to know that $(c_y)' \wedge i_y \neq 0$ but this is so since $c_y < i_y$.

$C24$. $a \leqq i_x$, $p_0 < i_y$ *implies* $(a')_x = i_x$.

For by $C16$, $a = a^*$, and $a_x = a$, $a_{xy} = p_0 = a_y$. $C24$ now follows from $C23$. (For $a = 0$, $C24$ is trivial).

$C25$. *If* $p_0 < i_y$ *then* $((i_x)')_x = i_x$, *and similarly for* i_y.

Let $a = i_x$ in $C24$.

$C26$. $p_0 = i_y$ *if and only if* $i = i_x$.

If $p_0 = i_y$, then $i = i_x \square i_y = i_x \square p_0 = i_x$ ($C10$, $P6$). If $i = i_x$, then $i_y = i_{xy} = p_0$.

$C27$. $((i_x)')_y = p'_0 \wedge i_y$.

If $p_0 = i_y$, by $C26$, $(i_x)')_y = ((i)')_y = 0_y = 0 = (i_y)' \wedge i_y = p'_0 \wedge i_y$. If $p_0 < i_y$, one has from $i_x = i_x \square p_0$ ($P6$) that $(i_x)' = (i_x \square p_0)' = i_x \square (p'_0 \wedge i_y)$ where $p'_0 \wedge i_y \neq 0$ ($C21$). Hence $((i_x)')_y = p'_0 \wedge i_y$ by $P5$.

$C28$. *The class of all x-projections is an ideal in* $\mathfrak{B}$ *and the correspondence* $a_x \rightarrow a_x \square i_y$ *is an isomorphism.*

The set of all a_x, $a \in \mathfrak{B}$ is exactly the set of all $a \leqq i_x$, by $C8$, and hence is an ideal, since the union of any two $a_1, a_2 \leqq i_x$ is $\leqq i_x$, as is $a_1 \wedge c$ for all

$c \in \mathfrak{B}$. Moreover the correspondence above preserves union ($P7$), intersection ($C11$), and complement ($C21$).

$C29$. *For* $c \in \mathfrak{B}$, $0 < a \leqq c_x$ *implies* $(a \square i_y) \wedge c > 0$.

Suppose $(a \square i_y) \wedge c = 0$. Then $c \leqq (a \square i_y)' = (a' \wedge i_x) \square i_y$. If now $a' \wedge i_x = 0$, $a = i_x$ and $(a \square i_y) \wedge c = i \wedge c = c = 0$. But then $c_x = 0$ and $a = 0$, which is a contradiction. If $a' \wedge i_x > 0$ then $a \leqq c_x \leqq a' \wedge i_x \leqq a'$ whence $a = 0$, also a contradiction.

$C30$. *If* p *is an atom in* $\mathfrak{B}$, *so are* p_x *and* p_y.

Suppose $0 < a < p_x$. By $C29$, $0 < (a \square i_y) \wedge p \leqq p$. Hence $(a \square i_y) \wedge p = p$ and $p \leqq a \square i_y$, $p_x \leqq a$, a contradiction.

$C31$. If $0 < c$, and $0 < p \leqq c_x$, where p is a point, then $p = ((p \square i_y) \wedge c)_x$. Moreover, if q is a point in $(p \square i_y) \wedge c$, then $q_x = p$.

This follows from $C29$, $P3$, and $C4$, thus: $0 < ((p \square i_y) \wedge c)_x \leqq p \wedge c_x = p$.

3. Completion of projective algebras. We shall prove the following general theorem:

THEOREM 1. *If* $\mathfrak{B}$ *is a projective algebra, then* $\mathfrak{B}$ *is embeddable with preservation of (unrestricted) union, intersection, complement, x- and y-projections, and* $\square$*-products, in a projective algebra* $\mathfrak{C}$ *which is complete-ordered in the sense that every set of elements of* $\mathfrak{C}$ *has a l. u. b. and a g. l. b.*

Let $\mathfrak{B}$ be a projective algebra with elements $0 \leqq a \leqq i$. We recall that if $U(A)$, $L(A)$ are the sets of *all* upper and *all* lower bounds, respectively, of *all* elements of $A \subset \mathfrak{B}$, then $LU(A)$ is a closure operation.[1] Specifically, $A \subset LU(A)$, $LULU(A) = LU(A)$, (indeed one has $LUL(A) = L(A)$ and $ULU(A) = U(A)$), and $A \subset B$ implies $LU(A) \subset LU(B)$. The class $\mathfrak{C}$ of all $LU(A)$, $A \subset \mathfrak{B}$, is a complete ordered lattice which embeds $\mathfrak{B}$ under the correspondence $a \rightarrow LU(a) = L(a)$, $a \in \mathfrak{B}$, with preservation of unrestricted union and intersection. For any class of elements $LU(A_\alpha)$ of $\mathfrak{C}$, set intersection $\cap LU(A_\alpha)$ is effective as g. l. b. $\wedge LU(A_\alpha)$, and *closure* of set union $LU(\cup LU(A_\alpha))$ is effective as l. u. b. $\vee LU(A_\alpha)$. Note that $LU(i) = L(i) = \mathfrak{B}$ contains all $LU(A) \in \mathfrak{C}$ and is thus the identity of $\mathfrak{C}$. Also $LU(0) = L(0) = (0)$ is the zero of $\mathfrak{C}$. It is emphasized that while the elements of $\mathfrak{C}$ are sets, and the order is that of set inclusion, the boolean algebra $\mathfrak{C}$ is not the ordinary one of subsets of a set, in particular, union is *not* set union, and the zero of $\mathfrak{C}$ is *not* the null set.

---

[1] G. Birkhoff, "Lattice theory," *American Mathematical Society Colloquium Publications*, vol. 25 (1940), p. 25; H. MacNeille, "Partially ordered sets," *Transactions of the American Mathematical Society*, vol. 42 (1937), pp. 416-460.

MacNeille [2] has shown that in $\mathfrak{C}$, $LU(A) \vee L(\text{all } a', a \in A) = LU(i)$, and $LU(A) \wedge L(\text{all } a') = LU(0)$, and that the correspondence $LU(A) \rightarrow L(\text{all } a')$ is a dual isomorphism of $\mathfrak{C}$ onto all of itself. From this it follows that $\mathfrak{C}$ is distributive and hence a boolean algebra, with $L(\text{all } a')$ effective as complement of $LU(A)$. The correspondence $a \rightarrow LU(a)$ obviously preserves complements, since $(LU(a))' = LU(a')$.

In this section we shall understand that if $A \subset \mathfrak{B}$, A_x is to mean the set of all a_x, $a \in A$. In $\mathfrak{C}$ define $(LU(A))_x = LU(A_x)$. We note that projection is well defined, i. e., $LU(A) = LU(B)$ implies $LU(A_x) = LU(B_x)$. It is sufficient to prove $A_x \subset LU(B_x)$. If $B = (0)$ we have $A \subset LU(A) = LU(0) = L(0) = (0)$; hence $A = (0)$, and $A_x = (0_x) = (0) \subset LU(B_x)$. Now let B contain at least one $b > 0$, and let $u \geqq B_x$. In particular, $u \geqq b_x > 0$, hence $u \wedge i_x \geqq b_x > 0$. Then $(u \wedge i_x) \square i_y) \geqq b_x \square b_y = b^* \geqq b$, all $b \in B$. But $U(A) = U(B)$, hence $(u \wedge i_x) \square i_y \geqq A$, and $u \geqq u \wedge i_x \geqq A_x$. Thus $A_x \subset LU(B_x)$. Similarly $B_x \subset LU(A_x)$. Moreover the correspondence $a \rightarrow LU(a)$ preserves projection, since $LU(a_x) = (LU(a))_x$.

Finally, in $\mathfrak{C}$ we define, for $(0) \neq LU(A) \subset LU(i_x)$, $(0) \neq LU(B) \subset LU(i_y)$, the direct product $LU(A) \square LU(B)$ as the (closed) set:

$$L(\text{all } u \square v;\ u \leqq i_x,\ v \leqq i_y,\ u \square v \geqq \text{all } \dot{a} \square \dot{b},\ \dot{a} \in LU(A),\ \dot{b} \in LU(B)).$$

The form of this definition is convenient in that it is clearly an element of $\mathfrak{C}$, since for any X, $L(X) = LUL(X)$; it is cumbersome, however, and we prove the following equivalences.

Lemma 1. *For $LU(A) \subset LU(i_x)$, $LU(A) = L_xU_x(A)$ where the sub-x means that the operator is restricted to the elements $\leqq i_x$, that is to the elements of the sub-boolean algebra of all x-projections.*

Clearly $U(A) \supset U_x(A)$; hence $LU(A) \subset LU_x(A)$. But $LU_x(A) = L_xU_x(A)$. For $LU_x(A) \supset L_xU_x(A)$, and if $l \leqq U_x(A)$, then $l \leqq i_x$ since $i_x \geqq (A)$; (recall that $LU(i_x) = L(i_x) \supset LU(A) \supset A$). Hence $LU(A) \subset L_xU_x(A)$. Now let $l_x \leqq U_x(A)$, $u \geqq A$; then i_x, $u \geqq u \wedge i_x \geqq A$, hence $l_x \leqq u \wedge i_x \leqq u$.

Lemma 2. $LU(A) \square LU(B) = L(\text{all } u \square v;\ u \in U_x(A),\ v \in (U_y(B)))$.

For in the $\square$-definition u ranges over all $U_xL_xU_x(A) = U_x(A)$.

Lemma 3. $LU(A) \square LU(B) = (\text{all } l;\ l \leqq \dot{a} \square \dot{b} \text{ for some } \dot{a} \in LU(A),\ \dot{b} \in LU(B))$.

For let $l \in L(u \square v;\ u \in U_x(A),\ v \in U_y(B))$. Then $l \leqq u \square v$, $l_x \leqq$ all u,

[2] H. MacNeille, *loc. cit.*

$l_y \leqq$ all v, $l_x \in L_xU_x(A)$, $l_y \in L_yU_y(B)$; hence $l_x = \dot{a}$, $l_y = \dot{b}$, and $l \leqq (l)^* = \dot{a} \Box \dot{b}$. Conversely, let $l \leqq \dot{a} \Box \dot{b}$, for $\dot{a} \in LU(A) = L_xU_x(A)$, $\dot{b} \in LU(B) = L_yU_y(B)$. Since $U_x(A) = U_xL_xU_x(A)$, $U_yL_yU_y(B)$, $l \leqq \dot{a} \Box \dot{b} \leqq$ all $u \Box v$. Thus $l \in L($all $u \Box v)$.

It is now clear that $a \to L(U(a))$ preserves $\Box$-product, for $LU(a) \Box LU(b) = LU(a \Box b) = L(a \Box b)$. For $LU(a) \Box LU(b) = (l; l \leqq$ some $\dot{a} \Box \dot{b}$, $\dot{a} \in LU(a)$, $\dot{b} \in LU(b)) = (l; l \leqq$ some $\dot{a} \Box \dot{b}$, $\dot{a} \leqq a$, $\dot{b} \leqq b)$.

We proceed now to verify $P1$-7 in $\mathfrak{C}$.

$P1$. $(LU(A) \dot{\vee} LU(B))_x = (LU(A))_x \dot{\vee} (LU(B))_x$.

If $A = (0)$ or $B = (0)$, $P1$ is trivial. Using the definitions, we reduce $P1$ to $LU((LU(A) \cup LU(B))_x) = LU(LU(A_x) \cup LU(B))$. Operating on both sides with U and recalling that $ULU(A) = U(A)$, we have $U((LU(A) \cup LU(B))_x) = U(A_x) \cap U(B_x)$. Since $A \subset LU(A)$, $B \subset LU(B)$, A_x and B_x are in $(LU(A) \cup LU(B))_x$ and the left member is contained in the right. Now let $u \geqq A_x$, B_x, and $l \in LU(A)$. Then $u \triangle i_x \geqq A_x$, B_x, and $(u \wedge i_x) \Box i_y \geqq a_x \Box a_y \geqq a$ for all $a \in A$. Since at least one $a > 0$, $u \wedge i_x > 0$. Thus $l \leqq (u \wedge i_x) \Box i_y$ and $l_x \leqq (u \wedge i_x) \leqq u$. Hence the left member is in the right.

$P2$. $(LU(i))_{xy} = (LU(i))_{yx}$ is a point in $\mathfrak{C}$; indeed both equal $LU(p_0) = (0, p_0)$.

For $(LU(i))_{xy} = LU(i_{xy}) = LU(p_0) = L(p_0) = (0, p_0)$, which is minimal over $LU(0)$.

$P3$. $(LU(A))_x = LU(0)$ if, and only if, $LU(A) = LU(0)$.

If $LU(A) = LU(0)$ clearly $(LU(A))_x = LU(0_x) = LU(0)$. Conversely if $LU(A_x) = (0) \supset A_x$, then $a_x = 0$ and $a = 0$ for all $a \in A$. Then $LU(A) = LU(0)$.

$P4$. $(LU(A))_{xx} = (LU(A))_x$.

The proof is trivial.

$P5$. For $(0) \neq LU(A) \subset LU(i_x)$ and $(0) \neq LU(B) \subset LU(i_y)$, the direct product has x-projection $LU(A)$, y-projection $LU(B)$, and is maximal in $\mathfrak{C}$ with these properties.

First, $LU((L(u \Box v; u \Box v \geqq$ all $\dot{a} \Box \dot{b}$, $\dot{a} \in LU(A)$, $\dot{b} \in LU(B))_x) = LU(A)$. Let $l \leqq$ all $u \Box v$, $u_0 \in U(A) = ULU(A)$, hence $u_0 \geqq$ all $\dot{a}$, and $u_0 \wedge i_x \geqq$ all $\dot{a}$, hence $u_0 \wedge i_x > 0$ since $LU(A) \neq (0)$. Then $(u_0 \wedge i_x) \Box i_y \geqq$ all $\dot{a} \Box \dot{b}$, and $l \leqq (u_0 \wedge i_x) \Box i_y$. Thus $l_x \leqq u_0 \wedge i_x \leqq u_0$. Hence the left side is in the right. Now let $0 < a_0 \in A$, $0 < b_0 \in B$; then $a_0 \Box b_0 \geqq$ all $u \Box v$, $a_0 = (a_0 \Box b_0)_x$, and $A \subset (L(u \Box v))_x$.

Second, suppose $(LU(S))_x = LU(A)$, $(LU(S))_y = LU(B)$, and let $s \epsilon S$. Then $s_x \epsilon LU(A)$, $s_y \epsilon LU(B)$, and $u \square v \geqq s_x \square s_y \geqq s$. Thus $S \subset L(u \square v) = LU(A) \square LU(B)$.

$P6$. $LU(i_x) \square LU(p_0) = LU(i_x)$.

This follows since $\square$-product is preserved under $a \rightarrow LU(a)$.

Before proving $P7$, we note the following:

LEMMA 4. *If* $S \subset LU(i_x)$, $u_1 \epsilon LU(S)$, $u_0 \geqq S \square i_y$, *then* $u_1 \square i_y \leqq u_0$.

First, if $i_x \epsilon S$, then $u_0 \geqq i_x \square i_y = i$, and the conclusion is trivial. Suppose $i_x \notin S$. We prove $u'_0 \leqq (u_1 \square i_y)' = (u'_1 \wedge i_x) \square i_y$. Since $u'_0 \leqq (u'_0)_x \square (u'_0)_y$, it is sufficient to prove $(u'_0)_x \leqq u'_1 \wedge i_x$, i. e., $(u'_0)_x \leqq u'_1$. But it is given that $u'_0 \leqq (S' \wedge i_x) \square i_y$; hence $(u'_0)_x \leqq S' \wedge i_x \leqq S'$. For if any $s' \wedge i_x = 0$, $s = i_x \epsilon S$ contrary to our present hypothesis. But it is given that $u_1 \leqq U(S)$; hence $u'_1 \geqq (U(S))' = L(S')$. Thus $u'_1 \geqq (u'_0)_x$.

$P7$. $(LU(A) \vee LU(B)) \square LU(i_y)$

$$= (LU(A) \square LU(i_y)) \vee (LU(B) \square LU(i_y)).$$

If either $A = (0)$ or $B = (0)$, $P8$ is trivially satisfied. Moreover, in any case, the right member is contained in the left by $C9$ which follows from $P1$-6.

We remark that if X and Y are any two subsets of $\mathfrak{B}$, then $LU(X \cup Y) = LU(\text{all } x \vee y;\ x \epsilon X, y \epsilon Y)$.

Now note the equalities: $(LU(A) \vee LU(B)) \square LU(i_y) = LU(LU(A) \cup LU(B)) \square L(i_y) = LU(\text{all } \dot{a} \vee \dot{b};\ \dot{a} \epsilon LU(A), \dot{b} \epsilon LU(B)) \square L(i_y) = (\text{all } l \leqq \text{some } u \square v;\ u \epsilon LU(\dot{a} \vee \dot{b}), v \epsilon L(i_y)) = (\text{all } l \leqq \text{some } u \square i_y;\ u \epsilon LU(\text{all } \dot{a} \vee \dot{b}))$.

We remark that $U(\text{all } l \leqq \text{some } \dot{a} \square i_y) = U(\text{all } \dot{a} \square i_y)$, and then note the equalities: $LU(A) \square LU(i_y) \vee (LU(B) \square LU(i_y)) = LU((\text{all } l \leqq \text{some } \dot{a} \square i_y) \cup (\text{all } l \leqq \text{some } \dot{b} \square i_y)) = L(U(l \leqq \text{some } \dot{a} \square i_y) \cap U(l \leqq \text{some } \dot{b} \square i_y)) = LU((\text{all } \dot{a} \square i_y) \cup (\text{all } \dot{b} \square i_y)) = LU(\text{all } a \square i_y \vee \dot{b} \square i_y) = LU(\text{all } (\dot{a} \vee \dot{b}) \square i_y)$.

We must therefore prove that $(\text{all } l \leqq \text{some } u \square i_y;\ u \epsilon LU(\text{all } \dot{a} \vee \dot{b})) \subset LU(\text{all } (\dot{a} \vee \dot{b}) \square i_y)$. Let $l \leqq \text{some } u_1 \square i_y$, $u_1 \epsilon LU(\text{all } \dot{a} \vee \dot{b})$, and $u_0 \geqq \text{all } (\dot{a} \vee \dot{b}) \square i_y$. Now use Lemma 4 with $S = (\text{all } \dot{a} \vee \dot{b})$. It follows that $u_1 \square i_y \leqq u_0$. Since $l \leqq u_1 \square i_y \leqq u_0$, the theorem is proved.

4. Representation theory. We shall prove a representation theorem for the case where $\mathfrak{B}$ is a projective algebra of *all* subsets of a set i of points p. We prove first two preliminary lemmas.

LEMMA 1. *If N is any cardinal number, there exists a (commutative) group containing exactly N elements.*

For a finite N we may use the cyclic group of order N. Now let N be infinite, and let V be a vector space of N basis elements over the galois field $GF(2)$. The elements of V may be considered as functions on a set of power N to the set $(0,1)$ where addition is component-wise, mod 2. Denote by W the subgroup of all elements of V with only a finite number of 1-components, together with the zero element. W consists of the mutually exclusive subsets W_j of elements containing exactly j 1-components. Thus $W = \sum_0^\infty W_j$. The power of W is therefore $1 + \sum_1^\infty N^j = N$.

LEMMA 2. *If i is any (additive) group of elements p, the correspondence $p \to p + q$, q fixed, p arbitrary in i, defines a permutation π_q (one-one transformation) of i onto all of i, and the correspondence $q \to \pi_q$ is an isomorphism of i onto a group Π of permutations of the elements of i. In particular* (a) *the powers of Π and of i are equal.* (b) *for every $p, q \in i$ there exists a $\pi_s \in \Pi$ such that $\pi_s(p) = q$,* (c) *if $\pi_p(s_0) = \pi_q(s_0)$ for some $s_0 \in i$, then $\pi_p = \pi_q$.*

This is only a statement of the familiar Cayley representation of a group by permutations.

Now let i be a projective algebra of *all* subsets of a set i. We recall that $p_0 = i_x \wedge i_y$ so that all points of i are in one of the following disjoint classes:

$$\text{(a) } p_0, \quad \text{(b) } p \in i_x,\ p \neq p_0, \quad \text{(c) } p \in i_y,\ p \neq p_0, \quad \text{(d) } p \notin i_x \vee i_y.$$

Moreover it is clear (C. 30) that the points in i_x are precisely all p_x where p ranges over i; similarly for i_y. Indeed, from C. 31, it follows that a_x is exactly the set of all p_x with p in a.

Let X be the class consisting of p_0 and of all pairs (p_x, q), p_x ranging over all points $\neq p_0$ in i_x, q over all points in i. Let Y be the class consisting of p_0 and all pairs (p_y, q), p_y ranging over all points $\neq p_0$ in i_y, q over all points in i.

Define the relation $p \sim q$ on points p, q of i to mean $p_x = q_x$, $p_y = q_y$. This is an equivalence relation splitting i into mutually exclusive sets (p) of points. In each class we distinguish a special representative $\bar{p}$ (Zermelo). The points in any class $(\bar{p})$ are thus all points of i of equal closure $\bar{p}_x \square \bar{p}_y$.

We define a group on the points of i, and set up the correspondence $q \to \pi_q$ of Lemma 2, and set up the following correspondence on points of i to subsets of $[X; Y]$.

(a) $p_0 \to [p_0; p_0]$.

(b) for $p \epsilon i_x$, $p \neq p_0$, $p \to \{[(p_x, q); p_0], q \epsilon i\}$.

(c) for $p \epsilon i_y$, $p \neq p_0$, $p \to \{[p_0; (p_y, q)], q \epsilon i\}$.

(d) for $p \notin i_x \vee i_y$, and $p \epsilon (\bar{p})$, if

$$p \neq \bar{p},\ p \to \{[(\bar{p}_x, q); (\bar{p}_y, \pi_p(q))], q \epsilon i\}.$$

But

$$\bar{p} \to \{[(\bar{p}_x, q); (\bar{p}_y, \pi_{\bar{p}}(q))], q \epsilon i\}$$

and

$\{[(\bar{p}_x, q); (\bar{p}_y, \pi_t(q))], q \epsilon i, t$ an arbitrary point in the i-complement of $(\bar{p})\}$.

For any $a \leqq i$, we let $a \to A$, where A is the set of all images of all points of a under (a-d).

LEMMA 3. *If* $s \to [x; y]$ *and* $t \to [x; y]$ *under* (a-d) *then* $s = t$.

Different points of the type (b) map into different $[x; y]$, since the point itself appears as one of the coordinates, and similarly for type (c). If $[(\bar{p}_x, q); (\bar{p}_y, \pi_s(q))]$ is the map of two different points, they would be points with the same closure, and hence in the same $(\bar{p})$. But the pair of elements q, $\pi_s(q)$ can arise only from a unique s (Lemma 2c), and the two points would then both be p or both be $\bar{p}$, according to (d).

LEMMA 4. *Every point* $[x; y]$ *of* $[X; Y]$ *occurs as an image of some point* p *of* i.

For the points of $[X; Y]$ fall into the following classes:

(a) $[x; y] = [p_0; p_0]$ which is the map of p.

(b) $[x; y] = [(p_x, q); p_0]$ which is the map of $p_x \leqq i_x$, $p_x \neq p_0$.

(c) $[x; y] = [p_0; (p_y, q)]$ which is the map of $p_y \leqq i_y$, $p_y \neq p_0$.

(d) $[x; y] = [(p_x, q); (r_y, s)]$ $p_x \neq p_0$, $r_y \neq p_0$. In i, form the product $p_x \square r_y \neq 0$. There exists at least one point $t \leqq p_x \square r_y$ and $0 < t_x \leqq p_x$, $0 < t_y \leqq r_y$. Since $t > 0$, $t_x = p_x$, $t_y = p_y$. Hence $[x; y] = [(\bar{t}_x, q); (\bar{t}_y, s)]$, $t_x \neq p_0$, $t_y \neq p_0$. But there exists an element $n \epsilon i$ (Lemma 2b) such that $\pi_n(q) = s$. Thus $[x; y] = [(\bar{t}_x, q); (\bar{t}_y, \pi_n(q))]$ is the map of some point in $(\bar{t})$.

THEOREM 2. *A projective algebra defined on all subsets of a set* i *of points* p *is isomorphic to a projective algebra of certain subsets of a direct product* $[X; Y]$, *with preservation of unit* $(i \to [X; Y])$, *union, intersection, complement, projection, and* $\square$*-product.*

The correspondence is that already defined. If now $a \to A$, $b \to B$,

$a \vee b \to S$, $a \wedge b \to P$, one readily verifies that $S = A \cup B$, $P = A \cap B$, since each member includes the other. For the latter equality one uses Lemma 3.

Now let $a \to A$, $a' \to B$. Then $B =$ complement of A in $[X; Y]$. Suppose $[x; y] \in B$, i. e., $s \to [x; y]$ where $s \in a'$. Then $[x; y] \in A'$, for if $[x; y] \in A$ one has $t \to [x; y]$, $t \in a$. By Lemma 3, $s = t$. Hence $B \subset A'$. Now let $[x; y] \in A'$. By Lemma 4, $s \to [x; y]$, for some $s \in i$. Hence $s \in a'$, and $[x; y] \in B$.

Next let $a \to A$, $a_x \to B$. Then $(A)_x = B$. Let $[x; p_0] \in A_x$. For some y, $[x; y] \in A$, hence $p \to [x; y]$ for $p \in a$. By examining the correspondences (a-d) one sees that $p \to [x; y]$ implies $p_x \to [x; p_0]$ in all cases. For example, if p is of the type (d), $p_x \in i_x$, and $p_x \neq p_0$ (for if so, $p \leqq i_y$, cf. (C. 18)). Thus $p_x \to [x; p_0]$ under (b). Hence in general $[x; p_0] \in B$. Similarly, if $a \to A$, $a_y \to C$, then $(A)_y = C$.

Finally, let $0 < a \leqq i_x$, $0 < b \leqq i_y$, and $a \to A$, $b \to B$, $a \square b \to C$. Then $A \square B = C$. For, let $[x; y] \in C$. Then $p \to [x; y]$, $p \in a \square b$, and $p_x \leqq a$, $p_y \leqq b$. But as pointed out above, $p_x \to [x; p_0]$ and $p_y \to [p_0; y]$, hence $[x; p_0] \in A$, $[p_0; y] \in B$. Thus $[x; y] \in A \square B$.

Conversely if $[x; y] \in A \square B$, $[x, p_0] \in A$, $[p_0; y] \in B$. Then for $p \leqq a$, $q \leqq b$, $p \to [x; p_0]$, $q \to [p_0; y]$. But then $0 < p \square q \leqq a \square b$, and there is at least one point $t \leqq p \square q \leqq a \square b$. Thus $t_x = p$, $t_y = q$, and $t \to [x; y]$. Hence $[x; y] \in C$.

UNIVERSITY OF WISCONSIN,
AND
UNIVERSITY OF SOUTHERN CALIFORNIA.

ANNALS OF MATHEMATICS
Vol. 48, No. 2, April, 1947

APPROXIMATE ISOMETRIES OF THE SPACE OF CONTINUOUS FUNCTIONS[1]

BY D. H. HYERS AND S. M. ULAM

(Received September 28, 1945—Revised January 7, 1946)

A transformation of a metric space which changes distances by less than a certain fixed positive number ϵ is called an approximate isometry, or an ϵ-isometry. In a preceding paper[2] it was shown that an ϵ-isometry taking Hilbert space into itself could be uniformly approximated within a distance of $k\epsilon$ by a true isometry, where k was a positive constant not greater than 10. The object of the present paper is to establish a similar "stability" theorem for transformations of the space of continuous functions over a compact metric space.

According to a theorem of Banach,[3] every isometry between the normed spaces E and E' of continuous, real valued functions over the compact metric spaces K and K', respectively, is generated by a homeomorphism between K and K', and conversely. This result suggests a method of attack on our stability problem. The first step is to prove the existence of an isometry between the spaces E and E', having given the approximate isometry between these spaces, and this can be done by showing that the underlying spaces K and K' are homeomorphic. However, the method used by Banach in establishing the correspondence between the points of K and K', which was based on a theory of peak functions, seemed rather difficult to apply to our problem. Therefore a different method, dealing with the hyperplane of all functions having the same value at a given point of K, is employed in the present paper.

The norms in the spaces E and E' will be defined in the usual way, as the maximum of the absolute value of the function in question. A transformation $T(f)$ taking E into E' will be called an ϵ-isometry if $| \, \| T(f) - T(g) \| - \| f - g \| \, | < \epsilon$ for all f and g in K.

THEOREM 1. *Let K and K' be compact metric spaces and let $T(f)$ be a continuous ϵ-isometry of E into E'. Put $M(p, b, h) = [\varphi; \varphi \in E, | \varphi(p) - b | \leqq h]$ and $M'(q, c, k) = [\psi; \psi \in E', | \psi(q) - c | \leqq k]$. Then for each p in K, each real number b and each $a \geqq 0$ there exists a point q of K' and a real number c such that $T(M(p\, b, a)) \subset M'(q, c, a + 3\epsilon/2)$.*

PROOF. Without loss of generality, we may assume that the metrics in K and K' have been chosen so that the diameter of each of these spaces is one. We shall assume for the present that $a > 0$, and consider the case $a = 0$ later. Put $f_n(x) = b + 3n - n\rho(x, p)$, $g_n(x) = b - 3n + n\rho(x, p)$, for $n = 1, 2, 3, \cdots$ and x in K, where $\rho(x, p)$ denotes the distance from x to p. Consider the spheres $S_n = [\varphi; \| \varphi - f_n \| \leqq 3n + a]$, $R_n = [\varphi; \| \varphi - g_n \| \leqq 3n + a]$ in the space E

[1] Presented to the American Mathematical Society in April, 1942.

[2] D. H. HYERS AND S. M. ULAM, "*On Approximate Isometries*", Bull. Amer. Math. Soc. 51 (1945) 288–292.

[3] S. BANACH, "*Operations Lineaires*", Warsaw, 1932, p. 170.

For commentary to this paper [33], see p. 686.

and their intersections $Q_n = R_n \cap S_n$. In order to obtain a metric characterization of the set $M(p, b, a)$ we shall show that $M(p, b, a) = \text{Cl}(\sum Q_n)$, where Cl (P) denotes the closure of the set P and where the set summation is extended over all positive integers. It is easily verified that if f is in Q_n then $|f(p) - b| \leqq a$. Therefore, since $M(p, b, a)$ is a closed set, $\text{Cl}(\sum Q_n) \subset M(p, b, a)$.

To prove that $M(p, b, a) \subset \text{Cl}(\sum Q_n)$, let P denote the set $[\varphi; |\varphi(p) - b| < a]$, so that $\text{Cl}(P) = M(p, b, a)$. For any given f in P, there exists $\delta > 0$ such that $|f(x) - b| < a$ for $\rho(x, p) < \delta$. If n is a positive integer greater than $\max[|f(x) - f(p)|/\rho(x, p); \rho(x, p) \geqq \delta]$, then f is in Q_n. For if $\rho(x, p) < \delta$, $|f(x) - f_n(x)| \leqq 3n - n\rho(x, p) + |b - f(x)| < 3n + a$, and similarly $|f(x) - g_n(x)| < 3n + a$. On the other hand, if $\rho(x, p) \geqq \delta$, $|f_n(x) - f(x)| \leqq |b - f(p)| + 3n - n\rho(x, p) + |f(p) - f(x)| \leqq a + 3n$, since $|f(p) - f(x)| \leqq n\rho(x, p)$ and $|b - f(p)| < a$. Similarly, $|f(x) - g_n(x)| \leqq a + 3n$ for all x in K. Hence $P \subset Q_n \subset \sum Q_m$. Taking closures, we see that $M(p, b, a) = \text{Cl}(P) \subset \text{Cl}(\sum Q_m)$. Therefore $M(p, b, a) = \text{Cl}(\sum Q_m)$.

Now the transformation T is an ϵ-isometry, so that $|\,\|T(\varphi) - T(f_n)\| - \|\varphi - f_n\|\,| < \epsilon$ and $|\,\|T(f_n) - T(g_n)\| - \|f_n - g_n\|\,| < \epsilon$, for all φ in K. Hence if φ is in Q_n, since $6n = \|f_n - g_n\|$, we have $\|T(\varphi) - T(f_n)\| \leqq \|\varphi - f_n\| + \epsilon \leqq 3n + a + \epsilon$, and $\|T(\varphi) - T(f_n)\| \leqq \frac{1}{2}\|f_n - g_n\| + a + \epsilon \leqq \frac{1}{2}\|T(f_n) - T(g_n)\| + a + 3\epsilon/2$. Similarly, $\|T(\varphi) - T(g_n)\| \leqq \frac{1}{2}\|T(f_n) - T(g_n)\| + a + 3\epsilon/2$.

Since K' is compact, there exists for each n a point y_n of K' such that $|f'_n(y_n) - g'_n(y_n)| = \|f'_n - g'_n\|$, where $f'_n = T(f_n)$ and $g'_n = T(g_n)$. Put $N = \frac{1}{2}\|f'_n - g'_n\|$, and assume for definiteness that $f'_n(y_n) \geqq g'_n(y_n)$ (in the contrary case f'_n and g'_n may be interchanged in the following argument). Let ψ be any element of $T(Q_n)$. By the inequalities developed in the preceding paragraph, $f'_n(y_n) - \psi(y_n) \leqq N + a + 3\epsilon/2$ and $\psi(y_n) - g'_n(y_n) \leqq N + a + 3\epsilon/2$. Since $f'_n(y_n) - g'_n(y_n) = 2N$, $f'_n(y_n) - \psi(y_n) = 2N - \{\psi(y_n) - g'_n(y_n)\}$, and hence $N + a + 3\epsilon/2 \geqq f'_n(y_n) - \psi(y_n) \geqq N - a - 3\epsilon/2$. Now $N = \frac{1}{2}f'_n(y_n) - \frac{1}{2}g'_n(y_n) = f'_n(y_n) - c_n$, where $c_n = \{f'_n(y_n) + g'_n(y_n)\}/2$. Therefore $a + 3\epsilon/2 \geqq c_n - \psi(y_n) \geqq -a - 3\epsilon/2$, or $|c_n - \psi(y_n)| \leqq a + 3\epsilon/2$, for all ψ in $T(Q_n)$.

Consider the sets $F_n = [y; |\psi_1(y) - \psi_2(y)| \leqq 2a + 3\epsilon, \psi_1, \psi_2, \epsilon\ T(Q_n)]$. Since by the results of the preceding paragraph, y_n is in F_n, each set F_n is non-empty. Clearly each F_n is closed. Since $Q_{n+1} \supset Q_n$, $T(Q_{n+1}) \supset T(Q_n)$, so that $F_{n+1} \subset F_n$. Therefore since K' is compact there exists at least one point q in the intersection of all the sets F_n, $n = 1, 2, \cdots$. It follows that

$$|\psi_1(q) - \psi_2(q)| \leqq 2a + 3\epsilon$$

for all functions ψ_1, ψ_2 contained in the set $\text{Cl}(\sum T(Q_n))$. By hypothesis T is continuous, so that $T(M(p, b, a)) = T(\text{Cl}(\sum Q_n)) \subset \text{Cl}(\sum T(Q_n))$. Therefore there exists a real number c such that

$$T(M(p, b, a)) \subset M'(q, c, a + 3\epsilon/2),$$

where a is any given positive number.

It remains to prove the theorem for the case where $a = 0$. Evidently $M(p, b, 0) \subset M(p, b, a)$ for every positive a. Hence $T(M(p, b, 0)) \subset T(M(p, b, 1/r)) \subset M'(q_r, c_r, 1/r + 3\epsilon/2)$, where $r = 1, 2, \cdots$, and the subscripts denote the possible dependence of c and q on r. Now the set of numbers c_r is bounded for $r = 1, 2, \cdots$, as can be seen by noting that the constant function $\varphi_0(x) \equiv b$ transforms into a function ψ_0 in $M(q_r, c_r, 1/r + 3\epsilon/2)$, and ψ_0 is evidently bounded. Hence, since K' is compact, there exist a subsequence σ of the sequence of natural numbers, a point q_0 and a number c_0 such that $\lim q_r = q_0$ and $\lim c_r = c_0$, where r tends to infinity over the sequence σ. If ψ is any element of the set $T(M(p; b, 0))$, then $| \psi(q_r) - c_r | \leqq 1/r + 3\epsilon/2$ for all r. Hence if r approaches infinity over the sequence σ the inequality $| \psi(q_0) - c_0 | \leqq 3\epsilon/2$ is true in the limit. Therefore $T(M(p, b, 0)) \subset M'(q_0, c_0, 3\epsilon/2)$.

COROLLARY. *Let T be a homeomorphism of E onto E' which satisfies the hypotheses of Theorem 1. Then the point q of K' is uniquely determined by the point p of K and is independent of the choice of the parameters a and b.*

PROOF. Given the real numbers b_1, b_2 and the positive real numbers a_1, a_2 and the point p of K, there exist points q and r of K' and real numbers c_1, c_2 such that $T(M(p, b_1, a_1)) \subset M'(q, c_1, a_1 + 3\epsilon/2)$ and $T(M(p, b_2, a_2)) \subset M'(r, c_2, a_2 + 3\epsilon/2)$, in accordance with Theorem 1. We shall prove that $q = r$. Since the inverse transformation T^{-1} is evidently an ϵ-isometry, we can apply Theorem 1 to obtain the relations $T^{-1}(M'(q, c_1, a_1 + 3\epsilon/2)) \subset M(p_1, b_{11}, a_1 + 3\epsilon)$ and $T^{-1}(M'(r, c_2, a_2 + 3\epsilon/2)) \subset M(p_2, b_{22}, a_2 + 3\epsilon)$, where p_1, p_2 are points of K, and the b's are real. Hence $M(p, b_1, a_1) \subset T^{-1}(M'(q, c_1, a_1 + 3\epsilon/2)) \subset M(p_1, b_{11}, a_1 + 3\epsilon)$. It follows that $p_1 = p$. Similarly $p_2 = p$. If we put $b_3 = \frac{1}{2}(b_{11} + b_{22})$ and $a_3 = | b_{22} - b_{11} |/2 + \max(a_1, a_2) + 3\epsilon$, then $M(p, b_{11}, a_1 + 3\epsilon) \subset M(p, b_3, a_3)$ and $M(p, b_{22}, a_2 + 3\epsilon) \subset M(p, b_3, a_3)$. Again by Theorem 1, there exists an element s of K' and a real number c_3 such that $T(M(p, b_3, a_3)) \subset M'(s, c_3, a_3 + 3\epsilon/2)$. Thus

$$M'(q, c_1, a_1 + 3\epsilon/2) \subset T(M(p, b_1, a_1 + 3\epsilon)) \subset T(M(p, b_3, a_3)) \subset M'(s, c_3, a_3 + 3\epsilon/2).$$

Similarly, $M'(r, c_2, a_2 + 3\epsilon/2) \subset M'(s, c_3, a_3 + 3\epsilon/2)$. It follows that $q = s$ and $r = s$, so that $q = r$.

THEOREM 2. *Let K and K' be compact metric spaces and let E and E' be the spaces of all real valued continuous functions on K and K', respectively. If $T(f)$ is a homeomorphism of E onto E' which is also an ϵ-isometry, then there exists an isometric transformation $U(f)$ of E onto E' such that $|| U(f) - T(f) || \leqq 21\epsilon$ for all f in E.*

PROOF. We may assume without loss of generality that $T(0) = 0$. Let $f(x)$ be any element of E and let p be any point of K. According to Theorem 1 there exists a point q of K' and a real number c such that $T(M(p, f(p), 0)) \subset M'(q, c, 3\epsilon/2)$. Since $T(0) = 0$, the constant function $\varphi(x) \equiv f(p)$, x in K, is transformed into a function $\varphi' = T(\varphi)$ satisfying the inequality $| \varphi'(y) | <$

$|f(p)| + \epsilon$ for all y in K'. Since $\varphi' \epsilon M'(q, c, 3\epsilon/2)$, it follows that $c - 3\epsilon/2 \leqq \varphi'(q) < |f(p)| + \epsilon$ and that

$$c - |f(p)| < 5\epsilon/2. \tag{1}$$

According to the corollary to Theorem 1, the point q is uniquely determined by the point p. Put $q = h(p)$. Similarly, if s is any point of K' there is by Theorem 1 and the corollary a unique point r of K and a real number b such that $T^{-1}(M'(s, c, 3\epsilon/2) \subset M(r, b, 3\epsilon)$. Put $r = h'(s)$. If we take $s = q = h(p)$ and put $p' = h'(q)$ then $M(p, f(p), 0) \subset T^{-1}(M'(q, c, 3\epsilon/2)) \subset M(p', b, 3\epsilon)$. It follows that $p' = p$ and that $|b - f(p)| \leqq 3\epsilon$. Hence $h'(h(p)) = p$ for all p in K. By reversing the roles of K and K' we can show similarly that $h(h'(s)) = s$ for all s in K'. It follows that h is one to one.

It has been shown incidentally that $T^{-1}(M'(q, c, 3\epsilon/2)) \subset M(p, b, 3\epsilon)$, where $|b - f(p)| \leqq 3\epsilon$. Assume first that $c > 0$. Put $\psi'(y) \equiv c - 3\epsilon/2$ for all y in K', and put $\psi = T^{-1}(\psi')$. Then $|\psi(p)| \leqq \|\psi\| < c - \epsilon/2$, since T^{-1} is an ϵ-isometry with $T^{-1}(0) = 0$. Since ψ' is in $M'(q, c, 3\epsilon/2)$, ψ is in $M(p, b, 3\epsilon)$, and $|b - \psi(p)| \leqq 3\epsilon$. Hence $|f(p)| \leqq b + 3\epsilon \leqq |\psi(p)| + 6\epsilon < c + 11\epsilon/2$. The last inequality together with inequality (1) shows that $||f(p)| - c| < 11\epsilon/2$. If $c < 0$, the same proof applies to show that $||f(p)| + c| < 11\epsilon/2$. Hence in all cases, we have

$$||f(p)| - |c|| < 11\epsilon/2. \tag{2}$$

In order to show that the transformation $q = h(p)$ is a homeomorphism, it remains to prove continuity. Put $f' = T(f)$. Since $T(M(p, f(p), 0)) \subset M'(q, c, 3\epsilon/2)$, it follows that $|f'(q) - c| \leqq 3\epsilon/2$, and that

$$||f'(q)| - |f(p)|| < 11\epsilon/2 + 3\epsilon/2 = 7\epsilon. \tag{3}$$

Now let x_n be any sequence of points converging to an arbitrary point p of K. Put $y_n = h(x_n)$, $q = h(p)$, and for any given positive number m, put $f'(y) = m\rho(y, q)$. Then by the inequality (3), $|m\rho(y_n, q) - |f(x_n)|| < 7\epsilon$, and $|m\rho(q, q) - |f(p)|| < 7\epsilon$, where $f = T^{-1}(f')$. By hypothesis T^{-1} is continuous, and $|f(x_n)| \to |f(p)|$, so that $\overline{\lim}_{n\to\infty} m\rho(y_n, q) \leqq 14\epsilon$. Since this inequality holds for arbitrarily large positive numbers m, it follows that $\lim_{n\to\infty} \rho(y_n, q) = 0$. Therefore $h(p)$ is continuous at each point of K, and hence is a homeomorphism of the compact metric space K into K'.

We shall now show that $q = h(p)$ generates an isometric transformation $g = U(f)$ of E into E', where $g(q) = s(q)f(h^{-1}(q))$, and where $s(q)$ is a function whose values are either 1 or -1, and which will be defined as follows. First take $b_2 > b_1 > 7\epsilon$, and consider the constant functions $\varphi_1(x) \equiv b_1$ and $\varphi_2(x) \equiv b_2$, for x in K. Then $\varphi_1' = T(\varphi_1)$ and $\varphi_2' = T(\varphi_2)$ will have the same sign at an arbitrary point q of K'. For if we choose a positive number $\delta < (b_1 - 7\epsilon)/2$, then by Theorem 1, $T[M(p, b_1, \delta)] \subset M'(p, c_1, \delta + 3\epsilon/2)$, and by inequality (2), $|c_1| - (\delta + 3\epsilon/2) > 0$. Hence all functions in the set $M'(q, c_1, \delta + 3\epsilon/2)$ have the same sign at the point q. By taking a finite chain of closed intervals, each of length 2δ, extending from b_1 to b_2, it is easily seen that $\varphi_1'(q)$ and $\varphi_2'(q)$ agree in sign. We now define $s(q) = \varphi'(q)/|\varphi'(q)|$, where $\varphi' = T(\varphi)$, and where $\varphi(x)$ is any given function having a constant value greater than 7ϵ throughout the space K. Since by inequality (3) φ' does not vanish at any point of K', it is clear that $s(q)$ is defined and continuous at every point of K'. It follows that $g(q) = s(q)f(h^{-1}(q))$ is also continuous, and that $g = U(f)$ is an isometry of E into E'.

Finally we must show that $T(f)$ is "close" to $U(f)$. Now whenever $|f(p)| > 7\epsilon$, the sign of $g(q)$ will agree with that of $f'(q)$, where $g = U(f)$, $f' = T(f)$ and $q = h(p)$, as can be seen by considering the constant function $\varphi(x) \equiv f(p)$, and using Theorem 1 together with inequality (2). Hence by inequality (3), $|f'(q) - g(q)| < 7\epsilon$, providing that $|f(p)| > 7\epsilon$. On the other hand if $|f(p)| \leqq 7\epsilon$, then again by inequality (3), $|f'(q) - g(q)| \leqq |f'(q)| + |f(p)| \leqq ||f'(q)| - |f(p)|| + 2|f(p)| < 21\epsilon$. Therefore $\| T(f) - U(f) \| < 21\epsilon$ for all f in E.

COROLLARY. *Under the hypotheses of the theorem, the underlying metric spaces K and K' are homeomorphic.*

THE UNIVERSITY OF SOUTHERN CALIFORNIA

Reprinted from the Proceedings of the NATIONAL ACADEMY OF SCIENCES,
Vol. 34, No. 8, pp. 403–405. August, 1948

MULTIPLICATIVE SYSTEMS, I

BY C. J. EVERETT AND S. ULAM

LOS ALAMOS SCIENTIFIC LABORATORY

Communicated by J. von Neumann, May 28, 1948

1. Introduction.—The purpose of this note is to introduce a formalism suitable for the description of systems which consist of particles of different types, each particle transmuting with given probability into a group of particles of these types, the number of particles in the system varying with time.[1, 2, 3] The type of a particle may refer to characteristics like position or momentum as well as to intrinsic properties.

The phase space in which we are interested here is that of all possible futures or genealogies of such systems, each such genealogy being considered as a point of the space. Given the elementary probabilities of transmutation, one can introduce a notion of probability or measure for certain subsets of the phase space.

The formalism is general enough to include as special cases the multiplication of bacteria, radioactive decay, cosmic ray showers, diffusion theory and the theory of trajectories in mechanical systems.

Detailed discussion of specific cases, at present in the form of reports, will be published elsewhere.

2. Remark on Measure.—It is convenient to have at hand some simple axioms on which the classical measure theory may be shown to rest. Suppose $\Gamma = \{\gamma\}$ is a point set, $I = \{i\}$ a class of distinguished subsets i of Γ called intervals, including the empty set θ and the entire set Γ. Denote by J the class of all subsets $S = \Sigma i$ which are sums of a finite or countable class of intervals (all sums hereafter are supposed to be over a finite or countable number of summands). Suppose that intervals satisfy the axioms:

I.1. Every set S of J can be represented as a sum of mutually disjoint intervals.

I.2. The complement $i' = \Gamma - i$ of an interval i is in J.

I.3. The set product ij of two intervals i, j is an interval.

Assume further

M.1. To every interval i is assigned a non-negative real number $m(i)$ called its measure.

M.2. $m(\Gamma) = 1$.

M.3. If $i = \Sigma i_\mu$, where the i, i_μ are intervals and the i_μ are pairwise disjoint, then $m(i) = \Sigma m(i_\mu)$.

An additive class C of subsets of Γ is one such that

C.1. All intervals belong to C.

C.2. If $A_1, A_2, \ldots$ are in C, so is ΣA_μ.

C.3. If A is in C, so is A'.

The intersection of all additive classes consists of the Borel sets. Thus all Borel sets are contained in any additive class.

If one defines outer measure $O(U) = glb\{m(S); U \subset S \epsilon J\}$ and inner measure $I(U) = 1 - O(U')$ as usual for an arbitrary subset U of Γ, one can show on the basis of the stated axioms that the class of all measurable sets M (for which $I(M) = O(M)$) is an additive class. Moreover, defining measure $m(M)$ for measurable sets to be the common value $I(M) = O(M)$, one can show $m(M) \geqq 0$; $m(\Sigma M) = \Sigma m(M)$, for disjoint measurable sets M; and if M is an interval, the assigned measure coincides with the original interval measure.

3. The set Γ_i of Graphs.—Consider t types of particles such that a particle of type i may produce, upon tranformation, $j_1 + \ldots + j_t \geqq 0$ of such particles, of which j_ν are of type ν, $\nu = 1, \ldots, t$. We suppose transformation times are the same for each type and hence that generations may be counted unambiguously. We agree to consider zero generation as consisting of one particle of a fixed type i. Then we consider the set Γ_i of all possible infinite histories or genealogies of such a particle, that is, the infinite records of the transformations of this particle and all its progeny through all generations $k = 0, 1, 2, \ldots$.

We may represent a genealogy in the plane if we make the following conventions:

(*a*) t different "colors" are assigned to the t types, a particle of type ν in the kth generation being represented by a dot of appropriate color.

(*b*) If a particle of type μ in generation k is transformed into no particle, i.e., if it dies or escapes, this is indicated by a sequence of Δ's (Δ = death) vertically below it, one in each succeeding generation.

(*c*) If a particle of type μ in generation k is transformed into $j_1 + \ldots + j_t > 0$ particles, j_ν of type ν, this is indicated by a branching from the corresponding dot of color μ in the kth generation to $j_1 + \ldots + j_t$ dots, j_ν of color ν, in the $k + 1$st generation, dots of the same color being placed consecutively and colors ranging in order $\nu = 1, \ldots, t$ from left to right.

Thus we may regard the set Γ_i of all such graphs γ in the plane. If γ is a graph, γ_n denotes the upper segment of γ from generation 0 through generation n. If $\gamma = \bar{\gamma}$, then $\gamma_n = \bar{\gamma}_n$ for all n. If $\gamma \neq \bar{\gamma}$, define $k(\gamma, \bar{\gamma})$ as the least integer n for which $\gamma_n \neq \bar{\gamma}_n$.

The set Γ_i has the natural metric: $d(\gamma, \bar{\gamma}) = 0$ for $\gamma = \bar{\gamma}$, $d(\gamma, \bar{\gamma}) = 1/k(\gamma, \bar{\gamma})$ for $\gamma \neq \bar{\gamma}$. The metric space Γ_i resulting is 0-dimensional, separable, and complete, but *not* locally compact.

A graph γ is said to terminate in case it contains no particles (only Δ's) in some generation. The set T of all terminating graphs may be split into the disjoint summands $T_0, T_1, \ldots,$ where T_n denotes the set of all terminating graphs $\tau^{(n)}$ which contain at least one particle in generation η, but none in the next.

By an interval of order n is meant the set $i(\tau^{(n)})$ of all graphs γ such that $\gamma_n = \tau_n^{(n)}$ where $\tau^{(n)}$ is a particular graph in T_n. The only interval of order 0 is Γ_i itself. We now define *interval* to be either θ, or any single graph γ, terminating or not, or any interval of order n, $n = 0, 1, \ldots$. It is easy to see that axioms I.1, 2, 3 are satisfied. Indeed we have the simple property that the intersection ij of two intervals is either θ or i or j.

4. *Measure in the Space of Graphs.*—While the notions of graph, distance and interval are geometric in character and depend only on the number t of types, measure may be introduced in various ways. One of the simplest, but by no means the only one to which the theory has been applied, may be imposed as follows. Suppose we assign a probability $p(i;\ j_1, \ldots, j_t)$ to the event that a particle of type i should produce, upon transformation, $j_1 + \ldots + j_t$ particles, j_ν of type ν. Then every segment γ_n of a graph γ has an associated probability $p(\gamma_n)$ that the event described by γ_n should occur. If we assign to intervals a measure by $m(\theta) = 0$ $m(\tau^{(n)}) = p(\tau_{n+1}^{(n)})$, $m(i(\tau^{(n)})) = p(\tau_n^{(n)})$, and $m(\gamma) = \lim p(\gamma_n)$ for γ non-terminating *it turns out that M.1, 2, 3 are satisfied* and thus Borel sets based on our intervals are measurable in the sense of §2. (Non-compactness of intervals makes M.3 non-trivial.)

The procedure applies in much more general systems where the transition probabilities are functions of the time.

[1] Hawkins, D., an Ulam, S., *Theory of Multiplicative Processes*, I. U.S.A.E.C., Oak Ridge MDDC 287, LADC 265. Declassified 1944.

[2] v. Neumann, J., and Ulam, S., "Random Ergodic Theorems," *Bulletin A. M. S.*, Abstract 51-9-165 (1945).

[3] v. Neumann, J., and Ulam, S., "On Combination of Stochastic and Deterministic Processes," *Ibid.*, Abstract 53-11-403 (1947).

APPROXIMATELY CONVEX FUNCTIONS

D. H. HYERS AND S. M. ULAM

In previous papers approximately linear functions [1] and approximately isometric transformations [2; 3; 4] have been studied.[1] In both cases it was shown that the properties of linearity and isometry are "stable" in a certain sense. For example, it was proved that if a function $f(x)$ satisfies the linear functional equation within an amount ϵ, that is, $|f(x+y)-f(x)-f(y)| \leqq \epsilon$, then there exists an actual solution $g(x)$ of the linear functional equation such that $|g(x)-f(x)| \leqq \epsilon$, where ϵ is a given positive number.

In the present paper we discuss a similar problem for the property of convexity. We consider real-valued functions defined on subsets of n-dimensional Euclidean space E_n. A function $f(x)$ defined on a convex subset S of E_n will be called *ϵ-convex* if $f(hx+(1-h)y) \leqq hf(x)+(1-h)\, f(y)+\epsilon$, for all x and y in S and for $0 \leqq h \leqq 1$. Here ϵ is a fixed positive number. Our object is to show that to an ϵ-convex function $f(x)$ there corresponds a convex function $g(x)$ such that $|f(x)-g(x)| \leqq k\epsilon$, for some constant k. In order to prove this we need some results on ϵ-convex functions and on approximating simplices given in the following four lemmas. The paper is self-contained.

LEMMA 1. *Let $f(x)$ be an ϵ-convex function defined on an n-dimensional simplex $S \subset E_n$. Let the vertices of the simplex be $p_0, p_1, \cdots, p_n$, then if $x=\sum_{i=0}^n \alpha_i p_i$, $\alpha_i>0$, $\sum_{i=0}^n \alpha_i=1$ is any point of S, we have*

$$f(x) \leqq \sum_{i=0}^{n} \alpha_i f(p_i) + 2k_n\epsilon, \tag{1}$$

where $k_n=(n^2+3n)/(4n+4)$.

PROOF. We prove the inequality by induction on n. For $n=1$, (1) reduces to the statement of ϵ-convexity, so it is true for $n=1$. We assume that (1) holds for n replaced by $n-1$, and prove it for n dimensions. The case in which some $\alpha_i=1$ is trivial, for in this case $x=p_i$, so we may assume that $\alpha_i<1$ for $i=1, \cdots, n+1$. For convenience we may also assume that $\alpha_n \geqq \alpha_j$, $j=0, \cdots, n-1$. Put $h=1-\alpha_n$, $a_j=\alpha_j/h$, $j=0, \cdots, n-1$, and $q=\sum_{j=0}^{n-1} a_j p_j$. Then $x =\sum_{i=0}^n \alpha_i p_i=hq+(1-h)p_n$, and since f is ϵ-convex,

$$f(x) \leqq hf(q) + (1-h)f(p_n) + \epsilon. \tag{2}$$

Presented to the Society, April 28, 1951; received by the editors February 6, 1952.

[1] For a discussion of these and other related questions, see [6].

Reprinted from Proceedings of the American Mathematical Society, vol. 3 (1952), pp. 821–8. Copyright 1952, American Mathematical Society. For commentary to this paper [46], see p. 686.

By the induction hypothesis,

$$f(q) \leqq \sum_{j=0}^{n-1} a_j f(p_j) + \frac{(n-1)(n+2)}{2n}\epsilon. \tag{3}$$

Substituting (3) into (2), we get

$$f(x) \leqq \sum_{i=0}^{n} \alpha_i f(p_i) + \left\{1 + \frac{h(n-1)(n+2)}{2n}\right\}\epsilon. \tag{4}$$

Since $\alpha_n \geqq \alpha_j$, $j=0, \cdots, n-1$, the minimum value which α_n can have is $1/(n+1)$, so the maximum value which h can have is $1-1/(n+1)=n/(n+1)$. Consequently an upper bound for the expression in brackets in inequality (4) is

$$1 + \frac{(n-1)(n+2)}{2(n+1)} = \frac{n^2+3n}{2n+2}.$$

Thus the lemma has been established.

LEMMA 2. *Let $f(x)$ be an ϵ-convex function defined on an open convex set $G \subset E_n$. Then on each closed bounded subset B of G, $f(x)$ is bounded.*

PROOF. f is bounded from above, since B may be covered with a finite number of n-dimensional simplices, each contained in G, and f is bounded on each simplex by Lemma 1.

To prove that f is bounded from below on B, let B be covered with a finite number of closed spheres S_i, such that each S_i is contained in G. Let x_i be the center of S_i, and let x_i+y be any point of the sphere S_i. Then by ϵ-convexity

$$f(x_i) \leqq 2^{-1}f(x_i+y) + 2^{-1}f(x_i-y) + \epsilon,$$

or

$$f(x_i+y) \geqq 2f(x_i) - f(x_i-y) - 2\epsilon.$$

Now x_i-y belongs to the sphere S_i, and since S_i is a closed subset of G, $f(x_i-y)$ is bounded from above as x_i+y varies over S_i. Hence $f(x_i+y)$ is bounded from below for $x_i+y \in S_i$, and it follows that f is bounded from below on B. The proof of the following two lemmas is left to the reader.

LEMMA 3. *Let x lie in an n-dimensional simplex with vertices $q_0, q_1, \cdots, q_n$, so that $x=\sum_{i=0}^n \alpha_i q_i$, $\alpha_i \geqq 0$, and $\sum_{i=0}^n \alpha_i = 1$. Suppose also that we have $n+1$ sequences $\{q_i^{(\nu)}\}$ $(i=0, \cdots, n; \nu=1, 2, 3, \cdots)$ of points such that $q_i^{(\nu)} \to q_i$ as $\nu \to \infty$, and that for each ν, x also lies in the n-dimensional simplex with vertices $q_i^{(\nu)}$ so that*

$$x = \sum_{i=0}^{n} \alpha_i^{(\nu)} q_i^{(\nu)},$$

where $\alpha_i^{(\nu)} \geqq 0$ *and* $\sum_{i=0}^{n} \alpha_i^{(\nu)} = 1$. *Then as* $\nu \to \infty$, $\alpha_i^{(\nu)} \to \alpha_i$.

LEMMA 4. *Suppose* x *is interior to an* n*-dimensional simplex in* E_n *whose vertices are* $q_0, q_1, \cdots, q_n$. *Then if* $q_i^{(\nu)} \to q_i$ *in* E_n *as* $\nu \to \infty$ $(i=0, \cdots, n)$, x *is also interior to the simplex* $S_n^{(\nu)}$ *whose vertices are* $q_i^{(\nu)}$ $(i=0, \cdots, n)$, *for sufficiently large* n.

THEOREM 1. *Let* $f(x)$ *be* ϵ*-convex on an open convex set* $G \subset E_n$, *and let* B *be any closed bounded convex subset of* G. *Then there exists a convex function* $\phi(x)$ *on* B *such that*

$$|\phi(x) - f(x)| \leqq k_n \epsilon, \qquad \text{for } x \in B,$$

where $k_n = (n^2+3n)/(4n+4)$.

PROOF. Let H be a bounded convex open subset of G such that $B \subset H$, and $\overline{H} \subset G$. Since B is a compact subset of the open convex set G, the existence of such an H is easily shown. Let K denote the convex hull of the closure of the graph of the function $f(x)$ for $x \in \overline{H}$, so that K is a convex set in E_{n+1}.

Define, for $x = (x_1, \cdots, x_n) \in \overline{H}$, $g(x) = \inf [y; (x_1, x_2, \cdots, x_n, y) \in K]$. Since $f(x)$ is bounded on $\overline{H}$ by Lemma 2, K is a compact set in E_{n+1} and $g(x)$ is well defined on $\overline{H}$. It is easily seen that $g(x)$ is a convex function, and that $g(x) \leqq f(x)$ for $x \in H$. Given a point $x \in B$, let p denote the point $(x_1, x_2, \cdots, x_n, g(x))$ in E_{n+1}. Now p evidently belongs to the boundary of K, and since K is closed, it also belongs to K. By a well known theorem,[2] p lies on an m-dimensional simplex S_m whose vertices are points or limit points of the graph of $f(x)$ for $x \in \overline{H}$, where $m \leqq n+1$. Notice that the assertion is actually true for some $m \leqq n$, for if p were in the interior of an $(n+1)$-dimensional simplex with vertices in K, then p would lie in the interior of K and not on its boundary.

There are three possible cases.

(i) p is a point of the graph of f.

(ii) p is a limit point of the graph of f.

(iii) p is "interior"[3] to some simplex S whose dimension is positive and less than or equal to m, and whose vertices are points or limit points of the graph of f.

In case (i), $f(x) = g(x)$, and there is nothing to prove. In case (ii)

[2] See [5, p. 9].

[3] A point will be called "interior" to a simplex S of dimension r if it belongs to S but not to any face of lower dimension than r.

it is convenient to translate the axes so that the origin of coordinates lies at the point x so that $x=0$. Then by hypothesis there exists a sequence of distinct points $x^{(\mu)} \in H \subset E_n$ tending to zero such that $\lim_{\mu\to\infty} f(x^{(\mu)}) = g(0)$. It is clear that an infinite number of these points must all lie in some one of the 2^n-tants determined by the coordinate hyperplanes. For definiteness, let us assume the first 2^n-tant contains an infinite number of these points. We denote them by $x^{(\nu)}$, so that all the coordinates of each $x^{(\nu)}$ may be assumed to be non-negative. Now choose on each coordinate axis a point p_j whose jth coordinate is negative, the others being zero, $j=1, 2, \cdots, n$, such that $p_j \in H$. Consider the simplex $S^{(\nu)}$ whose vertices are $p_1, p_2, \cdots, p_n$ and $x^{(\nu)}$. Then the origin belongs to this simplex, and there exist $\alpha_i^{(\nu)}$, $i=1, \cdots, n+1$, such that

$$\sum_{j=1}^{n} \alpha_j^{(\nu)} p_j + \alpha_{n+1}^{(\nu)} x^{(\nu)} = 0, \tag{5}$$

where $\alpha_j^{(\nu)} \geqq 0$, $\alpha_{n+1}^{(\nu)} > 0$, and $\sum_{i=1}^{n+1} \alpha_i^{(\nu)} = 1$.

To prove this, let p_{jj} be the jth coordinate of the point p_j and let $x_j^{(\nu)}$ be the jth coordinate of the point $x^{(\nu)}$. Then the "vector" equation (5) may be written in the form:

$$\alpha_j^{(\nu)} p_{jj} + \alpha_{n+1}^{(\nu)} x_j^{(\nu)} = 0, \qquad j = 1, \cdots, n, \tag{6}$$

where $p_{jj} < 0$, and $x_j^{(\nu)} \geqq 0$. Since $x^{(\nu)} \neq 0$, at least one of the $x_j^{(\nu)}$ must be positive. If $x_j^{(\nu)} = 0$, choose $\alpha_j^{(\nu)} = \rho_j^{(\nu)} = 0$. If $x_j^{(\nu)} \neq 0$ equation (6) determines the ratio $\rho_j^{(\nu)} = \alpha_j^{(\nu)}/\alpha_{n+1}^{(\nu)}$, which in this case is evidently positive. The value of $\alpha_{n+1}^{(\nu)}$ is then determined by the requirement that $\sum_{i=1}^{n+1} \alpha_i^{(\nu)} = (1 + \sum_{j=1}^{n} \rho_j^{(\nu)})\alpha_{n+1}^{(\nu)} = 1$. Thus relation (6) is established. By Lemma 1, it follows that

$$f(0) \leqq \sum_{j=1}^{n} \alpha_j^{(\nu)} f(p_j) + \alpha_{n+1}^{(\nu)} f(x^{(\nu)}) + 2k_n\epsilon. \tag{7}$$

Now as $\nu \to \infty$, $x^{(\nu)} \to 0$. Hence by (6), $\alpha_j^{(\nu)} \to 0$ for $j=1, \cdots, n$. It follows that $\alpha_{n+1}^{(\nu)} \to 1$. Since $f(x^{(\nu)}) \to g(0)$, we have $f(0) \leqq g(0) + 2k_n\epsilon$, or $f(x) \leqq g(x) + 2k_n\epsilon$.

We now turn to case (iii). Here p lies in the interior of an r-dimensional simplex S_r $(1 \leqq r \leqq n)$ whose vertices p_i $(i=0, 1, \cdots, r)$ are points or limit points of the graph of f.

Let π be a supporting hyperplane of $K \subset E_{n+1}$ through the point p. Now p is interior to at least one line segment S_1 belonging to S_r and hence to K. Any such line segment S_1 must lie in the hyperplane π, for otherwise S_1 would pierce the hyperplane π at p so that part of

S_1 would lie on one side of π and part on the other, which is impossible since all of K lies on one side of π. It follows that S_r, and hence its vertices p_i, lies in π, and the p_i are boundary points of K.

This supporting hyperplane π cannot be perpendicular to E_n, for in this case π would project (orthogonally) into a hyperplane in E_n which would be a supporting hyperplane of the projection of the convex set K and which would contain the point x. Thus x would be on the boundary of the projection of K. But the projection of K includes the open set H which by hypothesis contains x, so x cannot lie on the boundary of K's projection, and we have a contradiction.

Therefore the projection of S_r onto E_n is a simplex Σ_r of the same dimension r, and the interior of S_r projects into the interior of Σ_r, so that the point x which is the projection of p lies in the interior of Σ_r.

We use double subscripts to denote the coordinates of the vertices p_i of S_r, and we denote the projections of these vertices onto E_n by $q_0, q_1, \cdots, q_r$. Then by hypothesis there exist sequences $q_i^{(\nu)}$ such that $p_{i,n+1}=\lim_{\nu\to\infty} f(q_i^{(\nu)})$, where $\lim_{\nu\to\infty} q_i^{(\nu)}=q_i$, and $q_0, \cdots, q_r$ are the vertices of the r-dimensional simplex $\Sigma_r \subset E_n$, which contains the point x in its interior. Our object is to construct a simplex $S_n^{(\nu)}$ of dimension n in E_n such that x is interior to $S_n^{(\nu)}$, and such that r of its vertices are points $q_0^{(\nu)}, \cdots, q_r^{(\nu)}$. We can then apply Lemma 1 to this simplex and take the limit in the resulting inequality as $\nu\to\infty$.

Suppose first that $r=n$. In this case, x is interior to the n-dimensional simplex $\Sigma_n \subset E_n$, so that

$$x = \sum_{i=0}^{n} \alpha_i q_i, \qquad \alpha_i > 0, \qquad \sum_{i=0}^{n} \alpha_i = 1.$$

Since $q_i^{(\nu)}\to q_i$ in E_n as $\nu\to\infty$, it follows by Lemma 4 that $x = \sum_{i=0}^{n} \alpha_i^{(\nu)} q_i^{(\nu)}$, $\alpha_i^{(\nu)}>0$, $\sum_{i=0}^{n} \alpha_i^{(\nu)}=1$. Hence by Lemma 3, $\alpha_i^{(\nu)}\to\alpha_i$ as $\nu\to\infty$.

Now by Lemma 1, we have $f(x) \leqq \sum_{i=0}^{n} \alpha_i^{(\nu)} f(q_i^{(\nu)})+2k_n\epsilon$. By taking limits as $\nu\to\infty$ we get

$$f(x) \leqq \sum_{i=0}^{n} \alpha_i p_{i,n+1} + 2k_n\epsilon = g(x) + 2k_n\epsilon.$$

Now let us suppose that $1 \leqq r \leqq n$. Let F_r be the r-dimensional flat containing Σ_r. Now if for all but a finite number of ν's, the $q_i^{(\nu)}$, $i=0, \cdots, n$; $\nu=1, 2, 3, \cdots$, are contained in F_r, then $q_i^{(\nu)}\to q_i$ in F_r and one has essentially case (iiia) with r replacing n, so the proof follows as before.

Next suppose that an infinity of points $q_i^{(\nu)}$ for some i lie outside this flat. We may as well assume (by relabeling and suppressing a subsequence if necessary) that all of the $q_0^{(\nu)}$ lie outside F_r.

Let us choose a new coordinate system with origin at q_0 and with the first r axes belonging to F_r, so that the equations of F_r are $z_j=0$, $j=r+1, \cdots, n$. The last $n-r$ coordinates $q_{0,r+1}^{(\nu)}, \cdots, q_{0,n}^{(\nu)}$ of the point $q_0^{(\nu)}$ cannot all be zero for any ν. It follows that for some fixed j, $q_{0,j}^{(\nu)}\neq 0$, for all ν. We may without loss of generality assume that $q_{0,r+1}^{(\nu)}\neq 0$, for all ν. Now there must be an infinity of the numbers $q_{0,r+1}^{(\nu)}$ which are either all positive or all negative, and by reversing the $(r+1)$st coordinate axis if necessary, we may assume that $q_{0,r+1}^{(\nu)}>0$ for all ν.

Next, if $r+1<n$, we consider $q_{0,r+2}^{(\nu)}$. If $q_{0,r+2}^{(\nu)}=0$ for all but a finite number of ν's, we rotate the z_{r+1} and z_{r+2} axes through an acute angle, keeping all of the other axes fixed, in such a way that after the rotation $q_{0,r+1}^{(\nu)}$ will still be positive and $q_{0,r+2}^{(\nu)}$ will become positive for all but a finite number of ν's.

On the other hand if $q_{0,r+2}^{(\nu)}\neq 0$ for an infinite number of ν's, then for an infinite number of ν's, these numbers are all positive or all negative. By reversing the z_{r+2}-axis if necessary we have $q_{0,r+2}^{(\nu)}>0$ for an infinite number of ν's. Thus by suppressing a subsequence if necessary we can arrange matters so that $q_{0,r+1}^{(\nu)}>0$ and $q_{0,r+2}^{(\nu)}>0$ for all ν.

If $r+2<n$, we proceed in the same way, with $r+1$ replacing r, and so on. Thus, there will exist a coordinate system in E_n and sequences of points $q_i^{(\nu)}\to q_i$ $(i=0, 1, \cdots, r)$ such that the origin lies at the point q_0, and $q_{i,j}=0$, $q_{0,j}^{(\nu)}>0$ for $j=r+1, \cdots, n$, where $f(q_i^{(\nu)})\to p_{i,n+1}$, $x=\sum_{i=0}^r \alpha_i q_i$, $g(x)=\sum_{i=0}^r \alpha_i p_{i,n+1}$, $\sum_{i=0}^r \alpha_i=1, \alpha_i>0$.

Now let q_i $(i=r+1, \cdots, n)$ be a point in H whose $(r+1)$st coordinate is a negative number and whose other coordinates are all zero. We now show that x is interior to the n-dimensional simplex whose vertices are $q_0^{(\nu)}, q_1, q_2, \cdots, q_n$, for sufficiently large ν.

Thus we must show the existence of positive numbers β_i $(i=0, \cdots, n)$ with $\sum_{i=0}^n \beta_i=1$ such that $x=\sum_{i=0}^r \alpha_i q_i=\beta_0 q_0^{(\nu)} + \sum_{i=0}^n \beta_i q_i$. That is, the β_i are to satisfy the following system of $n+1$ linear equations:

$$\beta_0 q_{0,j}^{(\nu)} + \sum_{i=0}^{r} \beta_i q_{i,j} = \sum_{i=0}^{r} \alpha_i q_{i,j} \qquad (j = 1, \cdots, r),$$

$$\beta_0 q_{0,j}^{(\nu)} + \beta_j q_{j,j} = 0 \qquad (j = r+1, \cdots, n), \tag{8}$$

$$\sum_{i=0}^{n+1} \beta_i = 1.$$

Since $\alpha_i > 0$, $\sum_{i=0}^{r} \alpha_i = 1$, and $q_{0,j}^{(\nu)} \to q_{0,j} = 0$, it follows that for $0 < \beta_0 < 1$ there will exist a ν_0, independent of β_0, such that the first r equations of the system (8) have solutions for β_i, $i = 1, \cdots, r$, which are between zero and one, whenever $\nu \geqq \nu_0$. Since $q_{j,j}$ and $q_{0,j}^{(\nu)}$ are of opposite signs by construction for $j = r+1, \cdots, n$, it is clear that the next $n-r$ equations will also have solutions β_j, $j = r+1, \cdots, n$, which are between zero and one when β_0 is, and when ν is sufficiently large. With the help of the last equation all the β's may be determined, with $0 < \beta_i < 1$, $i = 0, \cdots, n$.

Next, for a given ν, so large that x is interior to the simplex with vertices $q_0^\nu, q_1, \cdots, q_n$, there will exist by Lemma 4 an index $\mu = \mu(\nu)$ such that x is also interior to the simplex with vertices $q_0^\nu, q_1^\mu, q_2^\mu, \cdots, q_r^\mu, q_{r+1}, \cdots, q_n$. Let one such index μ be determined for each ν and put $\bar{q}_i^\nu = q_i^{\mu(\nu)}$, $i = 1, \cdots, r$. For convenience we also put $\bar{q}_0^\nu = q_0^\nu$. Then there exist $\alpha_i > 0$, $i = 0, 1, \cdots, n$, such that $\sum_{i=0}^{n} \alpha_i^\nu = 1$ and $x = \sum_{i=0}^{r} \alpha_i q_i = \sum_{i=0}^{r} \alpha_i^\nu \bar{q}_i + \sum_{i=r+1}^{n} \alpha_i q_i$. By Lemma 1 we have

$$f(x) \leqq \sum_{i=0}^{r} \alpha_i f(\bar{q}_i^\nu) + \sum_{j=r+1}^{n} \alpha_j f(q_j) + 2k_n\epsilon.$$

Now as $\nu \to \infty$, $\bar{q}_i^\nu \to q_i$, $f(\bar{q}_i^\nu) \to p_{i,n+1}$, and, by Lemma 3, we know that $\alpha_i^\nu \to \alpha_i$ for $i = 0, 1, \cdots, r$, while $\alpha_j^\nu \to 0$, $j = r+1, \cdots, n$. Hence by letting $\nu \to \infty$ in the last inequality we get

$$f(x) \leqq \sum_{i=0}^{r} \alpha_i p_{i,n+1} + 2k_n\epsilon = g(x) + 2k_n\epsilon.$$

We have proved that for any point $x \in B$, $g(x) \leqq f(x) \leqq g(x) + 2k_n\epsilon$, where $g(x)$ is a convex function. Now define $\phi(x) = g(x) + k_n\epsilon$. Then $\phi(x)$ is convex and

$$|\phi(x) - f(x)| \leqq k_n\epsilon \qquad \text{for } x \in B.$$

This completes the proof of theorem 1.

THEOREM 2. *If $f(x)$ is an ϵ-convex function defined on a convex open subset of G of E_n, then there exists a convex function $\phi(x)$ defined on G such that $|f(x) - \phi(x)| \leqq k_n\epsilon$.*

PROOF. Let H_ν, $\nu = 1, 2, 3, \cdots$, be a sequence of convex, compact subsets of G such that $H_{\nu+1} \subset H_\nu$ and such that $G = \bigcup_{\nu=1}^{\infty} H_\nu$ (the existence of such a sequence is easily demonstrated). Then by Theorem 1, there exists for each ν a convex function $\phi_\nu(x)$ on H_ν such that $|\phi_\nu(x) - f(x)| \leqq k_n\epsilon$, for $x \in H_\nu$. For each fixed positive integer μ, the function $f(x)$ is bounded on H_μ by Lemma 2. Hence the sequence $\{\phi_\nu(x)\}$ is defined and uniformly bounded on H_μ for $\nu \geqq \mu$. By a well known selection theorem there exists a subsequence $\{\phi_{1p}(x)\}$ of the $\phi_\nu(x)$ which converges for all $x \in H_1$. Similarly there is a subsequence $\{\phi_{2p}(x)\}$ of the $\phi_{1p}(x)$ which is defined and convergent on H_2, and so on. Now consider the sequence $\{\phi_{pp}(x)\}$, $p=1, 2, 3, \cdots$. For any given $x \in G$, there exists a positive integer m so that $x \in H_m$. Hence for $p \geqq m$, the sequence $\{\phi_{pp}(x)\}$ is defined and converges to a limit $\phi(x)$. Thus $g(x)$ is defined, is convex, and satisfies the inequality $|\phi(x)-f(x)| \leqq k_n\epsilon$ for $x \in G$.

REFERENCES

1. D. H. Hyers, *On the stability of the linear functional equation*, Proc. Nat. Acad. Sci. U.S.A. vol. 27 (1941) pp. 222–224.

2. D. H. Hyers and S. M. Ulam, *On approximate isometries*, Bull. Amer. Math. Soc. vol. 51 (1945) pp. 288–292.

3. ———, *Approximate isometries of the space of continuous functions*, Ann. of Math. vol. 48 (1947) pp. 285–289.

4. D. G. Bourgin, *Approximate isometries*, Bull. Amer. Math. Soc. vol. 52 (1946) pp. 704–714.

5. T. Bonnesen and W. Fenchel, *Konvexe Körper*, New York, 1948.

6. D. G. Bourgin, *Classes of transformations and bordering transformations*, Bull. Amer. Math. Soc. vol. 57 (1951) pp. 223–237.

THE UNIVERSITY OF SOUTHERN CALIFORNIA AND
THE LOS ALAMOS SCIENTIFIC LABORATORY

ON THE STABILITY OF DIFFERENTIAL EXPRESSIONS

S. M. Ulam and D. H. Hyers

1. *Introduction*

Every student of calculus knows that two functions may differ uniformly by a small amount and yet their derivatives may differ widely. On the other hand, it is more or less obvious intuitively, and quite easily proved, that if a continuous function f on a finite closed interval has a proper maximum at a point $x=a$ then any continuous function g sufficiently close to f also has a maximum arbitrarily close to $x=a$. When f and g have first derivatives, this means that f' vanishes and changes sign at $x=a$ while g' vanishes at some point of a preassigned neighborhood of $x=a$ (see theorem 1 for the case $n=1$).

Thus, in spite of the first sentence of the above paragraph, there are certain cases in which the derivatives cannot "differ much" *providing* however that we are allowed to *shift* the point a little at which the derivative is taken. This paper gives a few elementary results along these lines. Later we plan to develop these ideas further and give var- is applications.

2. *Some Theorems for One Independent Variable*

In this section we shall use the term "neighborhood of a point" to mean an interval whose center is the point. Its "radius" is half the length of the interval.

Theorem 1: *Let $f(x)$ be a function having an nth derivative in a neighborhood N of the point $x=a$. If $f^{(n)}(a)=0$, and $f^{(n)}(x)$ changes sign at $x=a$, then corresponding to each $\varepsilon>0$ there exista a $\delta>0$ such that, for each function $g(x)$ having an nth derivative in N and satisfying the inequality $|g(x)-f(x)|<\delta$ in N, there exists a point $x=b$ such that $g^{(n)}(b)=0$ and $|b-a|<\varepsilon$.*

Proof: Let ϵ be chosen (without loss of generality) less than the radius of N, and assume that $f^{(n)}(x)$ is negative immediately to the left of $x=a$ and positive to the right (the opposite case is handled similarly). Then there exist points x_1 and x_2 such that $f^{(n)}(x_1)<0$, $f^{(n)}(x_2)>0$, $|x_1-a|<\varepsilon/2$, $|x_2-a|<\varepsilon/2$. Let $h>0$ be chosen so that $h<\varepsilon/2n$ and also so that $\Delta_h^n f(x_1)<0$ and $\Delta_h^n f(x_2)>0$, and keep h fixed. Let $\delta>0$ be chosen to be smaller than $(|\Delta_h^n f(x_1)|)/2^n$ and $(|\Delta_h^n f(x_2)|)/2^n$. Now, if $g(x)$ is any nth-differentiable function defined for $x\in N$ and such that $|g(x)-f(x)|<\delta$ for $x\in N$, we form the nth difference of g at x_1 and at x_2 and obtain:

For commentary to this paper [51], see p. 687.

$$|\Delta_h^n g(x_i) - \Delta_h^n f(x_i)| \le |g(x_i+nh) - f(x_i+nh)|$$
$$+ n|g(x_i+(n-1)h) - f(x_i+(n-1)h)|$$
$$+ \binom{n}{2}|g(x_i+(n-2)h) - f(x_i+(n-2)h)| + \cdots$$
$$+ |g(x_i) - f(x_i)|, \quad i=1,2.$$

By the choice of h and ε it follows that $x_i + rh \in N$, $r = 0,1,2, \cdots, n$, so that

$$|\Delta_h^n g(x_i) - \Delta_h^n f(x_i)| < \delta[1+\binom{n}{1}+\binom{n}{2}+\cdots+\binom{n}{n-1}+1] = 2^n\delta.$$

By the choice of δ it follows that

$$\Delta_h^n g(x_1) < 0 \quad \text{and} \quad \Delta_h^n g(x_2) > 0$$

Now by the law of the mean we have

$$\Delta_h^n g(x_1) = h^n g^{(n)}(x_1+n\theta_1 h), \quad \text{where} \quad 0<\theta_1<1$$
$$\Delta_h^n g(x_2) = h^n g^{(n)}(x_2+n\theta_2 h), \quad \text{where} \quad 0<\theta_2<1.$$

Since $g^{(n)}(x)$ is a derived function* it follows that $g^{(n)}(b)=0$ for some b between $x_1+n\theta_1 h$ and $x_2+n\theta_2 h$. Now $x_1 < b < x_2+nh < x_2+\varepsilon/2$, and $|x_i - a| < \varepsilon/2$, $i=1,2$, so that $|b-a| < \varepsilon$.

Theorem 2: *Let $F(x,y)$ be a continuous function of two variables in a region R containing the point (a,b), and let $f(x)$ be a function with a continuous derivative $f'(x)$ in the neighborhood N of the point a and such that $f'(x) - F(x,f(x))$ vanishes at a and changes sign in every neighborhood of the point a, where $b=f(a)$. Then corresponding to each $\varepsilon>0$ there is a $\delta>0$ such that, for each differentiable function $g(x)$ satisfying the inequality $|f(x)-g(x)|<\delta$, $(x\in N)$, there exists a point x_0 such that $g'(x_0) = F(x_0, g(x_0))$, and such that $|x_0 - a| < \varepsilon$.*

Proof: Choose x_1 and x_2, within N and also within the neighborhood $|x-a|<\varepsilon$, such that

$$\phi(x_1') = \alpha_1 = f'(x_1') - F(x_1', f(x_1')) < 0$$
$$\phi(x_2') = \alpha_2 = f'(x_2') - F(x_2' f(x_2')) > 0$$

Since the (proper, relative) maximum and minimum values of the continuous function $\phi(x)$ form a finite or enumerably infinite set**, it is possible

* See for example Franklin, A Treatise on Advanced Calculus, p. 117.
** Cf. Hobson, Theory of Functions of a Real Variable, vol. I, p. 329.

to select x_1', x_2' such that $\phi(x)$ does not have a maximum or minimum value at x_1' or at x_2'. Hence in every neighborhood of x_1' there are points ξ and η such that $\phi(\xi) > \phi(x_1') > \phi(\eta)$ and similarly for x_2'. Hence, if we put $C_1 = \alpha_1 + F(x_1', f(x_1'))$, the function $f'(x) - C_1$ vanishes at $x = x_1'$ and changes sign in every neighborhood of $x = x_1'$. Let η_1 be chosen so that the interval $|x - x_1'| < \eta_1$ is within N and also within the neighborhood $|x - a| < \varepsilon$, and also so that $|F(x', y') - F(x, y)| < |\alpha|/2$ for $|x' - x| < \eta_1$ and $|y' - y| < \eta_1$. Since $f(x)$ is continuous, there exists $\varepsilon_1 > 0$ such that $\varepsilon_1 < \eta_1$ and such that $|f(x') - f(x)| < \eta/2$ for $|x' - x| < \varepsilon_1$ (x and $x' \in N$). By theorem 1, there is a $\delta_1 > 0$, which may be chosen so that $\delta_1 < \eta_1/2$, so that for any differentiable function $g(x)$ defined on N, with $|g(x) - f(x)| < \delta_1$ for $x \in N$, there exists a point $x = x_1$ such that $|x_1' - x_1| < \varepsilon_1$ and such that $g'(x_1) = C_1 = \alpha_1 + F(x_1', f(x_1'))$. Hence

$$|g'(x_1) - F(x_1, g(x_1)) - \alpha_1| = |F(x_1', f(x_1')) - F(x_1, g(x_1))| < |\alpha|/2,$$

since $|x_1' - x_1| < \varepsilon_1 < \eta_1$ and $|f(x_1') - g(x_1)| \le |f(x_1') - f(x_1)| + |f(x_1) - g(x_1)| < \eta_1/2 + \delta_1 < \eta_1$. Therefore $g'(x_1) - F(x_1, g(x_1)) < \alpha_1/2 < 0$, where x_1 is a point in N such that $|x_1 - a| < \varepsilon$. Similarly, it can be shown that there exists a point x_2 in N such that $|x_2 - a| < \varepsilon$ and $g'(x_2) - F(x_2, g(x_2)) > \alpha_2/2 > 0$. Since $g'(x) - F(x, g(x))$ is a derived function it takes on the value 0 at some point x_0 between x_1 and x_2. Hence $g'(x_0) = F(x_0, g(x_0))$, where $|g(x) - f(x)| < \delta$ and $|x_0 - a| < \varepsilon$.

3. *A Two Dimensional Theorem.*

Theorem 3. *Let $F(x,y)$ be continuous with continuous second partial derivatives in a neighborhood* N of a point (x_0, y_0), and let us suppose that*

(i) $$F_x(x_0, y_0) = F_y(x_0, y_0) = 0$$

(ii) $$F_{yy}(x_0, y_0) \neq 0 \quad \text{and} \quad F_{xy}^2 - F_{xx}F_{yy} \neq 0$$

at (x_0, y_0). Then, given any $\varepsilon > 0$, there exists a $\delta > 0$ such that, if $G(x,y)$ is any function with continuous second partial derivatives and with $G_{yy} \neq 0$ on N, satisfying the inequality

(iii) $$|G(x,y) - F(x,y)| < \delta$$

on N, then there is a point (x_1, y_1) such that $G_x(x_1, y_1) = G_y(x_1, y_1) = 0$, where $|x_1 - x_0| < \varepsilon$, $|y_1 - y_0| < \varepsilon$.

Proof: Consider the equation

*Here N denotes a two dimensional neighborhood, e.g. a square centered at the point in question.

(1) $$F_y(x,y) = 0.$$

Since it is satisfied for $x = x_0$, $y = y_0$ and since $F_{yy} \neq 0$ at (x_0, y_0), there exists a solution $y = \varphi(x)$ of Eq. (1) passing through the point (x_0, y_0) and having a continuous derivative, for x in some neighborhood $|x - x_0| \leq a$ of the point x_0.

Given $\varepsilon > 0$, choose $\varepsilon_1 > 0$ so that $|\varphi(x) - \varphi(x_0)| < \varepsilon/2$ for $|x - x_0| < \varepsilon_1$, where $\varepsilon_1 < \varepsilon$.

Now consider the function

$$\Phi(x) = F(x, \varphi(x)),$$

which has continuous first and second derivatives

$$\Phi'(x) = F_x(x, \varphi(x))$$

$$\Phi''(x) = \left. \frac{F_{xx}F_{yy} - F_{xy}^2}{F_{yy}} \right|_{y = \varphi(x)}$$

By hypothesis, $\Phi''(x_0) \neq 0$. Hence $\Phi'(x)$ vanishes and changes sign at $x = x_0$, so by theorem 1, there exists $\delta_1 > 0$ such that, if $g(x)$ is any differentiable function with $|g(x) - \Phi(x)| < \delta_1$ for $|x - x_0| \leq a$, then $g'(x_1) = 0$ for some point x_1 satisfying $|x_1 - x_0| < \varepsilon_1$.

By hypothesis $F(x,y)$ is continuous so it is continuous in y, uniformly with respect to x for $|x - x_0| \leq a$. Let $\varepsilon_2 < \varepsilon/2$ be chosen so that

$$|F(x, y_1) - F(x, y_2)| < \delta_1/2$$

for $|y_1 - y_2| < \varepsilon_2$ and all x satisfying $|x - x_0| \leq a$, where it is understood that $(x, y_i) \in N$, $i = 1, 2$.

By theorem 1 it follows that for each fixed x in the interval I: $|x - x_0| \leq a$ there is a $\delta(x)$ such that for any differentiable function $h(y)$ satisfying $|h(y) - F(x,y)| < \delta(x)$, $y \in N$, there exists y_1 such that $|y_1 - \varphi(x)| < \varepsilon_2$ and $h'(y_1) = 0$. The closed interval I may be covered with intervals of length $\delta(x)$ and by the Heine-Borel theorem there exists a finite subcovering and hence a δ_I independent of x. Now let $G(x,y)$ be any function defined on N with continuous second derivatives and with $G_{yy} \neq 0$ on N, such that inequality (iii) is satisfied, where δ is the smaller of δ_I and $\delta_1/2$. Then there exists a function $y = \varphi_1(x)$ such that $|\varphi(x) - \varphi_1(x)| < \varepsilon_2$ and $G_y(x, \varphi_1(x)) = 0$ for $|x - x_0| \leq a$. Since $G_{yy} \neq 0$, $y = \varphi_1(x)$ is differentiable. Hence the function $\Psi(x) = G(x, \varphi_1(x))$ is differentiable and $\Psi'(x) = G_x(x, \varphi_1(x))$.

Now

$$|\Psi(x) - \Phi(x)| \leq |G(x, \varphi_1(x)) - F(x, \varphi_1(x))| + |F(x, \varphi_1(x)) - F(x, \varphi(x))|$$

The first term on the right is less than $\delta \leq \delta_1/2$, and since $|\varphi_1(x) - \varphi(x)| < \varepsilon_2$, for $|x - x_0| \leq a$ we have $|F(x, \varphi_1(x) - F(x, \varphi(x))| < \delta_1/2$. Therefore $|\Psi(x) - \Phi(x)| < \delta_1$ for $|x - x_0| \leq a$. Hence $\Psi(x)$ is a $g(x)$, and there exists a point x_1 with $|x_1 - x_0| < \varepsilon_1$ such that

$$\Psi'(x_1) = G_x(x_1, \varphi_1(x_1)) = 0.$$

Putting $y_1 = \varphi_1(x_1)$ we have $G_x(x_1, y_1) = 0$, $G_y(x_1, y_1) = 0$ and $|x_1 - x_0| < \varepsilon_1 < \varepsilon$, while

$$|y_1 - y_0| \leq |\varphi_1(x_1) - \varphi(x_1)| + |\varphi(x_1) - \varphi(x_0)| < \varepsilon_2 + \varepsilon/2 < \varepsilon,$$

so the theorem is proved.

4. *Some Counter-Examples.*

1. The following example is due to John von Neumann.

Take $F(x, y, y', y'') = y'^2 - yy'' - x = 0$, $f(x) \equiv 0$, $x_0 = 0$. Then $F(x, f(x), f'(x), f''(x)) = -x$ obviously changes sign as x passes through the value x_0. Now take $g(x) = \delta \sin(x/\delta)$ where δ is any positive number, and take $\varepsilon = \frac{1}{2}$. Obviously $|g(x) - f(x)| < \delta$ for all x. Since $g'(x) = \cos(x/\delta)$, $g''(x) = -(1/\delta)\sin(x/\delta)$, $F(x, g(x), g'(x), g''(x)) = \cos^2(x/\delta) + \sin^2(x/\delta) - x = 1 - x$. Thus the root of $F(x, g(x), g'(x), g''(x)) = 0$ is at $x = 1$, which is at a distance $1 > \frac{1}{2}$ from the root $x = 0$ of $F(x, f(x), f'(x), f''(x))$. The stability theorem fails in this case. One must be careful about formulating "implicit" stability theorems in derivatives.

2. In the two dimensional case, one has to be careful about formulating the condition of "change of sign", as the following example shows. Geometrically, it amounts essentially to taking a torus with axis OZ, and tipping it slightly.

Consider the function

$$z = F(x, y) = \sqrt{b^2 - a^2 + 2a\sqrt{x^2 + y^2} - x^2 - y^2}, \qquad (a > b)$$

in the neighborhood of the point $(a, 0)$. On differentiating we have:

$$F_x = \frac{x}{z}\left[\frac{a}{\sqrt{x^2 + y^2}} - 1\right]$$

$$F_y = \frac{y}{z}\left[\frac{a}{\sqrt{x^2 + y^2}} - 1\right]$$

Clearly F_x changes sign from + to − as x increases through $x = a$ and y

is held constant ($y=0$). Also F_y changes sign from + to − as y increases through $y=0$ and x is held fixed ($x=a$).

Now consider the following function

$$z = G(x,y) = \delta y + F(x,y),$$

where δ is a small positive number. Then for $|x-a| < b/2$ and $|y| < b/2$, $|G(x,y) - F(x,y)| < \delta$. We have

$$G_x = \frac{x}{F(x,y)}\left(\frac{a}{\sqrt{x^2+y^2}} - 1\right) = F_x$$

$$G_y = \delta + \frac{y}{F(x,y)}\left(\frac{a}{\sqrt{x^2+y^2}} - 1\right).$$

The locus of zeros of G_x in the neighborhood of $(a,0)$ is obviously the circle $x^2+y^2=a^2$. But along this circle, $G_y = \delta > 0$. Hence there is no point near $(a,0)$ where both $G_x = 0$ and $G_y = 0$.

Los Alamos Scientific Laboratory and
University of Southern California

Reprinted from JOURNAL OF COMBINATORIAL THEORY Vol. 1, No. 2, September 1966

On Some Possibilities of Generalizing the Lorentz Group in the Special Relativity Theory *

C. J. EVERETT AND S. M. ULAM

University of California, Los Alamos Scientific Laboratory, Los Alamos, New Mexico

The special Theory of Relativity deals with a four-dimensional space which is the (Euclidean)R_4 topologically and a Euclidean metric in it. This metric is not positive definite. The Lorentz group consists of linear transformations of this Minkowski space which leave invariant the "light cones," that is to say the sets of points such that the Euclidean norm of the space part of the vector is equal to the norm (i.e., the absolute value) of the time coordinate assuming the choice of units made so that the velocity of light $= 1$).

Starting with this definition of the Lorentz transformations, one might think of the problem of determining, analogously, the group of transformations in more general spaces. Such a space could be considered as a direct product of a purely spatial set S and a "time-space" T, i.e., $M = S \times T$. M would then be a more general "Minkowski space." If one assumes that S and T, in addition to being metric spaces, have an algebraic structure, e.g., they are groups, one could try to determine all automorphisms of M which leave invariant those sets consisting of all points of M for which the norm of the S-component is equal to the norm in T-component. One could, of course, consider a still more general problem when S and T are not necessarily provided with an algebraic structure but are merely metric spaces; consider their direct product suitably metrized and try to determine all the isometries of M onto itself which preserve the sets corresponding to the "light cones" which

* Work performed under the auspices of the U. S. Atomic Energy Commission.

For commentary to this paper [86], see p. 687.

would be then defined as the sets of all points for which the distance in $S =$ the distance in T, distance from any fixed point considered as the "vertex" of the cone.

Should there arise a need in some physical theories of having models of space-time which are not topologically Euclidean, the determination of groups of transformations like the above might be of interest. We are thinking here, primarily of global transformations of the whole space onto itself and not of only local such "coordinate" changes. Only a very special case of this general problem will be considered here—that of S being the p-adic vector space and T being the p-adic number system. The topology usually defined for these spaces is of course non-Euclidean. This topology gives in fact a totally discontinuous space which is 0-dimensional. It is conceivable that spaces of this sort might be useful in some future models of nuclear or subnuclear theories. They present frameworks which are less apt to lead to divergences in the computation of certain physical quantities.

Many attempts have been made to introduce in physical theories the idea of a minimal length λ. The main difficulty in doing so stems from the fact that it cannot be done in a Lorentz invariant way. Attractive as it would be to have a "quantum" of distance (which would presumably be of the order of 10^{-13} cm), one cannot reconcile the necessary relativistic handling of the nuclear or subnuclear phenomena with such a postulate. What we would like to suggest is, rather, the following possibility: Without assuming the existence of a minimal quantity for a distance in space, one could think of only a discrete set of possible distances which would, however, be infinite, with arbitrarily small values possible, e.g., a set of distances that are actually realizable forming a perfect nowhere dense set—topologically equivalent to the Cantor discontinuum. It is obviously necessary that in macroscopic phenomena, or even the atomic ones, the distances have to behave, metrically and algebraically, very much like the real number system. In the "very small," however, i.e., in dimensions of the order of 10^{-13} cm or less, it is conceivable that we could have a different topologic and algebraic model.

Mathematically, the possibilities of dealing with such a heterogeneous structure involve problems of stability of the notions of isomorphism and automorphism. We would like to insert here a brief discussion of the definitions and give examples of problems concerning such generalized stability.

A model of "space-time" for a physical theory could conceivably consist of a structure which in the small is quite different from the Euclidean one, but for distances exceeding a certain $\epsilon > 0$ would behave very much like the Euclidean space. In the large then, i.e., for distances greater than this fixed ϵ, the symmetry properties of space and space-time should be very much like the Euclidean or Lorentz ones. Some mathematical results are known on the behavior of transformations which are "ϵ-linear" or transformations which are "ϵ-isometries." The definition of the first class is as follows: a transformation $f(x)$ is called ϵ-linear on the vector space X, in case for all x, y

$$\| f(x+y) - f(x) - f(y) \| < \epsilon.$$

Some results of D. H. Hyers and one of us assert that such a transformation must of necessity be everywhere close to a linear one, i.e., that there exists a strictly linear transformation $l(x)$ such that, for all x,

$$\| l(x) - f(x) \| < K \cdot \epsilon$$

where K is a constant independent of ϵ and x (in case of one dimension K is actually 1). A transformation $T(p)$ of a metric space into itself is called an ϵ-isometry if for all p, q

$$| \varrho(p, q) - \varrho(T(p), T(q)) | < \epsilon.$$

Again it was proved that if the given metric space is a Euclidean space or even the Hilbert space, an analogous result holds, namely, there exists a transformation of a space in itself which is a strict isometry $I(p)$, such that for all p, $\varrho(I(p), T(p)) < K \cdot \epsilon$, where again K is independent of ϵ and p.

It would be of interest to prove analogous theorems for the Lorentz transformations, i.e., for transformations preserving the Minkowski (non-positive definite) metric. This group of transformations of the heterogeneous space would appear as the Lorentz group for all the pairs of distances exceeding a given $\epsilon > 0$—this ϵ should be some constant of the order of 10^{-13} cm and would mark, so to say, the boundary between the phenomena which are describable by the classical metric and the other ones taking place in the "very small." It is hoped to discuss these questions in a subsequent paper.

In the present paper, we consider an "event-space" $E = K^{n+1}$ over a non-Archimedean valued field K, and define the "light-cone" Γ as the set of events $\xi = (X, t)$ for which $|X| = v(t)$, where v is the valuation of K, and $|X| \equiv \max v(x_i)$ serves as a norm for the space K^n of "positions" X.

A Lorentz transformation is defined, in analogy with special relativity, as a non-singular linear transformation T such that $T\Gamma \subset \Gamma$, and T is said to be "proper" in case it maps the axis of Γ into its interior.

The set L^* of proper Lorentz transformations is a group, and contains as a subgroup the set S^* of "space-rotations" S of E. The group S^* defines in L^* an equivalence relation, $T' = S_1TS_2$, $S_1, S_2 \in S^*$, and for the corresponding equivalence classes we obtain a representation system F^*, i.e., a complete set of canonical forms.

The transformations of L^* are of a remarkably "classical" kind, being in a sense Galilean, and free of space-contraction and time-dilatation. The improper transformations, which here exist, have no analog in the real case, a brief summary of which is included for contrast.

An abstract formulation of our concretely defined, normed position-space is given in the final section.

1. THE VALUATED FIELD K. A real-valued function v defined on a field K is called a *valuation* in case

V1. $v(0) = 0$, $v(x) > 0$ for $x \neq 0$

V2. $v(xy) = v(x)v(y)$

V3. $v(x + y) \leqq v(x) + v(y)$

(hence, $v(\pm 1) = 1$, $v(-x) = v(x)$, and $v(x^{-1}) = 1/v(x)$ for $x \neq 0$). The fields of complex, real, and rational numbers have as a valuation the absolute value $|x|$.

Moreover, the rational field Q admits other valuations. If p is a fixed prime, then $v_p(0) = 0$, $v_p(x) = p^{-i}$ for $x = p^i a/b (a \neq 0,\ b > 0$ coprime, $p \nmid ab)$, defines the *p-adic* valuation of Q, which satisfies V3 in the stronger form

V4. $v(x + y) \leqq \max(v(x), v(y))$.

Such a valuation is said to be *non-Archimedean*, V4 being equivalent to the inequality $v(n) \leqq 1$ for all integers n. An important consequence of V4 is the property

V5. $v(x) > v(y)$ implies $v(x + y) = v(x)$.

A sequence $\{x_n\}$ of elements of a valuated field K is called *regular* in case $v(x_m - x_n) \to 0$, and K is *complete* if every regular sequence has a limit x in K, i.e., $v(x_n - x) \to 0$.

The method by which the real numbers are constructed from sequences $\{r_n\}$ of rationals, regular with respect to the valuation $v(x) = |x|$, is a quite general completion device for valuated fields. Applied to the valuation v_p of the rational field Q, it yields the field Q_p of *p-adic numbers*. Every such number $x \neq 0$ may be represented as a convergent power series

$$x = p^i(a_0 + a_1 p + a_2 p^2 + \ldots)$$

with integer coefficients, $0 \leqq a_j < p$, $a_0 \neq 0$. The function v defined by $v(0) = 0$, $v(x) = p^{-i}$ for $x \neq 0$, is a non-Archimedean valuation of Q_p, with respect to which Q_p is complete. The rational numbers are "contained" in Q_p as the series with (terminally) periodic a_j. Thus $-1 = 1 + 2 + 2^2 + \ldots$ in R_2. For further details, see [1, 3, 4].

2. THE POSITION-SPACE P. The vector-space $P = K^n$ of (column) vectors $X = [x_i]$ over a non-Archimedean valued field K admits a real-valued *norm*

$$|X| = \max(v(x_i); \quad i = 1, \ldots, n)$$

with the properties

N1. $|0| = 0$, $|X| > 0$ for $X \neq 0$

N2. $|aX| = v(a)|X|$. Hence $|X| = v(x) > 0$ implies

$|x^{-1}X| = 1$.

N3. $|X + Y| \leqq |X| + |Y|$, indeed

N4. $|X + Y| \leqq \max(|X|, |Y|)$, hence also

N5. $|X| > |Y|$ implies $|X + Y| = |X|$.

The set P of "positions" X is a metric space with distance function $d(X, Y) = |X - Y|$, complete when K is v-complete. In the p-adic case, $K = Q_p$, it is also separable and zero-dimensional, with closed compact *unit-sphere*

$$\sigma = (X; \mid X \mid = 1).$$

A *rotation* of P is a norm-preserving linear transformation, concretely, an $n \times n$ matrix $R = [r_{ij}]$ over K, such that $\mid RX \mid = \mid X \mid$ for all X. A permutation matrix P is an obvious example. Since $\mid Z_j \mid = 1$ for the *basic vector* Z_j (1 in row j, 0 elsewhere), the j-th column RZ_j of a rotation R has $\mid RZ_j \mid = 1$ also, and so $v(r_{ij}) \leqq 1$ for all i, j. In view of N2, the rotations are just those matrices for which $R\sigma \subset \sigma$.

THEOREM 2.1. *The set R^* of all rotations is a group, and, for every $R \in R^*$,*

$$\text{(a) } R\sigma = \sigma, \qquad \text{(b) } v(\det R) = 1, \qquad \text{(c) } R^\tau \equiv [r_{ji}] \in R^*.$$

PROOF. Obviously the *identity matrix* I_n is in R^*. If R_1, $R_2 \in R^*$, then $R_1R_2 \in R^*$ also, since $\mid R_1R_2X \mid = \mid R_2X \mid = \mid X \mid$. If $R \in R^*$, then: (1) R^{-1} exists, since $RX = 0$ implies $0 = \mid RX \mid = \mid X \mid$ and $X = 0$; also (2) $R^{-1} \in R^*$, since $\mid R^{-1}X \mid = \mid R(R^{-1}X) \mid = \mid X \mid$. Fix $R = [r_{ij}] \in R^*$. (a) If $\mid Y \mid = 1$, then $\mid R^{-1}Y \mid = 1$, and $R(R^{-1}Y) = Y$. Hence $\sigma \subset R\sigma \subset \sigma$. (b) Since $v(r_{ij}) \leqq 1$, we know from V4 and V2 that $d \equiv \det R$ has $v(d) \leqq 1$. Similarly, $v(d') \leqq 1$ for $d' = \det R^{-1}$. But $v(d)v(d') = v(dd') = 1$. (c) Let $\mid X \mid = v(x_m)$, $Y = R^\tau X$, $Z = R^{-1}Z_m$, where we know $\mid Z \mid = 1$. Since $v(r_{ij}) \leqq 1$, clearly $\mid Y \mid \leqq \mid X \mid$. On the other hand, $\mid X \mid = v(x_m) = v(X^\tau Z_m) = v(X^\tau R R^{-1} Z_m) = v(Y^\tau Z) \leqq \mid Y \mid \mid Z \mid = \mid Y \mid$.

THEOREM 2.2. (a) *If $\mid U \mid = 1 = v(u_1)$, the matrix R with columns $U, Z_2, \ldots, Z_n$ is a rotation.*

(b) *If $\mid X \mid = 1$, then $RZ_1 = X$ for some $R \in R^*$.*

(c) *If $\mid X \mid = \mid Y \mid$, there exists a rotation R such that $RX = Y$.*

PROOF. (a) Fix X with $\mid X \mid = 1$, and set $X' = RX$. Since all entries of R have $v \leqq 1$, we know $\mid X' \mid \leqq \mid X \mid = 1$. If $v(x_1) = 1$, then $v(x_1') = v(u_1)v(x_1) = 1$. If $v(x_1) < 1$, then $v(x_m) = 1$ for some $m > 1$, and $v(x_m') = v(u_m x_1 + x_m) = v(x_m) = 1$ by V5, since $v(u_m z_1) = v(u_m)v(x_1) \leqq v(x_1) < 1$. In either case, $\mid X' \mid = 1$.

(b) If $\mid X \mid = 1 = v(x_m)$, there is a permutation matrix $P \in R^*$ such that $PX \equiv U$ has $u_1 = x_m$. By (a) $RZ_1 = U = PX$ for $R \in R^*$. Hence $P^{-1}RZ_1 = X$, where $P^{-1}R \in R^*$.

(c) If $|X| = 0$, $R = I_n$ serves. Suppose $|X| = |Y| = v(x) > 0$. By V2, $X_1 \equiv x^{-1}X$ and $Y_1 \equiv x^{-1}Y$ have $|X_1| = 1 = |Y_1|$. By (b) there exist $R_1, R_2 \in R^*$ such that $R_1Z_1 = X_1$, $R_2Z_1 = Y_1$. Therefore, for $R \equiv R_2R_1^{-1} \in R^*$, $RX_1 = Y_1$ and $RX = Y$.

3. THE EVENT-SPACE E. For a non-Archimedean valued field K, we consider the space $E = K^{n+1}$ of "events" $\xi = \left|\begin{smallmatrix} X \\ t \end{smallmatrix}\right|$ where X is a "position" vector of the normed space $P = K^n$ ($n \geqq 2$!) and $t \in K$. The *light-cone* of E is the set of events

$$\Gamma = (\xi;\ |X| = v(t))$$

and a *Lorentz transformation* is a non-singular linear transformation T of E such that

$$T\ \Gamma \subset \Gamma.$$

Writing $T\xi = \xi'$, where

$$T = \begin{bmatrix} A & B \\ C^\tau & d \end{bmatrix}$$

is of order $n + 1$ over K, this condition reads

$$|X| = v(t) \text{ implies } |AX + Bt| = v(C^\tau X + dt)$$

or, equivalently,

$$|X| = 1 \text{ implies } |AX + B| = v(C^\tau X + d).$$

We need at once the trivial

LEMMA. *For $C \in K^n$, $d \in K$, with $|C| \geqq v(d)$, there exists an $X \in K^n$ such that $|X| = 1$ and $C^\tau X + d = 0$.*

PROOF. If $C = 0$, $X = Z_1$ serves. If $0 < |C| = v(c_1)$ (say), then $C^\tau X + d = 0$ for the vector X with $x_1 = -c_1^{-1}(c_2 + d)$; $x_2 = 1$; $x_i = 0$, $i > 2$; and $|X| = 1$ since $v(c_2 + d) \leqq \max(v(c_2), v(d)) \leqq v(c_1)$ implies $v(x_1) \leqq 1$.

THEOREM 3.1. *If T is a Lorentz transformation, then* (a) $|C| < v(d)$, $|X| = 1$ *implies* $|AX + B| = v(d)$, *and* (c) $|B| \leqq v(d)$. *Hence* $T = dT_1$, $d \neq 0$, *where*

$$T_1 = \begin{bmatrix} A_1 & B_1 \\ C_1{}^\tau & 1 \end{bmatrix}$$

has $|B_1| \leqq 1$, $|C_1| < 1$, *and* $|A_1X + B_1| = 1$ *when* $|X| = 1$.

PROOF. (a) If $|C| \geqq v(d)$, then for the X of the Lemma, $|X| = 1$, and hence $|AX + B| = v(C^\tau X + d) = 0$, violating the non-singularity of T.

(b) For $|X| = 1$, we must then have $|AX + B| = v(C^\tau X + d) = v(d)$, by V5, since $v(C^\tau X) \leqq |C||X| = |C| < v(d)$.

(c) Let $|B| = v(b_m)$. From the lemma, with $d = 0$ and the m-th row of A for C, we obtain an X with $|X| = 1$ such that $x_m' = \Sigma a_{mj} x_j + b_m = b_m$. Hence by (b), $v(d) = |AX + B| \geqq v(x_m') = v(b_m) = |B|$.

We say a Lorentz transformation T is *proper* in case $|B| < v(d)$, equivalently, for the event ξ with $X = 0$, $t = 1$, the event $\xi' = T\xi$ has $|X'| < v(t')$. Geometrically, this means that T maps the axis $X = 0$, $t = t$ of the cone Γ into a line $X' = Bt$, $t' = dt$ lying *inside* the cone, i.e., $|X'| < v(t')$ for $t' \neq 0$.

THEOREM 3.2. *Let* $T = dT_1$ *be a proper Lorentz transformation. Then*

(a) $|B_1| < 1$, $|C_1| < 1$, *and* A_1 *is a rotation of* K^n.

(b) *Writing* $X' = d(A_1X + B_1t)$

$$t' = d(C^\tau{}_1X + t)$$

we have the implications

1. $$|X| \leqq v(t) \Rightarrow v(t') = v(d)v(t)$$

$$|X| < v(t) \Rightarrow |X'| < v(t')$$

2. $$|X| \geqq v(t) \Rightarrow |X'| = v(d)|X|$$

$$|X| > v(t) \Rightarrow |X'| > v(t')$$

Hence T maps the interior and exterior of Γ *into themselves respectively*

(c) $T\Gamma = \Gamma$

(d) *The set* L^* *of all proper Lorentz transformations is a group.*

PROOF. (a) From Theorem 3.1, $|B_1|$, $|C_1| < 1$. To prove $A_1 \in R^*$, let $|X| = 1$. Since $|A_1X + B_1| = 1$ and $|B_1| < 1$, we cannot have $|A_1X| < 1$, by N4. Hence $|A_1X| \geqq 1 > |B_1|$ and so $1 = |A_1X + B_1| = |A_1X|$ by N5.

(b) Note first that the lst parts of 1,2 are trivial when $t = 0$, $X = 0$, respectively. Now

1. Let $|X| \leqq \nu(t), t \neq 0$. We know $\nu(t') = \nu(d)\nu(C_1{}^\tau X + t)$. Since $\nu(C_1{}^\tau X) \leqq |C_1||X| \leqq |C_1|\,\nu(t) < \nu(t)$, the result follows from V5. If $|X| < \nu(t)$, then $t \neq 0$. We know $|X'| = \nu(d)\,|A_1X + B_1t|$. Since $\nu(d) > 0$, and $\nu(d)\nu(t) = \nu(t')$ as just shown, it suffices to note $|A_1X + B_1t| \leqq \max\,(|A_1X|, |B_1t|) = \max\,(|X|, \nu(t)\,|B_1|) < \nu(t)$.

2. Let $|X| \geqq \nu(t), X \neq 0$. We know $|X'| = \nu(d)\,|A_1X + B_1t|$. Since $|B_1t| = \nu(t)\,|B_1| \leq |X||B_1| < |X| = |A_1X|$, the result follows from N5. If $|X| > \nu(t)$, then $X \neq 0$. We know $\nu(t') = \nu(d)\nu(C_1{}^\tau X + t)$. Since $\nu(d) > 0$ and $\nu(d)\,|X| = |X'|$, it suffices to note that $\nu(C_1{}^\tau X + t) \leqq \max\,(\nu(C_1{}^\tau X), \nu(t)) < |X|$.

(c) By definition, T^{-1} exists and $T\Gamma \subset \Gamma$. To prove $\Gamma \subset T\Gamma$, let $\eta \in \Gamma$. Then $T(T^{-1}\eta) = \eta$, and $T^{-1}\eta$ must be on Γ by the last remark of (*b*).

(*d*) Obviously $I_{n+1} \in L^*$. If $T_1, T_2 \in L^*$, then T_1T_2 is non-singular and $T_1T_2\Gamma \subset T_1\Gamma \subset \Gamma$. Setting $\xi' = (T_1T_2)\xi$ where $X = 0$, $t = 1$ in ξ, it follows from (b.1) that $|X'| < \nu(t')$, and T_1T_2 is proper. If $T \in L^*$, then T^{-1} is non-singular, and $T^{-1}\Gamma \subset \Gamma$ follows from $\Gamma \subset T\Gamma$. Setting $\xi' = T^{-1}\xi$ for the axial point $X = 0$, $t = 1$, we must have ξ' interior to Γ, otherwise $T\xi' = \xi$ would be on or outside Γ along with ξ'.

4. Lorentz Canonical Form. The set S^* of all *space-rotations*

$$S = \begin{bmatrix} R & 0 \\ 0^\tau & 1 \end{bmatrix}, \quad R \in R^*$$

of the event-space $E = K^{n+1}$ is a subgroup of the group L^* of proper Lorentz transformations. For $S \in S^*$, clearly $\nu(\det S) = 1$ and $S^\tau \in S^*$.

The relation $T' = S_1TS_2$ (for some $S_1, S_2 \in S^*$) is an equivalence relation ($T' \sim T$), which splits the group L^* into disjoint classes

$$\{T\} = (T';\, T' \sim T,\, T' \in L^*).$$

We now exhibit a *representation system* F^*, consisting of precisely one matrix F from each class, which may therefore be regarded as a complete set of "canonical forms" for L^*, every proper Lorentz transformation being equivalent to a unique F of F^*.

First we stipulate, for every real number $0 < r < 1$ which occurs as a value of K, a *unique* element $x_0 = f(r)$ of K for which $v(x_0) = r$, and denote by K_0 the set of all such x_0. For example, if $K = Q_p$, we might take $K_0 = (p^i; i \geqq 1)$, where $f(p^{-i}) = p^i$.

Now let F^* be the set of *all* matrices $F = dF_2$, $d \neq 0$,

$$F_2 = \begin{bmatrix} I_n & B_2 \\ C_2{}^\tau & 1 \end{bmatrix}$$

of the following *types*:

I. $B_2 = 0$, with

(A) $C_2{}^\tau = 0$

or (*B*) $C_2{}^\tau = (c_0, 0, \ldots, 0)$, $c_0 \in K_0$

II. $B_2 = b_0 Z_1$, $b_0 \in K_0$, with

(A) $C_2{}^\tau = 0$

or (B) $C_2{}^\tau = (c_1, 0, \ldots, 0)$, $c_1 \in K$, $0 < v(c_1) < 1$

or $C_2{}^\tau = (c_1, c_0, 0, \ldots, 0)$, $c_1 \in K$, $c_0 \in K_0$,

$0 \leqq v(c_1) < v(c_0)$.

This complexity, in sharp contrast to the real case (§7), is dictated by the following principal:

THEOREM 4.1. *The set F^* is a complete set of canonical forms for the group L^* of proper Lorentz transformations, i.e.,*

1. *No two distinct matrices F of F^* are equivalent.*
2. *Every $T \in L^*$ is equivalent to some $F \in F^*$.*
3. *Every F of F^* belongs to L^*.*

PROOF. 1 Suppose $S_1 F S_2 = F'$, where $S_1, S_2 \in S^*$ and $F = dF_2$, $F' = d'F_2' \in F^*$. This implies $d = d'$, and

$$R_1 R_2 = I_n, \quad R_1 B_2 = B_2', \quad C_2{}^\tau R_2 = C_2{}'^\tau; \quad \text{for } R_1, R_2 \in R^*.$$

Thus $|B_2| = |B_2'|$, $|C_2| = |C_2'|$, and we see at once that F of type

IA, B or IIA implies F' of the same type, and hence also $F = F'$, by the uniqueness of f: $v(x_0) = v(x_0')$, $x_0, x_0' \in K_0 \Rightarrow x_0 = x_0'$.

Let F be of type IIB or IIC. Then F' must be one of these types, and clearly $B_2' = B_2 = b_0 Z_1$. Hence $R_1 Z_1 = Z_1$, and $Z_1 = R_2 Z_1$ is the first column of R_2. But $C_2^\tau R_2 = C_2'^\tau$, so C_2 and C_2' have the same first component c_1. Thus $F = F'$ if both are of the same type IIB or IIC. But they cannot be of different types for this would imply two vectors, of form

$$(c_1, 0, \ldots, 0) \qquad \text{and} \qquad (c_1, c_0, 0, \ldots, 0)$$

having the same norm, and with $v(c_1) < v(c_0)$.

2. Given $T = dT_1 \in L^*$, $d \neq 0$, $|B_1| < 1$, $|C_1| < 1$, $A_1 \in R^*$ we have to produce $S_1, S_2 \in S^*$ such that $dS_1 T_1 S_2 = dF_2 \in F^*$ which reads, in terms of blocks,

$$R_1 A_1 R_2 = I_n, \qquad R_1 B_1 = B_2, \qquad C_1^\tau R_2 = C_2; \qquad R_1, R_2 \in R^*.$$

We consider T_1 under separate *cases*, which yield the corresponding *types* for F_2.

CASE IA. $|B_1| = 0 = |C_1|$. For $R_1 = I_n$, $R_2 = A_1^{-1}$, F_2 is of *type* IA. (This is the case of the space rotations $T_1 \in S^*$).

CASE IB. $|B_1| = 0 < |C_1| < 1$. Let $c_0 = f(|C_1|)$. Then $|c_0 Z_1| = v(c_0) = |C_1|$. Definining R_0 as a rotation such that $R_0 C_1 = c_0 Z_1$, $R_2 = R_0^\tau \in R^*$, and $R_1 = R_2^{-1} A_1^{-1}$, we obtain F_2 of *type* IB.

CASE II. (Preliminary). $0 < |B_1| < 1$. Let $b_0 = f(|B_1|)$, so that $|b_0 Z_1| = |B_1|$. Define $R_1' \in R^*$ so that $R_1' B_1 = b_0 Z_1$, and set $R_2' = A_1^{-1} R_1'^{-1}$. Using these rotations for S_1', S_2' yields an intermediate matrix $T_1' = S_1' T_1 S_2'$ of form

$$T_1' = \begin{bmatrix} I_n & b_0 Z_1 \\ Q^\tau & 1 \end{bmatrix}$$

where $Q^\tau = C_1^\tau R_2' = (q_1, \ldots, q_n)$ has $|Q| = |C_1|$.

CASE IIA. $|C_1| = 0$. Then $Q = 0$, and $F_2 = T_1'$ is of *type* IIA.

Finally, suppose $0 < |C_1| = |Q| < 1$. Using T_1' for T_1 in iteration of the original argument shows that we now require rotations R_1, R_2 such that

$$R_1 R_2 = I_n \qquad R_1 Z_1 = Z_1 \qquad Q^\tau R_2 = C_2$$

where C_2 has the form in *types* IIB or IIC. Our final cases are based on the nature of the intermediate Q where we recall $0 < | C_1 | = | Q | < 1$.

CASE IIB. $v(q_1) = | Q |$. Then $v(q_j) \leq v(q_1)$, $j \geq 1$, and

$$R_1 = \begin{bmatrix} 1 & q_2/q_1 & \cdots & q_n/q_1 \\ 0 & & I_{n-1} & \end{bmatrix}$$

is a rotation, such that $R_1 Z_1 = Z_1$, and $q_1 Z_1{}^\tau R_1 = Q^\tau$. Hence $Q^\tau R_1^{-1} = q_1 Z_1{}^\tau$, and defining $R_2 = R_1^{-1}$ yields an F_2 of type IIB.

CASE IIC. $v(q_1) < | Q | = v(q_m)$, $m > 1$. Let $P \in R^*$ be a permutation matrix ($P^{-1} = P^\tau$!) such that

$$Q^\tau P^{-1} = (q_1; q_2, \ldots, q_n) \begin{vmatrix} 1 & 0^\tau \\ 0 & P^\tau_{n-1} \end{vmatrix} = (q_1; q_m, \ldots, q_t).$$

Let $c_0 = f(| C_1 |)$, so that $v(c_0) = v(q_m)$, and hence $v(q_1) < v(c_0)$, $v(q_j/c_0) \leq 1$, $j \geq 2$. It follows that

$$R_{n-1} = \begin{vmatrix} q_m/c_0 & \cdots & q_t/c_0 \\ 0 & I_{n-2} & \end{vmatrix}$$

is a rotation of K^{n-1}, therefore

$$R_n = \begin{vmatrix} 1 & 0^\tau \\ 0 & R_{n-1} \end{vmatrix}$$

is a rotation of K^n. Moreover,

$$Q^\tau P^{-1} = (q_1; q_m, \ldots, q_t) = (q_1, c_0, 0, \ldots, 0) R_n.$$

Therefore $Q^\tau (R_n P)^{-1} = (q_1, c_0, 0, \ldots, 0)$ and $(R_n P) Z_1 = R_n Z_1 = Z_1$. If we set $R_1 = R_n P$ and $R_2 = R_1^{-1}$, we obtain an F_2 of *type* IIC.

3. If $F = dF_2 \in F^*$, we shall see that F_2 and hence F is in L^* by noting that F_2 is an element of the special subgroup G^* of L^* defined in

THEOREM 4.2. *Let*

$$G = \begin{bmatrix} A & B \\ C^\tau & d_1 \end{bmatrix}$$

be a matrix over K such that

1. *A is the identity I_n except for its first row, which is of the form $(a_1, a_2, 0, \ldots, 0) \equiv A_1{}^\tau$, where $| A_1 | = 1 = v(a_1)$.*

2. $B = b_1 Z_1$, *with* $|B| = \nu(b_1) < 1$.
3. $C^\tau = (c_1, c_2, 0, \ldots, 0)$, *where* $|C_1| < 1$.
4. $\nu(d_1) = 1$.

Then (a) $G \in L^*$, *and* $\nu(D) = 1$ *for* $D = \det G = a_1 d_1 - c_1 b_1$.
(b) *The set* G^* *of all such* G *is a group, with identity* I_{n+1}.

PROOF. (a) det $G = a_1 d_1 + (-1)^{n+2} c_1 (-1)^{n+1} b_1 = a_1 d_1 - c_1 b_1$ has $\nu = 1$ since $\nu(c_1 b_1) = \nu(c_1)\nu(b_1) < 1 = \nu(a_1 d_1)$. Hence G is non-singular. To prove $G\Gamma \subset \Gamma$, let $|X| = 1$. Then $|AX + B| = \nu(C^\tau X + d_1)$. Indeed, both are unity. For, A is a rotation, and $|B| < 1$, hence $|AX+B| = |AX| = |X| = 1$. Also, $\nu(C^\tau X + d) = \nu(d_1) = 1$, since $\nu(C^\tau X) \leq |C||X| < 1$. Finally $|B| < 1 = \nu(d_1)$ and G is proper.

(b) The matrix I_{n+1} is of the above form. Computing the product GG' of two such matrices shows it to be of the same structure, with first and last rows respectively:

$$(a_1 a_1' + b_1 c_1'),\quad (a_1 a_2' + a_2 + b_1 c_2'),\ 0, \ldots, 0,\ (a_1 b_1' + b_1 d_1')$$

$$(c_1 a_1' + d_1 c_1'),\quad (c_1 a_2' + c_2 + d_1 c_2'),\ 0, \ldots, 0,\ (c_1 b_1' + d_1 d_1').$$

That these have the required character is clear from V4, V5. From the product form, one easily infers that G has an inverse G' with first and last rows:

$$(d_1 D^{-1}),\ (-d_1 a_2 + b_1 c_2) D^{-1},\ 0, \ldots, 0,\ (-b_1 D^{-1})$$

$$(-c_1 D^{-1}),\ (c_1 a_2 - a_1 c_2) D^{-1},\ 0, \ldots, 0,\ (a_1 D^{-1})$$

where $D = \det G$ has $\nu(D) = 1$.

COROLLARY. *For* $T \in L^*$, $\nu(\det T) = \nu(d)$.

REMARK. The result of Theorem 4.2 is also true for the wider class of matrices with $A^\tau = (a_1, \ldots, a_n)$, $C^\tau = (c_1, \ldots, c_n)$ subject to the same conditions. G^* appears to be the smallest group containing all the canonical forms F_2.

5. SOME "PHYSICAL" ASPECTS. A Lorentz transformation $T\xi = \xi'$ may be regarded as a one-one mapping of *two* event-spaces E and E', corresponding vectors ξ, ξ' denoting the "same event" as defined in the

two frames. Then the basic property $T\Gamma \subset \Gamma'$ signifies that events ξ of E with $|X| = \nu(t)$ (light-front, speed of light $c = 1$) appear in E' as events ξ' with $|X'| = \nu(t')$ also.

The spatial origin $X = 0$ of E at time t appears in E' as the event $X' = Bt$, $t' = dt$. Hence $X' = B_1 t'$ is the trajectory of 0 in E', and B_1 is its velocity vector. The proper transformations $T \in L^*$ are those for which the relative speed $b_0 = |B_1|$ is less than 1, the speed of light. In contrast to the real case (§7), improper transformations T exist, and one can see "how the world looks" to a (p-adic!) photon (§6).

Since every $T \in L^*$ is of the form $T = S'^{-1}FS$, where $S, S' \in S^*$ and $F \in F^*$, we may write $S'\xi' = S'T\xi = FS\xi$, or simply $\eta' = F\eta$, where $\eta' = S'\xi'$, $\eta = S\xi$ involve only the spatial coordinate transformations $Y' = R'X'$, $Y = RX$ induced by the corresponding rotations of spatial axes, for example, by $[Z_1, \ldots, Z_n]R^{-1}$ in P.

It appeared already in Theorem 3.2(1) that $\nu(t') = \nu(d)\nu(t)$ for all events ξ in and on the cone Γ whenever $T \in L^*$. Since d appears as a uniform scaling factor for both X' and t' in E', and the canonical transformation $\eta' = F\eta$ is of form $y_i' = dy_i$, $i = 2, \ldots, n$, it is appropriate to suppose $d = 1$. It then turns out that the proper Lorentz transformations are essentially Galilean, with $\nu(t') = \nu(t)$, independent of position. Indeed, the surprisingly "classical" nature of the present pathological event-spaces becomes manifest when we consider the analogs of length-contraction and time-dilatation.

1. Let X_1, X_2 be two fixed positions in E, observed simultaneously at time t_0' in E' as X_1', X_2'. Then

$$X_1' = AX_1 + Bt_1 \qquad X_2' = AX_2 + Bt_2$$
$$C^\tau X_1 + dt_1 = t_0' = C^\tau X_2 + dt_2$$

Hence
$$\begin{aligned} X_2' - X_1' &= A(X_2 - X_1) + B(t_2 - t_1) \\ &= A(X_2 - X_1) - d^{-1}BC^\tau(X_2 - X_1) \\ &= d\{A_1(X_2 - X_1) - B_1C_1^\tau(X_2 - X_1)\} \end{aligned}$$

Since $|B_1| < 1$, $|C_1| < 1$, and $A_1 \in R^*$, for $T \in L^*$, it is clear from N5 that $|X_2' - X_1'| = \nu(d)\,|X_2 - X_1|$.

2. Let t_1, t_2 be two times at the same position X_0 of E, observed as t_1', t_2' in E'. Then

$$t_1' = C^\tau X_0 + dt_1 \qquad \text{and} \qquad t_2' = C^\tau X_0 + dt_2$$

whence $t_2' - t_1' = d(t_2 - t_1)$, and $\nu(t_2' - t_1') = \nu(d)\nu(t_2 - t_1)$.
Thus for $d = 1$, there is no effect in either case.

6. NOTE ON IMPROPER TRANSFORMATIONS. A Lorentz transformation $T = dT_1$ is *improper* in case $|B| = \nu(d)$, i.e., $|B_1| = 1$. Since such a T takes the axis of Γ onto Γ, T^{-1} is not a Lorentz transformation at all, and the set of Lorentz transformations is not a group. Moreover, A_1 is not a rotation, for if so, we should have $A_1^{-1}(-B_1) = X$ of norm 1, and hence $1 = |A_1X + B_1| = 0$. It *is* true that $|A_1X| \leqq 1$ for $|X| = 1$ (by N5), hence $|A_1X| \leqq |X|$ for *all* X. From this it follows that $|X| < \nu(t)$ implies $|A_1X + B_1t| = \nu(t) = \nu(C_1^{\tau}X + t)$, which shows that an improper T maps the entire interior of Γ onto Γ.

We make no attempt to extend Theorem 4.1, but give only a partial reduction in

THEOREM 6.1. *If $T = dT_1$ is improper, there exist $S_1, S_2 \in S^*$ such that*

$$S_1T_1S_2 = F_2 = \begin{vmatrix} a_1 & a_2 & \ldots & a_n & 1 \\ 0 & & A_{n-1} & & 0 \\ c_1 & c_2 & \ldots & c_n & 1 \end{vmatrix}$$

which has the invariant plane $x_1\zeta_1 + t\zeta_{n+1}$, where ζ_j denotes the j-th "1-spot" vector of E.

PROOF. First, let $T_1' = S_1T_1$, where S_1 involves a rotation R_1 such that $RB_1 = Z_1$. Then $T_1'\zeta_{n+1} = \zeta_1 + \zeta_{n+1}$, *and for some unique* ξ, $T_1'\xi = \zeta_{n+1}$. Since ζ_{n+1} and $\zeta_1 + \zeta_{n+1}$ are linearly independent, so are ξ and ζ_{n+1}, which shows that the spatial component X of ξ is not zero. Let $|X| = \nu(x) > 0$, and define S_2 as a space-rotation such that $R_2(xZ_1) = X$. Finally, set $F_2 = T_1'S_2$.

Then we find

$$F_2\zeta_{n+1} = T_1'S_2\zeta_{n+1} = T_1'\zeta_{n+1}, = \zeta_1 + \zeta_{n+1}, \text{ giving the last column of } F_2,$$

while

$$F_2\zeta_1 = T'_1S_2\zeta_1 = x^{-1}T_1' \begin{vmatrix} X \\ 0 \end{vmatrix}.$$

Since

$$\begin{vmatrix} X \\ 0 \end{vmatrix} = \xi - t\zeta_{n+1},$$

substitution leads to

$$F_2\zeta_1 = x^{-1}(\zeta_{n+1} - t(\zeta_1 + \zeta_{n+1})) = -x^{-1}t\zeta_1 + x^{-1}(1-t)\zeta_{n+1}.$$

The matrices (over Q_2)

$$T_1 = \begin{vmatrix} 2 & 0 & 1 \\ 0 & 1 & 0 \\ 0 & 0 & 1 \end{vmatrix} \text{ and } T_1 = \begin{vmatrix} 2 & 2 & 1 \\ 2 & 2 & 0 \\ 2 & 0 & 1 \end{vmatrix}$$

are improper Lorentz transformations, with A_1 non-singular, and singular, respectively.

7. Real Lorentz Transformations. For contrast, we include a sketch of the real case. Let $\mathscr{R}$ be the real field, with absolute value $|x| \geqq 0$, $\mathscr{R}^n$ the n-space ($n \geqq 2$) of vectors X, with norm $|X| = (X^\tau X)^{\frac{1}{2}} \geqq 0$. Call an $n \times n$ matrix R over $\mathscr{R}$ a *rotation* in case $|RX| = |X|$ for all X, equivalently (here) $R^\tau R = I_n$. The set R^* of all R is a group, and $|\det R| = 1$. The "Gram-Schmidt" process, employed in place of Theorem 2.2(a) implies its final conclusion (c). See [2, §24, 48, 59] for standard results.

Fix $c > 0$, and in the space $\mathscr{R}^{n+1}$ of events $\xi = \left[\begin{smallmatrix} X \\ t \end{smallmatrix}\right]$, $t \equiv x_{n+1}$, define the cone $\Gamma = \{\xi; |X| = c|t|\}$. A Lorentz transformation (L.T.) is a non-singular $(n+1)$-order matrix T such that $T\Gamma \subset \Gamma$. Defining

$$Q = \begin{bmatrix} I_n & 0 \\ 0 & -c^2 \end{bmatrix},$$

this condition reads:

$$\xi^\tau Q \xi = \Sigma_1{}^n x_j^2 - c^2 x^2_{n+1} = 0 \tag{1}$$

implies

$$\Sigma_1{}^n x_j'^2 - c^2 x_{n+1}'^2 = \xi'^\tau Q \xi' = \xi^\tau (T^\tau Q T)\xi = \xi^\tau P \xi = \Sigma_1^{n+1} x_i p_{ij} x_j = 0 \tag{2}$$

where $P = [p_{ij}] = T^\tau Q T$ is symmetric ($p_{ij} = p_{ji}$).

Theorem 7.1. *If T is a Lorentz transformation, then*

$$T^\tau Q T = qQ \tag{3}$$

and

$$\Sigma^n{}_1 x_j'^2 - c^2 t'^2 \equiv q(\Sigma^n{}_1 x_j^2 - c^2 t^2) \text{ for all } X, t, \tag{4}$$

where

$$q = d^2 - c^{-2}B^\tau B \neq 0 \tag{5}$$

Conversely, if $T^\tau QT = qQ$ where $q \neq 0$, then T is a L.T.

PROOF. Let T be a L.T. Since $X = \pm Z_i$, $x_{n+1} = c^{-1}$ satisfy (1), it follows from (2) that $p_{i,n+1} = 0$ and $p_{ii} = q \equiv - c^{-2}p_{n+1,n+1}$. Similarly, $X = 3Z_i + 4Z_j$ $(i \neq j)$, $x_{n+1} = 5c^{-1}$, leads to $p_{ij} = 0$, so (3) and (4) are true. Substitution of $X = 0$, $t = 1$, $X' = B$, $t' = d$ in (4) yields (5), where $q \neq 0$ is seen by taking determinants in (3). (It will appear that actually $q > 0$.) The converse statement is trivial. For, (3) with $q \neq 0$ implies det $T \neq 0$, and $T\Gamma \subset \Gamma$ is obvious since (1) clearly implies (2).

COROLLARY. *The set L^* of all Lorentz transformations is a group.*

PROOF. *This follows formally from the N. & S. condition $T^\tau QT = qQ$ for $q \neq 0$.*

THEOREM 7.2. *If $T \in L^*$, then*

(a) $d \neq 0$.

(*b*) *For* $T = dT_1$, $T_1 = \begin{vmatrix} A_1 & B_1 \\ C_1^\tau & 1 \end{vmatrix}$, *we have*

$$A_1^\tau A_1 - c^2C_1C_1^\tau = q_1I_n \tag{6}$$

$$A_1^\tau B_1 = c^2C_1 \tag{7}$$

where

$$q_1 = 1 - c^{-2}B_1^\tau B_1 = q/d^2 \neq 0. \tag{8}$$

(c) *$B = 0$ if and only if $C = 0$. In such a case, $q = d^2 > 0$, and $A_1 \in R^*$.*

PROOF. (a) If $d = 0$, let X be a vector (existence trivial!) such that $X \neq 0$ and $C^\tau X = 0$. Then for X, and $t = c^{-1}| X |$, $X' = 0$ in (4) since $t' = C^\tau X + 0 = 0$. But T is non-singular.

(b) is obtained by block multiplication from $T_1^\tau QT_1 = d^{-2}qQ$.

(c) If $B_1 = 0$, then $C_1 = 0$ in (7). If $C_1 = 0$, (6) and (8) imply A_1^τ non-singular, hence $B_1 = 0$ in (7). If $B_1 = 0 = C_1$, then $q_1 = 1$, $q = d^2 > 0$ in (8), and $A_1 \in R^*$ in (6).

The set S^* of all space-rotations S (definitions and notations as in §4) is a subgroup of the Lorentz group L^*, which defines equivalence classes in L^* as before. Let F^* be the set of all matrices $F = dF_2$, $d \neq 0$, where

$$F_2 \equiv \begin{vmatrix} A_2 & B_2 \\ C_2^\tau & 1 \end{vmatrix}$$

is of *type*

I. $\quad F_2 = I_{n+1}$

II. $\quad F_2 = \begin{vmatrix} 1 & 0^\tau & b_0 \\ 0 & \gamma_0^{-1} I_{n-1} & 0 \\ b_0/c^2 & 0^\tau & 1 \end{vmatrix}$

where $0 < b_0 < c$ and $\gamma_0 \equiv (1 - b_0^2 c^{-2})^{-\frac{1}{2}}$.

THEOREM 7.3. *The set F^* is a representation system for the equivalence classes of L^*.*

PROOF. 1. If $dS_1F_2S_2 = d'F_2'$, we infer $d = d'$, $R_1A_2R_2 = A_2'$, $R_1B_2 = B_2'$, hence $b_0 \equiv | B_2 | = | B_2' | \equiv b_0'$.

From this it is clear that F_2' is of the same type as F_2 and therefore $F_2' = F_2$.

2. Given $T = dT_1 \in L^*$, $d \neq 0$, as in Theorem 7.2, we must produce $S_1, S_2 \in S^*$ such that $S_1T_1S_2 = F_2$, or, in block form

$$R_1A_1R_2 = A_2, \qquad R_1B_1 = B_2, \qquad C_1^\tau R_2 = C_2^\tau.$$

There are now only two "cases" (cf. Theorem 7.2)

CASE I: $B_1 = 0 = C_1$. For $R_1 = I_n$, $R_2 = A_1^{-1}$, we obtain $F_2 = I_{n+1}$ of *type* I.

CASE II: $| B_1 | \equiv b_0 > 0$, $| C_1 | \equiv c_0 > 0$. Letting R_1', R_2' be preliminary rotations such that $R_1'B_1 = b_0Z_1$ and $C_1^\tau R_2 = c_0Z_1^\tau$, we obtain an intermediate matrix

$$S_1'T_1S_2' = T_1' \equiv \begin{vmatrix} A_3 & b_0Z_1 \\ c_0Z_1^\tau & 1 \end{vmatrix} \in L^*$$

which, by Theorem 7.2, satisfies

$$A_3^\tau A_3 - c^2c_0^2Z_1Z_1^\tau = q_1I_n \tag{9}$$

$$A_3^\tau Z_1 = c^2b_0^{-1}c_0Z_1 \tag{10}$$

where

$$q_1 = 1 - b_0^2c^{-2} \equiv q/d^2 \neq 0. \tag{11}$$

By (10) A_3 has the form

$$A_3 = \begin{vmatrix} c^2 b_0^{-1} c_0 & 0^\tau \\ U & A_4 \end{vmatrix}$$

From (9) we therefore conclude

$$c^4 b_0^{-2} c_0^2 + U^\tau U - c^2 c_0^2 = q_1, \quad (12)$$

$$A_4^\tau U = 0, \quad (13)$$

$$A_4^\tau A_4 = q_1 I_{n-1}. \quad (14)$$

Since $q_1 \neq 0$, we infer from (14) that $q_1 = 1 - b_0^2 c^{-2} > 0$ (hence $q = q_1 d^2 > 0$!) and therefore $0 < b_0 < c$. Defining $\gamma_0 = q_1^{-1/2} > 0$, we have from (14) that $\gamma_0 A_0$ is a rotation R_{n-1} of $\mathscr{R}^{n-1}$, from (13) that $U = 0$, and from (12) that $c_0 = b_0/c^2$.

Collecting these results, we see that

$$T_1' = \begin{vmatrix} A_3 & b_0 Z_1 \\ \frac{b_0}{c^2} Z_1^\tau & 1 \end{vmatrix}$$

where

$$A_3 = \begin{vmatrix} 1 & 0^\tau \\ 0 & \gamma_0^{-1} R_{n-1} \end{vmatrix}.$$

Clearly one further space-rotation S_3' suffices, with

$$R_3' = \begin{vmatrix} 1 & 0^\tau \\ 0 & R_{n-1}^{-1} \end{vmatrix}.$$

For then, $R_3' A_3 = \begin{vmatrix} 1 & 0^\tau \\ 0 & \gamma_0^{-1} I_{n-1} \end{vmatrix}$, $R_3' Z_1 = Z_1$, and hence

$$S'_3 T_1' = \begin{vmatrix} R_3 A_3 & b_0 Z_1 \\ \frac{b_0}{c^2} Z_1^\tau & 1 \end{vmatrix} \text{ of } \textit{type}\ \text{II}.$$

3. We omit the verification of $F^* \subset L^*$.

Corollary. *Every Lorentz transformation* $T = dT_1$ *is of form*

$$T = (d\gamma_0^{-1}) S_1 L S_2; \qquad S_1, S_2 \in S^*,$$

where

$$L = \begin{vmatrix} \gamma_0 & 0^\tau & \gamma_0 b_0 \\ 0 & I_{n-1} & 0 \\ \dfrac{\gamma_0 b_0}{c^2} & 0^\tau & \gamma_0 \end{vmatrix}.$$

$0 \leqq b_0 = | B_1 | < c$ *(hence T "proper"!)*, $\gamma_0 = (1 - b_0^2 c^{-2})^{-\frac{1}{2}}$, *det* $L = 1$.

Moreover in the relations (4), *and* (5) *of Theorem* 7.1, *one has*

$$q = d^2 \gamma_0^{-2} > 0.$$

8. Abstract Normed Spaces. In the theory of Euclidean spaces, one begins with a formal "inner-product" (X, Y) on a real n-space, defining $\| X \| = (X, X)^{\frac{1}{2}}$, and proves the existence of an *orthonormal* basis $\{N_i\}$ such that, for $X = \Sigma x_i N_i$, $Y = \Sigma y_i N_i$, these functions assume the simple forms $\Sigma x_i y_i$ and $(\Sigma x_i^2)^{\frac{1}{2}}$. The (linear) isometries then appear as the transformations relating such bases. This approach, which singles out no particular basis, is more general than that of initially defining $(X, Y) = \Sigma x_i y_i$ for X, Y expressed in terms of a stipulated basis.

The latter type of definition for $| X |$ was made in §2, for simplicity, and also because no result of comparable generality was known to us. If V_n is an n-space over a field K with a V1–3 valuation (cf. §1), and $\{B_j\}$ is any fixed basis, then $| X |_0 \equiv | \Sigma x_i B_i |_0 \equiv \max v(x_i)$ defines an N1–3 norm, and it *is* known that *every* N1–3 norm is "equivalent" to this $| X |_0$, in a topologic sense [1, Ch. IV]. And of course N4 for $| X |_0$ follows from V4, indeed this *is* the norm $| X |$ of §2. It is therefore clear that every *abstract* N1–4 norm, relative to a V1–4 valuation is "equivalent" to our norm $| X |$. (Note here that N2, 4 *imply* V4.)

Hereafter, let K be a field with V1–4 (hence V5) valuation, V_n a vector-space of order n over K, with an abstract norm $\| X \|$ satisfying N1–4 (hence N5).

Call $N_1, \ldots, N_m$; $m \leq n$, a *normal* set in case

$$\|\sum_1^m x_i N_i\| = \max v(x_i); \quad x_i \in K$$

It is apparent that:

(A) a set (N_j) is normal if and only if

$$\max v(u_i) = 1 \text{ implies } \| \Sigma u_i N_i \| = 1$$

(An equivalent condition is that $\| \Sigma u_i N_i \| = 1$ for every set of u_i, not all zero, with values 0,1 only.)

(B) a normal set is linearly independent, with every $\| N_i \| = 1$. We say V_n is an *N-space* in case it has a normal basis, and so characterize abstractly the position spaces of §2.

In analogy with the real case, it is true that, if $\{N_i\}$ is a normal basis and θ a linear transformation of V_n, then $\{\theta N_i\}$ is also a normal basis if and only if $\| \theta X \| = \| X \|$ for all X. However, a normal basis need not exist, for this surely implies:

N6. For every $X \in V_n$, there exists an $x \in K$ such that $\| X \| = v(x)$, which is hardly true for the norm $\| X \| = 3 \max v(x_i)$ on the space $V_n = Q_2^n$.

Our best result is contained in the final theorem, in which N6 is necessary, but V6 is not (witness the example $V_n = Q^n$). For the case $V_n = Q_p^n$ however, V6 holds, Q_p being complete, with compact unit sphere. We give a straightforward proof without reference to topology.

THEOREM 8.1. *If, in addition to* V1–5 *and* N1–5, *we have* N6, *and also* V6: *Every K-sequence* $\{s_h\}$ *with all* $v(s_h) \leqq 1$ *contains a convergent subsequence,*
then V_n *has a normal basis. Indeed, for every linearly independent set* $(B_1, \ldots, B_m)$, $m \leqq n$, *there exists a normal set* $(N_1, \ldots, N_m)$ *spanning the same sub-space.*

PROOF. It is clear from N6 and V2 that we may suppose all $\| B_i \| = 1$ for the given set $(B_1, \ldots, B_m)$. Defining $N_1 = B_1$, we have $\| x_1 N_1 \| = \max v(x_i)$, and the first step is complete.

Suppose a normal set $(N_1, \ldots, N_k)$, $1 \leqq k < m$ already defined, spanning the same sub-space as $(B_1, \ldots, B_k)$. Let

$$S^* = \{S = \Sigma_1^k s_i N_i + B_{k+1};\ v(s_i) \leqq 1\}$$

From linear independence and N4, we have

$$0 < \| S \| \leqq 1, \qquad S \in S^*$$

Define

$$n_0 = g \cdot l \cdot b \cdot \{\| S \|;\ \ S \in S^*\},\ \ 0 \leqq n_0 \leqq 1,$$

and let $\{S^h\}$ be a sequence with

$$S^h \equiv \Sigma_1{}^k s_i{}^h N_i + B_{k+1} \in S^*$$

such that

$$\| S^h \| \to n_0 .$$

Applying V6 successively to the components of the S^h, we obtain a subsequence $\{T^h\}$ of $\{S^h\}$, with

$$T^h \equiv \Sigma_1{}^k t_i{}^h N_i + B_{k+1} \in S^*$$

$$\| T^h \| \to n_0$$

such that $t_i{}^h \to l_i \in K$, $i = 1, \ldots, k$. Since

$$| v(t_i{}^h) - v(l_i) | \leqq v(t_i{}^h - l_i) \to 0$$

we know that $v(t_i{}^h) \to v(l_i) \leqq 1$. Therefore

$$L \equiv \Sigma_1{}^k l_i N_i + B_{k+1} \in S^*$$

and so,

$$0 < \| L \| \leqq 1.$$

Moreover,

$$| \, \| T^h \| - \| L \| \, | \leqq \| T^h - L \| = \max_i v(t_i{}^h - l_i) \to 0.$$

Since a sequential limit is unique, this shows that

$$n_0 = \| L \|, \quad L \in S^*.$$

Hence, by N6, we may write

$$0 < n_0 = \| L \| = v(l) \leqq 1.$$

Defining $N_{k+1} = l^{-1}L$, we have $\| N_{k+1} \| = 1$, and $(N_1, \ldots, N_{k+1})$ spans the same sub-space as $(B_1, \ldots, B_{k+1})$.

It now suffices to prove

(A) $$\| U \| \equiv \| \Sigma_1{}^k u_i N_i + u_{k+1} N_{k+1} \| = 1$$

provided $\max (v(u_1), \ldots, v(u_{k+1})) = 1$.

It is obvious by N4 that $\| U \| \leqq 1$. If $v(u_{k+1}) < 1$, then $\| U \| = 1$ by N5, since $(N_1, \ldots, N_k)$ is a normal set. For $v(u_{k+1}) = 1$, we have

$$1 \geqq \| U \| = v(u_{k+1}) \, \| \Sigma_1{}^k u_{k+1}^{-1} u_i N_i + N_{k+1} \|$$

$$= \| \Sigma_1{}^k(u_{k+1}^{-1}\, u_i + l^{-1}l_i)\, N_i + l^{-1}\, B_{k+1} \|$$
$$= v(l^{-1})\ \| \Sigma_1{}^k(l u_{k+1}^{-1} u_i + l_i)\, N_i + B_{k+1} \| .$$

Since $v(lu_{k+1}^{-1}u_i) = v(l)v(u_i) \leqq 1$ and $v(l_i) \leqq 1$ it is clear that the final vector is in S^*, hence has a norm $\geqq v(l)$, the $g \cdot l \cdot b \cdot$ of its norms $\| S \|$. Since $v(l^{-1})v(l) = 1$, $\| U \| = 1$ and the proof is complete.

References

1. G. Bachman, *An Introduction to p-adic Numbers and Valuation Theory*, Academic Press, New York, 1964.

2. P. R. Halmos, *Finite Dimensional Vector Spaces*, Princeton Univ. Press, Princeton, N. J., 1948.

3. C. C. Mac Dufeee, *An Introduction to Abstract Algebra*, Wiley, New York, 1940.

4. B. L. van der Waerden, *Modern Algebra*, Vol. I, Ungar, New York, 1949.

Reprinted from the PROCEEDINGS OF THE NATIONAL ACADEMY OF SCIENCES
Vol. 57, No. 5, pp. 1172–1174. May, 1967.

*ON VISUAL HULLS OF SETS**

BY G. H. MEISTERS AND S. M. ULAM

LOS ALAMOS SCIENTIFIC LABORATORY, LOS ALAMOS, NEW MEXICO

Communicated March 17, 1967

1. The problems treated in this note concern the possibilities of "reconstructing" a subset S of real Euclidean n-space E_n $(n \geq 2)$ or, more generally, a real Hilbert space H, from the knowledge of the orthogonal projections of S onto each member of a preselected family $\mathfrak{M}$ of proper subspaces of E_n (or of H). Obviously, in general, this is not possible: The surface of a sphere in E_3 has the same projections on each two-dimensional subspace as the solid sphere. We introduce, however, the notion of the "largest" set $\mathcal{P}\langle S\rangle$ which has the same projections as S (with respect to the given family $\mathfrak{M}$). We call this set $\mathcal{P}\langle S\rangle$ the *visual hull* of S, and identify by the equation $\mathcal{P}\langle S\rangle = S$ those subsets S of H which can be "reconstructed" from a knowledge of their orthogonal projections onto the members of $\mathfrak{M}$. Although there are many interesting choices for $\mathfrak{M}$, we mention three in particular:

(1) all one-dimensional subspaces of H,
(2) all two-dimensional subspaces of H (when dim $H \geq 3$),
(3) all hyperplanes (= closed maximal proper subspaces of H).

The visual hull $\mathcal{P}\langle S\rangle$ is related to the notion of the convex hull $\mathcal{C}\langle S\rangle$ in the following way. If $\mathfrak{M}$ is the family of all one-dimensional subspaces [case (1) above], then for every connected set S it turns out that $\mathcal{C}\langle S\rangle \subset \mathcal{P}\langle S\rangle \subset \overline{\mathcal{C}\langle S\rangle}$. However, if S is not connected, then $\mathcal{P}\langle S\rangle$ is only a subset of $\overline{\mathcal{C}\langle S\rangle}$. On the other hand, for other choices of $\mathfrak{M}$ [such as in cases (2) and (3) above] $\mathcal{P}\langle S\rangle$ need not be convex even when S is connected. For example, if S is an anchor-ring (solid torus) in E_3 and $\mathfrak{M}$ is as in case (2) above, then $\mathcal{P}\langle S\rangle$ is larger than S but still homeomorphic to S.

Case (2) above is obviously suited to problems of "photographic recognition" or "graphic representation."

In this note we present a precise formulation of these problems and some first results of a general nature.

2. Let H denote a real Hilbert space of any finite or transfinite dimension with inner product $\langle x,y\rangle$ and norm $||x|| = \langle x,x\rangle^{1/2}$. Given a family $\mathcal{P}$ of nonzero orthogonal projections (i.e., linear, symmetric, idempotent mappings $P:H \rightarrow H$), we say that two subsets A and B of H are *$\mathcal{P}$-equivalent*, written $A \overset{\mathcal{P}}{\sim} B$, iff $P(A) = P(B)$ for all $P \in \mathcal{P}$. This is obviously an equivalence relation for the subsets of H. For each subset S of H we define $\mathcal{P}\langle S\rangle$ to be the set-union of all subsets B of H satisfying $B \overset{\mathcal{P}}{\sim} S$. Obviously $S \subset \mathcal{P}\langle S\rangle$.

THEOREM 1. *$S \overset{\mathcal{P}}{\sim} \mathcal{P}\langle S\rangle$, $\mathcal{P}\langle\mathcal{P}\langle S\rangle\rangle = \mathcal{P}\langle S\rangle$, and $\mathcal{P}_1 \subset \mathcal{P}_2 \rightarrow \mathcal{P}_2\langle S\rangle \subset \mathcal{P}_1\langle S\rangle$.*

Proof: First, $S \subset \mathcal{P}\langle S\rangle$ implies $P(S) \subset P(\mathcal{P}\langle S\rangle)$ for all $P \in \mathcal{P}$. On the other hand, if $x \in P(\mathcal{P}\langle S\rangle)$, then $x = P(a)$ for some $a \in A$, where $A \overset{\mathcal{P}}{\sim} S$. But then $P(a) = P(s)$ for some $s \in S$, so that $x = P(s)$, or $x \in P(S)$. The second equation follows from the first and the definition of $\mathcal{P}\langle S\rangle$, since $\overset{\mathcal{P}}{\sim}$ is an equivalence relation. The last assertion of the theorem is just as obvious.

Thus for each subset S of H, *$\mathcal{P}\langle S\rangle$ is the largest subset of H which is $\mathcal{P}$-equivalent to S.* We are particularly interested in those subsets S of H satisfying $S = \mathcal{P}\langle S\rangle$, for it is these subsets of H which could be considered as "completely determined" or

For commentary to this paper [91], see p. 688.

"reconstructible" from a knowledge of their $\mathcal{P}$-projections. By "an *affine subspace* of H" we shall mean a translated subspace of the form $x + M$. If P is a projection, $\mathcal{R}(P)$ denotes its range. If $A \subset H$, $A^{\perp} = \{x \in H: \ \langle x,a \rangle = 0 \text{ for all } a \in A\}$.

THEOREM 2. *For each subset S of H, $x \in \mathcal{P}\langle S \rangle$ iff for every $P \in \mathcal{P}$ the affine subspace $x + \mathcal{R}(P)^{\perp}$ intersects S in at least one point.*

PROOF: $x \in \mathcal{P}\langle S \rangle$ iff $x \in A$ for some $A \overset{\mathcal{P}}{\sim} S$. The latter condition is equivalent to the statement that for every $P \in \mathcal{P}$, $P(x) = P(s)$ for some $s \in S$ [for then $A = \{x\} \cup S$ is $\mathcal{P}$-equivalent to S]. Now if $P(s) = P(x)$, then $s = x + (I - P)(s - x)$ which belongs to $x + \mathcal{R}(P)^{\perp}$. Conversely, if $s \in x + \mathcal{R}(P)^{\perp}$ for some $s \in S$, then $s = x + (I - P)(s - x)$ so that $P(s) = P(x)$.

COROLLARY 1. $S = \mathcal{P}\langle S \rangle$ *iff for every $x \notin S$ there exists a $P \in \mathcal{P}$ such that the affine subspace $x + \mathcal{R}(P)^{\perp}$ does not intersect S.*

A family $\mathcal{P}$ will be called *full* if every one-dimensional subspace of H is contained in the range of at least one member of $\mathcal{P}$. This property may be satisfied by $\mathcal{P}$ in many different ways. In particular, it is satisfied if $\mathcal{P}$ is the family of projections corresponding to any one of the three cases mentioned in the introduction.

COROLLARY 2. *If $\mathcal{P}$ is a full family of projections, then for every subset S of H, $\mathcal{P}\langle S \rangle$ is contained in the closed convex hull $\overline{\mathcal{C}\langle S \rangle}$ of S.*

Proof: If $x \notin \overline{\mathcal{C}\langle S \rangle}$, then by the Hahn-Banach theorem there exists a closed hyperplane M of H so that $x + M$ does not intersect $\overline{\mathcal{C}\langle S \rangle}$. *A fortiori*, $x + M$ does not intersect S. By hypothesis there exists a $P \in \mathcal{P}$ such that $M^{\perp} \subset (\mathcal{R}P)$. For this $P \in \mathcal{P}$, $x + \mathcal{R}(P)^{\perp}$ does not intersect S. Consequently, by Theorem 2, $x \notin \mathcal{P}\langle S \rangle$. Thus we have proved that $\mathcal{P}\langle S \rangle \subset \overline{\mathcal{C}\langle S \rangle}$.

COROLLARY 3. *If $\mathcal{P}$ is a full family of projections, then every closed convex subset S of H satisfies $S = \mathcal{P}\langle S \rangle$.*

COROLLARY 4. *If $\mathcal{P}$ is exactly the family of orthogonal projections onto all one-dimensional subspaces of H and if S is a connected subset of H, then $\mathcal{C}\langle S \rangle \subset \mathcal{P}\langle S \rangle \subset \overline{\mathcal{C}\langle S \rangle}$; in particular $\mathcal{P}\langle S \rangle$ is convex.*

Proof: $\mathcal{P}\langle S \rangle \subset \overline{\mathcal{C}\langle S \rangle}$ by Corollary 2. If $x \notin \mathcal{P}\langle S \rangle$, then there exists (by Theorem 2) a $P \in \mathcal{P}$ such that $x + \mathcal{R}(P)^{\perp}$ does not intersect S. But since $\mathcal{R}(P)$ is one-dimensional, $x + \mathcal{R}(P)^{\perp}$ is a closed affine hyperplane which separates H into two open half spaces H_+ and H_-. Since $S \subset H_+ \cup H_-$ and S is connected, it follows that S is completely contained in one of the half spaces or the other, say H_+. But these half spaces are convex so that $\mathcal{C}\langle S \rangle \subset H_+$. It follows now that $x \notin \mathcal{C}\langle S \rangle$. Therefore we have shown that $\mathcal{C}\langle S \rangle \subset \mathcal{P}\langle S \rangle$, and this completes the proof.

THEOREM 3. *If $S = A_1 \cup \cdots \cup A_m$ for some integer $m \geq 1$ where each A_k is convex and either open or closed (but not necessarily disjoint from the other A_k's) and if each m-dimensional subspace of H is contained in the range of at least one member of $\mathcal{P}$ (so that, in particular, we must assume $\dim H > m$), then $\mathcal{P}\langle S \rangle = S$.*

Proof: Suppose $x \notin S$. Then for each $k = 1, \ldots, m$ it follows from the Hahn-Banach theorem that there exists a closed hyperplane M_k such that $x + M_k$ does not intersect A_k. Let $L = \bigcap_{k=1}^{m} M_k$. Then L is a closed subspace of H, $L^{\perp}$ has dimension $\leq m$, and $x + L$ does not intersect S. By the hypothesis of the theorem there exists a projection $P \in \mathcal{P}$ such that $L^{\perp} \subset \mathcal{R}(P)$ and therefore $x + \mathcal{R}(P)^{\perp} \subset x + L$ so that $x + \mathcal{R}(P)^{\perp}$ does not intersect S. But then by Corollary 1 $\mathcal{P}\langle S \rangle = S$, as was to be shown.

If $S \subset E_2$, $\mathcal{P}$ is the family of projections onto all one-dimensional subspaces of E_2, and S is the union of two disjoint convex sets, then $\mathcal{P}\langle S\rangle$ is also the union of two convex sets which, however, may be larger than the two given sets. An analogous theorem is presumably valid in arbitrary E_n for n disjoint convex sets. We remark that given a set S, there exist besides $\mathcal{P}\langle S\rangle$ "smallest" (i.e., minimal) sets which have the same projections as the given set, but these are not unique and their existence depends on Zorn's lemma.

* This research was supported in part by the National Science Foundation, grant GP 5965.

Reprinted from JOURNAL OF COMBINATORIAL THEORY Vol. 4, No. 3, April 1968

Note on the Visual Hull of a Set*

W. A. BEYER AND S. M. ULAM

University of California, Los Alamos Scientific Laboratory, Los Alamos, New Mexico

ABSTRACT

The concept of "visual hull" of a set S in a linear topological space L (previously introduced by Meisters and Ulam) is discussed. The visual hull of a set S is the largest set containing S having the same projections as S on given subspaces of L. A "pathological" example is constructed showing that the visual hull of S less S itself is not necessarily topologically connected to S. Transfinite induction is used in the construction.

1. INTRODUCTION

The concept of a "visual hull of a set" in a Hilbert space H has been introduced by Meisters and Ulam [5]. For $S \subset H$, the visual hull $\mathscr{P}\langle S\rangle$ is the "largest" set containing S having the same projections as S with respect to a preselected family $\mathscr{M}$ of projections of H. One might conjecture from simple examples that $\mathscr{P}\langle S\rangle$ must be connected to S.

The purpose of this note is to construct a "pathological" example where this is false (Theorem 3). In a simple case this example reduces to the following. Take H to be the Euclidean plane and the projections $\mathscr{M}$ to be projections on all lines in all directions. For any set W interior to the unit circle of power less than that of the continuum there exists a set V on the circumference of the unit circle such that

$$\mathscr{P}\langle V\rangle = V \cup W.$$

In particular, there exists a set V on the circumference of a circle such that its visual hull is equal to V plus the center of the circle. Transfinite induction is used to construct V.

The definition of $\mathscr{P}\langle S\rangle$ is generalized slightly to permit construction of the example in a real linear locally convex space.

2. PRELIMINARIES

The concept of the visual hull of a subset S of a linear topological space L is discussed in this section. Two definitions are given for a visual hull:

* Work performed under the auspices of the U.S. Atomic Energy Commission.

For commentary to this paper [93], see p. 688.

the first is in terms of projections of S; the second is in terms of affine subspaces which do not intersect S. The relations between the different definitions are then discussed by two theorems. A few simple properties of such visual hulls are then stated. Other results similar to those obtained by Meisters and Ulam no doubt hold for these generalizations. Two lemmas needed for the construction of the "pathological" example are then given. For a definition of linear topological space, see Dunford and Schwartz [2, pp. 36, 49, 50].

A projection P in a linear topological space L is a linear continuous idempotent mapping of L into L [2, pp. 480–2]). Each P decomposes L into complementary subspaces L_1 and L_2 such that $L = L_1 + L_2$ (algebraic sum) and $L_1 \cap L_2 =$ null vector of L where $L_1 = P(L)$ and $L_2 = (I - P)(L)$. Conversely, to any complementary decomposition of L, $L = L_1 + L_2$, $L_1 \cap L_2 =$ null vector, there corresponds a projection P. For an arbitrary subspace L_1 there may not be a corresponding complementary decomposition of L. Every L_i of a complementary decomposition is closed. If L is a Banach space and L_1 is finite-dimensional, there exists such a decomposition [3, p. 49]. If L is a Hilbert space, there exists such a complementary decomposition for every subspace.

DEFINITION 1. Let $\mathscr{P}$ be a family of projections in L. Two subsets A and B of L are $\mathscr{P}$-equivalent, written $A \overset{\mathscr{P}}{\sim} B$, if $P(A) = P(B)$ for all $P \in \mathscr{P}$. Define $\mathscr{P}\langle s \rangle$ to be the set union of all subsets B of L satisfying $B \overset{\mathscr{P}}{\sim} S$.

DEFINITION 2. Let $\mathscr{M}^*$ be a family of subspaces of L. The visual hull $\mathscr{P}^*\langle S \rangle$ of a set $S \subset L$ is defined to be

$$\{x \mid x \in L, \exists\, m^* \in \mathscr{M}^* \ni (m^* + x) \cap S = \varphi\}^c$$

where $\{\ \}^c$ denotes complement and φ is the empty set.

THEOREM 1. *Let $\mathscr{P}$ be given. Let $\mathscr{M}^* = \{m^* \mid m^* = (I - P)(L), P \in \mathscr{P}\}$. Then for every $S \subset L$: $\mathscr{P}^*\langle S \rangle = \mathscr{P}\langle S \rangle$.*

PROOF: Suppose $x \in \mathscr{P}^*\langle S \rangle$. Then for every $m^* \in \mathscr{M}^*$, $(x + m^*) \cap S$ is nonempty. Choose $s \in (x + m^*) \cap S$. Choose P such that $m^* = (I - P)(L)$. Then $P(s) = P(x)$. Hence $x \in \mathscr{P}\langle S \rangle$.

Now suppose $x \in \mathscr{P}\langle S \rangle$. To each $P \in \mathscr{P}$, there exists a closed complementary space $m^* = (I - P)(L)$. $x \in \mathscr{P}\langle S \rangle$ implies that for every $P \in \mathscr{P}$ there exists $s \in S$ such that $P(x) = P(s)$. Now $(I - P)(s - x) = s - x$. Hence $s - x \in m^*$. Therefore $s \in x + m^*$. Hence $(x + m^*) \cap S$ is non-empty for every $m^* \in \mathscr{M}^*$ and $x \in \mathscr{P}^*\langle S \rangle$.

THEOREM 2. *Let $\mathscr{M}^*$ be given and suppose for each $m^* \in \mathscr{M}^*$ there exists a projection P such that $m^* = (I - P)(L)$. Put $\mathscr{P} = \{P \mid m^* = (I - P)(L), m^* \in \mathscr{M}^*\}$. Then for each $S \subset L$: $\mathscr{P}^*\langle S\rangle = \mathscr{P}\langle S\rangle$.*

The proof is almost the same as that of Theorem 1.

If L is a real Hilbert space and if we put $\mathscr{M} = \{m^{*\perp} \mid m^* \in \mathscr{M}^*\}$, then $\mathscr{P}^*\langle S\rangle$ defined by definition 2 is equivalent to the definition in Meisters and Ulam [5].

We now state some easy consequences of Definition 2.

LEMMA 1. $A \subset \mathscr{P}^*\langle A\rangle$.

LEMMA 2. *$A \subset B$ implies $\mathscr{P}^*\langle A\rangle \subset \mathscr{P}^*\langle B\rangle$.*

LEMMA 3. $\mathscr{P}^*\langle A\rangle \cup \mathscr{P}^*\langle B\rangle \subset \mathscr{P}^*\langle A \cup B\rangle$.

LEMMA 4. $\mathscr{P}^*\langle A \cap B\rangle \subset \mathscr{P}^*\langle A\rangle \cap \mathscr{P}^*\langle B\rangle$.

LEMMA 5. $\mathscr{P}^*\langle \mathscr{P}^*\langle A\rangle\rangle = \mathscr{P}^*\langle A\rangle$.

LEMMA 6. *If $\mathscr{M}_1^* \subset \mathscr{M}_2^*$, then $\mathscr{P}_2^*\langle A\rangle \subset \mathscr{P}_1^*\langle A\rangle$.*

LEMMA 7. *If S is connected and $\mathscr{M}^*$ is the family of all hyperplanes (maximal proper subspaces) of L, then $\mathscr{P}\langle S\rangle$ is convex.*

3. TWO LEMMAS

A set $S \subset L$ is convex if $a, b \in S$ implies the segment

$$\overline{ab} = \{ta + (1 - t)b \mid 0 < t < 1\} \subset S.$$

The interior $I(S)$ of a set S is the set of all points in S each of which has a neighborhood (with respect to L) contained in S. The boundary $B(S)$ of a set S is the set of all points in L such that each neighborhood of each point in $B(S)$ has points in S and points in $L \sim S = \{x \mid x \in L, x \notin S\}$. $S \subset L$ is strictly convex if S is convex and $\overline{ab} \not\subset B(S)$ for any $a \neq b$. If $S \subset L$ is convex and if $a \in B(S)$, then an antipodal point to a relative to $c \in I(S)$, called $(a)_c^*$, is a point in $B(S)$ such that $c \in \overline{a(a)_c^*}$. $\overline{\overline{W}}$ denotes the power (cardinality) of the point set W.

The following two lemmas, certainly well known, do not seem to be given in a convenient source. The plane case of Lemma 8 is given in Yaglom and Boltyanskii [7, p. 114].

LEMMA 8. *Suppose L is a real linear topological space and $S \subset L$ is convex. For any line l, if $l \cap I(S) \neq \varphi$ then $\overline{\overline{l \cap B(S)}} \leqslant 2$.*

PROOF. Suppose $x \in l \cap I(S)$, $x_0 \in l \cap B(S)$, $\bar{x} \in \overline{x_0 x}$, $\lim_{i \to \infty} x_i = \bar{x}$, and that $N \subset S$ is a neighborhood of x. For sufficiently large i one has

$$x^* = x + \frac{1}{1-t}(x_i - \bar{x}) \in N \subset S$$

where $t(0 < t < 1)$ is such that $tx_0 + (1-t)x = \bar{x}$. Then $x_i \in \overline{x_0 x^*}$; hence $x_i \in S$ and therefore $\bar{x} \in I(S)$. Thus for each $a \neq x$:

$$\overline{\overline{\{ta + (1-t)x \mid t > 0\} \cap B(S)}} \leqslant 1.$$

The conclusion of the lemma follows.

This lemma implies the uniqueness of the symbol $(a)_c^*$ used in the proof of Theorem 3.

LEMMA 9. *Let L be a real linear locally convex space of dimension greater than* 1. *If $S \subset L$ is convex, $\bar{\bar{S}} > 1$, $\bar{\bar{B}}(S) > 0$, then $\bar{\bar{B}}(S)$ is not finite.*

PROOF: Let $x \in B(S)$ and N_x be a convex neighborhood of x. Suppose $N_x \cap S \subset B(S)$. There must exist $\bar{x} \in N_x \cap S \sim x$ for otherwise $\bar{\bar{S}} = 1$. since S is convex. Then $\overline{\bar{x}x} \subset N_x \cap S \subset B(S)$ and $\bar{\bar{B}}(S)$ is not finite. Now suppose $\bar{x} \in I(S) \cap N$. By use of Lemma 8, for sufficiently small $\epsilon > 0$,

$$x_0 = (1+\epsilon)x - \epsilon\bar{x} \in N_x \cap \bar{S}^c (\bar{S} = \text{closure of } S).$$

x_0 has a convex neighborhood $N_{x_0} \subset \bar{S}^c$. N_{x_0} has dimension greater than 1. Choose $a \in N_{x_0}$ so that a is linearly independent of $x_0 - \bar{x}$. Then

$$\{\lambda x_0 + (1-\lambda)a \mid 0 < \lambda < 1\} \subset N_{x_0}$$

and

$$\overline{\overline{\{t\bar{x} + (1-t)[\lambda x_0 + (1-\lambda)a] \mid 0 < t < 1, 0 < \lambda < 1\} \cap B(S)}}$$

is not finite.

4. EXISTENCE OF A "PATHOLOGICAL" SET

For the remainder of the paper a visual hull is taken to be defined by Definition 2 above with $\mathscr{M}^*$ fixed as the family of all one-dimensional subspaces of L (called lines). The * in the notation for the corresponding visual hull will be suppressed.

THEOREM 3. *Suppose L is a real linear locally convex space of dimension >1 and $S \subset L$ is strictly convex. If $W \subset I(S)$ and $\bar{\bar{W}} < \overline{\overline{B(S)}} = \overline{\overline{I(S)}}$, then there exists $V \subset B(S)$ such that its visual hull $\mathscr{P}\langle V \rangle = V \cup W$.*

PROOF: Well-order the points of $I(S) \sim W$: $a_1, a_2, \ldots, a_\xi, \ldots$, so that for any initial segment $| a_\xi |$ one has $\overline{\overline{| a_\xi |}} < \overline{\overline{I(S) \sim W}}$. Call its order type μ. Let ξ be an ordinal $< \mu$. Put $B = B(S)$. Suppose that for each ordinal $\xi' < \xi$ a set $V_{\xi'}$ exists with the following four properties:

$$V_{\xi'} \subset V_{\xi''} \subset B \quad \text{if} \quad \xi'' < \xi', \tag{1}$$

$$\overline{\overline{B \sim V_{\xi'}}} \leqslant \overline{\overline{2\xi'}}, \tag{2}$$

$$W \subset \mathscr{P}\langle V_{\xi'} \rangle, \tag{3}$$

$$a_{\xi'} \notin \mathscr{P}\langle V_{\xi'} \rangle. \tag{4}$$

It follows from (2) that $\overline{\overline{V_{\xi'}}} = \overline{\overline{B}}$ since $\overline{\overline{B}}$ must not be finite (Lemma 9). Put $\bar{V}_\xi = \bigcap_{\xi' < \xi} V_{\xi'}$. If $a_\xi \notin \mathscr{P}\langle \bar{V}_\xi \rangle$, put $V_\xi = \bar{V}_\xi$. Otherwise put

$$T = \Big[\bigcup_{\substack{y \in B - \bar{V}_\xi \\ z \in W}} (y)_z^* \Big] \cup (B \sim \bar{V}_\xi),$$

where $(y)_z^*$ is the unique (Lemma 8) antipodal point to y with respect to z. Then $\overline{\overline{T}} < \overline{\overline{B}}$ and hence $\overline{\overline{B \sim T}} = \overline{\overline{B}}$.

Let $l(w, a_\xi)$ denote the line containing the points w and a_ξ. Put

$$*T = B \sim T \sim \bigcup_{y \in T} (y)_{a_\xi}^* \sim \bigcup_{w \in W} l(w, a_\xi) \cap B.$$

Then $\overline{\overline{*T}} < \overline{\overline{B}}$. Choose $d \in *T$. Put $V_\xi = \bar{V}_\xi \sim d \sim (d)_{a_\xi}^*$. V_ξ satisfies properties (1)–(4) if ξ replaces ξ' in the statement of these properties. By the principle of transfinite induction the sets V_ξ exist for all ordinals $\xi < \mu$ and have properties (1)–(4).

It will be shown that $V = \bigcap_{\xi < \mu} V_\xi$ satisfies the conclusion of the theorem. The sets $L \sim I(S) \sim B(S)$, V, $B(S) \sim V$, $I(S) \sim W$, and W are considered separately. Suppose $a \in L \sim \bar{S}$. Then a convex neighborhood of a can be separated from S by a hyperplane H [6, p. 27]. If $p \in H$, then for any $l \subset H + a - p$, $l \cap S = \varphi$; hence $a \notin \mathscr{P}\langle V \rangle$. Obviously $V \subset \mathscr{P}\langle V \rangle$.

Suppose $a \in B(S) \sim V$. Since S is strictly convex there exists l with $a \in l$ and $l \cap V = \varphi$. Hence $a \in \mathscr{P}\langle V \rangle$.

Suppose $a \in I(S) \sim W$. Let λ be the index of a in the well-ordering of $I(S) \sim W$. By (4) and (1), $a = a_\lambda \notin \mathscr{P}\langle V_\lambda \rangle \supset \mathscr{P}\langle V \rangle$. Hence $a_\lambda \notin \mathscr{P}\langle V \rangle$. Finally, since $W \subset \mathscr{P}\langle V_\xi \rangle$ for $\xi < \mu$ by (3) we have, using (1) to obtain the second inclusion:

$$W \subset \bigcap_{\xi < \mu} \mathscr{P}\langle V_\xi \rangle \subset \mathscr{P}\Big\langle \bigcap_{\xi < \mu} V_\xi \Big\rangle = \mathscr{P}\langle V \rangle.$$

This completes the proof of Theorem 3.

5. Remarks

a. Theorem 3 suggests the following problem: find simple non-trivial conditions on V under which $\mathscr{P}\langle V \rangle$ is connected to V.

b. Theorem 3 could be generalized to sets S having the property that for some finite k: if $l \cap I(S) \neq \varphi$ then $\overline{\overline{l \cap B(S)}} \leqslant k$.

c. In an entirely different line we shall discuss a conjectured separation property of the visual hull. As has been implied above, in a linear locally convex space of dimension > 1, if S is convex, then $\mathscr{P}\langle S \rangle \subset \bar{S}$. Suppose S_1 and S_2 are bounded strictly separated plane convex sets with interiors. Juel [4] and Brunn [1] have both shown that S_1 and S_2 have exactly four common support lines. In an obvious way by use of these 4 lines one can construct the visual hull of $S_1 \cup S_2$ and note that $\mathscr{P}\langle S_1 \cup S_2 \rangle$ has two components. Conjecture: in n-dimensional Euclidean space if $S_1, \ldots, S_n$ are convex, bounded, strictly separated bodies, then $\mathscr{P}\langle \bigcup_{i=1}^{n} S_i \rangle$ has n components.

The above construction of $\mathscr{P}\langle S_1 \cup S_2 \rangle$ gives the visual hull of a torus in Euclidean 3-space.

d. Let $\bar{\mathscr{H}}$ be a family of affine subspaces of a linear topologic space L. Put $\mathscr{H} = \{L \sim \bar{h} \mid \bar{h} \in \bar{\mathscr{H}}\}$. A set $K \subset L$ is called $\mathscr{H}$-convex if K is the intersection of a subfamily of $\mathscr{H}$. If $S \subset L$, then $\mathscr{P}^*\langle S \rangle = \bigcap_{K \supset S} K$ where K is $\mathscr{H}$-convex and $\mathscr{M}^* = \bar{\mathscr{H}}$. Thus the visual hull discussed above is an example of a generalized convex hull mentioned by Danzer, Grünbaum, and Klee [8, p. 156].

References

1. H. Brunn, Vom Normalenkegel der Zwischenebenen zweier getrennter Eikörper, *Bayer. Akad. Wiss. Math.-Natur. Kl. S.-B.* **1930**, 165–182.
2. N. Dunford and J. T. Schwartz, *Linear Operators, Part I: General Theory*, Interscience, New York, 1958.
3. S. Goldberg, *Unbounded Linear Operators*, McGraw-Hill, New York, 1960.
4. C. Juel, Indledning i Laeren om de grafiske Kurver, *Kongel. Danske Vid. Selsk. Skr.* 6, *Raekke, Math. Afd.* **X, 1** (1899), 1–90, with French summary (see page 78 in French summary).
5. G. H. Meisters and S. M. Ulam, On Visual Hulls of Sets, *Proc. Nat. Acad. .Sci U.S.A.*, **57** (1967), 1172–1174.
6. F. A. Valentine, *Convex Sets*, McGraw-Hill, New York, 1964.
7. I. M. Yaglom and V. A. Boltyanskii, *Convex Figures*, (English translation by P. J. Kelley and L. F. Walton), Holt, Rinehart and Winston, New York, 1961.
8. L. Danzer, B. Grünbaum and V. Klee, Helly's Theorem and Its Relatives, *Proc. Symp. Pure Math.* **7** (1963).

Reprinted from the PROCEEDINGS OF THE NATIONAL ACADEMY OF SCIENCES
Vol. 60, No. 4, pp. 1189–1195. August, 1968.

ON EQUATIONS WITH SETS AS UNKNOWNS

BY PAUL ERDÖS AND S. ULAM

DEPARTMENT OF MATHEMATICS, UNIVERSITY OF COLORADO, BOULDER

Communicated May 27, 1968

We shall present here a number of results in set theory concerning the decompositions of a set E in various ways as sum (union) of its subsets. These results have connection with problems on countably additive measure functions in abstract sets, but they may also bear on the problems of the axiomatics of set theory and generally on foundations of set theory itself. Some of these results employ the continuum hypothesis or the generalized continuum hypothesis. The several problems which will be presented also put these hypotheses in a certain limelight.

The impossibility of defining a countably additive measure for *all* subsets of a set of power of continuum (a measure which would vanish for subsets consisting of any single point) was first established with the use of the continuum hypothesis by Banach and Kuratowski.[1] Very shortly afterwards, one of us showed the impossibility of such a measure for subsets of a set of power $\aleph_1$ without the use of any hypothesis.[2] The same result was shown there to hold for sets of higher powers, in fact, for all the accessible alephs. More recently, these results have been extended to a large class of inaccessibles as well. These results show that this "problem of measure" is closely related to fundamental problems concerning the role of axioms of set theory. Recent developments have further clarified these relations. Important results have been obtained by Scott, Solovay, Martin, and others. The proofs of these relations make use of the methods introduced by Paul Cohen in proving the independence of the continuum hypothesis.

Both the results of Banach and Kuratowski and the stronger result of Ulam are obtained by exhibiting purely combinatorial schemata of decompositions of abstract sets with certain properties: $B\cdot$ and $K\cdot$ show a countable sequence of decompositions of a set of power of the continuum, each into countably many disjoint subsets so that, no matter how one takes a finite number of sets from each of these decompositions, the intersection of all these finite unions contains, at most, countably many points. Sierpinski[3] generalized the $B\cdot$ and $K\cdot$ schema in the following way. There exists a sequence of decompositions into aleph disjoint sets, each so that if one is selected from any countably many of these (not necessarily all), the union of the selected sets gives the whole of the space, except perhaps for countably many points. Decompositions given by $U\cdot$ show, without the use of the continuum hypothesis, the following phenomenon. A set E of power $\aleph_1$ can be decomposed countably many times into $\aleph_1$ disjoint sets in the following way:

A "matrix" of sets can be constructed such that we have countably many rows and noncountably many columns. Sets in each row are disjoint. The union of sets in any column gives the whole set E except for possibly countably many points. As is easy to see, the existence of such a decomposition (a sequence of decompositions, properly speaking) contradicts the possibility of defining a countably additive real-valued measure function.

In this paper we shall show various modifications of such constructions. In particular, we strengthen the result of Sierpinski. One can decompose E in the nth row into 2^{n+1} disjoint sets with the above property. It is clear that the impossibility of a measure function follows because if the measure for these sets existed, we could select a set of power less than $1/2^n$ and their union would have to measure less than 1. This is a contradiction, since the complement is countable. This construction uses the continuum hypothesis. Whether one can do it by using a weaker hypothesis remains an open problem. Should the number of sets in each row be finite and fixed, we show that one does not need the continuum hypothesis for this property to hold (but of course one does not get the impossibility of a measure function from such a "matrix"). We shall introduce a special symbol for decompositions of sets in such "matrix" patterns.

Finally, we would like to say that all the results and problems in this paper form only a special aspect of a more general problem which we formulate rather vaguely here: Given a class of Boolean relationships to be satisfied by unknown sets, all subsets of a given set, one wants to find or "construct" sets satisfying such relations, which may be countable or noncountable in number. We hope to attack this more general question in a paper to be published in the future.

THEOREM 1. *The real line (and in fact every set of power* c*) can be decomposed in infinitely many ways as the union of k disjoint sets*

$$S = \bigcup_{l=1}^{k} A_l^{(n)}, \quad n = 1,2,\ldots$$

so that

$$\left|S - \bigcup_{n=1}^{\infty} \cdot A_{l_n}^{(n)}\right| < k \tag{1}$$

for every choice of the sets $A_{l_n}^{(n)}$, $1 \leq l_n \leq k$.

We prove Theorem 1 without the axion of choice. Consider all sets of $k - 1$ disjoint rational intervals and write them in a sequence $\{I_n^{(k-1)}\}$, $n = 1,2\ldots.$ The first $k - 1$ sets $A_l^{(n)}$, $1 \leq l \leq k - 1$, of the nth row of our decomposition matrix are the $k - 1$ intervals of $I_n^{(k-1)}$. $A_k^{(n)}$ is the complement of the union of the intervals in $I_n^{(k-1)}$. Now let

$$\bigcup_{n=1}^{\infty} A_{l_n}^{(n)} = F$$

be a typical family of sets, one from each row. To prove (1) it suffices to show that if $x_1, x_2, \ldots x_k$ are any k real numbers, we must have $x_i \in F$ for at least one i, $1 \leq i \leq k$. To see this, observe that there is a set of $k - 1$ rational intervals, say $I_n^{(k-1)}$, which separates $x_1, x_2, \ldots x_k$. But then every $A_l^{(n)}$, $1 \leq l \leq k$ contains exactly one of the x_i or $x_i \in F$ for at least one i, as stated. This completes the proof of Theorem 1.

It is easy to see that (1) fails to hold with $k - 1$ instead of k, but we leave this to the reader.

Theorem 2. *Let* $|S| \geq \aleph_0$, $2 \leq k_1 \leq k_2 \leq, \ldots, k_n \to \infty$ *and let*

$$S = \bigcup_{l=1}^{k_n} A_l^{(n)}, \quad n = 1,2,\ldots$$

be a decomposition of S into k_n *disjoint sets. Then there is always an* l_n, $1 \leq l_n \leq k_n$, $n = 1,2, \ldots$ *so that*

$$|S - \bigcup_{n=1}^{\infty} A_{l_n}^{(n)}| \geq \aleph_0. \tag{2}$$

To prove Theorem 2 we will define elements x_u, $1 \leq u \leq \infty$ of S (the x_u's are not necessarily all different, but there are infinitely many different ones among them) and sets $A_{l_n}^{(n)}$ so that

$$x_u \not\subset A_{l_n}^{(n)}, \quad 1 \leq u < \infty, 1 \leq n < \infty, 1 \leq l_n \leq k_n \tag{3}$$

(3) will clearly imply Theorem 2.

We construct the x_n and the sets $A_{l_n}^{(n)}$ by induction with respect to n. Assume first that $2 = k_1 = ,\ldots = k_t < k_{t+1}$. Consider the 2^t sets

$$\bigcap_{n=1} A_l^{(n)} \quad l = 1 \text{ or } 2.$$

The union of these sets is S, thus at least one of them is infinite, say

$$|\bigcap_{n=1} A_{\epsilon_n}^{(n)}| = \aleph_0, \ (\epsilon_n = 1 \text{ or } 2).$$

Let x be an arbitrary element of $\bigcap_{i=1}^{t} A_{\epsilon n}^{(n)}$, put $x = x_1 = \ldots = x_t$ and $A_{l_n}^{(n)} = A_{3-\epsilon_n}^{(n)}$ for $n = 1,2,\ldots t$. Clearly the complement of $\bigcup_{n=1}^{\infty} A_{l_n}^{(n)} = \bigcup_{n=1}^{\infty} A_{3-\epsilon_n}^{(n)}$ (which equals $\bigcap_{n=1}^{t} A_{\epsilon n}^{(n)}$) is infinite.

Now assume that we have already succeeded in choosing elements $x_1, x_2, \ldots x_u$ and sets $A_{l_n}^{(n)}$, $1 \leq n \leq u$ having the following properties:

$$|S - \bigcup_{n=1}^{u} A_{l_n}^{(n)}| \geq \aleph_0, \qquad S_u = S - \bigcup_{n=1}^{u} A_{l_n}^{(n)} \tag{4}$$

$$x_s \not\subset A_{l_n}^{(n)}, \quad 1 \leq s \leq u; 1 \leq n \leq u \tag{5}$$

and finally there are at most $k_u - 2$ distinct elements among the x_i, $1 \leq i \leq u$. Now we construct x_{u+1} and $A_{l_{u+1}}^{(u+1)}$ so that (4) and (5) remain satisfied and so that there are at most $k_u - 2$ distinct elements among the $x, \ldots x_u, x_{u+1}$.

Assume first that $k_{u+1} = k_u$. Then there are at least two sets $A_{l_1}^{(u+1)}$ and $A_{l_2}^{(u+1)}$ which do not contain any of the elements $x_1, \ldots x_u$ (this follows from the fact that there are at most $k_u - 2$ distinct x's and the $A_l^{(u+1)}$'s are disjoint). At least one of these, say $A_{l_1}^{(u+1)}$, has an infinite complement in S_u (i.e., infinitely many elements of S_u do not belong to $A_{l_1}^{(u+1)}$). Put $A_{l_1}^{(u+1)} = A_{l_{u+1}}^{(u+1)}$ and clearly (4) and (5) are satisfied.

Assume next that $k_{u+1} > k_u$. Then there are at least three sets $A_{l_1}^{(u+1)}$.

$A_{l_2}{}^{(u+1)}$, $A_{l_3}{}^{(u+1)}$ which do not contain any of the x_i, $1 \leq i \leq u$. At least one of the sets, say $A_{l_1}{}^{(u+1)}$, has an infinite complement in S_u; we choose an arbitrary element x_{u+1} from this complement and put $A_{l_1}{}^{(u+1)} = A_{l_{u+1}}{}^{(u+1)}$. x_{u+1} is clearly different from $x_1, x_2, \ldots x_u$, and (4) and (5) are again satisfied. Thus we have constructed the infinite sequence $x_1, x_2, \ldots$ and the sets $A_{l_u}{}^{(u)}$. (3) is clearly satisfied and it is clear from our construction that there are infinitely many distinct elements among the x's, thus (2) is satisfied and Theorem 2 is proved.

Now we would like to state the following question which we cannot solve.

PROBLEM I. *Let* $\aleph_0 < |S| \leq c$. *Let* $2 \leq k_1 \leq k_2 \leq \ldots, k_n \rightarrow \infty$, *be any sequence of integers. Does there exist for every* n *a decomposition*

$$S = \bigcup_{l=1}^{k_n} A_l{}^{(n)}$$

into disjoint sets so that for every $1 \leq l_n \leq k_n$

$$|S - \bigcup_{n=1}^{\infty} A_{l_n}{}^{(n)}| \leq \aleph_0 \quad ?$$

If $c = \aleph_1$, we will see that the answer to our problem is affirmative; in fact very much more is true. But if we do not assume that $c = \aleph_1$, we cannot solve this question even if we assume that $|S| = \aleph_1$ and let our sequence k_n tend to infinity very slowly. If $|S| = c$ and $k_n > 2^n$ as stated in the introduction, we cannot expect a positive solution without some assumption on the power of the continuum since this would imply that there is no real-valued completely additive measure on the subsets of the reals where points measure 0.

It seems very likely that if we do not make some assumption about the power of the continuum, then we cannot obtain a solution to Problem I. It may even be possible to show that if Problem I has a solution for a fixed sequence $k_n \rightarrow \infty$, then a solution exists for every such sequence.

It will be convenient to introduce the symbol $m \rightarrow (p, q, r, s)$, which means that a set S of power m can be decomposed into the union of p disjoint sets in q ways:

$$S = \bigcup_{1 \leq \alpha \leq \omega_q} A_\alpha{}^{(\beta)}, \quad 1 \leq \beta \leq \omega_p$$

so that if we choose any one of the sets $A_{\alpha_i}{}^{(\beta_i)}$ for r different β_i, then

$$|S - \bigcup_i A_{\alpha_i}{}^{(\beta_i)}| < s, \tag{6}$$

$m \not\rightarrow (p, q, r, s)$ means that such a decomposition is impossible. Sierpinski proved[3] that

$$c \rightarrow (\aleph_1, \aleph_0, \aleph_0, \aleph_1) \tag{7}$$

is equivalent to the continuum hypothesis. Sierpinski's result implies that if we assume $c = \aleph_1$, the answer to Problem I is affirmative. We generalized this and proved other results on our symbol, but do not give these proofs since Hajnal observed that all our results follow from previous results of Erdös,

Hajnal, and Milner,[4] and Erdös, Hajnal, and Rado.[5] We studied the following symbol extensively:

$$\binom{m}{q} \rightarrow \binom{s\ s}{r\ r}. \tag{8}$$

The meaning of (8) is as follows: Let $|S_1| = m$, $|S_2| = q$; we split the pairs (x,y), $x \in S_1$, $y \in S_2$ into two classes. Then there is always a $U \subset S_1$, $V \subset S_2$, $|u| = s$, $|V| = r$ so that all the pairs (x,y), $x \in U$, $y \in V$ are in the same class. $\binom{m}{q} \not\rightarrow \binom{s\ s}{r\ r}$ means that there is a division of the pairs into two classes for which such sets U and V do not exist.

It is easy to see that

$$\binom{m}{q} \rightarrow \binom{s\ s}{r\ r} \tag{9}$$

is equivalent to

$$m \not\rightarrow (2, q, r, s). \tag{10}$$

First we show that (9) implies (10). Let $|S_1| = m$, $|S_2| = q$. Let $x \in S_1$, $\beta \in S_2$, $1 \leq \beta \leq \omega_q$. The pair (x,β) is in class I if $x \in A_1^{(\beta)}$ and in class II if $x \in A_2^{(\beta)}$. (9) implies that (6) is not always satisfied, hence (10) is proved.

Next we show that (10) implies (9). In fact we show that if (9) does not hold, then $m \rightarrow (2,q,r,s)$. Consider then a splitting of the pairs (x,β) into two classes so that if $U \subset S_1$, $V \subset S_2$, $|U| = s$, $(V) = s$, then there is always a pair (x_1,y_1) in class I and a pair (x_2,y_2) in class II where $x_1 \in U$, $x_2 \in U$; $y_1 \in V$, $y_2 \in V$.

Now we put x in $A_1^{(\beta)}$ if (x,β) is in class I, and in $A_2^{(\beta)}$ if (x,β) is in class II. Now we show that (6) is satisfied. Since $p = 2$ in (6), $\alpha_i = 1$ or $= 2$. Without loss of generality we can assume that for r values of i, $\alpha_i = 1$. But then if (6) is not satisfied, we would have

$$|S - \bigcup_i A_1^{(\beta_i)}| \geq s, \tag{11}$$

where β_i runs through a set of ordinals V of power r. (11) means that there is a set $U \subset S_1$, $|U| \geq s$ so that all the pairs (x,β_i), $\beta_i \in V$ are in class II, which contradicts our assumption that (9) is false; hence (6) and thus $m \rightarrow (2,q,r,s)$ is proved. Hence (10) implies (9) and thus the proof of the equivalence of (9) and (10) is complete.

Theorem 48 of Erdös-Rado states that

$$\binom{\aleph_1}{\aleph_0} \rightarrow \binom{\aleph_1\ \aleph_1}{\aleph_0\ \aleph_0},$$

thus clearly

$$\binom{\aleph_1}{\aleph_0} \rightarrow \binom{\aleph_0\ \aleph_0}{\aleph_0\ \aleph_0},$$

which by the equivalence of (9) and (10) implies

$$\aleph_1 \to (2, \aleph_0, \aleph_0, \aleph_0). \tag{12}$$

Perhaps

$$\aleph_2 \to (2, \aleph_r, \aleph_1, \aleph_1),$$

but this is undoubtedly very difficult since it is equivalent to one of the most difficult unsolved problems of Erdös-Hajnal-Rado (problem 12),

$$\begin{pmatrix}\aleph_2\\ \aleph_1\end{pmatrix} \to \begin{pmatrix}\aleph_1 & \aleph_1\\ \aleph_1 & \aleph_1\end{pmatrix}. \tag{13}$$

Perhaps $\aleph_1 \to (2, \aleph_2, \aleph_0, \aleph_1)$, but this is also very difficult, since (see problem 12)

$$\begin{pmatrix}\aleph_1\\ \aleph_2\end{pmatrix} \to \begin{pmatrix}\aleph_1 & \aleph_1\\ \aleph_0 & \aleph_0\end{pmatrix} \tag{14}$$

is also unsolved; (13) would imply (14), but (14) also seems very hard.

On the other hand, it follows from Theorem 33 of Erdös-Hajnal-Rado that

$$\aleph_2 \to (2, \aleph_1, \aleph_0, \aleph_1).$$

We will not discuss $m \to (2,q,r,s)$ further, but refer to reference 5.

We state another result which generalizes Sierpinski's result: Assume that $2^{\aleph_k} = \aleph_{\kappa+1}$, then

$$\aleph_{\kappa+1} \to (\aleph_{\kappa+1}, \aleph_{\kappa+1}, \aleph_\kappa, \aleph_{\kappa+1}). \tag{15}$$

(15) follows from Lemma 14.1 of Erdös-Hajnal-Milner.[4] A slightly weakened form of this lemma is stated as follows:

Assume that $2^{\aleph_k} = \aleph_{\kappa+1}$, $|S_1| = |S_2| = \aleph_{\kappa+1}$. The pairs (x,y), $x \in S$, $y \in S_2$ can be split into $\aleph_{\kappa+1}$ classes so that whenever $U \subset S_1$, $V \subset S_2$, $|U| = \aleph_{\kappa+1}$, $|V| = \aleph_\kappa$, there is an $x \in V$ so that every class is represented by the pairs (x,y), $y \in U$.

Now we deduce (15) from this lemma. The elements of S_1 are denoted by $\{x_\gamma\}$, $1 \leq \gamma \leq \omega_{\kappa+1}$, those of S_2 by $\{\beta\}$ $1 \leq \beta \leq \omega_{\kappa+1}$, and the classes into which the pairs are split are denoted by $\{\alpha\}$, $1 \leq \alpha \leq \omega_{\kappa+1}$.

We put $x_\gamma \in A_\gamma^{(\beta)}$ if the pair (x_γ,β) is in class α. A simple argument using Lemma 14.1 shows that (6) is satisfied with $r = \aleph_\kappa$, $s = \aleph_{\kappa+1}$, which proves (15).

Perhaps $\aleph_{\kappa+1} \to (\aleph_{\kappa+1}, \aleph_{\kappa+2}, \aleph_\kappa, \aleph_{\kappa+1})$ also holds, but as we already stated we could not even prove

$$\aleph_1 \to (2, \aleph_2, \aleph_0, \aleph_1).$$

Finally we state a trivial result. Let $m < \aleph_\kappa$ and assume $2^m < \aleph_{\kappa+1}$. Then $\aleph_{\kappa+1} \to (2, m, m, \aleph_{\kappa+1})$. In fact, the following stronger result holds. Let $|S| = \aleph_{\kappa+1}$. Put

$$S = A_1^{(\beta)} \cup A_2^{(\beta)}, \quad 1 \leq \beta \leq \omega_m.$$

Then for some $\beta_i = 1$ or $\beta_i = 2$, $\bigcup_\beta A_{\beta_i}^{(\beta)}$ has a complement of power $\aleph_{\kappa+1}$. We leave the simple proof to the reader.

[1] Banach and Kuratowski, *Fund. Math.*, **15** (1935).
[2] Ulam, S., "Uber die Masstheorie in der Allgemeinen Mengenlehre," *Fund. Math.*, **16** (1936).
[3] Sierpinski, W., *Hypothèse du Continu* (New York: Chelsea Publishing Company, 1956), 2d ed., pp. 53–55.
[4] Erdös, P., A. Hajnal, and E. Milner, "On the complete subgraphs of graphs defined by systems of sets," *Acta Math. Acad. Sci. Hung.*, **17,** 159–229 (1965).
[5] Erdös, P., A. Hajnal, and R. Rado, "Partition relations for cardinal numbers," *Acta Math. Acad. Sci. Hung.*, **16,** 93–196 (1965); see also the earlier paper by Erdös, P., and R. Rado, "A partition calculus in set theory," *Bull. Am. Math. Soc.*, **62,** 427–489 (1956).

Reprinted from JOURNAL OF COMBINATORIAL THEORY Vol. 6, No. 3, April 1969

On the Pairing Process and the Notion of Genealogical Distance

JAN MYCIELSKI AND S. M. ULAM

University of Colorado, Boulder, Colorado 80304

Communicated by Gian-Carlo Rota

Received March 1968

INTRODUCTION AND GENERAL FACTS

By a stochastic pairing process we mean one schematizing the life process of a species with sexual reproduction. A branching process [1] would describe the life history of a species with asexual reproduction (like some bacteria and viruses). Of course one could consider pairing-branching processes for some species like aphids and some bacteria which have both modes of reproduction. But our considerations will pertain only to a pairing process simplified almost to the point of triviality.

We assume that: (i) the whole population P of the species is divided into disjoint generations $A_i (i \in I)$, where I is the set of all integers; (ii) each generation A_i is divided at random into disjoint unordered couples whose union is all of A_i; (iii) each couple produces exactly two individuals; (iv) $a \in A_i$ if and only if it was produced by a couple in A_{i-1}; (v) all A_i have the same finite even number n of individuals. We could have supposed that each A_i is divided randomly into a set of males M_i and a set of females F_i, where $\operatorname{card} M_i = \operatorname{card} F_i = n/2$, and that each couple consists of a male and a female chosen at random, but this would make no difference in this paper (and would exclude species, like many plants, having bisexual individuals).

We have not engaged in a fuller investigation of such processes and this paper contains only three definitions of distances between individuals of P, and some properties of these distances. It would be interesting to relax in some natural way our (iii), (iv), and (v). We believe that our theorems remain true if each pair produces, with some probabilities, 0, 1, or 2 or more offspring with the *expected value* of 2 also for example the population may be variable in time but should not increase or decrease "too rapidly." Finally, the mating need not be random with uniform probability, but for the validity of our theorems it *may* suffice that it is not "too restricted"

For commentary to this paper [97], see p. 689.

to subgroups. But probably the most interesting problems pertain to the distribution and evolution of properties in a species. We have here the well-known laws of Mendel [2] and the theorem of Hardy [3]. Yet little is known about the rate of accumulation of mutations [4] or its relation to the value (i.e., impact on reproductivity) of those mutations. Some numerical experiments pertaining to this are actually being pursued by R. Schrandt and S. M. Ulam. But no full mathematical description of evolution has ever been made.

The relationship between two consecutive generations A_i and A_{i+1} is defined by a partition of A_i into couples, a partition of A_{i+1} into pairs produced by the same couple, and a one-to-one map of the first partition onto the second denoting which couple produced which pair. Since there are

$$\prod_{s=0}^{n/2-1} (n - (2s + 1)) = n!2^{-n/2}/(n/2)!$$

partitions of any A_i into pairs and $(n/2)!$ such maps we obtain the following

PROPOSITION. *There are $f(n)$,*

$$f(n) = (n/2)! \prod_{s=0}^{n/2-1} (2s + 1)^2 = (n!)^2 2^{-n}/(n/2)!$$

possible relationships between two consecutive generations.

A realization of our pairing stochastic process is any sequence $r = \langle r_i : i \in I \rangle$, where r_i denotes some possible relationship between A_i and A_{i+1}. R denotes the set of all realization r. We suppose that any two relationships are equally probable, i.e., each has probability $1/f(n)$. We introduce the corresponding probability measure μ over the space R of all realizations r, i.e. μ is the product measure.

1. THE METRIC d_r

For any $a \in P = \bigcup_{i \in I} A_i$ let $S(a)$ denote the set $A \cup \{a\}$, where A is the set of all ancestors of a, i.e., parents, grandparents, great grandparents, etc. Of course $S(a)$ depends on r.

One possible definition of a distance between two individuals $a, b \in P$ is:

$$d_r(a, b) = \text{card}(S(a) \triangle S(b)),$$

where $\triangle$ denotes the symmetric difference of sets, i.e.,

$$A \triangle B = (A \cup B) - (A \cap B).$$

This distance, which as we will show *is finite with probability* 1, could be interpreted as follows: Suppose that each $a \in P$ brings one new individual property (i.e., "mutation") and passes it on as well as all of its inherited properties to all its children. Then our distance d_r is a reasonable measure of the difference between individuals. (The reality is, of course, infinitely more complicated; inheritance is a kind of stochastic process, but we have not tried to include this in our framework.)

THEOREM 1. (i) *d_r is a metrization of the set P for almost all $r \in R$.*
(ii) *For every $a, b \in P$*

$$\int_R d_r(a, b)\, d\mu < \infty.$$

PROOF: (i) All the facts: $d_r(a, b) = 0$ iff $a = b$, $d_r \geqslant 0$,

$$d_r(a, b) + d_r(b, c) \geqslant d_r(a, c),$$

and $d_r < \infty$ for almost all r, which constitute the statement (i), are obvious.

(ii) Let $a \in A_i$ and $b \in B_j$. Clearly the number $p = \mu\{r : S(a) \neq A_k$ or $S(b) \neq A_k\}$ is less than 1 if $k \leqslant \min(i, j) - 2 \log n/\log 2$. Hence for some $p_0 < 1$ we have

$$\mu\{r : S(a) \neq A_k \text{ or } S(b) \neq A_k\} \leqslant p_0{}^s$$

for all

$$k \leqslant \min(i, j) - 2s \log n/\log 2.$$

Of course

$$\{r : d_r(a, b) \geqslant nt\} \subseteq \{r : S(a) \neq A_k \text{ or } S(b) \neq A_k\},$$

where $k = \max(i, j) - t$. Therefore

$$\mu\{r : d_r(a, b) \geqslant nt\} \leqslant p_0^{s(t)}$$

for all $t \geqslant \alpha = \max(i, j) - \min(i, j) + 2 \log n/\log 2$, where

$$s(t) = \frac{t + \min(i, j) - \max(i, j)}{2 \log n/\log 2}.$$

Hence $\mu\{r : n(t + 1) \geqslant d_r(a, b) \geqslant nt\} \leqslant p^{s(t)}$ for $t \geqslant \alpha$ and

$$\int_R d_r(a, b)\, d\mu \leqslant n\alpha + \sum n(t + 1)\, p_0^{s(t)} < \infty$$

2. The Metric ρ_r

To each $a \in P$ we will assign a vector

$$v(a) = \langle v_c(a) : c \in P \rangle,$$

where $v_c(a)$ are non-negative real numbers defined by the following recursive rules:

(i) $v_c(a) = 0$ if $c \neq a$ and c is not an ancestor of a,

(ii) $$v(a) = e(a) + \frac{v(a') + v(a'')}{2},$$

where $e(a) = \langle p_c : c \in P \rangle$, where $p_c = 0$ for $c \neq a$ and $p_a = 1$, and a', a'' are the parents of a (of course $+$ denotes vector-addition). It is easy to prove by induction that the above rules define uniquely the vectors $v(a)$ for all $a \in P$. Of course these vectors depend on r.

The vector $\langle v_c(a) : c \in A_i \rangle$ indicates the distribution of genes of the generation A_i in a. Of course

$$\sum_{c \in A_i} v_c(a) = 1.$$

Such vectors should appear natural to a genealogist or a lawyer.

Our second proposed description of distance is

$$\rho_r(a, b) = \sum_{c \in P} | v_c(a) - v_c(b) |.$$

It seems that d_r and ρ_r are two extremes in a family of possible distances. We do not have any absolute preference between them.

Theorem 2. (i) *ρ_r is a metrization of the set P for almost all $r \in R$.*

(ii) *For every $a, b \in P$*

$$\int_R \rho_r(a, b) \, d\mu < \infty.$$

Proof: (i) By the definition of ρ_r it remains only to show that $\rho_r(a, b) < \infty$ for almost all r. But this follows from (ii), which we will prove now.

(ii) Let

$$B_i = \bigcup_{i \geqslant j > -\infty} A_j$$

and $i(a)$ be defined by $a \in A_i$. Of course by the triangle inequality it will be enough to show

(1) $$\int_R \sum_{c \in B_{i(a)}} \left| v_c(a) - \frac{1}{n} \right| d\mu < \infty \quad \text{for every } a \in P.$$

We put

$$\alpha_c = v_c(a) - \frac{1}{n} \qquad \text{and} \qquad \delta_j = \operatorname{diam}\{\alpha_c : c \in A_j\}$$

(diam = max − min). Of course by the definition of $v(a)$ we have:

(2) $$\sum_{c \in A_j} \alpha_c = 0.$$

We will check that

(3) for each relationship between A_j and A_{j-1} we have

$$\max\{| \alpha_c | : c \in A_{j-1}\} \leqslant \max\{| \alpha_c | : c \in A_j\}.$$

Indeed for each $c \in A_{j-1}$ we have

$$\alpha_c = \tfrac{1}{2}(\alpha_{c_1} + \alpha_{c_2}),$$

where $c_1, c_2 \in A_j$ are the children of c. Hence

$$| \alpha_c | \leqslant \tfrac{1}{2}(| \alpha_{c_1} | + | \alpha_{c_2} |) \leqslant \max\{| \alpha_{c_1} |, | \alpha_{c_2} |\},$$

and (3) follows.

(4) For each $j < i(a)$, there exists a relationship between A_j and A_{j-1} such that

$$\delta_{j-1} \leqslant \tfrac{3}{4}\delta_j .$$

To prove this let us notice that, since $j < i(a)$, for each (married) couple $C = \{c_1, c_2\} \subseteq A_j$ we have $\alpha_{c_1} = \alpha_{c_2}$. Thus for each couple $C \subseteq B_{i(a)+1}$ we can put $\alpha(C) = \alpha_c$, where $c \in C$. Let $V = \{C_1^j, \ldots, C_{n/2}^j\}$ be the set of all couples in A_j. A relationship between A_j and A_{j-1} defines a graph over the set V, namely $C_{i_1}^j$ is joined by an edge to $C_{i_2}^j$ iff $i_1 \neq i_2$ and there exists a couple $C_i^{j-1} \subseteq A_{j-1}$, and $c_1 \in C_{i_1}^j$, and $c_1 \in C_{i_2}^j$ such that c_1, c_2 are the children of C_i^{j-1}. Clearly this graph is a disjoint union of closed circuits and isolated vertices, and of course each graph over the set of vertices V which consists of disjoint circuits and isolated vertices may result from a suitable relationship between A_j and A_{j-1}. We also obtain the relation

$$\alpha(C_i^{j-1}) = \tfrac{1}{2}(\alpha(C_{i_1}^j) + \alpha(C_{i_2}^j)),$$

i.e., the edges of the graph of couples of A_j correspond to couples of A_{j-1}, and the α-s corresponding to the edges are averages of the α-s corresponding to their vertices.

Now we consider two cases

(i) $\quad \operatorname{card}\{C \in V : \alpha(C) \leqslant \min\{\alpha(C) : C \in V\} + \delta_j/2\} \leqslant n/4;$

(ii) $\quad \operatorname{card}\{C \in V : \alpha(C) \geqslant \min\{\alpha(C) : C \in V\} + \delta_j/2\} \leqslant n/4.$

Clearly one of these cases must hold. Suppose (i). Then there exists a graph of the required kind over V such that every C_1 with

$$\alpha(C_1) \leqslant \min\{\alpha(C) : C \in V\} + \delta_j/2$$

is joined to C_2 and C_3 with

$$\alpha(C_2), \alpha(C_3) \geqslant \min\{\alpha(C) : C \in V\} + \delta_j/2.$$

Thus for every couple C^{j-1} of A_{j-1} we have

$$\min\{\alpha(C) : C \in V\} + \delta_j/4 \leqslant \alpha(C^{j-1}) \leqslant \max\{\alpha(C) : C \in V\},$$

and (4) follows. In case (ii) the proof is similar.

A relationship satisfying (4) will be called a reducing relationship.

Since the probability of each relationship between two consecutive generations is $p = 1/f(n)$ (see Proposition) then by the law of large numbers it follows that for almost all $r \in R$ there exists an integer $k \geqslant 0$ such that for every $j > k$ the sequence

$$r_{i(a)-1}, \qquad r_{i(a)-2}, \ldots, r_{i(a)-j}$$

contains more than $[pj/2]$ reducing relationships.

Let $k(r)$ denote the least such k. We will show that

$$(5) \qquad \sum_{c \in B_{i(a)}} |\alpha_c| < nk(r) + \beta,$$

where β is a constant independent from r.

Indeed we have of course $|\alpha_c| < 1$ and hence

$$\sum_{i(a) \geqslant s \geqslant i(a)-k(r)} \sum_{c \in A_s} |\alpha_c| < n(k(r) + 1).$$

By (2), (3), and (4) and the definition of $k(r)$ for every $j > k(r)$ and $c \in A_{i(a)-j}$ we have $|\alpha_c| \leqslant (\frac{3}{4})^{[pj/2]}$. Hence

$$\sum_{i(a)-k(r) > s > -\infty} \sum_{c \in A_s} |\alpha_c| \leqslant n \sum_{j=k(r)+1}^{\infty} (\tfrac{3}{4})^{[pj/2]}$$

and (5) follows. By (5)

$$\int_R \sum_{c \in B_{i(a)}} | \alpha_c | \, d\mu \leqslant n \int_R k(r) \, d\mu + \beta,$$

and hence to prove (1) we need only to show

$$(6) \qquad \int_R k(r) \, d\mu < \infty.$$

But to prove this we need first

(7) $\{r : k(r) = t\} \subseteq \{r :$ the sequence $r_{i(a)-1}, ..., r_{i(a)-t}$ contains exactly $[pt/2]$ reducing relationships$\}$.

Indeed if $r_{i(a)-1}, ..., r_{i(a)-t}$ contains more than $[pt/2]$ reducing relationships then $k(r) = t$ cannot hold. If it contains less than $[pt/2]$ reducing relationships, say u of them and $k(r) = t$ then $r_{i(a)-1}, ..., r_{i(a)-t}, r_{i(a)-t-1}$ must contain more than $[p(t + 1)/2]$ reducing relationships (since $t + 1 > k(r)$) and $r_{i(a)-t-1}$ is the only new one and hence $u + 1 > [p(t + 1)/2]$; but this is impossible since $u + 1 \leqslant [pt/2]$.

Now let us show (6). By (7), and the theorem of Bernoulli,

$$\mu\{r : k(r) = t\} \leqslant \binom{t}{[pt/2]} p^{[pt/2]}(1 - p)^{t-[pt/2]}$$

Thus we have

$$\int_R k(r) \, d\mu = \sum_{t=1}^{\infty} t\mu\{r : k(r) = t\} \leqslant \sum_{t=1}^{\infty} t \binom{t}{[pt/2]} p^{[pt/2]}(1 - p)^{t-[pt/2]}.$$

Using the formula of Stirling it is routine to check that the last series is convergent. Q.E.D.

3. The Metric $\rho_r^{(\epsilon)}$

Let us mention a third possible metric in P, which we denote by $\rho_r^{(\epsilon)}$, where $0 < \epsilon < 1$. $\rho_r^{(\epsilon)}$ is defined in the same way as ρ_r but rule (ii) of this definition is replaced by

$$(\text{ii})_\epsilon \qquad v(a) = \epsilon e(a) + (1 - \epsilon) \frac{v(a') + v(a'')}{2}.$$

The vector $v(a)$ defined in this way satisfies

$$\sum_{c \in P} v_c(a) = 1.$$

Therefore $\rho_r^{(\epsilon)}(a, b) \leqslant 2$ for all $r \in R$ and $a, b \in P$.

Perhaps a reasonable genetic interpretation of $\rho_r^{(\epsilon)}$ is possible.

References

1. T. E. Harris, *The Theory of Branching Processes*, Berlin, 1963.
2. See, e.g., J. D. Watson, *Molecular Biology of the Genes*, New York-Amsterdam, 1965.
3. See, e.g., J. H. B. Kemperman, On systems of mating. I,..., IV, *Indagationes Mathematicae* **29** (1967), 245–304.
4. H. J. Muller, Evolution by Mutation, *Bull. Amer. Math. Soc.* **64** (1958), 137–160.

ROCKY MOUNTAIN
JOURNAL OF MATHEMATICS
Volume 1, Number 4, Fall 1971

SOME PROBABILISTIC REMARKS ON FERMAT'S LAST THEOREM

P. ERDÖS AND S. ULAM

Let $a_1 < a_2 < \cdots$ be an infinite sequence of integers satisfying $a_n = (c + o(1))n^\alpha$ for some $\alpha > 1$. One can ask: Is it likely that $a_i + a_j = a_r$ or, more generally, $a_{i_1} + \cdots + a_{i_n} = a_{i_r}$ has infinitely many solutions. We will formulate this problem precisely and show that if $\alpha > 3$ then with probability 1, $a_i + a_j = a_r$ has only finitely many solutions, but for $\alpha \leqq 3$, $a_i + a_j = a_r$ has with probability 1 infinitely many solutions. Several related questions will also be discussed.

Following [1] we define a measure in the space of sequences of integers. Let $\alpha > 1$ be any real number. The measure of the set of sequences containing n has measure $c_1 n^{1/\alpha - 1}$ and the measure of the set of sequences not containing n has measure $1 - c_1 n^{1/\alpha - 1}$. It easily follows from the law of large numbers (see [1]) that for almost all sequences $A = \{a_1 < a_2 < \cdots\}$ ("almost all" of course, means that we neglect a set of sequences which has measure 0 in our measure) we have

$$A(x) = (1 + o(1))c_1 \sum_{n=1}^{x} \frac{1}{n^{1/\alpha - 1}} = (1 + o(1))c_1 \alpha x^{1/\alpha} \tag{1}$$

where $A(x) = \sum_{a_i < x} 1$. (1) implies that for almost all sequences A

$$a_n = (1 + o(1))(n/c_i\alpha)^\alpha. \tag{2}$$

Now we prove the following

THEOREM. *Let $\alpha > 3$. Then for almost all A*

$$a_i + a_j = a_r \tag{3}$$

has only a finite number of solutions. If $\alpha \leqq 3$, then for almost all A, (3) *has infinitely many solutions.*

It is well known that $x^3 + y^3 = z^3$ has no solutions, thus the sequence $\{n^3\}$ belongs to the exceptional set of measure 0.

Assume $\alpha > 3$. Denote by E_α the expected number of solutions of $a_i + a_j = a_r$. We show that E_α is finite and this will immediately

Received by the editors April 28, 1970.
AMS 1970 subject classifications. Primary 10K99, 10L10.

imply that for almost all sequences A, $a_i + a_j = a_r$ has only a finite number of solutions. Denote by $P(u)$ the probability (or measure) that u is in A. We evidently have

$$E_\alpha = \sum_{n=1}^{\infty} P(n) \sum_{u+v=n} P(u)P(v)$$

$$= c_1{}^3 \sum_{n=1}^{\infty} \frac{1}{n^{1-1/\alpha}} \sum_{u+v=n} \frac{1}{u^{1-1/\alpha}v^{1-1/\alpha}}$$

$$< c_2 \sum_{n=1}^{\infty} \frac{1}{n^{1-1/\alpha}} \frac{1}{n^{1-2/\alpha}} = c_2 \sum_{n=1}^{\infty} \frac{1}{n^{2-3/\alpha}} < c_3$$

which proves our theorem for $\alpha > 3$. One could calculate the probability that (3) has exactly r solutions ($r = 0, 1, \cdots$).

Let now $\alpha \leqq 3$. The case $\alpha = 3$ is the most interesting; the case $\alpha < 3$ can be dealt with similarly. Denote by $E_\alpha(x)$ the expected number of solutions of (3) if a_i, a_j and a_r are $\leqq x$. We have

$$E_3(x) = \sum_{n=1}^{x} P(n) \sum_{u+v=n} P(u)P(v) = c_1{}^3 \sum_{n=1}^{x} \frac{1}{n^{2/3}} \sum_{u+v=n} \frac{1}{(uv)^{2/3}}$$

(4)

$$= (1 + o(1))c_1{}^3 \sum_{n=1}^{x} \frac{1}{n^{2/3}} \frac{c_2}{n^{1/3}} = (1 + o(1))c_1{}^3 c_2 \log x.$$

By a little calculation, it would be easy to determine c_2 explicitly. Now we prove by a simple second moment argument that for almost all A the number of solutions $f_3(A, x)$ of $a_i + a_j = a_r$, $a_r \leqq x$ satisfies

(5) $\quad f_3(A, x) = (1 + o(1))c_1{}^3 c_2 \log x,$ that is $f_3(A, x)/E_3(x) \to 1.$

To prove (5) we first compute the expected value of $f_3(A, x)^2$.

The expected value of $f_3(A, x)$ was $E_3(x)$ which we computed in (4). Denote by $E_3{}^2(x)$ the expected value of $f_3(A, x)^2$. We evidently have

(6) $$E_3{}^2(x) = \sum_{1 \leqq n_1 \leqq x;\ 1 \leqq n_2 \leqq x} P(n_1)P(n_2) \sum_{u_1+v_1=n_1;\ u_2+v_2=n_2} P(u_1, u_2, v_1, v_2)$$

where $P(u_1, v_1, u_2, v_2)$ is the probability that u_1, v_1, u_2, v_2 occurs in our sequence. If these four numbers are distinct, then clearly $P(u_1, u_2, v_1, v_2) = P(u_1)P(u_2)P(v_1)P(v_2)$, but if say $u_1 = u_2$, the probability is larger. Hence $E_3{}^2(x) > (E_3(x))^2$ and to get the opposite inequality we have to add a term which takes into account that the four terms do not have to be distinct, or $n_1 < n_2$, $u_1 = u_2$.

$$E_3^2(x) < (E_3(x))^2$$

$$+ c \sum_{n_1=1}^{x} P(n_1)P(n_1 + v_2 - v_1) \sum_{u_1+v_1=n_1;\ v_2<x} P(u_1)P(v_1)P(v_2)$$

$$< (E_3(x))^2 + \sum_{n_1=1}^{x} \frac{c_1}{n_1} \sum_{v_2=1}^{x} P(v_2)P(n_1 + v_2 - v_1) \tag{7}$$

$$< (E_3(x))^2 + \sum_{n_1=1}^{x} \frac{c_1}{n_1} \sum_{v_2=1}^{\infty} P(v_2)^2 < (E_3(x))^2 + \sum_{n=1}^{x} \frac{c_2}{n}$$

$$< (E_3(x)^2) + c_3 \log x.$$

Thus

$$(E_3(x^2)) < E_3^2(x) < (E_3(x))^2 + c_3 \log x. \tag{8}$$

(8) implies by the Tchebycheff inequality that the measure of the set A for which

$$|f_3(A, x) - E_3(x)| > \epsilon \log x \tag{9}$$

is less than $c/\epsilon^2 \log x$. This easily implies that for almost all A

$$\lim_{x=\infty} f_3(A, x)/E_3(x) = 1. \tag{10}$$

To show (10) let $x_k = 2^{k(\log k)^2}$. From (9) and the Borel-Cantelli Lemma it follows that

$$\lim_{k=\infty} f_3(A, x)/E_3(x_k) = 1. \tag{11}$$

(11) now easily implies (10), $f_3(A, x)$ is a nondecreasing function of x, thus if $x_k < x < x_{k+1}$, $f_3(A, x_k) \leqq f_3(A, x) \leqq f_3(A, x_{k+1})$. Thus (11) follows from $E_3(x_n)/E_3(x_{k+1}) \to 1$.

By the same method we can prove that for $\alpha < 3$

$$\lim_{x=\infty} \frac{f_\alpha(A, x)}{E_\alpha(x)} \to 1.$$

Similarly we can investigate the equation

$$a_{c_r} = a_{c_1} + a_{c_2} + \cdots + a_{c_x}. \tag{12}$$

Here by the same method we can prove that for $\alpha > k + 1$ with probability 1, (12) has only a finite number of solutions and for $\alpha \leqq k + 1$ it has infinitely many solutions.

Euler conjectured that the sum of $k - 1$ (kth) powers is never a kth power. This is true for $k = 3$, unknown for $k = 4$ and has been recently disproved for $k = 5$ [2]. As far as we know it is possible that for every $k \geqq 3$ there are only a finite number of kth powers which are the sum of $k-1$ or fewer kth powers.

Let $\beta > 1$ be a rational number. One can ask whether $[n^\beta] + [m^\beta] = [l^\beta]$, has solutions in integers n, m, l. One would guess that for $\beta < 3$ the equation always has infinitely many solutions but that the measure of the set in β, $\beta > 3$, for which it has infinitely many solutions has measure 0, but it is not hard to prove that the β's for which it has infinitely many solutions is everywhere dense.

References

1. P. Erdös and A. Rényi, *Additive properties of random sequences of positive integers*, Acta Arith. **6** (1960), 83–110. MR **22** #10970; See also: H. Halberstam and K. F. Roth, *Sequences*, Vol. 1, Clarendon Press, Oxford, 1966. MR **35** #1565.

2. L. J. Lander and T. Parkin, *A counterexample to Euler's sum of powers conjecture*, Math. Comp. **21** (1967), 101–103. MR **36** #3721.

UNIVERSITY OF COLORADO, BOULDER, COLORADO 80302

Part II

Computations, Games, and Numbers

Reprinted from the Journal of the American Statistical Association
September, 19549, Vol. 44, pp. 335-341

THE MONTE CARLO METHOD

Nicholas Metropolis and S. Ulam
Los Alamos Laboratory

We shall present here the motivation and a general description of a method dealing with a class of problems in mathematical physics. The method is, essentially, a statistical approach to the study of differential equations, or more generally, of integro-differential equations that occur in various branches of the natural sciences.

Already in the nineteenth century a sharp distinction began to appear between two different mathematical methods of treating physical phenomena. Problems involving only a few particles were studied in classical mechanics, through the study of systems of ordinary differential equations. For the description of systems with very many particles, an entirely different technique was used, namely, the method of statistical mechanics. In this latter approach, one does not concentrate on the individual particles but studies the properties of *sets of particles*. In pure mathematics an intensive study of the properties of sets of points was the subject of a new field. This is the so-called theory of sets, the basic theory of integration, and the twentieth century development of the theory of probabilities prepared the formal apparatus for the use of such models in theoretical physics, i.e., description of properties of aggregates of points rather than of individual points and their coordinates.

Soon after the development of the calculus, the mathematical apparatus of partial differential equations was used for dealing with the problems of the physics of the continuum. Hydrodynamics is the most widely known field formulated in this fashion. A little later came the treatment of the problems of heat conduction and still later the field theories, like the electromagnetic theory of Maxwell. All this is very well known. It is of course important to remember that the study of the

For commentary to this paper [38], see p. 689.

physics of the continuum was paralleled through "kinetic theories." These consist in approximating the continuum by very large, but finite, numbers of interacting particles.

I

When a physical problem involves an intermediate situation, i.e., a system with a moderate number of parts, neither of the two approaches is very practical. The methods of analytical mechanics do not even give a qualitative survey of the behavior of a system of three mutually attractive bodies. Obviously the statistical-mechanical approach would also be unrealistic.

An analogous situation exists in problems of combinatorial analysis and of the theory of probabilities. To calculate the probability of a successful outcome of a game of solitaire (we understand here only such games where skill plays no role) is a completely intractable task. On the other hand, the laws of large numbers and the asymptotic theorems of the theory of probabilities will not throw much light even on qualitative questions concerning such probabilities. Obviously the practical procedure is to produce a large number of examples of any given game and then to examine the relative proportion of successes. The "solitaire" is meant here merely as an illustration for the whole class of combinatorial problems occurring in both pure mathematics and the applied sciences. We can see at once that the estimate will never be confined within given limits with certainty, but only—if the number of trials is great—with great probability. Even to establish this much we must have recourse to the laws of large numbers and other results of the theory of probabilities.

Another case illustrating this situation is as follows: Consider the problem of evaluating the volume of a region in, say, a twenty-dimensional space. The region is defined by a set of inequalities

$$f_1(x_1, x_2 \cdots x_{20}) < 0; f_2(x_1, x_2 \cdots x_{20}) < 0; \cdots f_{20}(x_1, x_2 \cdots x_{20}) < 0.$$

This means that we consider all points$(x_1, x_2, x_3, \cdots x_{20})$ satisfying the given inequalities. Suppose further that we know that the region is located in the unit cube and we know that its volume is not vanishingly small in general. The multiple integrals will be hardly evaluable. The procedure based on the definition of a volume or the definition of an integral, i.e., the subdivision of the whole unit cube, for example, each coordinate x_1 into ten parts, leads to an examination of 10^{20} lattice points in the unit cube. It is obviously impossible to count all of them. Here again the more sensible approach would be to take, say 10^4 points at random from this ensemble and examine those only; i.e., we should count how many of the selected points satisfy all the given inequalities. It follows from simple application of ergodic theorems that the estimate should be, *with great probability*, valid within a few per cent.

As another illustration, certain problems in the study of cosmic rays are of the following form. An incoming particle with great energy entering the atmosphere starts a whole chain of nuclear events. New particles are produced from the target nuclei, these in turn produce new reactions. This cascade process continues with more and more particles created until the available individual energies become too small to produce further nuclear events. The particles in question are protons, neutrons, electrons, gamma rays and mesons. The probability of producing a given particle with a given energy in any given collision is dependent on the energy of the incoming particle. A further complication is that there is a probability distribution for the direction of motions. Mathematically, this complicated process is an illustration of a so-called Markoff chain. The mathematical tool for the study of such chains is matrix theory. It is obvious that in order to obtain a mathematical analysis, one would have to multiply a large number of $(n \times n)$ matrices, where n is quite great.

Here again one might try to perform a finite number of "experiments" and obtain a class or sample of possible genealogies. These experiments will of course be performed not with any physical apparatus, but theoretically. If we assume that the probability of each possible event is given, we can then play a great number of games of chance, with chances corresponding to the assumed probability distributions. In this fashion one can study empirically the asymptotic properties of powers of matrices with positive coefficients, interpreted as transition probabilities.

II

Finally let us consider more generally the group of problems which gave rise to the development of the method to which this article is devoted. Imagine that we have a medium in which a nuclear particle is introduced, capable of producing other nuclear particles with a distribution of energy and direction of motion. Assume for simplicity that all particles are of the same nature. Their procreative powers depend, however, on their position in the medium and on their energy. The problem of the behavior of such a system is formulated by a set of integro-differential equations. Such equations are known in the kinetic theory of gases as the Boltzmann equations. In the theory of probabilities one has somewhat similar situations described by the Fokker-Planck equations. A very simplified version of such a problem would lead to the equation:

$$\frac{\partial u(x, y, z)}{\partial t} = a(x, y, z)\Delta u + b(x, y, z)u(x, y, z) \tag{1}$$

where $u(x, y, z)$ represents the density of the particles at the point (x, y, z). The Laplacian term, $a\Delta u$ on the right hand side corresponds to the diffusion of the particles, and bu to the particle procreation, or multiplication. [In reality, the equation describing the physical situation stated above is much more complicated. It involves more independent variables, inasmuch as one is interested in the density $w(x, y, z; v_x, v_y, v_z)$ of particles in phase space, v being the velocity vector.] The classical methods for dealing with these equations are extremely laborious and incomplete in the sense that solutions in "closed form" are unobtainable. The idea of using a statistical approach at which we hinted in the preceding examples is sometimes referred to as the Monte Carlo method.

The mathematical description is the study of a flow which consists of a mixture of deterministic and stochastic processes.[1] It requires its own laws of large numbers and asymptotic theorems, the study of which has only begun. The computational procedure looks in practice as follows: we imagine that we have an ensemble of particles each represented by a set of numbers. These numbers specify the time, components of position and velocity vectors, also an index identifying the nature of the particle. With each of these sets of numbers, random processes are initiated which lead to the determination of a new set of values. There exists indeed a set of probability distributions for the new values of the parameters after a specified time interval Δt. Imagine that we draw at random and *independently*, values from a prepared collection possessing such distributions. Here a distinction must be made between those parameters which we believe vary independently of each other, and those values which are strictly determined by the values of other parameters. To illustrate this point: assume for instance that in the fission process the direction of the emitted neutron is independent of its velocity. Or again, the direction of a neutron in a homogeneous medium does not influence the distance between its origin and the site of its first collision. On the other hand, having "drawn" from appropri-

[1] von Neumann, J., and Ulam, S., *Bulletin A.M.S.*, Abstract 51-9-165 (1945).

ate distributions the velocity of a new-born particle and the distance to its first collision, the time elapsed in travel is completely determined and has to be calculated accordingly. By considering a large number of particles with their corresponding sets of parameters we obtain in this fashion another collection of particles and a new class of sets of values of their parameters. The hope is, of course, that in this manner we obtain a good sample of the distributions at the time $t+\Delta t$. This procedure is repeated as many times as required for the duration of the real process or else, in problems where we believe a stationary distribution exists, until our "experimental" distributions do not show significant changes from one step to the next.

The essential feature of the process is that we avoid dealing with multiple integrations or multiplications of the probability matrices, but instead sample single chains of events. We obtain a sample of the set of all such possible chains, and on it we can make a statistical study of both the genealogical properties and various distributions at a given time.

III

We want now to point out that modern computing machines are extremely well suited to perform the procedures described. In practice, the set of values of parameters characterizing a particle is represented, for example, by a set of numbers punched on a card. We have at the outset a large number of particles (or cards) with parameters reflecting given initial distributions. The step in time consists in the production of a new such set of cards. The original set is processed one by one by a computing machine somewhat as follows: The machine has been set up in advance with a particular sequence of prescribed operations. These divide roughly into two classes: (1) production of "random" values with their frequency distribution equal to those which govern the change of each parameter, (2) calculation of the values of those parameters which are deterministic, i.e., obtained algebraically from the others. It may seem strange that the machine can simulate the production of a series of random numbers, but this is indeed possible. In fact, it suffices to produce a sequence of numbers between 0 and 1 which have a uniform distribution in this interval but otherwise are uncorrelated, i.e., pairs will have uniform distribution in the unit square, triplets uniformly distributed in the unit cube, etc., as far as practically feasible. This can be achieved with errors as small as desired or practical. What is more, it is not necessary to store a collection of such numbers in the machine itself, but paradoxically enough the machine can be made to produce numbers simulating the above properties by iterating a well-defined arithmetical operation.

Once a uniformly distributed random set is available, sets with a prescribed probability distribution $f(x)$ can be obtained from it by first drawing from a uniform uncorrelated distribution, and then using, instead of the number x which was drawn, another value $y=g(x)$ where $g(x)$ was computed in advance so that the values y possess the distribution $f(y)$.

Regarding the sequence of operations on a machine, more can be and has been done. The choice of the *kind* of step to be performed by the machine can be made to depend on the values of certain parameters just obtained. In this fashion even dependent probabilistic processes can be performed. Quite apart from mechanized computations, let us point out one feature of the method which makes it advantageous with, say, stepwise integration of differential equations. In order to find a particular solution, the usual method consists in iterating an algebraical step, which involves in the nth stage values obtained from the $(n-1)$th step. The procedure is thus serial, and in general one does not shorten the time required for a solution of the problem by the use of more than one computer. On the other hand, the statistical methods can be applied by many computers working in parallel and independently. Several such calculations have already been performed for problems of types discussed above.[2]

IV

Let us indicate now how other equations could be dealt with in a similar manner. The first, purely mathematical, step is to transform the given equation into an equivalent one, possessing the form of a diffusion equation with possible multiplication of the particles involved. For example as suggested by Fermi, the time-independent Schrödinger equation

$$\Delta\psi(x, y, z) = (E - V)\psi(x, y, z)$$

could be studied as follows. Re-introduce time dependence by considering

$$u(x, y, z, t) = \psi(x, y, z)e^{-Et}$$

u will obey the equation

$$\frac{\partial u}{\partial t} = \Delta u - Vu.$$

[2] Among others, problems of diffusion of neutrons, gamma rays, etc. To cite an example involving the study of matrices, there is a recent paper by Goldberger, *Phys. Rev.* 74, 1269 (1948), on the interaction of high energy neutrons with heavy nuclei.

This last equation can be interpreted however as describing the behavior of a system of particles each of which performs a random walk, i.e., diffuses isotropically and at the same time is subject to multiplication, which is determined by the value of the point function V. If the solution of the latter equation corresponds to a spatial mode multiplying exponentially in time, the examination of the spatial part will give the desired $\psi(x, y, z)$—corresponding to the lowest "eigenvalue" E.

The mathematical theory behind our computational method may be briefly sketched as follows: As mentioned above and indicated by the examples, the process is a combination of stochastic and deterministic flows.[1] In more technical terms, it consists of repeated applications of matrices—like in Markoff chains—and completely specified transformations, e.g., the transformation of phase space as given by the Hamilton differential equations.

One interesting feature of the method is that it allows one to obtain the values of certain given operators on functions obeying a differential equation, without the point-by-point knowledge of the functions which are solutions of the equation. Thus we can get directly the values of the first few moments of a distribution, or the first few coefficients in the expansion of a solution into, for example, a Fourier series without the necessity of first "obtaining" the function itself. "Symbolically" if one is interested in the value of $U(f)$ where U is a functional like the above, and f satisfies a certain operator equation $\psi(f)=0$, we can in many cases obtain an idea of the value of $U(f)$ directly, without "knowing" f at each point.

The asymptotic theorems so far established provide the analogues of the laws of large numbers, such as the generalizations of the weak and strong theorems of Bernoulli, Cantelli-Borel.[3] The more precise information corresponding to that given in the Laplace-Liapounoff theory of additive processes has not yet been obtained for our more general case. In particular it seems very difficult to estimate in a precise fashion the probability of the error due to the finiteness of the sample. This estimate would be of great practical importance, since it alone would allow us to suit the size of the sample to the desired accuracy.

The "space" in which our process takes place is the collection of all possible chains of events, or infinite branching graphs.[4] The general properties of such a phase space have been considered but much work remains to be done on the specific properties of such spaces, each corresponding to a given physical problem.

---

[3] Everett, C. J. and Ulam, S., U.S.A.E.C., Los Alamos reports LADC-533 and LADC-534. Declassified, 1948.

[4] Everett, C. J. and Ulam, S., *Proc. Nat. Acad. Sciences*, 34,403 (1948).

Reprinted from Vol. II, PROCEEDINGS OF THE
INTERNATIONAL CONGRESS OF MATHEMATICIANS, 1950

RANDOM PROCESSES AND TRANSFORMATIONS

S. ULAM

I

It is intended to present here a general point of view and specific problems connected with the relation between descriptions of physical phenomena by random processes and the theory of probabilities on the one hand, and the deterministic descriptions by methods of classical analysis in mathematical physics, on the other. We shall attempt to formulate procedures of random processes which will permit heuristic and also quantitative evaluations of the solutions of differential or integral-differential equations. Broadly speaking, such methods will amount to construction of statistical models of given physical situations and statistical *experiments* designed to evaluate the behavior of the physical quantities involved.

The role of probability theory in physics is really manifold. In classical theories the role of *initial conditions* is consciously idealized. In reality these initial conditions are known only within certain ranges of values. One could say that probability distributions are given for the actual values of initial parameters. The influence of the variation of initial conditions, subject to "small" fluctuations, on the properties of solutions has been studied in numerous cases and forms one subject of the theories of "stability."

In a more general way, not only the initial constants, but even the operators describing the behavior of a given physical situation may not be known exactly. We might assume that, for example, the forces acting on a given mechanical system are known only within certain limits. They might depend, for example, to some extent on certain "hidden" parameters and we might again study the influence of random terms in these forces on the given system.

In quantum theory, of course, the role of a stochastic point of view is even more fundamental. The variables describing a physical system are of higher mathematical type. They are sets of points or sets of numbers (real, complex, or still more general) rather than the numbers themselves. The probability distributions enter from the beginning as the primitive notions of the theory. The observable or measurable quantities are values of certain functionals or eigenvalues of operators acting on these distributions. Again, in addition to this fundamental role of the probabilities formulation, there will enter the fact that the nature of forces or conditions may not be known correctly or exactly, but the operators corresponding to them will depend on "hidden" parameters in a fashion similar to that in classical physics. In fact, at the present time considerable latitude exists in the choice of operators corresponding to "forces" in nuclear physics.

There is, in addition, another reason for the recourse to descriptions in the spirit of the theory of probabilities which permit from the beginning, a flexibility and, therefore, greater generality of formulations. It is obvious that a general mathematical formalism for dealing with "complications" in models of reality

For commentary to this paper [48], **see p. 689.**

is needed already on a heuristic level. This need is mainly due to the lack of simplicity in the presently employed models for the behavior of matter and radiation. The combinatorial complexity alone, present in such diverse problems as hydrodynamics, the theory of cosmic rays, the theory of nuclear reaction in heterogenous media, is very great. One has to remember that even in the present theories of so-called elementary particles themselves one employs rather complicated models for each of these particles and their interactions. Often the complications relate already to the qualitative topological and algebraic structure even before one attempts to pursue analysis of these models. One reason for these complications is that such problems involve a considerable number of independent variables. The infinitesimal analysis, i.e., the methods of calculus, become, for the case of many variables, unwieldy and often only purely symbolic. The class of "elementary" functions within which the operators of the calculus act in an algebraically tolerable fashion is restricted in the main to functions of one variable (real or complex). Mathematical physics deals with this increasing complexity in two opposite limiting methods. The first is the study of systems of differential or integral-differential equations describing in detail the behavior of each element of the system under consideration. The second, an opposite extreme in treatment, is found in theories like statistical mechanics dealing with only a few total or integral properties of systems which consist of enormous numbers of objects. There we resign ourselves to the study of only a few functionals or operators on such ensembles.

Systems involving, so to say, an intermediate situation have been becoming, in recent years, more and more important in both theory and practice. A mechanical problem of a system of N bodies with forces acting between them (we think here of N as having a value like, say, 10 or 20) would present an example of this kind. Similarly one can think of a continuum, say a fluid subject to given forces in which, however, we are interested in the values of N quantities describing the whole continuum of the fluid. Neither of the two extreme approaches which we mentioned is very practical in such cases. It will be impractical to try to solve exactly the deterministic equations. The purely statistical study of the system, in the spirit of thermodynamics, will not be detailed enough. The approach should be rather a combination of the two extreme points of view, the classical one of following step by step in time and space the action of differential and integral operators and the stochastic method of averaging over whole classes of initial conditions, relations, and interactions. We propose a way to combine the deterministic and probability method by some general mathematical algorithms.

In mathematics itself combinatorial analysis lacks general methods, and methodologically resembles an experimental science. Its problems are suggested by arrangements and combinations of physically existing situations and each requires for solution specific ingenuity. In analysis the subject of functional equations is in a similar position. There is a variety of special cases, each treated by special methods. According to Poincaré it is even impossible to define, in general, functional equations.

We shall now give examples of heuristic approaches all based on the same principle: of an equivalent random process through which one can examine the various problems of mathematical physics alluded to above.

One should remember that mathematical logic itself or the study of mathematics as a formal system can be considered a branch of combinatorial analysis. Metamathematics introduces a class of games—"solitaires"—to be played with symbols according to given rules. One sense of Gödel's theorem is that some properties of these games can be ascertained only by playing them.

From the practical point of view, investigation of random processes by playing the corresponding games is facilitated by the electronic computing machines. (In this connection: a simple computational device for production of a sequence of numbers with certain properties of randomness is desirable. By iterating the function $x' = 4x(1 - x)$ one obtains, for almost all x, the same ergodic distribution of iterates in $(0,1)$ [10; 12].)

II

One should remember that the distinction between a probabilistic and deterministic point of view lies often only in the interpretation and not in the mathematical treatment itself. A well-known example of this is the comparison of two problems, (1) Borel's law of large numbers for the sequence of the throws of a coin, and (2) a simple version of the ergodic theorem of Birkhoff: if one applies this ergodic theorem to a very special situation, namely, the system of real numbers in a binary expansion, the transformation T of this set on itself being a shift of the binary development by 1, one will realize that the theorems of Borel and Birkhoff assert in this case the same thing (this was noticed first, independently, by Doob, E. Hopf, and Khintchine.) In this case a formulation of the theory of probability and a deterministic one of iterating a well-defined transformation are mathematically equivalent.

In simple situations one might combine the two points of view: the one of probability theories, the other of iterating given transformations as follows. Given is a space E; given also are several measure preserving transformations $T_1, T_2, \cdots, T_n$. We start with a point p and apply to it in *turn* at *random* the given transformations. Assume for simplicity that at each time each of the N given transformations has an equal chance $= 1/N$ of being applied. It was proved by von Neumann and the author that the ergodic theorem still holds in the following version: for almost every sequence of choices of these transformations and for almost every point p the ergodic limit will exist [10; 12]. The proof consists in the use of the ergodic theorem of Birkhoff in a suitably defined space embodying, as it were, the space of all choices of the given transformations over the space E. The question of metric transitivity of a transformation, i.e., the question whether the limit in time is equal to the space average, can be similarly generalized from the iteration of a given transformation to the situation dealt with above; that is, the behavior of a sequence of points obtained by using several transformations at random. One can again show, similarly to the case of one transformation [11], that metric transitivity obtains in very general cases.

III

A very simple practical illustration of a statistical approach to a mathematically well-defined problem is the evaluation of integrals by a sampling *procedure*: suppose R is a region in *a* k-dimensional space defined by the inequalities:

$$f_1(x_1, \cdots, x_k) < 0$$

$$f_2(x_1, \cdots, x_k) < 0$$

$$\cdots\cdots\cdots\cdots\cdots$$

$$f_l(x_1, \cdots, x_k) < 0.$$

The region is contained, say, in the unit cube. The problem is to evaluate the volume of this region. The most direct approximation is from the definition of the integral: one divides each of the k axes into a number N of, say, equidistant points. We obtain in our cube, N^k lattice points and by counting the fraction of those which do belong to the given region we obtain an approximate value of its volume. An alternative procedure would be to produce, at random, with uniform probability a number M of points in the unit cube and count again the fraction of those belonging to the given region. From Bernoulli's law of large numbers it follows that as M tends to infinity this fraction will, with probability 1, tend to the value of the volume in question. It is clear from the practical point of view that for large values of k, the second procedure will be, in general, more economical. We know the probability of an error in M tries and given the error, the necessary value of M will be for large k much smaller than N^k. Thus it can be seen in this simple problem that by playing a game of chance (producing the points at random) we may obtain quantitative estimates of numbers defined by strictly deterministic rule. Analogously, one can evaluate by such statistical procedures, integrals occurring in more general problems of "geometric probabilities."

IV

Statistical models, that is, the random processes equivalent to the deterministic transformations, are obvious in the case of physical processes described by differential diffusion equations or by integral differential equations of the Boltzmann type. These processes are, of course, the corresponding "random walks". One finds in extensive literature dealing with stochastic processes the foundations for construction and study of such models, at least for simple problems of the above type. It is known that limiting distributions resulting from such processes obey certain partial differential equations. Our aim is to invert the usual procedure. Given a partial differential equation, we construct models of suitable games and obtain distributions or solutions of the corresponding equations by playing these games, i.e., by experiment. As an illustration consider the problem of description of large cosmic ray showers. It can be schematized as follows:

An incoming particle produces with certain probabilities new particles; each of these new particles, which are of several kinds, is, moreover, characterized by additional indices giving its momenta and energies. These particles can further multiply into new ones until the energies in the last generation fall under certain given limits. The problem is first: to predict, from the given probabilities of reactions, the statistical properties of the shower; secondly, a more difficult one, the inverse problem, where the elementary probabilities of transformation are not known but statistics of the showers are available, to estimate these probabilities from the properties of the shower. Mathematically, the problem is described by a system of ordinary differential equations or by a matrix of transitions, which has to be iterated.

A way to get the necessary statistics may be, of course, to "produce" a large number of these showers by playing a game of chance with assumed probabilities and examine the resulting distributions statistically. This may be more economical than the actual computation of the powers of the matrices describing the transition and transmutation probabilities: the multiplication of matrices corresponds to evaluation of all contingencies at each stage, whereas by playing a game of chance we select at each stage only *one* of the alternatives.

Another example: given is a medium consisting of several nuclearly different materials, one of which is uranium. One introduces one or several neutrons which will cause the generation of more neutrons through fissions in uranium. We introduce types, i.e., indices of particles corresponding to different kinds of nuclei present. In addition, the positions and velocities of particles of each type can be also characterized by additional indices of the particle so that these continuous variables are also, approximately, represented by a finite class of discrete indices. The given geometrical properties of the whole assembly and nuclear constants corresponding to the probabilities of reaction of particles (they are, in general, functions of velocities) would give us a matrix of transitions and transmutations. Assuming that time proceeds by discrete fixed intervals, we can then study the powers of the matrix. These will give us the state of the system at the nth interval of time. It is important to remember that the Markoff process involved here has infinitely many states because the numbers of particles of each type are not a priori bounded. A very schematized mathematical treatment would be given by the partial differential equation

$$\frac{\partial w}{\partial t} = a\Delta w + b(x)w.$$

This equation describes the behavior of a diffusing and multiplicative system of particles of one type, x denoting the "index" of position. For a mathematical description of this system it is preferable, instead of picturing it as an infinite-dimensional Markoff process, to treat it as an iteration of a transformation of a space given by the generating functions [2; 3; 5; 6; 9]. (Considerable work has been done on a theory of such processes also by Russian mathematicians [8].) The transformation T, given by the generating functions which is of the form $X_i' = f_i(x_1, \cdots, x_n)$, $i = 1, \cdots, n$, where the f_i are power series with non-negative coefficients, will define a linear transformation A whose terms a_{ij} will be the *expected values* of the numbers of particles of type j produced by starting with a particle of type i. Ordinarily, to interpret a matrix by a probabilistic game, one should have all of the terms non-negative, and the sum of each row should be equal to 1. One can generalize the interpretation of matrices, however, by playing a probability game, considering the terms not as transition probabilities but rather as the first moments or expected values of the numbers of particles of type j produced by one particle of type i. (The probabilities, of course, can be fixed in many different ways so as to yield the same given values of the moments.) One can go still further. Multiplication of matrices with arbitrary real coefficients can be studied by playing a probability game if we interpret the real numbers in each term as matrices with non-negative coefficients over two symbols:

$$1 \approx \begin{vmatrix} 1 & 0 \\ 0 & 1 \end{vmatrix} \qquad -1 \approx \begin{vmatrix} 0 & 1 \\ 1 & 0 \end{vmatrix}.$$

The negative and positive numbers require then each its own "particles" with separate indices. This correspondence preserves, of course, both addition and multiplication on matrices. Obviously, more general matrices with complex numbers as general terms admit, therefore, also of analogous probabilistic interpretation, each complex number requiring 4 types of "particles" in this correspondence [4].

The following theorem provides one mathematical relation between the properties of the iterates of the transformation given by generating functions and the iterates of the associated linear transformation (given by the expected values): With probability 1 the ratios of the numbers of particles of any two types will approach the ratios defined by the direction of the *invariant* vector given by Frobenius' theorem for the linear matrix [2; 3; 5; 6; 9].

It is possible to interpret the "particles" in a rather general and abstract fashion. Thus, for example, one may introduce an auxiliary particle whose role is that of a clock [2, part 2]. A distribution in the 4-dimensional time-space continuum can be investigated by an iteration of transition and transmutation matrices. The parameter of iteration will then be a purely mathematical variable τ, having no direct physical meaning since physical time is now one of the dependent variables.

V

In some cases one could deal with a partial differential equation as follows.

First, purely formally, we transform it into an equation of the diffusion-multiplication type. We then interpret this equation as describing the behavior of a system consisting of a large number of particles of various types which diffuse and transmute into each other. Finally we study the behavior of such a system empirically by playing a game with these particles according to prescribed chances of transitions. Suppose, for example, we have the time independent Schrödinger equation:

$$a\Delta\psi + (E - V(x, y, z))\psi = 0.$$

By introducing a new variable τ, and the function

$$u = \psi e^{-E\tau},$$

we shall obtain the equation

$$\frac{\partial u}{\partial t} = a\Delta u - Vu.$$

This latter is of the desired type. The potential $V(x, y, z)$ plays the role of expected value of the multiplication factor at the position given by the vector x [1]. Dirac's equation can also be treated in a similar fashion. (We have to introduce at least 4 types of particles since the description is not by means of real numbers but through Dirac's matrices. Again the parameter τ, as in Schrödinger's equation, is a purely auxiliary variable not interpretable as time.) Such probability models certainly have heuristic value in cases where no analytical methods are readily applicable to obtain solutions of the corresponding equations in closed form. This is, for example, the case when the potential function is not of simple enough type or in problems dealing with three or more particles. The result of a probability game will, of course, never give us the desired quantities accurately but could only allow the following possible interpretation: Given $\epsilon > 0$, $\eta > 0$, with probability $1 - \eta$, the values of quantities which we try to compute lie within ϵ of the constants obtained by our random process for sufficiently great number n of the sampling population.

One should remember that in reality the integral or partial differential equations often describe only the behavior of averages or expected values of physical quantities. Thus, for example, if one assumes as fundamental a model of the fluid as does the kinetic theory, the equations of hydrodynamics will describe the behavior of average quantities; velocities, pressures, etc., are defined by averaging these over very large numbers of atoms near a given position. The results of a probability game will reflect, to some extent, the deviation of such quantities from their average values. That is to say, the fluctuations unavoidably present as a result of the random processes performed may not be purely mathematical but may reflect, to some extent, the physical reality.

VI

One economy of a statistical formulation is this: often, in a physical problem, one is merely interested in finding the values of only a few *functionals* of an unknown distribution. Thus, for example, in a hydrodynamic problem we would like to know, say, the average velocity and the average pressure in a certain region of the fluid. In order to compute these one has to know, in an analytic formulation of the problem, the positions of all the particles of the fluid. One needs then the knowledge of the functions for all values of the independent variables. In an abstract formulation the situation is this: given is an operator equation $U(f) = 0$ where f is a function of k variables; what we want to know is the value of several given functionals $G_1(f), G_2(f), \cdots, G_l(f)$. (Sometimes, of course, even the existence of a solution of the equation $U(f) = 0$ or, which is the same, of the equation $V(f) = U(f) + f = f$, that is, the fixed point of the operator $V(f)$, is not a priori guaranteed.) The physical problem, however, consists merely in finding the values of $G_i(f)$. Mathematically it amounts to looking for functions f for which $G_i(V(f)) = G_i(f)$. We might call such f *quasi-fixed* points of the transformation V (with respect to the given functionals G_i). Obviously, the existence of quasi-fixed points is, a priori, easier to establish than the existence of a solution in the strict sense. A simple mathematical illustration follows: let T be a continuous transformation of the plane onto itself given by $x' = f(x, y)$; $y' = g(x, y)$. There need not, of course, exist a fixed point. There will always exist a point (x_0, y_0) such that $|x'_0| = |x_0|$; $|y'_0| = |y_0|$, analogously in n dimensions. Similar theorems in function spaces would permit one to assert the existence of quasi-solutions of operator equations $V(f) = f$. A quasi-solution (for given functionals) is then a function which possesses the same first n moments or the same first n coefficients in its Fourier series as its transform under V. For each n there should exist such quasi-solutions.

In a random process "equivalent" to a given equation, the values of *functionals* of the desired solution or, more generally, quasi-solutions, are obtained quite automatically as the process proceeds. The convergence in probability of the data, obtained during the process, to their true value may, in some cases, be much more rapid than the convergence of the data describing the functions themselves. This will be in general the case for functionals which have the form of integrals over the distributions.

VII

The role of "small" variations introduced in the operators which describe physical processes is discussed in elementary cases in the theories of stability. In the simplest cases one deals with the influence which variations of constants have on the behavior of solutions, say, of linear differential equations. In many purely mathematical theories one can conceive the problem of stability in a very general way. One can, for example, study instead of functional *equations*, functional *inequalities* and ask the question whether the solutions of these inequalities are, of necessity, close to the solutions of the corresponding equations. Perhaps the simplest example would be given by the equation

$$T(x + y) = T(x) + T(y)$$

for all x, y which are elements of a vector space E, and the corresponding functional inequality:

$$\| S(x + y) - S(x) - S(y) \| < \epsilon$$

for all x, y.

A result of Hyers is that there exists a T satisfying the equation such that for all x, we have then

$$\| T(x) - S(x) \| < \epsilon.$$

Or, more generally, one could ask the question: given an ϵ-isomorphism F of a metric group, is there always an actual isomorphism G within, say, k times ϵ of the given F. Another example is the question of ϵ-isometric transformations T, i.e., transformations T such that for all p, q:

$$| \rho(p, q) - \rho(T(p), T(q)) | < \epsilon.$$

Here again one can show that such T differ only by $k \cdot \epsilon$ from strictly isometric transformations. To give still another example one can introduce a notion of almost convex functions and almost convex sets. Again it is possible to show that such objects differ little from strictly convex bodies which, one proves, will exist in their vicinity.

All this is mentioned here because, in order to establish rigorously the comparison between random process models of physical problems and their classical descriptions by analysis, mathematical theorems will be needed which will allow us to estimate more precisely the influence of variations not merely of constants but of the operators themselves.

In many mathematical theories it is natural to subject the definitions themselves to ϵ-variations. Thus, for example, the notion of the homeomorphic transformation can be replaced by a notion of a continuous transformation which is up to ϵ a one-to-one transformation. Again one finds that many theorems about one-to-one transformations can be generalized to hold for the almost one-to-one case.

Little is known at present about solutions of functional inequalities. One needs, of course, beyond theorems on stability, more precise information on the rapidity of the convergence in probability.

VIII

In theories which would deal with *actually* infinite assemblies of points—the probability point of view can become axiomatic and more fundamental rather than only of the approximative character evident in the previous discussion. Let us indicate as an example a purely schematic set-up of this sort. We want to treat a dynamic system of an infinite number of mass points interacting with each other. Imagine that on the infinite real axis we have put, with probability equal to $\frac{1}{2}$, on each of the integer points a material point of mass 1. That is to say, for each integer we decided by a throw of a coin whether or not to put such a mass point on it. Having made infinitely many such decisions, we shall obtain a distribution of points on the line. It can be denoted by a real number in binary development, e.g., the indices corresponding to *ones* give us, say, for odd places, the non-negative integers where mass points are located, for the even indices of ones, we obtain the location of the mass points on the negative part of the line. Imagine that this binary number represents our system at the time $T = 0$. Assume further that the mass points attract each other with forces proportional to the inverse squares of the distances. (It is obvious that forces on each point are well-defined at all times since the sum of the inverse squares of integers converges absolutely.) Motions will now ensue. We propose to study properties of the motion common to almost all initial conditions, or theorems valid for almost all binary sequences (normal numbers in the sense of Borel). As representing initial conditions one may make the assumption that as the two points collide they will from then on stay together and form a point with a greater mass whose motion will be determined by the preservation of the momentum. It is interesting to note here that, because the total mass of the system is infinite, the various formulations of mechanics which are equivalent to each other in the case of finite systems cease to be so in this case. One can use, however, Newton's equations quite legitimately in our case. The interesting thing to notice is that the behavior of our infinite system will not be obtainable as a limiting case of the behavior of very large but finite systems approximating it. One shows, for example, that the average density of the system will remain constant equal to $\frac{1}{2}$ for all time. One can prove that collisions will lead to formations or condensations of arbitrarily high orders. For all time T there will be particles which have not yet collided with another particle. On the other hand, given a particle, the probability that it will collide at some time tends to 1. We might add that one could treat similarly systems of points distributed on integer-valued lattice points in the plane or in 3-dimensional space. The forces will not be determined any more by absolute convergence, but in 2 and 3 dimensions one can show that if we sum over squares or spheres the forces acting on a point from all the other points in the spheres whose radii tend to infinity, the limits will exist for each point with probability 1. That is, for almost every initial condition of the whole system the force is defined everywhere. In a problem of this sort it is obvious that the role of probability formulation is fundamental. Actually infinite systems of this kind may be thought of, however, as a new kind of idealization of systems already considered in present theories. This is so if we allow in advance for an infinity of hidden parameters present in the physical system, and which are not so far treated explicitly in the model. An important case in which the idealization to an actual infinity of many degrees of freedom interacting with each other seems to be useful is the recent theory of turbulence of Kolmogoroff, Onsaeger, and Heisenberg.

An interesting field of application for models consisting of an infinite number of interacting elements may exist in the recent theories of automata.[1] A general model, considered by von Neumann and the author, would be of the following sort:

Given is an infinite lattice or graph of points, each with a finite number of connections to certain of its "neighbors." Each point is capable of a finite number of "states." The states of neighbors at time t_n induce, in a specified manner, the state of the point at time t_{n+1}. This rule of transition is fixed deterministically or, more generally, may involve partly "random" decisions.

One can define now closed finite subsystems to be called *automata* or *organisms*. They will be characterized by a periodic or almost periodic sequence of their states as function of time and by the following "spatial" character: the state of the neighbors of the "organism" has only a "weak" influence on the state of the elements of the organism; the organism can, on the contrary, influence with full generality the states of the neighboring points which are not part of other organisms.

One aim of the theory is to establish the existence of subsystems which are able to multiply, i.e., create in time other systems identical ("congruent") to themselves.

As time proceeds, by discrete intervals, one will generate, starting from a finite "activated" region, organisms of different types. One problem is again to find the equilibrium ratios of the numbers of individual species, similarly to the situation described in §IV. The generalization of Frobenius' theorem mentioned there gives one basis for the existence of limits of the ratios.

The existence of finite universal organisms forms one of the first problems of such theory. These would be closed systems able to generate arbitrarily large (or "complicated") closed systems.

One should perhaps notice that any metamathematical theory has, to some extent, formally a character of the above sort: one generates, by given rules, from given classes of symbols, new such classes.

Mathematically, the simplest versions of such schemes would consist simply of the study of iterates of infinite matrices, having nonzero elements in only a finite number of terms in each row. The problems consist of finding the properties of the finite submatrices appearing along the diagonal, as one iterates the matrix.

References

1. M. D. Donsker and M. Kac, *A sampling method for determining the lowest eigenvalue and the principal eigenfunction of Schrödinger's equation*, Journal of Research of the National Bureau of Standards vol. 44 (1950) pp. 551–557.

2. C. J. Everett and S. Ulam, Los Alamos Report L. A. D. C. 533, 534, and 2532.

3. ———, *Multiplicative systems*, I, Proc. Nat. Acad. Sci. U. S. A. vol. 34 (1948) pp. 403–405.

4. ———, *On an application of a correspondence between matrices over real algebras and matrices of positive real numbers*, Bull. Amer. Math. Soc. Abstract 56-1-96.

5. T. E. Harris, *Some mathematical models for branching processes*, RAND Corp. Report, September, 1950.

[1] J. von Neumann lectures at the University of Illinois, December, 1949.

6. D. Hawkins and S. Ulam, Los Alamos Report L. A. D. C. 265, 1944.

7. G. W. King, *Stochastic methods in quantum mechanics*, A. D. Little Co. Report, February, 1950.

8. A. N. Kolmogoroff and M. A. Sevatyanov, *The calculation of final probabilities for branching random processes*, C. R. (Doklady) Acad. Sci. URSS (N.S.) vol. 56 (1947) pp. 783–786.

9. N. Metropolis and S. Ulam, *The Monte Carlo method*, Journal of the American Statistical Association vol. 44 (1949) pp. 335–341.

10. J. von Neumann and S. Ulam, *Random ergodic theorems*, Bull. Amer. Math. Soc. Abstract 51-9-165.

11. J. C. Oxtoby and S. Ulam, *Measure preserving homeomorphisms and metrical transitivity*, Ann. of Math. vol. 42 (1941) pp. 874–920.

12. S. M. Ulam and J. von Neumann, *On combination of stochastic and deterministic processes*, Bull. Amer. Math. Soc. Abstract 53-11-403.

Los Alamos Scientific Laboratory,
Los Alamos, N. M., U. S. A.

Reprinted from the AMERICAN MATHEMATICAL MONTHLY
Vol. LX, No. 4, April, 1953

A PROPERTY OF RANDOMNESS OF AN ARITHMETICAL FUNCTION

N. METROPOLIS and S. ULAM, Los Alamos Scientific Laboratory

Let $f(x)$ be a transformation of a set E into itself. One can decompose E into minimal invariant subsets or "trees" by considering for each point x the smallest set X with the following properties: 1. $x \epsilon X$, 2. If $y \epsilon X$ then $f(y) \epsilon X$, 3. If $y \epsilon X$ all the points x such that $f(x) = y$ also belong to X. One will obtain a decomposition of $E = X_1 + \cdots + X_\xi + \cdots$ into disjoint trees X_ξ. These characterize $f(x)$ up to a conjugating transformation h; *i.e.*, all $h(f(h^{-1}))$ where h is an arbitrary one-one transformation of E into itself.*

Suppose that $f(x)$ is a random function defined on E with values in E, *i.e.*, for each x a value y called $f(x)$ was chosen from E, say with a uniform distribution of probability in E (assumed to be a measure space).§ In case of E finite, one could ask about the "probable" or expected number of trees in the decomposition given by f. This expected number, not easy to estimate, is likely of the order of $\log n$; n is the number of points of E. For the case where f is postulated as a one to one transformation, *i.e.*, a permutation, the trees become cycles. It is well known that the expected number of cycles is $\log n$.

We examined the number of trees for some specific *not a priori* random functions. One function, often used in various Monte Carlo problems to produce random *digits* (by iteration), is the following: let x be an integer ranging from 0 to $2^k - 1$ written in the binary development $x = \alpha_0 2^0 + \cdots \alpha_{k-1} 2^{k-1}$. Let $f(x)$ be defined as follows: we take x^2 written again in the binary notation and define $f(x)$ as the number given by the *middle* k (out of $2k$) digits of x^2. It will, of course, again range between 0 and $2^k - 1$.

With this $f(x)$, for $k = 12$, *i.e.*, x ranging from 0 to 4095, the number of trees is 7.

Imagine that, as mentioned before, for each element x of a set E, one performs a selection of an element from E, with, say, uniform probability, *i.e.*, each time, every element has an equal chance ($= 1/n$) of being selected. One can ask of such a "random function," or transformation, for the probable number λ_0 of points y in the set E which are *not* of the form $y = f(x)$. Also one can ask for the number λ_1 of points y for which there is exactly *one* x such that $y = f(x)$—more generally, λ_i of points for which there are exactly i distinct

* See S. Ulam, Bull. Am. Math. Soc., 1943, Abstract, p. 49.

§ We consider the product space E^E. Logically, the study of a random function is, of course, the study of the probability distribution in E^E, the totality of all $f(x)$ on E into E.

For commentary to this paper [50], see p. 690.

values of x so that $f(x_1) = \cdots = f(x_i) = y$. For large n, the numbers λ_i should have a Poisson distribution, *i.e.*, $\lambda_i = ne^{-1}/i!$ (*e.g.*, λ_0 is the number of points not selected. But, given a number, at each choice, the probability of it *not* being selected is $1-1/n$, and this experiment is repeated n times independently—similarly for $i=1, 2, \cdots$.)

It is perhaps amusing that for some *specific* functions, for instance $f(x)$ as defined above by the "middle of the square," the actual distribution of the λ_i is extremely close to the one expected for a "random transformation." Sample values are given below:

For $k=12$; *i.e.* $n=4096$

i	λ_i/n observed	λ_i/n Poisson
0	.366	.368
1	.377	.368
2	.178	.184
3	.061	.061
4	.012	.015
5	.004	.003

For $k=16$, $n=65,536$

i	λ_i/n observed	λ_i/n Poisson
0	.370	.368
1	.367	.368
2	.183	.184
3	.062	.061
4	.015	.015
5	.003	.003
6	.0006	.0006

The only sizable deviation from what one would expect in a "random" f are, of course the sets $f^{-1}(0)$ and $f^{-1}(2^{k/2})$.

The problem of enumeration for the case $k=12$ was done with the aid of a desk calculator. The larger problem of $k=16$, $n=65,536$ was performed on the recently completed Los Alamos electronic computer. The computing problem consisted of generating the particular set $f(E)$ and of counting how many times each y occurred.

It was possible to keep track in the "memory" of 4096 numbers at any one time, so that it was necessary to regenerate the whole set E sixteen times to get a complete enumeration and check. The total computing time was one and one-half hours.

We propose sometime to examine, empirically on the computer, certain other arithmetical functions for their "tree" distribution and other combinatorial properties. We express our thanks to Miss Lois Cook for aid in doing both the hand computation and "coding" for the automatic calculation.

ON CERTAIN SEQUENCES OF INTEGERS DEFINED BY SIEVES

Verna Gardiner; R. Lazarus,
N. Metropolis and S. Ulam

The sequence of primes can be defined by the sieve of Eratosthenes. One can think of variations in the definition of a sieve. The following problem was studied by means of actual calculation on an electronic computing machine (partly with the aim of trying to develop "coding procedures" which would obviate the need for large memories which at first appear necessary for problems of this sort). Consider the sequence of all positive integers, 1, 2, 3, ... We shall now strike out from this sequence every *second* term by counting from 1. The odd integers will be left. We shall now strike out every *third* integer in the remaining sequence, again starting to count from 1, but considering only the remaining integers. We shall obtain a second sequence of integers. The next step is to strike out every *fourth* integer counting only the remaining ones and we obtain another subsequence. We can continue this process indefinitely. It is obvious that infinitely many integers will remain after we have completed the process. This sieve is different from that of Eratosthenes since in striking out all multiples of successive integers we count off only among the remaining ones. It could perhaps be called a sieve of Josephus Flavius. The result of this sieve is a sequence of integers of a density much smaller than that of the primes.

Another sieve could be the following: consider again the sequence of integers starting with 1. We shall strike out from it every second term. Apart from 1, the first integer which remains is 3; now in the remaining sequence we shall strike off every term whose index is a multiple of 3. In the sequence which remains now and which consists of 1, 3, 7, 9, 13, 15, 19, ..., the first term which has not been used already is 7. We shall, therefore, strike off every term in this sequence whose index is a multiple of 7; that is to say, the 7th (which is 19), 14th, 21st, etc., term in this sequence. In the remaining sequence we shall look up the first term which has not been used (it is 9) and again strike off terms whose indices are multiples of it. We continue this process indefinitely; it is obvious again that infinitely many terms remain. They may be called the result of our sieve.

The aim of our exercise was to consider certain asymptotic properties of this latter sequence of numbers, let us call them for brevity "lucky numbers." All lucky numbers up to 48,000 were quickly computed on the machine and the following data about them were obtained. (For the first few see Table I.)

For commentary to this paper [53], see p. 690.

1) The number of lucky numbers between 1 and n seems to compare quite well with the number of primes from 1 to n. We append a short table of their frequencies (Table II).

2) We noticed the number of luckies of the form $4n + 1$ and of the form $4n + 3$. This ratio seems to tend to 1 with a preponderance, at first, of the luckies of the form $4n + 3$. More generally we looked at the number of luckies of the form, say, $5n + \alpha$, $\alpha = 0, 1, \ldots 4$, etc. It is obvious that there are no luckies of the form $3n + 2$.

4) More generally we looked at gaps of a given length between successive luckies. It seems that as far as we went the number of gaps $N(k)$ of a given length k corresponds to the number of gaps $P(k)$ of the same length between successive primes.

5) Every even integer between 1 and 100,000 is a sum of two luckies.

6) There are 715 numbers which are simultaneously prime and lucky between 1 and 48,600.

More detailed tables exist on magnetic tapes.

The similarity between the behavior of the lucky numbers and that of primes seems to be rather striking in the range of the integers which we have considered. Obviously it is rather difficult to prove general theorems. For example, the question of whether there are infinitely many primes among the luckies is certainly difficult to answer. The sieve defining the lucky numbers can be written down as follows: let us define a sequence of sequence of integers

$$a_n^{(1)} = n; \; a_n^{(2)} = \{a_{nv}^{(1)}\} - \{a_{2v}^1\}; \; a_n^3 = \{a_n^{(2)}\} - \{a_{v \cdot a_2^2}^2\}$$

$$a_n^l = \{a_n^{l-1}\} - \{a_{v \cdot a_l^l}^{l-1}\}, \; v = 1, 2, 3, \ldots$$

(All our sequences on the left hand side consist of the sets as given by the right hand side in natural order.) The lucky numbers are simply $a_k^{(k)}$. One could consider other variations of the sieve. It is proposed to make a few more experiments on numbers resulting from such sieves.*

* Some years ago one of us (S.U.) has discussed sieves of this kind with P. Erdös, Professor D. H. Lehmer has mentioned that Erdös recently obtained some general results on a sieve of this sort.

TABLE I

FIRST 45 LUCKY NUMBERS

1	3	7	9	13
15	21	25	31	33
37	43	49	51	63
67	69	73	75	79
87	93	99	105	111
115	127	129	133	135
141	151	159	163	169
171	189	193	195	201
205	211	219	223	231

Between 1 and 48,600 there are 715 numbers which are simultaneously prime and lucky.

TABLE II

N	LUCKIES IN INTERVAL N		PRIMES IN INTERVAL N	
	No. in Interval	Total thru N	No. in Interval	Total thru N
1-2,000	276	276	304	304
2,000-4,000	227	503	247	551
4,000-6,000	213	716	233	784
6,000-8,000	204	920	224	1008
8,000-10,000	198	1118	222	1230
10,000-12.000	195	1313	209	1439
12,000-14,000	188	1501	214	1653
14,000-16,000	196	1697	210	1863
16,000-18,000	183	1880	202	2065
18,000-20,000	186	2066	198	2263
20,000-22,000	186	2252	202	2465
22,000-24,000	181	2433	204	2669
24,000-26,000	179	2612	192	2861
26,000-28,000	178	2790	195	3056
28,000-30,000	180	2970	190	3246
30,000-32,000	173	3143	187	3433
32,000-34,000	177	3320	206	3639
34,000-36,000	168	3488	186	3825
36,000-38,000	179	3667	193	4018
38,000-40,000	173	3840	186	4204
40,000-42,000	170	4010	189	4393
42,000-44,000	178	4188	187	4580
44,000-46,000	160	4348	182	4762
46,000-48,000	175	4523	185	4947
48,000-48,600	48	4571	53	5000

TABLE III

NUMBER OF GAPS OF LENGTH K

K	$N(k)$	$P(k)$
2	647	680
4	621	677
6	824	1075
8	351	411
10	361	478
12	509	517
14	184	238
16	172	168
18	267	253
20	106	105
22	112	101
24	130	77
26	32	34
28	51	38
30	66	65
32	21	12
34	24	15
36	33	20
38	13	4
40	9	7
42	10	5
44	4	3
46	5	0
48	1	1
50	2	2
52	6	4
54	2	3
56	4	0
58	1	1
60	0	1
62	0	1
64	1	0
66	0	0
68	0	0
70	0	0
72	0	1

TABLE IV

	LUCKIES OF THE FORM $4n+1$, $4n+3$		PRIMES OF THE FORM $4n+1$, $4n+3$	
n	A_1^4	A_3^4	A_1^4	A_3^4
1-2,000	133	143	148	155
2,000-4,000	102	125	122	125
4,000-6,000	108	105	114	119
6,000-8,000	97	107	116	108
8,000-10,000	99	99	110	112
10,000-12,000	93	102	98	111
12,000-14,000	96	92	111	103
14,000-16,000	102	94	102	108
16,000-18,000	96	87	97	105
18,000-20,000	88	98	108	90
20,000-22,000	95	91	98	104
22,000-24,000	96	85	104	100
14,000-16,000	99	80	96	96
26,000-28,000	95	83	99	96
28,000-30 000	98	82	89	101
30,000-32,000	92	81	94	93
32,000-34,000	109	68	105	101
34,000-36,000	95	73	86	100
36,000-38,000	89	90	99	94
38,000-40,000	90	83	90	96
40,000-42,000	90	80	92	97
42,000-44,000	82	96	97	90
44,000-46,000	75	85	93	89
46,000-48,000	81	94	92	93
48,000-48,600	19	29	27	26
Total	2319	2252	2487	2512

TABLE V

n	LUCKIES OF THE FORM $5n+1$, $5n+2$, $5n+3$, $5n+4$, $5n+5$					PRIMES OF THE FORM $5n+1$, $5n+2$, $5n+3$, $5n+4$, $5n+5$				
	A_1^5	A_2^5	A_3^5	A_4^5	A_5^5	A_1^5	AA_2^5 A	A_3^5A	A_4^5	A_5^5
1-2,000	60	50	56	60	50	74	78	78	73	1
2,000-4,000	47	44	52	39	45	61	64	60	62	
4,000-6,000	42	41	49	50	31	60	57	61	55	
6,000-8,000	51	37	40	41	35	52	58	57	57	
8,000-10,000	44	42	42	34	36	60	52	54	56	
10,000-12,000	45	39	40	38	33	47	54	53	55	
12,000-14,000	36	45	39	41	27	52	55	54	53	
14,000-16,000	38	43	40	38	37	55	55	1	49	
16,000-18,000	35	44	36	28	40	51	52	49	50	
18,000-20,000	35	37	38	34	42	52	45	52	49	
20,000-22,000	39	35	37	36	39	50	52	50	50	
22,000-24,000	43	36	30	39	33	52	51	49	52	
24,000-26,000	37	38	52	33	29	45	45	52	50	
26,000-28,000	36	35	36	30	41	52	54	47	42	
28,000-30,000	36	33	35	36	40	46	47	48	49	
39,000-32,000	38	34	36	35	30	46	49	46	46	
32,000-34,000	30	37	39	31	40	49	47	56	54	
34,000-36,000	40	38	41	24	25	50	44	46	46	
36,000-38,000	30	34	34	40	41	44	55	49	45	
38,000-40,000	43	31	32	30	37	47	38	49	52	
40,000-42,000	39	30	36	31	34	43	45	51	50	
42,000-44,000	35	26	40	42	35	47	47	51	42	
44,000-46,000	43	20	31	35	31	44	47	46	45	
46,000-48,000	34	35	29	37	40	51	49	36	49	
48,000-48,600	9	8	6	13	12	13	12	14	14	
Total	965	892	936	895	883	1243	1252	1259	1245	1

Los Alamos Scientific Laboratory

Reprinted from JOURNAL OF THE ASSOCIATION FOR COMPUTING MACHINERY
Volume 4, Number 2, April 1957

Experiments in Chess*

J. KISTER, P. STEIN, S. ULAM, W. WALDEN, M. WELLS
Los Alamos Scientific Laboratory, Los Alamos, New Mex.

The aim of this article is to report briefly on some experiments performed on a fast computing machine (MANIAC I—Los Alamos) on the coding of computers to play the game of chess. The idea that machines could be made to play this game is very old and there is considerable literature devoted to this problem which has a long and amusing history, including the well known specious machines which were operated by humans hidden inside the alleged machinery. We shall mention merely an article by C. Shannon on the problem of coding for machines which could play chess. There is an account by Turing of a computing machine which actually played a game of chess. One game played by his computer is given in the book *Faster than Thought* by Bowden. Recently, an article in the Russian newspaper *Pravda* mentioned the fact that a computing machine in Moscow had been coded to play the game of chess. No detailed account was given either of the method by which this game was coded nor of the results of such play beyond a statement that a fair chess player is able to beat the machine.

In Los Alamos recently an attempt was made to try out the ideas which some of us had advanced during the course of the last few years by putting them to test on one of the Los Alamos computing machines. The problem of coding the rules of chess, i.e., of making individual moves, is of course very simple. The problem of finding methods of play which would enable the machine to produce anything even faintly comparable with the result of human planning is extremely difficult. This difficulty is evident at every step of the game. Before a decision can be made to make a move, an enormous number of positions has to be surveyed if one wants to look "ahead" for even a moderate number of moves. If we assume that the average number of legal moves at any given stage is of order 20, then to look 3 moves ahead, i.e., 3 by white and 3 by black, would involve consideration of 20^6 *chains* of moves. At the end of each of such chains, an evaluation must be made according to some prescribed rule which eventually singles out one move in preference to others.

The game which was played on the machine is not really chess but rather, so to speak, a miniature of it; we play on a 6 x 6 board, omitting the bishops, and with 6 pawns on each side. (For the first move we allow the pawn to move only one square ahead.) The game retains much of the flavor of real chess but is very much simpler. Castling is not permitted; promotion of a pawn was allowed to take place as usual. The reason for the reduction of chess to this mutilated version was in order to allow the machine to look two moves ahead, i.e., 2 by white and 2 by black. The time taken to consider all chains came out on the average to about 12 minutes per move; thus it was possible to play a game in the course of

*Presented at the meeting of the Association, August 27–29, 1956.

some hours. In real chess the number of possible moves at each stage would be almost twice as great. This factor has to be raised to the fourth power for our 4 chain survey and that means that the time taken for a single move would be roughly a factor of 10 times longer, which would have been prohibitive. We might add that on a machine like the 704 essentially everything we shall report in this paper could be performed for the ordinary 8 x 8 chess in comparable times to those required for our 6 x 6 game. In analyzing the position after each possible chain of 4 moves we use first of all the criterion of *material* advantage, that is to say, captures and re-captures are considered with the resulting evaluation of the material. For the relative values of pawns and pieces we took numbers proportional to the well known ones more or less agreed upon by chess players in the practice of the game. In addition to the material, the criterion employed *mobility*, defined by the number of legal moves available to the player following the first, second and third moves of the chain. More specifically, the machine chose as the move to be made, say by white, that move which produced

$$\max_{\text{1st moves}} \quad \min_{\text{2nd moves}} \quad \{\max_{\text{3rd moves}} \quad [\min_{\text{4th moves}} \quad (E - N_4) + N_3] \quad - N_2\}.$$

N_j is the number of legal moves at the jth step of the chain and E is the value of the white pieces minus the value of the black pieces left on the board at the end of the chain. The moves by black are, of course, evaluated by the same rule with the sign of E reversed. In case two or more moves give equal evaluation, the machine chooses the one which comes first in its survey of chains. We might add that the comparison between mobility and material value chosen for the first game played was such that one pawn was equivalent to 8 legal moves in a given position. These criteria for evaluation seemed to us at the time extremely crude. The value of the material is certainly the overruling consideration in a chess player's decisions, except for "sacrifices." But mobility, it seems, is only *one* of a number of general considerations which eventually result in the player's choice of a specific move. These other considerations are hard to define even in a chess player's language; to formulate them in computing machine language is incomparably more difficult. To anticipate the results of the few games played, we might say that our simple criteria turned out surprisingly well.

The first game matched the computer against itself; i.e., both black and white moves were chosen by the above rule. Like any game between beginners it contained weak moves, but in general we were very pleased with the quality of the play. Several changes in the code were then made to fix the most obvious of the remedial weaknesses. For example, the machine seemed to have a mortal fear of checks, since its freedom after check was nearly nil and it tended to sacrifice material to avoid checks.

With an improved code, the second game matched the machine against a human. Dr. M. Kruskal, a strong player from Princeton, offered to play the machine giving it odds of a queen and playing white. This seemed a fair start for such a contest. The game took many hours to play and drew wide local interest. After about 15 moves Kruskal had made no gain and had even started calling his opponent "he" instead of "it." As the game went on it appeared Kruskal might even lose, although a draw seemed the most probable result. At about move 19, however, the machine chose a weak continuation which enabled its opponent to lay a 3 move mating trap. The machine's only way to delay checkmate was to sacrifice its queen which it did somewhat to the sadness of the authors (and all onlookers but one). After the forced exchange of its queen for a pawn, an end-game resulted in which the machine had no chance; the finish came at move 38. It was generally conceded that the changes made in the code improved, on the whole, the play of the computer although some additional changes which could have been made at small cost in time and complexity of code would have enabled the machine to sacrifice a knight for a pawn and a check and an almost certain victory.

The third game played by the machine matched it against a young lady member of the laboratory staff who had learned the game a week before with the express purpose of playing the machine. She was coached during the week in elements of play and the elementary combinations, and had played several practice games with players of average strength. This is the way the game proceeded:

MOVE	WHITE (MANIAC)	BLACK
1.	P-K3	P-QN3
2.	N-KR3	P-K3
3.	P-QN3	P-N3
4.	N-N	P-QR3
5.	P x RP[1]	N x P
6.	K-K2[2]	N-Q4[3]
7.	N x N	NP x N ch[4]
8.	K-K	P-R3
9.	P-QR3	R-N[5]
10.	P-R4	R-R
11.	P-R5	K-K2
12.	Q-R3	Q-N2
13.	Q-R2 ch	K-N2
14.	R-N[6]	R x P
15.	R x Q	R x Q
16.	R-N	R-QR2[7]
17.	P-R3	R-R3[8]
18.	RP x P	P-Q3[9]
19.	N-R3 ch	K-K
20.	P-N5 ch	K-K2
21.	P x R(Q)	N-Q2
22.	Q x P ch	K-Q
23.	N-N5 mate	

(1) A strategic error; isolating White's QRP and allowing Black's Kt out could prove fatal. N-K2 would be better.

(2) Could lead to a lost game if Black should play P-Q3.

(3) Whew!

(4) A weak move giving White a passed pawn. KP x N ch is forced.

(5) Pointless, as Black takes it back on the next move.

(6) Needlessly giving up pawn. MANIAC growing overconfident?

(7) A timid move. Black should hold on to the fifth rank.
(8) Compounding the error. White gets a pawn free. P-N4 was imperative.
(9) Allows a quick finish, as MANIAC mercilessly demonstrates.

Some of the authors of this article are chess players, and it is their feeling that the play of the machine with the above rules for the selection of moves could be compared to that of one who has average aptitude for the game and experience amounting to 20 or so full games played.

Changes which can be made in the code immediately suggest themselves: For example the chains involving exchanges of pawns and pieces can be pursued further than our limit of 4 moves. This will not materially increase the time per move. Another change suggested is to evaluate the mobility in a somewhat more sophisticated manner. For example, weighting moves into the opponent's territory somewhat more heavily than those moves on one's own territory. This should make the machine more aggressive. It is well known that knights, especially, increase in value as they penetrate more deeply into the enemy's camp.

The machine has an addition time of 90 microseconds. This is also the time taken for the execution of individual orders employed in determining the legality of the move, etc. In the middle game when the position was opened up and many pieces still remained, the number of chains to be considered was of the order of 400,000. These took about 20 minutes to survey and therefore each move at that stage took as much. At one point in the third game, just before the machine traded queens, it very humanly deliberated a full half hour.

What can one say about future developments in this direction? It is clear that even much faster machines which will appear in the course of the next few years having, say, *one* microsecond order times will not enable one to look more than 3 moves ahead (3 by white and 3 by black). Most of the weak moves or blunders committed by our machine can be attributed to its inability to look far enough ahead in certain continuations. Additional criteria for selection of moves and restriction to some chains only undoubtedly will be found, and various pragmatic rules for evaluation of positions will be utilized. Some features of the feeling for analogy in positions can be incorporated into the *memory* of the machine. This will involve increasingly large demands on the storage capacity.

It is not our belief that a machine will be made in the near future which could be coded to beat a strong player. The value of experiments of the kind here discussed lies in another direction: It would seem that playing such games as chess on a machine serve to illuminate that mechanism by which the human brain operates, gathering at a glance a great amount of information and then evaluating it by as yet incomprehensible criteria. It is clear, for instance, that the human brain operates for varied purposes in many parallel channels in contrast to the machine which, although very fast, does one simple thing at a time before going on to the next. A chess player's feeling for position and proper evaluation of it is a prime example of this remarkable faculty. As more and more penetrating studies of games of strategy are made perhaps new insights will be gained into a significant area of knowledge: The organization of thought.

INFINITE MODELS IN PHYSICS

BY

S. M. ULAM

The purpose of this paper is to present certain mathematical models of physical systems with the intention of indicating a whole class of interesting possibilities not normally apparent in the orthodox formulation. These models differ from the usual ones in two ways. First, they employ an actual infinity of points or objects present from the start; second, instead of using exclusively the Euclidean continuum (topologically) and the real-number system (algebraically), they will start with somewhat more general schemes. (The mathematical description of the physical world need not be specialized to the usually made assumptions that for a discrete description we use a finite system of mathematical points and for a continuum the familiar Euclidean n-dimensional space.) In the first part of this paper we shall discuss in a very special case the behavior of a system consisting of infinitely many discrete points and concentrate on some properties of such systems which cannot be obtained by a passage to the limit $n \to \infty$ of very large finite systems of n points. In the second part there will be a proposal of study of physical systems in which the number of points is of the power of the continuum and which are not Euclidean in character, but topologically these systems will be zero-dimensional, nowhere-dense perfect sets (Cantor sets).

PART I

Let us imagine the following "physical" system. On the real axis, in a one-dimensional space, we are given at time $t = 0$ an infinite collection of mass points distributed at random, on the points of the axis with integer coordinates. We assume that all the masses are equal and the random distribution is defined by the following procedure. For each integer value of x we have a priori probability $\frac{1}{2}$ that a mass point will occupy the position. Any such distribution of points may be denoted by a real number between zero and 1 written in the binary notation. We can, for example, reserve the even indices for the nonnegative integers, the odd ones for the negative ones, writing a 1 in the $2n$th place if there is a point in the place $x = -n$, zero otherwise, 1 in the position $2n + 1$ if there is a mass point in the position $x = +n$, otherwise zero. In this fashion every initial distribution will correspond to a unique real number. The assumption of equal probability for zero and 1 leads to the ordinary Lebesgue measure on the interval.

We shall now assume that our mass points exert forces on each other. In the simplest case, we may consider attractions with the magnitude of the force proportional to the inverse square of the distance. Assuming an initial distribution as given at time $t = 0$, we want to study the properties of such a system for subsequent times and establish some theorems about the behavior of such systems. It is important to notice that we introduce a probabilistic point of view from the beginning. A meaningful statement on properties of motion of such a system will concern "almost all" such initial distributions. An exceptional set of initial distributions, *i.e.*, a set such that the representative set of real numbers has measure zero in the sense of Lebesgue, will be of no concern to us. The statements made from now on about systems of this sort should be understood to be valid for almost all such systems. For simplicity we consider the objects of this sort located in a one-dimensional space. It is possible to make analogous constructions in three or n dimensions. Our mass points can then be thought of as located on the lattice points with all coordinates integer-valued in the (x,y,z)-space. (The representation of any such system of mass points could be, again, a point with all coordinates written in a binary system and, say, located in the unit cube.) Given such a system of points, we may also assume forces between any pair of these mass points. The first question that arises is whether the resultant total force acting on any point is well defined. In one dimension there is no problem since the sum of inverse squares of the distances converges absolutely. One can show, using rather elementary forms of the law of large numbers about sums of random vectors, that in two and even in three dimensions the resultant force due to attractions between points with an inverse-square law exists, with probability 1 for every point of our system, if this resultant is defined as a limit of resultant forces on a given point from all points located in spheres with the given point as a center and with radii tending to infinity. (In four dimensions this would no longer be true; it is the inverse square in the expression for force which ensures that convergence obtains in three dimensions.) The force on each point being well defined, we can write down an infinite system of total differential equations (Newton's equations of motion) defining the motion of each point of our system. One has to show, and this can be done, that the force is well defined not only for $t = 0$ but for positive values of time. We should notice here that a system of Newtonian equations is meaningful for such an infinite collection of points whereas some of the more general formulations of analytical mechanics, *e.g.*, a variational principle, can be defined only with difficulty. For example, a quantity like the total energy of a system is not defined since the total mass is infinite; the total kinetic energy may be infinite for $t > 0$, and so on. We hope to present a discussion of the question of how to formulate the more general principles in such infinite systems in another paper.

A convention is necessary to describe the process of collision between two mass points. Since we deal here, in our idealization, with finite masses located at mathematical points, the velocity of approach will become infinite before collision, and a special convention is necessary for describing the result of such a collision between a pair of points. These collisions are certain to happen in a one-dimensional system. Let us assume here that when two points collide, the momentum will be preserved. If masses m_1 and m_2 collide, a new mass

$m_1 + m_2$ is formed with a velocity given by the condition of preservation of momentum. The energy, then, will not be preserved, but we could imagine that it disappears in some internal degrees of freedom (of the point!), *e.g.*, as heat. It is clear that in a one-dimensional problem, collisions will lead to the appearance of ever-increasing masses. One is interested in the behavior of our system as a function of time. Some of the properties of (almost all) such systems can be established rather easily:

1. The masses appearing in the course of time will be unbounded. In other words, for almost every initial condition of our system there will exist for every M a time t such that a mass greater than M will appear after this time.

2. There will always exist single particles. In other words, for almost every system and for every t there will exist in the system a point with unit mass.

3. The asymptotic density of our system remains constant and equal to the original density. We define the asymptotic density as follows. Consider the totality of particles contained in an interval from $-N$ to $+N$, and denote by M_N the total mass of all particles in this interval. The $\lim_{N\to\infty} M_N/2N$ shall be called the asymptotic density if it exists. With our initial masses equal to 1, and the random placing of these masses on integer points, this limit (from Bernoulli's theorem) is equal to $\frac{1}{2}$. It is easy to see that this limit will exist and be equal to $\frac{1}{2}$ for all t. This is simply due to the fact that, given any t, the displacement of each particle will be bounded. If we take a sufficiently large interval, the flux across its ends will constitute an arbitrarily small fraction of the total number of particles, and our assertion follows.

4. Arbitrarily large "holes" will appear in our system; *i.e.*, for almost every system and for all d there will exist a time t so that there will be infinitely many mass points separated by intervals larger than d. Moreover, for all greater times these long empty intervals will continue to exist.

All this is very easy to establish in one dimension. In two or more dimensions, we shall *not* have, in general, collision between point masses, and we should instead have to define captures, *i.e.*, formation of double or multiple systems. The corresponding theorems on the existence of stable or semistable captures are much harder to prove. Probably an easier way to deal with an analogue of our system would be to give to each point a finite size and then consider, say, certain collisions as leading to formations of larger masses. Properties 2 and 3 should be easy to establish.

More interesting would be a quantitative description of the behavior of such a system. For example, it is interesting to calculate in one dimension the average mass of a particle in our infinite system for any time t and to determine the shape of the distribution of masses as a function of time. Another set of data would concern the velocity distribution of our particles as a function of time. (To define a meaningful average velocity one would have to introduce a cutoff in the distance between two particles which approach each other just before a collision; otherwise, in the system consisting of mathematical points these velocities become arbitrarily large during the collision process.)

If we define a cutoff distance, then the average of our velocity will have a meaning for all t, and the question arises: What is the "temperature" of the system as a function of time?

In order to study these and other questions, a series of experiments was performed on a computing machine by John Pasta and the author. We tried to imitate our infinite system by a finite one composed of a great number of masses placed on points of a regular subdivision of an interval with a random decision made for having or not having points in successive positions. In order to "approximate" an infinite system somewhat realistically, one has to imagine the two end points of the interval on which the points are located as coinciding; *i.e.*, we have a finite system of points on the circumference of a circle. This attenuates the end effects. It is clear that a finite system will imitate an infinite one only for times less than a certain value. Given a finite system, it is certain that it will ultimately collapse to a single point, whereas in the infinite one, we know that the asymptotic density will remain constant for all times. Therefore, in interpreting the results of a calculation for a finite system, one has to carry them only up to a value of time T where the system still consists of "many" points. In order to make a rigorous analysis, for the study of any given functional of the distribution of distances, masses, velocities, etc., of our system, one would have to give a priori inequalities for T as a function of N and a $\delta > 0$; that is, given only a finite system of N points, T such that the functional under consideration computed up to time T will differ in value by less than δ from the value of this functional for an infinite system. Many finite systems were computed, *i.e.*, we started with many distributions of mass points given by our initial random procedure, and the subsequent motions of each point were calculated in each case with the hope of obtaining heuristic results on the behavior of such systems. The number of points initially taken was of the order of 1,000. Among the quantities studied were the distribution of masses at any time t, the distribution of distances, etc. The results of these experiments will be reported on in a separate paper. I shall indicate here very briefly some of the qualitative facts about them:

a. The average mass of a particle appears to be increasing linearly with time.

b. There is a suspicion, at least, that if one considers the distribution of masses of particles in the units of the average mass at time t, this distribution tends to a fixed function. In other words, a steady state may establish itself.

c. A quantity which was called hierarchy was studied. This is defined by induction as follows: the original particles have by definition a hierarchy rank zero. When two particles of rank m and n collide, they form a particle whose hierarchy rank is equal to the greater of the two numbers m, n. In case $m = n$, the rank of the new particles is $m + 1$. This index gives an idea of the degree or order of conglomeration as distinguished from a mere increase of mass by accretion.

d. The average kinetic energy was studied as a function of time. We used in the computation a cutoff in the distance of approach in order to eliminate the arbitrarily high velocity just before collapse. The shape of this function cannot yet be discussed in detail, but it is obvious that this average energy increases with time and then starts decreasing again, which, of course, is due to the fact that our system ultimately will end up as one big particle at rest. Presumably, our finite systems imitate the infinite one only up to the time when this average energy stops increasing. Nothing as yet, therefore, can be stated about the behavior of "temperature" in time.

It should be pointed out here that the type of distribution which we have introduced in our numerical work (à la Bernoulli) could equally well be chosen differently. For example, we could assume that there is a fixed probability $\alpha \cdot dx$ of finding a unit mass in the interval dx; that is, the initial randomness could be chosen à la Poisson. It is easy to see that the convergence of the force on each particle of the system would be equally valid for almost all initial conditions with this setup.

One could postulate a finite interval with infinitely many points of various masses distributed in this interval at time $t = 0$ by a given recipe and then discuss the ensuing motions. In this connection it is amusing to notice that there exists a sequence of numbers x_i, all between zero and 1, and a sequence of masses m_i, such that $\sum_{i=1}^{\infty} m_i = 1$, with the following property. If we place the mass m_i at the position x_i and let any two of these masses attract each other with a force inversely proportional to the square of their distance, the whole system will be in static equilibrium! Similarly with repulsive forces.

One could assume, of course, not only that the positions of our points on the line or in space are given at random but also that there exist, already at time $t = 0$, velocities of each of these points, distributed in a random fashion. It is proposed to study, heuristically, the behavior of such systems by performing computations on a large finite system.

Part II

We have briefly discussed an infinite but countable system of material points subject to forces acting on each of them. We shall now outline some models of mass distributions which combine such a discrete character with certain properties of a continuum. The most elementary schema of this sort would perhaps be the following.

Imagine a mass point with a mass equal to unity, located in the middle of the interval 0,1. This mass point can now, with a probability p_1, remain forever in its original position or, with a probability $p_2 = 1 - p_1$, split into two parts, each of mass $\frac{1}{2}$, which will be located in the positions $\frac{1}{4}$ and $\frac{3}{4}$, respectively. If the latter eventuality has occurred, we shall assume again that either each of the two masses can stay where it is or each can independently split into two masses (equal to $\frac{1}{4}$ each) which will be located at $\frac{1}{8}$ and $\frac{3}{8}$ for the first point or $\frac{5}{8}$ and $\frac{7}{8}$ for the second point. This process is to continue indefinitely. We imagine that each of the points can split into two equal ones, which will then be located to the left and right of it, with a probability p_2, or stay "dead" forever. We have then a branching or multiplicative process which will define, at each stage, a possible distribution of masses on dyadically rational points of the interval. The process is defined by the two constants p_1, p_2. If p_2 should be equal to 1 and $p_1 = 0$, we would have a certainty of splitting every time, and the process would lead to all rational binaries, each having in the limit mass zero. The closure of our set would be the full continuum of real numbers between zero and 1. If p_2 is less than 1, we shall get as a result of our continuing branching process a countable set of points whose closure will be, *with probability* 1, a nowhere-dense set consisting of some isolated points and a perfect nowhere-dense set of the kind first defined by Cantor. There are three cases to distinguish, a subcritical, just critical, and supercritical system. In our simple setup, these correspond to the cases $p_2 < \frac{1}{2}$, $p_2 = \frac{1}{2}$, $p_2 > \frac{1}{2}$, respectively. In the last case there is a finite probability of the process never ending, and as a result, in addition to having a finite collection of points which are isolated, we shall obtain a perfect set on the interval as a closure of the "unending" part of the process. A detailed discussion of such branching can be found, *e.g.*, in [1]. We have to make here more precise the sense in which we speak of a result "of one such process." What is meant, of course, is that one considers all possible outcomes of such a branching process. There exists a measure in the space of all possible branching processes defined in a rather natural way [2]. When we speak of the process leading with probability 1 to a set with given properties, we mean that the subset of all outcomes with these properties has measure 1 in the space of all possible ones.

One can look upon the sets of points obtained by the above construction as describing "virtual positions of a physical object" or upon the space itself as being a collection of virtual symbols. These then will not, in general, form Euclidean continua. Neither will they consist of a discrete set of points. It is obvious that an analogous procedure can be effected in three dimensions. One could, for example, perform our branching independently in all three dimensions, or one could imagine the following single process. A particle with mass 1, located at the center of a unit sphere splits with probability p_2 into two particles, each with mass $\frac{1}{2}$, and located on the opposite ends of an interval with length α_1. The direction of this interval can be obtained by a random process, say with isotropic distribution in space. Each of the ends with mass equal $\frac{1}{2}$ can again be subject to the same splitting possibility, say, with the same probability p_2 independently, and then split into two particles, each with mass equal to $\frac{1}{4}$, located at the ends of an isotropically chosen interval of length α_2. If the process continues indefinitely, we shall obtain a three-dimensional analogue of the sets above. It might be of interest to add here that it is by no means easy to determine the topological character of the resulting set. It has been known (since Antoine) that some perfect nowhere-dense sets in 3-space are equivalent to sets located in one dimension under a homeomorphism of the whole three-dimension space and others are not. The question which of the two is more likely under our process is not immediately answered.

This way of considering a space of symbols to correspond to a model of a physical situation is perhaps the simplest one of its kind and, so to say, applies to the "configuration space" of a particle. An analogous construction should be thought of as proceeding in the phase space. Not merely the positions but the momenta or velocities of a particle could be thought of as generated by a process like the above, leading to a Cantor-set structure for all possible values of the physical quantities in which we are interested. In this paper we shall restrict ourselves to the configuration space alone in order to bring out the features of the proposal more clearly. In addition, we shall confine ourselves here to very classical, *i.e.*, nonrelativistic and non-quantum-theoretical, features of such models. Any pretended attempt to take more seriously the implications of such models for physics would have to consider such constructions in Minkowski space.

We mentioned above some of the properties of sets obtained by our branching process; *e.g.*, the set will be, with probability 1, nowhere dense. To obtain quantitative data about this set seems quite difficult. Again exploratory numerical work seemed of some value, and a series of computations on an electronic machine was undertaken as follows. Starting with the original point in the middle of the interval, the process was continued by the use of random numbers. That is, a great number of finite sets was determined, the process being stopped each time after a certain number of "generations." Given the probabilities p_1, p_2, one may obtain a sufficient number N of sets produced by the splitting process for a statistical study. If the number of generations k for the splitting process is kept constant, we shall obtain a variable number of points in the set. One can then compute the average value of any functional of such a set. The integration in the space of all possible branching processes is then replaced by averaging on the N sets actually produced. To justify the averaging as an approximation to integration is very easy for many functionals. The quantities studied are as follows:

1. Given a set of mass points at generation time k, one can compute its moment of inertia I. (The center of gravity of each system, as is obvious from its definition, is located in the original position of the mass point since our splitting preserves the center of mass.) If we average the value of the moment of inertia over all N sets which we have manufactured, we obtain the value of the average moment of inertia of the infinite set with good approximation.

2. If we imagine that any two points of our set attract each other with a force proportional to the masses and inversely to the square of the distance, we can investigate the question of the value of the "gravitational" self-energy. Here again if we compute this quantity for each of our N sets and take the average, we shall get an approximation to the average or expected value of this quantity in our infinite process. This value exists in the three-dimensional case.

In addition to computing the average, one can get a fair idea of the distribution of the value of this self-energy. In order to ascertain it with any accuracy, a very great number N of sets has to be manufactured, of course.

3. One could ask about the attraction of two systems of the above sort, located on an interval of length 1, but apart from each other, if we assumed that any element of one set attracts any element of the other set with a force proportional to the product of the masses and inversely proportional to the square of the distance. In other words, the resultant of these forces can again be computed approximately by considering a number N^2 of such pairs. This resultant will, of course, be a function of the distance d between the centers of the two systems. As should be obvious, this resultant varies approximately as 1 divided by d^2 for large d but grows to infinity more rapidly than that when d is comparable to 1 and when the two systems overlap.

A detailed account of these experiments will be presented separately, together with the account of computations mentioned in Part I.

The motivation for a calculation of quantities like self-energy is, of course, that it is clear a priori that the values of these will be less than the values for corresponding Euclidean continuum models and may have finite values when these latter diverge. It is proposed to calculate in the near future a mass-splitting process more general than the one outlined above; *viz.*, the probabilities for splitting which we have assumed to be constant from generation to generation and from point to point could be assumed to depend on the existence of another object of the above sort in proximity to our given system. Indeed, if we assume that a function $V(x)$ is given and the probability of multiplication is proportional to this function, we obtain a distribution u as a result of our splitting process which will obey an equation similar to the Schrödinger equation

$$\Delta u + (E - V)u = 0.$$

In addition, the process, as its stands, has the unsatisfactory feature that it still leaves mathematical points with finite masses. If one wanted to insist on all mathematical points having masses zero, the following iteration of our procedure should be considered: let ω correspond to the first passage to the limit of our process as defined above. We shall now iterate it in this fashion. Each of the points which remained with finite mass will be again subject to a splitting procedure, say, with the same probabilities, into two equal masses but this time located on the opposite ends of an interval shorter than the one in the first process by a fixed ratio R. For example, let $R = \frac{1}{100}$; then if it happened that the first mass point located at $\frac{1}{2}$ has not split during the first process, it will have a probability p_2 of splitting into two masses equal to $\frac{1}{2}$, but located at positions $\frac{1}{2} - \frac{1}{100} \cdot \frac{1}{4}$ and at $\frac{1}{2} + \frac{1}{100} \cdot \frac{1}{4}$. Again, for example, if the point located at $\frac{3}{4}$ has not split in the first process, let it have probability p_2 of splitting into two masses each equal to $\frac{1}{4}$, and located at $\frac{3}{4} - \frac{1}{100} \cdot \frac{1}{8}$, and $\frac{3}{4} + \frac{1}{100} \cdot \frac{1}{8}$. If this continued, we would have a second limiting set 2ω. If at the end of this procedure some points still have finite masses, we continue, assuming now another still smaller ratio R for the splitting distance and repeat this splitting. It is obvious that, with probability 1, a sequence of these processes will lead to a Cantor set of points so that at ordinal ω^2 all points will have mass zero, and we shall obtain a distribution of density without finite values at any point.

Another generalization seems strongly suggested. In the processes described, the objects which are obtained by our branching are still algebraically of the nature of real numbers. It would seem that this is too special. The formalism of the new quantum theory would suggest considering each of the "manufactured" points as having spinlike properties; *i.e.*, there should be several kinds $x_1, x_2, \cdots, x_n$ of these points. This is, in fact, already the case for the branching process as we described it in three dimensions.

The aim in presenting this collection of problems about distributions of infinite sets of points endowed with masses and exerting "forces" on each other is twofold. In the first place, the modern theories of physics could very well use (in addition to their present point of view, oscillating as it does between the field and the discrete-particle description) something that combines from the beginning the two aspects and is not necessarily restricted to Euclidean topology. In the second place, it appears possible that a geometrization of physics, if it ever takes place, will not be effected through merely generalizing the Euclidean character of the distance in the large, which is the content of Riemannian geometry. Since physical phenomena change radically in nature in the small, it is clear that a differentiable metric will never bring out this feature. A more radical change, perhaps even that of local topology, will be necessary. Finally, the recent evidence for frequent transmutation in nature of what at each stage of the historical development was considered an elementary particle and the multiplicative character of phenomena in the small may make it appear amusing, at least for a mathematician, to consider models of this sort.

Bibliography

1. C. J. Everett and S. Ulam, *Multiplicative systems in several variables*, I, AECD-2164 (1948); II, AECD-2165 (1948); III AECD-2532 (1948), Technical Information Branch, Oak Ridge, Tenn.

2. ——— and ———, *Multiplicative systems*, I, Proc. Nat. Acad. Sci. U.S.A. vol. 34 (1948) pp. 403–405.

Los Alamos Scientific Laboratory,
Los Alamos, N. Mex.

Reprinted from MATHEMATICAL TABLES AND OTHER AIDS TO COMPUTATION
Vol. XIII, No. 65, January, 1959

Heuristic Numerical Work in Some Problems of Hydrodynamics

By John R. Pasta and S. Ulam

It is well known that in many problems of hydrodynamics one cannot, at the present time, obtain valid solutions in a closed form. The asymptotic behavior of the solutions and even more generally their qualitative properties are barely obtainable through analytical work alone. This is even true of some problems which are already highly schematized by neglecting various physical parameters, like the change in the equation of state in a fluid or gas during its motion. One proceeds in numerical computations with the continuum of the fluid replaced by a finite network of discrete points, and thus replaces the partial differential equations by a system of difference equations. The time variable, too, is replaced by a discrete succession of steps in time. This is the usual procedure in solving initial value problems which do not yield to analytical methods. In recent years the advent of electronic computing machines introduced the possibility of large scale experimentation in calculation of problems in more than one dimension.

It is the purpose of the following discussion to outline some general properties of such numerical work and to propose several different methods for numerical computations. Some numerical work already performed will be discussed in the sequel; it dates from 1952, when the authors first applied such computations on electronic computing machines. More recently, F. Harlow [1] has applied similar methods to calculations performed on electronic machines in Los Alamos. With the constant improvement in electronic computers, both as regards their speed and the size of the memory, it will be possible to perform more ambitious calculations; both the variety and the magnitude of the problems which can be handled will increase. Such calculations can play a role analogous to that of experiments in physics and may suggest new theoretical lines of attack.

There are, broadly speaking, at least two different ways to approach numerically the problems dealing with the dynamical behavior of continua. The kinetic theory of gases assumes the physical reality of the discontinuum. One calculates the properties in the *large* of the motion of N "points"—atoms or molecules which have statistically given velocity distributions and are subject to incessant collisions with each other. There are forces acting between these points, e.g., deriving from potentials, but whose form is only imperfectly known. Many, but not all, properties of the macroscopic motions are largely independent of the exact form of the interactions between these particles. This is not the place to enter into a description of the way to derive the hydrodynamical equations from the Boltzmann integral-differential equations describing the microscopic behavior of such systems. Suffice it to say that, e.g., Navier-Stokes equations can be so derived and they will describe the behavior of certain statistical averages (or functionals on the $6N$ dimensional space) which are interpreted as macroscopic quantities like density, pressure and velocity, of a point in space (three dimensions) as functions of time.

Received 12 May, 1958. This work was performed under the auspices of the U. S Atomic Energy Commission.

What we prefer to discuss here briefly is the numerical procedures which are suggested by this model. The number N is, in problems of hydrodynamics, enormous. N is greater than, say, 10^{22} for 1cc of liquid. Is it possible to scale down this number to a value practical for numerical computations and still be able to observe meaningfully the behavior of the functionals in which we are interested?

The answer, it seems to us, is in the negative at the present time. If one takes the Boltzmann equations literally and considers the individual points of calculation as representing atoms, then to obtain the "average" velocity of a point of the gas one would require, say, k particles per cell in a spatial resolution in which we are interested. Let the number of cells be l. (In practice, the resolution equals d on a linear scale and l equals d^r; where r is the number of dimensions, and equals 1, 2, 3 (in problems in one, two, or three dimensions without special symmetries reducing the number of independent space parameters).) The total number of points is then $\vartheta = kd^r$. If we want a statistical error of the order of 1%, then $k \sim 10^4$. If the linear resolution is to be of the order of 5%, say, then d is 20. Even in one dimension, we would have to consider 200,000 points. This is much too high for present computers and should make it clear that in a numerical calculation, the "points" have to be thought of as representing not individual atoms but rather large aggregates of atoms. The behavior of each point has to be schematized so as to represent a statistical average of a great number of atoms. The numerical work will not reflect the Boltzmann equations, but the simpler equations which are its consequences; each point of our calculation represents, for example, the center of mass of a collection of atoms. The implicit assumption in this set-up is, therefore, that such a mass remains coherent during the entire course of the problem. That is to say, the molecules initially close to each other remain so throughout the problem and do not diffuse too much with respect to each other—the small globule of the fluid does not distort too much.

Parenthetically, we may add that this question is connected with the whole complex of difficulties which one encounters even in the purely mathematical studies in the theory of functions of several real variables. The so-called density theorems and the notion of set-derivation can serve as examples: The *density* of a set of points at a given point has to be defined through a sequence of sets such that not merely their diameters shrink to zero, but the sets have to be "not too thin"—for example, in case of rectangles enclosing the given point, the ratio of the sides has to remain bounded. This sort of assumption is necessary for the validity of Vitali's theorem, etc. Physically it is clear that in our case we have to assume that the surface-to-volume ratio for the globules has to remain bounded for any a priori evaluation of the pressure, which acts on its center of mass. One has to introduce pressure, and there seem to be at least two obvious ways to do it:

a) Assuming the knowledge of the equation of state, there is the dependence of the pressure on local density (we assume as given the thermodynamic nature of the process). In this outline we may specialize, say, to either an *isothermal* or *adiabatic* process, that is, $p = f(\rho)$. The pressure gradients which are evaluable through gradients of density will be calculated then by estimating these gradients through the *geometry* of the instantaneous appearance of our system of points which represent, we emphasize, the positions of centers of mass of globules. Our problem, then is that of finding a rule or recipe to estimate the density at a point of space given only a finite system of points (rather widely separated!). The limiting case of an enormous number of points presents no difficulties, for one could simply count the number of points in a square of a fixed mesh, and this number will be proportional to the density. In practice, since we are limited to a moderate number of points for the whole fluid, the question is to estimate the density in the most reliable way. Consider the case of two dimensions. We can think of the points located initially in a regular array, for example, on the vertices of a rectangular division, or better, points of a triangular subdivision of space. Of course, after some "cycles" of the computation, the geometry of the system will change and one could proceed to estimate the density as follows: Enclose each point by triangles with the closest points as vertices. The smallest (in area) triangles in whose interior a given point is located would give, through the ratio of its area to the area of the original triangle, at least an idea of the change in density at that point.

This procedure suffers from several drawbacks. There is the question of the computational stability of the calculation. The selection of the *nearest* points leads to discontinuity, in time, of the area. In a rectangular subdivision the more "classical" definition of density through the Jacobian in a finite approximation requires the knowledge of the position of four nearby points $x \pm h$, $y \pm k$. In the equations of motion for each of our points, we need the gradients of the density in the Lagrangian coordinates (the "a,b,c" not the "x,y,z"). In our crude way of calculating the densities themselves, the computation of *differences* in fixed *directions* in space at a given time, the nearby points for increasing a,b,c may not be sufficiently close, and the gradients may be very inaccurately estimated. In the case of points in the boundary of the fluid, say with vacuum, one needs special prescriptions. All this introduces even more serious errors.

b) There is another way to introduce, numerically, the forces due to pressure gradients. We could imagine repulsive forces acting between any pair of our points. They would in simple cases depend on the distance alone (the forces should derive from potentials if we assume scalar pressure—no tensor forces—no viscosity, etc.). The form of the potential will, of course, depend on the equation of state. In a one-dimensional problem the nature of this correspondence is clear—it suffices to have forces between neighboring points only. The continuity of motion guarantees the permanence, in time, of the relation of neighborhood. The situation is however completely different in two or more space dimensions. (A general remark here: In one-dimensional problems the points of our calculation can best represent the boundaries of zones into which our fluid or gas is sub-divided. These keep their coherence and shape, and the only meaningful parameter for the spatial distribution, the density, is inversely proportional to the width of the zones. In two or more dimensions the boundaries of zones are not easily describable by points or systems of a few points, and these points, we will repeat once more, correspond to centers of mass of volume.) We have to define the neighbors of a given point. One can do it in reference to the original geometry of a system of points that is to keep the relation which existed initially. But this is not good except in the case of infinitesimal or small deformation. The neighbors of a given point will change in the course of time. If one tries to calculate the resultant force on a point by calculating it from *all* the points in the mesh, not merely the neighbors, one should remember that the number of calculations increases with the square of the number of points. A "cut-off" for the force is necessary at some distance and the force is not calculated if the distance between two points exceeds a certain constant. One has to remember that in the initial position, that is to say, in the regular lattice in which the points are arrayed, the cut-off has to be selected so that with it we obtain the initial distribution of pressure. The pressure gradient is given directly as the resultant of all the forces acting on a point from its "neighbors" and depends on the actual position of the point without reference to the ("forgotten" by the system) initial configuration. We will discuss in the sequel, for some concrete problems, how such calculating schemata operated.

To summarize briefly the above computational scheme: The particles represent small parts of the fluid. The forces due to pressure gradients are introduced directly by imagining that neighboring or "close" points repel each other. The dependence of this force on the distance between points is so chosen that in the limiting case of very many points, it would represent correctly the equation of state. That this is possible, in principle, is clear a priori: the density is inversely proportional in the limit of a very large number of points to the square (in two dimensions) or cube (in three dimensions) of the average distance between them. The pressure is a function of density, and this being a function of the distances, we obtain an analogue of the equation of state by choosing a suitable distance dependence of the force. The Lagrangian particles are at P_1:(x_1, y_1, z_1), P_2:(x_2, y_2, z_2), $\cdots$, P_N: (x_N, y_N, z_N). The forces (repulsive) between any two are given by $F_{i,j} = F(d(P_i, P_j))$ where $d(P_i, P_j)$ is the distance between the two points.

The average value of d at a point of the fluid is, in the limit of $N \to \infty$, a function of the local density: $\bar{d} \simeq \rho^{-1/3}$.

The pressure p is a function of ρ alone in, say, isothermal or adiabatic problems. The pressure gradients give then the expression for our F.

There is so far no general theory and the convergence of such finite approximations to the hydrodynamical equations remains to be proved, but even more important than that would be an estimate of the speed of convergence.

One could assume, as a starting point for a numerical calculation, instead of the partial differential equations of hydrodynamics, e.g., the equations of Euler-Lagrange, a mathematical description which is somewhat more general: the "points" of our calculation need not correspond to the material points of a fluid, but instead may represent—more generally—some other parameters of the problem.

After all, in many problems one is not interested in the positions of every given particle of the fluid, but rather in the behavior, in time, of a few functionals of the motion; for example, if a,b,c are the "laboratory" or Lagrangian coordinates in fixed space in the classical formulation, one deals with the functions $x(a,b,c,t)$, $y(a,b,c,t)$, $z(a,b,c,t)$ which are interpreted as a position at time t of the point which at time $t = 0$ occupies the point (a,b,c). $\rho(x,y,z,t)$ is the local density which is computed by differentiation from the knowledge of x,y,z and the derivatives

$$\frac{\partial x}{\partial a}, \frac{\partial x}{\partial b}, \frac{\partial x}{\partial c}, \frac{\partial y}{\partial a} \cdots \text{etc.}$$

The pressure p is, for our purpose, computable from ρ. It is always possible to think of the functions x,y,z (which satisfy the Euler-Lagrange equations) as developed into series:

$$x = \sum_{i,j,k} \alpha_{ijk}(t)\psi_{ijk}(a,b,c) \qquad z = \sum_{i,j,k} \gamma_{ijk}(t)\psi_{ijk}(a,b,c)$$

$$y = \sum_{i,j,k} \beta_{ijk}(t)\psi_{ijk}(a,b,c) \qquad i,j,k = 1, 2, \cdots$$

Here the ψ's are fixed functions of a,b,c alone. The functions need not form orthogonal systems but, to fix the ideas, we might think of them as being terms of the Fourier series of Rademacher functions, for instance. The α,β,γ are functions of time alone and can be treated as abstract Lagrangian particles. The partial differential equations for x,y,z are replaced by a system of infinitely many (a discrete infinity) of ordinary differential equations describing the change in time of the α,β,γ. To see the validity of such an approach, we shall illustrate this proposal in an example.

In a one-dimensional problem, we have to find the function $x(a,t)$ satisfying the equation

$$\ddot{x} = -\frac{\partial p}{\partial x} - F(x); \quad p = f(\rho); \quad \rho = \frac{\rho_0}{\dfrac{\partial x}{\partial a}}$$

with a given initial distribution of density and velocity of the fluid. The ordinary numerical procedure for a solution consists in the replacement of the continuous variable x by a discrete one, that is, $x(a)$ is replaced by $x_i \quad i = 1, 2, \cdots N$. Each x_i obeys a Newtonian equation of the second order for the $x_i(t)$ where in practice we also replace t by a discrete sequence of times and obtain a system of difference equations. Our introduction of the fixed functions of space $\psi(x)$ and the α_i's amounts to a Lagrangian change of variables where instead of the x_i which are the actual "points" of our fluid, we introduce new variables (q_i) in Lagrangian notation:

$$q_1 = q_1(x_1 \cdots x_n), \qquad q_2 = q_2(x_1 \cdots x_n) \cdots q_k = q_k(x_1 \cdots x_n)$$

where the functions are "holonomic", that is, they do not involve the velocities $\dot{x}_i$ in a non-integrable form. For example, the analogue of the Rademacher functions would be: If n is of the form $n = 2^k$

$$q_1 = \sum_{i=1}^{n/2} x_i - \sum_{i=(n/2)+1}^{n} x_i$$

$$q_2 = \sum_{i=1}^{n/4} x_i - \sum_{i=(n/4)+1}^{n/2} x_i + \sum_{i=(n/2)+1}^{(3/4)n} x_i - \sum_{i=(3/4)n+1}^{n} x_i$$

. . . .

The differential equations for the x_i will be replaced by a system of equations for the q_i; the forces which are given directly for the x variable by pressure gradients will be replaced by "generalized" forces which are functions of the q_i and their derivatives. In cases where the ψ form an orthogonal system, the kinetic energy will still be a quadratic function of q_i and the α_i. This procedure is, of course, strictly legitimate in the case where we consider the x variables being expressible by the q_i.

A general question arises: Which, in a given problem, are the most convenient variables (q)? It is clear that in many problems of hydrodynamics one is interested mostly in certain over-all properties of the motion of the fluid—one wants to know the behavior of certain given *functionals.*

In the classical approach one calculates first the x,y,z for each of the points of the space a,b,c. In practice, one is limited to a finite number of these points, and it is perhaps plausible that in many cases it would be "better" to know, instead, an equal number of coefficients α_{ijk} in the development of $x(a,b,c,t)$ in a given (Fourier) series than to know the value of x in a corresponding number of points in the space of the α_{ijk}'s. This is especially clear if one should know, a priori, from the physical nature of the problem that the functions x,y,z are reasonably smooth (e.g., in the sense that the absolute value of the partial derivatives remains bounded). In the terminology of the α_{ijk}'s, this smoothness amounts to the knowledge that the coupling between the α's which have high indices and those with small values is small. That is to say, the series converge rapidly. Physically speaking, it means that the high frequencies are less excited.

Such reduction of a continuum to a discrete countable infinity is, of course, very familiar in some problems of quantum theory—the radiation field, etc. The problems that have been dealt with in quantum theory have been mostly linear; there are no forces between the coefficients and each mode or "particle" represented by the coefficients behaves independently of all the others.

For the case of our hydrodynamical problems, the forces due to pressure gradient are functions of all the coefficients; that is, we have a true n-body problem, and the "quantization" is justified practically only if a cut-off at a finite index (i) is permissible together with an estimate, in advance, of the error for all t under consideration—that is to say, if the high modes do not become increasingly important. When the high modes do acquire more energy as time goes on, the classical approach becomes difficult also. The onset of turbulence or of the positional mixing of the fluid renders the classical treatment (partial differential equations) illusory. The absolute values of spatial derivatives increase and even their existence for finite times cannot be guaranteed and a statistical approach is indicated. It would seem that in problems where this behavior is expected, the approach through a study of a finite model involving perhaps a change of coordinates, like the one proposed, may be of possible utility. If one wants to study the *rate* of development of instabilities or the *rate* at which mixing proceeds, then the flow of energy from the low modes (that is to say, from the α's with small indices to those with high indices) will show just the rate of increase in mixing positionally or, in the derivatives, the rate of increase in vorticity.

We would like to mention another possible advantage in the use of general coordinates in numerical work. A description of the motion of the fluid through the partial differential equations of hydrodynamics postulates the existence of the partial differential expressions. It is well known that even in comparatively simple problems these derivatives exist only up to a finite value of time, after which discontinuities develop, e.g., in pressure and in density (that is to say, shocks), and one has to use different methods to treat such discontinuities. On the other hand, a Rademacher series or Fourier series may very well represent a step function, more generally functions without derivatives at some points. The α_i might then continue to be used as dynamical variables even after discontinuities occur in de-

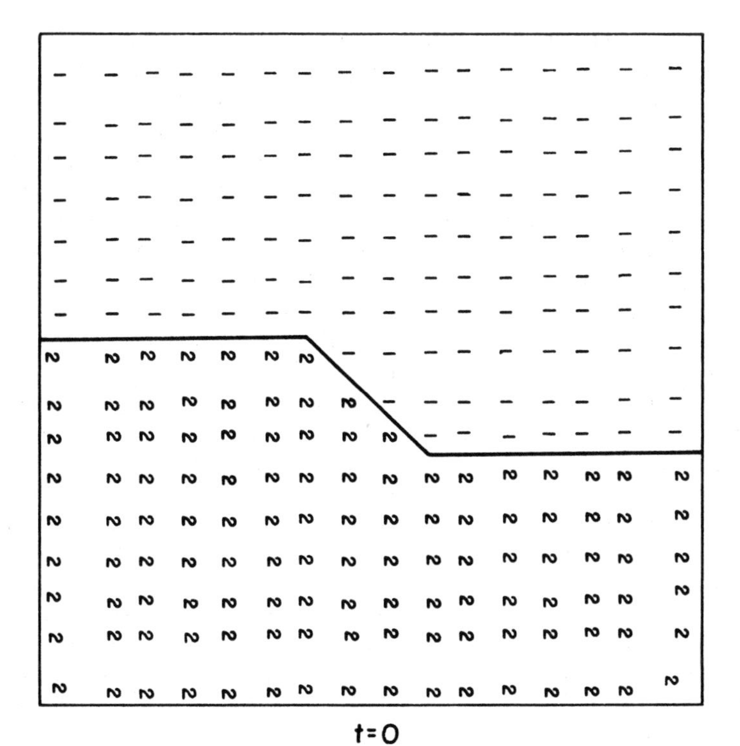

FIG. 1.—Initial configuration, $t = 0$

rivatives like $\partial x/\partial a$ making the differential expressions in the Euler-Lagrange equations unmanageable.

In order to test some of these general speculations on actual problems, we have run some numerical computations on an electronic machine, the MANIAC, in the Los Alamos Scientific Laboratory and on its prototype at the Institute for Advanced Study in Princeton. The main purpose of these calculations was exploratory, and the feasibility of using certain numerical schemes on the machine was considered of more interest than the precision of the results. The main point of interest was the amount of time necessary in order to compute on the machine the time behavior of certain functionals of our systems. The problems were mostly of the initial value type, the integration was in time, and the calculation ran in cycles, for which we decided that about five minutes would be allowed. The nature of the problems and the characteristics of the machine with this requirement fixed the maximum number of mass points at about 256. The first problem studied involved the motion of a heavy fluid on top of a lighter one—usually known as the Taylor instability configuration.

The initial configuration was the 16 × 16 array shown in Fig. 1. The particles are of two types, one having a mass which was double the mass of the other. The force was the same between any two particles and was chosen to be inversely proportional to the distance of separation. This algebraically simple choice was made mainly to keep the computation time to a minimum. The whole system of particles is enclosed in a unit square with the heavier particles on top and all particles subject to an external gravitational field, an unstable configuration. With our simple force law the computation for each pair of particles took about 30 milliseconds. With our 256 particles, or over 32,000 pairs, the computation of resultant forces would have taken 15 minutes. This time can be reduced by introducing a cut-off in the force so that for a particle lying outside such range of influence, the only computation necessary is that of a separation between the particles. In order to avoid even such computations of distance which involve the long multiplication time of the machine, we arranged the cut-off in terms not of the Euclidean metric, but in terms of another one (Minkowski): the sum of the horizontal and vertical distances of separation thereby reducing the time of ascertaining the distance in such case to about 3 milliseconds for these pairs. With this, the total computation time (for a cycle) was about 5 minutes, and a typical problem would run 150 cycles —all together, with printing of results, somewhat more than 10 hours.

When a particle approached the side of the container closer than a cut-off distance, a special situation arose. The most convenient way of treating the wall was to create a virtual particle located on the wall at the same horizontal or vertical position as required to contain the real particle. Again we should emphasize that for our first experiments, this recipe was chosen for computational conveniences rather than as a mathematical or physical requirement. Thus, also, all the parameters for the problem were chosen to be simple powers of 2 so as to be able to use the fast operation of the left and right binary shift, rather than the slower operation of direct multiplication.

With our grid of points, crudely as it was chosen, it hardly can be expected that the details of the motion will be exactly described. A typical configuration is shown in Fig. 2; a much later cycle (later time) configuration is shown in Fig. 3 and the formation of the "atmosphere" is apparent. What can be hoped, however, is that some *functionals* of the motion will be more accurately depicted. We are interested in the transition phase and in the time rates of mixing on a large scale between the two fluids. One of the functionals which we observed and plotted as a function of time is the total kinetic energy of the particles divided into two parts: the kinetic energy of the horizontal and vertical motions separately (these are shown on Fig. 4). Even though the motion itself is very irregular, these quantities seem to present rather smooth functions of time. In the case of a stable configuration with the lighter fluid on top and the heavy fluid on the bottom, there would be only a periodic interchange of kinetic and potential energies.* In the unstable case, there should be an increase of kinetic energy which persists for a considerable time. The hope is that with many more calculations one could try to guess from such numerical results at least the form of an empirical law for the increase of this quantity as a function of initial parameters.

Another quantity of interest is the spectrum of angular momentum which may be defined, for example, in the following way: we draw around each particle a circle of fixed radius and calculate the angular momentum in each such region. We then

* See Fig. 5, showing the configuration at a time *later* than the one in Fig. 2.

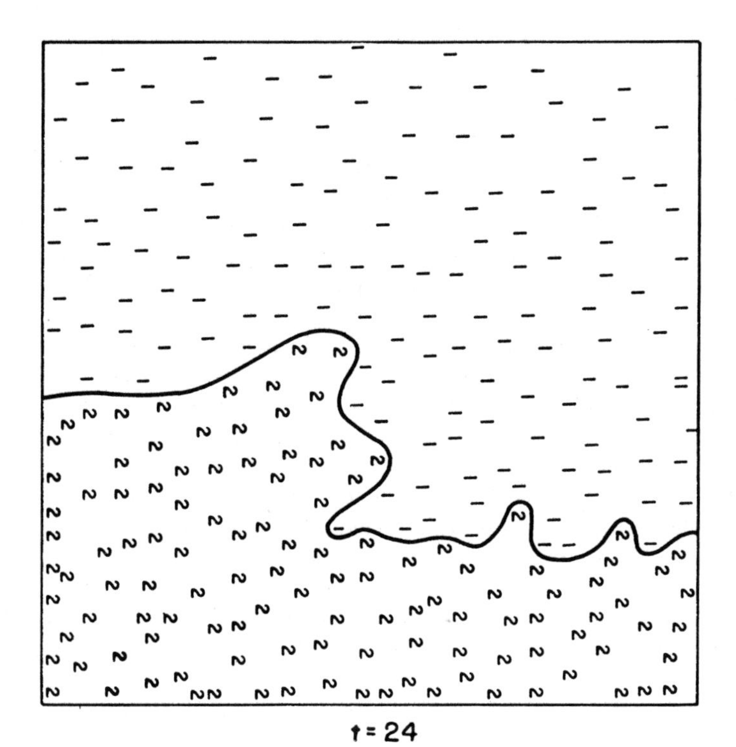

Fig. 2.—Typical configuration, $t = 24$

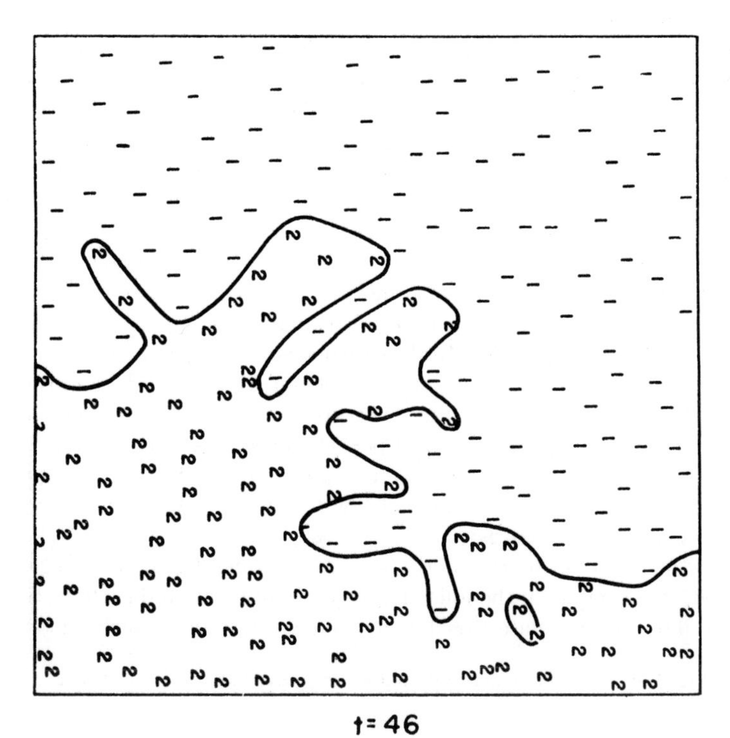

Fig. 3.—Typical configuration, $t = 46$

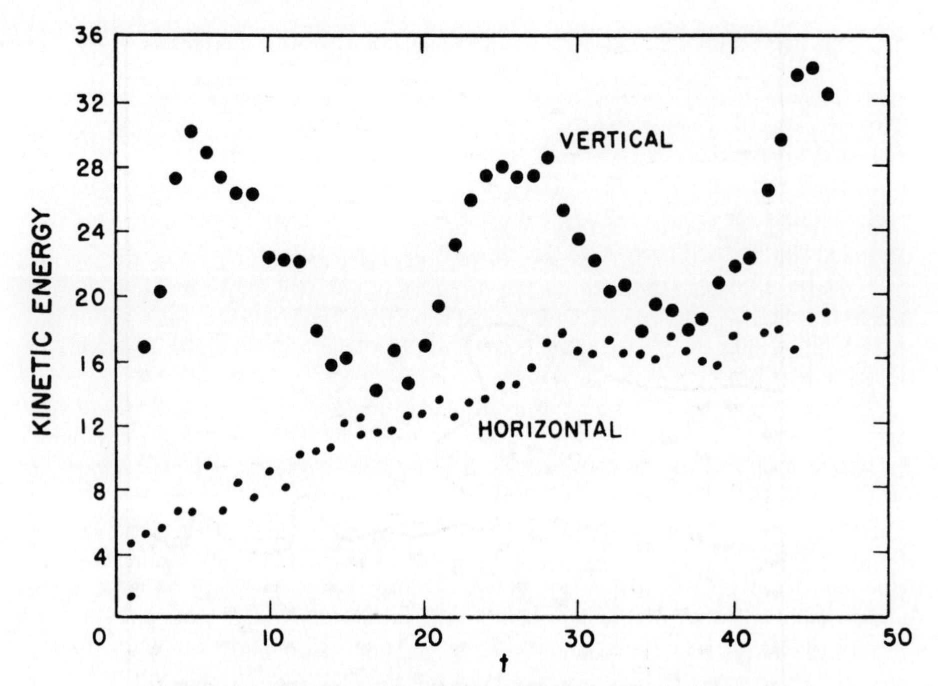

FIG. 4.—Kinetic energies of horizontal and vertical motions

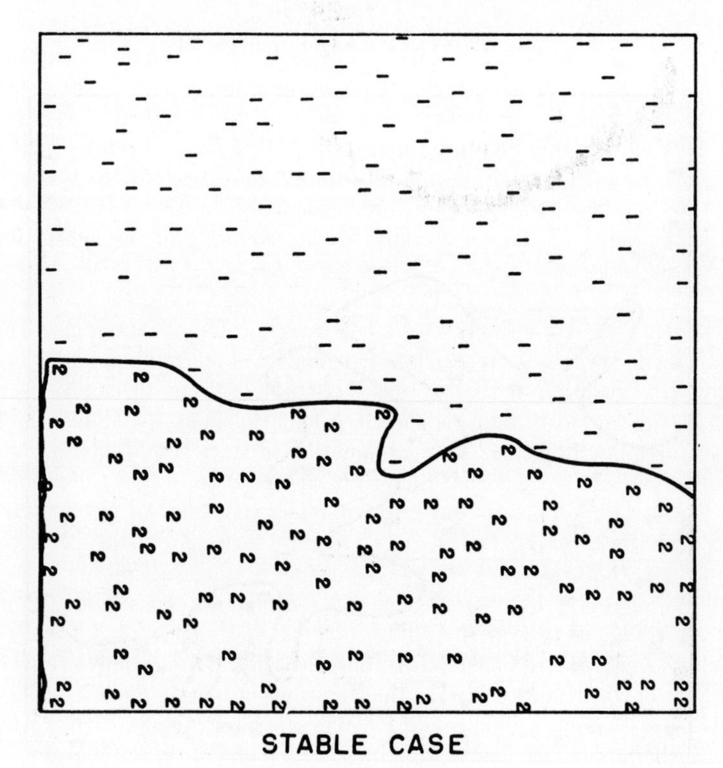

FIG. 5.—Typical configuration, stable case

10

take the sum of the absolute values or the sum of the squares of these quantities. This gives us an over-all measure of angular momentum on a scale given by the radius. Varying the radius, these numbers may then be studied as a function of both time and the radius. It is interesting to consider the rate at which the large scale angular momentum is transferred to small scale eddies—also, the spectrum in the asymptotic state, if it exists. All this, of course, as a function of the parameters of the problem, the ratio of the densities of the two fluids and the external force. It is clear that very many numerical experiments would have to be performed before one would trust such pragmatic "laws" for the time behavior of mixing or increase in vorticity.

Another reason for selection of our problem was an interest in the degree of spatial mixing of two fluids, starting from an unstable equilibrium. To measure quantitatively such mixing in configuration space, a functional was adopted similar to the one just mentioned for the spectrum of angular momentum. At a given time t, a circle is drawn around each point P. In each circle, we look at the ratio r of the number of heavy particles to the total number of particles in this region. We take the quantity $\mu(P,t) = 4r(1 - r)$. This gives an index of mixing of the fluid in the circle, being equal to zero if only one fluid is present and equal to one if particles of both fluids are equally present in it. We average this quantity over all the circles. Initially this average is very close to zero, the only region where it differs from zero being around the interface. As the mixing proceeds, our average measure of mixing increases with time. We have plotted this quantity for several different radii. These functions are again comparatively smooth even though the interface between the two fluids becomes highly convoluted. Again, one might expect that after a sufficient number of numerical experiments, one would obtain a hint of the time behavior of this "mixing functional", or at least an idea of the time T taken for the over-all mixing (which is initially close to zero) to become of the order of $\frac{1}{2}$ or $1/e$, say. This time, for dimensional reasons, must depend on $\sqrt{L/a}$ where L is the depth of the vessel and a the acceleration of the lighter fluid into the heavier one.

The problem of Taylor's instability in its initial stages was also examined in Los Alamos experimentally for the case of a heavy gas on top of a light gas. The interface between the two gases involved an irregularity. Photographs taken at various times show configuration of the two gases not unlike those developing in our calculation. If one wanted to put credence in calculations like the above, the form of the proper force law imitating the real equation of state would have to be chosen very carefully. Also one would have to study the question of how the accumulation of error due to the finiteness of our grid will effect the behavior of our functionals. We have treated compressible gases. An attempt to treat numerically incompressible fluids in a similar way encounters serious additional difficulties:

In order to preserve the volume of each fluid element, one would have to keep constant the areas, in two-dimensional problems, of triangles or other elementary figures whose vertices are occupied by our particles. This means that there is a great number of constraints added to the equations of motion for the repelling mass points. This was computationally not feasible on the available machines for problems where the number of points was of the order of 100 or more.

Another way of calculating in the above spirit would have been to postulate additional forces, doing no work but strongly tending to preserve the elementary area defined by our mass points. This also turned out to be impractical.

Atomic Energy Commission, Washington, D. C. and
Los Alamos Scientific Laboratory, Los Alamos, New Mexico

1. M. W. Evans & F. H. Harlow, *The Particle-in-Cell Method for Hydrodynamic Calculations*, Los Alamos Report 2139, 1957.

ON SOME STATISTICAL PROPERTIES OF DYNAMICAL SYSTEMS

S. M. ULAM
UNIVERSITY OF CALIFORNIA
LOS ALAMOS SCIENTIFIC LABORATORY

1. Introduction

It is intended to present in this paper a number of problems and a brief summary of some numerical computations on the asymptotic behavior of certain simple dynamical systems.

These problems refer to the behavior of a few mass points with given mutual interactions and concern the ergodic properties of the system. Broadly speaking, the questions pertain to the time rates with which a statistical or thermodynamical equilibrium-like states, might be attained. That the approach to equilibria as postulated in statistical dynamics might be extremely slow as compared to times obtained by phase-space volumes or relaxation estimates was indicated in some computations performed a number of years ago by J. Pasta, E. Fermi, and the author [1]. This problem dealt with the long time range behavior of a vibrating string with nonlinear forces added to the usual linear ones. In reality, the problem concerned a dynamical system of a finite number (for example, 64) of particles and was pursued numerically over hundreds of cycles, each corresponding to a would-be period, that is, times corresponding to a full period of the purely linear part of the problem. The results were somewhat surprising in that no tendency towards equilibrization of energy between all the possible modes was noted. Instead, these results showed a transfer of energy between the first few modes of oscillation of the string. The high modes (say from number 5 on up to the last), even in their totality do not acquire more than a few per cent of the total potential plus kinetic energy. Ultimately, the system came back practically to the initial condition. An account of this work is also given in my book [2].

Imagine, quite generally, a system of particles with different masses, all considered as mass points which attract each other according to a given law, say with inverse square forces. Let us assume, furthermore, that the system is in a quasi equilibrium in the sense that most of the particles will stay within a certain bounded distance from each other for a time long compared to the time it takes the radius vector of each particle to describe a full rotation through 2π. One question is, how long will it take for the velocities of the particles to be distributed approximately in accordance with the equilibrium law of statistical mechanics, that is, to have Maxwellian distribution? For a bounded system—if we assume that there exist walls confining the whole system, or else if there exist, for example, outside magnetic constraints to help confine charged particles—one could rely on the ergodic theorem and the metric transitivity to bring about an approach to such an equilibrium.

It is known that in a bounded phase-space the continuous ergodic (metrically transitive) transformations are everywhere dense in the space of all continuous measure-preserving transformations. What is more, they form, in a topological sense, the "bulk" of the whole space [3]. A corresponding theorem stating an analogous property of a real dynamical system of n bodies has not yet been obtained.

The discussion which follows is occasioned by speculations contained in a Los Alamos Report [4]. This deals with the following situation: assume a system composed of two or more stellar bodies and a vehicle which, as an additional body of mass infinitesimally small compared to the celestial objects, forms a many (for example, three) body system. Assume furthermore that the "rocket" describes a trajectory under the action of gravitational forces due to the two large masses, but also still has a certain amount of reserve energy available for steering, that is to say, changing its course by suitably emitted impulses. This available energy is roughly of the order of the kinetic energy which the vehicle already possesses at time $t = 0$. The problem is whether one can use this reserve energy in such a way as to obtain, by suitable near collisions with one or the other of the celestial bodies, much more kinetic energy than that at time $t = 0$, perhaps more by an order of magnitude. As an illustration, assume that the vehicle is between two members of a double star system, it is describing a trajectory in between the two. The question is whether by planning the orbit and changing it suitably one could acquire in leaving the system many times the velocity initially present. It is clear that in a two-body system such possibilities do not exist. The orbit, unperturbed by further impulses, would be a Keplerian ellipse and obviously no multiple "collisions" are possible. One might expect that in a double star system such possibilities do exist. Obviously, in a triple star system the chances of finding suitable orbits and suitable maneuvering seem greater. In an n-body system, say of equal masses, and a rocket whose mass is infinitesimally small by comparison, we will approach the situation of a volume of gas containing both heavy and light atoms, where in equilibrium the velocities of the light particles are greater. From the ergodic theorem, at least applied to a bounded system, it would also follow that the light particle will require very high velocities. The ergodic behavior guarantees that arbitrarily near the given dynamical motion there exists one which will make the rocket approach as close to a small sphere surrounding any of the given heavy mass points as we please, which in particular implies high velocities. The question of whether such motions can be obtained by small changes effected by impulses emitted from the rocket is not answered by the general theorem, but this seems, in view of the prevalence of ergodic motions near given ones, extremely likely. Nothing precise is known, however, about the *times* necessary for obtaining such motions. They might be of super-astronomical lengths. One could say that our question is that of the existence of a Maxwellian Demon in a restricted and, so to say, more modest sense: is it possible, by using "intelligently" a small amount of available energy, to shorten the times for near-equilibrization by large factors?

2. Dr. Kenneth Ford has studied, with the author [6], a specific version of this problem in the summer of 1959. A rocket, whose mass is negligible, is navigating between two heavy bodies. The question is whether trajectories within such stellar systems can be so arranged that the rocket would finally acquire a velocity which is many times greater than the velocity of the heavy bodies. It is easily seen that the change in its speed after a single collision cannot be greater than twice the speed of the heavy body. The problem has to be also considered with the limitation that the radii of the stellar bodies are finite, which makes it harder to arrange trajectories for repeated collisions which would result in considerable gain of velocity of the rocket. Instead of the speed of the heavy body, the escape velocity from its surface becomes a limit for an additional increase. The model specifically studied, both analytically and numerically, in some detail by K. Ford assumed two heavy bodies of equal mass executing a circular motion about the center of mass. In spite of its very specialized form, this problem is already very complicated, since there is a variety of weird rocket orbits. Ford first finds, in the rotating frame of reference, the properties of a continuously infinite set of periodic solutions and examines solutions slightly perturbed from these looking for net energy changes of the rocket in the laboratory frame. Several orbits are found in which, for example, the rocket arrives with negligible velocity from infinity, is captured in a large orbit of low energy, then eased with judicially applied power into an orbit which loops both stars. At the point of nearest approach to one of these, a downward thrust may be applied giving an orbit in which the kinetic energy increases, and then goes off into infinity with a velocity many times that of the star.

Some of the orbits which are periodic are stable against perturbations. In numerical tries, the greatest final velocity of the rocket was about 3.71 times the velocity of the star.

3. A still simpler model to illustrate the general problem will now be considered. A material point (of mass 1) is confined, in one dimension, on a unit interval between two heavy oscillating walls whose mass is infinite. We assume, for example, a harmonic oscillation of the two confining walls. The point is thrown into the interval with the initial velocity, say, equal 1. Assume furthermore that the collisions are always elastic. In succession they will lead to changes in the velocity of our point. In a head-on collision, the point will gain twice the speed of the wall. In a collision which overtakes the wall, the result will be a loss of speed for our point. The maximum velocity of the wall may be assumed to be, say, also equal to 1, and the problem is to study the behavior of the velocity of our point after many collisions. One expects that, after sufficiently long times, the average velocity of the point will become very large, if we look upon the problem as one of statistical mechanics. The tendency towards equipartition of energy would imply this. Mathematically, the problem involves also difficulties of diophantine analysis. The successive collisions take place at times increasingly difficult to compute precisely, and small changes in these lead later on to widely different patterns. Since the collisions which are head-on are slightly more frequent than the unfavorable ones, one might expect a gradual increase in velocity on the average, and the question is to compute or estimate the rate of this increase. One would like such estimates for "almost every" initial position and initial speed of the small particle.

A numerical study of this problem was undertaken with Mark Wells in the Los Alamos Scientific Laboratory. To simplify the computations, the motion of the wall was not assumed to be harmonic, but in a form of a tooth-shaped, that is, broken linear, displacement in time. The problem was studied in two versions, one with velocity of the wall oscillating linearly from 0 to 1, the other with the velocity constant and reversing with a fixed period.

Several thousand successive collisions were computed and great care was taken to examine the influence of the roundoff of errors on the behavior of successive collisions. The results showed a rather surprising behavior. Instead of the expected—perhaps somewhat erratic but, on the average steady increase of speed of particle—enormous fluctuations were observed. With initial speed = 1 of the particle, and speed = 1 of the wall, the velocity of the particle obtained during several thousand collisions sometimes varied between 3 and 4, but the periods of time when the velocity was high were followed by longer periods when it dropped back to 1 or below. It was not possible to conclude from these computations whether the long time average of the energy would increase linearly or with a smaller power of time.

The numerical work was performed for the case where the two walls were moving in phase. (Computationally, it is sufficient, of course, to assume just one heavy wall.) The numerical results, such as they were, would rather indicate a very slow approach to situations envisaged in a statistical mechanics picture, and definitely large fluctuations which seemed to be increasing with time.

Since even this deterministic problem shows unexpected difficulties, it seems futile to superimpose on it the original question of whether, by suitably planned additional small impulses, the rate of increase of speed might be greatly accelerated. Such a restricted Maxwellian Demon would have to be in possession of a super-computer and solve, in addition to the diophantine problem, a game theoretic question. From the greatly fluctuating nature of the nonperturbed motion, it is clear that no local recipe of the kind used in problems of the calculus of variations would be suitable for optimization. It may be that, on the contrary, before certain collisions, occasional "sacrifices," in the sense of the term as used in the game of chess, might be necessary for an overall optimum, that is to say, occasionally a few collisions should be planned which might lead to a lower value of speed so as to have favorable collisions later on.

It might be that the restriction to one dimension makes the problem less typical of situations usually dealt with in statistical mechanics. One could imagine an analogous system in two-space dimension, say a ball colliding with pulsating walls on a billiard table. The additional parameter of the angle of impact might introduce more random-like properties, and the asymptotic behavior of the speed of the particle would be perhaps a less fluctuating one. This is by no means certain, however. It is only plausible that a large number of particles or constraints in a dynamical system is required to insure a thermodynamic-like behavior of the system.

4. A more realistic problem would assume, still in its simplest version, instead of an infinite, a large but finite mass of particles corresponding to the walls. The asymptotic properties of this system might be different from the ones in the previous discussion.

Here, an idealized mathematical model, still in one-space dimension but able to test some of the schemes above, could be as follows:

Imagine, on an infinite line, masses put in on every point with integer coordinates. These masses are either 1 or 2, and for each integer value of the coordinate we decide by the throw of a coin which one of the two values of the mass to locate there. In addition, we give to each of these points a velocity of $+1$ or -1, again deciding independently with probability $1/2$ which one to use. All this is done at time $t = 0$. We can represent the initial state oi such a system by, say, two real numbers, ξ and η. One may characterize symbolically the distribution of masses by using the symbol 0 on the nth binary of ξ if the mass is 1 at the point $x = n$, and symbol 1 if the mass is 2. The other number η, similarly, will contain all the information about the velocities of the system by using symbol 0 in η_n if the velocity at $x = n$ is -1 and symbol 1 if the velocity is $+1$. In this fashion it will be possible to talk about "almost every initial distribution" (in the sense of Lebesgue measure). The mass points will start colliding and, assuming collisions to be perfectly elastic, new velocities will appear and new sets of collisions will ensue.

A whole set of problems now arises. Will the distribution of velocities, which initially was random in the sense of Bernouilli, tend to a distribution more resembling that of a gas? In one dimension high values of velocities will not be established, but the question of the rate of approach to an equilibrium-like situation is of interest. Or perhaps the fluctuations will continue indefinitely on a large scale. The proper way to consider limits is obviously to take any point, an interval of length $2N$ around it, compute the functional in question, and examine the limit as $N \to \infty$.

This model can be varied, of course. One could have, instead of giving each point a velocity ± 1, say, a continuous distribution of velocities to start with, and so forth.

A similar problem was considered previously [5]. The point of view adopted there was different. The collisions, on the contrary, were assumed to be totally inelastic. The points were not put on *every* integer valued coordinate, but only with probability 1/2. The question studied there was that of formation of condensations and superclustering. Needless to say, in our problem above there will be initially, with probability 1, arbitrarily large but increasingly rarely-spaced groups of points with velocities in the same direction. Therefore, the spatial distribution of points after some time will be quite nonuniform. Whether the asymptotic density will remain constant is not obvious a priori.

Coming back to our problem of a light particle colliding with heavy ones: the indications given by numerical tries reported above are then that the rate of energy increase is both slow and irregular. If also true for a random distribution of heavy masses in three dimensions, this would have some consequences for models of mechanisms by which cosmic ray particles acquire very high energies. One such model, considered in literature, postulates charged particles colliding with magnetic fields of stars. These stars move at random and would ultimately transfer some of their energy to the elementary particle. Another model is of a particle moving in continuous and varying magnetic fields in interstellar space. This latter is more difficult to schematize as simply as our model above, but could perhaps provide a more efficient way for endowing the particle with very high energy.

REFERENCES

[1] E. Fermi, J. Pasta, and S. M. Ulam, "Studies of nonlinear problems," LA-1940, Office of Technical Services, U.S. Dept. of Commerce, Washington, D. C.

[2] S. M. Ulam, *A Collection of Mathematical Problems*, New York, Interscience, 1960, p. 109.

[3] J. C. Oxtoby, and S. M. Ulam, "Measure preserving homeomorphisms and metric transitivity," *Ann. of Math.*, Vol. 42 (1941), pp. 874–920.

[4] S. Ulam, "On the possibility of extracting energy from gravitational systems by navigating space vehicles," LAMS-2219 (1958).

[5] ———, "Random processes and transformations," *Proceedings of the International Congress of Mathematics*, Cambridge, Mass., 1950, Vol. 2.

[6] K. W. Ford, "Transfer of energy from astronomical bodies to space vehicles," T-Division Report, Los Alamos, 1959.

Some Properties of Certain Non-Linear Transformations

STANISLAW M. ULAM
University of California
Los Alamos Scientific Laboratory

Several of the papers presented at this meeting dealt with new developments in linear theories in quantum theory. Professor Dyson's lecture especially showed how far one can push the ideas of linear representations—the applications he made of the theory of operators constitute a real triumph of the economy of these ideas. In several other talks there were, however, allusions to non-linear transformations which may be necessary in a future fundamental physical theory.

Very little is known about non-linear transformations, even simple ones, e.g., quadratic transformations in vector spaces, even in the finite dimensional ones. In this talk a brief account will be given of some experimental work concerning certain special nonlinear transformations and their asymptotic or ergodic properties. This work was done jointly with Paul Stein and was possible because of the use of the fast computing machines and I shall present some of the observations made in these studies. Mathematics is not really an observational science and not even an experimental one. Nevertheless, the computations which were performed were useful in establishing some rather curious facts about simple mathematical objects.

To plunge into medias res: The transformations studied were of the type illustrated by the example

$$X' = 2XY + 2XZ + 2YZ$$
$$Y' = X^2 + Z^2$$
$$Z' = Y^2.$$

Here X, Y, Z are real numbers between 0 and 1 and their sum is 1. The new variables X', Y', and Z' will be also in this range and their sum will be also 1. This transformation, in the three-dimensional space, of a convex set into itself is an example of a more general type: we expand $(X + Y + Z)^2$, divide the six binomial terms into three classes—arbitrarily—and have, it turns out, 91 different types of quadratic transformations. (These are all distinct even under permutation of letters.) The interpretation or motivation for this special type of transformations is, for example, the following: Suppose X, Y, Z denote fractions, in a population which is assumed to be very large, of three different types of particles. We might call them blue, red, and green particles. We assume that these mate in pairs at random and each pair produces two offspring whose color is determined by the colors of the parents. For example, in the transformation given above, the X-type together with Y-type give an X-type, so do X with Z and Y with Z. Mating X with its own type and mating Z with its type give the Y type. Mating Y with itself gives type Z. The expected number of particles of each color in the next generation will be then given by the formulae above. If one wants to compute the fractions of the population of each of the three types in the next generation, one will iterate the transformation.

In general, as time goes on by discrete generations, the fractions will be given by the successive iterates of our transformation. One may be interested, among other properties, in the asymptotic behavior of the fractions of the population of each type. It turns out that for some transformations of this type the iterates of almost every point converge to a unique fixed point of the transformation. This means that asymptotically the population becomes stable and a "steady state" is approached. For certain other transformations, still of this type, the fixed point is repulsive and instead of converging to a fixed point, the iterates of a point might converge to a periodic configuration. The period might be for example of order 3; for a transformation of this type in four variables, we found convergence to a period of order 12. For some other transformations but still of this type, the asymptotic or limiting configuration might not be a finite set of points and this limiting set can have a very peculiar structure. A full account of this work is given in a

Los Alamos Report No. LA-2305 where the case of three variables was studied exhaustively.

The main part of this talk is concerned with cubic transformations in three variables. (Strictly speaking, since the sum of the variables is equal to 1, the transformation is then in two variables.) There the examples studied are still of this form: we take $(X + Y + Z)^3$. We obtain 10 binomial terms which we divide as before into three classes. Setting X' equal to one group of terms, Y' to the other, and Z' to the remaining, we obtain an example of the transformations we have studied. Here the limiting set may exhibit a

Figure A. The limiting set of points under the iteration of the transformation:

$$x_1' = x_1^2 + x_3^2 + 2x_1x_2$$

$$x_2' = 2x_1x_3 + 2x_2x_4 + 2x_3x_4$$

$$x_3' = x_2^2 + 2x_2x_3$$

$$x_4' = x_4^2 + 2x_1x_4$$

$$x_1' + x_2' + x_3' + x_4' = x_1 + x_2 + x_3 + x_4 = 1$$

The view is a projection of the 3-dimensional set, the three straight line segments serve merely as an indication of the orientation of the coordinate axes in space.

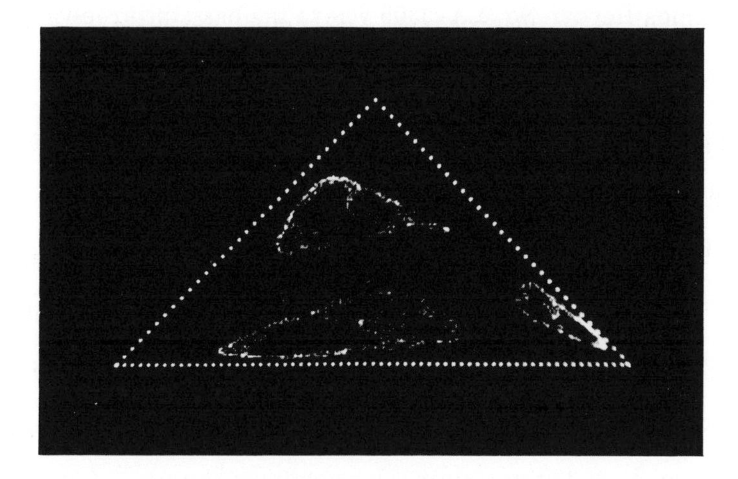

Figure B. The limiting set of points for the transformation:

$$x_1' = x_2^3 + 3x_1x_2^2 + 3x_2x_1^2 + 3x_3x_2^2$$

$$x_2' = 3x_1x_2^2 + 3x_3x_1^2 + 6x_1x_2x_3$$

$$x_3' = x_1^3 + x_3^3 \qquad (x_1 + x_2 + x_3 = x_1' + x_2' + x_3' = 1)$$

The triangle shown is the boundary of the region which is being transformed into itself.

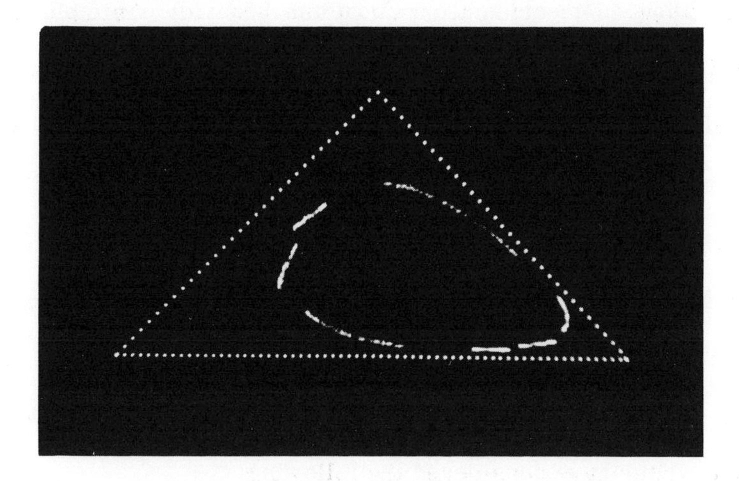

Figure C. The limiting set of points for the transformation:

$$x_1' = x_3^3 + 3x_1x_2^2 + 3x_2x_3^2 + 3x_3x_2^2 + 6x_1x_2x_3$$

$$x_2' = 3x_1x_3^2 + 3x_3x_1^2$$

$$x_3' = x_1^3 + x_2^3$$

$$(x_1 + x_2 + x_3 = x_1' + x_2' + x_3' = 1)$$

great pathology in many cases. In some it consists of a set of points which is infinite but nowhere dense. In other cases it gives the appearance of a curve, in some others still, finite periods of very high order seem to be the limiting configuration. For these cubic transformations, however, it is also true that in the majority of cases there is a convergence to a fixed point. If this is not the case, then the limiting set may depend on the initial starting point but there are regions within the domain of definition such that no matter with which point in the region one starts, the limiting set will be the same.

The computing work was facilitated by a device which displayed visually the iterated points—hundreds or thousands of them—on a screen and photographed the result. One might look upon our study as providing some orientation about the behavior of simple algebraic non-linear transformations which occur in problems analogous to the one studied by Volterra on symbiosis of several types of species and, more generally, in biological—genetic problems. A detailed account of this work with Paul Stein will appear as a Los Alamos Report. In a different context, the study of such transformations can serve as a preparation for the study of some non-linear generalizations of the Schrödinger equation. C. J. Everett and the writer have considered the following:

In the most simple and naive case consider the time independent Schrödinger equation:

$$\Delta\psi + [E - V(x, y, z)]\psi = 0$$

where V is a given function of position (defined on the configuration space). E is an eigenvalue to be determined. One requires the solution to be integrable over the whole space of its definition. There might be boundary conditions other than square integrability; for example, the function sought should vanish at prescribed places, etc.

This equation can be studied in the form

$$\frac{\partial u}{\partial t} = \Delta u - V \cdot u$$

where u(x, y, z, t) of four variables and t is a re-introduced parameter enabling one to omit an explicit mention of E: for large values of t the solution u, one can prove, will have the asymptotic form

$$u(x, y, z, t) = e^{-Et}\psi(x, y, z)$$

where ψ is independent of t and will satisfy the first equation.

In problems of nuclear physics and in the attempted theories of elementary particles, V may not be easily determined or "given." Some time ago, it was proposed to consider—somewhat in the spirit of field theories with "self-consistent" equations—instead of a given potential another unknown function whose square, that is to say its probability density, would serve as a potential for the function ψ. This unknown function would itself obey a differential equation similar to the first equation above. We would then have a system of linked equations, the solution of each serving as a "potential" for the other.

$$(1)\quad \begin{aligned} \frac{\partial u}{\partial t} &= \Delta u + a_1(v^2) \cdot u + b_1(u^2)u + \cdots \\ \frac{\partial v}{\partial t} &= \Delta v + a_2(u^2) \cdot v + b_2(v^2)v + \cdots \end{aligned}$$

We have considered as the most elementary case a system of equations

$$\frac{\partial u}{\partial t} = \Delta u + v^2 u$$

(2)

$$\frac{\partial v}{\partial t} = \Delta v + u^2 v$$

or, for a "steady state"

$$\Delta u + v^2 u = E_1 u$$

(3)

$$\Delta v + u^2 v = E_2 v$$

In <u>one</u> dimension a solution of the system (3) above with $u = v$ is

$$u = \frac{\sqrt{2}}{2} \frac{e^{-|x|/4}}{1 + e^{-|x|/2}} \tag{4}$$

if one requires the function to be square integrable and positive. For the case where one requires the function to be defined on a finite interval only—or, say, vanishing at two points on the line—one obtains elliptic functions.

The equation above is non-linear; in fact, in the way we have written it, it is a cubic in the unknown function. It seems to bear a formal similarity to the equations devised by Heisenberg. These are written as follows:

$$\gamma_\nu \frac{\partial}{\partial x_\nu} \psi \pm l^2 \gamma_\mu \gamma_5 \psi (\psi^+ \gamma_\mu \gamma_5 \psi) = 0 \tag{5}$$

It is a single equation (instead of a system of several linked ones) and the unknown function is defined not for scalar entities but for hyper-complex entities. To my knowledge there is no known exact solution for these but, using the Tamm-Dankoff method, Heisenberg claims to have found approximate solutions which seem to give constants resembling ratios of masses for some mesons.

A fuller treatment of a model given above of our linked equations would be

$$(6) \qquad \frac{\partial u_i}{\partial t} = \Delta u_i + \sum_{j,k} (a^i_{jk} u_j u_k) u_i \qquad i,j,k = 1 \ldots n.$$

The various coefficients now determine the various algebraic relations between the distributions whose mutual interactions and coexistence define the "elementary" objects.

The equations may be looked upon as generalizing, as a partial differential equation the equations of Volterra (total differential equations) on the coexistence of biological species.

The numerical work, undertaken to get at least an approximate idea of what the solutions of our system might be like, gave the following results and equations:

In one dimension, starting with a variety of initial u(x), v(x) at time t = 0 and integrating (2) one obtains very rapid convergence of u(x,t) and v(x,t) to the form given in (4), for example:

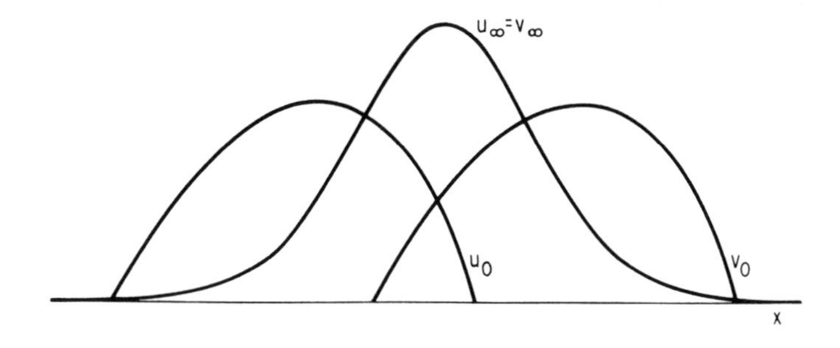

The solution (4) seems to be a strongly attracting fixed point in function space, at least in starting with positive u_0 and v. However, for certain pairs of non-positive initial u_0, v_0 there is convergence to a pair u_∞, v_∞ where u_∞ is the function given by (4) and $v_\infty = -u_\infty$.

In three dimensions, the story seems very involved; apparently the solution is not so strongly attractive.

It should be pointed out that if one requires in (3) the solution to have integral of its square equal to 1 then E_1 and E_2 can not be arbitrary, but, for positive u,v have to be $E_1 = E_2 = \frac{1}{16}$.

The numerical work in three dimensions is proceeding. There seems to be some evidence pointing to the possibility of several topologically distinct pairs of solutions.

The constants E_1 and E_2 can be interpreted as being "rest masses" of the system. That is because, in natural units, in the equations (6) and indeed in (3) they have dimensions of length $\underline{\underline{1}}$, (not length square as in Heisenberg (5)).

Given a length 1 one can use, as a first approximation, $m = \hbar/lc$ to obtain the "spectrum" of m.

Starting in (2) with two functions u_0 and v_0 of the form shown by

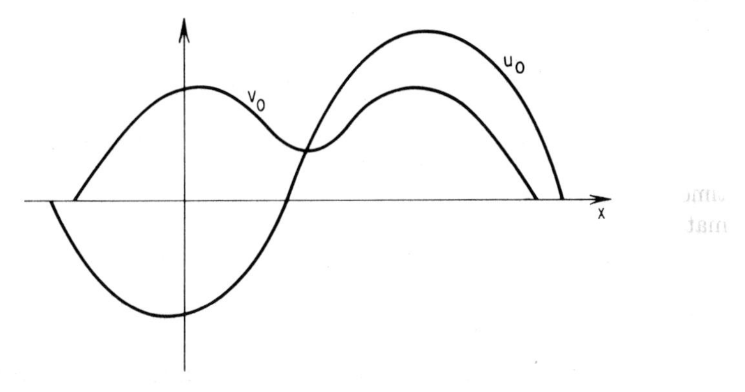

one obtains convergence to:

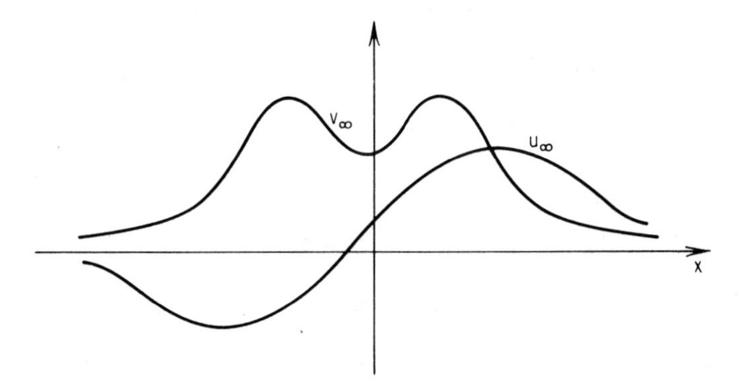

SIAM Review
Vol. 6, No. 4, October, 1964

COMBINATORIAL ANALYSIS IN INFINITE SETS AND SOME PHYSICAL THEORIES*

S. ULAM†

I intend to discuss my subject by giving first a very brief perspective on the relevant historical developments which have taken place in foundations of mathematics and in the fundamentals of the picture in physical theories during the time which has passed since the early work of von Neumann and the present—more than a third of a century.

Von Neumann's work in the late twenties has affected profoundly both the developments in mathematical logic and in the mathematical foundations of quantum theory. But, what a change in outlook in both of these and what revolutions intervened after the completely unexpected turn of events in experimental discoveries!

The program of Hilbert's, which was a Credo of most mathematicians at that time and which von Neumann tried to establish and realize in foundations of mathematics, had to be abandoned or perhaps completely reinterpreted as a result of the work of Gödel. The quantum theory of the atom turned out to be inadequate as a general schema for a description of nuclear phenomena and the physics of elementary particles. These events had already at that time cast an ominous shadow ahead of them. The first problem of mathematics—on Hilbert's list—the continuum hypothesis—remained unsolved and a number of problems involving simple combinatorial properties of infinite sets began to rival Cantor's problem on the continuum—in their simplicity and intractability. In theoretical physics there already existed examples of divergences to infinity of some of the most important quantities—e.g., the self-energies of the electron.

By now, in 1963, the bewildering mass of new *facts* discovered in physics, in the astronomical universe and in mathematics (sic!) has rendered illusory the hopes, so much entertained through the 19th century and at the beginning of the present one, of ultimate simplicity of the foundations of science.

Perhaps it would not be exaggerated to say that the situation confronting us now would appear as perplexing to Poincaré and Hilbert as it would have to Eratosthenes and Appolonius were they transported to the end of the last century!

1. The concept of infinity played several roles in pure mathematics and in its applications to the physical sciences. It is not a priori obvious why the infinite algorithms prove themselves so enormously useful in simplifying mathematical constructions. The calculus is so much more maniable than operations with finite differences. In addition, even in combinatorial studies, the use of continuous variables, for example, in the employ of generating or of characteristic functions

* The fourth John von Neumann Lecture delivered at the Summer meeting of SIAM on August 27, 1963, in Boulder, Colorado. Received by the editors February 3, 1964.

† Los Alamos Scientific Laboratory, P.O. Box 1663, Los Alamos, New Mexico.

allows very often a very great insight in the behavior of finite quantities. In probability theories and in statistical mechanics, the limit theorems and the ergodic theorems show a simplicity which is not inherent until one goes to asymptotic problems.

All this is perhaps partly due to the fact that one is content to examine only the simplest or *linear properties*. The simplicity is apparent and concerns essentially only the *first means* or the averages. Indeed, the enormous size of the literature dealing with linear spaces and operators contrasts very strongly with the meager amount of available information about nonlinear phenomena.

We shall try now to show, in set theory, that already for a long time there were lying, immediately under the surface, so to say, very simple unresolved questions, some of which may perhaps now be considered as independent axioms.

As a first example, we shall take Suslin's hypothesis. It can be stated as follows:

Given is a class $\mathfrak{M}$ of sets of points (they may be thought of as sets of points on the real axis) with the following properties:

1. For any two sets A, B in $\mathfrak{M}$ either $A \supset B$ or $B \supset A$ or $A \cdot B = 0$.
2. Any subclass of $\mathfrak{M}$ with the property that any two sets of the subclass are disjoint is countable.
3. Any subclass of $\mathfrak{M}$ of sets such that no two sets in it are disjoint is countable.

The problem: Is the whole class $\mathfrak{M}$ countable?

Despite all the efforts of set theoreticians, apart from some partial results the problem remains unsolved.

Another problem [17, p. 15] is as follows.

Given is a class $\mathfrak{N}$ of sets of points on the interval with the following properties:

1. All the subintervals belong to this class.
2. The class forms a Borel field; that is to say, given any set in it, its complement (with respect to the whole interval) also belongs to it. Given any countable sequence of sets $A_1, A_2, \cdots, A_n, \cdots$ belonging to the class, their sum $\sum_{n=1}^{\infty} A_n$ also belongs to the class $\mathfrak{N}$.
3. Given any decomposition of the whole interval into disjoint sets, containing at least two elements each—the decomposition may be into noncountably many sets—there exists in the class $\mathfrak{N}$ a set selecting exactly one element from each of the sets of the decomposition.

Is the class $\mathfrak{N}$ necessarily the class of all subsets of the interval?

This problem is typical of a class which concerns the general metamathematical question of what are the sets that are definable in a given set of axioms of set theory. This is, of course, fundamental to the results à la Gödel on consistency of the axiom of choice and the continuum hypothesis. There are several other rather simply formulated questions of the type: Given is a subclass of subsets of a space (e.g., the interval or the set of all ordinals up to Ω—the first noncountable ordinal). We allow the formation of new sets and certain operations, e.g., Borel operations of complementation and countable summation or even projective operations. The question then is: do we obtain *all* subsets of the given space?

Problems of combinatorial nature, dealing with infinite sets, arise rather naturally in the theory of abstract measures. Since some of these have a direct bearing on the problem of continuum and on inaccessible alephs, we shall present a few of these [18]:

Given a (noncountable) set E, we are interested in defining for its subsets A a measure function $m(A)$ with the following properties: $m(A) \geqq 0$; $m(E) = 1$; $m(p) = 0$ for any set consisting of a single point p; if A_i, $i = 1, 2, 3, \cdots$, is a sequence of disjoint sets ($A_i \cdot A_j = 0$ for $i \neq j$), then

$$m\left(\sum_{i=1}^{\infty} A_i\right) = \sum_{i=1}^{\infty} m(A_i).$$

The first question that arises is: is it possible to define such a measure function for all subsets of the continuum; that is to say, in the case when E is the interval $(0, 1)$? Using the continuum hypothesis Banach and Kuratowski have proved that this is impossible. Assuming this hypothesis they prove therefore also that it is impossible in the case when E is a set of power $\aleph_1$. The writer has succeeded in showing, without any hypotheses, that a measure function of the above sort is impossible to define for all subsets of a set of power $\aleph_1$. This result holds also for sets of power $\aleph_2, \cdots, \aleph_\omega$, etc.; in fact, for all known alephs. The question remains unresolved only for sets whose power is an inaccessible $\aleph$.

The impossibility of such a measure function results directly from an existence of a class of decompositions of E forming an infinite matrix in certain purely combinatorial set-theoretical properties:

$$\begin{aligned}
E &= A_1^1 + A_2^1 + \cdots + A_n^1 + \cdots + A_\xi^1 + \cdots \\
E &= A_1^2 + A_2^2 + \cdots + A_n^2 + \cdots + A_\xi^2 + \cdots \\
&\cdots\cdots\cdots\cdots\cdots\cdots\cdots\cdots\cdots\cdots \\
&\cdots\cdots\cdots\cdots\cdots\cdots\cdots\cdots\cdots\cdots \\
E &= A_1^n + A_2^n + \cdots + A_n^n + \cdots + A_\xi^n + \cdots \\
&\cdots\cdots\cdots\cdots\cdots\cdots\cdots\cdots\cdots\cdots \\
&\cdots\cdots\cdots\cdots\cdots\cdots\cdots\cdots\cdots\cdots,
\end{aligned}$$

where all the sets in each row are mutually disjoint:

$$A_\xi^n \cdot A_\eta^n = 0, \qquad \text{for all } n \text{ and for } \xi \neq \eta.$$

The sum of every column gives the whole set E, with an exception of at most countably many points. The index n runs through all integers from 1 to infinity; the index ξ runs through all the trans-finite ordinals up to Ω. In other words, there are countably many rows and noncountably many columns. Granting the existence of such a schema of sets, the impossibility of a measure function for all the sets in it follows immediately. One reasons as follows:

Since the sum of each column gives a set of measure 1, there must exist in it a set of positive measure. There are noncountably many columns and only countably many rows. Therefore, there would exist a row in which there would be noncountably many sets of positive measure. Since the sets in each row are mutually disjoint, we get a contradiction.

Using the axiom of choice, it is possible to construct a table of decomposition as above as follows. Well order all points of the sets E as follows:

$$p_1, p_2, \cdots, p_n, \cdots, p_\xi, \cdots, \qquad \xi < \Omega.$$

Consider the sets A_c^n as empty boxes into which we shall put successively the points P in this manner: the point P_ξ will be put in every row once and in every column up to the column whose index is ξ. This is possible because ξ is a countable ordinal and we can make a one-to-one correspondence between the countable infinity of rows and the countable infinity of columns up to and including the ξth. Since the point appears in every row just once, the sets in each row will be disjoint. The sum of every column will give the whole set E with exception of at most countably many points because all points, whose index is larger than ξ, will be in it from our construction.

In this fashion we established a property of sets of power $\aleph_1$ without the use of any hypotheses. One can generalize this construction for higher alephs very easily. The difficulty arises only for the inaccessible alephs.

The matrix of sets established above shows only one example of combinatorial behavior of the algebra of infinite sets. There are a number of other problems on the decompositions of sets, connected with the existence of measures, assuming only two values 0 and 1, which were introduced in the paper quoted above. Recently some interesting partial results were obtained in connection with these abstract measures. It should be quoted here that the work of Gödel [7] allows us to assert that it is free of contradiction to assume that the sets of our decomposition A_c^n are projective sets (in the sense of Luzin) of real numbers, E being the interval $0 \leqq x \leqq 1$.

Other examples of combinatorial problems involving infinite sets may be formulated about properties of infinite games.

It was S. Mazur in Poland who first defined an infinite game: given is a subset X of points of the interval $(0, 1)$. Two players A and B play in this manner: the first player selects a subinterval I_1. Then it is player B's turn; he has to select a subinterval I_2 of I_1 in any manner he likes. Then A plays again by selecting a subinterval of I_2, and so on. If the intersection of all the intervals I_n contains a point in common with the set X, it is the first player who wins. If the intersection is void, it is B who wins the game. The question is, for what set X is there a method of winning, say, for the first player? It is immediately obvious that if X is a set of first category in the sense of Baire, then player B can win against any play of A. It was Banach who first proved the theorem that if the set X does not have the property of Baire in any interval, that is to say, neither X nor its complement is a first category in any interval, there is no method of winning for either player. A number of similar games are then considered by Mazur, Banach, and the author. These games are not purely set-theoretical, since they involve definitions of topology.

Recently the writer considered a purely set-theoretical game. This starts from an abstract set E—it could be the continuum or a set of, say, power $\aleph_1$. Again there are two players A and B. The first player divides E into two disjoint sets.

Player B now has to *choose* one of these and then can divide it, in any way he pleases, into two disjoint subsets. Next A has to choose one of these, divide it again into two disjoint subsets as he likes, and present the choice to B again, etc. If the intersection of all the chosen sets is empty, we say that A wins. Is there a way for A to win this game no matter how B would play?[1] This problem is obviously one of pure set theory. It seems to be still unresolved even in the case where X is a set of power $\aleph_1$. The important fact to be stressed is that it is not too difficult to devise a number of other questions of this sort, all rather easy to state and, therefore, apparently going, so to say, to the core of properties of abstract sets. As we shall see later, the present indications are that many of these may possibly be *independent* of the usual axioms of set theory. Gödel's work and the very recent (still unpublished!) most important results of Paul Cohen[2] give such indications. This presents a most startling situation: as strange as if, say, in elementary geometry of triangles one had found simple statements which would turn out to be independent of the axioms of Euclid!

The yet unpublished papers by Paul Cohen appear to contain the result that Cantor's continuum hypothesis is independent of the axioms of set theory in their present form—as used by Gödel, for example.[3] Does it mean that it is "natural" to assume the falsity of the statement that continuum is $\aleph_1$? It seems to the present writer that as with the axiom of choice whose validity is more natural to postulate, this would be the case. If one could draw the analogy with a more special mathematical theory—the non-Euclidean geometries are "true" in a philosophical sense. The result of Cohen may open, so to say, a large class of "non-Cantorian" set theories. The freedom of construction in mathematics, proclaimed by Cantor and defended by Hilbert, would certainly include the validity of the more general foundations of mathematics, where the continuum would be beyond the first aleph.

At the moment it may be premature to say whether it is equally sensible to put the power of continuum as an inaccessible aleph. The above-mentioned theorems on additive set functions (measures) for all subsets of the alephs could justify such postulates—especially since it may very well turn out that the existence of a completely additive set function (in our sense) for all subsets of the continuum may also be an assertion independent of the classical axioms for set theory.

It seems very interesting to note that after his early work in foundations of mathematics and logic, von Neumann turned much later, towards the end of his life, towards the study of computing machines and a theory of automata, where he made some of his most important and lasting contributions. Turing's work has also proceeded from the more classical type of investigations in foundations to this seemingly more concrete domain.

[1] See S. Ulam, *An open problem*, Some Recent Advances in Games Theory, Princeton University Conference, October 4–6, 1961; Princeton, New Jersey, 1962.

[2] Communications by Paul Erdös and other friends from preprints of the as yet unpublished paper by P. Cohen.

[3] Note added in proof: see [22].

This seems to be both historically and epistemologically significant. Some of the oldest mathematical recursive definitions, those of prime integers, rely on the recursive functions of the simplest, i.e., first type. Almost immediately one has an object—a sequence of primes, some of the simplest properties of which are beyond the present power of mathematics to establish. It is the contention of the author that a similar situation may appear now in metamathematics: the very set of axioms themselves could, in the future, be thought of as defined, not by a finite collection given in advance, but by a recursive operation. Speaking somewhat vaguely, we would have, starting from a finite number of expressions—which include the usual axioms—new ones appearing genetically by rules of procedure which, in addition to the pre-accepted rules of inference, etc., contain some other operations leading to new expressions which we also treat as "axioms." This would be in analogy to some special constructions of which I will give one or two examples.

One can consider a rule for growth of patterns—in one dimension it would be merely a rule for obtaining successive integers. We mentioned already the primes which could be thought of as obtainable by a sieve. The so-called lucky numbers are obtained by a different sieve [6]. In both cases simple questions that come to mind about the properties of a sequence of integers thus obtained are notoriously hard to answer. Another one-dimensional sequence was studied recently. This one was defined purely additively: one starts, say, with integers 1, 2 and considers in turn all integers which are obtainable as the sum of two different ones previously defined but only if they are so expressible in a unique way. The sequence would start as follows:

$$1, 2, 3, 4, 6, 8, 11, 13, \cdots;$$

even sequences as simple as that present problems! For two dimensions one can imagine the lattice of all integral valued points or the division of the plane into equilateral triangles (the hexagonal division). Starting with one or a finite number of points of such a subdivision, one can "grow" new points defined recursively by, for example, including a new point if it forms with two previously defined points the third vertex of a triangle, but only doing it in the case where it is uniquely so related to a previous pair; in other words, we don't grow a point if it should be a vertex of two triangles with two different pairs previously taken. Apparently the properties of the figure growing by this definition are difficult to ascertain. For example, with the definition above, it is not easy to decide whether or not there will be infinitely long side branches coming off the "stems." It will be a much more complicated matter to define, in analogy with these simple constructions, a genetic or successive construction of axioms considered as a finite collection of symbols again "growing" from an initial set.

One of the rules of putting down new expressions would then involve taking *some* which are undecidable, in the ordinary sense, from the previously constructed ones. (Of the two statements: the one coming up "first" and its negation, the first one would be taken as a new axiom.)

A procedure of this sort would still not obviate existence of undecidable statements for such an expanded system—in fact, by Gödel's method there would still be some in the form of Diophantine problems but these presumably would be less elementary.

2. We shall now present some comments on the role of infinite constructions in physical theories. The idea of a universe which is actually infinite is very old. The actual speculations go back to antiquity and were pursued by many philosophers later, e.g., by Giordano Bruno. It is interesting to note that Cantor himself was very deeply concerned with this problem (see, e.g., [3, pp. 166–183 et al.]).

Perhaps the most elementary consideration involving "paradoxes" of actual physical infinity can be illustrated by the so-called Olbers' paradox. This is simply the following. If, in euclidean space, the stars existed in infinite number and if they were approximately uniformly distributed, then the sky would appear to be infinitely bright—since the apparent luminosity of a star diminishes inversely proportionally to the square of the distance but the number of stars increases in proportion to the square of the distance. Since the luminosity of the sky is finite, one has to conclude either that space is not infinite or else it is not uniformly populated with equally luminous objects. One way out of this difficulty, still in the framework of classical conceptions, was found by Charlier. He proposed a possible model for the stellar universe, still in a sense isotropic and homogeneous, by using an infinite hierarchy of groupings. One can imagine a cluster of stars separated by distances large compared to the distances between individual stars in a single cluster. These clusters again are not uniformly spread out but are grouped in aggregates: super-clusters, with distances between super-clusters much larger than the distances between the clusters themselves in a single super-cluster, etc. If the distances between members of successive hierarchies increase relatively to each other, say, by a fixed ratio, the total luminosity will converge in every point of space—so will the total gravitational force on any point, etc. One could arrange the geometry in such a way that it would be isotropic in the sense that no special point of space is singled out. The geometry of the universe in the large would be somewhat like that of the well-known Cantor discontinuum (a perfect nowhere dense set) but "turned inside out."

Einstein's general relativity theory introduced a variety of new possible models for the physical universe. One can arrive at a resolution of Olbers' paradox, already in the special theory, in a purely kinematic model, still with the euclidean geometry of space but considering the relativistic time-space continuum. This can be done, for example, by postulating an expansion of space according to Hubble's law: the velocity of recession between any two points is proportional to their distance. The velocity of light being the upper limit of relative velocities, one arrives at a finiteness of the "decidable" universe; that is to say, at any given *time* the number of points which can physically influence any given point is finite.

As is well known in the *general* theory of relativity, the metric and even the topology of space is determined by the distribution of masses and may be such as to yield a finite, i.e. compact, space for the physical universe. This is of course not at all necessary, in fact some models of the curved four-dimensional time-space have negative curvature, i.e., the metric is hyperbolic and the mathematical model of the universe becomes noncompact. It is, however, still the finite value for all physically meaningful velocities of propagation of effects which determines convergence of, e.g., the total forces acting on a point, the total luminosity on any point due to radiation from other sources, etc.

Really, strictly speaking, the question of convergence should be kept distinct from the problem of determinacy of the lifelines of physical points. If in the time direction the four-dimensional continuum is not bounded, then events of the arbitrarily remote past continue to influence the course of "events."

We can not enter here into any detailed discussion of these problems, which are bound with the definition and meaning of natural constants, but one should be aware of their existence—later we will discuss the question of decidability.

Now we will formulate on a very elementary model, still within classical physics, some problems involving actual infinities in physics.

Strictly speaking of course, all the problems in mechanics concerning a continuous mass distribution involve passages to the limit and, therefore, assume a certain topology for a mathematical model of any physical theory. This has almost invariably been the euclidean continuum. Whether in very classical physics we use the real number system exclusively or in continuum theory the primitive notion is that of a measure function, i.e., probability distribution, it is always over sets of real numbers or systems of real numbers. In other words, the Hilbert spaces are constructed over locally euclidean manifolds. There is no certitude that in the future this will always be the case. On the contrary, one can discern indications that something like the p-adic number system or the Cantor discontinuum will be found as a useful basis for describing phenomena in the very small.[4]

Imagine in an infinite set S of mass points distributed at random a euclidean space on the lattice points of a given subdivision, say on the points with integer coordinates. In the simplest problem of this type, we can postulate that each such point of space is occupied by a mass point with probability one-half. Assume now that the mass points attract each other with newtonian forces, i.e., inversely proportional to the square of the distances. It is easy to show that, *with probability* 1, the total force is well defined on every point. Assume now, say, that at time $t = 0$ all the points were at rest, motions will ensue for positive t as a result of our postulated forces and the question is to describe properties of such a system whose total mass is infinite but with motions well defined dynamically.

One merely has to make a rule or convention governing collisions between mass points. In one dimension, for instance, we could require that points after colliding stay together as a single point (with mass equal to the sum of masses of the colliding particles) moving in such a way as to preserve the momentum (and not the total kinetic energy which one could imagine is converted into some form involving "internal" degrees of freedom).

The behavior of such a truly infinite system is not obtainable as a limit of the

[4] See in this connection a discussion in [17, pp. 89–104].

behavior of a finite approximating configuration. By this I mean that the asymptotic properties of certain functionals of S are not "continuous upper limits" of these for systems S_n consisting of N points. This is very obvious; under attracting forces a finite system starting from rest will, under our conventions in one dimension, come to a single point. For the infinite system, on the contrary, it is easy to show [19] that the asymptotic density of the points on the whole infinite line will remain constant despite the fact that the individual masses become larger and the distances between neighboring points will, on the average, increase with time. Some elementary properties of such an infinite system can be demonstrated. It is easy though to ask questions that are hard to answer. For example, the question of the "temperature" of the system, i.e., the average value of the kinetic energy of the individual points—will it remain finite and bounded for all time? Will the distribution of masses expressed in terms of the average mass tend to a fixed function, etc.? It is conceivable that some problems of this sort, i.e., intuitively interpreted as having "physical" meaning, may be undecidable.

In analogy to models involving an infinity in the large of physical or pseudophysical points, one can imagine problems with a quasi-physical interpretation and dealing with the "infinitesimally small." This can be done in the framework of classical mechanics by imagining an infinity of mass points located in a bounded part of space with total mass finite. The kinematics and dynamics of motion of such systems may involve difficult or perhaps undecidable problems. Again I shall refer to the discussion mentioned above (cf. [17, pp. 95 ff.]).

One may consider now analogues of the Olbers' paradox but, so to say, in the infinitely small rather than in the infinity of stars in the large as in the original one. Indeed, one may consider the divergence of self-energy of a charged continuous euclidean sphere as a case in point. This difficulty was realized long ago, as soon as the electron was discovered. As is well known in quantum theory in modern physics of elementary particles, the divergences, if anything, are even more serious. They affect the field theories and in addition to making some physical quantities *prima facie* divergent—that is, infinite—it appears that the logical schemata themselves exhibit an infinity of alternatives each contributing to a physical quantity in such a way as to preclude convergence without arbitrary cut-offs or "renormalizations." I am speaking here of the problems involving virtual processes and particles in arbitrarily high multiplicity—that is to say, an infinity of Feynman diagrams, in certain cases, all on equal footing contributing equally to the divergence of what should be observable quantities.

As in the elementary paradox due to uniform infinity of stars, there are at least two ways open to resolve this—one is to abrogate the infinity by assuming a finite space-time and a finite number of points not further analyzable. This is extremely hard to accomplish successfully, since no theory with a minimal length seems to be reconcilable with Lorenz invariance. The other way would be to relinquish the postulate or belief in uniformity or homogeneity in the euclidean sense. This would amount to postulating a geometry for physics which would result in a structure in the small obtained by taking the Charlier type distribution in infinite space and, by inverse radii, transforming it into the small. (This should not be taken literally; what one has in mind is a mathematical substratum for configuration and momenta space describable by, e.g., Cantor-set discontinua; there is no minimal length; arbitrarily small distances occur but not *all* the real numbers are now allowable as physically meaningful quantities.)

Certainly, the experimental discoveries and observations of the last decade indicate that something of this sort will perhaps be necessary. In particular, Hofstadter's studies of the electromagnetic structure of single nucleons reveal a nonuniform type of geometry. The charge distribution of protons and neutrons (sic!) seems to present a shell-like form. But in the field of elementary particles in general the combinatorial complexity is ever greater. To quote from a recent publication: "Scarcely more than a decade ago it appeared to many of us that the study of the physics of elementary particles was entering a period of refinement and consolidation. The paths of study of the weak interactions seemed well determined, the pi-meson appeared to have all of the qualities necessary to account for the strong interactions, ... Now after more than ten years progress we see again that God is more subtle than we would suppose. Our simplicities were but the shadows of a more complex reality. . . ." (Preface in *Strange Particles*, by R. K. Adair and E. C. Fowler, Interscience Publishers, New York, 1963.) Such sentiments are to be found periodically in the literature of the last fifteen or twenty years!

One could think of "experimenta crucis" to determine whether there is an asymptotic uniformity in the very small—or in the large. The counts of number of galaxies as a function of the distance (apparent magnitude), either visually or by radio signals, could decide between an elliptic or hyperbolic (that is to say, compact or noncompact) nature of the universe. There can, of course, never be a final decision but it would make it philosophically more satisfactory to build plausible theories on an "as if" basis. One could try to attach such interpretations to some experiments in the very small again to "decide" whether the complexity varies and increases indefinitely or whether there is any asymptotic uniformity. The indications are in this case, however, that there are no "atoms of simplicity" and, which is most strange, one would almost be tempted to say that in the physical world the set-theory axiom of regularity (that is to say, that every set contains a minimal element with respect to the relation of "belonging to a set") *does not hold*!!

Assuming now that indeed the physical universe presents an infinite complexity in the small and the astronomical universe an infinite variety in the large, the question arises: will there be meaningful undecidable statements in physics and in cosmology? It seems to me that the answer should be in the affirmative. By this is meant that no finite number of axioms will suffice to deduce statements summarizing satisfactorily the physical world. The crux of this belief lies in the proper definition or interpretation of the word "meaningful." To many physicists, the original undecidable propositions of Gödel would appear either intuitively true or else perhaps without too much real import. The more recent results of P. Cohen should certainly dispel such feelings. Statements like the continuum hypothesis of Cantor are certainly meaningful to most scientists. If it will turn out, as I consider it likely, that the other examples of set-theoretical propositions form a set of statements independent from the existing axioms, one should be prepared for a similar situation in the more physical theories, also. It might be that such concise and simple, therefore meaningful, propositions are not at all rare in the space of all statements.

Many theories deal with infinity in *time* with a class of problems concerning the ergodic or more generally asymptotic behavior of physical systems. One is interested in time averages of some functionals of the system and the average is really taken for $t \to \infty$. The statements asserting the existence of first means form the substance of the well-known ergodic theorems. Other, in some cases more delicate, properties of behavior and time are more difficult to prove and it is quite possible to have undecidable statements of this sort.

Nonlinear systems pose interesting questions concerning their long time behavior. One of the most elementary examples is given by a nonlinear vibrating string. We were interested (see [4]) in the distribution of energy, which was initially present (for example) only in the first mode, among the higher modes or in the rate of "thermalization" for increasing values of time. An exploratory numerical work has shown rather unexpected properties of this system. In particular, recurrences in phase space were taking place much sooner than expected from general Poincaré type of arguments. This work was followed by analytical studies by several authors [5, 9, 21]. We can not here enter even in the formulation of these asymptotic problems but merely want to stress the point that already in such a simple physical system (e.g., the one-dimensional string) the behavior of certain functionals of its motion for infinite time presents mysteries.

More generally, the problems of metric transitivity of dynamical systems are in most cases unsolved. It is interesting to note that, even though for most measure preserving flows on manifolds the metric transitivity holds [14], as a general case there are very few actual mechanical systems for which this has been proved. The recent work of Kolmogoroff, Arnold and J. Moser produced most interesting examples and also general theorems for certain classes of such systems.

It turns out that in certain dynamical systems there is no metric transitivity—there exist, for example, invariant closed curves in the phase space. These correspond to almost periodic solutions of the differential equations. This has bearing on the problem of stability of periodic solutions of systems of ordinary differential equations which are nonlinear but conservative [1, 2, 8, 11, 12, 13].

We mention these results because they indicate the complex variety of behavior of classical systems for infinite time. The asymptotic properties of motions, e.g., the question of stability, might be conceivably undecidable in some cases. Even when one stylizes into simple form the transformation of a phase-space—considering now transformations which are *not* one-one—for example, in *quadratic* transformations of a euclidean simplex into itself, there appears an enormous complexity in the behavior of the iterates of such [15, 16]. E. Lorenz succeeded in stylizing some models of flows of the atmosphere into forms resembling such nonlinear transformations [10]. Noting the notorious "unpredictability" of atmospheric phenomena, we see again that already very simple algorithms (iterations of simple algebraic transformations) will after a long time exhibit a behavior which is hardly decidable.

Problems in dynamics of a finite system of particles, one of which may be thought of as a vehicle with some internal energy available for changing its motion, lead to still more complicated problems concerning long-time behavior. The question of whether one can have a "strategy of motion" to utilize the gravitational energy of a many-body system has been studied [20]. Again there appear very difficult questions of asymptotic nonlinear Diophantine analysis.

3. The above examples of problems in physical situations will perhaps suggest that, as in questions of pure mathematics, no finite system of axioms may suffice to render all the "natural" questions decidable. Perhaps, as suggested above for the mathematical problems, a genetic or organic production of new natural postulates from a number of given "laws" will be the procedure. It is possible that in the game of trying to obtain knowledge and understanding of natural phenomena one will not be able to proceed from a finite system with the hope to obtain all the answers but—instead—the axioms which one will adjoin will themselves be dictated by the development of this knowledge. In analogy to a game like chess, one could paraphrase this possibility by stating that there is no "forced win", a winning strategy of moves which can be announced ahead of time independently of the moves of the opponent, but only that there exists a winning sequence for every choice of the opposing strategy. The newly developing ideas in the theories of automata and organisms, which owe so much to the work of von Neumann, will in the future bring important results.

REFERENCES

[1] V. I. Arnold, *Generation of quasi-periodic motion from a family of periodic motions*, Dokl. Akad. Nauk SSSR, 138 (1961), pp. 13–15.

[2] ———, *Small denominators. I. Mapping the circle onto itself*, Izv. Akad. Nauk SSSR Ser. Mat., 25 (1961), pp. 21–86.

[3] Georg Cantor, *Gesammelte Abhandlungen; Mathematischen und Philosophischen Inhalts*, George Olms Verlagsbuchhandlung, Hildesheim, 1962.

[4] E. Fermi, J. R. Pasta, and S. Ulam, *Studies of Nonlinear Problems. I*, Los Alamos Scientific Report LA-1940, 1955.

[5] Joseph Ford, *Equipartition of energy for nonlinear systems*, J. Mathematical Phys., 2 (1961), pp. 387–393.

[6] V. Gardiner, R. Lazarus, N. Metropolis, and S. Ulam, *On certain sequences of integers defined by sieves*, Math. Mag., 29 (1956), pp. 117–122.

[7] K. Gödel, *Consistency proof for the generalized continuum hypothesis*, Proc. Nat. Acad. Sci. U. S. A., 25 (1939), pp. 220–224.

[8] A. N. Kolmogoroff, *General theory of dynamical systems and classical mechanics*, Proceedings of the International Congress of Mathematicians 1954, vol. 1, Amsterdam, 1957, pp. 315–333.

[9] Martin Kruskal, *Asymptotic theory of Hamiltonian and other systems with all solutions nearly periodic*, J. Mathematical Phys., 3 (1962), pp. 806–828.

[10] E. Lorenz, *Simplified dynamical equations and their use in the study of atmospheric prediction*, Report, Air Force Aerospace Center, Bedford, Massachusetts.

[11] J. Moser, *Perturbation theory for almost periodic solutions for undamped nonlinear dif-*

ferential equations, presented at the Symposium on Nonlinear Differential Equations, Colorado Springs, August 1961, to appear.

[12] ———, *Stability and nonlinear character of ordinary differential equations*, to appear.

[13] ———, *A new technique for the construction of solutions of nonlinear differential equations*, Proc. Nat. Acad. Sci. U. S. A., 47 (1961), pp. 1824–1831.

[14] J. C. OXTOBY AND S. M. ULAM, *Measure preserving homeomorphisms and metric transitivity*, Ann. of Math., 42 (1941), pp. 874–920.

[15] P. STEIN AND S. ULAM, *Quadratic Transformations, Part I*, Los Alamos Scientific Report LA-2305, May 1959.

[16] ——— AND ———, *Nonlinear transformation studies on electronic computers*, to appear in Rozprawy Mat.

[17] S. ULAM, *A Collection of Mathematical Problems*, Interscience, New York, 1960.

[18] ———, *Zur Masstheorie in der allgemein Mengenlehre*, Fund. Math., 16 (1930), pp. 140–150.

[19] ———, *Infinite models in physics*, Applied Probability, Proceedings of Symposia in Applied Mathematics, Vol. VII, McGraw-Hill, New York, 1957, pp. 51–66.

[20] ———, *On some statistical properties of dynamical systems*, Proceedings of 4th Berkeley Symposium, vol. 3, University of California Press, Berkeley, 1961, pp. 315–321.

[21] NORMAN J. ZABRUSKY, *Exact solution for the vibrations of a nonlinear continuous model string*, J. Mathematical Phys., 3 (1962), pp. 1028–1039.

[22] PAUL J. COHEN, *The independence of the continuum hypothesis*, Proc. Nat. Acad. Sci. U.S.A., 50 (1963), pp. 1143–1148; 51 (1964), pp. 105–110.

INSTYTUT MATEMATYCZNY POLSKIEJ AKADEMII NAUK

ROZPRAWY MATEMATYCZNE

XXXIX

P. R. STEIN and S. M. ULAM

Non-linear transformation studies on electronic computers

WARSZAWA 1964
PAŃSTWOWE WYDAWNICTWO NAUKOWE

For commentary to this paper [81], see p. 697.

INTRODUCTION

This paper will deal with properties of certain non-linear transformations in Euclidean spaces—mostly in two or three dimensions. In the main they will be of very special and simple algebraic form. We shall be principally interested in the iteration of such transformations and in the asymptotic and ergodic properties of the sequence of iterated points. Very little seems to be known, even on the purely topological level, about the properties of specific non-linear transformations, even when these are bounded and continuous or analytic. The transformations we study in this paper are in fact bounded and continuous, but in general many-to-one, i.e., not necessarily homeomorphisms. In one dimension such transformations are simply functions with values lying in the domain of definition; for example, if $f(x)$ is continuous and non-negative in the interval $[0, 1]$ and $\max[f(x)] \leqslant 1$, then $x' = f(x)$ is a transformation (¹) of the type considered. Even in one dimension, however, nothing resembling a complete theory of the ergodic properties of the iterated transformation exists. On the algebraic side, we study in this paper the invariant points (fixed points), finite sets (periods) and invariant subsets (*curves*) of these transformations—together with the means of obtaining them constructively. The topological properties of two (not necessarily one-dimensional) transformations $S(p)$, $T(p)$ are identical under a homeomorphism H: that is, when $S(p) = H[T[H^{-1}(p)]]$. When S and T are themselves homeomorphisms—and for *one dimension*—necessary and sufficient conditions for conjugacy are known (²). When S and T are one-dimensional, but not necessarily one-to-one, it is possible to give a set of necessary conditions for conjugacy; no meaningful sufficient conditions, however, are known.

For example, the set of fixed points of S has to be topologically equivalent to those of T. The same must hold for the set of periodic points, i.e., points such that the nth power of the transformation returns the point to its original position. The

(¹) Here and throughout the paper a primed variable always represents the value obtained on the next iterative step. In a more explicit notation, the above equation would read: $x^{(n+1)} = f(x^{(n)})$.

(²) J. Schreier und S. Ulam, *Eine Bemerkung über die Gruppe der topologischen Abbildungen der Kreislinie auf sich selbst*, Studia Math. 5 (1935).

attractive and *repellent* fixed points must correspond, etc. These conditions are known from the corresponding study of homeomorphisms. For many-to-one transformations one may generalize these conditions by considering the *tree* of a point. For a given transformation T we define the tree of a point P as the smallest set Z of points such that:

a) P belongs to Z.

b) If a point Q belongs to Z, then $T(Q)$ belongs to Z.

c) If Q belongs to Z, then all points of the form $T^{-1}(Q)$ belong to Z.

Obviously, for two transformations to be conjugate, the trees of corresponding points must be combinatorially equivalent and, in addition, their topological interrelations must be the same [3].

The present study was initiated several years ago [4] with the consideration of certain homogeneous, quadratic transformations which we called *binary reaction systems*. A typical example is the following:

$$\begin{aligned} x_1' &= x_2^2 + x_3^2 + 2x_1x_2 , \\ x_2' &= 2x_1x_3 + 2x_2x_3 , \\ x_3' &= x_1^2 , \end{aligned} \tag{1}$$

where we consider initial points P with coordinates x_1, x_2, x_3 satisfying:

$$0 \leqslant x_i \leqslant 1, \quad x_1 + x_2 + x_3 = 1 . \tag{2}$$

Since

$$x_1' + x_2' + x_3' = (x_1 + x_2 + x_3)^2 ,$$

the transformation (1) maps the two-dimensional region (2) into some sub-region of itself. The choice of these transformations was motivated by certain physical and biophysical considerations. For example, the set of equations (1) could be interpreted as determining the composition of a hypothetical population whose individuals are of three *types*, conventionally labeled 1, 2 and 3. The x_i would then represent the fraction af the total population which consists of individuals of type "i". The transformation can be thought of as mathematical transcription of the *mating rule*:

(3)

type 2 and type 2 produce type 1,
type 3 and type 3 produce type 1,
type 1 and type 2 produce type 1,
type 1 and type 3 produce type 2,
type 2 and type 3 produce type 2,
type 1 and type 1 produce type 3.

---

[3] One-dimensional transformations are considered in more detail in appendix I.

[4] Menzel, Stein and Ulam, *Quadratic Transformations*, Part I, Los Alamos Report LA-2305, May, 1959 (Available from the Office of Technical Service, U. S. Dept. of Commerce, Washington 25, D. C.).

For any assigned initial composition, i.e., any initial vector (x_1, x_2, x_3) satisfying (2), we may then ask: What is the final (or limiting) composition of the population after infinitely many "generations", that is, after infinitely many matings according to the scheme (3)?

In the present context, a *mating rule* can be defined as a system of three non-linear first-order difference equations of the form:

$$(4)\qquad \begin{aligned} x_1' &= f_1(x_1, x_2, x_3)\,, \\ x_2' &= f_2(x_1, x_2, x_3)\,, \\ x_3' &= f_3(x_1, x_2, x_3)\,, \end{aligned}$$

where each f_i is the sum of some subset of the six homogeneous monomials $x_1^2, x_2^2, x_3^2, 2x_1x_2, 2x_1x_3, 2x_2x_3$, and each such term must belong to one and only one f_i. Two transformations are called equivalent if they are conjugate under the (linear) transformation defined by a given permutation of the indices $1, 2, 3$. (This is the only linear homeomorphism which preserves the homogeneous quadratic character of the transformation.) Under this definition of equivalence, it turns out that there are 97 inequivalent transformations of the above type. It quickly becomes apparent that, despite their formal simplicity, these transformations are very difficult to study analytically, particularly if one is interested in their iterative properties. For example, for most initial points in the region of definition, the sequence of iterates generated by repeated application of the transformation given by equation (1) converges to a set of three points:

$$(5)\qquad \begin{aligned} p_2 &= T(p_1)\,, \\ p_3 &= T(p_2)\,, \\ p_1 &= T(p_3)\,. \end{aligned}$$

Using a standard terminology to be explained in detail below, we say that the "limit set" is a "period of order three". It is clear by inspection of transformation (1) that another limit set exists; if we write

$$p_1 = (x_1 = 1, x_2 = x_3 = 0), \qquad p_2 = (x_1 = x_2 = 0, x_3 = 1)\,,$$

then

$$(6)\qquad p_2 = T(p_1), \quad p_1 = T(p_2)\,.$$

In addition there is the algebraic fixed point of (1):

$$(7)\qquad p = T(p)\,.$$

The general initial vector, however, always leads to (5). Certain other quadratic transformations show an even more complicated behavior.

An example is the transformation:

$$(8)\qquad x_1' = x_2^2 + x_3^2 + 2x_1x_2, \quad x_2' = x_1^2 + 2x_2x_3, \quad x_3' = 2x_1x_3\,.$$

This bears a close formal relationship to (1); in fact, they differ only by the exchange of a single term. The limit sets, however, are quite different. Transformation (8) has an attractive fixed point with coordinates:

$$x_1 = \frac{1}{2}, \quad x_2 = \frac{\sqrt{2}}{4}, \quad x_3 = \frac{2-\sqrt{2}}{4}. \tag{9}$$

It also has a limit set of the type (6) with $p_1 = (1, 0, 0)$, $p_2 = (0, 1, 0)$. In this case, both limit sets are observed. It is found experimentally that the set of initial points leading to (9) is separated from those leading to the oscillatory limiting behavior (6) by a closed curve surrounding the fixed point (figure 2 of the reference in footnote 4). The analytical nature of this boundary curve remains unknown.

In view of the complicated behavior exhibited by these examples, we felt it would be useful to study these transformations numerically, making use of the powerful computational aid afforded by electronic computing machines. From one point of view our present paper may be looked on as an introduction, through our special problems, to modern techniques in *experimental* mathematics with the electronic computer as an essential tool. Over the past decade these machines have been extensively employed in solution of otherwise intractable problems arising in the physical sciences. In addition to solving the particular practical problem under consideration, this work has in some cases resulted in significant theoretical advances. Correspondingly, attempts to solve difficult physical problems have led to considerable improvements in the logical and technical design of computers themselves. In contrast, the use of electronic computers in pure mathematics has been relatively rare (5). This may be partly due to a certain natural conservatism; in our opinion, however, the neglect of this important new research tool by many mathematicians is due simply to lack of information. In other words, the average mathematician does not yet realize what computers can do. It is our hope that the present paper will help to demonstrate the effectiveness of high-speed computational techniques in dealing with at least one class of difficult mathematical problems. With this end in mind, we have devoted the first section of our paper to a brief discussion of how computing machines can be used to study problems in pure mathematics. Much of this section is introductory in character, and is meant primarily for those readers who have had no first-hand experience in the use of computers. It also includes, however, a description of the numerical techniques used in this study; these may be of interest even to seasoned practitioners.

---

(5) Perhaps the greatest computational effort has been expended on problems in number theory. See section I, footnote 2.

After our study of quadratic transformations in three variables [6], we decided to investigate the iterative properties of other classes of polynomial transformations. As a natural generalization of the quadratics described above, we consider transformations of the form:

$$x'_i = f_i(x_1, \dots, x_k) \qquad (i = 1 \text{ to } k)\,, \tag{10}$$

where the f_i are disjoint sums of the homogeneous monomials which arise on expanding the expression:

$$F = \Big(\sum_{i=1}^{k} x_i\Big)^m . \tag{11}$$

The number of such terms—each taken with its full multinomial coefficient—is

$$N_k^m = \binom{m+k-1}{k-1} . \tag{12}$$

By construction, $\sum_{i=l}^{k} f_i = F$, so that if we take

$$\sum_{i=l}^{k} x_i = 1, \qquad x_i \geqslant 0\,, \tag{13}$$

the (additive) normalization of the x_i is preserved. We are then dealing with positive transformations in a bounded portion of the Euclidean space of $k-1$ dimensions, i.e., just the hyperplane defined by (13). If $m = 2$, $k = 3$, these transformations are the 97 quadratics in three variables introduced above. The bulk of the present paper is devoted to the case $m = k = 3$, i.e., cubic transformations in three variables; there are 9370 independent transformations of this form. We have also examined the 34337 quadratic transformations in four variables, but our analysis of the results is not yet complete (January, 1963); for this case ($m = 2$, $k = 4$) we include only some statistical observations and a few interesting examples. These three cases — $m = 2$, $k = 3$; $m = 2$, $k = 4$; $m = 3$, $k = 3$ — are the only ones for which an exhaustive survey is at present feasible. For other values of m and k the number of transformations to be studied is much too large.

The determination of the exact number T_k^m of inequivalent transformations for arbitrary m and k is an unsolved combinatorial problem. It can, of course, be reduced to enumerating those transformations which are invariant under one or more opera-

---

(6) We shall not discuss this work here. Full details are contained in the reference given in footnote 4. That report contains, in addition, some fragmentary results on a few particular quadratic transformations in higherdimensional spaces.

tions of the symmetric group on the k indices, but no convenient way of doing this is known. The problem, however, is not of much practical significance. A *lover limit* T^{*m}_k to the number T^m_k of inequivalent transformations is given by:

$$T^{*m}_k = S_k^{N^m_k}, \tag{14}$$

where S^i_j is the Stirling number of the second kind. S^i_j is also the number of ways of putting i objects in j identical boxes, no box being left empty. This underestimates T^m_k by assuming in effect that each transformation has $k!$ non-identical copies, i.e., that no transformation is invariant under any permutation (except the identity). The following table illustrates the trend:

TABLE I

	N^m_k	T^{*m}_k	T^m_k
$m = 2,\ k = 3$	6	90	97
$m = 3,\ k = 3$	10	9 330	9 370
$m = 2,\ k = 4$	10	34 105	34 337
$m = 4,\ k = 3$	15	2 375 101	—
$m = 2,\ k = 5$	15	210 766 920	—
$m = 3,\ k = 4$	20	45 232 115 901	—

The T^m_k were obtained by direct enumeration—using, of course, all known short-cuts. For $m = 2$, $k = 4$, this enumeration was actually performed on a computer. In view of the huge values of the T^{*m}_k in the lower half of this table, it is unlikely that anyone will be interested in attempting a comprehensive numerical study of these transformations for values of m and k larger than those we have considered.

A general discussion of our results for the cubics in three variables and the quadratics in four variables is given in section II; the reader will also find there formal definitions of a few basic concepts and an explanation of the special terminology employed throughout the paper. Perhaps the most interesting result of this study is our discovery of limit sets of an extremely "pathological" appearance. The existence of such limit sets was quite unexpected (7)—and is indeed rather surprising in view of the essential simplicity of the generating transformations. Sections III and IV are concerned with the effect—on the iterative properties of our transformations—of two types of structural generalization. Specifically, in section III we consider the one-parameter generalization—called by us the "Δt-modification"—which consists in replacing equation (10) by:

$$x'_i = (1 - \Delta t)x_i + \Delta t f_i(x_1, \ldots, x_k) \quad (i = 1 \text{ to } k), \qquad 0 < \Delta t \leqslant 1. \tag{12a}$$

(7) Quadratic transformations in three variables apparently do not exhibit similar pathologies.

This generalization has the special property of leaving the fixed points of the transformation invariant, although their character—i.e., whether they are attractive or repellent—may be altered. The detailed discussion of the behavior of such transformations under variations of the parameter Δt is limited to the cubic case.

Section IV describes the result of introducing small variations in the coefficients of the monomials which make up the various f_i. Again we deal only with the cubic case, and indeed only with a few interesting examples chosen from our basic set of 9370 transformations. Let us denote the N_k^m monomials (e.g., x_1^3, $3x_2x_3^2$, $6x_1x_2x_3$, ...) in the expansion of (11) by the symbol M_j, $j = 1$ to N_k^m. The assignment of a particular index to a particular monomial is arbitrary.

Then we have

(13a) $$f_i = \sum_{j=1}^{N_k^m} d_{ij} M_j\,, \qquad 1 \leqslant i \leqslant k\,,$$

with

(14a) $$d_{ij} = 1 \text{ or } 0\,,$$

(15) $$\sum_{i=1}^{k} d_{ij} = 1\,.$$

The generalization then consists in relaxing the restriction (14). If this were done subject only to the condition that the d_{ij} all be non-negative, we should be dealing with a $(k-1)N_k^m$ parameter family of positive, bounded, homogeneous polynomial transformations. At present nothing significant can be said about this class as a whole. As explained in section IV, our procedure has been to study one-parameter families of transformations which are in a certain sense "close" to some particular transformation of our original set [8].

In section V we give a brief, heuristic discussion of the connection between our transformations—which are really first-order non-linear difference equations—and differential equations in the plane. Our conclusion is that the connection is not, in fact, very close, and that the techniques so far developed for treating non-linear differential equations do not seem suitable for handling the problems discussed in this paper.

---

[8] Some analogous but rather unsystematic investigations were carried out on quadratics in three variables, and are contained in the report cited in footnote 4. Subsequent to the appearance of that report we made some studies (unpublished) on quadratics with randomly chosen positive coefficients satisfying (15). For quadratics (at least in three variables), the conclusion seems to be that such randomly chosen transformations are most likely to lead under iteration to simple convergence for almost all initial points.

The final section of our main text—section VI—contains a description of a class of piece-wise linear transformations on the unit square. These transformations exhibit interesting analogies with our polynomial transformations in three variables. Relatively little work has been done on this "two-dimensional broken linear" case, but the preliminary results we report seem to indicate that a detailed study might prove worthwhile.

There are two appendices: Appendix I is largely devoted to an extended discussion of certain non-linear transformations in one dimension, on the unit interval. Some of these are special cases of our cubics in three variables; others originated independently of our principal study. It is perhaps rather surprising how little can be said theoretically even about this simple one dimensional case. It turns out that some of the same phenomena are observed in one dimension as are found in the plane—e.g., the apparently discontinuous behavior of limit sets as a function of a monotonically varying parameter. Of course, the repeated iteration of a one-dimensional transformation is a much simpler matter than the corresponding process in several dimensions. However, as we soon discovered, great care must be taken to avoid the phenomenon of "spurious convergence". This point is discussed in some detail and a few—rather alarming—examples are given.

Appendix II contains the bulk of the photographic evidence—including the "pathology" of the limit sets—on which the discussion of sections III and IV is based. These pictures, together with others scattered throughout the main body of the text, constitute in a sense the unique contribution of this paper. In retrospect, it seems unlikely that our investigation could have been successfully carried out without the visual aid afforded by the oscilloscope and the polaroid camera. Put in the simplest terms, unless one knows precisely what one is looking for, mere lists of numbers are essentially useless. Automatic plotting devices however,—such as the oscilloscope—allow one to tell at a glance what is happening. Very often the picture itself will suggest some change in the course of the investigation—for example, the variation of some hitherto neglected parameter. The indicated modification can often be effected in a few seconds and the result observed on the spot (9).

Visual display is of very great value when one is in effect studying sets of points in the plane; when one passes to three dimensions automatic plotting ceases to be merely a convenience and becomes essential.

---

(9) This interaction of man and computing machine has sometimes been referred to as "synergesis". See for example, S. M. Ulam, *On some new possibilities ... computing machines*, I. B. M. Research Report 68, 1957. I. B. M. Corporation, Yorktown Heights, New York.

A glance at our pictures of three-dimensional limit sets—the result of iterating certain quadratic transformations in four variables—should convince even the most skeptical reader. In our opinion, it would be virtually impossible to make sense out of a mere numerical listing of coordinates of the points plotted in these photographs.

Of the many who have helped with this work, there are three to whom we are particularly indebted: Cerda Evans, Verna Gardiner, and Dorothy Williamson. These ladies did the actual coding and supervised all the machine calculations. Without their help this paper could not have been written.

I. THE ROLE OF THE COMPUTING MACHINE

1. The use of electronic computers for the solution of complicated or tedious problems—usually of practical origin—is by now familiar. Typical computer tasks are: the evaluation of integrals, the solution of large systems of linear equations, the solution of minimax problems (linear programming), the treatment of complicated boundary value or initial value problems, etc. One of the more impressive jobs that computers have done is to calculate the time history of immensely complicated physical systems (e.g. involving hydrodynamical motions, magnetic fields, etc.). Recently there has been considerable interest in using computers to attack problems of a less applied nature, for example those arising in combinatorial analysis (10) and number theory (11). This work often takes on an experimental flavor; such experimentation has led to results of considerable interest, for example, the construction of certain types of mutually orthogonal latin squares (12). Computers can also be used to investigate formal mathematical systems (13), to reduc symbolic expressions (14), and—with less success—to study games of "skill" like chess (15).

The use of computing machines that we describe in the present paper differs in two respects from the examples just cited. On the one hand, our study is not essentially combinatorial in character, but falls rather in the domain of algebra and real variable function theory. On

---

(10) See, e.g., Proceedings of Symposia on Applied Mathematics, Vol. X, American Mathematical Society (1958), or the article by Marshall Hall, Jr. in Surveys of Applied Mathematics, Vol. IV, 1958.

(11) In the absence of a comprehensive reference, we refer the reader to the recent ssues of Mathematics of Computation (1960-1962). See e.g. Vol. 16, No. 80, October 1962, especially the article by D. H. Lehmer, et al.

(12) See the article by E. T. Parker in Proceedings of Symposia in Pure Mathematics, Vol. VI, American Mathematical Society (1962). Parker's original construction supplied the final step in the disproof of a famous conjecture by Euler.

(13) See H. Gelernter et al., Proceedings of the Western Joint Computer Con ference, San Francisco, May 1960, pp. 143-149.

(14) M. B. Wells, Proceedings of the IFIP Congress (1962) (to be published.

(15) C. Shannon, Philosophical Magazine 41, March, 1950. — Kister, Stein, Ulam, Walden and Wells, Journal of the Association for Computing Machinery, Vol. 4, Number 2, April 1957.

the other hand, we are not attempting to "solve" some well-defined problem; instead we investigate via repeated trials the asymptotic properties of certain non-linear transformations, usually without any advance knowledge of what we may find in a given case. Even "after the fact", so to speak, it is difficult to classify these asymptotic properties in a meaningful fashion; the broadness of the categories we employ for this classification ([16]) is merely a measure of our lack of insight into the structure of the observed limit sets.

Faced with this situation, one may ask the question: how does one recognize "convergence"—i.e., the existence of an invariant set—when one has no a priori numerical criteria to apply? We can only supply a partial answer to this question, but that answer has the advantage of simplicity, viz: "use your eyes". The practical application of this "technique" involves, of course ([17]), the use of automatic plotting devices.

2. Roughly speaking, computing machines are devices which perform the four elementary arithmetic operations on numbers in a certain—not necessarily simple—sequence. This sequence of operations is called the "program", and consists of a set of logical commands, both of the sequential ("do this and then do that") and of the branching ("if this holds, then do that") type. The program is composed by an investigator (the "programmer") and must therefore reflect his own limitations. Nevertheless, the machine may easily produce results quite unanticipated by the programmer, even if the program is essentially deterministic in nature ([18]). A classic example—which happens to be relevant here—is the step-by-step application of some recurrence relation which generates a sequence whose trend the programmer cannot determine in advance. As an example, we may cite the following one-step recursion in a single variable:

$$y_{n+1} = w_n(3 - 3w_n + \sigma w_n^2)\,, \qquad w_n \equiv 3y_n(1 - y_n)\,. \tag{1}$$

Given some initial $0 < y_0 < 1$, we may ask: what is the result of applying the rule (1) N times, where N is some larger number, say 10^5? This particular transformation is discussed in detail in appendix I; here we quote three examples for the purpose of illustration.

(a) If $\sigma = 0.99004$, then for almost all y_0 the sequence of iterates produced by (1) converges (in $<10^5$ steps) to a period of order 14.

([16]) See section II.

([17]) Hand plotting is in general highly impractical, and clearly relinquishes the principal advantage of machine computation: SPEED.

([18]) Strictly speaking, all programs used on digital computers are deterministic in nature: even when random numbers are employed, these are generated according to some fixed algorithm so that the sequence is in principle known.

(b) If $\sigma = 0.99005$, the corresponding limit set is a period of order 28 ([19]).

(c) If $\sigma = 0.99008$, no finite period is observed after $N = 5 \times 10^5$ steps.

So far as we are aware, this behavior could not be predicated by current analytical or algebraic techniques. Such phenomena are easy to study on a computer, however, because of the great speed with which it can carry out the (relatively simple) operations implied by an expression such as (1) above. In fact, 200000 iterations of this transformation takes slightly less than one minute on a really fast computer ([20]).

3. As we mentioned in the introduction, the principal content of this paper is the study of the asymptotic properties of certain non-linear transformations of relatively simple form. This means that, if T is such a transformation, we examine the sequence

$$T(p),\ T^2(p),\ T^3(p),\ \ldots$$

for various initial points p lying in the domain of T. The mathematical object of interest to us is the set (or sets) of points to which these sequences converge. In the absence of any general analytical technique for calculating these "limit sets", we must have recourse to "brute-force" methods.

Some non-linear transformations which appear morphologically similar to those considered here can in fact be completely analyzed by elementary methods. We discovered one such case in the course of some earlier work on biological systems. It is described in our report on quadratics in three variables (see the appendix of the reference cited in footnote (4) in the introduction). We restate these special results here:

Let

$$C_i' = \sum_{k,m=1}^{N} \gamma_i^{km} C_k C_m\,, \qquad 1 \leqslant i \leqslant N\,, \tag{2}$$

with coefficients satisfying

$$\begin{aligned} \gamma_i^{km} &= \gamma_i^{mk} > 0, \quad \min(k, m) \leqslant i \leqslant \max(k, m)\,, \\ \gamma_i^{km} &= \gamma_i^{mk} = 0, \quad \text{otherwise}\,, \end{aligned} \tag{3}$$

$$\sum_{i=m}^{k} \gamma_i^{km} = 1\,, \tag{4}$$

$$\sum_{i=m}^{k} i\gamma_i^{km} = \frac{m+k}{2}\,. \tag{5}$$

([19]) These results were found by using the IBM "STRETCH" computer. The periods are exact to within the accuracy of that machine, i.e., 48 binary digits (~15 decimals). See further in appendix I.

([20]) This figure applies to "STRETCH" and includes all additional "diagnostic" operations such as checking for "convergence", etc.

We normalize the C_i by

$$(6) \qquad 0 \leqslant C_i \leqslant 1, \quad \text{all } i, \quad \sum_{i=1}^{N} C_i = 1\,.$$

This property is clearly preserved under iteration. With the coefficients defined as above, there exists a linear invariant:

$$(7) \qquad \sigma \equiv \sum_{i=1}^{N-1} (N-i)\,C_i = \sum_{i=1}^{N-1} (N-i)\,C_i'\,.$$

Given an initial vector $(C_1^{(0)}, C_2^{(0)}, \ldots, C_N^{(0)})$ whose coordinates satisfy (6), σ is explicitly determined. It can then be shown that every initial vector satisfying (6) converges to a definite fixed point which is determined as follows. For the given value of σ, there is one value of the index j such that

$$(8) \qquad N-j \geqslant \sigma > N-j-1\,.$$

The fixed point is then explicitly given by

$$(9) \qquad C_j = \sigma-(N-j-1), \quad C_{j+1} = N-j-\sigma, \quad \text{all other} \quad C_i = 0\,.$$

Note that the fixed point is independent of the values of the coefficients γ_i^{km}.

As simple examples of coefficients satisfying (3), (4), and (5), we may mention

$$(10) \qquad \gamma_i^{km} = \frac{1}{2^{|k-m|}} \binom{|k-m|}{i-\min(k,m)}$$

and

$$(11) \qquad \gamma_i^{km} = \begin{cases} \dfrac{1}{|k-m|+1}, & \text{if } \min(k,m) \leqslant i \leqslant \max(k,m)\,, \\ 0, & \text{otherwise.} \end{cases}$$

For a fuller discussion of this transformation and its possible applications, we refer the reader to the original report.

The term "brute-force" refers to the fact that, in order to determine the convergence properties of some transformation T belonging to our class, we must in general actually evaluate $T^k(p)$ for $k = 1, 2, \ldots, N$, where N is likely to be quite large—sufficiently large, that is, so that we can observe convergence [21] to the limit set. To make matters clear, let us consider a specific example. We choose the cubic transformation

$$(12) \qquad \begin{aligned} x_1' &= x_3^3 + 3x_1x_3^2 + 3x_3x_1^2 + 6x_1x_2x_3\,, \\ x_2' &= x_1^3 + 3x_2x_3^2 + 3x_3x_2^2\,. \\ x_3' &= x_2^3 + 3x_1x_2^2 + 3x_2x_1^2\,. \end{aligned}$$

We taken some initial point $p = (x_1, x_2, x_3)$ whose coordinates satisfy:

$$(13) \qquad x_1 + x_2 + x_3 = 1, \quad 0 \leqslant x_i \leqslant 1, \ i = 1, 2, 3\,.$$

[21] "Convergence" must be of course understood in some approximate numerical sense. Our usual criteria are set forth in the next sub-section.

The program then instructs the computing machine to evaluate the right hand side of (12), thus producing a new point $p' = (x_1', x_2', x_3')$; the coordinates of p' again of course satisfy (13). p' is then set to p, and the process is repeated. The iteration proceeds in this fashion until either some finite limit set is found ([22]) or an invariant set—presumably infinite—is "observed". The observation consists in looking at successive groups of consecutive iterates—in practice we have usually taken 900 points at a time—until no qualitative visual change is noted over a sample of several successive such groups of points. Since the transformation (12) is really two-dimensional, we may plot the successive points p in the plane. Accordingly, we define new coordinates S, a by the linear transformation ([23])

$$S = \frac{1 + x_1 - x_3}{2}, \qquad a = \frac{x_2}{2}. \tag{14}$$

The domain of the transformation is then the 45° isosceles triangle:

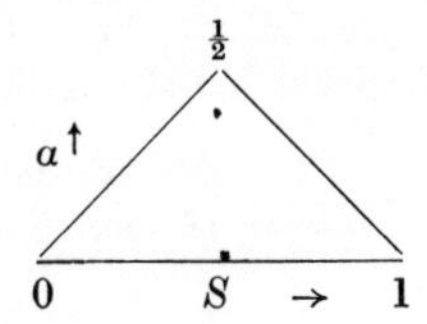

In terms of these new variables, (12) takes the form:

$$\begin{aligned} S' &= -\tfrac{1}{2} S^3 - \tfrac{15}{2} S^2 a - \tfrac{3}{2} S a^2 + \tfrac{3}{2} a^3 + 6Sa - 3a + 1 \equiv F(S, a), \\ a' &= \tfrac{1}{2} S^3 + \tfrac{3}{2} S^2 a + \tfrac{3}{2} S a^2 - \tfrac{7}{2} a^3 - 6Sa + 3a \equiv G(S, a). \end{aligned} \tag{15}$$

The computer is instructed to store 900 successive points

$$p(S^{(n)}, a^{(n)}),\ p(S^{(n+1)}, a^{(n+1)}), \ldots,$$

and, when the last point has been calculated, to plot all 900 points on our oscilloscope screen ([24]). If we choose, we may then photograph the resulting pattern with a polaroid camera. Such a photograph is shown in figure 1. Here one sees 900 successive high-order iterates ($n = 2700$ to 3600) of the initial point, $S = \frac{1}{2}$, $a = 0.17$. For convenience, the triangle of reference is also shown.

([22]) See the next sub-section.

([23]) These are the coordinates employed in our earlier work on quadratics in three variables; we have retained them more for historical reasons than for any particular advantage they may possess.

([24]) The points are actually plotted in the order in which they are calculated, the whole pattern being replotted as many times as we wish. Actually, the plotting of 900 points is effectively instantaneous so far as the human eye is concerned. If we wish to *see* the points plotted in succession, we must introduce artificial time delays between the plotting of successive points.

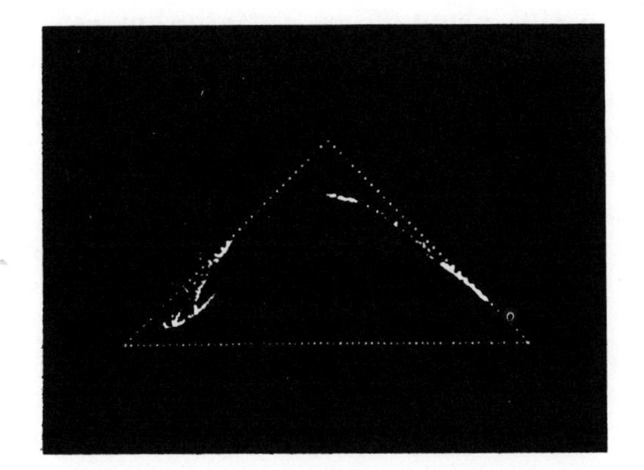

Fig. 1

This calculation—as well as all others which produced the photographs in this paper—was performed on the Los Alamos Laboratory's MANIAC II computer ([25]). MANIAC II requires about 15 seconds to calculate 900 iterates of a point by repeated applications of a cubic transformation like (12) above. This figure includes the time spent in examining the successive points for simple convergence, as well as other "diagnostic" operations ([26]). The actual numerical values of the coordinates may be printed out whenever desired by simply flipping a switch. On MANIAC II a decimal number is normally limited to eight significant figures. In the present paper, when there is occasion to quote numerical values obtained from MANIAC II print-outs, we shall generally reproduce them to seven figures without further specifying their accuracy.

Computer programs are, of course, not limited to generating sequences of numbers from an iterative formula such as (12). A considerable amount of sophistication can be incorporated into such a program so as to allow the machine to make "decisions" in the course of the calculation. It can, in fact, examine any property or any functional of the data that the programmer can describe in appropriate terms. One problem that is met with frequently in this work is to determine the points in a sequence of iterates that lie closest to some chosen point, say within some chosen angle or set of angles. This sort of experiment is frequently of help in elucidating the local structure of a complicated limit set. Then again we may want to determine the average values of S and a, i.e. ergodic means, taken over the sequence. To achieve any sort of accuracy in such problems ([27]) we may be required to go to 50000 or even 100000 iterations. One saving feature is that several such diagnostic experiments can be carried out simultaneously. There are, however, special questions that must be dealt with by special programs. One such question arises in connection with our illustrative transformation (12). The complicated limit set shown in figure 1 is not the only one observed. This transformation has an attractive fixed point at:

$$(16) \qquad S = F(S, a) = 0.6259977, \qquad a = G(S, a) = 0.1107896;$$

indeed, the eigenvalues of the jacobian matrix ([28]) evaluated at this point are complex, with $|\lambda|^2 = 0.4366967$. Consequently, there must be a neighborhood of this point in which all sequences will converge to it. The only

([25]) For the use of other computing machines in this work, see the next sub-section.

([26]) For reasons of accuracy, the calculations are performed in the x_i coordinates; the transformation (14) to the S, a coordinates is carried out only for plotting purposes.

([27]) More properly, to have any confidence in the results. The accuracy cannot always be satisfactorily estimated.

([28]) The criterion for the nature of a fixed point is discussed in Section III.

way to find the boundary of this neighborhood is by trial and error. This is a time-consuming job, even for an electronic computer; if one picks a point close to the boundary of the region of convergence, several hundred—or even several thousand—iterations may be required before one can tell whether the chosen point lies inside or outside the region. Figure 2 (see page 24, section II) shows the approximate boundary for the present case, drawn through 107 experimentally determined points. One of these is known to one part in 10^7, while the others have been determined only to 1 part in 10^4 [29].

4. General procedure

a. *Cubic transformations in three variables.* Enough has been said above to make clear the necessity of using an electronic computer in such investigations. We must now say something about the systematic aspects of the study. All 9370 cubic transformations were initially studied on an IBM 7090 [30]. First a complete list of inequivalent cubics was prepared—this was also done on the 7090, incidentally serving to check our original pencil-and-paper enumeration. Then by a completely automatic procedure, each transformation was taken in turn and four randomly-generated initial points were each taken as the start of an iterative sequence. For each point the iteration was continued until either convergence to a finite set of points was "observed" or 10000 iterations had been performed. By "observed" we mean that the machine sensed convergence to a fixed point or to a finite period of order $\leqslant 300$. More precisely, the computer was programmed to test whether the following conditions was satisfied

$$|x_i^{(n_1)} - x_i^{(n)}| < 10^{-7}, \qquad i = 1, 2, 3. \tag{17}$$

If (17) is satisfied, a finite limit set has been reached to within the indicated accuracy. For $n = n_1 + 1$ this means convergence to a fixed point ("simple convergence"). Otherwise, the limit set is a period of order $n - n_1$. In practice, values of the x_i were stored at fixed time steps $n_1 = 300$, 600, ..., the test (17) being performed on each step. If "convergence" was found, the appropriate values of the x_i were printed out and the next random initial point was used, etc. If no such convergence was found after 10000 steps, the values of the iterates for the last few steps were printed, and the computer proceeded as before.

When all the cubic transformations had been studied in this fashion, the "interesting" cases—i.e., those in which no convergence was ob-

[29] The point $S = \frac{1}{2}$, $a = 0.2952833$ lies in the region of convergence, while the point $S = \frac{1}{2}$, $a = 0.2952834$ gives rise to a sequence which converges to the class IV limit set (see definition in section II).

[30] This computer is approximately 5 times as fast as MANIAC II.

served—were examined one by one on MANIAC II, where the visual oscilloscope display could be consulted. Many cases of apparent non-convergence turned out in fact to be convergent with the iteration carried further. It should be stressed that the restriction to 10000 iterations, which we imposed in the course of the systematic, fully automatic survey of all cubic transformations, was merely one of convenience; without some such reasonable limitation, the automatic survey would have taken too long. The same remark applies to the decision that only four randomly-generated initial points be taken for each case. Past experience has shown that this last restriction is not unreasonable when a complete survey of transformations is contemplated. By this we mean that the behavior of an arbitrary transformation of our class is "likely" to be defined even if iterates of only four random points are studied. To be sure, in some cases the limit set depends in a very complicated way on the initial point; for such a transformation this crude sampling technique is not adequate. In these cases, however, the four random trials are likely to produce two different limit sets; this in itself is an indication that the transformation in question should be studied in more detail.

For the detailed examination of a given transformation, many relatively sophisticated MANIAC programs are available. We may, in effect, study any properties of the transformation that seem of interest. Typically, these may include:

1. Determination of non-attractive fixed points (see section III).

2. Checking for periodicity.

3. Exhibiting some qualitative properties of the mapping, e.g., by showing the images under the transformation of a family of lines.

4. Determining the dimensions of the limit set.

5. Verifying that low-order periods are attractive (see section III).

6. Examining the dependence of the limit set on the initial point.

We cannot expatiate here on the actual procedures involved; sufficient to say that the use of visual display (i.e., the oscilloscope plot) is an essential tool in all this analysis.

b. *Quadratic in four variables.* All (34337) inequivalent transformations of this class were studied by the same fully automatic method as that used to study the cubics. For this purpose a faster machine than the IBM 7090 was clearly required; we were fortunate enough to have access to the IBM 7030 "STRETCH" computer, which is approximately 4 times as fast as the 7090 and 20 times as fast as MANIAC II. Only partial results are reported in section II, since our analysis of the STRETCH print-outs is not complete.

The detailed study of a given quadratic in four variables is more difficult than the corresponding analysis for the three variable cubics: the domain is three-dimensional, being in fact the tetrahedron defined by

$$\sum_{i=1}^{4} x_i = 1, \quad 0 \leqslant x_i \leqslant 1,\ 1 \leqslant i \leqslant 4.$$

Thus a meaningful visual display involves plotting some properly chosen projection of the three-dimensional limit set. In some cases it may require several trials before an appropriately "revealing" viewing angle is found; consequently it was not feasible to plot every potentially "interesting" limit set in this fashion, and some sort of selective procedure had to be resorted to. The method we chose was to look at three plane projections first—e.g., x_1 versus x_2, x_1 versus x_3, and x_2 versus x_3. It turns out that one soon develops a feeling for the "interesting" case even without being able to build up an image of the actual three-dimensional configuration from the plane "slices". More serious than this purely technical difficulty is that resulting from the generally more complicated dependence of the limit set on the initial point: it turns out that in these transformations one is much more likely to miss something by restricting one's self to a few randomly-generated initial points. At the present time, lacking any local or structural criteria for the prediction of asymptotic behavior, we see no way to overcome this difficulty.

II. LIMIT SETS

1. Abbreviated notation for transformations. In order to have a convenient way of referring to a particular transformation without having to reproduce its explicit form, we introduce at this point a simple shorthand notation. As already noted in the introduction, our cubic transformations in three variables may be written in the form:

$$x_i' = \sum_{j=i}^{10} d_{ij} M_j, \quad i = 1, 2, 3, \tag{1}$$

with

$$d_{ij} = 0 \text{ or } 1, \quad \text{all } i, j, \tag{2}$$

and

$$\sum_{i=1}^{3} d_{ij} = 1, \quad \text{all } j, \tag{3}$$

where the M_j are the separate terms in the expansion of $(x_1+x_2+x_3)^3$. We now choose the following conventional ordering of the M_j:

$$\begin{aligned} &M_1 = x_1^3, \quad M_2 = x_2^3, \quad M_3 = x_3^3, \quad M_4 = 3x_1x_2^2, \quad M_5 = 3x_1x_3^2, \\ &M_6 = 3x_2x_1^2, \quad M_7 = 3x_2x_3^2, \quad M_8 = 3x_3x_1^2, \quad M_9 = 3x_3x_2^2, \quad M_{10} = 6x_1x_2x_3. \end{aligned} \tag{4}$$

Any cubic transformation of our class is then completely determined by specifying which terms M_j—or, equivalently, which indices j—appear in the first two lines of the schema (1). Let us call the set of indices belonging to the first line C_1 and those belonging to the second line C_2; C_3 is of course the complement of C_1+C_2 with respect to the full set $\{1, 2, \ldots, 10\}$ and need not be written down. Thus, for example, the transformation:

$$\begin{aligned} x_1' &= x_3^3 + 3x_1x_2^2 + 3x_1x_3^2 + 3x_2x_3^2 + 3x_3x_2^2 + 6x_1x_2x_3, \\ x_2' &= x_1^3 + 3x_3x_1^2, \\ x_3' &= x_2^3 + 3x_2x_1^2 \end{aligned} \tag{5}$$

would appear in the form:

$$T_{C_1C_2}: \quad \begin{aligned} C_1 &= \{3, 4, 5, 7, 9, 10\}, \\ C_2 &= \{1, 8\}. \end{aligned} \tag{6}$$

An analogous notation may be adopted—mutatis mutandis—for quadratics in four variables. Any such transformation can be written in the form:

$$(7)\qquad \begin{aligned} x'_i &= \sum_{j=1}^{10} d_{ij} F_j\,, \qquad i = 1, 2, 3, 4\,, \\ d_{ij} &= 0 \text{ or } 1, \qquad \text{all } i, j\,, \\ \sum_{i=1}^{4} d_{ij} &= 1, \qquad \text{all } j\,. \end{aligned}$$

Our conventional assignment of indices to the F_j is as follows:

$$(8)\qquad \begin{matrix} F_1 = x_1^2, & F_2 = x_2^2, & F_3 = x_3^2, & F_4 = x_4^2, & F_5 = 2x_1x_2, \\ F_6 = 2x_1x_3, & F_7 = 2x_1x_4, & F_8 = 2x_2x_3, & F_9 = 2x_2x_4, & F_{10} = 2x_3x_4\,. \end{matrix}$$

Let Q_k denote the set of indices belonging to the kth line of the schema (7). Then any such transformation is specified by writing down three of the four Q_k thus:

$$(9)\qquad T_{Q_1Q_2Q_3}:\quad \begin{aligned} Q_1 &= \{2, 8, 9\}\,, \\ Q_2 &= \{3, 7, 10\}, \\ Q_3 &= \{5, 6\} \end{aligned}$$

represents the transformation

$$(10)\qquad \begin{aligned} x'_1 &= x_2^2 + 2x_2x_3 + 2x_2x_4\,, \\ x'_2 &= x_3^2 + 2x_1x_4 + 2x_3x_4\,, \\ x'_3 &= 2x_1x_2 + 2x_1x_3\,, \\ x'_4 &= x_1^2 + x_4^2\,. \end{aligned}$$

This notation will be used extensively throughout the paper.

2. Limit set terminology. By a *limit set* $L_p(T)$ we shall mean the set of all points of the region of definition ([31]) which are limit points, in the ordinary sense, of the set $T^n(p)$, $n = 1, 2, \ldots$, for fixed p. It may happen that $L_p(T)$ is independent of the initial point p; $L_p(T) \equiv L(T)$ could then be called the limit set of T. In general, $L(T)$ will only be defined for interior points p, since points on the boundary frequently ([32]) behave in a rather special way.

([31]) For cubics this is the S, a triangle introduced in section I; for quadratics it is the tetrahedron

$$\sum_{i=1}^{4} x_i = 1, \qquad x_i \geqslant 0\,.$$

([32]) I.e. often enough to make it worthwhile excluding them in the definition of $L(T)$.

Thus, for example, if p_0 is a unique fixed point of the transformation: $T(p_0) = p_0$, and if the iterated images $T^n(p)$ of all interior points p converge to p_0, then $L(T) \equiv \{p_0\}$. If $p_0, p_1, \ldots, p_{k-1}$ form a system S of k points such that $T(p_i) = p_{i+1}$, $i = 1, 2, \ldots \pmod{k}$, and if for all interior points p, $\lim_{n\to\infty} T^{nk}(p)$ is one of these points, then $L_p(T) \equiv S$.

It might happen that the interior points divide into a finite number of classes $C_1, C_2, \ldots, C_r$ such that for all points p belonging to the same class $L_p(T)$ forms the same set; we should then have a finite number of limit sets $L_1, L_2, \ldots, L_r$. Some of these may contain a finite number of points, others may be infinite. For convenience we shall usually refer to a finite limit set containing k distinct points as a *period of order* k.

Although a given finite limit set belonging to some transformation T may legitimately be considered a "property" of that transformation, it is in no sense characteristic; many different transformations of our type may possess the same limit set, even for the same set of initial points. It should also be stressed that not every set of points $S \equiv \{p_1, \ldots, p_k\}$ such that $T(p_i) = p_{i+1}$ ($i \bmod k$) is properly a limit set. Such a set of points—each of which is a solution of the equation $T^k(p) = p$—must have the additional property that there exists a set of initial points whose iterated images converge to S. Finite sets S which have this property are conventionally termed *attractive*. Thus, we should properly refer to a finite limit set of k points as an *attractive period of order* k. In the sequel we shall usually omit the word *attractive* when the context makes it clear that this is what is meant.

There is, of course, no *structure problem* so far as finite limit sets are concerned; they are completely described by giving the coordinates of their constituent points. For infinite limit sets, the situation is different. On the basis of our numerical work alone, we cannot say with certainty that our transformations have such limit sets; the sets may in fact be finite (with an enormous number of points in them), but the presumption that they are infinite is very strong. For any observed *infinite* limit set we can at most say that it is not a period of order less than some very large k. Granting, however, that we are dealing with infinite sets, and that we may infer some of their properties by examining a sufficiently large finite sub-set [33], we may attempt to classify them according to their macroscopic morphological properties.

3. Infinite limit sets for cubics in three variables. On the basis of our empirical study of cubic transformations, we may make a rough division of infinite limit sets into four classes:

---

[33] This assumption underlies all our numerical work.

Class I. This includes all limit sets that appear to have the form of one or more closed curves. Figures 3 through 6 will serve as examples of this class. The detailed structure of these "curves" has been studied numerically in some cases, but there are as yet no theoretical arguments to the effect that these are really one-dimensional continua.

To illustrate one type of numerical study that we have carried out on these limit sets, we cite the case of figure 3. This shows the "infinite" limit set $L(T)$ belonging to the transformation:

$$T_{C_1C_2}: \quad \begin{aligned} C_1 &= \{2, 5, 7, 9, 10\}, \\ C_2 &= \{1, 3, 6, 8\}. \end{aligned} \tag{11}$$

In the S, a coordinates, this takes the form:

$$\begin{aligned} S' &= \tfrac{3}{2}S^3 - \tfrac{3}{2}S^2a - \tfrac{15}{2}Sa^2 + \tfrac{23}{2}a^3 - 3S^2 - 3a^2 + \tfrac{3}{2}S + \tfrac{3}{2}a + \tfrac{1}{2}, \\ a' &= -\tfrac{3}{2}S^3 + \tfrac{3}{2}S^2a - \tfrac{9}{2}Sa^2 + \tfrac{1}{2}a^3 + 3S^2 + 3a^2 - \tfrac{3}{2}S - \tfrac{3}{2}a + \tfrac{1}{2}. \end{aligned} \tag{12}$$

There is a (repellent) fixed point at:

$$S_0 = 0.6149341, \quad a_0 = 0.1943821. \tag{13}$$

To six decimal places, the overall bounds on the curve are [34]

$$\begin{aligned} S_{\max} &= 0.816878 \quad \text{at} \quad a = 0.058022, \\ S_{\min} &= 0.411270 \quad \text{at} \quad a = 0.204391, \\ a_{\max} &= 0.435861 \quad \text{at} \quad S = 0.552246, \\ a_{\min} &= 0.017750 \quad \text{at} \quad S = 0.728386. \end{aligned} \tag{14}$$

To five decimal places, the average value of coordinates is found to be [34]

$$\bar{S} \equiv \frac{1}{N}\sum_{i=1}^{N} S^{(i)} = 0.62231, \quad \bar{a} \equiv \frac{1}{N}\sum_{i=1}^{N} a^{(i)} = 0.20772. \tag{15}$$

This set $L(T)$ is the only infinite limit set the transformation seems to possess (the pair $(S = 1, a = 0)$, $(S = \frac{1}{2}, a = \frac{1}{2})$ turns out to be an attractive period for this transformation). For "most" initial points, the sequence of iterates converges to $L(T)$. If we choose as our initial point some $p \subset L(T)$, the curve will be traced out by successive images of p, though not in a continuous fashion. If, however, we look only at successive iterates of the 71st power of T: $T^{(71)}$, the curve is indeed generated in a relatively continuous fashion; the successive points $T^{(71n)}(p)$, $n = 1, 2, \ldots$, lie close to each other and trace the curve in a clockwise sense. This is illustrated in figure 7, where 246 successive values of $T^{(71)}(p)$ are plotted. It is striking that the non-uniform density of points along the curve—as shown in figure 3—is reproduced by this sequence of iterates. We are thus led to the conjecture that $L(T)$ and $L(T^{(71)})$ coincide. It is, of course, by no means generally true that $T^{(k)}$ and T will have the same limit set for an arbitrary T of our type (cf. the case of periodic limit sets where k is a multiple of the period). Further experiments have convinced us that $L(T) \equiv L(T^{(k)})$ for all k in this case. If this is so, the set is certainly infinite. That it is a continuum is also very probable.

---

[34] These results were obtained by carrying out $N = 96000$ iterations, starting rom a point on the curve with coordinates: $S = 0.5841326$, $a = 0.4125823$.

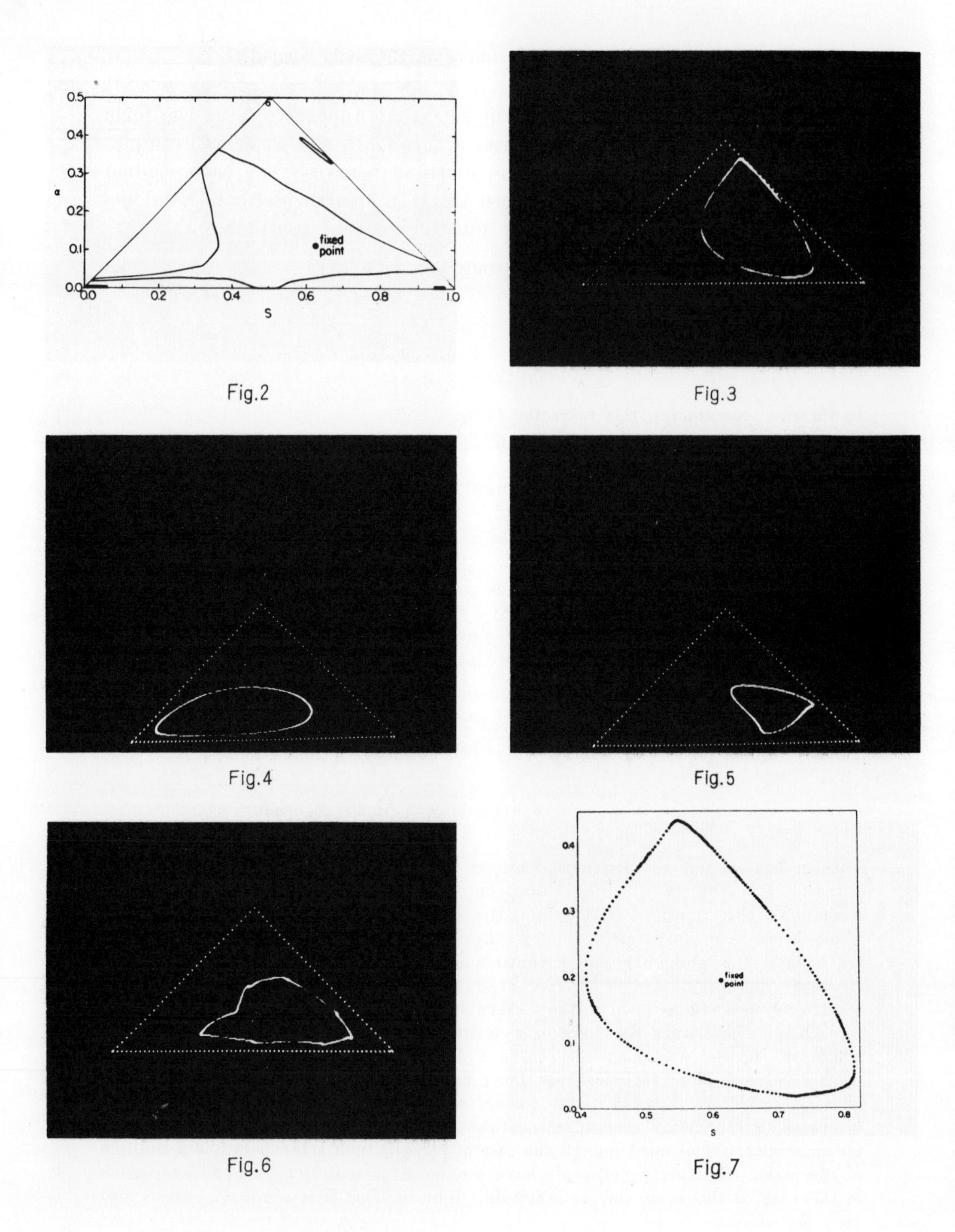

Fig.2

Fig.3

Fig.4

Fig.5

Fig.6

Fig.7

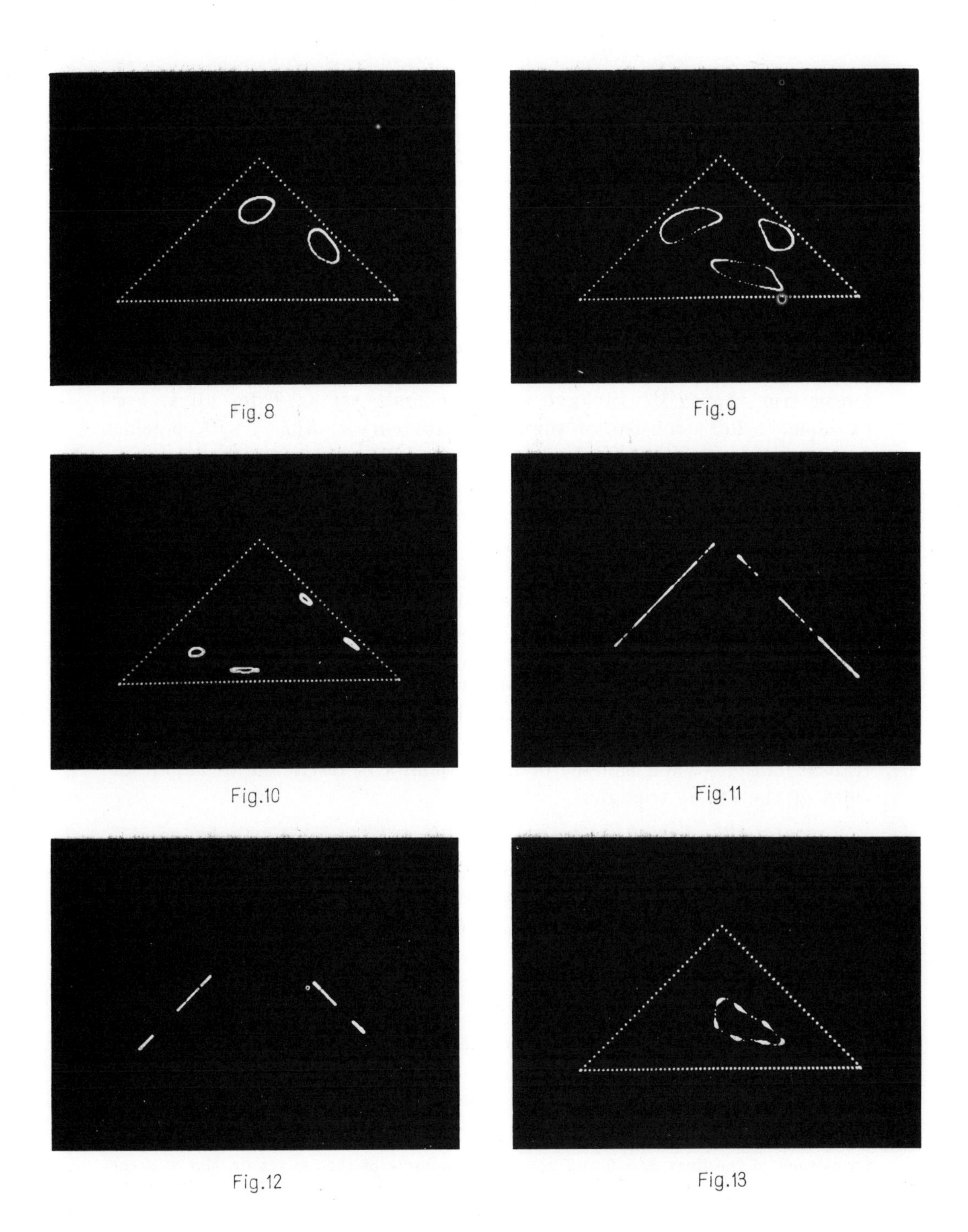

Fig. 8

Fig. 9

Fig. 10

Fig. 11

Fig. 12

Fig. 13

The presumption that $L(T)$ is one-dimensional is supported by the following experiment. We choose a point p_0 which seems to lie, with all available precision, on a convex portion [35] of the curve, and obtain 100000 iterated images of it, keeping track of those iterates which lie closest to p_0. We find that the two points p_1 and p_2 of closest approach lie in opposite quadrants with respect to p_0, and that the slopes of the two line segments (p_1, p_0) and (p_2, p_0) are the same to within a fraction of a percent. This suggests (1) that the limit set is a curve, and (2) that the curve probably has a continuous derivative at p_0 [36].

Limit sets consisting of several separate curves (figures 8, 9, 10) may in principle be treated in the same manner, although it is then no longer true that $T^{(k)}$ will have the same limit set as T for all k. For example, if $L(T)$ consists of three separate curves, $L(T^{(3)})$ will coincide with only one of these—which curve depends, of course, on the initial point.

Class II. This class consists of those infinite limit sets all points of which lie on a pair of boundaries of the (S, a) triangle. Alternate iterates lie on alternate sides, hence the square of the transformation will have a limit set confined to one side of the triangle. $T^{(2)}(p)$ is then strictly one-dimensional for all p situated on one or the other of the two sides in question. The situation is illustrated in figures 11 and 12. There seem to be only a few such one-dimensional limit sets possible within our class of cubic transformations. Correspondingly, many different cubic transformations lead to the same pair of one-dimensional transformations when the set of initial points is restricted to a pair of sides of the (S, a) triangle.

For example, every transformation of the form:

$$
(16)\qquad
\begin{aligned}
x_1' &= x_3^3 + 3a_1x_1x_3^2 + 3x_2x_3^2 + 3b_1x_3x_1^2 + 3x_3x_2^2 + 6c_1x_1x_2x_3\,,\\
x_2' &= x_1^3 + x_2^3 + 3a_2x_1x_3^2 + 3b_2x_3x_1^2 + 6c_2x_1x_2x_3\,,\\
x_3' &= 3x_1x_2^2 + 3a_3x_1x_3^2 + 3x_2x_1^2 + 3b_3x_3x_1^2 + 6c_3x_1x_2x_3
\end{aligned}
$$

with non-negative a_i, b_i, c_i satisfying:

$$
(17)\qquad \sum a_i = \sum b_i = \sum c_i = 1
$$

will lead to the pair of one-variable polynomial transformations of 6th order:

$$
(18)\qquad y' = w[3 - 3w + w^2]\,, \quad w \equiv 3y(1-y)\,,
$$

$$
(19)\qquad u' = 3v(1-v)\,, \quad v \equiv u[3 - 3u + u^2]\,.
$$

In other words, transformations T of the form (16) have the property that $T^{(2)}$ transforms each of the lines $x_3 = 0$ and $x_1 = 0$ into subsets of themselves (in the S, a coordinates, these lines are respectively the boundaries $S + a = 1$ and $S = a$). The study

[35] Overall convexity is rarely, if ever, a property of these limit sets.

[36] We do not conjecture that the derivative exists at every point, but we think it likely that the number of points where the derivative does not exist is at most a set of measure zero.

of such one-dimensional transformations is much easier than that of the original plane transformations, but there are certain serious computational pitfalls connected with high-order iteration (see appendix I).

Class III. The limit sets comprising this class will be referred to as *pseudo-periods*. They consist of relatively dense clusters of points localized at a finite set of *centers*, with a few scattered points in between (figure 13). Such limit sets have not been observed for our original cubic transformations with integer coefficients; they are, however, a prominent feature of the more general transformations discussed in sections III and IV.

Class IV. In this class we place all infinite limit sets not included in the first three classes. Viewed on the oscilloscope they appear as very complicated distributions of points with no recognizable orderly structure. Some examples are shown in figures 14 through 17. A few other examples will be discussed in detail in the following sections. For illustrative purposes, however, we include here a few remarks about figure 17.

This limit set belongs to the transformation:

$$(20) \qquad T_{C_1C_2}: \quad \begin{array}{l} C_1 = \{3, 4, 6, 7, 9, 10\}, \\ C_2 = \{5, 8\}. \end{array}$$

As is evident from the photograph, it consists of seven separated pieces; each of these is invariant under the 7th power of the transformation. Extensive experimentation indicates that the gaps are *really there*. There appears to be no orderly structure within the separate pieces; in figure 18 we show about 385 consecutive images $T^{(7)}(p)$ (in the upper left-hand piece of the limit set) of some p lying in this sub-set.

4. Statistical observations.

a. A large majority of our 9370 cubic transformations in three variables—some 75 per cent—exhibited what might be called *simple* convergence for all initial points tried. For these the limit sets consist of a single point; i.e., a fixed point of the transformation. In many cases there are two such attractive fixed points, but we have not found a case in which both such points are interior to the (S, a) triangle.

We exclude here a few trivial cases such as the following. Consider

$$(21) \qquad T_{C_1C_2}: \quad \begin{array}{l} C_1 = \{1, 2, 3, 4, 5, 10\}, \\ C_2 = \{6, 9\}. \end{array}$$

Explicitly, the second and third line read:

$$(22) \qquad \begin{array}{l} x_2' = 3x_2(x_1^2 + x_2x_3), \\ x_3' = 3x_3(x_1^2 + x_2x_3). \end{array}$$

Thus

$$(23) \qquad \frac{x_2'}{x_3'} = \frac{x_2}{x_3},$$

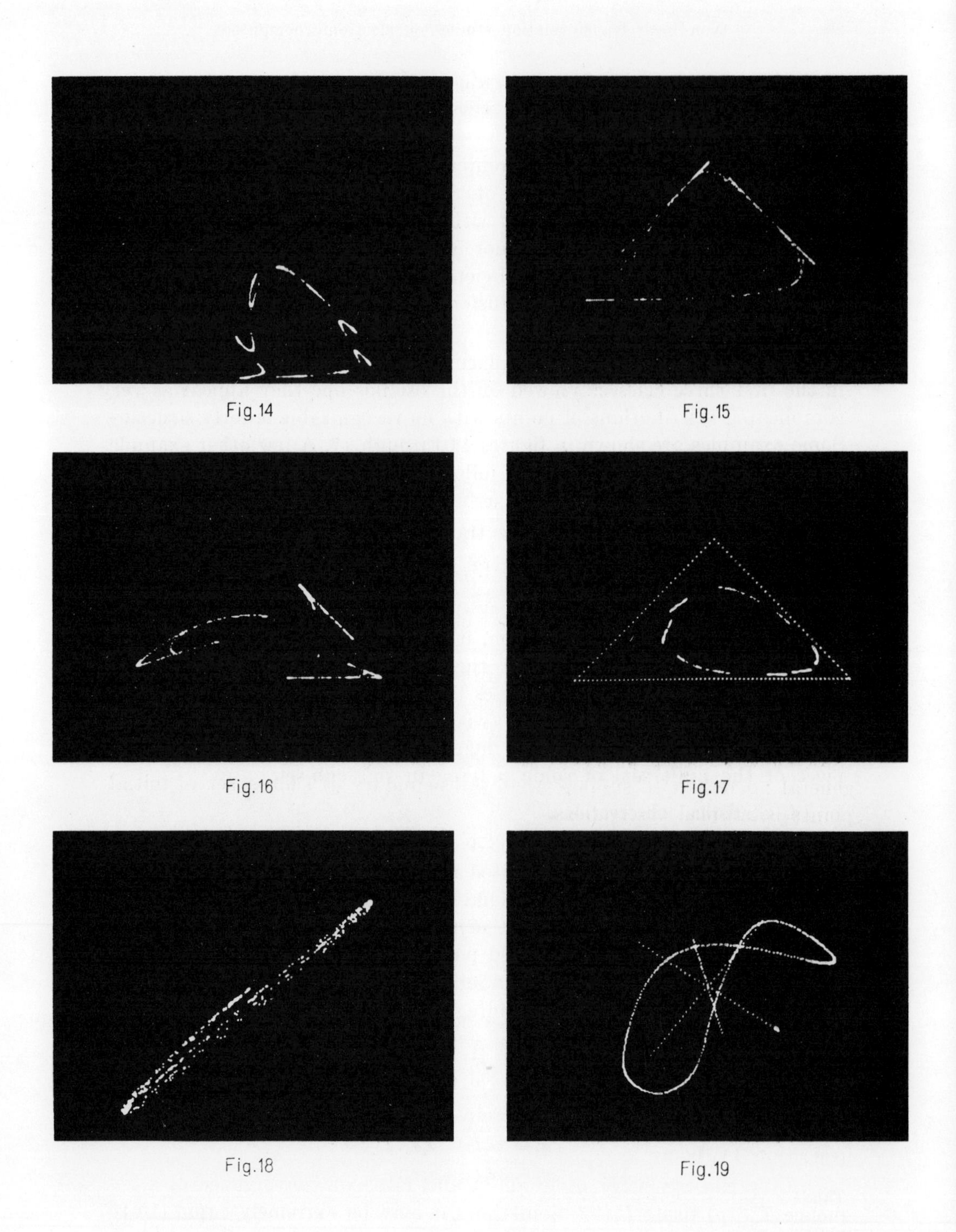

Fig.14

Fig.15

Fig.16

Fig.17

Fig.18

Fig.19

so that this ratio is fixed by the initial value, and we have a continuum of fixed points. Setting $x_2/x_3 \equiv r$, we find that the fixed point is given by

$$x_3 = \frac{1+r-\sqrt{(1+r^2)/3}}{1+3r+r^2} \tag{24}$$

with, of course,

$$x_2 = rx_3, \quad x_1 = 1-(1+r)x_3. \tag{25}$$

If we consider the transformation derived from the above by interchanging the right-hand sides of (22), we shall have:

$$\frac{x_2''}{x_3''} = \frac{x_3'}{x_2'} = \frac{x_2}{x_3}, \tag{26}$$

yielding a corresponding continuum of limit sets which are periods of order two.

b. About 16.5 percent of the transformations seem to have only finite (periodic) limit sets; not surprisingly, most of these are of order two. More than half of the latter are of a *trivial* nature, that is, two vertices of the triangle permute under T. Less than 20 cases were found for which the limit set was a period of order $k > 3$. High-order periods are, however, frequently encountered in the study of the generalized transformations discussed in sections III and IV.

c. Some 5 percent of the cases were found to have several (i.e., two, rarely three) distinct finite limit sets of the types described above. For a given transformation it would in principle be possible to determine numerically the set of initial points whose iterated images converge to a particular one of the several limit sets; lack of time has prevented us from doing this except in a few cases. We only remark that there is in general no reason to suppose that the boundary of such a set of initial points is simple.

d. The remaining 3.5 percent—some 334 transformations—possess infinite limit sets. Most of these—roughly three-quarters of them—belong to class I, that is they look like closed curves. Perhaps 5 percent of the rest belong to class II, the 20 percent residuum being of class IV type. As mentioned above, no examples of class III (pseudo-periods) limit sets were encountered in the study of our original group of cubic transformations (i.e., those with integer coefficients 1, 3 or 6).

e. No case has been found in which a transformation has two distinct class IV limit sets, although there are cases where one of several limit sets was of class IV type. One such has already been described in section I (page 16); a more complicated example will be mentioned in section IV below.

f. We can say very little about the rate of convergence of a sequence $T^{(n)}(p)$ to its $L_p(T)$. Sometimes it may be extremely rapid (10 to 20 iterations); in other cases many thousands of iterates may be required.

If $L_p(T)$ consists of a single point, $L_p(T) = \{p_0\}$, this rate can, of course, be calculated (for points sufficiently close to p_0) by solving the approximate, linear difference equations explicitly.

This is, however, not always sufficient. If the jacobian matrix, evaluated at p_0, has complex roots, and $|\lambda^2| = 1$, the linear difference equations may generate an invariant ellipse. Such a case was found in our of quadratics in three variables, and is discussed in our report on that work. In the S, a coordinates, this transformation is

$$S' = 1 - 4a + 4a^2 + 2aS, \quad a' = 2aS \tag{27}$$

with fixed point at

$$S = \tfrac{1}{2}, \quad a = \tfrac{1}{4}. \tag{28}$$

Letting

$$x = S - \tfrac{1}{2}, \quad y = a - \tfrac{1}{4}, \tag{29}$$

the linear approximation is

$$x' = -y + \tfrac{1}{2}x, \quad y' = y + \tfrac{1}{2}x. \tag{30}$$

This then generates the invariant ellipse:

$$x'^2 + x'y' + 2y'^2 = x^2 + xy + 2y^2. \tag{31}$$

In fact, however, for the full (non-linear) transformation, the fixed point is attractive.

5. Limit sets for quadratic transformations in four variables. All 34337 distinct systems of this type have been investigated on the "STRETCH" computer, as described in section I above ([37]). A preliminary survey of the results indicates that only about two percent of these transformations possess infinite limit sets. The finite limit sets need no special comment; they are of the same sort as those found in cubics in three variables—except, of course, that they are not in general plane sets. A few periods of rather high-order (more than 100 points!) were found, as well as a fair number of cases with 10 to 80 points. This probably should be expected in view of the greater variety of possible algebraic structures.

We are not yet in a position to classify the infinite limit sets as we have in the case of the cubics. Perhaps the closest analogy to sets of the class I type are those which appear to be closed curves in space. These are illustrated in figures 19 through 23. They are shown in convenient projections; the "coordinate system" in the center of the picture merely indicates the orientation relative to the viewer, who is conceived

([37]) The computing time required for the whole study was only a fraction of what one would predict on the basis of 7 seconds for 10^4 iterations—an average for these recurrence relations as actually coded—because a large majority of cases "converged" in a few ($\sim$50) steps.

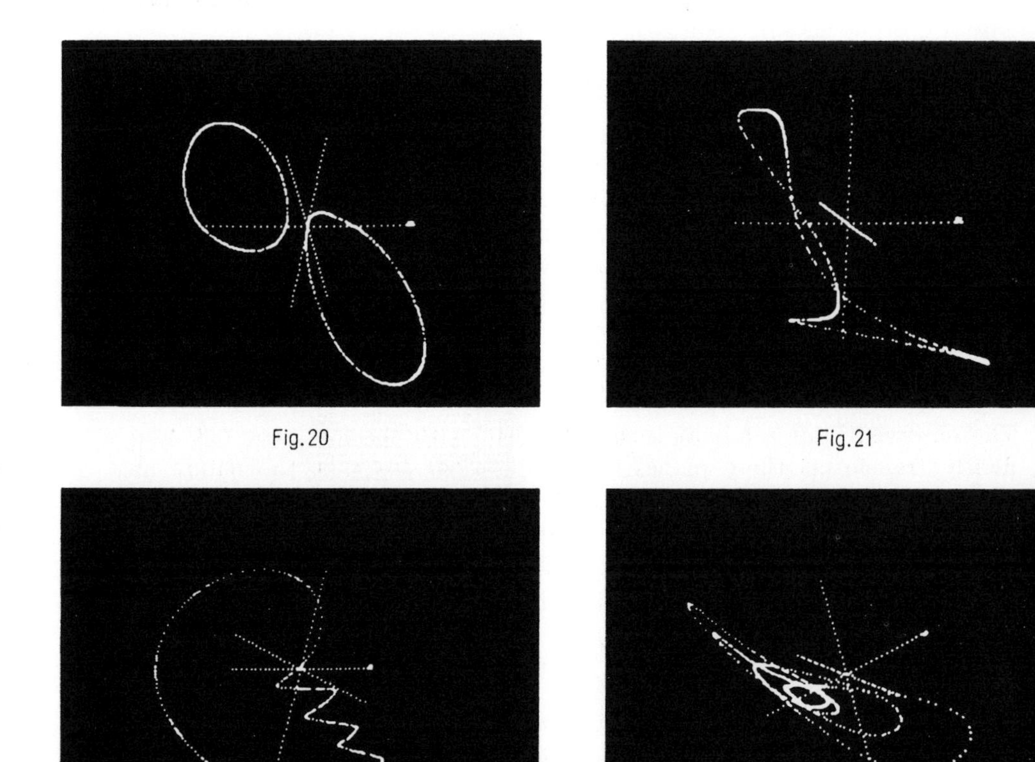

Fig. 20

Fig. 21

Fig. 22

Fig. 23

of as stationed at a certain distance from the origin along the y axis ([38]). Figure 19 (see page 26) shows a limit set belonging to the transformation:

$$(32) \qquad T_{Q_1Q_2Q_3}: \quad \begin{aligned} Q_1 &= \{1, 3, 4\}, \\ Q_2 &= \{5, 6, 8\}, \\ Q_3 &= \{7, 10\}. \end{aligned}$$

Presumably what one sees is a twisted space curve.

In figure 20, the limit set consists of two plane curves, one of which lies in the (x_1, x_3) plane, the other lying in a plane inclined at 45° with respect to the first. The corresponding transformation is

$$(33) \qquad T_{Q_1Q_2Q_3}: \quad \begin{aligned} Q_1 &= \{1, 2, 9\}, \\ Q_2 &= \{4, 7, 10\}, \\ Q_3 &= \{3, 5, 6\}. \end{aligned}$$

The observed limit set is at least consistent with the fact that (33) evidently transforms these planes into each other (so that the points lie alternately on the separate curves). More complicated twisted curves are possible (figure 21). We have also found quite implausible looking limit sets like that shown in figure 22. As a final example, we cite the transformation:

$$(34) \qquad T_{Q_1Q_2Q_3}: \quad \begin{aligned} Q_1 &= \{1, 7, 9\}, \\ Q_2 &= \{3, 4, 8\}, \\ Q_3 &= \{2, 5, 10\}. \end{aligned}$$

This has at least two infinite limit sets, one of which may be of class IV type (not shown); the other (figure 23) is a "curve" of unknown structure.

At the time of this writing (January, 1963) we are unable to say anything more specific about the limit sets for quadratics in four variables; to date, less than one-third of the seven hundred or so potentially "interesting" cases have been looked at on the oscilloscope.

([38]) The "reference system" (x, y, z) is parallel to, but displaced relative to, the actual coordinate system (x_1, x_2, x_3). The origin of the (x, y, z) system is in the (approximate) center of the picture; that of the (x_1, x_2, x_3) system is in the lower left-hand corner.

III. THE "Δt-MODIFICATION"

1. We discuss here a particular one-parameter generalization of our original cubic transformations in three variables which we have called the *Δt-modification* ([39]). It consists in replacing the usual difference equations ([40]):

$$(1) \qquad x_i' = \sum_{j=1}^{10} d_{ij} M_j , \qquad i = 1, 2, 3 ,$$

by

$$(2) \qquad x_i' = (1 - \Delta t) x_i + \Delta t \sum_{j=1}^{10} d_{ij} M_j ,$$

with

$$(3) \qquad 0 < \Delta t \leqslant 1 .$$

If $\Delta t = 1$, we recover the original set (1); $\Delta t = 0$ is excluded, since the equations then become the trivial identity transformation.

The abbreviated notation of section II is extended in an obvious manner to cover this case. Thus, if (1) is represented symbolically by $T_{C_1C_2}$, (2) may be symbolized by $T_{C_1C_2(\Delta t)}$. For a given p, the limit set will correspondingly be denoted by $L_p(T)$, $L_{p(\Delta t)}(T)$.

In the S, a coordinates, (1) appears in the form

$$(4) \qquad S' = F(S, a), \qquad a' = G(S, a) .$$

Correspondingly, (2) reads

$$(5) \qquad S' = (1 - \Delta t) S + \Delta t F(S, a), \qquad a' = (1 - \Delta t) a + \Delta t G(S, a) .$$

It is clear that the fixed points of (5) coincide with those of (4). As we shall see below, this fact enables us to find these fixed points by simple

---

([39]) This has already been mentioned in the introduction, equation (12). The modification can, of course, be introduced for the general case (equations (10) through (13) of the introduction).

([40]) The M_j are defined in section II, sub-section 1, equation (4). Unless otherwise stated, equations (2) and (3) of section II are assumed to hold, as well as the condition

$$\sum x_i = 1, \; x_i \geqslant 0, \quad \text{for all initial points} \quad p = (x_1, x_2, x_3) .$$

iteration, thus avoiding the unpleasant algebra involved in eliminating one variable from the pair of general cubics

$$S = F(S, a), \qquad a = G(S, a), \tag{6}$$

and then solving for the roots of the resulting high-order ($\leqslant 9$) polynomial. One can look upon (5) at the simplest (and most naive) finite difference scheme for approximating the first-order differential system

$$\frac{dS}{dt} = -S + F(S, a), \qquad \frac{da}{dt} = -a + G(S, a). \tag{7}$$

The analogy between (5) and (7) is not, however, very close ([41]); consequently it is better to discuss (5) on its own merits. The effect of setting $\Delta t < 1$ (but > 0) on a single iteration is easy to see. Let us take a particular point S, a; the image produced by (4) will be denoted, as usual, by S', a', while we shall call the corresponding image under (5) S'mod, a'mod. Then

$$S'\text{mod} - S' = \Delta t(S' - S), \qquad a'\text{mod} - a' = \Delta t(a' - a). \tag{8}$$

In other words, the length of the iterative step is altered, while the direction remains the same. What happens on repeated iteration is, however, not all obvious. One expects that the limit set $L_{p(\Delta t)}(T)$ will in general have smaller diameter as we decrease Δt, but we cannot at present predict its structure as a function of Δt, even relative to the (observed) structure of $L_p(T)$. It is worthwhile illustrating this in a particular case. Consider the transformation:

$$T_{C_1C_2}: \quad \begin{aligned} C_1 &= \{3, 5, 7, 9, 10\}, \\ C_2 &= \{1, 2, 8\}. \end{aligned} \tag{9}$$

In the S, a coordinates, this reads explicitly:

$$T_A: \quad \begin{aligned} S' &= S^3 - 6S^2a - 3Sa^2 + 4a^3 - \tfrac{3}{2}S^2 + 3Sa - \tfrac{3}{2}a^2 + 1. \\ a' &= -S^3 + 3Sa^2 + 2a^3 + \tfrac{3}{2}S^2 - 3Sa + \tfrac{3}{2}a^2. \end{aligned} \tag{A}$$

Since we shall refer to this transformation quite often in the sequel, we have given it the distinctive label (A). T_A has one interior ([42]) fixed point (repellent), whose coordinates are:

$$S_0 = 0.5885696, \qquad a_0 = 0.1388662. \tag{10}$$

There are two infinite limit sets; these are shown in figure 24 and in figure 11 of section II. At the moment, we shall not be concerned with

([41]) For further discussion on this point see section V.

([42]) There is another repellent fixed point at a vertex of the triangle, namely $S = a = \frac{1}{2}$.

the limit set shown in figure 11; this is evidently of class II type and can therefore be studied in one-dimensional form [43]. The limit set shown in figure 24—which we shall henceforth refer to as $L(T_A)$—appears as an irregular pattern surrounding the fixed point (shown superimposed on the picture). Figure 25 again shows $L(T_A)$—this time enlarged by a factor 3, while figure 26 shows a portion of the upper left-hand corner [44] enlarged about 14 times. Figure 24 shows 900 consecutive iterates, while figure 25 shows these same 900 points plus 1800 more. For comparison, in figure 27 we plot just 50 consecutive iterates. The approximate outer dimensions of $L(T_A)$ are [45]

$$(11)\quad \begin{aligned} S_{\text{max}} &= 0.754696 \quad \text{at} \quad \alpha = 0.077251\,, \\ S_{\text{min}} &= 0.443911 \quad \text{at} \quad \alpha = 0.204610\,, \\ \alpha_{\text{max}} &= 0.277406 \quad \text{at} \quad S = 0.491266\,, \\ \alpha_{\text{min}} &= 0.071196 \quad \text{at} \quad S = 0.739170\,. \end{aligned}$$

We now contrast with $L(T_A)$ the limit sets $L_{(\Delta t)}(T_A)$ belonging to the generalized transformation $T_{A(\Delta t)}$. If we set $\Delta t = 0.9931$, we get a limit set entirely different from $L(T_A)$ (from the same initial point). This is shown in figure 28. It exhibits what we have called "pseudo-periodic" structure, that is, almost all the iterated images of the initial point p are concentrated in the neighborhood of seven distin·t "centers"—an example of a class III limit set [46].

With a very small change in Δt—namely, by setting $\Delta t = 0.9930$ — we find instead a period of order 7. This is shown in figure 29. As we decrease Δt in small steps down to $\Delta t = 0.9772$ (figure 30), the corresponding $L_{(\Delta t)}(T_A)$ remains a period of order 7; the coordinates of the individual points appear to change continuously with Δt. For $\Delta t = 0.97713$, $L_{(\Delta t)}(T_A)$ is again a pseudo-period, and this character persists down to $\Delta t = 0.9770$ (figure 13 of section II). Below [47] $\Delta t = 0.9770$ $L_{(\Delta t)}(T_A)$ is a closed curve [48] around the fixed point which shrinks in more or less continuous fashion as Δt is decreased. Figures 31, 32 and 33 illustrate, respectively, the limit sets for $\Delta t = 0.97$, 0.94 and 0.92. Finally, for $\Delta t < 0.9180154$ (see below), the limit set consists of a single point—the fixed point (10) [49]. This peculiar behavior of $L_{(\Delta t)}(T_A)$ as a func-

---

[43] See section II.

[44] This is the region $0.455 \leqslant S \leqslant 0.525$, $0.225 \leqslant \alpha \leqslant 0.278$.

[45] These results are based on a calculation with $N = 96000$ iterations.

[46] See section II for this classification scheme.

[47] We have not attempted to find the *critical* values of Δt with greater precision, though this could in principle be done to, say, 7 decimal places on MANIAC II.

[48] This is, of course, only a conjecture. See the discussion in section II.

[49] In the language of functional analysis, $T_{A(\Delta t)}$ is a *shrinking operator* in this range of Δt values.

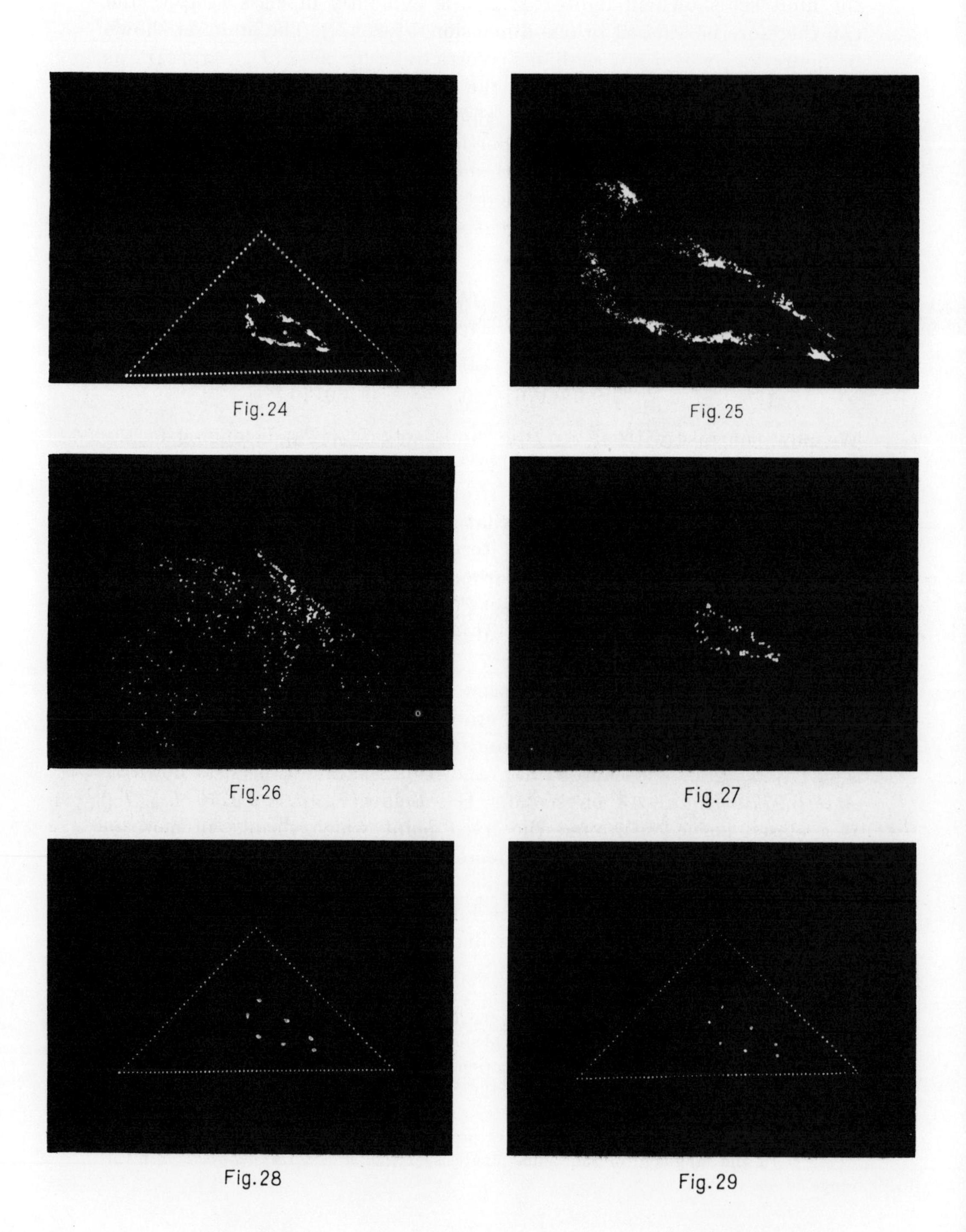

Fig. 24

Fig. 25

Fig. 26

Fig. 27

Fig. 28

Fig. 29

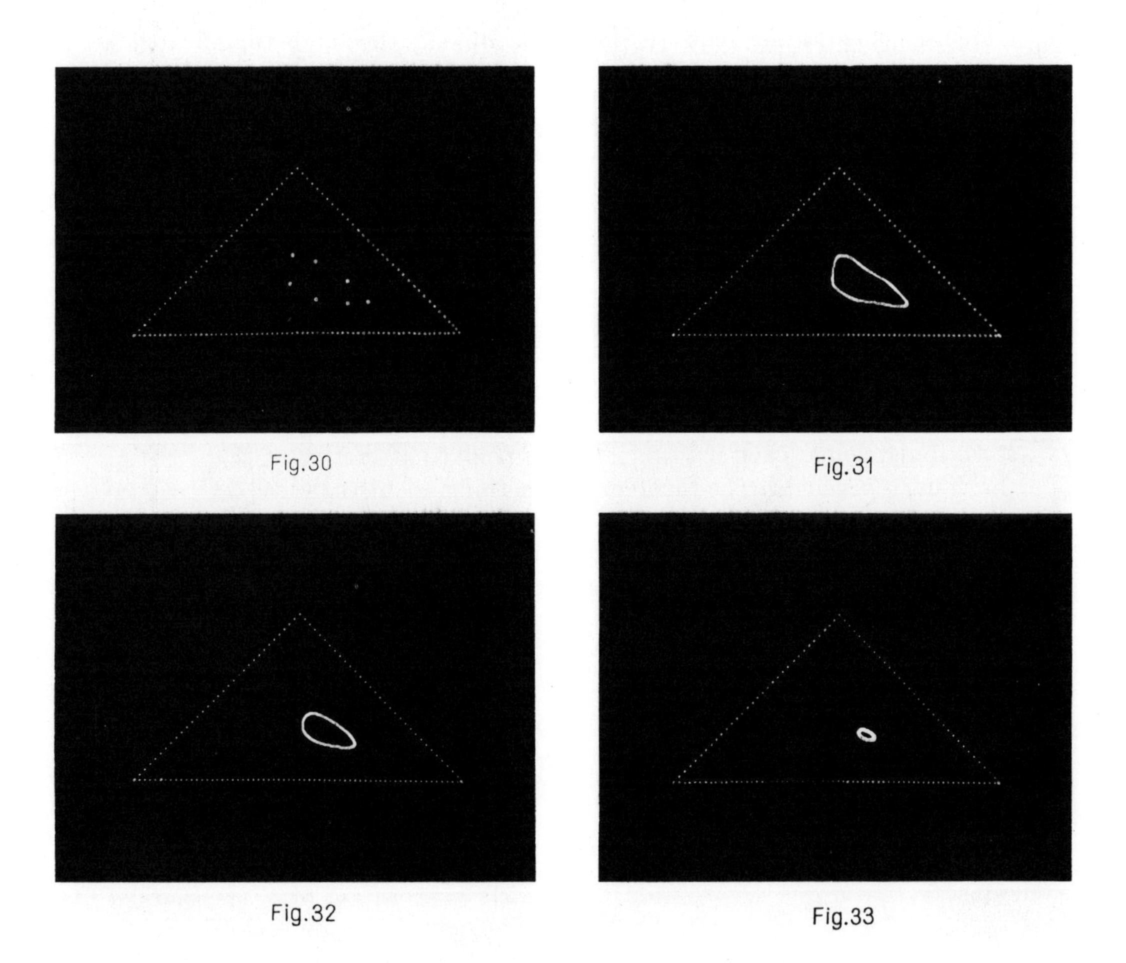

Fig.30

Fig.31

Fig.32

Fig.33

tion of Δt is not an isolated instance, nor is it by any means among the most extreme examples we have encountered (see section IV for a considerably more "pathological" case). Within the class of cubic transformations we have studied, it seems to be an empirical rule that the more pathological the limit set looks for $\Delta t = 1$, the more complicated will be its behavior as Δt is decreased.

2. Attractive and repellent fixed points. The fixed points of a cubic transformation in the standard form (4) are those real roots of the algebraic system (6) which lie in or on the boundary of the S, a triangle ([50]). We are interested both in finding the values of the coordinates of these fixed points and in determining whether the points are attractive or repellent. By attractive we mean as always that, for any point p in a sufficiently small neighborhood of the fixed point p_0, the sequence $T^{(n)}(p)$ will converge to p_0. A general criterion for the attractiveness of a fixed point has been given by Ostrowski ([51]), viz: let $|\lambda_{\max}|$ be the largest eigenvalue in absolute value of the jacobian matrix evaluated at the fixed point. Then if $|\lambda_{\max}| < 1$, the point is attractive; if $|\lambda_{\max}| > 1$, the point is repellent. The theorem says nothing about the case $|\lambda_{\max}| = 1$, nor does it yield a method for determining theoretically the appropriate neighborhood. For the two-variable transformation (4), we may give the eigenvalues of the jacobian matrix explicitly:

(12) $$\lambda = \frac{T_0 \pm \sqrt{T_0^2 - 4J_0}}{2},$$

where T_0 is the trace:

(13) $$T_0 = \frac{\partial F}{\partial S} + \frac{\partial G}{\partial a} + \Big|_{\substack{S=S_0\\a=a_0}},$$

and J_0 is the jacobian:

(14) $$J_0 = \frac{\partial F}{\partial S}\frac{\partial G}{\partial a} - \frac{\partial G}{\partial S}\frac{\partial F}{\partial a}\Big|_{\substack{S=S_0\\a=a_0}}.$$

For the modified system (5), we find correspondingly:

(15) $$\lambda_{\text{mod}} = 1 - \Delta t + \frac{\Delta t}{2}\left(T_0 \pm \sqrt{T_0^2 - 4J_0}\right).$$

If the roots are complex, i.e., if $T_0^2 < 4J_0$, we have

(16) $$|\lambda_{\text{mod}}|^2 = 1 - \Delta t(2 - T_0) + \Delta t^2(1 - T_0 + J_0).$$

([50]) Brouwer's theorem assures us that there is at least one fixed point.

([51]) A. Ostrowski, *Solution of Equations and Systems of Equations*, Chapter 18, (New York 1960).

Defining Δt_{lim} as the value of Δt for which $|\lambda_{\text{mod}}|^2 = 1$, we obtain

$$\Delta t_{\text{lim}} = \frac{2 - T_0}{1 - T_0 + J_0}. \tag{17}$$

Thus, for the case of complex roots, we may make the fixed point (S_0, a_0) attractive by choosing Δt such that

$$0 < \Delta t \leqslant \Delta t_{\text{lim}}. \tag{18}$$

Similarly, if $\lambda_{\max}$ is real and negative, and $|\lambda_{\max}| > 1$,

$$\Delta t_{\text{lim}} = \frac{2}{1 + |\lambda_{\max}|}. \tag{19}$$

It is clear that this artifice will not work if $\lambda_{\max} \geqslant 1$. Such a situation arises in the one-dimensional case:

$$x_1' = x_1^3 + 3x_1^2 x_2, \quad x_2' = x_2^3 + 3x_2^2 x_1, \quad x_1 + x_2 = 1, \tag{20}$$

that is

$$x_1' = x_1^2\,(3 - 2x_1). \tag{21}$$

The fixed points are $x_1 = 0, \frac{1}{2}, 1$; at these points the derivative dx_1'/dx_1 has the values $0, \frac{3}{2}, 0$. Clearly both $x_1 = 0$ and $x_1 = 1$ are attractive fixed points; for all $x_1^{(0)} < \frac{1}{2}$, $x_1^{(n)} \to 0$, while for all $x_1^{(0)} > \frac{1}{2}$, $x_1^{(n)} \to 1$. The interior fixed point $x_1 = \frac{1}{2}$ is repellent and cannot be made attractive by using the Δt-modification. The corresponding situation does not seem to occur for any of our cubic transformations in the plane.

In practice, all one has to do to obtain the numerical value of a repellent fixed point is to choose a sufficiently small Δt and iterate; on a computer, this calculation requires only a few seconds ([52]).

3. Attractive periods. The set of points constituting a period of order k are fixed points of $T^{(k)}$. Thus one may test whether a periodic limit set is attractive by applying Ostrowski's criterion to $T^{(k)}$. Let

$$J^{(n)} \equiv \begin{bmatrix} \dfrac{\partial S^{(n)}}{\partial S} & \dfrac{\partial S^{(n)}}{\partial a} \\[2ex] \dfrac{\partial a^{(n)}}{\partial S} & \dfrac{\partial a^{(n)}}{\partial a} \end{bmatrix}_{(S^0,\, a^0)}, \tag{22}$$

$$J_{n-1} \equiv \begin{bmatrix} \dfrac{\partial S^{(n)}}{\partial S^{(n-1)}} & \dfrac{\partial S^{(n)}}{\partial a^{(n-1)}} \\[2ex] \dfrac{\partial a^{(n)}}{\partial S^{(n-1)}} & \dfrac{\partial a^{(n)}}{\partial a^{(n-1)}} \end{bmatrix}_{(S^{(n-1)},\, a^{(n-1)})}, \tag{23}$$

where e.g. (S^0, a^0) is a fixed point of $T^{(k)}$.

([52]) Early in this investigation we made the "mistake" of taking $\Delta t > \Delta t_{\text{lim}}$ in a few cases, and thereby discovered the interesting limit sets $L_{p(\Delta t)}(T)$.

Then, by the chain rule:

$$J^{(k)} = J_{k-1} \times J_{k-2} \times \dots \times J_0 . \tag{24}$$

Thus $J^{(k)}$ is easily obtained by evaluating (24) over the periodic set in question; the application of Ostrowski's criterion is then immediate. We have often used this technique to convince ourselves that the periods are really limit sets and not the result of spurious or accidental convergence (53).

---

(53) This technique has actually been used for periods with orders k as large as 148. For very large k the method might fail owing to round-off errors or other numerical inaccuracies.

IV. MODIFICATION OF THE COEFFICIENTS

1. In this section we present some result on the effect of modifying the original integer-valued coefficients of our cubic transformations in three variables. That is, we consider, as before, transformations of the form:

$$x_i' = \sum_{j=1}^{10} d_{ij} M_j , \tag{1}$$

$$\sum_{i=1}^{3} d_{ij} = 1 , \tag{2}$$

but we no longer require that the d_{ij} all be 1 or 0. As already remarked in the introduction, if we impose on the d_{ij} only the additional condition:

$$0 \leqslant d_{ij} \leqslant 1 , \tag{3}$$

then (1), (2), and (3) define a class of cubic transformations depending on 20 parameters, e.g., d_{1j}, d_{2j} $(j = 1$ to $10)$ ([54]). Since we are unable to formulate a complete theory for the finite sub-class of transformations characterized by the restrictions: $d_{ij} = 0$ or 1, all i, j, it is clear a fortiori that we do not have a theory for the infinite class.

In this paper we limit ourselves to showing how an experimental study of some special cases can help to throw light on the properties of our original cubic transformations.

In effect, what we do is study certain transformations which are "close to" some particular transformations of our basic type. A natural way to define a transformation close to some given $T_{C_1C_2}$ would be to choose its coefficients as follows:

$$\begin{aligned} d_{ij} &= 1 - \varepsilon_{ij}, \qquad j \subset C_i , \\ d_{ij} &= \varepsilon_{ij}, \qquad j \not\subset C_i , \end{aligned} \tag{4}$$

([54]) For the definition of the cubic monomials M_j, see equation (4) of section II. The domain of this class of transformations is again the region

$$\sum_{i=1}^{3} x_i = 1, \quad 0 \leqslant x_i \leqslant 1 .$$

where the d_{ij} must satisfy (2) and (3), and the ε_{ij} are small. This class of transformations, defined with respect to some $T_{C_1C_2}$ is still too extensive to study, even if the various ε_{ij} are restricted to a few discrete values. What we have actually done is to consider 20 such transformations, each of which depends on a single parameter ε. We denote these by the symbol:

$$T_{(r,s)\varepsilon}\,, \quad 1 \leqslant r \leqslant 10\,, \quad 0 \leqslant \varepsilon \leqslant 1\,, \quad s = 0, 1\,. \tag{5}$$

It is understood that these transformations are only defined relative to some $T_{C_1C_2}$. For convenience we shall generally refer to the transformations $T_{(r,s)\varepsilon}$ defined relative to some $T_{C_1C_2}$ as *associated* transformations. The coefficients of the $T_{(r,s)\varepsilon}$ are specified as follows:

For $j \neq r$:

$$\begin{aligned} d_{ij} &= 1, & j &\subset C_i\,, \\ d_{ij} &= 0, & j &\not\subset C_i\,. \end{aligned} \tag{6}$$

For $j = r$:

$$\begin{aligned} d_{1r} &= 1 - \varepsilon, & r &\subset C_1\,, \\ d_{1r} &= (1-s)\,\varepsilon, & r &\not\subset C_1\,; \\ d_{2r} &= 1 - \varepsilon, & r &\subset C_2\,, \\ d_{2r} &= s\varepsilon, & r &\not\subset C_2\,. \end{aligned} \tag{7}$$

In words: $T_{(r,s)\varepsilon}$ is formed from $T_{C_1C_2}$ by the replacement $M_r \to (1-\varepsilon) M_r$ wherever the term M_r occurs, and by adding εM_r to one of the other two lines of the three-line schema. As an example, consider the transformation T_A introduced in section III:

$$T_A\colon \quad \begin{aligned} C_1 &= \{3, 5, 7, 9, 10\}\,, \\ C_2 &= \{1, 2, 8\}\,. \end{aligned} \tag{8}$$

Relative to T_A, $T_{(5,1)\varepsilon}$ would read:

$$\begin{aligned} x_1' &= M_3 + (1-\varepsilon) M_5 + M_7 + M_9 + M_{10}\,, \\ x_2' &= M_1 + M_2 + \varepsilon M_5 + M_8\,, \\ x_3' &= M_4 + M_6\,, \end{aligned} \tag{9}$$

while $T_{(4,0)\varepsilon}$ would take the form:

$$\begin{aligned} x_1' &= M_3 + \varepsilon M_4 + M_5 + M_7 + M_9 + M_{10}\,, \\ x_2' &= M_1 + M_2 + M_8\,, \\ x_3' &= (1-\varepsilon) M_4 + M_6\,. \end{aligned} \tag{10}$$

In the S, a coordinates, the $T_{(r,s)\varepsilon}$ can be written:

$$S' = F(S, a) + \varepsilon f_{rs}(S, a), \quad a' = G(S, a) + \varepsilon g_{rs}(S, a); \tag{11}$$

the original $T_{C_1C_2}$ is obtained from (11) by setting $\varepsilon = 0$. For the two examples given above, we have:

$$(12) \qquad T_{(5,1)\varepsilon}: \quad \begin{aligned} f_{rs} &= f_{51} = -\tfrac{3}{2}(S-a)(1-S-a)^2, \\ g_{51} &= -f_{51}\,; \end{aligned}$$

$$(13) \qquad T_{(4,0)\varepsilon}: \quad \begin{aligned} f_{rs} &= f_{40} = 12a^2(S-a)\,, \\ g_{40} &= 0\,. \end{aligned}$$

It turns out that for these one-term modifications $T_{(r,s)\varepsilon}$ we always have $g_{rs} = \pm f_{rs}$ or 0. f_{rs} can further be factored into a numerical coefficient c_{rs} and a function $M_r(S, a)$; the M_r are of course just the original cubic monomials expressed in terms of S and a. The c_{rs} and g_{rs} are determined as follows:

For $r \subset C_1$:

$$(14) \qquad \begin{aligned} s = 0&: \quad g_{rs} = 0, \qquad c_{rs} = -1\,, \\ s = 1&: \quad g_{rs} = -f_{rs}, \; c_{rs} = -\tfrac{1}{2}\,; \end{aligned}$$

for $r \subset C_2$:

$$(15) \qquad \begin{aligned} s = 0&: \quad g_{rs} = -f_{rs}, \; c_{rs} = \tfrac{1}{2}\,, \\ s = 1&: \quad g_{rs} = f_{rs}, \qquad c_{rs} = -\tfrac{1}{2}\,; \end{aligned}$$

for $r \not\subset C_1$, $r \not\subset C_2$:

$$(16) \qquad \begin{aligned} s = 0&: \quad g_{rs} = 0, \qquad c_{rs} = 1\,, \\ s = 1&: \quad g_{rs} = f_{rs}, \qquad c_{rs} = \tfrac{1}{2}\,. \end{aligned}$$

2. We have studied the modified transformations $T_{(r,s)\varepsilon}$ for a variety of our original cubic $T_{C_1C_2}$ that happen to have infinite limit sets. Our usual procedure has been to vary ε in steps of $\frac{1}{100}$ in the range $\frac{1}{100} \leqslant \varepsilon \leqslant \frac{1}{10}$ for a given $T_{(r,s)\varepsilon}$ relative to a given $T_{C_1C_2}$, although on occasion intermediate values of ε have been used. Only for the transformation T_A (equation (8) above) have we looked at all 20 modified transformations. For a few other $T_{C_1C_2}$ we have limited ourselves to selecting certain of the associated $T_{(r,s)\varepsilon}$ for detailed study. Since this selection has generally been made on intuitive grounds, we cannot claim that the most "interesting" modifications of the original transformations have always been considered. Nevertheless, this part of our study has proved most revealing, especially as regards the structure of class IV limit sets.

Before describing the results, we insert a few remarks on the difference between the two types of generalizations we have considered—the *Δt-modification* of section III and the *associated* transformations $T_{(r,s)\varepsilon}$. The Δt-modification is essentially nothing but the application of a technique frequently employed in the practical solution of non-linear equations by iterative methods; it is, in fact, one way—perhaps the simplest—of introducing a linear convergence factor. Apart from our use of this device for obtaining the coordinates of the fixed point, our principal interest is in small convergence factors (Δt close to unity)—too small, in fact, to produce convergence to the fixed point. In view of the fact that the Δt-modified transformation $T_{C_1C_2(\Delta t)}$ has precisely the same fixed point as the original transformation $T_{C_1C_2}$, one might expect that there exists a close relationship between the corresponding limit sets $L_{p(\Delta t)}(T)$ and $L_p(T)$. In some sense this is true, as the examples given in section III show (see also below, sub-section 4). We may express this more formally as follows:

We define a sequence of transformations $T_{C_1C_2(\Delta t_i)} \equiv T_{\Delta t_i}$ with corresponding limit sets $L_{p(\Delta t_i)}(T)$ by some convenient rule:

$$\Delta t_0 = \Delta t_{\text{lim}}\,, \qquad \Delta t_i = 1 - \frac{1-\Delta t_0}{2^i}\,. \tag{17}$$

The sequence $T_{\Delta t_i}$, $i = 1, 2, \ldots$, clearly converges to $T_{C_1C_2}$.

We then formulate the following conjecture:

Given a $T_{C_1C_2}$ and a $\delta > 0$, then, for all p in the triangle, there exists an $N_{(p)}$ such that, for $i > N_{(p)}$ and for all $x \subset L_p(T)$, there exists a $y \subset \subset L_{p(\Delta t_i)}(T)$ satisfying $|y - x| < \delta$.

The modification of $T_{C_1C_2}$ defined by the associated transformations $T_{(r,s)\varepsilon}$ differs from the Δt-modification in several respects. In the first place, the perturbation introduced is not linear. Furthermore, the fixed points of $T_{(r,s)\varepsilon}$ are in general not the same as those of $T_{C_1C_2}$ (fixed points on the boundary of the triangle may, of course, be common to $T_{(r,s)\varepsilon}$ and $T_{C_1C_2}$ for some pairs r, s) ([55]). Finally, each pair r, s must be treated separately; for fixed ε, perturbations of different terms of $T_{C_1C_2}$ may lead to quite different limit sets. Nevertheless, a conjecture analogous to that formulated for the sequence $T_{\Delta t_i}$ would most probably turn out to be correct.

3. Limits sets of the transformations $T_{(r,s)\varepsilon}$ associated with T_A. Since we usually deal with values of ε of the form:

$$\varepsilon_i = \frac{i}{100}\,, \qquad 1 \leqslant i \leqslant 10\,, \tag{18}$$

we introduce a symbol to denote a set of such values:

$$I(i, j) \equiv \{\varepsilon_n\}\,, \qquad i \leqslant n \leqslant j\,. \tag{19}$$

In addition, $R^k_{(r,s)}$ will denote the closed interval of ε:

$$R^k_{(r,s)} = [\,{}^-R^k_{(r,s)}\,,\; {}^+R^k_{(r,s)}]\,, \qquad {}^-R^k_{(r,s)} \leqslant \varepsilon \leqslant {}^+R^k_{(r,s)} \tag{20}$$

for which the limit sets $L_{(r,s)\varepsilon}$ of $T_{(r,s)\varepsilon}$ are periods of order k. The photographs illustrating the examples that follow will be found, suitably labelled, in sub-section 2 of appendix II.

([55]) The new fixed point $S_{(r,s)\varepsilon} = S + \Delta s$, $a_{(4,s)\varepsilon} = a + \Delta a$, calculated to first order in $(\Delta S, \Delta a)$, has both ΔS and Δa proportional to ε. The ratio $\Delta S/\Delta a$ in this order is therefore independent of ε, though not of r, s. There are, in fact, 6 possible directions of displacement, two for each of the three cases: $g_{rs} = f_{rs}$, $g_{rs} = -f_{rs}$, $g_{rs} = 0$, cf. equations (14), (15), and (16).

There is one significant feature common to all the $T_{(r,s)\varepsilon}$ associated with T_A; for every pair r, s at least one periodic limit set—that is, a period of order $k > 1$—was found in the range $I(1, 10)$. The order of periodicity of most frequent occurrence was $k = 7$. Thus, for example, for $(r, s) = (10, 0)$, we found periodic limit sets with $k = 7$ over the range $I(3, 10)$, and the case $(r, s) = (6, 1)$ behaves in the same fashion over the same range. For both series of associated transformations, the limit sets for $\varepsilon = 0.01$ are of class IV type and closely resemble $L(T_A)$. At $\varepsilon = 0.02$, bright spots show up in the pattern (figure A-1); this usually indicates that one is *near* a period, i.e., that a relatively small change in ε will yield a transformation having a finite limit set. In the notation (20), this would be written: $^{-}R^{7}_{(r,s)} - 0.02 \ll 1$. In these two examples it happens that a period of order 7 is observed over the range $I(3, 10)$, that is: $I(3, 10) \subset R^{7}_{(r,s)}$. This is not generally the case. Thus for the case $(r, s) = (9, 0)$, $I(2, 9) \subset R^{7}_{(9,0)}$, whereas $L_{(9,0)0.01}$ and $L_{(9,0)0.10}$ are of class IV type and are morphologically similar to $L(T_A)$. It may be recalled (section III) that an analogous behavior was observed for the Δt-modified transformations $T_{A(\Delta t)}$, namely that $L_{\Delta t}(T_A)$ was found to be a period of order 7 for a particular range of values of Δt $(0.9930 \leqslant \Delta t \leqslant 0.9772)$, and different in character (actually, of class III type) outside the range on both sides.

Periodic limit sets of order 7 have been found for some range $I(i, j)$ of ε in 9 out of the 20 possible cases. For one of these, $(r, s) = (2, 1)$, $I(4, 7) \subset R^{7}_{(2,1)}$, while $L_{(2,1)0.10}$ is periodic with $k = 28$ (figures A-2, A-3). In the transition region, i.e., for $^{+}R^{7}_{(2,1)} < \varepsilon < {}^{-}R^{28}_{(2,1)}$, the limit sets are infinite. These are shown in figures A-4 and A-5 for the range $I(8, 9)$. They look like pseudo-periods, but, when suitably enlarged, they are seen to be of class I type (figures A-6, A-7). In these pictures one clearly sees with increasing ε the onset of instability—to use an expression from mechanics—and the eventual attainment of a different stable state. The transition region at the lower end of the range also contains infinite limit sets. Figures A-8 and A-9 show $L_{(2,1)0.03}$, first to normal scale, then enlarged. It is manifestly a class III limit set.

For other $T_{(r,s)\varepsilon}$ periods of order $k > 7$ are found for certain ranges of the parameter ε, viz: $k = 9, 16, 23, 30, 37, 46, 62, 148$. In two cases, two periods of relatively prime order are found in different sub-ranges of $I(1, 10)$. Thus $T_{(5,1)\varepsilon}$ has two periodic limit sets, one with $k_1 = 23$ for $\varepsilon = 0.01$ and one with $k_2 = 16$ over $I(9, 10)$. Similarly, $T_{(1,1)\varepsilon}$ has a periodic limit set with $k_1 = 16$ for $\varepsilon = 0.01$, and one with $k_2 = 9$ for $\varepsilon = 0.10$. In these cases the dependence of the limit set on ε in the transition region $^{+}R^{k_1}_{(r,s)} < \varepsilon < {}^{-}R^{k_2}_{(r,s)}$ is more complicated than that described above. For ε-values in this region and sufficiently close to the end-points we observe the expected pseudo-periodic limit sets. For values of ε not too close to either boundary the limit set may be either of class IV or of class I type. Figure A-10 shows $L_{(5,1)0.04}$ to normal scale; in figure A-11, a portion of the limit set is shown enlarged.

We conclude this sub-section with two further examples. These illustrate a phenomenon previously mentioned in our general description of limit sets (section II), namely the coexistence of finite periods and class IV sets. Figures A-12 and A-13 show two distinct limit sets belonging to $T_{(3,1)0.01}$. One is a period of order $k = 23$, while the other is a class IV set closely resembling $L(T_A)$. The same phenomenon is perhaps more strikingly illustrated by the case of $T_{(5,0)0.02}$. Here we find both a class IV limit set and a period of order $k = 148$ (figures A-14 and A-15). We can say virtually nothing in this case about the dependence of the limit set on the initial point. Current computing facilities and techniques are not sufficiently powerful to effect an acceptably accurate determination of the respective regions of convergence without using prohibitive amounts of computing time. We have, however, carried out a few numerical experiments, the results of which certainly confirm our first impression that the geometrical structure of these regions is immensely complicated.

4. Study of the associated transformations for other $T_{C_1C_2}$. In this sub-section we discuss a few additional examples to illustrate the dependence of infinite limit sets on the parameter ε. The relevant photographs and tables will be found in appendix II.

For our first example, we choose the transformation:

$$T_{C_1C_2} \equiv T_B\colon \quad \begin{aligned} C_1 &= \{2, 4, 6, 7, 9\}, \\ C_2 &= \{5, 8, 10\}. \end{aligned} \tag{21}$$

The class IV limit set $L(T_B)$ belonging to this transformation is shown in figure B-1. As is evident, it consists of three separate pieces. Each of these is, of course, a limit set for $T_B^{(3)}$. It is instructive to compare the limit sets $L_{\Delta t}(T_B)$ with those belonging to certain of the associated $T_{(r,s)\varepsilon}$. In appendix II we list the results for only one case: $(r, s) = (1, 0)$. The limit sets $L_{\Delta t}(T_B)$ and $L_{(1,0)\varepsilon}$ are described in table B. There are (at least) three ranges of Δt values for which $L_{\Delta t}(T_B)$ is periodic; for Δt close to unity the behavior of $L_{\Delta t}(T_B)$ as a function of Δt is rather *wild*. As Δt approaches $\Delta t_{\text{lim}} = 0.854320$ the (class I) limit set shrinks in a continuous manner. The behavior of $L_{(1,0)\varepsilon}$ as ε is varied over $I(1, 10)$ is, if anything, more "pathological"; there are at least 6 different intervals $R_k^{(1,0)}$ for which the limit set is periodic, and each period has a different order. Note the similarity in appearance between the two class IV limit sets: $L_{\Delta t}(T_B)$ $(\Delta t = 0.994)$ and $L_{(1,0)0.01}$.

The next two examples may be taken together:

$$(22) \qquad T_D: \quad \begin{aligned} C_1 &= \{2, 7, 8, 9, 10\}, \\ C_2 &= \{4, 5, 6\}; \end{aligned}$$

$$(23) \qquad T_E: \quad \begin{aligned} C_1 &= \{2, 5, 7, 8, 9\}, \\ C_2 &= \{4, 6, 10\}. \end{aligned}$$

The basic class IV limit sets $L(T_D)$ and $L(T_E)$ are shown in figures D.1 and E.1; their morphological resemblance is apparent. The behavior of the $L_{\Delta t}$ and $L_{(1,0)\varepsilon}$ for these two cases is set forth in the tables and photographs of appendix II. Detailed comment is perhaps superfluous at this state of our knowledge; we limit ourselves to drawing attention to the following comparisons:

1. Compare $L_{\Delta t}(T_D)$ ($\Delta t = 0.97$) with $L_{(1,0)0.10}(T_D)$.

2. Compare $L_{\Delta t}(T_E)$ ($\Delta t = 0.97$ and $\Delta t = 0.96$) with $L_{(1,0)0.09}(T_E)$ and $L_{(1,0)0.10}(T_E)$.

5. The original transformations T_B, T_D, T_E are closely related from the point of view of formal structure. T_D and T_E differ by exchange of a single term between the defining sets C_1 and C_2, while each of these goes over into T_B under the simultaneous interchange of two terms between C_1 and C_2. A comparison of the associated limit sets for T_D and T_E shows that the initial similarity of $L(T_D)$ to $L(T_E)$ is roughly preserved under perturbation. This suggests the possibility that some meaningful classification based on algebraic form might be devised ([56]). Of even greater interest is the correspondence, in these examples, between the $L_{\Delta t}$ and the $L_{(1,0)\varepsilon}$ for some ranges of the respective parameters. We are not at present in a position to draw any significant conclusions from the existence of this correspondence; it seems likely, however, that a closer study of these examples would yield criteria enabling one to predict such behavior.

6. There is one property of these transformations which may safely be inferred from the data, namely, that they are close to transformations having periodic limit sets (for some common set of initial points), where *close* is to be interpreted with reference to some appropriate parameter space—e.g., a range of ε values of Δt values. Their limit sets are "close" to periods, not in the sense that pseudo-periods are, but rather

---

([56]) The difference in behavior of $L_{\Delta t}(T_B)$ on the one hand and $L_{\Delta t}(T_D)$, $L_{\Delta t}(T_E)$ on the other is undoubtedly due in part to the fact that in the first case the jacobian matrix has complex eigenvalues at the fixed point, while for T_D and T_E the eigenvalues are real; this is probably sufficient to explain the qualitative difference of behavior of the corresponding $L_{\Delta t}$ as $\Delta t \to \Delta t_{\lim}$ for $\Delta t - \Delta t_{\lim}$ sufficiently small.

by virtue of the fact that they contain points which lie close—perhaps arbitrarily close—to a set of algebraic solutions of $T^k(p) = p$. In other words, the Hausdorff distance between the set of period points and the limit set L is small. In this connection, the following piece of evidence may be presented. Consider the transformation $T_{(1,0)0.01}$ associated with T_E, for which we have observed that the sequence $T^{(n)}_{(1,0)0.01}(p)$, $n = 1, 2, ...$, converges to a period of order $k = 10$ for almost all p. Let us choose a p close to the fixed point. If we then examine the sequence for $n = 1, 2, ..., N$, where N is sufficiently large, we find that the images $T^{(n)}(p)$ of p have traced out a pattern which closely resembles the original class IV limit set $L(T_E)$ of figure E.1. This is shown in figure E.2. The bright spots are the points belonging to the periodic limit set $L_{(1,0)0.01}$. Presumably this means that the effect of introducing a small perturbation into T_E, of the form specified by $T_{(1,0)0.01}$ ([57]) is to make the limit set $L(T_E)$ contract to 10 points. Alternatively, we could say that, as $\varepsilon \to 0$, the periodic limit set $L_{(1,0)\varepsilon}(T_E)$ *spreads out* until it becomes the class IV limit set $L(T_E)$.

This and other similar examples suggest that it might be useful to consider the period limit sets as fundamental, the hope being that one could develop an appropriate perturbation method, taking these periods as the *unperturbed states*. The effect of a small change of a parameter (in the direction of *instability*) is then simply to make the period non-attractive. This can in principle be studied by purely algebraic methods. Determining the structure of the resulting limit set—the *perturbed state*—is of course a more difficult matter.

In some cases this may amount to nothing more than the development of improved techniques for handling algebraic expressions of very high order. To clarify this statement, we offer one further example. Consider the Δt-modified transformations $T_{\Delta t}(T_A)$, where T_A is the transformation introduced in sub-section 2 above. For $\Delta t = 0.99300$, $L_{\Delta t}(T_A)$ is a period of order 7. With a very small change in Δt—namely, for $\Delta t = 0.99301$—the limit set is of class III type, a pseudo-period. Rather than investigating $T_{A(\Delta t)}$ let us turn our attention to the seventh power of the modified transformation, $T^7_{A(\Delta t)}(p)$ ($\Delta t = 0.99301$). If we choose our initial point p sufficiently close to one of the (repellent) fixed points of $T^7_{A(\Delta t)}$ ([58]), we find that the first 516 iterated images of p, $T^{(7n)}_{A(\Delta t)}(p)$, $n = 1, 2, ..., 516$, lie on an almost exact straight line in the S, a triangle. This is shown in figure A-16. The initial point p is at the lower

---

([57]) If T_E is written in the form: $S' = F(S, a)$, $a' = G(S, a)$, then $T_{(1,0)\varepsilon}$ is $S' = F(S, a) + \varepsilon(S - a)^3$, $a' = G(S, a)$.

([58]) The actual values are not known: we have not yet developed good techniques for finding the coordinates of the points of a non-attractive period. The initial point for this example was taken as: $S = 0.7034477$, $a = 0.1159449$, chosen on the basis of some simple numerical experimentation. It is close to one of the periodic points belonging to the limit set $L_{A(\Delta t)}$ ($\Delta t = 0.99300$), viz.: $S = 0.7037400$, $a = 0.1157123$.

right, and the successive images trace out the line continuously from right to left [59]. If we continue the iteration, we find that the later images deviate from the straight line, then oscillate in position, and finally settle down to generate another straight line with a different end-point—presumably very close to another fixed point of $T^{(7)}_{A(\Delta t)}$. It is clear that if one had powerful enough algebraic tools, one could calculate this linear behavior.

7. We close this section with two remarks:

1) A study of the $T_{\Delta t}$ and $T_{(r,s)\varepsilon}$ associated with those $T_{C_1C_2}$ which have only class I limit sets indicates that the latter are much more stable with respect to these one-parameter modifications than are the limit sets discussed above.

2) Even these *unstable* limit sets appear to be stable with respect to some one-parameter perturbations of the corresponding transformations. Thus the transformations $T_{(3,0)\varepsilon}$ associated with T_E have limit sets visually identical with $L(T_E)$ over the whole range $I(1, 10)$. Anomalies such as these make general pronouncements about absolute stability (or instability) impossible.

To illustrate: One might be tempted to explain the observed stability in this case as follows:

Explicitly, $T_{(3,0)\varepsilon}$ has the form:

$$S' = F(S, a) + \varepsilon(1 - S - a)^3, \qquad a' = G(S, a). \tag{24}$$

Now the density of $L(T_E)$ is relatively large near the right-hand boundary of the triangle, $S + a = 1$. The perturbing terms, however, vanishes on this line. Thus the transformation is *on the average* very little altered by the perturbation. But this "explanation" becomes less convincing when one looks at other transformations associated with T_E. $T_{(2,1)\varepsilon}$, for example, has the form:

$$S' = F(S, a) + \tfrac{1}{2}\varepsilon(1 - S - a)^3, \qquad a' = G(S, a) + \tfrac{1}{2}\varepsilon(1 - S - a)^3. \tag{25}$$

One would expect the same argument to apply here, but in fact the limit sets only resemble $L(T_E)$ over the two ranges $I(1, 2)$ and $I(6, 10)$. In between, we get the familiar periodic and pseudo-periodic behavior.

---

[59] The final point plotted has coordinates: $S = 0.7030206$, $a = 0.11628136$, so the slope of the line is roughly $\Delta a / \Delta S \simeq -0.713$. For this photograph, the scaling factor is approximately 2340.

V. RELATION TO THE THEORY OF DIFFERENTIAL EQUATIONS

1. As we remarked in section III, the non-linear transformations discussed in this paper exhibit certain analogies with systems of differential equations. In the following we confine ourselves to discussing the plane case.

An important study in the theory of differential equations, particularly as applied to non-linear mechanics, is that of so-called autonomous systems ([60], [61], [62]):

$$\frac{dx}{dt} = P(x, y), \qquad \frac{dy}{dt} = Q(x, y). \tag{1}$$

The theory, initiated by Poincaré, seeks to determine the properties of the solutions of (1) under very general conditions, and to deduce such properties for particular cases without actually solving the equations explicitly (i.e., obtaining the general integral). In particular, the trajectories, given parametrically as a function of t:

$$x = x(t), \qquad y = y(t) \tag{2}$$

are investigated from a topological point of view. Fundamental is the classification of the singular points of the system (1), that is, the points x, y, where $P(x, y) = Q(x, y) = 0$. The behavior of trajectories in the neighborhood of singular points can be found by consideration of the linear approximation to (1); the real object of the theory, however, is to characterize and, where possible, predict behavior in the large. One of the most interesting phenomena connected with behaviour in the large is the existence of closed trajectories, or *limit cycles*. The theorem of Poincaré and Bendixson ([63]) gives sufficient conditions for the existence of such. Unfortunately, the fulfillment of these conditions in particular

---

([60]) A general reference is N. Minorsky, *Non-linear Oscillations*, Princeton 1962. Full references are given here. More detail on theoretical points can be found in:

([61]) S. Lefschetz, *Differential Equations: Geometric Theory*, New York 1957. See also:

([62]) Nemitsky and Stepanov, *Qualitative Theory of Differential Equations*, Princeton 1960.

([63]) See reference in footnote 60.

cases is often hard to verify; to date no satisfactory theoretical method for dealing with an arbitrary given system has been found [64].

2. If we write our general two-dimensional system of non-linear difference equations in the form:

$$
(3)\qquad
\begin{aligned}
\frac{S^{(n)} - S^{(n-1)}}{\Delta t} &= -S^{(n-1)} + F(S^{(n-1)}, a^{(n-1)}),\\
\frac{a^{(n)} - a^{(n-1)}}{\Delta t} &= -a^{(n-1)} + G(S^{(n-1)}, a^{(n-1)}),
\end{aligned}
$$

the analogy with (1) is evident. The fixed points of (3) correspond to the singular points of (1), and the behavior of solutions in the neighborhood of a fixed point can be investigated via the linear approximation; this procedure, in fact, yields Ostrowski's criterion (see section III). If the fixed point is attractive, the asymptotic solution in its neighborhood can of course be obtained. In the case of repellent fixed points (or if the initial point is outside the region of attraction of all attractive fixed points), the sequence of iterates sometimes converges to a limit set which appears to resemble a Poincaré limit cycle, i.e. a closed curve. In other cases, finite limit sets (periods) are obtained; on the other hand, one may observe limit sets of quite ambiguous geometrical, not to say topological, structure. These last two alternatives have no analogues in the case of differential equations.

In fact, the analogy between (3) and (1) is more apparent than real. The significant distinction lies, perhaps, in the fact that for our difference equations there is nothing corresponding to the trajectories of (1); successive iterates do not in general lie close to each other. This fact makes it difficult to use topological arguments to determine the character of the limit set. For sufficiently small Δt the sequence of iterates may resemble a trajectory to some extent, but the limit as $\Delta t \to 0$ is almost certain to be a single point [65].

---

[64] See reference in footnote 62, appendix I. The practical applications are largely confined to stability theory. See reference in footnote 60 and the literature there cited.

[65] It may happen that some power $T^{(n)}$ of a transformation more closely resembles a trajectory; cf. the example cited in section II.

VI. BROKEN-LINEAR TRANSFORMATIONS IN TWO DIMENSIONS

1. For certain special quadratic transformations in one dimension one can give an almost complete discussion of the iterative properties; this is possible because these transformations are conjugate to piecewise linear (*broken-linear*) mappings of the interval into itself. For example, the transformation: $x' = g(x)$, where $g(x) = 2x$, $0 \leqslant x \leqslant \frac{1}{2}$; $g(x) = 2 - 2x$, $\frac{1}{2} \leqslant x \leqslant 1$ (66). The iterative properties of the latter can be obtained from a study of the law of large numbers for the elementary case of Bernoulli. Stated differently, the behavior of iterates of this simple quadratic transformation turns out to depend on combinatorial rather than analytic properties of the function. With this in mind, we tried to see whether an analogous situation would obtain in two dimensions. Our non-linear, polynomial transformations of a triangle into itself might, we thought, be similar to suitably chosen broken-linear mappings of a square into itself, at least as regards their asymptotic behavior.

One simple generalization to two dimensions of broken-linear transformations in one variable is a mapping:

$$(1) \qquad x' = f(x, y), \qquad y' = g(x, y),$$

where each of the functions f and g is linear in regions of the plane. In other words, the graphs of these functions consist of planes fitted together to form pyramidal surfaces. The motivation for studying such transformations is the hope that their iterative properties will turn out to depend only on the folding of the plane along straight lines or, more specifically, on the combinatorics of the overlap of the various linear regions which is generated by the mapping. The simplest non-trivial case to investigate consists in taking $f(x, y)$ as a function defined by choosing a point in the square and making f maximum at this point, the function being linear in the four triangles into which the square is divided. $g(x, y)$ is defined in an analogous manner.

Each of the functions $f(x, y)$, $g(x, y)$ is thus made to depend on three parameters. Thus for f we choose a point x_1, y_1 in the square and erect a perpendicular of height

(66) See further in appendix I.

$0 < d_1 \leqslant 1$ at this point; this defines a surface consisting of 4 intersecting planes. The transformation can then be given explicitly as follows:

$$
(2) \qquad
\begin{aligned}
&\text{I:} && x' = \frac{d_1}{x_1} x , \\
&\text{II:} && x' = \frac{d_1}{1-y_1}(1-y) , \\
&\text{III:} && x' = \frac{d_1}{1-x_1}(1-x) , \\
&\text{IV:} && x' = \frac{d_1}{y_1} y ,
\end{aligned}
$$

where the regions I to IV are specified by the bounding lines:

$$
(3) \qquad
\begin{aligned}
&L_1: && y = \frac{y_1}{x_1} x , \\
&L_2: && y = \frac{y_1 - 1}{x_1} x + 1 , \\
&L_3: && y = \frac{1-y_1}{1-x_1} x + \frac{y_1 - x_1}{1-x_1} , \\
&L_4: && y = \frac{y_1}{1-x_1}(1-x) ,
\end{aligned}
$$

then:

$$
(4) \qquad
\begin{aligned}
&\text{region I is bounded by } L_1,\ L_2 \text{ and } x = 0 , \\
&\text{region II is bounded by } L_2,\ L_3 \text{ and } y = 1 , \\
&\text{region III is bounded by } L_3,\ L_4 \text{ and } x = 1 , \\
&\text{region IV is bounded by } L_4,\ L_1 \text{ and } y = 0 .
\end{aligned}
$$

Analogous equations hold for $y' = g(x, y)$, with parameters x_2, y_2, d_2.

Of the several transformations of this type that we have studied numerically we mention only the following:

$$
(5) \qquad T_1: \quad
\begin{aligned}
&x_1 = \tfrac{1}{3}, \quad y_1 = \tfrac{1}{3}, \quad d_1 = 0.95 , \\
&x_2 = 0.6, \quad y_2 = 0.5, \quad d_2 = 0.95 ,
\end{aligned}
$$

$$
(6) \qquad T_2: \quad
\begin{aligned}
&x_1 = 0.5, \quad y_1 = 0.9, \quad d_1 = 1 , \\
&x_2 = 0.3, \quad y_2 = 0.7, \quad d_2 = 0.8 ,
\end{aligned}
$$

and the one-parameter family:

$$
(7) \qquad T_z: \quad
\begin{aligned}
&x_1 = y_1 = z, \qquad d_1 = 1 , \\
&x_2 = y_2 = 1 - z, \quad d_2 = 1 .
\end{aligned}
$$

The limit sets are shown in figures H-1, H-2 and H-3 through H-17 of appendix II.

$L(T_2)$ (figure H-2) is, in a sense, analogous to the class I limit sets we observed for some of our cubic transformations in three variables. In contrast, $L(T_1)$ (figure H-1) represents a new phenomenon—a connected "curve" (in this case, a collection of line segments) that does not close. More interesting, however, is the behavior of the limit sets $L(T_z)$ as z varies from 0.49 down to 0.01 ([67]). Initially $L(T_z)$ is an *open cycle* like $L(T_1)$. With decreasing z, the limit set becomes more complex, until it resembles a class IV limit set (e.g., $\hat{z} = 0.27$). With further decrease of z, the limit set appears to contract; the *tails* get shorter, and the points cover some sub-region of the square more and more densely. One interesting question—unanswered at the time of this writting—is: does T_z become ergodic for some range of z values? In figure H-10 we see 1000 (consecutive) points belonging to $L(T_z)$ for $z = \frac{1}{4}$. Figure H-11 shows these same 1000 points together with the next 2000 points. It is evident that the region containing $L(T_z)$ is *filling in*. In our opinion, this is a strong indication of ergodicity.

For lower values of z, the same *ergodic* behavior is observed, until, at $z = 0.15$, the limit set splits into four disconnected pieces (figure H-15). As z is further decreased, these four pieces shrink; by the time we reach $z = 0.01$, the limit set is *nearly* a single point.

2. It is clear that one can devise broken-linear transformations that are dense in the unit square; one may take, for example, product transformations with *independent* coefficients and use the one-dimensional result for each factor. For transformations of the type considered in this section, however, it is not easy to determine a priori what the limit set will be. It should be emphasized that there is no hope of demonstrating that our polynomial transformations in three variables are exactly c o n j u g a t e to some two-dimensional broken-linear transformations. Presumably there is no such conjugacy. Nevertheless, one might hope that a somewhat weaker notion of equivalence than that of strict conjugacy could be introduced.

One suggestion along these lines is the following: define two transformations T and S to be asymptotically similar if for almost every initial point p the limit set $L_p(T)$ is topologically equivalent to $L_{p'}(S)$ for some suitably chosen p', and vice versa. Thus, for example: if for any two transformations T, S, the sets of iterates $T^{(n)}, S^{(n)}$ are dense in the (common) domain of definition for almost every initial

---

([67]) The case $z = \frac{1}{2}$ has not been studied. The reason for this is technical. Straightforward iteration will always produce sequences which degenerate to zero in a finite number of steps, owing to the fact that every iteration involves multiplication by 2. In a binary machine, this operation is a "shift" to the left. A sufficiently long chain of such left shifts will always result in zero. If one wants to study this case, one must replace multiplication by 2 by some arithmetically equivalent operation, e.g., multiplication by $C/(\frac{1}{2}C)$, where C is not a power of 2.

For this case ($z = \frac{1}{2}$), the problem becomes, of course, one-dimensional. The iterates remain on the line $x = y$, and the limit set is identical with that of the transformation $x' = g(x)$ introduced in sub-section 1 above.

point, then T and S are asymptotically similar in the above sense. Another special case in which two transformations, T and S, are asymtotically similar is when each transformation possesses just one attractive fixed point, the region of attraction being, in both cases, the whole space.

As we remarked above, in the case of broken-linear transformations the asymptotic behavior of iterates depends only on the combinatorial structure of the subdivisions of the fundamental regions (triangles) under repeated folding. Just how complicated this can be is shown by the behavior of the limit sets $L(T_z)$ belonging to the one-parameter family T_z discussed above. To date we have not managed to devise any good method for handling the Boolean algebra of these iterated intersections.

APPENDIX I

In this appendix we collect some general remarks about the process of iterating transformations, particularly in one dimension. We also discuss, in some detail, a few special one-dimensional transformations which we have had occasion to study.

1. One of the first, simple, transformations whose iterative properties were established is the following:

$$x' = f(x) \equiv 4x(1-x) . \tag{1}$$

To obtain these properties we consider, instead of (1), the broken-linear transformation (68):

$$x' = g(x), \quad g(x) = \begin{cases} 2x & \text{for } x \leqslant \frac{1}{2}, \\ 2(1-x) & \text{for } x > \frac{1}{2}. \end{cases} \tag{2}$$

The study of the iterates of this transformation is equivalent to investigating the iterates of a function $S(x)$ defined as follows:

(3) if $x = 0.a_1 a_2 a_3 \ldots a_n \ldots ,$ where the a_i are either 0 or 1,

(4) then $S(x) = 0.a_2 a_3 a_4 \ldots$

In other words, $S(x)$ is merely a *left shift* of the binary word x by one place. The iterative properties of $S(x)$ are in turn deducible from the law of large numbers in the case of Bernoulli. In effect, $S^{(i)}(x)$ falls into the first half of the interval if and only if $a_i = 0$. The *ergodic average*

$$\frac{1}{N} \sum_{i=1}^{N} F_{(I)}[S^{(i)}(x)]$$

is therefore the same as the fraction of ones among the a_i for $1 \leqslant i \leqslant N$. $F_{(I)}$ is the characteristic function of the interval $[0, \frac{1}{2}]$.

The relation between (1) and (2) is that of conjugacy: there is a homeomorphism $h(x)$ of the interval $[0, 1]$ with itself such that

$$g(x) = h\left[f[h^{-1}(x)]\right] . \tag{5}$$

(68) This transformation has already been mentioned in section VI.

Thus the study of the iterates of the quadratic transformation (1) reduces to the corresponding study for the broken-linear transformation (2). In this case, $h(x)$ can be written down explicitly [69]:

$$h(x) = \frac{2}{\pi} \sin^{-1}(\sqrt{x}). \tag{6}$$

2. The set of exceptional points. In the case of the function $f(x) = 4x(1-x)$, it is true, then, for almost every [70] initial point, that the sequence of iterated images will be everywhere dense in the interval, and what is more, the ergodic limit can be explicitly computed; it is positive for every sub-interval.

There exist, however, initial points x such that the sequence x, $f(x)$, $f[f(x)]$, ... is not dense in the whole interval $[0, 1]$. Obviously, all periodic points, i.e., points such that, for some n, $f^{(n)}(x) = x$, are of this sort. It is interesting to notice, however, that there exist points x for which the sequence $f^{(i)}(x)$, $i = 1, 2, \ldots$, is infinite without being dense; there are, in fact, non-countably many such points. To show this we consider the equivalent problem of exhibiting such points for the function $S(x)$ introduced above. The construction then proceeds as follows. Consider a point $x = 0.a_1 a_2 \ldots a_n \ldots$ We define a set Z consisting of all those x's which have $a_n = a_{n+1} = a_{n+2}$ for all n of the form $n = 3i$. In other words, the set Z consists of points which have every binary digit repeated three times, the sequence being otherwise arbitrary. Consider now the transformation $x' = S(x)$, where $S(x)$, as defined in (4) above, is a shift of x one index to the left. We now look at the sub-interval from 0.010 to 0.011. Starting with any point in Z, it is clear that no iterated image will fall in this sub-interval; no three successive binary digits of a point in Z are of the form (010). It is easy to see that Z contains non-countably many points; in particular, it contains non-periodic points, so that the set of images $S^{(i)}(x)$, $i = 1, 2, \ldots$, is infinite, but not dense in $[0, 1]$.

Presumably, one can find points in Z for which the ergodic limit exists. The measure of the set Z is zero, but, relative to Z, the set S of those points for which the *sojourn time* exists still form a *majority*. We may define *majority* either in the sense of Baire category or as follows. Take points $na \pmod 1$, $n = 1, 2, \ldots$, where a is an irrational constant. Consider the set N_1 of those n's for which $na \pmod 1$ belongs to Z, and also the set N_2 of those n's for which $na \pmod 1$ belongs to S. We then say that the points belonging to S form a *majority* of those points belonging to Z if the relative frequency of N_2 in N_1 is one.

---

[69] This result was first published by S. Ulam and J. von Neumann, American Mathematical Society Bulletin, Vol. 53 (1947), p. 1120, Abstract 403. See also the work of O. Rechard, Duke Mathematical Journal, Vol. 23 (1956), pp. 477-488.

[70] *Almost every* is to be understood in the sense of Lebesgue measure.

The behavior exhibited by the points belonging to Z is more general, in the sense of measure, for some other transformations of the interval $[0, 1]$. It is possible to give examples of continuous functions such that, for almost every point, the iterated sequence will be nowhere dense in the interval, although the sequence does not converge to a fixed point.

3. A remark on conjugacy.

THEOREM. *Let $g(x)$ be the broken-linear function of equation* (2) *above, i.e.,*

$$g(x) = \begin{cases} 2x, & 0 \leqslant x \leqslant \frac{1}{2}, \\ 2(1-x), & \frac{1}{2} > x > 1. \end{cases} \tag{2}$$

Let $t(x)$ be a convex function on $[0, 1]$ which transforms the interval into itself, and such that $t(0) = t(1) = 0$. For some p in the interval, we must have $t(p) = 1$; by convexity, there is only one such point. Consider the lower tree (71) *generated by the point* 1. *The necessary and sufficient condition that $t(x)$ be conjugate to $g(x)$ is that this tree be combinatorially the same as that generated by* 1 *under $g(x)$, and that the closure of this set of points be the whole interval, i.e., that the tree be dense in* $[0, 1]$.

The condition is obviously necessary, since the point 1 generates a tree under $g(x)$ which is simply the set of binary rational points. Under any homeomorphism $h(x)$ which has to effect the conjugacy, the point $\frac{1}{2}$ must go over into p, and our assertion follows.

To prove sufficiency, we construct $h(x)$ in the following manner.

We take $h(\frac{1}{2}) = p$ by definition. We next choose $h(\frac{1}{4})$ to be the smaller of the two values of $t^{-1}(p)$; $h(\frac{3}{4})$ is then by definition the larger of these two values. We then take $h(\frac{3}{8})$ to be the smaller of the two values of $t^{-1}[h(\frac{3}{4})]$, and so on. Proceeding in this fashion, we thus construct a function $h(x)$ defined for all binary rationals. It remains to prove that we can define it for all x by passage to the limit. This, however, follows from the assumption that these points are dense in $[0, 1]$ and that their order is preserved. The function $h(x)$ will obviously be monotonic, and, being continuous, will possess an inverse $h^{-1}(x)$. From our construction it then follows that $h[g(x)] = t[h(x)]$.

4. Broken-linear transformations. In one dimension these are functions $f(x)$ that are continuous on $[0, 1]$ and linear in sub-intervals of $[0, 1]$. We assume that the graph of the function has a finite number of vertices, i.e., that $f(x)$ consists of a finite number of lines fitted

(71) By the *lower tree* of p (under $f(x)$) we understand the smallest set of all points with the following properties:

(a) The set contains the given point p.

(b) If a point belongs to the set, then so do all its counter-images under f.

together continuously. For these broken-linear transformations one certainly expects that the ergodic limit exists for almost every point. For example: if one considers the sequence of iterated images $T^{(n)}(p)$, then, for almost every initial point p, the time of sojourn should exist for all sub-intervals, i.e.,

$$\lim_{N\to\infty} \frac{1}{N} \sum_{i=1}^{N} f_R[T^{(i)}(p)]$$

should exist for almost all p and all measurable sets R; here f_R is the characteristic function of the set R ([72]). The value of this limit may indeed depend on the initial point p; it is likely, however, that all the points of the interval can be divided into a finite number of classes such that, within each class, the value of the limit is the same ([73]).

There is another *finiteness* property that these transformations may possess. Given n, consider all broken-linear transformations which have at most n pieces (i.e., the space divides into n regions, in each of which the transformation is linear). Then it may be conjectured that there are only a finite number of different types of such transformations, where any two transformations of the same type are asymptotically similar (in the sense defined above, section VI). In other words, according to this conjecture, the *type* (or class) that a given transformation belongs to does not depend on the precise numerical values of the coordinates of those points where the derivative is undefined (*corner points*), but is determined only by the combinatorial structure of the subdivision of space into linear domains. In one dimension this means that the *type* of a transformation is determined by the number and inter-relation of the nodes in the graph of the function, and not by their precise location.

([72]) We should perhaps mention here a more general conjecture. Suppose T is a polynomial transformation of the sort described by equations (10) to (13) of the Introduction. We then conjecture that the sequence of iterated images $T^{(n)}(p)$ has the following property: Let C be any cone of directions in n-space, and let $f_C(p)$ be the characteristic function of this cone, i.e., $f_C(p) = 0$ if p does not belong to C, $f_C(p) = 1$ otherwise. Then, for almost every p,

$$\lim_{N\to\infty} \frac{1}{N} \sum_{i=1}^{N} f_C[T^{(i)}(p)]$$

exists. Cf. the article by S. Ulam in *Modern Mathematics for the Engineer*, second series, edited by E. F. Beckenbach, New York 1961, p. 280.

([73]) See pages 71, 72 of S. Ulam, *A Collection of Mathematical Problems*, New York 1960. An analogous conjecture can be made concerning the actual limit sets $L_p(T)$ for our cubics in three variables. Cf. the discussion in sub-section 2 of section II of this paper.

5. Numerical accuracy. The machines we use to compute the iteration process work with a fixed number of significant digits [74]; in MANIAC II, for example, this number is 8. It is therefore clear that any direct, single-step iterative process carried out on this computer will exhibit a period in not more than 10^8 steps. Given an algorithm which is iterative and of first order (i.e., the nth step depends only on the $(n-1)^{\text{st}}$), the process will, with great probability, exhibit a period which is much shorter. Statistically, one can reason as follows. If we assume a random distribution of, say, the last 4 digits of all computed numbers, then the probability that the cycle will close long before the full theoretical run of 10^8 steps is extremely close to one. Indeed, after producing numbers $A_1, A_2, \ldots, A_k$, the chance that A_{k+1} will be equal to one of the preceding numbers is of the order of $k/10^8$. If we continue the calculations up to A_{2k}, the chance that at least two numbers in the chain will coincide is approximately $1-\left(1-\frac{k}{10^8}\right)^k$. This is practically equal to $1-\frac{1}{e}$ if $k \sim 10^4$. Clearly, going to $3k, 4k, \ldots$, the probability that the chain will be cyclic gets very close to one. The situation is quite different in an iterative process involving two or more variables, e.g., computing (A_{k+1}, B_{k+1}) from (A_k, B_k). On a probabilistic basis alone, one expects to encounter periods of length $\sim 10^8$ (the maximum possible being 10^{16}). In practice, this means that in two dimensions one does not expect to encounter accidental or *false* periodicity unless one generates very long sequences.

As the argument above shows, fortuitous periodicity can be a very real danger in one-dimensional iterative calculations. Indeed, we came across a striking example in the course of studying the asymptotic properties of the transformation

$$y' = \sin \pi y, \quad 0 \leqslant y \leqslant 1\,. \tag{7}$$

This classical transformation is symmetric about $y = \frac{1}{2}$, and maps the unit interval into itself. Furthermore, both fixed points — $y = 0$ and $y = 0.7364845$ — are repellent. Now a simple argument shows that for such a function there cannot be any attractive periods of any finite order, that is, the derivative

$$\frac{d}{dy} T^{(k)}(y)\big|_{y=y_p}$$

[74] For any particular machine one can increase the accuracy by resorting to so-called *multiprecision arithmetic*. This can generally be done only at the cost of considerable loss in speed.

evaluated at any periodic point $T^{(k)}(y_p) = y_p$ is always greater than one in absolute value ([75]). Nevertheless, when we performed the iteration on MANIAC II, the limit set we observed was invariably one of two finite periods, the first of order 1578, the second of order 6168. These periods were exact to the last available binary digit! This *spurious convergence* is undoubtedly produced by the complicated interaction of several factors, e.g., the particular machine algorithms for multiplication and round-off, our choice of finite approximation to the function $\sin \pi y$, and so forth. When we repeated the calculation on the "STRETCH" computer (which works with about 15 significant figures), the periods were no longer observed; even after a million or so iterations, there was no observable tendency toward convergence to such periods.

If one is interested in obtaining the asymptotic distribution of iterates under some transformation like (7), one must either provide for more significant figures or resort to ingenious devices. One such *trick*—which appears to work well and is relatively convenient—consists essentially in computing the inverse transformation. Since the functions under consideration are two-valued, this involves introducing a random choice at each step. Specifically, taking some initial point p_0, we compute the sequence

$$p_0, f^{-1}(p_0), f^{-2}(p_0), \dots$$

Here the symbol $f^{-k}(p_0)$ implicitly contains the prescription that we choose one of the two values of the true inverse at random; thus the sequence

$$f^{-i}(p) = f[f^{-(i-1)}(p)], \quad i = 1, 2, \dots,$$

implies a sequence of random decisions as to which counter-image to choose at each step. If the calculation is carried out in this fashion, the chance of falling into an exact period is vanishingly small until we reach a chain length in the neighborhood of the theoretical maximum. Once having obtained our inverse sequence, we can conceptually invert it,

([75]) It is essential for this argument that the transformation maps the whole interval into itself. The number of distinct periods of order k can then be easily enumerated. The conclusion follows on noting that

$$\frac{d}{dy} T^{(k)}(y)$$

must be the same for all points y belonging to the same period. If, however, the maximum value of the function is less than one, $T^{(k)}(y)$ can have relative minima above the y axis; then, indeed, there may exist attractive periods, i.e., the line $y' \equiv T^{(k)}(y) = y$ may intersect the curve sufficiently close to an extremum so that the kth order fixed point is attractive.

that is pretend that we started with the last point and proceeded to p_0 by direct iteration ([76], [77]).

6. In this sub-section we present some results of a study of a particular one-dimensional transformation ([78]):

$$(8)\qquad T_\sigma:\quad y' = W(3-3W+\sigma W^2),\qquad W \equiv 3y(1-y)\,.$$

This arises in a natural manner from a certain sub-class of our generalized cubic transformations in three variables:

$$(9)\qquad x'_i = \sum_{j=1}^{10} d_{ij} M_j\,,$$

$$(10)\qquad \sum_{i=1}^{3} d_{ij} = 1,\ \text{all } j,\qquad 0 \leqslant d_{ij} \leqslant 1\,.$$

Namely, we choose certain of the d_{ij} as follows:

$$d_{13} = \sigma,\qquad d_{17} = d_{19} = d_{34} = d_{36} = 1,\qquad d_{31} = 0\,.$$

The rest are arbitrary, except that they must, of course satisfy (10).

If we restrict ourselves to the sub-class of initial points such that $x_3 = 0$ (i.e., the side $S + a = 1$ of our reference triangle), the second power of the transformation can be written in the form (8).

T_σ is symmetric about $y = \frac{1}{2}$, but it does not map the whole interval $[0, 1]$ into itself; its maximum value is

$$(11)\qquad y'_{\max} = T_\sigma(\tfrac{1}{2}) = \frac{9}{16}\left(1+\frac{3\sigma}{4}\right)$$

so this is the right-hand boundary of the invariant sub-region. The left-hand boundary is then the image of $y'_{\max}$. In the range $1 \geqslant \sigma \geqslant 0.9$,

---

([76]) One cannot, of course, actually reverse the calculation and expect to reproduce the sequence. If one could, there would be no need for this stochastic device.

([77]) This procedure presupposes a good method for generating random numbers. There are several of these which are well suited for use in automatic digital computers. One of the most common—and, in fact, the method used by us—is to generate the numbers by the chain:

$$R_{n+1} \equiv R_0 R_n\ (\mathrm{mod}\ 2^k)\,,$$

where k is the binary word length of the machine, and R_0 is some properly chosen constant. On MANIAC II, $R_0 = 5^{13}$. This chain closes, but its length is greater than 10^{12}. The lowest order binary digits are themselves not random, but this makes no difference in practice.

([78]) This transformation was introduced by the way of illustration in section I.

the fixed point varies from $y_0 = 0.8224922$ at $\sigma = 1$ down to $y_0 = 0.8193719$ at $\sigma = 0.90$. Over this range, the derivative at y_0 is negative and greater than 1 in absolute value; thus the fixed point is repellent.

We have studied this transformation as a function of the two parameters—the initial point y_0 and the coefficient σ—on the "STRETCH" computer. To study the asymptotic distribution, we divide the interval into 10000 equal parts and have the machine keep track of the number of points which fall into each sub-interval over the iteration history. Such a history was usually taken to be a sequence of length $n \times 10^5$, with $4 \leqslant n \leqslant 9$ (79). Should a period (of order $k \leqslant 3 \times 10^5$) be detected by the machine, the calculation is automatically terminated. If the order of the period is not too great ($k \leqslant 500$), the values of the periodic points are printed out, and the value of $\frac{d}{dy} T_\sigma^{(k)}(y)$ is calculated over the period; if $\left| \frac{d}{dy} T_\sigma^{(k)}(y) \right| < 1$, this fact is strong evidence that the observed period is actually a limit set.

Our numerical investigation of (8) has been mostly restricted to the range $0.98 \leqslant \sigma \leqslant 1$ (80). Even in this restricted parameter range, the observed asymptotic behavior is of bewildering complexity. For $\sigma = 1$, the distribution of iterates in the interval is extremely non-uniform. There are large peaks at the end-points of the allowed sub-interval, with much complicated *fine-structure* in between, i.e., many relative maxima as well as sizeable intervals in which the distribution is locally uniform. This general behavior persists as σ is decreased down to $\sigma = 0.9902$. At $\sigma = 0.9901$, however, a dramatic change takes place; most of the points concentrate in a few small sub-intervals. The limit set is, in fact, a pseudo-period (81). At $\sigma = 0.99009$, the pseudo-periodic behavior is still evident, but the occupied sub-intervals are larger. They contract, however, for $\sigma = 0.99008$, and at $\sigma = 0.990079$ a period of order $k = 42$ is found. Actually, for this particular value of the parameter σ, there appear to be two possible periodic limit sets, with $k = 42$ and $k = 84$ respectively. The dependence on the initial point is, of course, quite complicated[82]. For example, as y_0 is varied over the values $y_0 = 0.11$,

(79) As we have already remarked (section I), the calculation of 2×10^5 iterates of this transformation requires about 1 minute on "STRETCH".

(80) Below $\sigma = 0.98$, various periodic limit sets of order $k \leqslant 24$ were found on MANIAC II. We have not studied this parameter range on "STRETCH".

(81) It may actually be a period with order $k = 63049$. This is the result indicated by the machine. At present we have no way of verifying this.

(82) Subsequent numerical work has shown that this y_0 dependence is spurious. The calculations were performed with multi-precision arithmetic; in some cases as many as its decimal places were retained! (Note added in print.).

0.12, 0.13, ..., 0.21, the period with $k = 42$ is found in all except three cases, namely $y_0 = 0.11, 0.16, 0.18$, which apparently lead to the periodic limit set with $k = 84$. Both periods are numerically well-attested; they are exact to the last binary digit and have $\left| \frac{d}{dy} T_\sigma^{(k)}(y) \right| < 1$ (83).

As we continue down in σ, the limit sets are pseudo-periodic (84) until we reach $\sigma = 0.99007$, for which a period of order $k = 112$ is found. This appears to be the approximate right-hand boundary of a *periodic belt*; that is, in the range $0.98990 \leqslant \sigma \leqslant 0.99007$ there are only periodic limit sets for this transformation. All these have orders which are multiples of 14 (with $k \leqslant 112$), except for the (approximate) left-hand boundary of the σ-range ($\sigma = 0.9899$) where a period of order 7 exists. The dependence of these limit sets on the initial point is complicated, and will not be reproduced here.

At $\sigma = 0.98988$ there is another large discontinuity in behavior; we again find an asymptotic distribution which covers the whole allowable sub-interval in non-uniform fashion. No more periodic limit sets are found as σ is decreased to $\sigma = 0.9800$. The only phenomenon of note is the splitting of the limit set into two parts; this occurs somewhere between $\sigma = 0.986$ and $\sigma = 0.985$. The resulting gap—which contains the fixed point—continues to widen as $\sigma \to 0.980$ (85).

The results of this investigation—which is rather incidental and sub-ordinate to the larger study reported in the present paper—clearly show that there is a great deal to be learned about the asymptotics of

---

(83) It is difficult to decide whether this behavior is real or not. Some supporting evidence for its *reality* is the observed behavior for values of σ very close to this *critical* value $\sigma = 0.990079$. We find that, with $\sigma = 0.9900789$, only a period with $k = 42$ is obtained, while on the other side, $\sigma = 0.9900791$, the limit set is always a period with $k = 84$.

(84) In this range, the machine *detected* some exact periods of huge order—e.g., $k = 295148$ ($\sigma = 0.990079$)! There seems to be no compelling reason to take this at face value.

(85) Such gaps have been observed in other cases. One interesting example is the transformation:

$$y' = W^2(3-2W), \qquad W \equiv 3y(1-y),$$

which is also a special case of one of our cubics in three variables. For this transformation, the limit set consists of 4 separate pieces, as follows: (I) $0.3455435 < y < 0.4086018$, (II) $0.4296986 < y < 0.5791385$, (III) $0.7562830 < y < 0.8146459$, (IV) $0.8220964 < y \leqslant 0.84375$.

In the present case, i.e. that of T_σ, the existence of the gap can easily be predicted. Let $y_0(\sigma)$ be the fixed point. There will clearly be a gap providing $T_\sigma^2(y_{\max}) > y_0(\sigma)$, since in this case only one of the two inverses of y_0 lies in the allowed interval. Of course, the determination of the critical value σ_c of σ from the equation $T_\sigma^2(y_{\max}) = y_0(\sigma)$ would in any event have to be carried out numerically.

iterative processes, even in one dimension. It seems that "pathological" behavior is not a property of higher-dimensional systems alone. With regard to our study of the periodic limit sets, it may be argued that what we have really done is to investigate, in a rather indirect manner, the behavior of the roots of high-order algebraic equations as a function of their coefficients. This is certainly true, at least in part. It is therefore of interest to observe that the *iterative* method seems at present to be the only effective tool for treating this purely algebraic problem.

APPENDIX II

1. In this appendix we collect the photographs and tables illustrating the phenomena discussed in section IV and VI. The notation used has been described in sections II, III, IV and VI. For convenience, some of the transformations are written out explicitly.

2. Modifications of the transformation T_A. In shorthand notation, this transformation is

$$T_A: \quad \begin{aligned} C_1 &= \{3, 5, 7, 9, 10\}, \\ C_2 &= \{1, 2, 8\}. \end{aligned} \tag{1}$$

In the S, a coordinates, this reads:

$$\begin{aligned} S' &= F(S, a) \equiv S^3 - 6S^2a - 3Sa^2 + 4a^3 - \tfrac{3}{2}S^2 + 3Sa - \tfrac{3}{2}a^2 + 1, \\ a' &= G(S, a) \equiv -S^3 + 3Sa^2 + 2a^3 + \tfrac{3}{2}S^2 - 3Sa + \tfrac{3}{2}a^2. \end{aligned} \tag{2}$$

The (repellent) fixed point has coordinates:

$$\begin{aligned} S_0 &= 0.5885696, \\ a_0 &= 0.1388662. \end{aligned} \tag{3}$$

The *generalized* transformations based on T_A may be written:

$$\begin{aligned} S' &= (1 - \Delta t)S + \Delta t\{F(S, a) + \varepsilon f_{rs}(S, a)\}, \\ a' &= (1 - \Delta t)a + \Delta t\{G(S, a) + \varepsilon g_{rs}(S, a)\}; \end{aligned} \tag{4}$$

the original transformation is recovered by setting $\Delta t = 1$, $\varepsilon = 0$.

TABLE A

Figure Number	Δt	ε	r, s	Comments
A-1	1	0.02	6, 1	Class IV limit set
A-2	1	0.07	2, 1	Period: $k = 7$
A-3	1	0.10	2, 1	Period: $k = 28$
A-4	1	0.08	2, 1	Class I limit set
A-5	1	0.09	2, 1	Class I limit set
A-6	1	0.08	2, 1	Scaled plot of part of A-4. Scaling factor ~77 ($0.472 \leqslant S \leqslant 0.485$, $0.125 \leqslant a \leqslant 0.135$)
A-7	1	0.09	2, 1	Scaled plot of part of A-5. Scaling factor ~37 ($0.465 \leqslant S \leqslant 0.492$, $0.120 \leqslant a \leqslant 0.140$)

TABLE A (continued)

Figure Number	Δt	ε	r, s	Comments
A-8	1	0.03	2, 1	Pseudo-period
A-9	1	0.03	2, 1	Scaled plot of part of A-8. Scaling factor $\sim$5.6 ($0.440 \leqslant S \leqslant 0.620$, $0.175 \leqslant a \leqslant 0.265$)
A-10	1	0.04	5, 1	Class IV limit set
A-11	1	0.04	5, 1	Scaled plot of part of A-10. Scaling factor $\sim$5 ($0.44 \leqslant S \leqslant 0.64$, $0.20 \leqslant a \leqslant 0.30$)
A-12	1	0.01	3, 1	Period: $k = 23$
A-13	1	0.01	3, 1	Class IV limit set
A-14	1	0.02	5, 0	Period: k = 148
A-15	1	0.02	5, 0	Class IV limit set
A-16	0.99301	1	—	514 successive iterates of $T^{(7)}_{A(\Delta t)}(p)$. Initial point: $S_0 = 0.7034477$, $a_0 = 0.1159449$. Scaling factor $\sim$909.

3. Modifications of T_B. In shorthand notation, this transformation is

$$T_B: \quad \begin{aligned} C_1 &= \{2, 4, 6, 7, 9\}, \\ C_2 &= \{5, 8, 10\}. \end{aligned} \tag{5}$$

In the S, a coordinates, it takes the form

$$\begin{aligned} S' &= F(S, a) \equiv 9S^2a - a^3 - \tfrac{3}{2}S^2 - 9Sa - \tfrac{3}{2}a^2 + \tfrac{3}{2}S + \tfrac{9}{2}a, \\ a' &= G(S, a) \equiv -3S^2a + 3a^3 - \tfrac{3}{2}S^2 + 3Sa - \tfrac{3}{2}a^2 + \tfrac{3}{2}S - \tfrac{3}{2}a. \end{aligned} \tag{6}$$

The coordinates of the fixed point are

$$\begin{aligned} S_0 &= 0.6887703, \\ a_0 &= 0.1592083. \end{aligned} \tag{7}$$

The only associated $T_{(r,s)\varepsilon}$ discussed is the case $(r, s) = (1, 0)$; for this case, the generalized transformation, written in the form of equation (4), has

$$\begin{aligned} f_{rs} &= f_{10} = (S - a)^3, \\ g_{rs} &= g_{10} = 0. \end{aligned} \tag{8}$$

In table B below, all scaled plots show the region: $0.30 \leqslant S \leqslant 0.65$, $0.20 \leqslant a \leqslant 0.36$, i.e., the upper left-hand piece of the complete limit set (shown in figure B-1 for the unmodified transformation). The scale factor is $\sim$2.9.

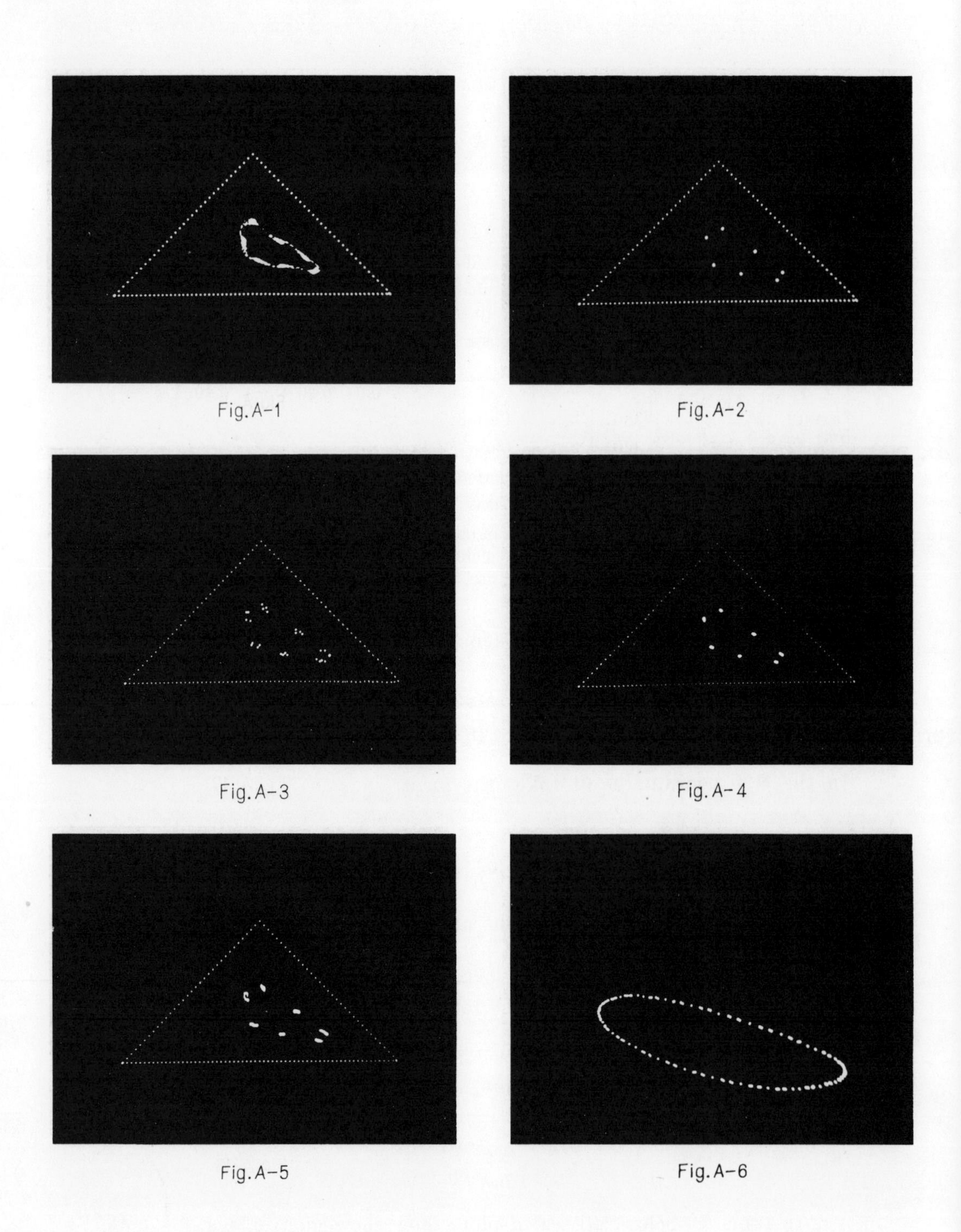
Fig. A-1
Fig. A-2
Fig. A-3
Fig. A-4
Fig. A-5
Fig. A-6

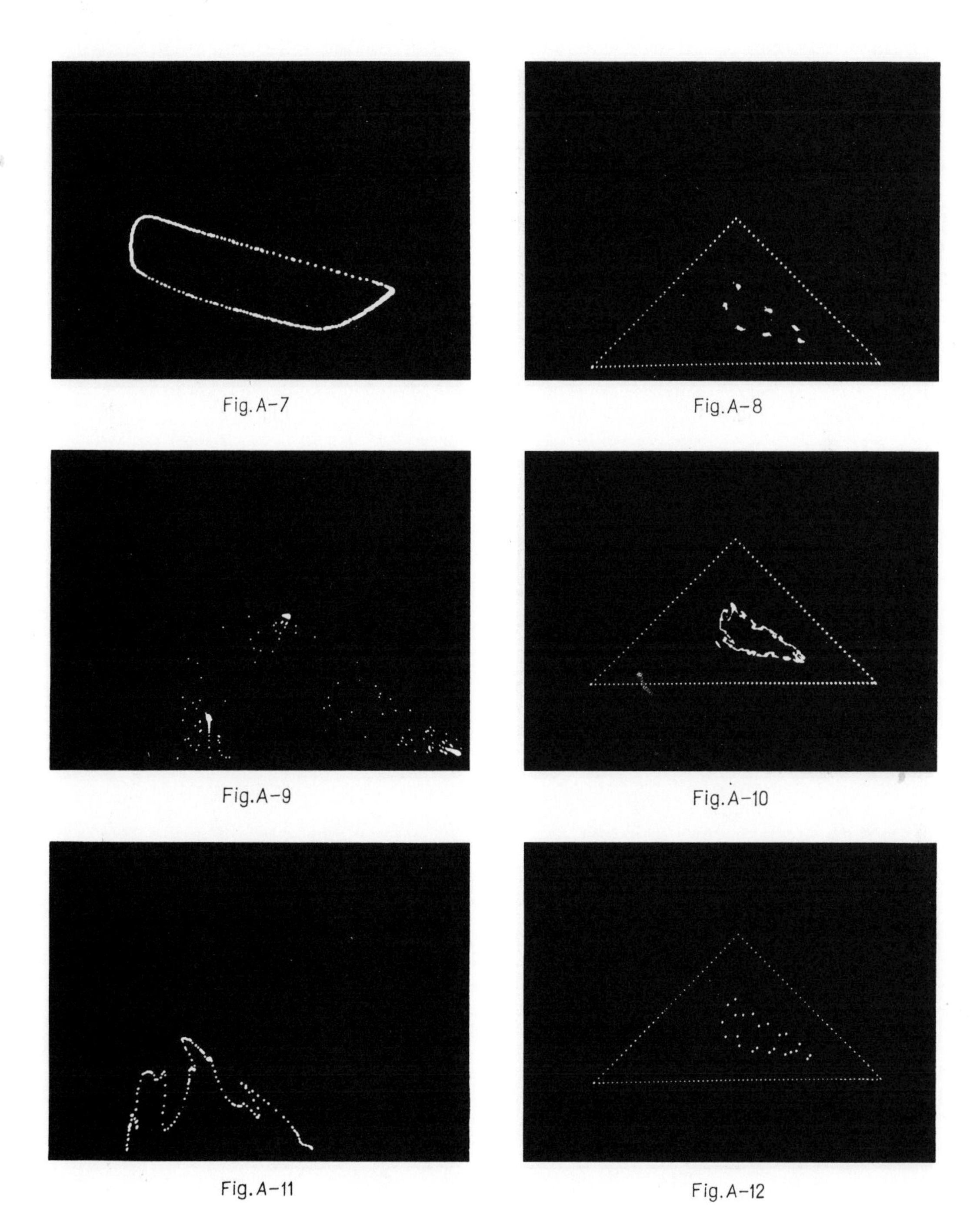

Fig. A-7

Fig. A-8

Fig. A-9

Fig. A-10

Fig. A-11

Fig. A-12

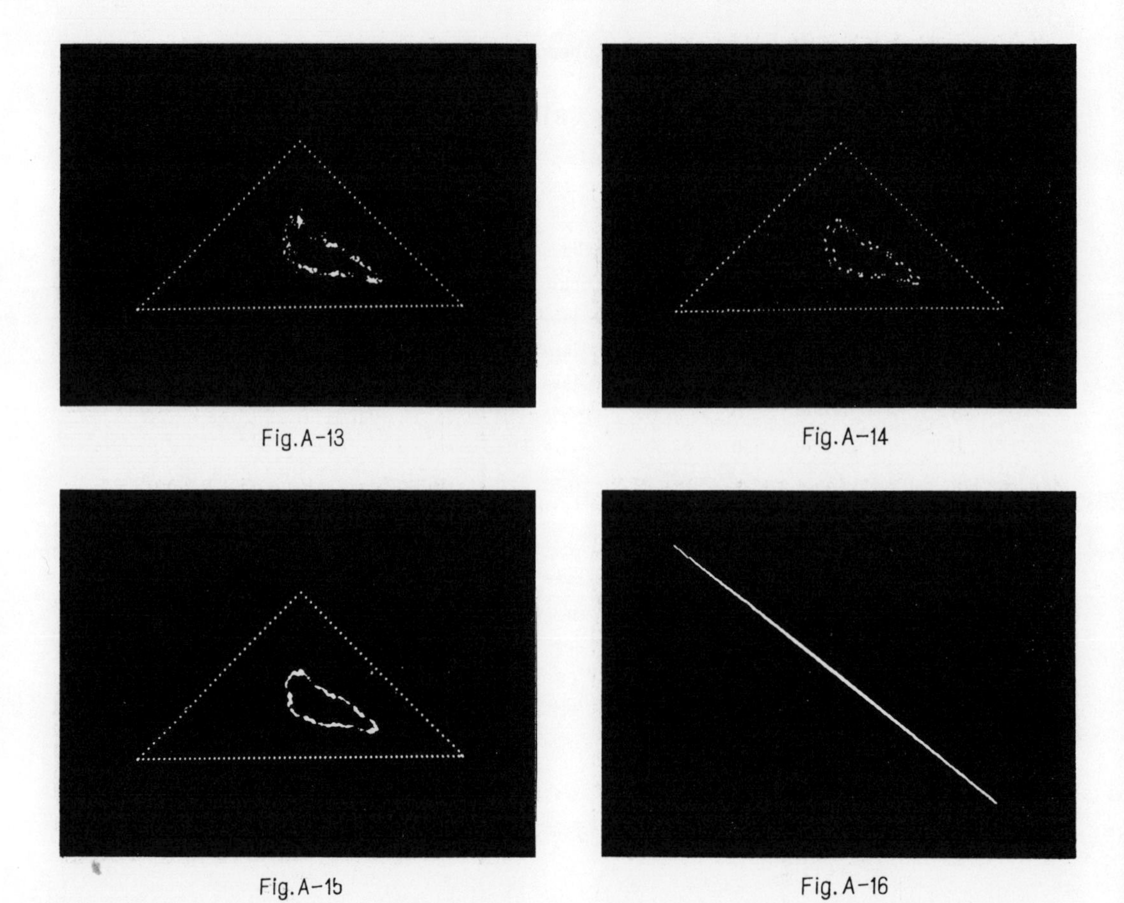

Fig. A-13

Fig. A-14

Fig. A-15

Fig. A-16

TABLE B

Figure Number	Δt	ε	r, s	Comments
B-1	1	0	—	Class IV limit set (2700 points)
B-2	1	0	—	Scaled plot of part of figure B-1.
B-3	0.996	0	—	Class IV
B-4	0.994	0	—	Class IV. Compare figure B-8.
B-5	0.992	0	—	Class IV
B-6	0.990	0	—	Period: $k = 78$ (26 points in this piece)
B-7	0.980	0	—	Class IV
B-8	1	0.01	1, 0	Class IV: 12 separate pieces (4 shown here). Compare figure B-4.
B-9	1	0.02	1, 0	Period: $k = 24$ (8 points shown here)
B-10	1	0.04	1, 0	Class IV
B-11	1	0.06	1, 0	Period: $k = 84$ (28 points shown here)
B-12	1	0.08	1, 0	Class IV
B-13	1	0.09	1, 0	Period: $k = 102$ (34 points shown here)
B-14	1	0.10	1, 0	Period: $k = 30$ (10 points shown here)
B-15	0.91	0	—	Class IV; shows "transition" from period with $k = 3$ $(\Delta t = 0.92)$

4. Modifications of T_D and T_E. These transformations are given by the schemes:

$$(9)\qquad T_D:\quad \begin{aligned} C_1 &= \{2, 7, 8, 9, 10\}, \\ C_2 &= \{4, 5, 6\}, \end{aligned}$$

$$(10)\qquad T_E:\quad \begin{aligned} C_1 &= \{2, 5, 7, 8, 9\}, \\ C_2 &= \{4, 6, 10\}, \end{aligned}$$

In the S, a coordinates, these read explicitly:

$$(11)\qquad T_D:\quad \begin{aligned} S' &= -\tfrac{3}{2}S^3 + \tfrac{3}{2}S^2a + \tfrac{3}{2}Sa^2 + \tfrac{13}{2}a^3 - 6Sa - 6a^2 + \tfrac{3}{2}S + \tfrac{9}{2}a, \\ a' &= \tfrac{3}{2}S^3 + \tfrac{9}{2}S^2a - \tfrac{3}{2}Sa^2 - \tfrac{9}{2}a^3 - 3S^2 + 3a^2 + \tfrac{3}{2}S - \tfrac{3}{2}a, \end{aligned}$$

with fixed point:

$$(12)\qquad \begin{aligned} S_0 &= 0.6525211, \\ a_0 &= 0.3056821; \end{aligned}$$

$$(13)\qquad T_E:\quad \begin{aligned} S' &= 9S^2a - a^3 - 3S^2 - 12Sa + 3a^2 + 3S + 3a, \\ a' &= -3S^2a + 3a^3 + 6Sa - 6a^2, \end{aligned}$$

with fixed point:

$$(14)\qquad \begin{aligned} S_0 &= 0.6444612, \\ a_0 &= 0.3219578. \end{aligned}$$

TABLE D

Figure number	Δt	ε	r, s	Comments
D-1	1	0	—	
D-2	0.97	0	—	Compare figure D-6
D-3	0.96, 1.0	0	—	Limit set for $\Delta t = 0.96$ superimposed on $L(T_D)$
D-4	0.83	0	—	
D-5	1	0.07	1, 0	
D-6	1	0.10	1, 0	Compare figure D-2

TABLE E

Figure number	Δt	ε	r, s	Comments
E-1	1	0	—	
E-2	1	0.01	1, 0	Shows convergence to periodic limit set ($k = 10$) from initial point close to fixed point
E-3	0.97	0	—	Compare with E-5, E-6
E-4	0.96	0	—	Compare with E-5, E-6
E-5	1	0.09	1, 0	
E-6	1	0.10	1, 0	

5. Broken linear transformations. These are described in section VI. Figures H-1 and H-2 show, respectively, 1000 points in the limit sets of T_1 and T_2. The latter are specified as follows:

$$T_1: \quad \begin{array}{lll} x_1 = \frac{1}{3}, & y_1 = \frac{1}{3}, & d_1 = 0.95, \\ x_2 = 0.6, & y_2 = 0.5, & d_2 = 0.95, \end{array} \tag{15}$$

with fixed point:

$$x_0 = \tfrac{1}{2}, \quad y_0 = \tfrac{2}{3}, \tag{16}$$

and

$$T_2: \quad \begin{array}{lll} x_1 = 0.5, & y_1 = 0.9, & d_1 = 1, \\ x_2 = 0.3, & y_2 = 0.7, & d_2 = 0.8, \end{array} \tag{17}$$

with fixed point:

$$x_0 = \tfrac{80}{143}, \quad y_0 = \tfrac{72}{143}. \tag{18}$$

The remaining figures, H-3 through H-17, show the limit sets belonging to the one-parameter family T_z:

$$T_z: \quad \begin{array}{ll} x_1 = y_1 = z, & d_1 = 1 \\ x_2 = y_2 = 1 - z, & d_2 = 1. \end{array} \tag{19}$$

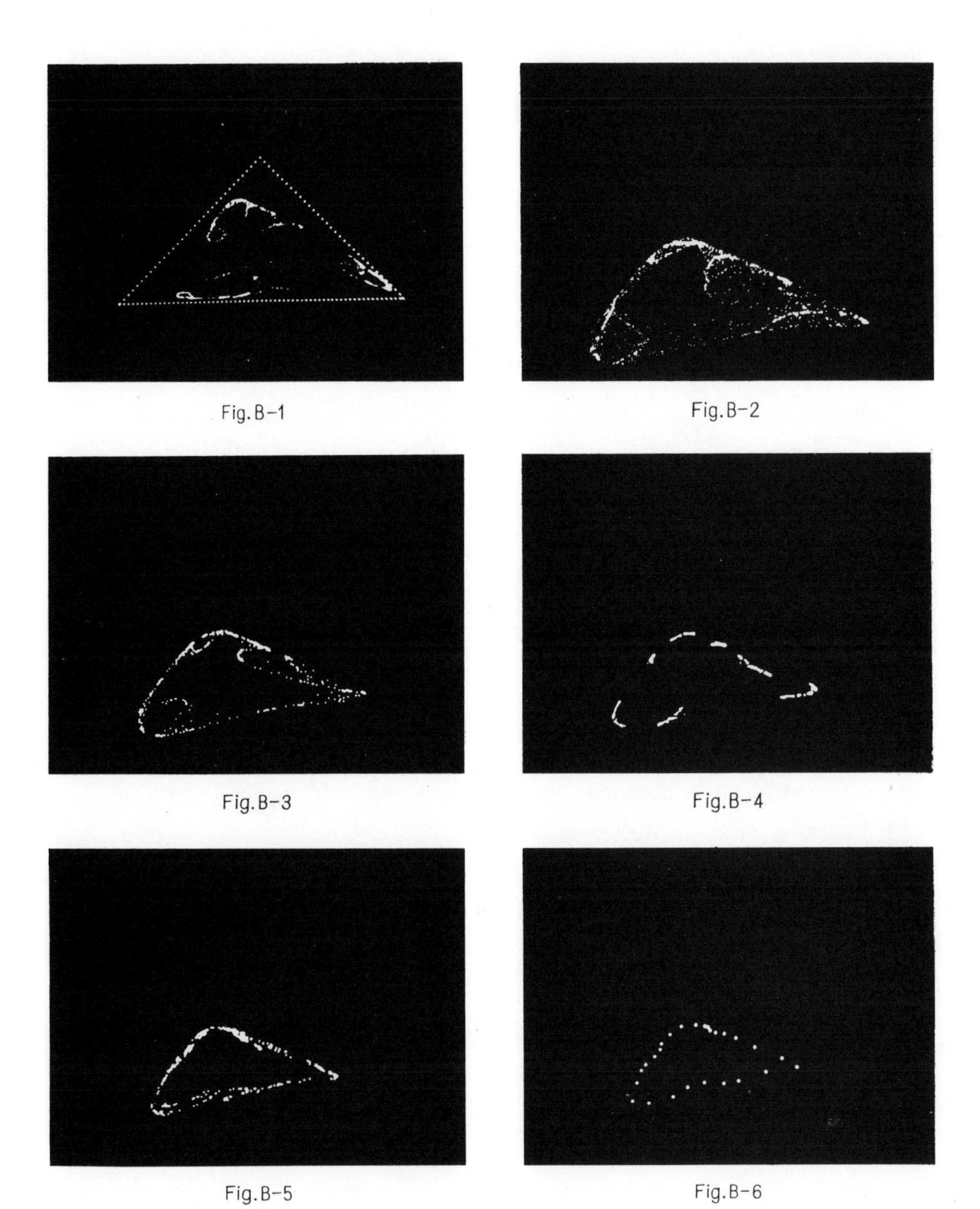

Fig. B-1

Fig. B-2

Fig. B-3

Fig. B-4

Fig. B-5

Fig. B-6

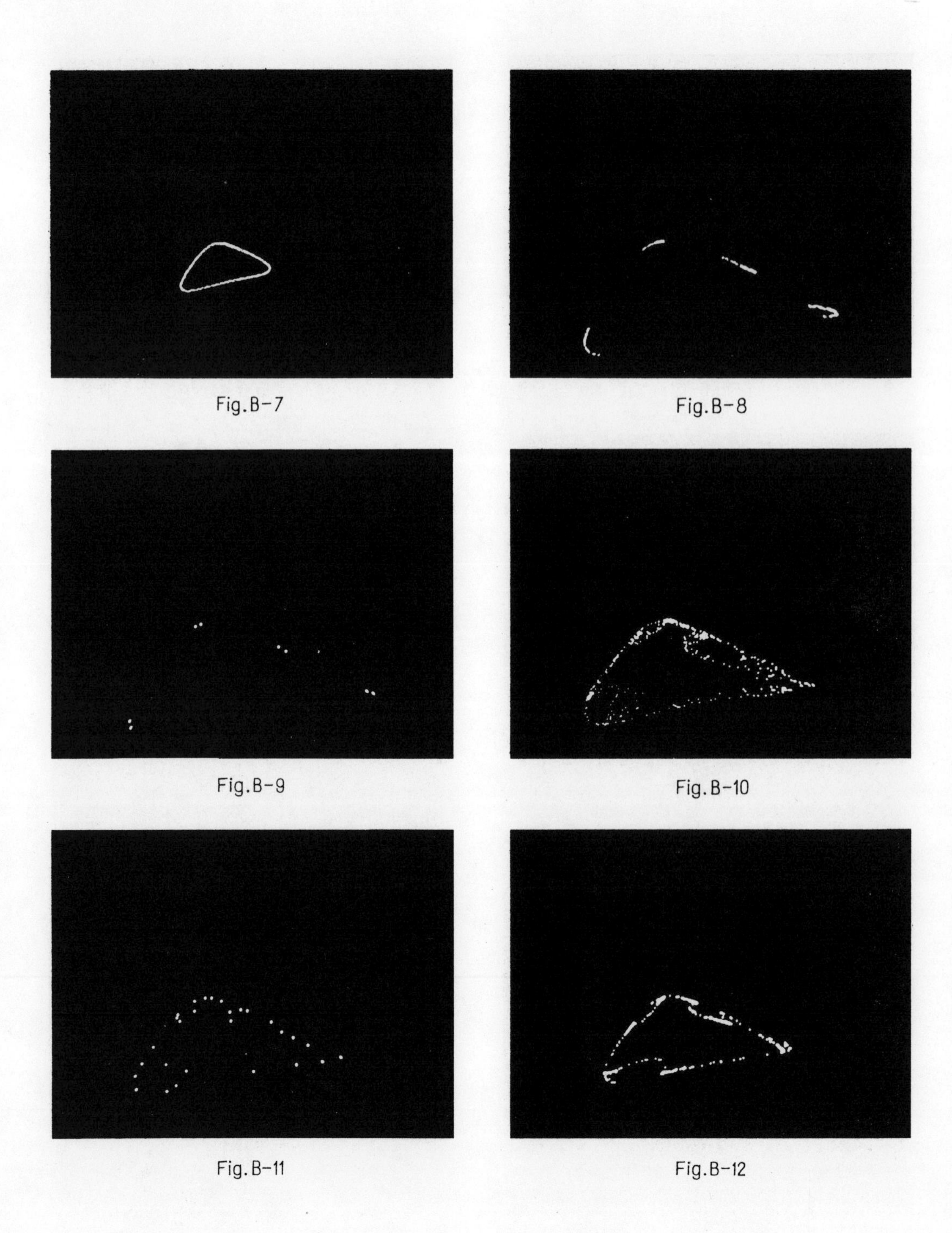

Fig. B-7

Fig. B-8

Fig. B-9

Fig. B-10

Fig. B-11

Fig. B-12

Fig. B-13

Fig. B-14

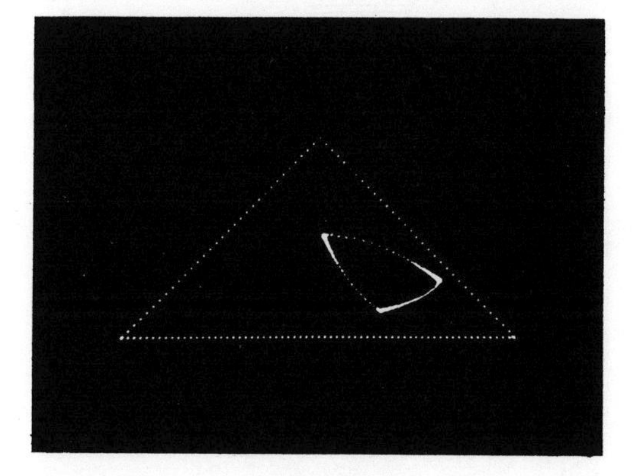

Fig. B-15

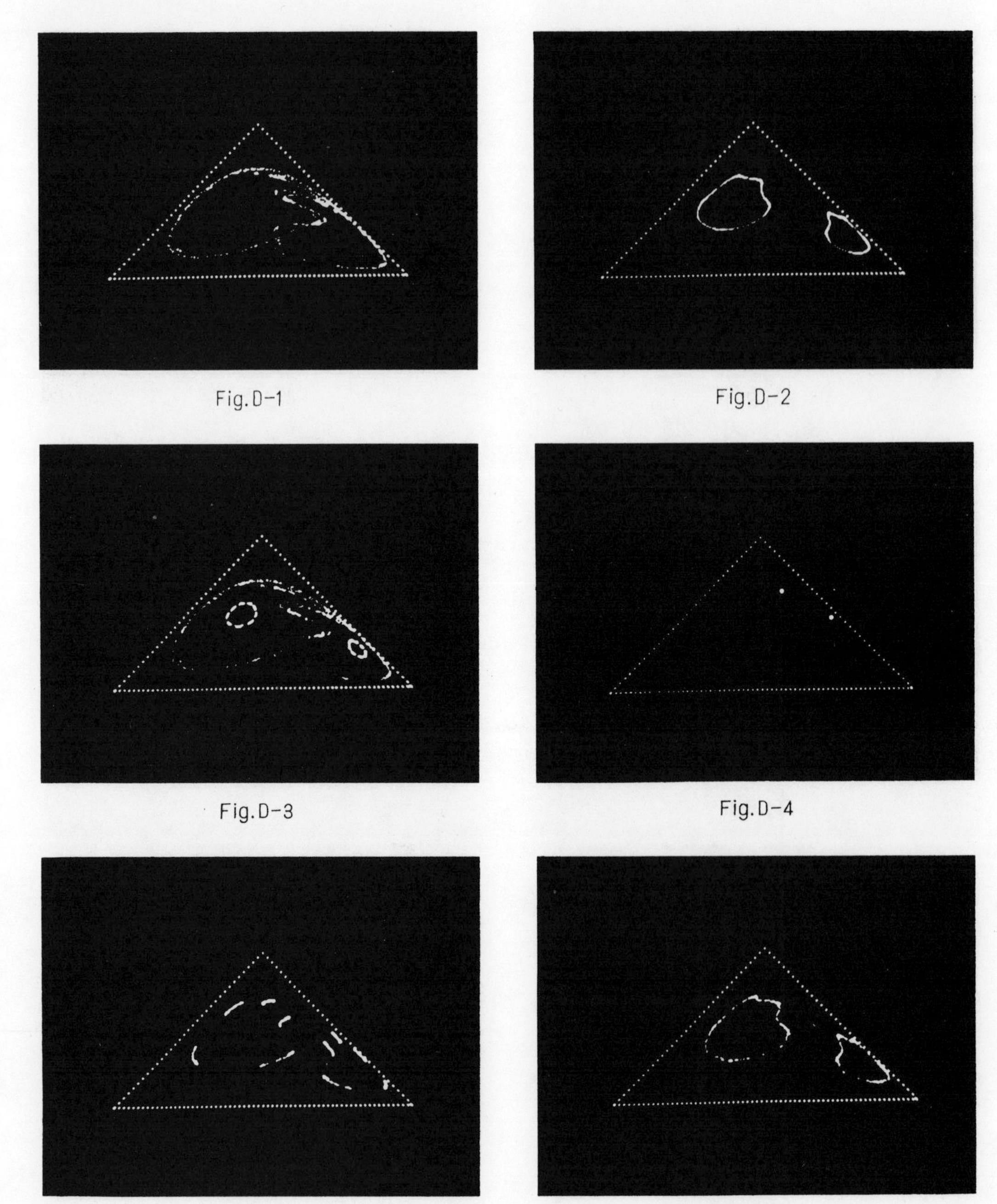
Fig. D-1 Fig. D-2 Fig. D-3 Fig. D-4 Fig. D-5 Fig. D-6

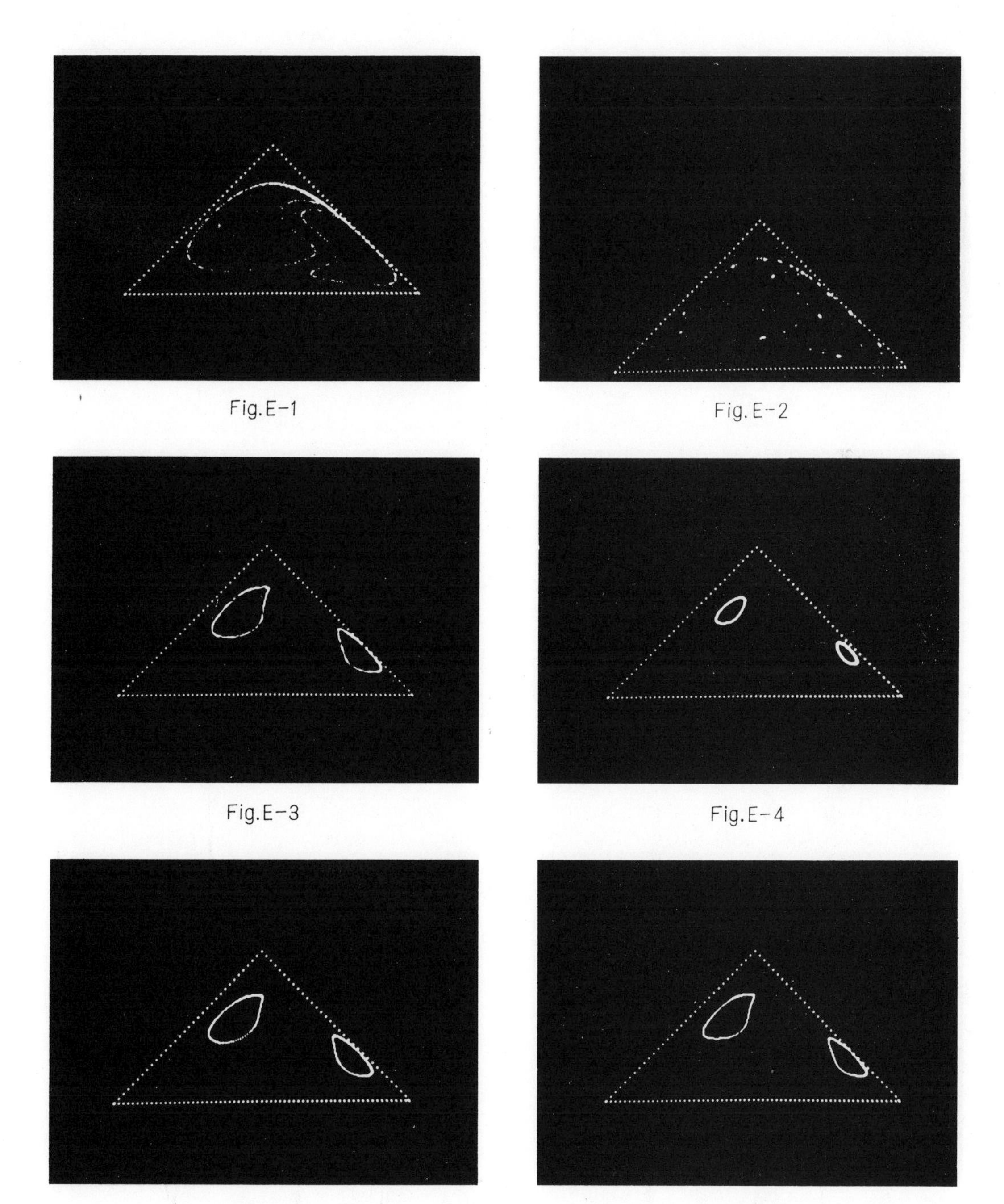
Fig. E-1
Fig. E-2
Fig. E-3
Fig. E-4
Fig. E-5
Fig. E-6

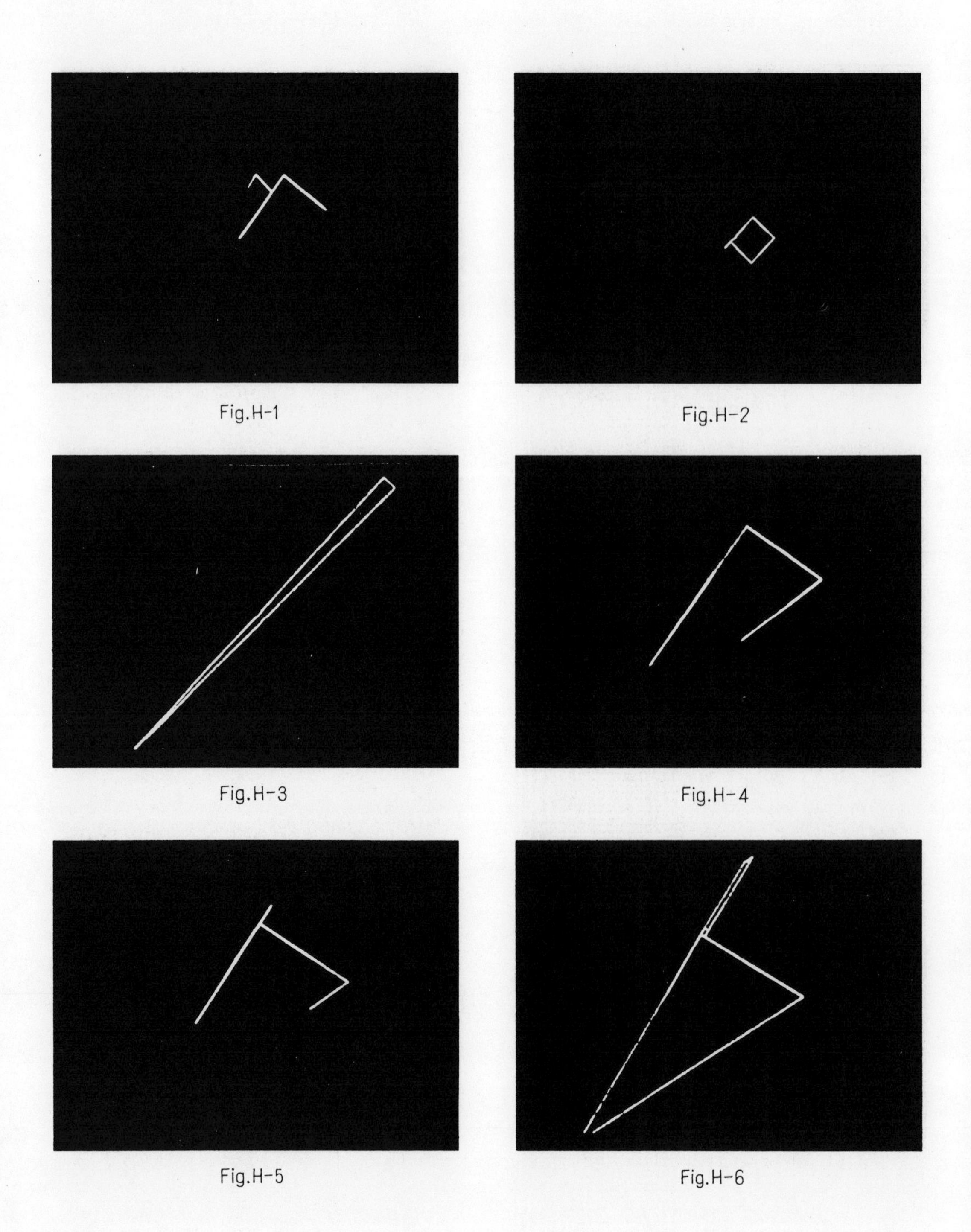

Fig.H-1

Fig.H-2

Fig.H-3

Fig.H-4

Fig.H-5

Fig.H-6

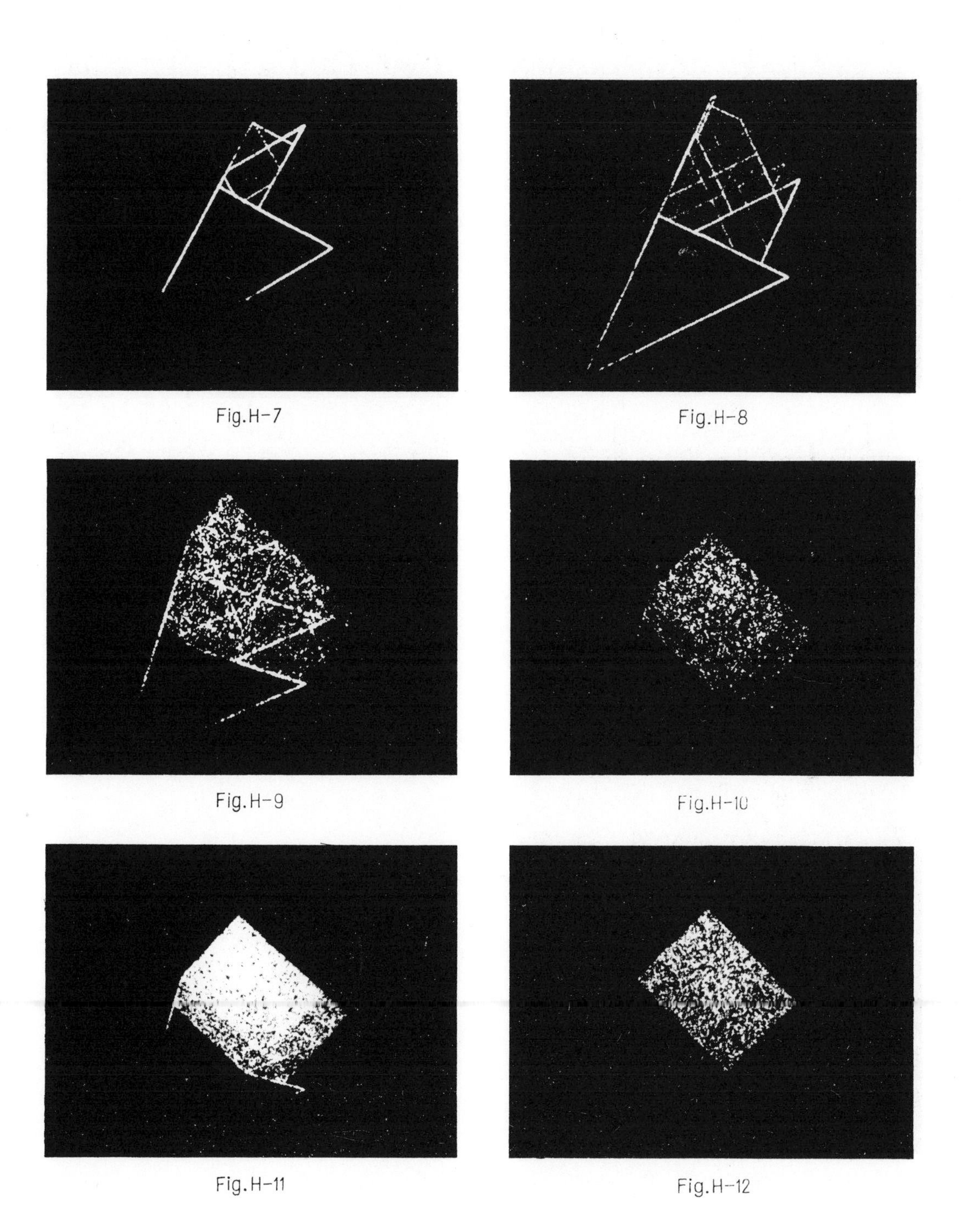

Fig.H-7

Fig.H-8

Fig.H-9

Fig.H-10

Fig.H-11

Fig.H-12

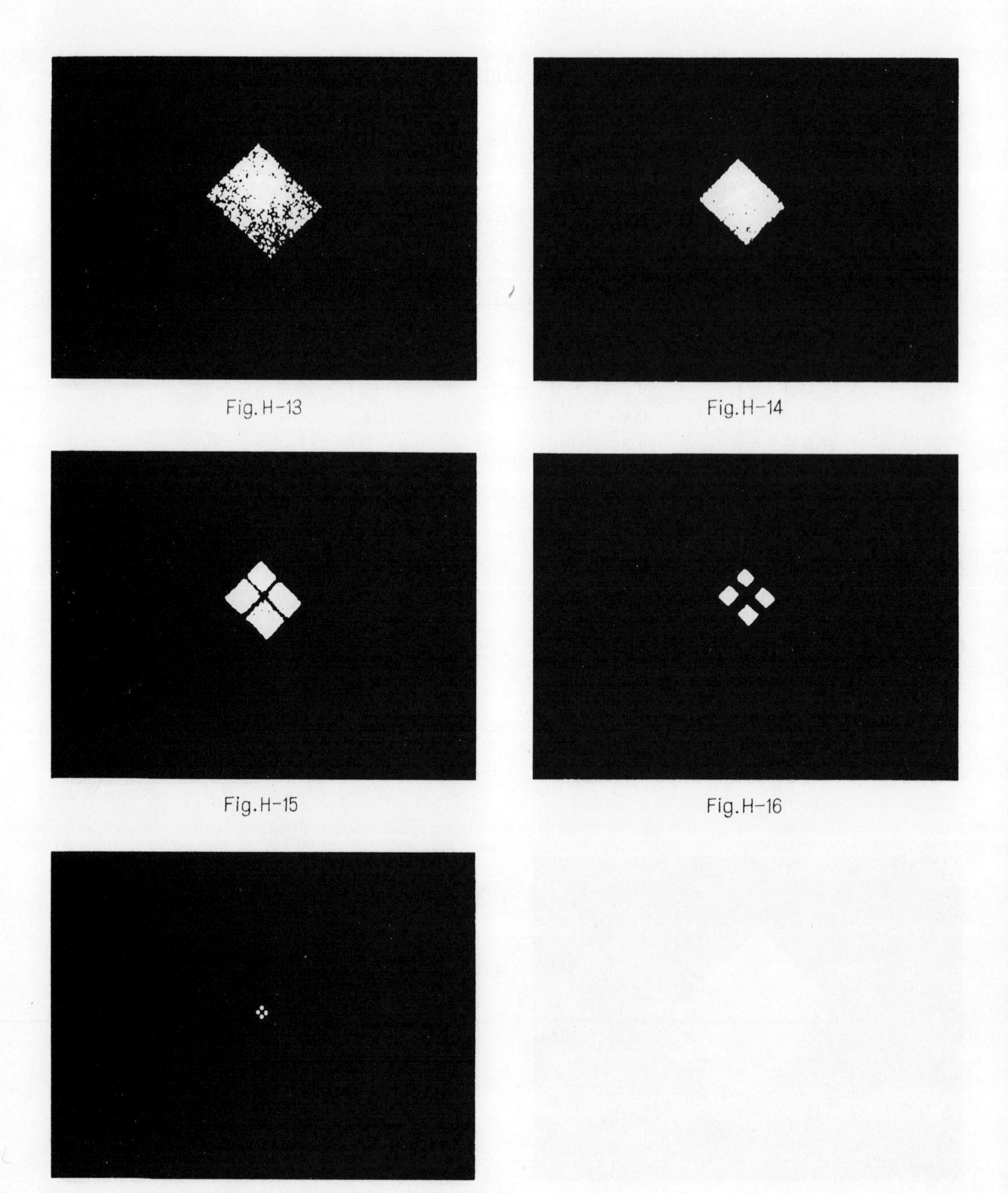
Fig. H-13

Fig. H-14

Fig. H-15

Fig. H-16

Fig. H-17

The identification is given by the following table (all figures except H-11 show 1000 points):

TABLE H

Figure number	z	Comments
H-3	0.49	For $z = 0.5$, see the remarks in section VI, note 2
H-4	0.42	
H-5	0.40	Compare figure H-1
H-6	0.38	
H-7	0.36	
H-8	0.32	Like "class IV" limit set
H-9	0.27	Like "class IV" limit set
H-10	0.25	
H-11	0.25	Points of H-10 plus the next 2000 consecutive iterates
H-12	0.23	
H-13	0.19	
H-14	0.16	
H-15	0.15	
H-16	0.12	
H-17	0.03	Fixed point becomes attractive below $z = 0.01$

CONTENTS

A VISUAL DISPLAY OF SOME PROPERTIES OF THE DISTRIBUTION OF PRIMES

M. L. Stein, S. M. Ulam, and M. B. Wells, University of California,
Los Alamos Scientific Laboratory, Los Alamos, New Mexico

Suppose we number the lattice points in the plane by a single sequence, e.g. Fig. 1 by starting at (0, 0) and proceeding counterclockwise in a spiral so that (0, 0)→1, (1, 0)→2, (1, 1)→3, (0, 1)→4, (−1, 1)→5, (−1, 0)→6, (−1, −1)→7, (0, −1)→8, (1, −1)→9, (2, −1)→10, (2, 0)→11, (2, 1)→12, (2, 2)→13, etc.

For commentary to this paper [83], see p. 698.

Consider the set P of those lattice points whose single index becomes a prime. Under our correspondence, points of P located on straight lines have indices which ultimately consist of values of a quadratic form. This is easily seen because the third differences between neighboring points on a straight line are 0 and after a finite number of indices which vary linearly have been passed, the progression becomes truly quadratic.

FIG. 1

The set P appears to exhibit a strongly nonrandom appearance (i.e. a different appearance from randomly chosen sets whose densities are like those of primes; that is, asymptotically $\log n/n$). This is due, of course, to the fact that some lines corresponding to quadratic forms which are factorable are devoid of primes; some other quadratic forms are rich in primes. A glance at a picture showing the set P reveals many such lines. It is a property of the visual brain which allows one to discover such lines at once and also notice many other peculiarities of distribution of points in two dimensions. In a visualization of a one-dimensional sequence this is not so much the case. (Perhaps an acoustic interpretation would be more suggestive?)

In addition to the well-known Euler form: y^2+y+41, one could observe instantly many other prime-rich forms. One line rather prominent in Fig. 1 in the lower half of the picture has numbers of the form $4x^2+170x+1847$; as pointed out by the referee, this is reducible into Euler's form by putting $y=2x+42$. (Under the enumeration above, the horizontal, vertical or diagonal straight lines correspond to quadratic forms, which have a leading term $4x^2$.) We have tried other "Peano numberings" of the lattice points. For example, (Fig. 2) for lattice points in the positive quadrant:

$$(0, 0) \to 1,\ (0, 1) \to 2,\ (1, 1) \to 3,\ (1, 0) \to 4$$
$$(0, 2) \to 5,\ (1, 2) \to 6;\ (2, 2) \to 7,\ (2, 1) \to 8,\ (2, 0) \to 9, \text{ etc.}$$

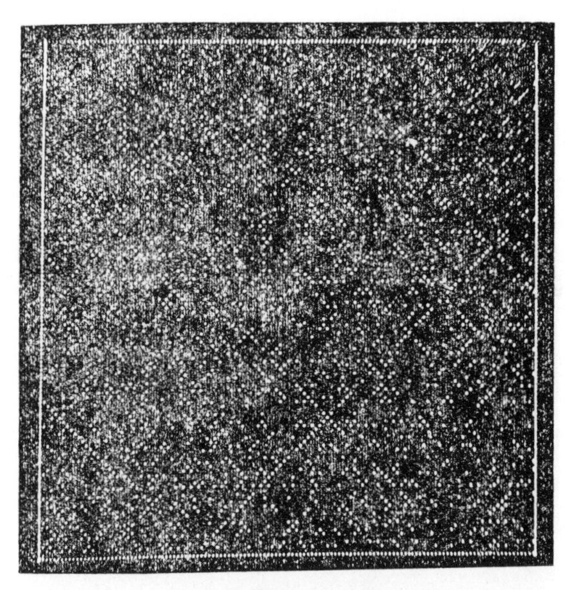

FIG. 2

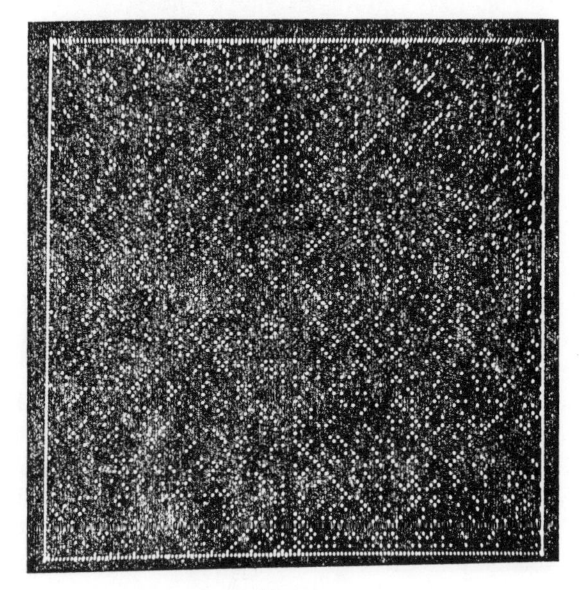

FIG. 3

The successive points on the principal lines will ultimately have coordinates given by quadratic forms with leading term $1 \cdot x^2$. Or, (Fig. 3) for points in the upper half plane:

$$(0, 0) \to 1,\ (-1, 0) \to 2,\ (-1, 1) \to 3,\ (0, 1) \to 4$$
$$(1, 1) \to 5,\ (1, 0) \to 6,\ (-2, 0) \to 7,\ (-2, 1) \to 8$$
$$(-2, 2) \to 9,\ (-1, 2) \to 10,\ (0, 2) \to 11,\ (1, 2) \to 12$$
$$(2, 2) \to 13,\ (2,1) \to 14,\ (2, 0) \to 15 \text{ and so on.}$$

There, the principal straight lines correspond to forms with the term $2x^2$.

The obvious questions, e.g.: Is the distribution of points of P asymptotically symmetric in every angle from the origin? Are there lines containing infinitely many primes? What is the asymptotic density of points on lines (e.g. are there pairs of nonfactorable lines with different asymptotic densities)? etc. seem to be hardly answerable with the present knowledge of the distribution of primes.

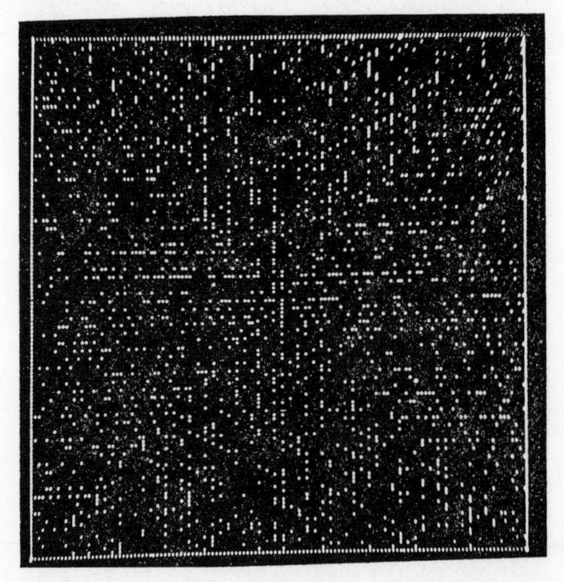

FIG. 4

We have observed many nonfactorable but, so to say, "almost factorable" lines, i.e. lines extremely poor in primes. We should add, as a curiosity, that as we displayed similarly the set L of lucky numbers (see [1]), in (Fig. 4) (the numbering of lattice points in this case is by a "discontinuous spiral," i.e. as in (Fig. 3) but going through both half planes); it appears again that there is a great deal of "structure"; in particular, some of the principal lines are manifest. This is much more surprising, of course, since there is no obvious multiplicative property of the set of luckies and no relation which is rigorous between the divisibility of quadratic forms and the definition of the sieve determining the lucky numbers.

The first observation of the properties of the P set was made on a few hundred points by hand. On the electronic computing machine "Maniac II" in Los Alamos we have been able to use a scope attached to the machine, which can display up to 65,000 points obtained as a result of calculation. This is then photographed and our pictures show a few of the results. We have magnetic tapes containing tables of primes up to ninety million. After discovering the quadratic forms which seem to be rich in primes up to $n=100{,}000$ or so, we then investigated primes up to ten million for such forms (see [2]). A few of the statistics are given below:

For primes in the Euler form $n=x^2+x+41$ we found the ratio r of these to all numbers of this form n up to 10,000,000 to be $r=.475\cdots$.

(1) For the form $n=4x^2+170x+1847$ there are 727 primes in the first 1560 for numbers of this form $(1\leqq n\leqq 10{,}000{,}000)(r=.466)$.

(2) For $n=4x^2+4x+59$ yields $r=.437\cdots$.

(3) For $n=2x^2+4x+117$ (the "rare" form) $r=.050\cdots$.
(A reason for rarity is that for no prime $p<29$ is divisibility by p excluded)

(4) The quadratic form $n=x^2+x+1$ is rich in "luckies." Up to numbers $n\cong 300{,}000$ $r=.29\cdots$.

This work was performed under the auspices of the U. S. Atomic Energy Commission.

References

1. V. Gardiner, R. Lazarus, N. Metropolis, and S. Ulam, On certain sequences of integers defined by sieves, Math. Mag., 29 (1956) 117–122.

2. Cf. Beeger, Nieuw Arch. Wisk., 20 (1939) 50.

Reprinted from the AMERICAN MATHEMATICAL MONTHLY
Vol. 71, No. 5, May, 1964

N° 266.

After the war, during one of his frequent summer visits to Los Alamos, Fermi became interested in the development and potentialities of the electronic computing machines. He held many discussions with me on the kind of future problems which could be studied through the use of such machines. We decided to try a selection of problems for heuristic work where in absence of closed analytic solutions experimental work on a computing machine would perhaps contribute to the understanding of properties of solutions. This could be particularly fruitful for problems involving the asymptotic—long time or "in the large" behavior of non-linear physical systems. In addition, such experiments on computing machines would have at least the virtue of having the postulates clearly stated. This is not always the case in an actual physical object or model where all the assumptions are not perhaps explicitly recognized.

Fermi expressed often a belief that future fundamental theories in physics may involve non-linear operators and equations, and that it would be useful to attempt practice in the mathematics needed for the understanding of non-linear systems. The plan was then to start with the possibly simplest such physical model and to study the results of the calculation of its long-time behavior. Then one would gradually increase the generality and the complexity of the problem calculated on the machine. The Los Alamos report LA–1940 (paper N° 266) presents the results of the very first such attempt. We had planned the work in the summer of 1952 and performed the calculations the following summer. In the discussions preceding the setting up and running of the problem on the machine we had envisaged as the next problem a two-dimensional version of the first one. Then perhaps problems of pure kinematics e.g., the motion of a chain of points subject only to constraints but no external forces, moving on a smooth plane convoluting and knotting itself indefinitely. These were to be studied preliminary to setting up ultimate models for motions of system where "mixing" and "turbulence" would be observed. The motivation then was to observe the *rates* of mixing and "thermalization" with the hope that the calculational results would provide hints for a future theory. One could venture a guess that one motive in the selection of problems could be traced to Fermi's early interest in the ergodic theory. In fact, his early paper (N° 11 *a*) presents an important contribution to this theory.

It should be stated here that during one summer Fermi learned very rapidly how to *program* problems for the electronic computers and he not only could plan the general outline and construct the so-called flow diagram but would work out himself the actual *coding* of the whole problem in detail.

The results of the calculations (performed on the old MANIAC machine) were interesting and quite surprising to Fermi. He expressed to me the opinion that they really constituted a little discovery in providing intimations that the prevalent beliefs in the universality of "mixing and thermalization" in non-linear systems may not be always justified.

A few words about the subsequent history of this non-linear problem. A number of other examples of such physical systems were examined by calculations on the electronic computing machines in 1956 and 1957. I presented the results of the original paper on several occasions at scientific meetings; they seemed to have aroused considerable interest among mathematicians and physicists and there is by now a small literature dealing with this problem. The most recent results are due to N. J. Zabusky. (1) His analytical work shows, by the way, a good agreement of the numerical computations with the continuous solution up to a point where a discontinuity developed in the derivatives and the analytical work had to be modified. One obtains from it another indication that the phenomenon discovered

(1) Exact Solutions for the Vibrations of a non-linear continuous string. A. E. C. Research and Development Report. MATT–102, Plasma Physics Laboratory, Princeton University, October 1961.

is not due to numerical accidents of the algorithm of the computing machine, but seems to constitute a real property of the dynamical system.

In 1961, on more modern and faster machines, the original problem was considered for still longer periods of time. It was found by J. Tuck and M. Menzel that after one continues the calculations from the first "return" of the system to its original condition the return is not complete. The total energy is concentrated again essentially in the first Fourier mode, but the remaining one or two percent of the total energy is in higher modes. If one continues the calculation, at the end of the next great cycle the error (deviation from the original initial condition) is greater and amounts to perhaps three percent. Continuing again one finds the deviation increasing—after eight great cycles the deviation amounts to some eight percent; but from that time on an opposite development takes place! After eight more i.e., sixteen great cycles altogether, the system gets very close—better than within one percent to the original state! This supercycle constitutes another surprising property of our non-linear system.

Paper N° 266 is not the only work that Fermi and I did together. In the summer of 1950 we made a study of the behavior of the thermonuclear reaction in a mass of deuterium and wrote a report, LA-1158, which is still classified. The problem is of enormous mathematical complexity, involving the hydrodynamics of the motion of the material, the hydrodynamics of radiation energy, all interwoven with the processes of the various reactions between the nuclei whose probabilities and properties depend i.a., on temperature, density, and the changing geometry of the materials. The aim of this work was to obtain, by a schematized but still elaborate picture of the evolution of all these physical processes, an idea of the propagation of such a reaction. This was to complement a previous work by Everett and myself, dealing with the problem of ignition of a mass deuterium. Assuming an ignition somehow started in a large volume, one wanted to evaluate the prospects of propagation of the reactions already started. Many ingenious schematizations and simplifications had to be introduced in order to describe the process, without the possibility of calculating in exact detail the innumerable geometrical and thermodynamical factors. The results of our computations on the chances of propagation were negative and the report played an important role in channeling imagination and energies towards a search for a different scheme for a successful hydrogen reaction. This was indeed found later on on a different basis. All the calculations on which the work of the report is based were performed on desk computers and slide rules. The subsequent massive and lengthy work on the electronic computer machines (organized and performed by von Neumann, F. and C. Evans and others) confirmed in large lines, qualitatively and to a good degree quantitatively the behavior of the system as estimated and predicted in our report—with its combination of intuitive evaluations, schematized equations and hand calculations.

S. M. ULAM.

266.

STUDIES OF NON LINEAR PROBLEMS

E. FERMI, J. PASTA, and S. ULAM
Document LA-1940 (May 1955).

ABSTRACT.

A one-dimensional dynamical system of 64 particles with forces between neighbors containing nonlinear terms has been studied on the Los Alamos computer MANIAC I. The nonlinear terms considered are quadratic, cubic, and broken linear types. The results are analyzed into Fourier components and plotted as a function of time.

The results show very little, if any, tendency toward equipartition of energy among the degrees of freedom.

The last few examples were calculated in 1955. After the untimely death of Professor E. Fermi in November, 1954, the calculations were continued in Los Alamos.

This report is intended to be the first one of a series dealing with the behavior of certain nonlinear physical systems where the non-linearity is introduced as a perturbation to a primarily linear problem. The behavior of the systems is to be studied for times which are long compared to the characteristic periods of the corresponding linear problems.

The problems in question do not seem to admit of analytic solutions in closed form, and heuristic work was performed numerically on a fast electronic computing machine (MANIAC I at Los Alamos). [1] The ergodic behavior of such systems was studied with the primary aim of establishing, experimentally, the rate of approach to the equipartition of energy among the various degrees of freedom of the system. Several problems will be considered in order of increasing complexity. This paper is devoted to the first one only.

We imagine a one-dimensional continuum with the ends kept fixed and with forces acting on the elements of this string. In addition to the usual linear term expressing the dependence of the force on the displacement of the element, this force contains higher order terms. For the purposes of numerical work this continuum is replaced by a finite number of points (at most 64 in our actual computation) so that the partial differential equation defining the motion of this string is replaced by a finite number of total differential equations. We have, therefore, a dynamical system of 64 particles with forces acting between neighbors with fixed end points. If x_i denotes the displacement of the i–th point from its original position, and α denotes the coefficient of the quadratic term in the force between the neighboring mass points and β that of the cubic term, the equations were either

$$x_i = (x_{i+1} + x_{i-1} - 2\,x_i) + \alpha\,[(x_{i+1} - x_i)^2 - (x_i - x_{i-1})^2] \qquad (i = 1\,,\,2\,,\cdots,64), \tag{1}$$

or

$$x_i = (x_{i+1} + x_{i-1} - 2\,x_i) + \beta\,[(x_{i+1} - x_i)^3 - (x_i - x_{i-1})^3] \qquad (i = 1\,,\,2\,,\cdots,64). \tag{2}$$

α and β were chosen so that at the maximum displacement the nonlinear term was small, e.g., of the order of one-tenth of the linear term. The corresponding partial differential equation obtained by letting the number of particles become infinite is the usual wave equation plus non-linear terms of a complicated nature.

Another case studied recently was

$$\ddot{x}_i = \delta_1\,(x_{i+1} - x_i) - \delta_2\,(x_i - x_{i-1}) + c \tag{3}$$

(1) We thank Miss Mary Tsingou for efficient coding of the problems and for running the computations on the Los Alamos MANIAC machine.

where the parameters δ_1, δ_2, c were not constant but assumed different values depending on whether or not the quantities in parentheses were less than or greater than a certain value fixed in advance. This prescription amounts to assuming the force as a broken linear function of the displacement. This broken linear function imitates to some extent a cubic dependence. We show the graphs representing the force as a function of displacement in three cases.

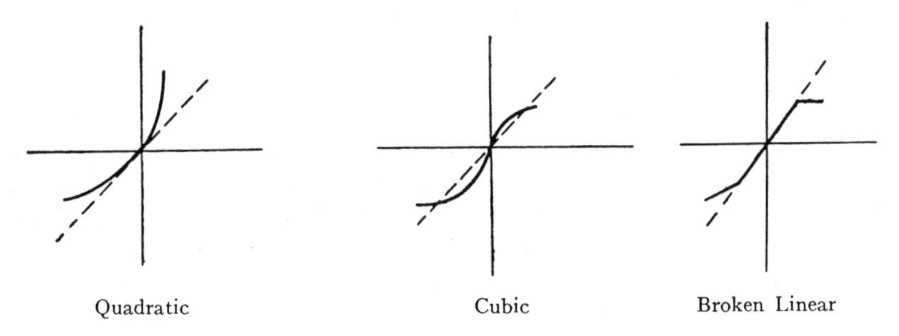

Quadratic Cubic Broken Linear

The solution to the corresponding linear problem is a periodic vibration of the string. If the initial position of the string is, say, a single sine wave, the string will oscillate in this mode indefinitely. Starting with the string in a simple configuration, for example in the first mode (or in other problems, starting with a combination of a few low modes), the purpose of our computations was to see how, due to nonlinear forces perturbing the periodic linear solution, the string would assume more and more complicated shapes, and, for t tending to infinity, would get into states where all the Fourier modes acquire increasing importance. In order to see this, the shape of the string, that is to say, x as a function of i and the kinetic energy as a function i were analyzed periodically in Fourier series. Since the problem can be considered one of dynamics, this analysis amounts to a Lagrangian change of variables: instead of the original $\dot{x}_i$ and x_i, $i = 1, 2, \cdots, 64$, we may introduce a and $\dot{a}_k$, $k = 1, 2, \cdots, 64$, where

$$a_k = \Sigma x_i \sin \frac{ik\pi}{64} . \tag{4}$$

The sum of kinetic and potential energies in the problem with a quadratic force is

$$E^{\text{kin}}_{x_i} + E^{\text{pot}}_{x_i} = \frac{1}{2} \dot{x}_i^2 + \frac{(x_{i+1} - x_i)^2 + (x_i - x_{i-1})^2}{2} \tag{5 a}$$

$$E^{\text{kin}}_{a_k} + E^{\text{pot}}_{a_k} = \frac{1}{2} \dot{a}_k^2 + 2 a_k^2 \sin^2 \frac{\pi k}{128} \tag{5 a}$$

if we neglect the contributions to potential energy from the quadratic or higher terms in the force. This amounts in our case to at most a few percent.

The calculation of the motion was performed in the x variables, and every few hundred cycles the quantities referring to the a variables were computed by the above formulas. It should be noted here that the calculation of the motion could be performed directly in a_k and $\dot{a}_k$. The formulas, however become unwieldy and the computation, even on an electronic computer, would take a long time. The computation in the a_k variables could have been more instructive for the purpose of observing directly the interaction between the a_k's. It is proposed to do a few such calculations in the near future to observe more directly the properties of the equations for $\ddot{a}_k$.

Let us say here that the results of our computations show features which were, from the beginning, surprising to us. Instead of a gradual, continuous flow of energy from the first mode to the higher modes, all of the problems show an entirely different behavior. Starting in one problem with a quadratic force and a pure sine wave as the initial position of the string, we indeed observe initially a gradual increase of energy in the higher modes as predicted (e.g., by Rayleigh in an infinitesimal analysis). Mode 2 starts increasing first, followed by mode 3, and so on. Later on, however, this gradual sharing of energy among successive modes ceases. Instead, it is one or the other mode that predominates. For example, mode 2 decides, as it were, to increase rather rapidly at the cost of all other modes and becomes predominant. At one time, it has more energy than all the others put together! Then mode 3 undertakes this role. It is only the first few modes which exchange energy among themselves and they do this in a rather regular fashion. Finally, at a later time mode 1 comes back to within one percent of its initial value so that the system seems to be almost periodic. All our problems have at least this one feature in common. Instead of gradual increase of all the higher modes, the energy is exchanged, essentially, among only a certain few. It is, therefore, very hard to observe the rate of " thermalization " or mixing in our problem, and this was the initial purpose of the calculation.

If one should look at the problem from the point of view of statistical mechanics, the situation could be described as follows: the phase space of a point representing our entire system has a great number of dimensions. Only a very small part of its volume is represented by the regions where only one or a few out of all possible Fourier modes have divided among themselves almost all the available energy. If our system with nonlinear forces acting between the neighboring points should serve as a good example of a transformation of the phase space which is ergodic or metrically transitive, then the trajectory of almost every point should be everywhere dense in the whole phase space. With overwhelming probability this should also be true of the point which at time $t = 0$ represents our initial configuration, and this point should spend most of its time in regions corresponding to the equipartition of energy among various degrees of freedom. As will be seen from the results this seems hardly the case. We have plotted (figs. 1 to 7) the ergodic sojourn times in certain subsets of our phase space. These may show a tendency to approach limits as guaranteed by the ergodic theorem. These limits, however, do not seem to correspond to equipartition even in the time average. Certainly, there seems to be very little, if any, tendency towards equipartition

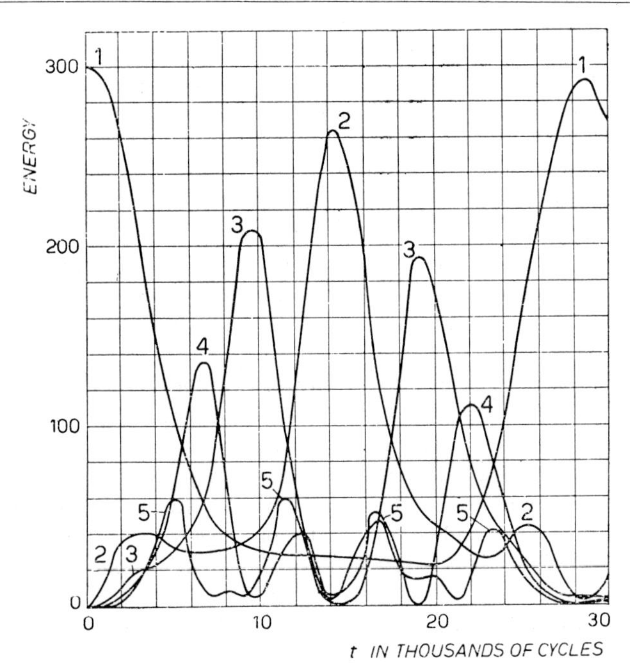

Fig. 1. – The quantity plotted is the energy (kinetic plus potential in each of the first five modes). The units for energy are arbitrary. $N = 32$; $\alpha = 1/4$; $\delta t^2 = 1/8$. The initial form of the string was a single sine wave. The higher modes never exceeded in energy 20 of our units. About 30,000 computation cycles were calculated.

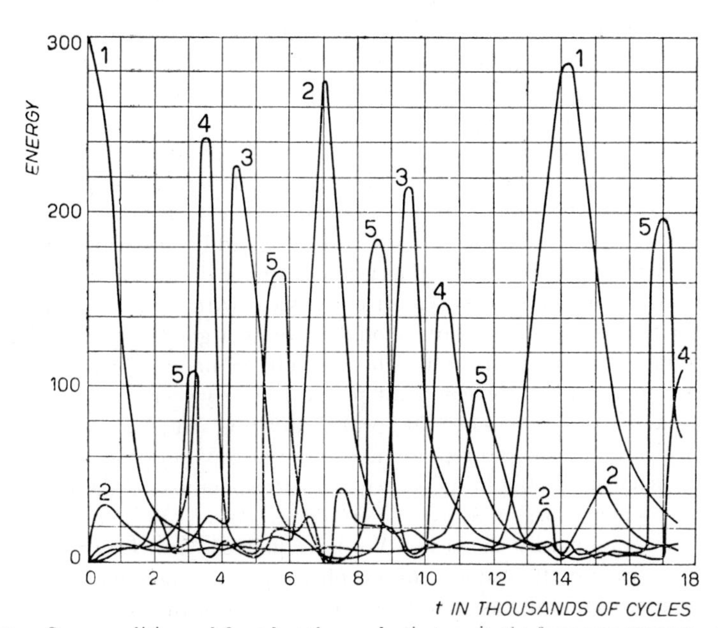

Fig. 2. – Same conditions ad fig. 1 but the quadratic term in the force was stronger. $\alpha = 1$. About 14,000 cycles were computed.

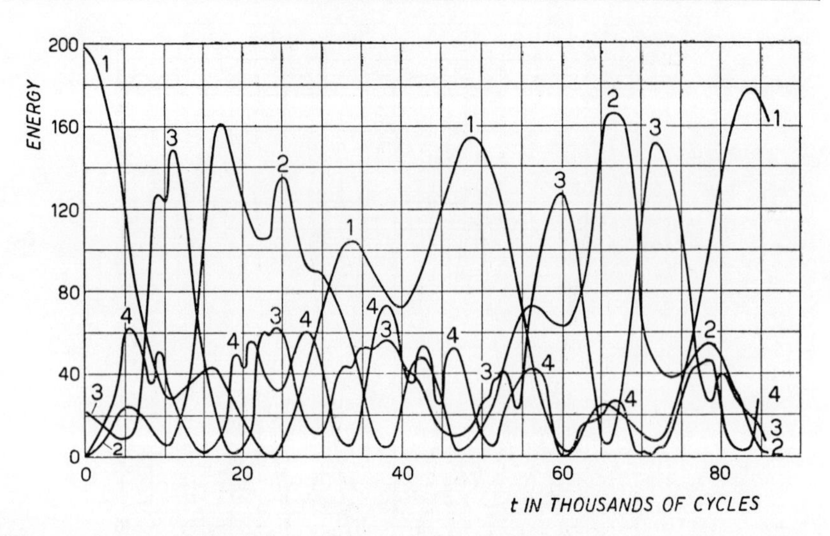

Fig. 3. – Same conditions as in fig. 1, but the initial configuration of the string was a " saw-tooth " triangular-shaped wave. Already at $t = 0$, therefore, energy was present in some modes other than 1. However, modes 5 and higher never exceeded 40 of our units.

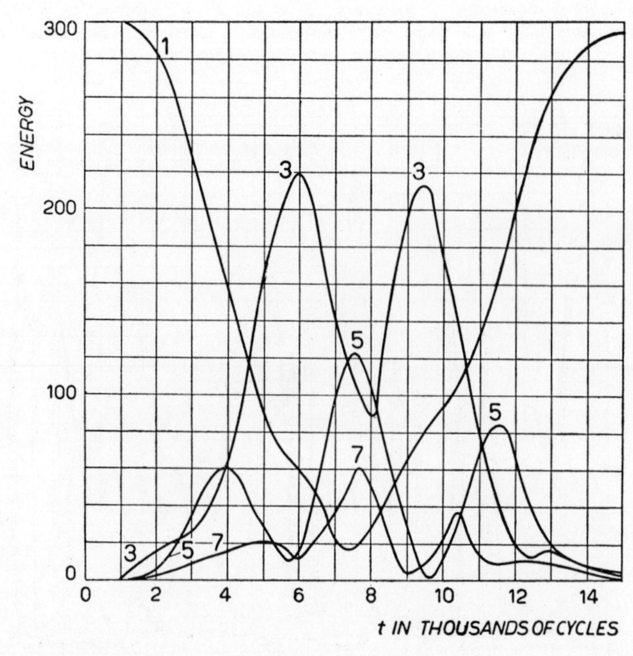

Fig. 4. – The initial configuration assumed was a single sine wave; the force had a cubic term with $\beta = 8$ and $\delta t^2 = 1/8$. Since a cubic force acts symmetrically (in contrast to a quadratic force), the string will forever keep its symmetry and the effective number of particles for the computation is $N = 16$. The even modes will have energy 0.

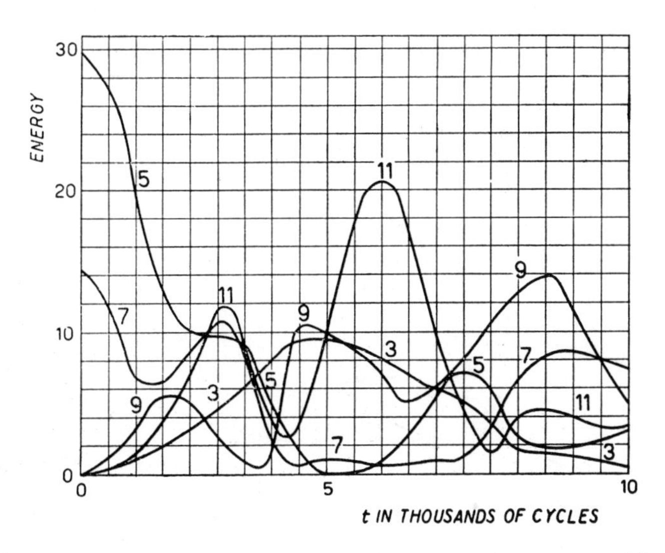

Fig. 5. – $N = 32$; $\delta t^2 = 1/64$; $\beta = 1/16$. The initial configuration was a combination of 2 modes. The initial energy was chosen to be 2/3 in mode 5 and 1/3 in mode 7.

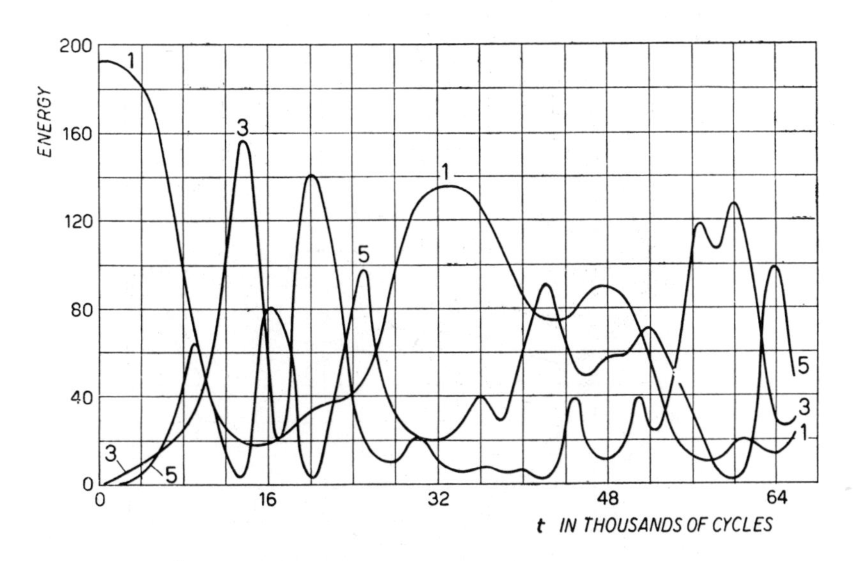

Fig. 6. – $\delta t^2 = 2^{-6}$. The force was taken as a broken linear function of displacement. The amplitude at which the slope changes was taken as $2^{-5} + 2^{-7}$ of the maximum amplitude. After this cut-off value, the force was assumed still linear but the slope increased by 25 percent. The effective $N = 16$.

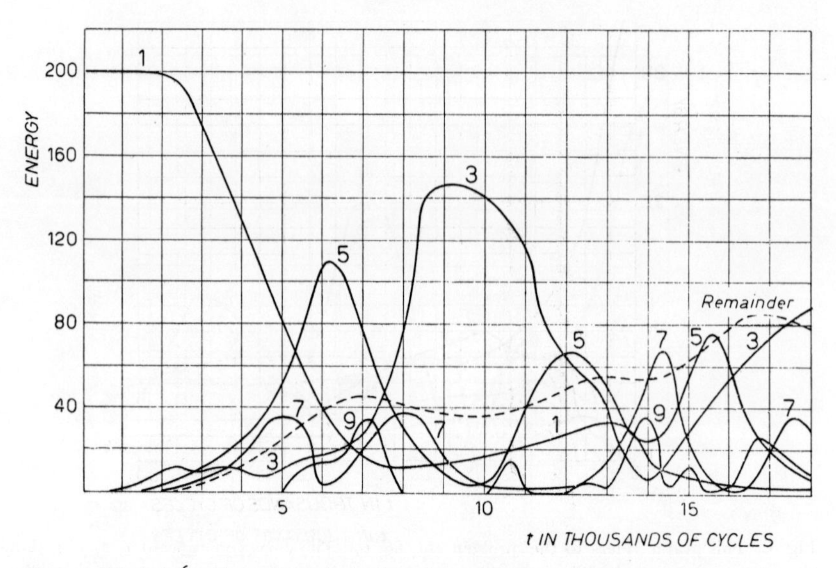

Fig. 7. – $\delta t^2 = 2^{-6}$. Force is again broken linear function with the same cut-off, but the slopes after that increased by 50 percent instead of the 25 percent charge as in problem 6. The effective N = 16.

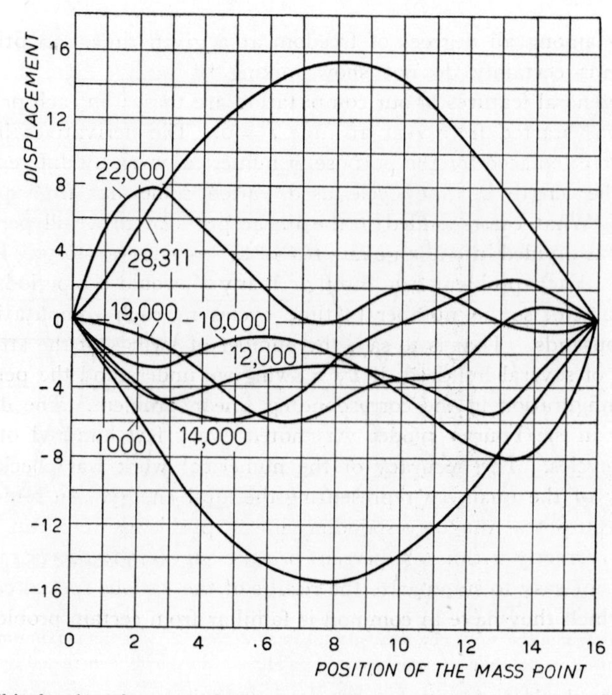

Fig. 8. – This drawing shows not the energy but the actual *shapes*, i.e., the displacement of the string at various times (in cycles) indicated on each curve. The problem is that of fig. 1.

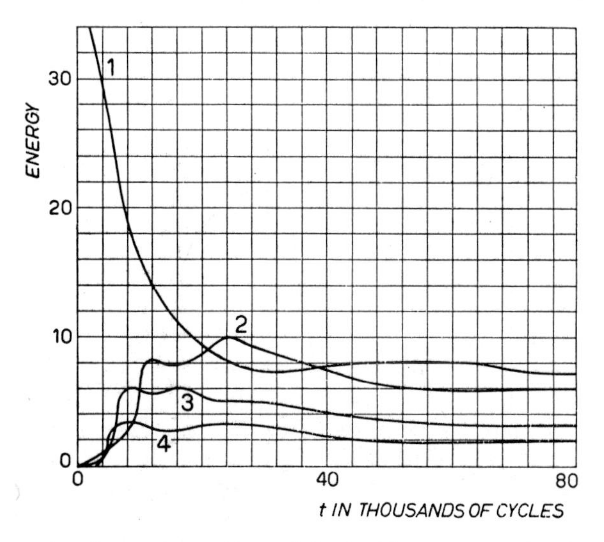

Fig. 9. This graph refers to the problem of fig. 6. The curves, numbered 1, 2, 3, 4, show the time averages of the kinetic energy contained in the first 4 modes as a function of time. In other words, the quantity is $\frac{1}{\nu}\sum_{i=1}^{\nu} T^i_{a_k}$. ν is the cycle no., $k = 1, 3, 5, 7$.

of energy among all degrees of freedom at a given time. In other words, the systems certainly do not show mixing. (2)

The general features of our computation are these: in each problem, the system was started from rest at time $t = 0$. The derivatives in time, of course, were replaced for the purpose of numerical work by difference expressions. The length of time cycle used varied somewhat from problem to problem. What corresponded in the linear problem to a full period of the motion was divided into a large number of time cycles (up to 500) in the computation. Each problem ran through many "would-be periods" of the linear problem, so the number of time cycles in each computation ran to many thousands. That is to say, the number of swings of the string was of the order of several hundred, if by a swing we understand the period of the initial configuration in the corresponding linear problem. The distribution of energy in the Fourier modes was noted every few hundred of the computation cycles. The accuracy of the numerical work was checked by the constancy of the quantity representing the total energy. In some cases, for checking purposes, the corresponding linear problems were run and these behaved correctly within one percent or so, even after 10,000 or more cycles.

It is not easy to summarize the results of the various special cases. One feature which they have in common is familiar from certain problems in me-

(2) One should distinguish between metric transitivity or ergodic behavior and the stronger property of mixing.

chanics of systems with a few degrees of freedom. In the compound pendulum problem one has a transformation of energy from one degree of freedom to another and back again, and not a continually increasing sharing of energy between the two. What is perhaps surprising in our problem is that this kind of behavior still appears in systems with, say, 16 or more degrees of freedom.

What is suggested by these special results is that in certain problems which are approximately linear, the existence of quasi-states may be conjectured.

In a linear problem the tendency of the system to approach a fixed "state" amounts, mathematically, to convergence of iterates of a transformation in accordance with an algebraic theorem due to Frobenius and Perron. This theorem may be stated roughly in the following way. Let A be a matrix with positive elements. Consider the linear transformation of the n-dimensional space defined by this matrix. One can assert that if $\bar{x}$ is any vector with all of its components positive, and if A is applied repeatedly to this vector, the directions of the vectors $\bar{x}$, $A(\bar{x}), \cdots, A^i(\bar{x}), \cdots$, will approach that of a fixed vector $\bar{x}_0$ in such a way that $A(\bar{x}_0) = \lambda(\bar{x}_0)$. This eigenvector is unique among all vectors with all their components non-negative. If we consider a linear problem and apply this theorem, we shall expect the system to approach a steady state described by the invariant vector. Such behavior is in a sense diametrically opposite to an ergodic motion and is due to a very special character, linearity of the transformations of the phase space. The results of our calculation on the nonlinear vibrating string suggest that in the case of transformations which are approximately linear, differing from linear ones by terms which are very simple in the algebraic sense (quadratic or cubic in our case), something analogous to the convergence to eigenstates may obtain.

One could perhaps conjecture a corresponding theorem. Let Q be a transformation of a n-dimensional space which is nonlinear but is still rather simple algebraically (let us say, quadratic in all the coordinates). Consider any vector $\bar{x}$ and the iterates of the transformation Q acting on the vector $\bar{x}$. In general, there will be no question of convergence of these vectors $Q^n(\bar{x})$ to a fixed direction.

But a weaker statement is perhaps true. The directions of the vectors $Q^n(\bar{x})$ sweep out certain cones C_α or solid angles in space in such a fashion that the time averages, i.e., the time spent by $Q^n(\bar{x})$ in C_α, exist for $n \to \infty$. These time averages may depend on the initial $\bar{x}$ but are able to assume only a finite number of different values, given C_α. In other words, the space of all direction divides into a finite number of regions R_i, $i = 1, \cdots, k$, such that for vectors $\bar{x}$ taken from any one of these regions the percentage of time spent by images of $\bar{x}$ under the Q^n are the same in any C_α.

The graphs fig. 1–9 show the behavior of the energy residing in various modes as a function of time; for example, in fig. 1 the energy content of each of the first 5 modes is plotted. The abscissa is time measured in computational cycles, δt, although figure captions give δt^2 since this is the term involved directly in the computation of the acceleration of each point.

In all problems the mass of each point is assumed to be unity; the amplitude of the displacement of each point is normalized to a maximum of 1. N denotes the number of points and therefore the number of modes present in the calculation. α denotes the coefficient of the quadratic term and β that of the cubic term in the force between neighboring mass points.

We repeat that in all our problems we started the calculation from the string at rest at $t = 0$. The ends of the string are kept fixed.

AN OBSERVATION ON THE DISTRIBUTION OF PRIMES

M. STEIN AND S. M. ULAM, University of California, Los Alamos Scientific Laboratory

Primes up to 100,000,000 have been stored on a rather fast access memory disc connected with a computer, MANIAC II, in our Laboratory. We have looked at some properties of the distribution of primes when the sequence is "compressed" by dividing each prime by its logarithm and taking the nearest integer to this quotient. Of several properties which this new sequence seems to show, the following is perhaps most striking.

It can be stated as follows: For each integer k consider the interval between k and $k+1+E(\log k)$. For some k's there will be no prime in this interval. For some others there will be exactly one prime in this range. For other k's there will be exactly 2 primes, etc. We consider the frequencies of these classes. Let $\gamma_0(n)$ be, therefore, the ratio to n of the number of all k's from 1 to n, such that there is no prime in the range between k and $k+E(\log k)+1$. Actually the interval examined by the machine was from k to $k+E(\log^2 k/\log k-1)$. Similarly we define $\gamma_1(n)$, $\gamma_2(n)$, etc. We have tabulated the γ's in intervals of 10,000 up to $n=100{,}000{,}000$.

It seems that they approach definite limits.

n	γ_0	γ_1	γ_2	γ_3	γ_4	γ_5	γ_6	γ_7
10,000	.2502..	.4681..	.2415..	.0385..	.0017..	0.0000	0.0000	0
50,000	.2693..	.4520..	.2350..	.0421..	.0014..	0.0000	0.0000	0
500,000	.2780..	.4456..	.2300..	.0443..	.0025..	0.0000	0.0000	0
1,000,000	.2851..	.4407..	.2249..	.0459..	.0033..	0.0000	0.0000	0
2,000,000	.2853..	.4379..	.2269..	.0466..	.0033..	0.0000	0.0000	0
5,000,000	.2902..	.4335..	.2242..	.0484..	.0037..	0.0001..	0.0000	0
10,000,000	.2960..	.4293..	.2209..	.0492..	.0045..	0.0002..	0.0000	0
20,000,000	.3017..	.4242..	.2182..	.0504..	.0052..	0.0002..	0.0000	0
40,000,000	.3023..	.4226..	.2180..	.0513..	.0056..	0.0002..	0.0000	0
80,000,000	.3048..	.4211..	.2163..	.0516..	.0059..	0.0003..	0.0000	0
100,000,000	.3052..	.4202..	.2162..	.0521..	.0060..	0.0003..	0.0000	0

The figures in the Table are rounded to four decimals. We have accumulated our statistics by counting in contiguous intervals of length given above.

These numbers do not appear, at least as yet, to approach a Poisson distribution. γ_0 is smaller and γ_1 is larger than the Poisson value and the tail of the distribution seems to go down to zero faster. There is no proof, of course, of the *existence* of the limit of the γ's, but the impression which the full data give is certainly that they do exist. The first time when six primes appear in our range is somewhere after 22579983 and before 22589996. Up to our $n=10^8$ there is not a single case of seven primes in our intervals. Presumably, for every ℓ, there will exist ℓ primes in this range for sufficiently great n, but the frequency of these in the set of all integers goes to zero extremely fast with increasing ℓ.

We are planning to obtain other statistical data about the distribution of these "compressed" or scaled primes.

This work was performed under the auspices of the U. S. Atomic Energy Commission.

Reprinted from the AMERICAN MATHEMATICAL MONTHLY
Vol. 74, No. 1, Part I, January, 1967

Part III

A Collection of Mathematical Problems

PROBLEMS IN MODERN MATHEMATICS

First published under the title

A COLLECTION OF MATHEMATICAL PROBLEMS

S. M. ULAM

Los Alamos Scientific Laboratories, Los Alamos, New Mexico

In memory of
J. SCHREIER

In memory of
J. SCHREIER

PREFACE TO THE PAPERBACK EDITION

In the small space allotted for a preface to this book in its paperback edition, I have had to restrict myself to just the briefest comments.

In my opinion by far the most important mathematical event of the last four years is the discovery by Paul Cohen.[1] The continuum hypothesis is independent of the axioms of set theory as usually presented. This means that one could, if one wanted to, assume that there are intermediate powers between $\aleph_1$ and C. Cohen's result opens a way to a class of large and "free," so to say, non-Cantorian, set theories. (See in this connection p. 1, line 8 from the bottom, *et. seq.*)

It may be that other unsolved problems of set theory have a similar status. Here perhaps is a new question in pure set theory, formulated in terms of infinite games (see pp. 23 and 24). Given is an abstract noncountable set E. Two mathematicians—A and B—play as follows: A can divide E in any way he pleases into two disjoint sets. B has to choose one of these sets and divide it into two disjoint sets in any manner he likes. Then it is A's turn. He also has to choose a set, divide it again, present to B for choice, and so on. Let us say that A wins if the intersection of all chosen sets is vacuous. Is there a strategy for one or the other player to win? (See An Open Problem, by S. Ulam in *Recent Advances in Game Theory,* published by Princeton University Conference in 1961, p. 223.) Possibly this problem and others of its sort may also be undecidable in the system of axioms usually employed for set theory.

Several of the problems mentioned in the original edition have been solved. To mention a few in particular, the problem in the game

[1] P. Cohen, The Independence of the Continuum Hypothesis, *Proc. Nat. Acad. Sci., U.S.*, 50, No. 6 (December 1963), pp. 1143–1148; and The Independence of the Continuum Hypothesis, II, *ibid.*, 51, No. 1 (January 1964), pp. 105–110.

of bridge (p. 36) was solved by several mathematicians—there does exist a distribution of hands in which a grand slam is possible against the best defense in every suit but not even a small slam is possible in no trump. The question now is, can a five-trump contract be made under these assumptions?

Problem 10 in Chapter II has been solved. A positive answer follows from results by Buchsbaum [2] and Samuel.[3] The problem on p. 32 has not really been well formulated. As stated, it was solved by H. Tverberg. The problem starting on p. 38, line 13 from the bottom, has been solved by H. Steinhaus.[4] The problem of mapping a disk onto a torus has been solved by Fort.[5]

Some enumeration problems connected with the operation of the direct product have been solved by G. M. Bergman.

The work discussed in Chapter VII, paragraph 8, has been followed by a number of investigations. Analytical results by J. Ford,[6] M. Kruskal,[7] and N. J. Zabusky [8] have brought some clarification to this intriguing situation. (See, for literature, M. Kruskal and N. J. Zabusky, to appear in the *Journal of Mathematical Physics.*)

In connection with problems discussed in Chapter VI, paragraph 1, Stability, I would like to note in particular the results of F. John on deformations of surfaces and manifolds.

It is impossible to resist the temptation to mention another problem or two. In the *Proceedings of the 1963 Number Theory Conference* held at the University of Colorado in Boulder there are two rather simple problems of mine that I should like to reproduce here:

[2] D. A. Buchsbaum, Some Remarks on Factorization in Power Series Rings, *J. Math. Mech.,* 10 (1961), pp. 749–753.

[3] P. Samuel, On Unique Factorization Domains, *Ill. J. Math.,* 5 (1961), pp. 1–17.

[4] H. Steinhaus, A Note on Astatic Equilibrium, *Bull. acad. polon. sci. sér. sci. math. astronom. phys.* 11 (1963), pp. 173–174. *Math. Rev.* 27 (1964) Rev. No. 2143.

[5] M. K. Fort, Jr., ϵ-Mappings of a Disk onto a Torus, *Bull. acad. polon. sci. sér. math. astron. phys.,* 7 (1959) pp. 51–54. *Math. Rev.* 211 (1960), Rev. No. 325.

[6] J. Ford, *J. Math. Phys.* 2 (1961), p. 387.

[7] M. Kruskal, *J. Math. Phys.* 3 (1962), p. 806.

[8] N. J. Zabusky, in *Mathematical Models in Physical Sciences,* edited by Stefan Drobot, Prentice-Hall, Englewood Cliffs, N. J., 1963.

Let p be a prime.

If a_1, a_2, b_1, b_2 are integers less than p, the minimum of $d_2 = \left|\frac{a_1}{b_1} \cdot \frac{a_2}{b_2} - p\right|$ is obtained for $a_1 = a_2 = p - 1$, $b_1 = 1$, $b_2 = p - 2$. Suppose we wanted to minimize $d_3 = \left|\frac{a_1}{b_1} \cdot \frac{a_2}{b_2} \cdot \frac{a_3}{b_3} - p\right|$ for a_i and b_i all less than p; e.g., if $p = 13, a_1 = a_2 = a_3 = 10, b_1 = 1, b_2 = 7$, $b_3 = 11$. Conjecture: min $d_3 \to 1/p^2$ if $p \to \infty$. *Note.* This problem was investigated numerically for all $p < 100{,}000$ on an electronic computing machine in Los Alamos several years ago. The results seem to support the conjecture. In every case min d_3 was of the form $1/N$ (N an integer). I cannot even establish this for every p.

Let a_1, a_2 be given integers; we construct a sequence of further integers by adding in increasing order those integers that can be represented in one way only as a sum of two different preceding integers in this sequence. So, if $a_1 = 1$, $a_2 = 2$, the sequence is 1, 2, 3, 4, 6, 8, 11, 13, 16, ... Determine the asymptotic density of this sequence. *Note.* Analogously constructed sequences with three integers given initially may have high (positive) densities.

It is planned in the near future to have a supplementary volume, containing more problems, as an addition to the present collection.

S. M. ULAM

May, 1964

PREFACE

In introducing the collection of problems forming the substance of this work, it is perhaps necessary to offer more explanations than is usually the case for a mathematical monograph. The problems listed are regarded as unsolved in the sense that the author does not know the answers. In this sense the structure of this small collection differs inherently from that of the well-known collection of problems by Pólya and Szegö [1]. The questions, drawn from several fields of mathematics, are by no means chosen to represent the central problems of these fields, but rather reflect the personal interests of the author. For the main part, the motif of the collection is a set-theoretical point of view and a combinatorial approach to problems in point set topology, some elementary parts of algebra, and the theory of functions of a real variable.

In spirit, the questions considered in the first part of this collection belong to a complex of problems represented in the *Scottish Book*. This was a list of problems compiled by mathematicians of Lwów in Poland before World War II, also containing problems written down by visiting mathematicians from other cities in Poland and from other countries. The author has recently translated this document into English and distributed it privately; the interest shown by some mathematicians in this collection encouraged him to prepare the present tract for publication. Many of the problems contained here were indeed first inscribed in the *Scottish Book*, but the greater part of the material is of later origin beginning with the years spent at Harvard (1936–1940) and a large proportion stems from recent years, appearing here for the first time. Many of the problems originated through conversations with others and were stimulated by the transitory interests of the moment in various mathematical centers. In addition, several problems were communicated by friends for inclusion in this collection. The last few chapters have a different character: the stress is on computations on calculating machines with examples of problems whose study through the use of this modern tool would have, in the author's opinion, great heuristic value.

Most of the problems were seriously considered and worked upon, but with different degrees of attention and time spent on attempts to solve them. Some have been studied by other mathematicians to whom they were communicated orally but others have not been thoroughly investigated and it would not surprise the author if a number admitted trivial solutions. Most of the problems are, so to say, of medium difficulty. A majority of them should definitely not fall into the category of mere exercises to be solved by routine applications of known lemmas and theorems. In fact one of the aims was a selection of "simple" questions in various domains of mathematics; simple, for example, in the sense that no elaborate definitions beyond those used in general courses on set theory, analysis, and algebra would be necessary for their understanding. The author believes that, on a purely heuristic level, a survey of this sort, if properly enlarged and deepened by others, could bring out the possible general and typical common "reasons" for the difficulties encountered in quite diverse branches of mathematics.

The present situation in mathematical research is perhaps different from that of previous epochs in its very great degree of specialization. The connections between different fields are growing more tenuous, or else so general and purely formal, that they become illusory. It has been said that unsolved problems form the very life of mathematics; certainly they can illuminate and, in the best cases, crystallize and summarize the essence of the difficulties inherent in various fields. The very existence of mathematics can be considered as fruitful only because it produces simple and concise statements whose proofs are much more complicated in comparison. Moreover, Gödel's discovery [1] of the existence of undecidable propositions in every consistent system of mathematics, including arithmetic, renders the "probably true" propositions all the more precious. The intriguing possibility which now exists *a priori* of undecidability lends an additional flavor, to some at least, of the unsolved mathematical problems (cf. Weyl [1]).

The separation between mathematical research stimulated by pure mathematics alone and the ideas stemming from theoretical physics has been increasing in the development of these fields during the last few decades. This may seem at first sight surprising, since the ideas and models of reality employed nowadays in physics tend toward increasing abstractness. However, it appears that on the whole, applied mathematics, so-called, deals at the present time in the majority of cases, with questions of classical physics—or else, when it concerns itself with the new theories, its role is restricted to a purely technical intervention. On the conceptual level one does not have enough, it seems, of cross fertilization of ideas! In the author's opinion it appears likely that in the near future the large class of concepts which have their origin in Cantor's set theory [1], which have influenced so many of the purely mathematical disciplines, will play a role in physical theory. The difficulties of the phenomena of divergence in present formulations of field theory may indicate the need for a type of mathematics capable of dealing with physical problems employing actual infinities *ab initio*. Several elementary problems are included here which are intended to indicate the nature of such possible formulations and the kind of mathematical schemes which may be of use in some future physical theories.

The set-theoretical motivation underlying the selection of questions in the various fields to which the problems refer influenced the choice of the more elementary problems and made the illustration of the more sophisticated ideas of recent years, in topology or algebra for example, impractical.

It is impossible to give detailed credit to all who have indirectly contributed to the set of ideas illustrated in the list of problems, but I would like to acknowledge in particular the pleasure of past collaboration with Banach, Borsuk, Kuratowski, Schreier, and

Mazur in Poland, and John von Neumann, Garrett Birkhoff, J. C. Oxtoby, P. Erdös, and C. Everett in this country. Thanks are due to Mrs. Lois Iles and Miss Marie Odell for their work in preparing the manuscript for publication.

S. M. ULAM

Fall, 1959

CONTENTS

CHAPTER I

Set Theory

CHAPTER II

Algebraic Problems

CHAPTER III

Metric Spaces

Chapter IV

Topological Spaces

Chapter V

Topological Groups

Chapter VI

Some Questions in Analysis

Chapter VII

Physical Systems

Chapter VIII

Computing Machines as a Heuristic Aid

CHAPTER I

Set Theory

1. Introductory remarks

The outstanding unsolved problem of set theory is the well-known continuum hypothesis of Cantor which asserts that the power $c = 2^{\aleph_0}$ of the set of all real numbers is equal to the power $\aleph_1$, the common power of all well-ordered noncountable sets, all of whose segments are finite or countable. We shall not discuss it here; Sierpiński's book [1] in the Polish Monograph Collection deals extensively with various formulations of this hypothesis and with problems which, on the surface, appear more "concrete," but are equivalent or logically related to it. Gödel [2] has investigated the question from the point of view of certain special axiomatizations of set theory. The principal result is that in many such systems, Cantor's hypothesis is either true or else forms an independent statement. Also true or independent is the proposition that certain subsets of the real number system with "paradoxical" properties are projective sets in the sense of Lusin [1]. The problem of the continuum cannot yet be regarded as settled, since none of the axiomatic formulations of set theory can be considered as "definitive" or all-comprehensive, and it is at present impossible to assert that the "naive" set theory, or the intuitive conception of what set theory should be, has found a definitive axiomatic formulation. Apparently it is Gödel's present impression that in a suitable large and "free" axiomatic system for set theory the continuum hypothesis is false. This feeling, based on indications provided by results on projective sets and the abstract theory of measure, has been shared by the author for many years.

The weaker hypothesis: c is less than the first inaccessible aleph, suffices to establish certain results, e.g., in measure theory, valid under the assumption of the continuum hypothesis. For example, the *nonexistence* of a completely additive measure vanishing for sets consisting of a single point and defined for *all* subsets of the interval follows from this hypothesis (Ulam [1]). It is the author's feeling that, in "reasonable" systems of axioms for set theory, even this weaker hypothesis may be false.

Another, perhaps less well known, problem in the general theory of sets is the following question of Suslin: Let C be a class of sets such that every two sets of this class are either disjoint or else one is contained in the other. Every subclass of C consisting only of mutually disjoint sets is countable, as is every subclass such that for every pair of its sets one contains the other. Is C a countable class?

There are many equivalent formulations and some interesting partial results, but the problem must be regarded as unsolved (cf. G. Birkhoff [1], p. 47). The present state of Suslin's and related problems is thus one more of many indications that abstract set theory is far from forming a complete or "dead" field. On the contrary, *the combinatorics of the infinite*, abounding in problems, lead to a vast study which now seems only in its beginnings and is not even systematically formulated in a general form. Indeed, it is possible to generalize and reformulate many problems such as one finds in Netto's book [1] on combinatorics or MacMahon's treatise [1] so as to obtain nontrivial problems about infinite sets. Coming back to Suslin's problem, one can formulate it equivalently in an abstract Boolean algebra or more generally in terms of lattice operations (cf. G. Birkhoff [1]).

Difficult problems arise already in rather simple commutative structures, e.g., in the study of infinite (say countable) Boolean algebras, the equivalent of the propositional calculus, but especially in the more general algebras in which projection operators corresponding to the logical quantifiers are introduced in addition to Boolean operations (cf. Everett and Ulam [1]). Such algebras present ever deeper questions which can be still regarded as concerning pure set theory (Halmos [1]).

Going still further, one employs in set theory operations going beyond all the above, namely, a passage to variables of higher type, i.e., the formation of classes of sets, classes of classes, etc., so that it is possible to treat algebraically mathematical systems of still greater generality. One can formulate problems in naive set theory dealing with properties of *repeated* passages to *variables* of *higher type*, studied *per se*. We shall content ourselves in stating one:

A "super-class" K of objects is imagined which is closed with respect to the operation of forming the class of all subsets. Starting, say, with the set S_0 of integers, one forms the class of all subsets of this set. This is a set which we denote by S_1; the class of all subsets of S_1 will be denoted by S_2, etc. In addition, the class K is closed with respect to the following construction: if S_i, $i = 1, 2, \ldots$, is any countable collection of classes of sets belonging to K, we form the class of all possible sequences of sets $s_i \in S_i$; we postulate that this class Σ of all such sequences $\{S_i\}$ also belongs to K. Imagine K is the smallest class closed under the two operations. The question now arises of classifying the objects of K by means of transfinite ordinals and of determining the powers of sets forming the elements of K.

2. *The operation of direct product*

One deals with the operation of the direct product, in a more or less explicit form, in every mathematical theory involving more than one variable. It is used quite explicitly in topology, group theory, measure theory, in the theory of metric spaces, and it also occurs in one form or another in many algebraic theories. It seems, however, that a *general* investigation of the properties of this operation for its own sake, on a set theoretical basis, has not been undertaken in spite of the many common features, presented by problems of "many variables," apparent in the various theories referred to.

The notion of *phase space* in mechanics is essentially that of a product space. The state of a system of particles is represented by a point in a product space, namely a direct product of spaces each describing the state of one particle. The component spaces themselves may be infinitely dimensional, as in quantum theory, where the state of a single particle requires a function for its description. In dealing with infinitely many particles, as in the physics of continua, it is necessary to introduce a direct product of an infinity of component spaces. A somewhat different type of operation, the "symmetric product," also arises in physics in connection with the Fermi-Dirac statistics, and requires a notion of the direct product as a basic substructure.

But in the foundations of mathematics itself the direct product enters implicitly in every theory involving the logical quantifiers (the "there exists" and "for all" expressions; cf. the work of Kuratowski and Tarski [1]). A mathematical interpretation of the existential quantifier is the operation of *projection* of a set, located in a product space, on one of the component spaces. The theory of projective sets due to Suslin and Lusin [1], and Sierpiński [2] exhibits some of the difficulties of this operation in problems of point set theory, the sets being considered in a topological space. It seems, however, that the real causes of the difficulties in the theory of projective sets have their origin already in general set-theory including the *general* theory of the operation of direct product, rather than in the topology of the real line or the Euclidean space where the projective sets were originally defined.

The importance for mathematical logic of the study of the direct product and the contingent operations of projection in a purely algebraic spirit is manifest. Just as the study of the algebraic properties of Boolean algebras, their structure isomorphisms and representations give a mathematical counterpart of the elementary logic of the calculus of propositions, the theory of such algebras, widened to include the direct product and projection operators, may provide a mathematical representation of logical systems where quantifiers are admitted, and thus afford an adequate algebraic structure for "constructive" mathematical theories.

We now formulate a few definitions and problems concerning the direct product of sets.

3. Product-isomorphisms and some generalizations

The direct product $A \times B$ of two sets A and B is the set of all ordered pairs (a, b) with a in A and b in B. Analogously the product ΠA_i is the set of all sequences $\{a_1, a_2, \ldots\}$ with a_i in A_i. In case all $A_i = A$ and $i = 1, \ldots, n$, we shall write $\Pi A_i = A^n$.

Two subsets A and B of a product E^2 are said to be product-isomorphic in case there exists a one-one transformation $f(x)$ on E to all of E such that the resulting transformation

$$(x, y) \to (f(x), f(y))$$

of E^2 to itself takes A into all of B. The relation of product-isomorphism is reflexive, symmetric, and transitive, and thus constitutes an equivalence relation on subsets of E^2 which divides the class of all such subsets into mutually disjoint subclasses of sets, product-isomorphic among themselves.

The first questions that arise in connection with this relation concern enumeration properties. It is obvious that sets of different cardinal numbers cannot be product-isomorphic.

What is the power of the equivalence class of all subsets of E^2 product-isomorphic to a given subset A of E^2 (in case E has, for example, the power c of the continuum)? In general, it is 2^c; of course it cannot be greater, since the power of the class of all subsets of E^2 is 2^c. In special cases it will be smaller; for example, if A should consist of only one point (a, a) on the "diagonal" of E^2, then every product-isomorphic set also consists of a single point of this diagonal, and the number of such sets is c. Moreover, if $A = E^2$, then A is product-isomorphic only with itself.

A product-isomorphism of a subset A with itself is called a product-automorphism. The number of product-automorphisms of a subset A of E^2, different on A, is in general 2^c when E has power c; this is true, for example, when $A = E^2$. One easily constructs examples of sets A which have only a finite number of product-automorphisms, in particular, some which admit only the identity as such an automorphism. Does there exist, for every n, a set having exactly n product-automorphisms?

Consider now the class K of equivalence classes of product-isomorphic subsets of E^n, where $n \geqq 2$ and product-isomorphism is defined in the obvious way by generalization from the case $n = 2$. The power of K is 2^e when the power e of the set E is infinite, a result which follows from a theorem on the power of nonisomorphic relation sets. What is the power of K when E is finite?

Most of the enumeration questions are difficult in such cases. Specifically, what is the power of: (a) the class of subsets of E^n product-isomorphic to a given one, (b) the class K of equivalence classes of product-isomorphic sets, (c) the set of product-automorphisms of a set with itself? Even good inequalities on such powers should be of interest. Of course, the power of (a) above cannot exceed $e!$ nor can the power of K be less than that of e^n, but the "best possible" bounds may not be easy to find.

The concept of product-isomorphism bears an interesting relation to that of isomorphism for various mathematical structures. Suppose that under an operation O the elements of an abstract set E form a group G. In the set E^3 consider the set $\mathfrak{G}$ of all points (x, y, z) such that $xOy = z$. We may call $\mathfrak{G}$ the "representation" of G in E^3. If H is also a group defined on the elements of E with representation $\mathfrak{H}$, then G and H are group isomorphic if and only if their representations are product-isomorphic in E^3.

The wide applicability of the notion of product-isomorphism is obvious since the definition of isomorphism of mathematical structures depends only on the number and kind of operations and not on their special properties. Thus, if G is a partially ordered set defined on the elements of an abstract set E by an order relation $(<)$, we may take as its representation $\mathfrak{G}$ the set of all pairs (x, y) in E^2 for which $x < y$, and state that two partially ordered sets G, H over E are order-isomorphic if and only if their representations in E^2 are product-isomorphic.

A greater complexity of the system naturally demands higher exponents n of the basic set E for its "representation." Thus a "ring isomorphism = product isomorphism" statement similar to the one above for groups and relation sets obviously holds when a "representation" is constructed in E^6 (the idea of a ring involving *two* binary operations). The question of the minimal dimension n of the product space E^n necessary to afford a precise representation $\mathfrak{G}$ of a system G is a fundamental one. For example: Can one "represent" a group already in G^2, that is to say, attach *effectively* to every group of the same cardinal power as G a subset in G^2, so that the "representations" would be product-isomorphic if and only if the groups are isomorphic in the usual sense?

Topological systems may be characterized by representations in E^∞. Suppose, for example, that G is a Fréchet space defined on a set E, with $a_0 = \lim a_\nu$ defined for certain sequences $\{a_\nu;\ \nu = 1, 2, \ldots\}$. We may define the representation $\mathfrak{G}$ of G in E as the set of all points $\{a_0, a_1, a_2, \ldots\}$ where $a_0 = \lim a_\nu$. Two Fréchet spaces G and H defined on E are homeomorphic if and only if their representations are product-isomorphic in the obvious sense. The extension of such procedures to combinations of algebraic and analytic structures like topological groups is manifest.

Questions concerning the representations of mathematical systems arise immediately. For example, suppose G is a group defined on the unit interval $E = [0, 1]$. (All this means, of course, is that G is a set of the power of the continuum.) Its representation $\mathfrak{G}$ is then a subset Z in the unit cube E^3. Subsets of E^3 can be classified as follows: A sequence $\{A_n\}$, $n = 1, 2, 3, \ldots$, of sets in E is given; one considers sets B belonging to the Borel field over $\{A_n\}$. (On the line this sequence is usually taken as the sequence of rational intervals.) The simplest sets in E^3 are subproducts, i.e., sets Z of the form $Z = B_1 \times B_2 \times B_3$ where B_1, B_2, B_3 are B-subsets of E. One can then consider sets which are complements of these. All these will be called of "class 0." The next class 1 of subsets would be countable sums of sets of class 0 and their complements. Class 2: again sums of sets of class 1, also their complements, and so on. We get an analogue of the Borel classification for subsets of E^3. One problem is of existence of algebraic structures whose representation $\mathfrak{G}$ is of minimal Borel class α, α given—for *any* choice of the sequence A_n, $n = 1, 2, \ldots$. (For example: Does there exist a group G of power of continuum, whose representation $\mathfrak{G}$ would be of class $\alpha > 3$, for any choice of a countable sequence—of the "elementary" sets?).

Starting with a given sequence $\{A_n\}$ of the "elementary" sets in E one can of course go beyond the Borel classification and define "projective sets" in E^3 in a manner completely analogous to that of defining the familiar projective sets of Lusin.

The definition of product-isomorphism of two sets A and B in E^n suggests a generalization which leads to interesting questions about the abstract systems G and H. Let us say that the two sets A and B are *weakly* product-isomorphic if there exist biunique transformations $f_i(x)$ on E to all of E such that the induced transformation

$$(x_1, \ldots, x_n) \rightarrow (f_1(x_1), \ldots, f_n(x_n))$$

maps A onto all of B. One may then define the *weak* isomorphism of two abstract mathematical structures G and H on the elements of E as the weak product-isomorphism of their representations $\mathfrak{G}$ and $\mathfrak{H}$ in E^n. Thus the weak isomorphism of two groups G and H over E is tantamount to the existence of three biunique correspondences $x = u(a)$, $y = v(b)$, $z = w(c)$ on G to H such that $c = ab$ in G implies $xy = z$ for the corresponding elements in H.

A different generalization may be obtained by defining two subsets A and B of E^n as (weakly) product-isomorphic under decomposition in case there exist decompositions

$$A = A_1 + \ldots + A_m, \quad B = B_1 + \ldots + B_m, \quad A_i A_j = 0 = B_i B_j \quad (i \neq j)$$

such that A_i and B_i, $i = 1, \ldots, m$, are (weakly) product-isomorphic. This leads naturally to the concept of "the weak isomorphism of two structures G and H under decomposition" defined in terms of the (weak) product-isomorphism of their representations $\mathfrak{G}$ and $\mathfrak{H}$ under decomposition. For example, are the groups S (all permutations of the set of integers) and H (all homeomorphisms of the unit interval of real members) isomorphic under decomposition?

Given a "representative" set Z of an algebraic structure (over a set of power c) one can ask whether it is Borelian or projective—in the sense of definitions given in the last paragraph. These refer to a given basic sequence of sets A_n in E. We call a sequence of abstract sets A_n in a set E *measurable* if it is possible to define for all sets S of the Borel field over $\{A_n\}$ a real-valued measure function $m(S)$ with the following properties:

1. $m(E) = 1$, $m(S) = 0$ if S consists of a single point.

2. $m(\sum_{i=1}^{\infty} S_i) = \sum_{i=1}^{\infty} m(S_i)$ if $S_i \cdot S_j = 0$ for $i \neq j$.

If the set Z is Borelian or projective with respect to a measurable sequence A_n we call the given algebraic structure abstractly Borelian or abstractly projective.

Among the first problems that arise is that of *existence* of a group defined on a set of power c whose representation would *not* be Borelian. More generally, for other algebraic structures—e.g., lattices or rings—how far can sets, closed under these operations, still exhibit a set-theoretical "pathology" of their representative set $\mathfrak{G}$ of n-tuples?

The motivation behind the above definitions is to have provision for a connection between the purely algebraic properties on one hand and the topological or "analytic" properties on the other—of structures which are given combinatorially, not dependent on a given topology of the given group (or relation algebra, etc.).

4. Generalized projective sets

Let A_n be a class of subsets of a set E, the latter having the power of the continuum. The projections on E of the sets of Borel class *over* the "rectangles" $A_m \times A_n$ in E^2 constitute the projective sets of class $k = 1$ over $\{A_n\}$. Continuing inductively, one defines the projective sets of class $k = 2, 3, \ldots$ over A_n. The problems of this section are based upon this definition (cf. Sierpiński [2]).

Is it true that, for every countable sequence of sets A_n in E, there exists a countable sequence of sets B_n such that the Borel class over B_n contains all projective sets over A_n?

Does there exist a sequence of sets A_n in E with the properties:

(a) the Boolean algebra generated by the sets A_n contains a noncountable set of atoms, and (b) all projective sets over A_n are $G_{\delta\sigma}$ sets relative to the sets $\{A_n\}$?

Does there exist a sequence of sets A_n in E with the properties: (a) the Borel class over A_n contains sets of arbitrarily high (Borel) class number, and (b) all projective sets over $\{A_n\}$ are Borel sets over the $\{A_n\}$?

More specifically, it is true that, for every positive integer k, there exists a sequence of sets A_n with the property (a) of the preceding problem, and the property (b) *all* projective sets over the A_n are sets of projective class k?

Given a sequence of sets A_n in E, and a transformation f on E to E, we shall say that f is a Borel transformation relative to $\{A_n\}$ in case the counter-image of every Borel set over the A_n is again such a Borel set. Does the product-isomorphism of two Borel sets over the class of rectangles $A_m \times A_n$ in E imply their product-isomorphism under a Borel transformation relative to A_n?

Given an arbitrary sequence of sets A_n in E, does there exist a one-to-one mapping of E into E^2 such that the Borel sets over A_n in E go into Borel sets over $A_m \times A_n$ in E^2 and conversely?

In the following problem, the term analytic has its classical connotation. Can every analytic subset of the unit square be obtained by Borel operations from "rectangles" $A \times B$ where A, B are analytic subsets of the unit interval?

The motivation for investigating the Borel operations and, beyond it, the projective operations when one starts with a *general* sequence of sets A_n—instead of the usual one which is the sequence of rational intervals or binary intervals—lies in the following possibility. There might exist a sequence of sets such that the number of its *atoms* is noncountable (i.e., still "nontrivial") and yet such that the projective class over this sequence is "*simpler*" than the "classical" projective class. For example, a sequence such that one could define a completely additive measure function for all sets of this projective class—this is impossible, according to a result of Gödel, for the familiar projective sets: i.e., it is free from contradiction in certain systems of axioms to assume that there exist projective sets which are nonmeasurable in the sense of Lebesgue. Even more generally, one can extend this result to show that *no* completely additive measure is possible for all projective sets (by a measure we understand a set function with the properties: (*1*) $m(E) = 1$; $m(p) = 0$ where E is the whole space, (p) is a set composed of any single point).

$$(2) \qquad m\left(\sum_{i=1}^{\infty} A_i\right) = \sum_{i=1}^{\infty} m(A_i) \text{ if } A_i \cdot A_j = 0 \text{ for } i \neq j.$$

Paradoxically enough, it is conceivable that a measure function like the above could exist, if one starts with a sufficiently "wild" sequence of sets A_n in a class of projective sets "over" this sequence. Possibly all such sets could have the Baire property, that is each set of the class would be of first category—or complement of such?

4a. Relations between products of different orders

Consider an infinite set E and the sets E^n, E^m, $n \neq m$. In each of these we have a special class of subsets, the "R-set" subproducts, i.e., sets of the form $A_1 \times A_2 \times \ldots A_m$ and $A_1 \times A_2 \times \ldots A_n$, respectively where the A's are arbitrary subsets of E. What is a one-one mapping of E^m into E^n such that the R-sets in E^n become sets of lowest possible Borel class over the R-sets in E^m? If the power of the set E is greater than c (of continuum) does there exist such a mapping so that the R-sets in E^n go over into Borel sets over R-sets in E^m?

In the above question the A's are arbitrary subsets of E. An analogous problem exists for the case where the sets A are restricted to the Borel sets over a sequence of sets S_i, $i = 1, 2, \ldots$, given once for all in E.

The same problem for E^n, E^∞.

We mentioned before the "representations" of various algebraic structures through subsets of E^k (the index k being $= n + 1$ for an "algebra" involving a n-ary operation). The question arises whether such structures can be "represented" by sets in E^m with $m < n + 1$ (cf. section 5) and still "effectively" up to an isomorphism; in particular, e.g., whether the *groups* requiring offhand a representation in E^3 can be *effectively* put in a one-one correspondence with subsets of E^2 in such a way that isomorphic groups would have product-isomorphic sets in E^2 attached to them. Obviously the problem of existence of such a correspondence is trivial if we do not require *effectiveness* or *constructiveness*—one can, using the axiom of choice, map all groups, isomorphic to each other into the same subset of E^2. However, if one requires that groups whose representation in E^3 is a Borel set (with respect to a given sequence of sets in E^1) correspond to sets in E^2 which are related to these sets in E^3 through one "Borelian" mapping, the relation with the problems above becomes obvious.

Questions of this sort are related in spirit to the recent results of Kolmogoroff [1], on the reduction of functions of n variables to superposition of functions of 2 variables, reducing transformations of E^m space into itself to superposition of transformations of a space E^n into itself, with $n < m$.

5. Projective algebras

Projective algebras form a generalization of Boolean algebras, and will permit, to a certain extent, an algebraic treatment of the logical quantifiers. For our present purposes it is sufficient to consider a representation of a projective algebra and, for simplicity, we shall restrict ourselves to the two-dimensional case, although the problems formulated below are meaningful in n-dimensions, $n > 2$.

Assume then that we have a class of sets situated in the Euclidean plane, closed under Boolean operations and under projection onto either axis, and containing the direct product $A \times B$ whenever A and B belong to the class and are situated on the X and Y axes respectively. Such a class constitutes the simplest example of a projective algebra (cf. Everett, Ulam [1]; MacKinsey [1]).

Given a countable class of sets in the plane, does there exist a finite number of sets which generate a projective algebra containing all sets of this countable class? Another statement might make this assertion for a countable class of sets given in E^m with the generating sets required to be in some E^n with $n < m$.

That this may be possible is shown by an example due to David Nelson in which two plane sets generate an infinite projective algebra. Note that the situation here is radically different from that of Boolean algebras, in which case one can obtain at most 2^n elements from n generating sets.

Does there exist a universal countable projective algebra, i.e., a countable projective algebra such that every countable projective algebra is isomorphic to some subalgebra of it?

Is it true that, for every positive integer k, there exists a projective algebra generated by k sets in the plane and which is free in the sense that no relations exist between the generated sets except those that are true in every projective algebra? Can every projective algebra be obtained by a homomorphism of a free projective algebra?

How many nonisomorphic projective algebras exist with k generators?

Many theorems in mathematics amount to stating that two sets, obtained by different sequences of Boolean operations and quantifications operating on a finite number of given sets are identical. It is, therefore, desirable to establish in some projective algebras a theorem to the effect that for two identical sets their identity can always be established by the rules of formal projective algebra. We are asking in particular whether there exists a countable sequence of sets in E^m such that the projective algebra generated by them is free; that is to say, whenever two sets, constructed from the generators by formal operations of projective algebra are identical, then they may be demonstrated to be so by formal projective-algebraic operations.

6. Generalized logic

The attempts to exhibit the nature of the essential difficulties in the foundations of mathematical logic in purely algebraic schemes have a long history. We will be concerned in this section with a type of problem which embodies some of the desired features of such a program, and at the same time seems to admit less familiar models, among them, what might be described as a system with infinitely many quantifiers.

One of the most striking weaknesses of projective algebra, as one naturally conceives it in generalizations from the plane to E^n, $n > 2$, is its limitation to a number of quantifiers bounded in advance. Ordinary logic, while it makes statements about only a finite number of variables at a time, suffers no such restriction. Moreover, it is apparent that the postulation of n projection operators along different axes is cumbersome and unnecessary. Instead, we might postulate *one* projection operator and one transformation of variables (say in E^3 $x' = y$, $y' = z$, $z' = x$). These two together will generate all projection operators.

Let us therefore define a more general type of "projective algebra" as a class R of subsets of a set E closed under Boolean operations, and under two operators P and T which we conceive of as a "projection" operator on the class of subsets to itself corresponding to projection of E^n along one "axis", and a one-to-one point-to-point transformation of E to E, corresponding to the permutation of the axes, respectively.

It seems possible to formulate a system of postulates in terms of such operators which, when applied to the special set E of all two-way infinite sequences $(\ldots, x_{-1}, x_0, x_1, x_2, \ldots)$, where x_i are real numbers, would properly contain ordinary logic. However, the possibilities, of *extending* the formalism and still using this special set E, seem to include a logic of propositions with infinitely many variables. For an example of a statement involving infinitely many quantifiers see the infinite games of Chapter I, Section 11.

The formal structure would be the following: A class R of subsets of E is closed with respect to the Boolean operations and the operators T and P. In addition, we can require countable additivity in R.

Any such class could be considered as an infinitely dimensional projective algebra. It is then a class of sets contained in E such that:

1. If $Z \subset R$, then $(E - Z) \in R$; if $Z_1 \in R$ and $Z_2 \in R$, then $(Z_1 + Z_2) \in R$.

2. If $Z \in R$, then $T(Z) \in R$ and $P(Z) \in R$; if $Z \subset R$, then $T^{-1}(Z) \in R$.

3. If $Z_i \in R$ for $i = 1, 2, \ldots$, then $\sum_{i=1}^{\infty} Z_i \in R$.

It is in this way, by using *3*, that one will obtain sets defined by an infinite class of "quantifier" operations or sets of an infinite projective class.

This could be of course achieved through a more orthodox procedure, using a finite number of quantifier operators only, but then we would have to operate with spaces involving additional variables.

Perhaps the set-up above would have a greater "algebraic" homogeneity.

The first problems would involve a representation theorem — then the possibility of generating countable projective algebras of the above type from a finite number of sets, etc. — similarly to problems on the two-dimensional projective algebras.

7. Some problems on infinite sets

1. Let A and B be infinite sets which admit a transfinite sequence of point transformations $t_\xi(a) \in B$, $a \in A$, with the properties: (*1*) $t_\xi(X) \cdot t_\xi(Y) = 0$ for $X \subset A$, $Y \subset A$, and some ξ implies $t_\eta(X) \cdot t_\eta(Y) = 0$ for all $\eta > \xi$; (*2*) for every infinite subset $X \subset A$ there exists a ξ such that $t_\xi(X)$ contains at least two distinct points; (3) $X \cdot Y = 0$ for finite X, Y implies existence of η such that $t_\eta(X) \cdot t_\eta(Y) = 0$.

Is the power of A necessarily less than or equal to that of B?

2. Let C be a class of subsets of the interval $(0, 1)$ with the following properties: (*1*) C contains the Borel sets (in the usual sense); (*2*) C is closed under complementation and countable unions; (*3*) For every decomposition of $(0, 1)$ into disjoint sets, each containing at least two points, there exists a set in the class C which has exactly one point in common with every set of the decomposition.

Is the class C necessarily the class of all subsets of $(0, 1)$?

3. The power of the class of all additive subgroups of the real field R is 2^c, namely the power of the class C of *all* subsets of R. Is the Borel class of sets over the sets which are arbitrary subgroups R identical with the whole class C? In the event of a negative answer, one may ask a similar question about the analytic class (of all sets generated by analytic operations from sets which are subgroups) or the kth projective class. That is, is every set of real numbers obtainable by projective operations from sets which are subgroups? For partial results see Erdös, Kakutani [1]

8. Measure in abstract sets

It is known (cf. S. Ulam [1]) that no countably additive measure function $m(A)$ exists, defined for *all* subsets A of a set E of power $\aleph_1$, which vanishes for all subsets consisting of single points and for which $m(E) = 1$. Does there exist a class of measure functions, $m_\xi(A)$, ξ in a set of lower power than $\aleph_1$, such that every subset of E is measurable in at least one of these measures? This is a problem of Erdös and the author. Partial results have been obtained by Alaoglu and Erdös.

Let E denote the set of transfinite ordinals less than Ω, the first ordinal corresponding to a noncountable power. Thus E has the power $\aleph_1$. Is it possible to define a countably additive measure function such that all sets of the Borel field over the subsets of E which form arithmetic progressions shall be measurable? (The subclass of arithmetic progressions form an analogue in this set of the class of binary intervals for the set of real numbers). Similar questions may be asked concerning the measure for (a) analytic class over the set of arithmetical progressions in E, and (b) "projective" subsets of E (over the same class).

If a set E is of a power which is an inaccessible aleph $\aleph$, does there exist a countably additive measure for *all* subsets of E with $m(E) = 1$ and $m(p) = 0$ for all points p?

It is known that the results on the impossibility of defining a completely additive measure for *all* subsets of a set E hold true for all sets whose powers are *accessible* cardinals. It seems rather likely that at any rate the existence of a two-valued measure, countably additive and defined for all subsets of a set of an *inaccessible* power does not contradict the axioms of set theory. Indeed, probably a stronger additivity for all $\aleph < \overline{\aleph}$ can be obtained where $\overline{\aleph}$ denotes the power of the first inaccessible cardinal.

9. Nonmeasurable projective sets

Gödel has proved that the existence of projective sets which are nonmeasurable in the sense of Lebesgue does not contradict the axioms of set theory; that is to say, the statement that such sets exist is either true, within certain axioms of set theory, e.g., those of von Neumann, or is independent of these axioms. It would be interesting to prove that the existence of such sets follows from the continuum hypothesis.

In the sequel we shall sketch some possible lines of attack on this problem through the use of certain constructions in the product space.

Gödel's result shows really more: it is free of contradiction in such axiomatic treatments of set theory to state that no countably additive measure, vanishing on sets consisting of a single point, is possible for all projective sets.

A similar problem can be formulated about the existence (assuming the continuum hypothesis) of projective sets with other "paradoxical" properties, e.g., sets not satisfying the Baire property (a set satisfies this property if it or its complement is on every interval a sum of countably many nowhere dense sets).

We shall indicate how, given an effective or constructive decomposition of the interval into $\aleph_1$ sets each of Lebesgue measure 0, one can define constructively, i.e., projectively, a nonmeasurable set consisting of a sum of the sets used in this decomposition. A construction is given in the paper referred to (Ulam [1]) which amounts, in essence, to the following:

One can construct a doubly infinite matrix whose elements are subsets of a set of a power $\aleph_1$ with the following properties. Each row of the matrix represents a decomposition of E into $\aleph_1$ disjoint sets. There are countably many rows and noncountably many columns. The sum of the sets in each column gives the whole set E with the exception of, at most, countably many elements (which would be the original sets each of measure 0). The sum of each column, therefore, would be a set of measure 1. So in each row there would exist at least one set of positive measure. This leads to a contradiction because there are only countably many rows and noncountably many columns, there would exist at least one row with noncountably many sets, each of positive measure and all disjoint, which is impossible.

If we can, therefore, establish the existence of a constructive decomposition of the interval into $\aleph_1$ projective sets each of measure 0, the construction used in l.c. would turn out, as we shall prove, also projective and nonmeasurable.

Here is a suggestion of a possible way to obtain such a decomposition:

Consider a set X of measure 0 on the interval, and such that under a Peano mapping T of the interval into the square which preserves measure (i.e., linear measure in the interval = area of the image in the square and *vice versa*) its image is contained in the set X^2. Assume furthermore that $T(X)$ is a proper subset of the set X^2. We want to show the existence of a constructive set S_1 still of measure 0 and containing X as a proper subset.

Take $T^{-1}(X^2)$, and define X_1 as $X+T^{-1}(X^2)$. Since T is measure-preserving, $T^{-1}(X^2)$ is of measure 0 and X^{-1} is of measure 0 too.

Consider now X_1^2; this set contains $T(X_1)$. More generally, suppose that sets X were defined for all $\alpha < \eta$ and that X_α^2 contained $T(X_\alpha)$. Take X_η as $\sum X_\alpha$ for all α, then X_η^2 contains X_α^2 for all α and, therefore, also $T(X_\alpha)$ for all α. Take $T^{-1}(X_\eta^2)$. This will be a set containing X_η. Let us call it $X_{\eta+1}$. Proceeding in this fashion we shall obtain a well-ordered sequence of sets that are increasing and are all of measure 0.

Under the assumption that the power of continuum is that of $\aleph_1$ this process will stop at a transfinite ordinal of at most third class. We obtain thus the desired decomposition of the interval into a well-ordered sequence of projective sets all of measure 0.

We can use now the construction described in the paper in

Fund. Math. Vol. 16, by using instead of points, the sets of our sequence. The sets of our matrix would all be projective sets. At least one of these sets must be nonmeasurable. This proves our theorem, the existence of a projective set not measurable in the sense of Lebesgue.

REMARK 1. This method really could prove more. For every measure that is completely additive and such that there will exist a mapping of the interval into the square preserving the measure in the square, and such that Fubini's theorem holds for the measure, there will exist sets that are projective and not measurable in the sense of this measure.

REMARK 2. The same argument would yield the existence of projective sets which do not possess the Baire property, i.e., that there exist projective sets which are not of first category on any perfect set, and their complements are not of first category on any perfect set.

REMARK 3. One could possibly extend such a method to obtain the existence of a projective set which does not possess the λ-property of Lusin.

All these results would be valid under the assumption of the continuum hypothesis. Presumably they would still hold under the assumption of a weaker hypothesis, namely, that the power of continuum is less than that of the first inaccessible aleph (greater than $\aleph_1$).

PROBLEM. Does there exist a two-valued measure for all subsets of a set whose power is the first inaccessible aleph?

We shall summarize again our approach to the question of existence of projective sets, nonmeasurable in the sense of Lebesgue, and will sketch an alternative approach to establish the lemma on "projective" decompositions of the interval into $\aleph_1$, sets of measure 0. The basic role is played by the theorem, mentioned above:

There does not exist a completely additive measure function defined for all subsets of a set Z of power $\aleph_1$ which would assume the value 0 *for sets consisting of a single point and equal to* 1 *for the whole set Z* (cf. Ulam [1]).

The proof depends on the existence of a decomposition of the set Z into a doubly infinite matrix of sets (all subsets of Z) which would apply equally well if the set Z was of the power of continuum, but could be decomposed into $\aleph_1$ sets all of measure 0. The proof of the nonexistence of measure is contructive if the decomposition used in the paper quoted above could be assumed constructive.

It would be interesting to strengthen Gödel's result by proving that the existence of projective nonLebesgue measurable sets follows from the assumption of the continuum hypothesis or even, the weaker hypothesis, namely, that the power of continuum is less than that of the first inaccessible $\overline{\aleph}$. Let us note that it is sufficient to show the existence of a decomposition of the interval 0,1 into $\aleph$ sets all of Lebesgue measure 0, where the $\aleph$ is a cardinal number smaller than the first inaccessible $\aleph$, and the decomposition would be constructive in the following sense: the sum of a subsequence of the sets of the decomposition corresponding to a constructive class of ordinals is understood here in the sense of, e.g., Kuratowski [2, 3] as "constructive". Given such a decomposition, the construction given in the paper by Ulam [1] yields a set which is obtained by projective operations from the given sets and, as shown, would be nonmeasurable.

The crucial point, therefore, is to show that one can decompose the interval into, for example $\aleph_1$ sets all of measure 0 and so that the decomposition itself would be "projective." An approach to such a construction, different from the one above, could be as follows:

Let us start with the well-known decomposition of Lebesgue of the interval into $\aleph_1$ disjoined sets of Borel sets and all having, in addition, the property that if a number z belongs to a set A_ξ, $\xi < \Omega$, then any number of the form $z + r$, where r is any rational number, belongs also to A_ξ. In virtue of this property, the sets A_ξ must all be of Lebesgue measure either 0 or 1. If all were of measure 0 we would have the desired decomposition so we can assume that one of these sets will have measure 1. The operations which we shall perform on sets from now on will lead always to sets that are invariant with respect to addition of a rational number so that these sets will also be of measure either 0 or 1 or else nonmeasurable. Since the sets will also be projective, we may assume that they are all measurable, otherwise our result would be already proved.

One needs the following general lemma:

Given a projective set that is uncountable, we shall attach to it "projectively" a proper subset of it. This will replace for our purposes the necessity of using the axiom of choice which, of course, in general destroys the constructive character of the set. Let $T(p)$ be a one-to-one transformation of the interval into the square. This $T(p)$ can be chosen as Borelian transformation, in fact, one of second Borel class. Consider a set Z contained in the interval. Consider the set Z^2 and the set $T(Z)$. We may consider the division of the unit square into the four sets Z^2, $C \times Z$, where CZ is the complement of Z to the whole interval, $(CZ)^2$ and $CZ \times Z$. If $T(Z)$ is not contained in any one of these sets, then the counter image of the two parts of $T(Z)$ lying in two different sets of this decomposition will give us a decomposition of Z into two nonvoid parts, which proves our assertion that there exists a projective proper subset of Z. Suppose then, that $T(Z)$ is wholly contained in one of the four sets mentioned above. We may assume that if $T(Z)$ is a subset of Z^2 it does not contain any points of the diagonal D of the square, because we could, by subtracting these points, obtain a proper subset of Z. Consider the part of the diagonal that corresponds to the set Z. This part, called Z_1, is contained in $T(CZ)$ in virtue of the remark that $T(Z) \cdot D = 0$. Consider, therefore, $T^{-1}(Z_1)$. If this set has a constructive proper subset we shall be able to define a constructive proper subset of the set Z itself, because of the existence of a constructive mapping of Z into Z_1

We may assume that Z_1 is contained wholly in just one of the nine sets into which the square is decomposed, by taking the decomposition of the interval into Z, $CZ - Z_1$ and Z_1 and multiplying it by itself which gives us the sets Z^2, $(CZ - Z_1)^2$, Z_1^2 and the six cross products of the three sets. Consider the set on the diagonal D corresponding to the set Z_1, call it Z_2. Any projective proper subset of Z_2 would give us immediately a projective subset of Z which would prove our theorem. It suffices, therefore, to find such a subset of Z_2. We can repeat the reasoning given before, and we shall arrive at a set Z_3. One can by transfinite induction prolong this construction for any ordinal of the second or third class. If we make the assumption that the power of continuum is only $\aleph_1$ this chain of sets Z must stop, which gives us a proper subset of the set Z given in the beginning. Our lemma is, therefore, proved.

Consider now for a given set Z its proper subset which we shall denote by $U(Z) = Z'$. Of the sets Z' and $Z - Z'$, one and one alone must have measure 1. We shall apply to it the lemma and obtain sets Z^2 and its complement. In this fashion we can define sets Z^n all of measure 1; that is to say, if they are measurable. The intersection of these sets would be a set still of measure 1, because of the additivity property of the measure function. Using our lemma for this set, which we shall call Z^ω, we shall obtain a set, still of measure 1, $Z^{\omega+1}$. We can continue our construction by transfinite induction remembering, however, that the axiom of choice is not used since we always select the proper subset effectively by taking the one which has measure 1. We can assume that the intersection of $\aleph_1$ of such sets is still a set of measure 1, otherwise we would have the desired decomposition into $\aleph_1$ sets of measure 0 by using the complements of these sets. Therefore, our construction can be prolonged to transfinite ordinals of class three. However, if we assume the continuum hypothesis, we must arrive at a vacuous set for some ordinal of the third class. Therefore, there must be a sequence of length $\aleph_1$ of sets that have measure 1 with a void intersection which proves our assertion.

We repeat that all the steps of our construction would be effective, i.e., the axiom of choice was not used, so the decomposition of the interval into $\aleph_1$ sets of measure 0 is *projective,* at least in the *wider sense* of the word.

10. Infinite games

The following combinatorial scheme was first proposed by S. Mazur around 1928. Imagine a set of points M contained in the unit interval (0, 1) and two indefatigable mathematicians A and B who will play the following game. Each will define an interval in turn, the first interval being given by A, and each successive interval being a subinterval of the preceding one but otherwise arbitrary. The game is won by A if the intersection of all intervals contains a point of M, otherwise by B.

In case the set M is a residual set in some interval, the player A can always win with the following strategy. The set M having a complement with respect to a certain interval which is of first category, A chooses this interval initially, and, regardless of the choices of B, always selects, at his nth turn, a subinterval disjoint with the nth nowhere dense set which figures in the decomposition of his opponent's set into a sum of countably many nowhere dense sets. Obviously A wins!

It is interesting that, using the axiom of choice, one can construct sets M such that for every subinterval neither M nor M' is of 1st category with respect to this subinterval. These are sets which do *not* have the so-called property of Baire. Banach has proved (unpublished) that for such a set M there exists no method of winning for either player.

One can generalize and vary Mazur's game in many directions. The problem occurring in all such games involving a subset M is: what is the class of sets M for which no method of winning for either player exists? It should be pointed out that, in all known games of this sort, a method of winning exists for either A or B when the subset M is *effectively defined.* For example, for the original game, up to the present no *effectively* constructed set is known which does not possess Baire's property. All known proofs of the existence of such sets utilize the axiom of choice.

A result of Gödel from which it follows that it is safe to assume, in certain systems of axioms for set theory, that such sets exist and are projective yields an interesting interpretation of Banach's result!

It is also interesting to note that games of this kind cannot easily be defined between three players without a trivial reduction to a game between two of the three players. The dichotomy inherent in such infinite constructions is to be observed in all constructive parts of set theory. Thus, for example, a Lebesgue measurable set possesses at almost all of its points a density which is either zero or one. A Baire set is of 1st category or a complement of such in all the points, etc.

The game first mentioned can also be specialized in various ways. Thus the choice of intervals open to A and B can be restricted by some rule. The simplest example might be one such that, given a set M, the players produce successive digits of a real number x in binary expansion, player A trying to make x belong to M, while B tries to make x belong to the complement of M. An analogue of Banach's result holds for these more special games. Compare Gale, Stewart [1]; Mycielski, Swierczkowski, Zieba [1]; Mycielski, Zieba [1]. See also the *Scottish Book* for original version of Mazur and modifications of Banach and Ulam.

An interesting variation of the latter situation would be provided by allowing a random element to enter: For a given subsequence n_i the play n_i is to be 0 or 1 determined by chance. A method of winning would be understood here as a strategy by means of which A succeeds in defining x in M for almost every sequence of 0's and 1's on the plays n_i.

11. Situations involving many quantifiers

It is in statements on existence of a winning strategy that a use of a large number of quantifiers appears most natural. In games between two persons they have, e.g., this form. For every move α_1 (of the player A), there exists a move α_2 for player B so that for every move α_3 there exists a move α_4 and so on so that after α_{2n} the situation is a win for B. In a notation of Kuratowski-Tarski:

$$\prod_{\alpha_1} \sum_{\alpha_2} \prod_{\alpha_3} \cdots \sum_{\alpha_{2n}} W(\alpha_1 \ldots \alpha_{2n})$$

where $W(\alpha_1 \ldots \alpha_{2n})$ is a "Boolean expression" describing a winning position.

Thus a mate announced in 5 moves involves, in this notation, 10 quantifiers. This has to be compared with the usual mathematical definitions, e.g., of uniform convergence, or the definition of an almost periodic function, etc. These, granting already the customary elementary mathematical notions and abbreviations, get by with 3–5 quantifiers.

In a study of the formal properties of repeated applications of quantifiers treated as mathematical operations (in the theory of projective sets of Luzin — or in projective algebras — cf. Section 5) the number of times one employs these operators is arbitrary. In the chapter on computing machines we shall mention the possibility of a heuristic investigation of combinatorial problems arising in this connection.

12. Some problems of P. Erdös

Several problems pertain to the theory of graphs. They have equivalent formulations in our terminology of product operation and can be stated as problems on subsets of a set E^2 through the obvious correspondence of considering pairs of points which are joined by a segment. When one considers vertices of triangles in a graph, one can consider the corresponding set of triplets in E^3 etc.

Some other problems belong to number theory but are of combinatorial character and we include them in this section.

1. Let S be a set of power $\aleph$; to each finite subset A of S there corresponds an element $f(A)$ of S and $f(A) \notin A$. A subset $S_1 \subset S$ is called independent if for every $A \subset S_1$, $f(A) \notin S_1$. Does there always exist an infinite independent subset of S? If S has power $\aleph_n$ this is false (a result of Erdös-Hajnal) but it may be true for $\aleph_\omega$.

2. A problem on infinite graphs (Erdös-Rado): Suppose an infinite graph is given whose vertices form an ordered set of type ω^2. If this graph does not contain a triangle, then its vertices have a subset of type ω^2, no two vertices of which are connected by an edge. This is denoted symbolically as follows: $\omega^2 \rightarrow (\omega^2, 3)^2$. Specker (*Comm. Helv.* 1957) showed that $\omega^n \rightarrow (\omega^n, 3)^2$ is false for $n > 2$. Does $\omega^\omega \rightarrow (\omega^\omega, 3)^2$ hold?

2a. Erdös-Rado: Let S be a set of power greater than c. All the finite subsets of S are divided into two classes (in an arbitrary way). Does there exist an infinite subset $S_1 \subset S$ so that for every integer k all subsets of S_1 having k elements belong to the same class (but the class can depend on k)? If the power of S is $\leqq c$ this is not true (Erdös, Hajnal).

3. Let g be a complete graph of power $\aleph_1$, $g = g_1 + g_2 + g_3$ and we assume that $g - g_i$, $i = 1, 2, 3$, does not contain a complete graph of power $\aleph_1$. Does there then exist a triangle, each edge of which is in a different g_i^2? Another problem: Let g be a complete graph of power $\aleph_1$, $g = \sum_\alpha g_\alpha$, $1 \leqq \alpha < \Omega_1$, so that $g - g_\alpha$ does not contain a complete graph of power $\aleph_1$ $1 \leqq \alpha < \Omega_1$ (such a decomposition is possible (Erdös-Rado)). Does there exist a complete subgraph g' of g of power $\aleph_0$, each edge of which is in a different g^2?

4. Erdös-Turan: Let $a_1 < a_2 < \ldots$ be an infinite sequence of integers. Denote by $f(n)$ the number of solutions of $n = a_i + a_j$. Assume that $f(n) > 0$ for $n > n_0$. Then $\limsup_{n=\infty} f(n) = \infty$. A still sharper conjecture is: let $a_k < ck^2$, $1 \leqq k \leqq \infty$, then $\limsup_{n=\infty} f(n) = \infty$.

5. Erdös-Turan: Denote by $r_k(n)$ the maximum number of integers not exceeding n in a set which does not contain an arithmetic progression of k terms. How large can $r_k(n)$ be? $r_k(n) < \frac{1}{2}n$ would prove van der Waerden's theorem according to which if one splits the set of integers into two groups, at least one of them contains an arbitrarily long arithmetic progression. (The best results so far are:

$$r_3(n) < cn/\log \log n$$

$$r_3 > n^{1 - c/\frac{1}{2} \log n}$$

$$r_3(n) > n^{1 - c/\log \log n}$$

Compare Roth [1] and references given there.)

6. Let $f(n) = \pm 1$, chosen arbitrarily. Prove that to every c there exists an m and a d so that

$$\left| \sum_{k=1}^{m} f(dk) \right| > c$$

This conjecture has connections with van der Waerden's theorem (cf. Khinchin [2]). Also it would imply that if $g(n) = \pm 1$, $g(n)$ multiplicative, then

$$\limsup_{n=\infty} \left| \sum_{k=1}^{n} g(k) \right| = \infty$$

7. Let $a_1, a_2, \ldots, a_n$ be n elements. $A_1, A_2, \ldots, A_k$ are sets formed of the a's. Assume that no A_i contains any A_j. Then $k \leqq {}^nC_{n/2}$. This is a theorem of Sperner [1]. Let now $B_1, B_2, \ldots, B_l$ be sets formed from the a's so that no B is the union of any two other B's distinct from it. What is the maximum l? Quite possibly $l < c\, {}^nC_{n/2}$. Erdös states that he cannot even prove that $l = 0(2^n)$.

8. Let $a_k \geqq 1$, $1 \leqq k \leqq n$. Consider all sums of the form

$$\sum_{k=1}^{n} \varepsilon_k a_k \text{ where } \varepsilon_k = \pm 1.$$

Erdös proves (sharpening a previous result of Littlewood and Offord), using the above result of Sperner, that the number of these sums falling in the interior of an interval of length 2 is $\leqq {}^nC_{n/2}$; equality for $a_k = 1$, $1 \leqq k \leqq n$. The conjecture is that the same holds if the a's are complex numbers of absolute value $\leqq 1$ and the interval is replaced by a circle of radius 1. The same result may even hold if the a's are vectors in a Hilbert (Banach) space.

9. Let $a_1 < a_2 < \ldots < a_k \leqq n$; $b_1 < b_2 < \ldots < b_l \leqq n$ be two sequences of integers such that all the products $a_i \cdot b_j$, $1 \leqq i \leqq k$, $1 \leqq j \leqq l$, are different. Prove that

$$k \cdot l < c\,(n^2/\log n).$$

If true this is the best possible result.

10. How many distinct residues $a_1, a_2, \ldots, a_k$ can one give (mod p) so that none of the $2^k - 1$ sums $a_{i_1} + a_{i_2} + \ldots + a_{i_j}$ should be $\not\equiv 0 \pmod p$? Clearly one can give $[\sqrt{(2p)}]$ such residues ($a_i = i$, $1 \leqq i \leqq [\sqrt{(2p)}]$). No decent upper bound is known.

11. (Erdös and the writer). Let I be any finitely additive ideal in the set of all integers. Consider the Boolean algebra of subsets of integers mod I. Does one obtain $2c$ nonisomorphic Boolean algebras in this way?

CHAPTER II

Algebraic Problems

1. An inductive lemma in combinatorial analysis

We shall illustrate this lemma first on structures with a given binary relation. Suppose that in two sets A and B, each of n elements, there is defined a distance function ρ for every pair of distinct points, with values either 1 or 2, and $\rho(p, p) = 0$. Assume that for every subset of $n - 1$ points of A, there exists an isometric system of $n - 1$ points of B, and that the number of distinct subsets isometric to any given subset of $n - 1$ points is the same in A as in B. Are A and B isometric? This assertion is true for $n \leqq 6$, as has been shown by P. Kelly [1] by examination of all possible cases.

Clearly the metric formulation is equivalent to a similar question about sets with a binary relation pRq, holding if and only if $\rho(p, q) = 1$. Can one infer the relational isomorphism of A and B from $n - 1$ level relational isomorphisms of subsets in the manner indicated?

Similar problems may be formulated in other algebraic systems. Specifically, suppose that G and H are groups of order n. We shall say two subsets $G_k \subset G$, $H_k \subset H$ of k elements each are conditionally isomorphic if there exists a one-to-one mapping f on G_k to H_k, such that whenever a, b and $c = ab$ are in G_k, then $f(c) = f(a) \cdot f(b)$. What is the minimum number $k(n)$ such that the conditional isomorphism of every G_k to some H_k implies the isomorphism of G and H? One might include the stronger hypothesis that if $\{G_k\}$ is any class of l distinct subsets G_k conditionally isomorphic to each other, then there are also l distinct subsets H_k of H, each conditionally isomorphic to the G_k sets.

2. A problem on matrices arising in the theory of automata

The theory of automata leads to some interesting questions which in the simplest case reduce to matrix theory formulations. Suppose one has an infinite regular system of lattice points in E^n, each capable of existing in various states $S_1, \ldots, S_k$. Each lattice point has a well defined system of m neighbors, and it is assumed that the state of each point at time $t+1$ is uniquely determined by the states of all its neighbors at time t. Assuming that at time t only a finite set of points are active, one wants to know how the activation will spread. In particular, do there exist "universal" systems which are capable of generating arbitrary systems of states. Do there exist subsystems which are able to "reproduce," i.e., to produce other subsystems like the initial ones? In a simple case, one would ask: Does there exist an infinite matrix $A = [a_{ij}]$ of zeros and ones with $\sum_j a_{ij} < B$ for all rows i, such that *every* possible finite matrix of zeros and ones will appear as a main-diagonal submatrix of some power A^p of A? A positive result would provide a simple example of a "universal" and "reproducing" system (in a very limited sense only).

More generally, an analogous question may be asked about matrices whose elements are integers modulo p.

A similar inquiry is pertinent in case of the "recursive functions." Can one obtain all recursive functions by a *prescribed* algorithm operating on a finite set of such functions? More generally, are all expressions in Gödel's system obtainable from a finite system of such expressions and a finite number of rules of composition performed in a prescribed order? That is to say, for example, application of two operations, applied in turn in an order given by one sequence of two symbols.

Perhaps there exists a logical analogue of our universal matrix model.

3. A fundamental transformation in the "theory of equations"

The transformation

$$T_n:\quad x_1' = -\sigma_1(x_1, \ldots, x_n),$$
$$x_n' = (-1)^n \sigma_n(x_1, \ldots, x_n),$$

where σ_j is the jth elementary symmetric function, gives the coefficients x'_i of the equation

$$z^n + x'_1 z^{n-1} + \ldots + x'_n = 0$$

in terms of the roots x_i. The inverse T_n^{-1} of the transformation T_n "solves" the equation of n-th degree. This transformation can be considered operating on the n-dimensional real or on the n-dimensional complex space.

Many of the statements about algebraic equations are translatable into the elementary properties of this mapping. Thus Gauss' theorem on the existence of roots is simply the statement that T_n is a mapping (many-one) on E^n to *all* of E^n, where E is the complex plane. The points "constructable by ruler and compass" are related to those resulting from iteration of the inverse transformation T_2^{-1} where $n = 2$.

However, the topological nature of this transformation does not seem to have been very thoroughly investigated. For example, what are the nontrivial fixed points $p = T_n(p)$? The origin is always a fixed point, but there are others, e.g., $p = (1, -2)$ when $n = 2$. What are the invariant analytic manifolds $M = T_n(M)$? What points are periodic under T_n?

The impossibility of solving the general equation of degree $n \geqq 5$ "by radicals" means that the corresponding T_n^{-1} is not a transformation involving only field operations and extraction of roots. Does there, however, exist a homeomorphism H of E^n such that the inverse of $S = H^{-1} T_n H$ would involve only such operations?

The solution of the equation of fifth degree may be made to depend on elliptic functions (Hermite) and such methods were generalized by Poincaré. Can one show that for any $n \geqq 6$ the transformation T_n^{-1} can be obtained by composition of such transformations T_m^{-1} of lower degree m operating on suitable subspaces of E^n?

Can one show that T_n itself is a composite of a finite number of mappings, each of which is a conjugate HT_mH^{-1} of some T_m, $m < n$, operating on a suitable subspace of E^n?

4. A problem on Peano mappings

Let R be the set of positive rational integers with the usual operations $a + b = s(a, b)$ and $a \cdot b \equiv m(a, b)$. Every one-to-one (Peano) mapping $c = p(a, b)$ on $R \times R$ to all of R may serve so associate with $s(a, b)$ and $m(a, b)$ two functions σ and μ on R to R by the definitions $\sigma(c) = \sigma(p(a, b)) = s(a, b)$, and $\mu(c) = \mu(p(a, b)) \equiv m(a, b)$. Does there exist a Peano mapping $p(a, b)$ such that "addition commutes with multiplication" in the sense that $\sigma(\mu(c)) \equiv \mu(\sigma(c))$ for all c of R? To illustrate, we note that the well-known Peano mapping $c = p(a, b) \equiv 2^{a-1}(2b-1)$ fails. For, $\sigma(\mu(14)) = \sigma(\mu(2^{2-1} \cdot [2.4 - 1])) = \sigma(8) = \sigma(2^{4-1} \cdot [2.1 - 1]) = 5$, while $\mu(\sigma(14)) = \mu(\sigma(2^{2-1} \cdot [2.4 - 1])) = \mu(6) = \mu(2^{2-1} \cdot [2.2 - 1]) = 4$.

5. The determination of a mathematical structure from a given set of endomorphisms

One of the fundamental tasks of abstract algebra is the determination of the automorphisms or homomorphisms into itself (endomorphisms) of a given algebraic structure. The inverse problem, though not as familiar, presents many features of interest (cf. Everett, Ulam [4]).

Suppose that we are given the operation of ordinary *multiplication* on the rational integers $R = 0, \pm 1, \pm 2, \ldots$. What are all the possible operations of "addition" definable on the set R which, with the given multiplication, will yield a ring? It is easy to show that the characteristic of such a ring must be 0 or 3.

Less specifically, what are all possible rings with identity, countably many primes, and unique factorization up to units?

Given the class of homeomorphisms of a topological space, what other topologies exist on the same set which have these mappings as the class of all their homeomorphisms?

6. A problem on continued fractions

Apparently the explicit form of the simple continued fraction corresponding to a real algebraic number of degree exceeding two is not known in any individual case. The following special questions may, however, be more tractable. Does there exist an algebraic number of degree > 2 in whose continued fraction $n_1 + 1/n_2 + + 1/n_3 + \ldots$ the sequence n_i is not bounded? (Consider in particular the number ξ defined by $\xi = 1/(\xi + y)$ where $y = 1/(1 + y)$.) Or is it perhaps true that *every* real algebraic number of degree > 2 has an unbounded sequence $\{n_i\}$ for its continued fraction?

(The set of real numbers for which the sequence $\{n_i\}$ is bounded has, as is well known, measure zero (and is of first category), so that one might say that the *a priori* chance of a number x having the sequence $\{n_i\}$ unbounded is 1.)

7. Some questions about groups

Is every separable continuous group, considered solely as an abstract group, isomorphic to a subgroup of the group S_∞ of all permutations of the integers? It is obvious that S_∞ is "universal" for all *countable* groups in this sense (that is: every countable group is isomorphic to a subgroup of it), but one can also show that *some* groups of power c, like the additive group R of real numbers is isomorphic to a subgroup of S_∞. The proof is based on the fact that R is a rational vector space with a (Hamel) basis having the power of the continuum, and that S_∞ contains a free product of continuum many groups isomorphic to the rational numbers under addition.

Let G be a subgroup of S_∞ with the property that for every two sets of integers of the same power whose complements are also of the same power, there exists a permutation g of G which transforms one set into the other. Is $G = S_\infty$ (Chevalley, von Neumann, *et al.*)?

If in the symmetric group S_n on n integers two pairs of elements a, b and α, β are simultaneously conjugate, i.e., there exists an element x of S_n such that $\alpha = x^{-1}ax$ and $\beta = x^{-1}bx$, then obviously every element generated by a and b is conjugate to the corresponding element generated by α and β. Is the converse true? That is, if every combination of a and b is conjugate to the corresponding combination of α and β (through perhaps a variable

x depending on this combination) are then a, b simultaneously conjugate to α, β?

The following question is due to H. Auerbach. Let G be a group of $n \times n$ matrices g such that every cyclic subgroup of G: $\ldots g^{-2}, g^{-1}, e, g, g^2, \ldots$ is bounded. Is G bounded? The answer is affirmative for $n = 2$.

We shall raise here a question on some purely group-theoretic properties of certain important infinitely-dimensional continuous groups. Later in the discussion of topological groups we shall refer to the simplicity of the group of all homeomorphisms of the circumference of the circle — a result of J. Schreier and the author [2] — to the corresponding result of von Neumann and the author on the group of all homeomorphisms of the surface of the sphere and to the recent results of Anderson. This question, whether the group under consideration possesses no invariant subgroups (except the identity element) is of interest for the groups of isometric transformations of Banach spaces onto themselves.

Is the group of all measure preserving transformations of the interval $(0, 1)$ a simple group?

Does there exist a universal constant c (independent of dimension) such that, for *every* irreducible group G of orthogonal $n \times n$ matrices g, there is a vector u of unit length, *some* n of whose images $g_1 u, \ldots, g_n u$ under G have separation c from each other, i.e.,

$$| g_i u - g_j u | \geqq c, \quad i \neq j, \; i, \; j = 1, \ldots, n$$

This constant could be greater than $\frac{1}{2}$. In fact, the group of rotations of a pentagon in the plane very likely has the minimum value of c.

The affirmative statement, if true, could serve as an important lemma in a geometric approach to Hilbert's problem [1] on the introduction of analytic parameters in a continuous group, recently solved by Gleason and Montgomery, cf. Montgomery [1].

8. Semi-groups

Let G be a semi-group (associative multiplicative system with identity). A semi-group H of G is *normal* if it has the property that, whenever $h = h_1 g_1 h_2 g_2 \ldots h_k g_k h_{k+1}$, with $h, h_1, \ldots, h_{k+1}$ in H and $g_1, \ldots, g_k$ in G, then $g = g_1 g_2 \ldots g_k$ is in H. If G and H are groups, normality of H in this sense coincides with the usual definition. Two elements a and b of G are said to be congruent mod H (for H normal) if there exist elements $a_1, \ldots, a_l, b_1, \ldots, b_l$ in G and $h_1, \ldots, h_l, h'_1, \ldots, h'_l$ in H such that $a = a_1 \ldots a_l$, $b = b_1 \ldots b_l$ and $h_1 a_1, \ldots, h_l a_l = h'_1 b_1, \ldots, h'_l b_l$. This again coincides with the usual congruence in the group case. It would be of great interest to establish the analogues of the classical chain theorems culminating in the Jordan-Holder theorem. We mention the following statements, some of which may be proved easily.

The group S_∞ of all one-to-one transformations of the integers (permutations) is not a normal semi-group of the semi-group of T_∞: the set of all transformations of the integers into themselves, while S_m *is* a normal subsemi-group of T_n, the latter referring to the corresponding set of operators on the set of integers $1, \ldots, n$. In T_∞, the semi-group F consisting of all mappings $f(n)$ such that $f(n) \neq n$ for only a finite number of integers n is a normal subsemi-group. If N is a normal subsemi-group of T_n, which contains an element not in F, then $N = T$.

The homeomorphisms of the line form a normal subsemi-group of the semi-group of all continuous functions.

8a. Topological semi-groups

Problems of A. D. Wallace:

Let S be a compact, connected semi-group.

1. If S is finite dimensional, homogeneous and has a unit, is S a group? (Yes, if $\dim S = 1$.)

2. If S has a zero and a unit, does S have the fixed point property?

3. If S has a zero and if $S^2 = S$, can S be homeomorphic with an n-sphere? (No, if $\dim S = 1$.)

9. A problem in the game of bridge

Many of the problems of combinatorial analysis, especially those of the theory of probability, derive from situations arising in various games of chance or even games of "skill". The majority of such problems refer to given or fixed situations. We give here an example of a problem in the game of bridge involving, so to say, one more existensional quantifier than the usual problems of the game.

Does there exist an initial distribution of hands with the following properties? (a) East and West can make, against best defense, a grand slam (all 13 tricks) in every suit if this suit were trumps. (b) In a no trump contract, however, against a good defense, East and West are unable to make even a small slam.

More specifically, what is the greatest number of tricks that East and West can always make, even against the best defense, assuming property (a)? It seems almost certain, that a hand distribution with the property (a) guarantees at least 5 tricks in "no trumps." (J. Schreier and the author have found an example of a distribution with the property (a) such that a *grand slam* in "no trumps" cannot be made.)

10. A problem on arithmetic functions

The set of integer valued arithmetic functions $\alpha(n)$, $n = 1, 2, 3, \ldots$ forms a domain of integrity under ordinary addition, and multiplication:

$$\alpha\beta(n) = \sum_{\alpha/n} \alpha(d)\beta(n/d).$$

Is this ring a unique-factorization domain? (E. D. Cashwell, C. J. Everett, who have proved unique-factorization in case of functions $\alpha(n)$ on integers to a *field*.)

CHAPTER III

Metric Spaces

1. Invariant properties of trajectories observed from moving coordinate systems

Suppose that we have, given in a fixed cartesian coordinate system (x, y, z), n moving points describing given curves, $x_i(t)$, $y_i(t)$, $z_i(t)$, $i = 1, 2, \ldots, n$. Suppose now that we have another cartesian coordinate system x', y', z', which is in motion relative to the given system. The given curves will appear differently in the moving system. The motion of the second system with respect to the first one is a general rigid motion, that is to say, the origin of the coordinates moves on an arbitrary curve, and the *rotation* of the system x', y', z' with respect to x, y, z is quite arbitrary as a function of time.

The question arises: what are the invariants of the given system of trajectories in respect to the arbitrarily moving observer? It is clear that for just one trajectory nothing can be said. In a suitably moving system, this trajectory will appear as a stationary point. It suffices to put the origin of the coordinate system onto the moving point. For two given trajectories it is clear that we can move the system x', y', z' in such a way that, for example, one of the points will appear to be stationary, say, again the origin, while the second point is moving on a straight line, say, the x-axis. Likewise, for three points, the invariants with respect to arbitrarily moving coordinate systems are trivial. It is clear that the invariants are functions of the mutual distances between the moving points at any given time.

If, however, $m \geqq 4$, some more interesting questions begin to arise. For example, given arbitrary continuous motions of four points, can one move the system of coordinates in such a way that to an observer of this moving system, the given trajectories will all appear as convex plane curves or perhaps as conics? If we have a sufficient number k of moving points whose trajectories are enlaced, is it true that in any moving coordinate system at least some two of them will appear enlaced?

Analogous questions of invariants of systems of trajectories (or, for that matter, more general parametrically represented surfaces, etc.) could be studied for a given class of topological transformations of space, more general than the rigid motions of the coordinate system.

2. Problems on convex bodies

(Mazur): In the three-dimensional, Euclidean space there is given a convex surface W and a point in its interior. Consider the set V of all points P defined by: the length of the interval OP is equal to the area of the plane section of W through O and perpendicular to OP. Is the set V convex?

A solid S of uniform density ρ has the property that it will float in equilibrium (without turning) in water in every given orientation. Must S be a sphere? (In a two-dimensional version of this problem, H. Auerbach [1] found shapes other than the circle with the desired property.) In the limit ($\rho \to 0$) one obtains the following problem: If a body rests in equilibrium in every position on a flat horizontal surface, is it a sphere?

Let C be a star-shaped closed plane curve, i.e., a polar curve given by $\rho = \rho(\theta)$, and suppose that $\rho(\theta)$ has a continuous derivative except possibly at a finite number of points. It can be shown that there exists a constant $k > 0$ such that the curve given by $\rho = \rho(\theta) + k$ is convex. An analogous remark applies to surfaces in n-dimensions. Suppose C is a curve $(\theta(t), \phi(t))$ in three-space contained in the surface S of a star-shaped region including the origin. Under what conditions is it true that the surface S can be expanded by adding a constant to each radius so that the curve which results from the given curve can be obtained as an intersection of convex surfaces?

3. Some problems on isometry

If A and B are metric spaces, then A^2 and B^2 may also be regarded as metric spaces, the metric of a product space A^2 being defined, for example, by

$$\rho((a_1, a_2), (a_3, a_4)) = [\rho^2(a_1, a_3) + \rho^2(a_2, a_4)]^{\frac{1}{2}}.$$

Does isometry of A^2 and B^2 imply that of A and B? By an isometry between two metric spaces is meant a bi-unique transformation of one space onto all of the other which preserves all distances. A similar question may be asked for other metrizations of the product space — instead of the "Euclidean" formula above, one may use the formula:

$$\rho((a_1, a_2), (a_3, a_4)) = \max [\rho(a_1, a_3), \rho(a_2, a_4)]$$

or another "Minkowski" gauge function. This is a metric version of problems concerning the "extraction of the square root" in algebraic structures, e.g., if the groups A^2 and B^2 are assumed to be isomorphic, does it follow that A is isomorphic to B? (Cf. Fox [1].)

Is Hilbert space characterized metrically among Banach spaces by the fact that its group of isometries is transitive on the unit sphere (Mazur)?

4. Systems of vectors

Let $V_1, \ldots V_n$ be a system of n vectors in k-dimensional space. We are interested here in "bound" vectors, that is to say, a vector V is defined by an ordered pair of points (A, B) in E^k. We allow three types of operations on vectors, namely, (a) replacement of a vector $V = (A, B)$ by a vector $V' = (A', B')$ obtained from it by a translation T along the line through A, B, i.e., $A' = T(A)$, $B' = T(B)$; (b) replacement of a pair of vectors $V = (A, B)$ and $V' = (A, B')$ with common origin by their sum $V'' = (A, B - A + B' - A)$; (c) the inverse operation to (b), i.e., the splitting of a vector V into any two vectors with the same origin as V whose sum is V. Any two systems of vectors obtainable one from the other by a finite number of such operations are said to be equivalent.

It has been shown that, if σ_k is an arbitrary k-simplex in E^k, every finite system of vectors is equivalent to a set of at most $k + 1$ vectors lying on the edges of σ_k and the latter system is uniquely determined by the original one. (A result of L. W. Cohen [1] and the author).

It would be interesting to prove an analogous representation for arbitrary countable systems of vectors in Hilbert space, allowing countably many operations and infinite summations in (b) and (c).

5. Other problems on metrics

Characterize subsets of the plane such that the distance between any two of their points has a rational value. (Can such a set be dense?)

A problem is mentioned elsewhere in this collection on introducing a metric in an abstract algebraic structure (e.g., group) in such a way that the group operations would be continuous in the metric and the topology resulting from the introduction of such a metric would be of a specified type. We shall raise here the vague question: given a metric space, can one introduce a *metric* in it which would lead to the given topology, the *metric* being the "most natural one" among all metrics giving this topology. One can try to formulate precise questions which would attempt to define concretely some aspects of the phrase "most natural." For example, given a topological space, can one find a metric in it so that the group of all isometric transformations under this metric would be maximal in the following sense: for no other metric (leading to the same topology) would the group of isometries contain this group as a proper subgroup? In particular, is the Euclidean metric defined on the surface of the n-dimensional sphere maximal in this sense? The same question for the Hilbert space sphere in the usual metric. Obviously, in general, a topological space will possess many different maximal metrics in the above sense. One could perhaps consider a metric introduced in a topological space as stable if transformations which are "almost isometric" must, of necessity, be near to strictly isometric transformations (cf. Chapter VI, Section 1). The question is now for which topological spaces can one introduce stable metrics in the above sense? One obviously would want such metrics to be also maximal. Without that requirement, the problem would not have much sense since, in general, one can find metrics for which only the identity would be the isometric transformation.

It is not without interest to consider, in certain algebraic structures, an introduction of metrics such that the algebraic automorphisms would be isometric transformations, but we shall not go into this subject.

CHAPTER IV

Topological Spaces

1. A problem on measure

Let E be a compact metric space. Does there exist a finitely additive measure $m(A)$ defined for at least all the Borel subsets A of E, such that $m(E) = 1$, $m(p) = 0$ for all points p of E, and such that congruent sets have equal measure (Banach-Ulam)?

Two sets A and B are called *congruent* if there exists an isometry between A and B alone, not necessarily a congruence under an isometry of the whole space E taking A into B. (If one postulates this latter more restricted notion of congruence, then such a measure is known to exist.) It is also clear that the term "finitely" cannot be replaced by "countably" for an affirmative solution. (Consider the set of points (x, y) in the plane where $0 \leqq y \leqq 1$, $x = 0, 1/n$; $n = 1, 2, 3, \ldots$.)

2. Approximation of homeomorphisms of E^n

Let E^2 be the Euclidean plane and G the group of biunique, bicontinuous transformations of E^2 to E^2 generated by the set of all such correspondences of the form:

$$X: \begin{cases} x' = f(x, y) \\ y' = y \end{cases} \qquad Y: \begin{cases} x' = x \\ y' = g(x, y). \end{cases}$$

An arbitrary homeomorphism can be approximated, arbitrarily closely (uniformly in every bounded part of the plane) by transformations belonging to G (cf. Eggleston [1]).

Similar questions may be posed for Euclidean n-space in various ways depending on the type of generators allowed for the group G. For example, in three dimensions we may permit generators of the form X: $x' = f(x, y, z)$, $y' = y$, $z' = z$ and its two analogues, or again, let G be generated by all homeomorphisms of the type

$x' = f(y, z)$, $y' = g(x, z)$, $z' = h(x, y)$. The question still open is: Are arbitrary homeomorphisms of E^n approximable by transformations of the above type?

The possibility of such approximations for the general case of E^n would be of considerable importance for topology, providing an inductive procedure for proving various topological theorems; for example, the famous conjecture of Alexander [1] on the approximability of arbitrary homeomorphisms by differentiable mappings, which in turn implies the possibility of triangulation of any topological manifold. The basic lemma here would be the Alexander conjecture for the plane (proved by N. Wiener and P. Franklin). Quite recently Moïse [1]) proved that three-dimensional homeomorphisms are approximable by simplicial homeomorphisms.

The problem of this section, even for $n = 2$, i.e., E^2, is open for the case where E is a more general topological space. Moreover, in the formulation as it stands, X, Y, etc., need not be restricted to homeomorphisms, and the approximability is of interest when the class of transformations is widened to include general, continuous, or even, say, Borelian mappings.

It would be interesting to attempt to utilize the recent results of Kolmogoroff [1] and Arnold [1] on representation of *functions* of any number of variables by composition of functions of two variables to obtain such results for topological, that is to say, one-to-one transformations. In other words, even theorems allowing only approximation, if not exact representation of homeomorphisms of n-dimensional spaces by compositions of homeomorphisms involving only two dimensions at a time, would be extremely valuable.

2a. On the approximability of transformations in three dimensions by compositions of cylindrical mappings

Suppose we consider the smallest group G of transformations containing the transformations of the form:

$$\begin{cases} (x + iy)' = W(x + iy) \\ z' = z \end{cases}$$

where W is an analytic function, and all rotations of the three-dimensional space. Can one approximate arbitrarily closely, by transformations from the group G, a mapping of a sphere onto a cube? Somewhat more generally, can one by means of such transformations obtain approximate mappings of a polyhedron to any other, topologically equivalent polyhedron, by compositions of two fixed transformations?

3. A problem on the invariance of dimension

There may be a possibility of strengthening Brouwer's theorem (cf. Hurewicz, Wallman [1]) on the invariance of dimension in the following way:

Does there exist for every integer $m > 1$ a one-one continuous mapping T on E^m to E^m, (E is the real line) such that for every biunique Borel transformation U of E^m to all of E^n, $n < m$, the transformation UTU^{-1} of E^n to E^n is *discontinuous*?

An affirmative answer would imply Brouwer's theorem on the nonhomeomorphism of E^n and E^m, $m \neq n$. For, if H should be a homeomorphism of E^m to E^n, $n < m$, then H is trivially Borelian and HTH^{-1} would be continuous. One may also consider the above questions with T continuous and (possibly) many-one. The special case $m = 2$, $n = 1$ is trivially verified.

As often in our problems, the statement conjectured above could have a wider applicability: for a general space E such that E^n and E^m are *not homeomorphic* for any pair $n \neq m$.

We may add here parenthetically that it suffices to show that E is not homeomorphic to E^n, $n > 1$, and that if E^{2n} is not homeomorphic to E^{2m}, then E^n is not homeomorphic to E^m. With this we may conclude that for all $n \neq m$, E^n and E^m are not homeomorphic. This is actually a purely arithmetical fact: If K is a collection of pairs of integers such that

1. from $(a, b) \in K$ and $(b, c) \in K$ it follows that $(a, c) \in K$,

2. from $(a, b) \in K$ it follows that $(a + 1, b + 1) \in K$,

3. from $(2a, 2b) \in K$ it follows that $(a, b) \in K$,

then, if K contains any pair (n, m), $n \neq m$, K must contain a pair $(1, n)$, $n > 1$.

4. *Homeomorphisms of the sphere*

The group of homeomorphisms of the surface of the sphere S in three dimensions has two components. The component of the identity forms a simple group G (a result of von Neumann and Ulam [1]). In fact, a stronger theorem holds: for every two homeomorphisms A and B of S different from the identity there exists a fixed number n of conjugates of A, i.e., $H_1AH_1^{-1}, \ldots, H_nAH_n^{-1}$, H_i in S, whose product is B. This number n does not exceed 23; to determine the minimum number seems very difficult.

Analogous theorems for the k-sphere, $k > 2$, are as yet unproved, as are theorems on the simplicity of groups of homeomorphisms (forming the component of the identity) of manifolds, other than the sphere.

Very recent work by R. D. Anderson [1] generalizes these results to groups of homeomorphisms of sufficiently homogeneous (setwise) spaces: In particular, the group of all homeomorphisms of the Cantor ternary set, the universal curve, the set of all rational and the set of all irrational numbers are simple, also some interesting partial results on the group of all (orientation preserving) homeomorphisms of S^n --- the n-sphere in n-space.

A question considered by Borsuk and the writer [2] is: Given an arbitrary closed subset C of the surface S of the sphere in n-space, does there exist a sequence of homeomorphisms H_n of S to all of itself such that $\lim H_n(E) = C$, that is to say, such that for every p of S, $\lim H_n(p)$ exists and is in C, and every point of C is such a limit.

5. *Some topological invariants*

No algorithm has yet been found which would permit one to decide whether two given curves in three-dimensional space are mutually enlaced. A sufficient condition for enlacement is that the Gaussian integral over the two curves is different from zero (cf. Alexandroff, Hopf [1]). Elementary examples show that the condition is not necessary. (We consider here two curves as not enlaced if there exists a homeomorphism of the whole space under which the images of the two curves are contained in disjoint geometrical spheres.)

Let us draw from each point of one curve a vector to every point of the other curve. If these vectors are all referred to a common origin and are normed to one, we shall have a mapping of the torus, which is the direct product of the two curves, onto the unit sphere. Does enlacement of the curves imply that the vectors cover the surface of the sphere essentially, that is to say, the mapping of the torus to the sphere is not retractable to a mapping into a single point?

Suppose that a system of n-vectors in 3-space forms a closed polygon. The vector sum is zero. If we consider the total moment vector relative to some point, it is curious that this may be zero as well. In a plane, for a polygon forming a simple Jordan curve, the total outer moment of these vectors can never be zero; indeed its magnitude is twice the area enclosed by the polygon. However, it is easy to find a hexagon in 3-space whose total moment with respect to any point is zero. This points to a possibility that some topological invariants may be derived from the moments of the various polygons formed by edges of a complex, or from even more general tensorial expressions. So, for example: We have discussed (Chapter III, Section 4) an equivalence of a system of vectors in n-dimensional space to a unique system of vectors located on the edges of a fixed simplex in n-space. Consider now a simplicial subdivision of a complex C located in the n-dimensional space. The system of its edges, properly oriented, will form a system of vectors which we will "represent" on the edges of a fixed (but arbitrary) simplex σ. The question arises as to what properties of this representation remain invariant under a subdivision C' of C for the corresponding representation of the resulting systems of vectors of the edges of C' on σ.

Another construction which may lead to significant topological invariants is the following: Let E be a topological space, and f a real-valued continuous function $f(p)$, defined for all points p of E. Let $G(f, E)$ be the group of all homeomorphisms A of E into itself such that $f(A(p)) = f(p)$ for all p. There is a possibility of distinguishing between two nonhomeomorphic spaces E and F by producing a function f on E, whose group $G(f, E)$ is not isomorphic to $G(g, F)$ for *any* function g on F. For example, if E is the circumference of a circle, and F is the unit interval, one can easily define a function f on E (say of period $2\pi/3$) whose group $G(f, E)$, the cyclic group of order 3, is not isomorphic to any group of homeomorphisms of F, since the interval F admits no homeomorphism of order 3.

One can, of course, employ mappings f on E to other spaces X than the real line. A general conjecture would be: If E and F are two nonhomeomorphic manifolds, there exists a space X and a mapping f of E on X so that $G(f, E)$ is not isomorphic to any $G(g, E)$ for any mapping g of F on X.

Among the simplest questions are the following: What abstract groups can be realized as groups $G(f, E)$ for a given space E? For instance, can every finite group be realized as a group $G(f, E)$ where E is the plane? What are all *countable* groups $G(f, E)$ which can result for functions f on Euclidean n-space?

Let f and g be two commuting transformations of I into itself (I is the closed interval): $f(g) = g(f)$. Does there exist a common fixed point $p_0 = f(p_0) = g(p_0)$? Communicated by A. L. Shields — originally raised by E. Dyer. The answer is not known even for $n = 1$. There are interesting partial results of Shields, J. R. Isbell, R. E. Chamberlain, and others.

6. Quasi-fixed points

Let E be a topological space, $T(p)$ a continuous mapping of E into itself, and $\phi_1(p), \ldots, \phi_k(p)$ a set of k real-valued continuous functions defined on E. We shall say p_0 is a quasi-fixed point of T relative to the ϕ_i in case

$$\phi_i(T(p_0)) = \phi_i(p_0), \qquad i = 1, \ldots, k$$

As an example, let E be the Euclidean plane, $T(p)$ a continuous transformation of E onto (into the whole of) E, $\phi_1(x, y) = |x|$, $\phi_2(x,y) = |y|$. One can show then the existence of a quasi-fixed point (x_0, y_0) of T relative to ϕ_1, ϕ_2.

Surely one will have to restrict either T or the ϕ_i for a significant theorem. Obviously for some T there may exist points quasi-fixed relative to every set of ϕ_i, in particular if T has a fixed point $p_0 = T(p_0)$.

Does an orthogonal transformation T of the unit sphere E in n-space always possess a point p_0 quasi-fixed relative to a set of $n - 1$ (arbitrary) continuous functions? This is of interset, of course, only in case the determinant of T is -1, and if true would constitute a generalization of the "Antipodensatz" of Borsuk [1] and the author.

It would be worthwhile to obtain results on the existence of quasi-fixed points in case T is a transformation of a function space E into itself. For, suppose that U is a functional operator and we are interested in the existence of a solution f_0 of the equation $U(f) = 0$. This is equivalent to finding a fixed point f_0 of the transformation $T(f) = U(f) + f$. Now in cases where the existence of a fixed point is difficult or impossible to establish, we may be satisfied with the knowledge that $\phi_i(T(f_0)) = \phi_i(f_0)$ for some or all sets of k continuous functions ϕ_i on E. These functionals ϕ_i might be, for example, the first k coefficients of f in its Fourier or power series development, or the first k moments of f, etc. Of course, the f_0 will, in general, depend on the ϕ_i in case there is no true fixed point of T. Nevertheless it may be useful in some applications to know that a function f_0 exists which has the same ϕ_i-values as its transform $T(f_0)$.

If the $f_i(p)$ are *linear* real valued functions defined on the Hilbert space S, and T is a continuous transformation of S into itself, does there exist a point p_0 such that $f_i(\Gamma(p_0)) = f_i(p_0)$?

Let $T(p)$ be a homeomorphism of Euclidean n-space. Suppose that for every point p, the set of iterates $p, T(p), T^2(p), \ldots$ is a set of finite diameter d_p and the d_p are bounded: $d_p < B$ for all p. Does $T(p)$ possess a fixed point $p_0 = T(p_0)$?

Does there exist for every manifold M a constant B such that every continuous transformation T of M into part of itself, having the property

$$| T^\nu(x) - x | < B$$

for all iterates ν and all x of M, must have a fixed point $x_0 = T(x_0)$? More generally, does such a constant exist for every locally-connected continuum?

7. Connectedness questions

Suppose that $T(p)$ is a differentiable transformation of the plane. If for some point p, the closure C of the set of all iterates $p, T(p), T^2(p), \ldots$ is connected, is the set C necessarily locally connected (Borsuk)?

Let S_1 and S_2 be two topological spherical surfaces (i.e., sets homeomorphic to the surface of the geometric sphere) in Euclidean 3-space, with S_2 contained in the interior of S_1. According to the Jordan-Brouwer theorem, S_2 decomposes space into two regions, and we assume S_1 is contained in one of them. Does there exist a surface S_3, topologically spherical, containing S_2 in its interior, and contained in the interior of S_1. The problem is not trivial because of the well-known examples of Alexander which show that the interior of a topological sphere need not be homeomorphic to that of a geometrical sphere. Analogous problems exist for higher dimensions (Schreier-Ulam [1]).

One may generalize the situation: Given are three topological spherical surfaces S_1, S_2, S_3 so that S_3 is contained in the interior of S_2, which in turn is contained in the interior of S_1. Can one find a topological sphere S_4 *either* contained in the interior of S_1 and containing in its interior S_2, *or* contained in the interior of S_2 and having S_3 in its interior?

8. Two problems about the disk

Suppose that $T(p)$ is a homeomorphism of the disk D (all (x, y) with $x^2 + y^2 \leqq 1$) onto all of itself. Do there exist arbitrarily small "triangles," i.e., triplets of points p_1, p_2, p_3 congruent to the triangles formed by the images $T(p_1)$, $T(p_2)$, $T(p_3)$? Do such triangles exist with prescribed angles?

Given a metric space A, a topological space B, and a (many-one) continuous mapping T on A to all of B, we consider the diameter d_b of the set $T^{-1}(b)$ for every b in B, and denote by

η_T the least upper bound of all d_b. Does there exist, for every $e > 0$, a continuous mapping T of the disk D onto the surface B of the torus with $\eta_T < e$? *

9. Approximation of continua by polyhedra

Let C be a simple closed curve in Euclidean 3-space which is non-knotted, that is to say, there exists a homeomorphism of the whole space which transforms C into the circumference of a circle. Can C be approximated arbitrarily closely by nonknotted polygons? This problem, considered by Borsuk and the author in 1930, has connection with the problem (of Alexander) of approximability of arbitrary homeomorphisms of the n-dimensional space by simplicial ones. An affirmative answer can be obtained from a theorem recently established by Moïse [1]: *vice versa* from a positive answer to the above, i.e., from an approximability of nonknotted curves by nonknotted polygons, the approximability of homeomorphisms follows. In the spaces of higher dimension than three, an analogous situation may obtain, namely, approximability of nonknotted spheres by nonknotted polyhedra may be sufficient to prove the approximability of general one-one continuous transformations by one-one differential transformations.

A problem of Borsuk connected with the above concerns the possibility of approximation of a unicoherent continuum in 3-space by unicoherent polyhedra. (A unicoherent continuum is a continuum E such that for every decomposition $E = A + B$ into two continua, $A \cdot B$ is a continuum. The disk is unicoherent, the circumference of the circle is not.)

There are a number of unsolved problems dealing with the approximability of continua with various given properties by polyhedra having the same properties.

10. The symmetric product

By the symmetric product E_s^n of a set E with itself is meant the class of all subsets of at most n distinct elements of E. Thus

* *Note added in proof*: A negative answer has been demonstrated by M. K. Fort, Jr.

E_s^n may be obtained from E^n by identifying all n-tuples $(p_1, \ldots p_n)$ whose component points form the same *set*. The importance of this "phase space" lies in the fact that some quantum statistics, for example, the Fermi-Dirac statistics, operates in similar spaces, just as the Maxwell-Boltzmann statistics operates in the direct product E^n. Thus the phase space of a system in which certain particles are indistinguishable is a symmetric product of the spaces corresponding to the particles.

If E is a topological space, a metric may be introduced into E_s^n using the Hausdorff distance between two sets of points. The properties of symmetric products are less well known than those of the direct product of spaces. It has been shown that, whereas the direct product of n circles results in the n-dimensional torus, their symmetric product, for $n = 2$, is the Möbius band. Moreover, if E is a real interval, E_s^n, for $n = 2$, 3, and 4, is the corresponding E^n, whereas for $n \geq 5$ this is not the case (Borsuk, Ulam [1]).

The metrized symmetric product E_s^n forms an nth order approximation to the space of all closed subsets of a compact space E (with Hausdorff distance as the metric for the latter space).

The exact topological structure and, in some cases, even some very general topological properties of E_s^n remain unknown (even when E is the interval and $n > 5$). For example, the existence of fixed points (for arbitrary continuous mappings) has not been established.

Can the quasi-fixed point (see Section 6) theorem be proved for E_s^n, for $k < n$ arbitrary real valued continuous functions, where E is the surface of the n-dimensional sphere? (Compare Bott [1].)

The symmetric product as defined above is only one of many different constructions possible on E^n. Products based on other rules for identifications of certain sets of n-tuples in E^n may lead to interesting spaces. Such possibilities have not been studied systematically.

While the formation of a topological space E_s^n from a topological space E is readily accomplished, it is not easy to see a possible symmetrization in the case of algebraic structures. There seems to be no simple way to obtain a symmetric product of a group with itself. As a matter of fact, the symmetric product of a group space forms often (as in the case of the Möbius band) a topological space which will not support any continuous group operation. In some cases, however, e.g., the 2nd and 3rd symmetric product of the line with itself, E_s^2 and E_s^3 are homeomorphic to E^2 and E^3, respectively. This means that the analogue of vector addition can be continuously defined here. This leads, however, to an algebraically rather artificial rule for "adding" elements in E^2 or E^3.

11. A method of proof based on Baire category of sets

The theorem stating that a complete metric space is not a set of first category (sum of countably many nowhere dense sets) has been used to advantage for obtaining *existence* proofs in modern mathematics, notably in the theory of functions of a real variable, and in topology. Thus, for example, in order to show the existence of continuous functions without derivatives at any point, one may prove that the set of such functions forms a residual set (complement of a set of first category) in the space of all continuous functions. Again, in order to show the existence of metrically transitive measure-preserving transformations, one proves that the set of all such mappings is residual in the space of all measure-preserving transformations. To show the existence of a homeomorphic image of an arbitrary n-dimensional set Z_n contained in the $2n+1$-dimensional Euclidean space E^{2n+1}, it is quite simple to show that, in the space of all continuous (possibly many-one) mappings of the set Z_n into E^{2n+1}, the one-one mappings form a residual set. There are many other examples.

A possibility exists of using another theorem on residual sets (or, more generally, of second category) to prove, instead of existence theorems, propositions involving the complementary quantifier, i.e., the "for all" theorems. Thus, a well-known theorem asserts that if G is a connected topological group and H is a subgroup of it which is not a set of first category with respect to G, then H contains all the elements of G.

Similarly, if G is such a group with a finite Haar measure defined for its subsets, and if H is a subgroup of measure > 0, then $H = G$.

To illustrate this possibility let us consider as an example the well-known theorem of van der Waerden which asserts that every partition of the set of all positive integers into two subsets N_0 and N_1 has the property (P): at least one of the sets N_0 and N_1 contains finite arithmetic progressions of unbounded length. Suppose that we make correspond to every partition a real number x, (regarded as a point on the circle of the circumference 1)

$$x = a_1 2^{-1} + a_2 2^{-2} + \dots$$

where $a_n = 0$ if n is in N_0 and $a_n = 1$ if n is in N_1. It is immediately apparent that the set H of all x for which the corresponding partition has property (P) is a set of measure 1 in the continuous group G of all x (under addition modulo 1). Since van der Waerden's theorem is true, it is clear *a fortiori* that $H = G$. It would be interesting if it could be shown by a relatively simple argument that our set H is a group, and thus that $H = G$.

There are quite a few combinatorial theorems and problems still unsolved, where such an approach could be tried.

12. Quasi-homeomorphisms

Let A and B be two manifolds (i.e., topological spaces such that neighborhoods of points are homeomorphic to n-dimensional Euclidean spheres); we suppose them to be metrized. We define A and B to be quasi-homeomorphic if for every $e > 0$ there exists a continuous mapping T_e of A onto B (all of B) such that for every a, a' in A whose distance exceeds e, $T_e(a) \neq T_e(a')$ in B, and there exist similar transformations S_e of B onto A. Are A and B necessarily homeomorphic?

This problem was proposed in a paper by C. Kuratowski and the author [1]. It would even be useful to show that some general topological invariants remain unchanged, in case of manifolds, for which there exist ε-mappings of both A into B and B into A.

The property of possessing a fixed point under every continuous mapping into itself could be such an invariant. Similarly the "quasi-fixed point" property (for any finite number k of arbitrary real valued continuous functions) may perhaps be shown to be invariant.

13. Some problems of Borsuk

In these problems AR-sets denote compact absolute retracts, that is to say, sets such that in every containing space they can be obtained as continuous images of the space by transformations which are equal to identity on these sets. ANR-sets denote compact absolute retracts of their neighborhood.

1. Are all homology dimensions (in the sense of Alexandroff) for ANR-sets identical with the usual dimension?

2. Is it true that every n-dimensional compactum is homeomorphic with a subset of an $(n+1)$-dimensional AR-set?

3. Is it true that every n-dimensional ANR-set contains an $(n-1)$-dimensional ANR-set?

4. Is it true that every $(2n+1)$-dimensional ANR-set contains topologically every compactum of dimension $\leq n$?

5. In the three-dimensional Euclidean space, are the AR-sets the same as acyclic locally contractible compacta?

6. Does there exist for every ANR-set a polytope having the same homotopy type (in the sense of Hurewicz)?

7. Is the factorization of a continuum into one-dimensional factors unique? (Factorization = decomposition into Cartesian product.)

Note added in proof: Compare the beautiful recent example of Bing [1].

CHAPTER V

Topological Groups

1. *Metrization questions*

One of the general problems of topological algebra is to determine all possible topological groups which are definable for a given abstract group G. This means the characterization of all topologies on the set G in which the group operation will be continuous. There are many special known results related to this problem. For example, it has been proved (Ulam, von Neumann) that there exists an abstract group of the power of the continuum such that no metrization, for which the group operation is continuous, will make this group separable. Such a group can even be chosen to be abelian (commutative). We may mention here a few open questions:

Can the group S_∞ of all permutations of the set of all integers be metrized in such a way that the group operation is continuous in the metric, and the group becomes a locally compact space? J. Schreier and the author have proved that no metrization is possible which would make it compact. This is due to the following purely group-theoretical property: Starting with any truly infinite permutation p, that is to say, one which changes the positions of infinitely many integers, one can obtain, by multiplying a fixed number N of suitable conjugates of this permutation, an arbitrary infinite permutation. In a compact topological group, given any integer N, one can find a neighborhood of the identity so small that the products of at most N conjugates of the elements of this neighborhood will still form a set which is not the whole group; in fact, by choosing ε sufficiently small, one can insure that elements of this form will be confined to another neighborhood of the identity. Now any neighborhood of the identity in this group would have to contain, were it compact, a truly infinite permutation.

There are the four obvious metrizations of the group S_∞ based on its invariant subgroups: (a) the identity permutation, (b) the finite even permutations, (c) the finite permutations, and (d) the whole group S_∞. The "natural" topology in the whole group S_∞ is the following one. A sequence of permutations converges to a fixed permutation if it converges weakly, that is to say, the images of every integer in these permutations become, after a while, equal to the image of the limiting permutation and stay constant. The same should be required about the inverses of these permutations. This topology is easily defined by means of a metric. The topology thus obtained leads obviously to a nonlocally compact space. An analogous topology introduced in the invariant subgroups will also lead to nonlocally compact spaces since it amounts to having the space of co-sets discrete.

Are there any continuous metrizations of S_∞ different from these four? Since all four lead to nonlocally compact topologies, a theorem stating that no other continuous topologies are possible would, combined with certain results of A. Weil [1], provide definite examples of groups in which no invariant measure (Haar-Weil measure) is possible.

2. *Universal groups*

The theorems of von Neumann, Pontrjagin *et al.* [1], on the representation of continuous groups imply that every compact topological group G is continuously isomorphic to some subgroup of the direct product of all finite dimensional rotation groups.

Does there exist a universal compact semi-group, i.e., a semi-group U such that every compact topological semi-group is continuously isomorphic to a subsemi-group of it?

Can one show that the group R of all rotations in the three-dimensional space is isomorphic (as an abstract group, not continuously, of course) to a subgroup of the group S_∞ of all permutations of integers? Or, perhaps quite generally: is every Lie group isomorphic (as an abstract group) to a subgroup of the group S_∞?

Does there exist a separable, locally-compact group U such that every locally compact group is isomorphic to a subgroup of U? The results of Gleason, Montgomery [1], and Zippin generalize those of von Neumann and Pontrjagin to *locally* compact groups.

A separable continuous group U, "universal" for all separable continuous groups G (in the sense that every such G be continuously isomorphic to a subgroup of U) may not exist.

3. Basis problems

A finite set of elements of a topological group which generate a subgroup dense in the whole space is called a finite basis. There exist isolated results on the existence of such bases, but there seems to be no systematic investigation of their existence and properties for general continuous groups. Nor has the minimal number of elements in a basis been determined in most cases where a finite basis had been shown to exist. We mention here a few of the known results.

Let S_∞ be the group of all permutations of the integers, with the "natural" topology, i.e., a sequence of permutations converges to a limit if the convergence is termwise, and the sequence of inverse permutations converges in the same sense to the inverse of the limit. There exist two elements in S_∞ such that the group generated by them is dense in the whole group. In fact, a specific set of products of powers of these two elements can be shown to form a dense set in S_∞ (J. Schreier, Ulam [4]). There is no scarcity of such pairs of permutations; almost every pair (in the sense of category) will serve as a basis.

An analogous situation obtains in the group of homeomorphisms of n-dimensional Euclidean space (J. Schreier, Ulam [1]). Does this group possess a basis consisting of differentiable transformations?

In the case of finite dimensional topological groups, one knows, for example, that the group of rotations contains pairs of elements generating dense subgroups.

In every semi-simple connected Lie group there exists a basis of four elements. It is not known whether this is the minimal number (Schreier, Ulam [3]).

One can find a pair of rotations of Hilbert space which generates a dense subgroup (in a weak sense) in the group of all such rotations (J. Schreier [1]).

Less is known about topological semi-groups, but the general situation seems to be quite similar to the one for groups, to judge from the evidence we have so far. Thus the class T_∞ of all (many-one) transformations of the set of integers into itself has a natural topology. There exists a finite set of elements generating a subset, dense in this topology, in the semigroup T_∞. A similar result is known for the semi-group of continuous transformations of the sphere into itself (Schreier, Ulam [1]).

Let H_n denote the group of all homeomorphisms of the n-cube I^n onto all of itself, metrized by means of the distance function:

$$\rho(h_1, h_2) = \max_{p \in I^n} \rho(h_1(p), h_2(p)) + \max_{p \in I^n} \rho(h_1^{-1}(p), h_2^{-1}(p))$$

the latter distances referring to the Euclidean metric in E^n. Is it true that almost every pair of homeomorphisms generate a group dense in H_n? ("Almost every" is used here in the sense of category, i.e., do these pairs form a residual set in the space of all pairs?)

Again, let H_n^k denote the kth direct product of the space H_n with itself, with the usual extension of the above H_n metric to a product space metric. Does almost every element $(h_1, \ldots, h_k)$ of H_n^k provide a set $h_1, \ldots, h_k$ which generate a free group of k generators in H_n? The same question may be asked about the semi-group G of all continuous, not necessarily one-one, transformations of n-space.

Given a pair of measure preserving transformations S, T of a space E into itself one may consider transformations such as S, $T(S)$, $T(T(S))$, $S(T(T(S)))$ ... etc. Specifically, to every real number $0 \leqq x \leqq 1$ in binary expansion $x = \alpha_1 \alpha_2 \ldots \alpha_n \ldots$, $\alpha_i = 0$ or 1, we may make correspond a sequence of transformations interpreting the symbol 0 as applying S and 1 as applying T in the order indicated. Is it true that, for almost every point p of the space E and for almost all x, the resulting sequence of images: p, $S(p)$, $T(S(p))$, $T(T(S(p))$, $S(T(T(S(p)))$ etc. will be *uniformly* dense?

The problem is: is this sequence uniformly dense for almost every (in the sense of category) pair of measure preserving transformations of E_n?

Is this true for a Haar group when S, T are transformations obtained by left multiplying the group by two elements s, t (if the Haar group is simple)?

It appears likely that there exist finite sets $T_1, T_2, \ldots, T_k$ ($k = 2$?) of measure-preserving transformations of the Euclidean cube generating a group dense in the space Σ of *all* continuous measure preserving transformations. If this were so, one could define a measure function in the space of all incompressible flows of say the three-dimensional space into itself. This measure would be obtained by considering the Lebesgue measure in the space of the x's corresponding to the transformations, as above.

4. Conditionally convergent sequences

Let V_n, $n = 1, 2, \ldots$, be a conditionally convergent series of vectors. It is well known that if we rearrange the terms in all possible ways and form their sums, these sums will form a linear manifold in the vector space. Garrett Birkhoff and the author have noticed that this well known theorem can be generalized if one considers the V_n to be elements of a compact group G; then the "sums" of all possible rearrangements of the sequence form a *coset* modulo a certain subgroup of G. Does a similar result hold for more general, noncompact topological groups?

CHAPTER VI

Some Questions in Analysis

1. Stability

We intend now to discuss, by means of a few examples, the notion of the stability of mathematical theorems considered from a rather general point of view: When is it true that by changing "a little" the hypotheses of a theorem one can still assert that the thesis of the theorem remains true or "approximately" true? The notion of stability arose naturally in problems of mechanics. It involves there, mathematically speaking, the continuity of the solution of a problem in its dependence on initial parameters. This continuity may be defined in various ways. Often it is sufficient to prove the boundedness of the solutions for arbitrarily long times, e.g., the boundedness of the distance between the point representing the system at any time from the initial point, etc. Needless to say, problems of stability occur in other branches of physics and, in a way, also, even in pure mathematics.

We shall not try to formulate a generally applicable definition of stability. One could attempt to do this by introducing suitable function spaces for physical theories and various metrics in them, but we shall be content instead to indicate some of the salient features of this concept as it appears in purely mathematical formulations. In particular we shall formulate some problems concerning the stability of solutions of functional equations.

For very general functional equations one can ask the following question. When is it true that the solution of an equation differing slightly from a given one, must of necessity be close to the solution of the given equation? Similarly, if we replace a given functional equation by a functional inequality, when can one assert that the solutions of the inequality lie near to the solutions of the strict equation? Instead of trying to define a general class of functional equations (one recalls a dictum of Poincaré's: "On ne peut guère définir les équations fonctionnelles en général") for this purpose, let us restrict ourselves to examples. A good illustration, in an elementary case, is provided by a result of Hyers' [1], resolving a problem of the author:

If $f(x)$ is a measurable real-valued function defined for all real x satisfying the inequality

$$|f(x+y) - (f(x) + f(y))| < \varepsilon$$

everywhere, one can show that there exists a function $l(x) = ax$ such that

$$l(x+y) = l(x) + l(y) \text{ and } | l(x) - f(x) | \leqq \varepsilon$$

everywhere. We say then that the functional equation of linearity

$$f(x+y) = f(x) + f(y)$$

is stable with respect to a change into an inequality. (By the way, even if $f(x)$ is not measurable, one can assert that the solution of the inequality is close to some — nonmeasurable perhaps — solution of the strict equation.)

One can ask, much more generally, for what metric groups G is it true that an ε-automorphism of G is necessarily near to a strict automorphism? (An ε-automorphism means a transformation f of G into itself such that $\rho(f(x \cdot y), f(x) \cdot f(y)) < \varepsilon$ for all x, y in G.) There should exist then a constant k, depending only on G and not on f, and an $a(x)$ depending of course on f, with $a(x \cdot y) = a(x) \cdot a(y)$ such that $\rho(a(x), f(x)) < k\varepsilon$ for all x. We require this to hold e.g., for all continuous or measurable, f. The above result of Hyers' answers the question when G is the group of real numbers relative to addition. Hyers also obtained results in the case when G is a more general vector space. Another paper, by Hyers and the author, answers the question, in an affirmative sense, for some infinite-dimensional vector spaces. In this and other examples, it should be pointed out that a formulation of stability "in the large" requires a metric in the space of functions. This is, of course, true in all the classical studies of the problem of stability in differential equations or systems of differential equations. One could ask for weak convergence of solutions — pointwise — or more strongly for uniform convergence in the norm (cf. Hyers, Ulam [1, 2]).

It is interesting that the notion of stability in the above sense can be introduced even for discrete structures; thus, for example, Shapiro [1] has answered a question of the author by the following result. If $t(a)$ is an automorphism "up to an integer k" of the set of all the integers mod p where p is a "large" prime and k is an integer "much smaller" than p, then under suitable conditions $t(a)$ is a strict automorphism. One could also ask for stability of configurations or constructions even in elementary geometry. Thus, to give an elementary and *ad hoc* example, and to depart from linear formulations and illustrate what we mean by a "quadratic" problem — if the constructions of the Pascal and Brianchon theorems for arbitrary sextuplets of points located on a continuous curve always result in points which are almost collinear — or in lines which are almost concurrent — is the given curve approximately a conic?

Still another illustration in metric geometry: D. Hyers and the author [1, 2] have shown that a transformation of Euclidean space which changes all distances by at most $\varepsilon > 0$ is of necessity near to an isometry, that is, a transformation *strictly* preserving all distances. It is not known for what general metric spaces the above statement remains true.

Many questions about transformations suggest themselves in the same vein. It can be shown that a transformation which very nearly preserves measure in a Euclidean space is close to a transformation which strictly preserves the measure of all subsets. Can one prove it in more general measure spaces? Is a transformation which is nearly laminar of necessity close to one which is strictly so? What can be said about transformations which are almost irrotational? Questions like the above on stability of properties involving differential flows lead to general problems about the stability of differential expressions. The following theorem was also noticed by D. Hyers and the author [4].

Let $f(x)$ be a real function on the line with its nth derivative $f^{(n)}(x_0) = 0$ for a certain n, and suppose that $f^{(n)}(x)$ changes sign in the neighborhood of x_0. For every $\varepsilon > 0$ there exists a $\delta > 0$ such that for every function $g(x)$ of class $C^{(n)}$ satisfying $|f(x) - g(x)| < \delta$, there is a point x_1 such that $g^{(n)}(x_1) = 0$ and $|x_1 - x_0| < \varepsilon$. It is perhaps remarkable that the hypothesis involves only the nearness of the functions f and g themselves, and the number δ is dependent on ε alone.

One can obtain a partial generalization of this result. Let $F(x, f(x), f'(x))$ be a continuous function of the three arguments, with $F(x_0, f(x_0), f'(x_0)) = 0$ for some x_0 and suppose that F changes sign in a neighborhood of x_0. Then again for every $\varepsilon > 0$ a $\delta > 0$ exists such that for every $g(x)$ of class $C^{(1)}$ closer than δ to $f(x)$ there is a point x_1 close to x_0 such that $F(x_1, g(x_1), g'(x_1)) = 0$.

Under what conditions can the previous theorem be generalized to $F(x, f(x), \ldots, f^{(n)}(x))$ or to $F(f(x, y); \partial f/\partial x, \partial f/\partial y, \ldots)$? The most interesting question of this kind concerning functions of two variables is the simultaneous vanishing of several partial derivatives at a point (x_0, y_0). Here one has to be careful about the meaning of "change of signs" in the neighborhood of a given point. It seems necessary to assume at least that all combinations of sign occur. Very little is known about such questions in the n-dimensional case.

A theory of the Calculus of Variations in the Large developed by Marston Morse [1] operates with qualitative or topological definitions of critical points of functions of several variables. This theory provides a general qualitative basis for the phenomena implied by the vanishing of first derivatives. Our remarks would seem to indicate that there may exist topological definitions relative to expressions involving higher derivatives.

We have dealt above with stability of solutions of some functional equations — when we change these equations into inequalities. One might study this question when an inequality is given *ab initio*. For example, there is the result of Hyers and the author [3] that a function which is almost convex lies near to a strictly convex function. In the most elementary case, it states that a solution of the functional inequality

$$f(x + y) - f(x) - f(y) < \varepsilon$$

is a function differing everywhere from a convex function by at most a fixed multiple of ε.

The subject of small deformations is treated explicitly in topology which, as once defined by Poincaré, is the study of those properties of figures which remain true even when the figure is drawn by unskilled draftsmen. This definition demands more than the invariance of properties of sets under 1—1 continuous transformations or homeomorphisms. It would seem to require such invariants under more general ε-deformations. Many topological properties are known to be invariant under such transformations. For example, two manifolds that are ε-deformable into each other for every $\varepsilon > 0$ have the same Betti numbers and other homology invariants. However, there are many topological properties for which this more general invariance is not yet demonstrated.

It is of interest not only to prove that for sufficiently small ε certain topological properties remain invariant under ε-deformations but actually to find the maximal value of this ε. More precisely, we have a given metric set and we consider all continuous mappings of this set into another set such that no two points distant by more that ε have the same image point. In other words, the transformation does not coalesce any pair of points whose distance is ε or more. Suppose, for example, the given set is an S_n (surface of a sphere in n dimensions with radius one). A theorem of Borsuk and the author [2] asserts that if this set S_n is mapped into a set in the Euclidean space E_n in such a way that no two points whose distance is greater than a certain l_n coalesce, then the image still cuts the Euclidean space into at least two regions. Determination of the number l_n which is the best (largest) constant has been made in this case. An analogous determination of such constants would be of some interest in the case of other topological properties. For example, suppose that T_n is a torus given metrically in the Euclidean space E_n as a product of two circumferences of the circle of radius 1. One should determine the maximum k_n such that if this set T is mapped into a Euclidean space in such a way that no two points more distant than k_n coincide, then the Betti numbers of the image must be greater than zero. (Compare our problems on quasi-fixed points in the previous chapter.)

The theorems and problems concerning the stability (of the vanishing) of differential expressions for functions of n variables, discussed above, can be considered for functionals. If we consider a typical elementary problem, that of finding an extremum y_0 of

$$I(y) = \int_a^b F(x, y(x), y'(x))\, dx$$

an analogous question to the one about derivatives of functions of a finite number of variables arises, namely, the conditions which guarantee that for every $\varepsilon > 0$ there exists a $d > 0$ such that for all sufficiently "regular" $G(x, y, z)$ with $|F - G| < d$, there exists a minimum $y_1(x)$ for

$$J(y) = \int_a^b G(x, y(x), y'(x))\, dx$$

where $|y_1 - y_0| < \varepsilon$. We assume here merely the proximity of F and G, and nothing is assumed about the proximity of their partial derivatives, occurring in the Lagrangian equations. Speaking descriptively, the question is: when is it true that solutions of two problems in the calculus of variations which correspond to "close" physical data must be close to each other? Affirmative theorems of this sort would ensure the stability of the solutions of physical problems even with respect to the introduction of small additional "hidden" parameters. Since this situation regarding the stability is obscure at present even for total differential equations, it is perhaps futile at this stage to speculate about more general formulations involving partial differential equations from this point of view. However, the following remark should perhaps be made. It seems desirable in many mathematical formulations of physical problems to add still another requirement to the well-known desiderata of Hadamard of existence, uniqueness, and continuity of the dependence of solutions on the initial parameters. Specifically one should have a stability in the stronger sense illustrated by us above: the solutions should vary continuously even when the operator itself is subject to "small" variations.

But even in problems of discrete mathematical structures, the notion of stability can be introduced quite generally. One could even speculate about defining a *distance* between statements in some formal system of mathematics in such a way that definitions of sets which are "close" to each other, say in the sense of a Hausdorff distance, would correspond to points with a small separation.

2. Conjugate functions

We call two transformations f and g of a space E into itself conjugate if there exists a biunique h such that $f = hgh^{-1}$. It is well known that two one-one functions f and g on an abstract set E to itself are conjugate if and only if the two decompositions of E into *cycles* under f and under g are similar. This means that the number of f-cycles of length l is the same as the number of g-cycles of length l for every cardinal l. In the case of functions f and g which are many-one a similar theorem may be proved, generalizing the concept of cycle to that of a "tree". By a tree is meant a minimal set T containing the image $f(x)$ and the complete counter-image $f^{-1}(x)$ for every point x in T. A tree may be represented as a graph containing at most one closed cycle. Different trees are disjoint. It is obvious what is meant by two trees being of the same type. A necessary and sufficient condition for two many-one functions on an abstract set E to itself to be conjugate is that the number of trees of each type be the same under f as under g.

A general investigation of conditions for conjugacy in case E is a given space and f, g, and h are of restricted character, e.g., where continuity for h is required, seems lacking. In particular we may ask the following questions:

If two transformations f, g of n-dimensional Euclidean space E^n are each given by polynomial forms and are conjugate under a continuous h, are they then conjugate under a linear transformation h_1 of E^n?

If two continuous transformations f, g are conjugate under a Borel transformation h, are they then necessarily conjugate under a continuous transformation h_1?

Under what conditions is a homeomorphism of E^n conjugate to a uniformly continuous transformation?

Restricting ourselves to one dimension: is every "smooth" function $f(x)$ on (0, 1) to (0, 1) (e.g., every such polynomial) conjugate to a suitable piecewise linear function? For example the parabola $f(x) = 4x(1 - x)$ is conjugate to the function defined on (0, 1) by the "broken line"

$$g(x) = 2x, \qquad 0 \leqq x \leqq \tfrac{1}{2}$$
$$g(x) = 2(1 - x), \quad \tfrac{1}{2} \leqq x \leqq 1$$

under the biunique transformation

$$h(x) = 2/\pi \sin^{-1} \sqrt{x}$$

An affirmative answer to the above question would reduce the study of the iteration of such functions $f(x)$ to a purely combinatorial investigation of the properties of "broken line" functions.

It may be advantageous to consider a type of conjugacy (at least formally) weaker than the one defined. Let us say that two functions $f(x)$ and $g(x)$ are asymptotically conjugate if the behavior of iterates of points under f and under g is similar in the following sense: there exists a biunique function $h(x)$ on (0, 1) to itself, such that for almost every a on (0, 1), $h(R_a) = S_a$, where R_a is the set of all points which have identical sojourn time in $(0, a)$ under iteration of f, and S_a is the corresponding set for g.

Is it then true that every polynomial $f(x)$ is asymptotically conjugate to a broken-line function $g(x)$? These problems are, of course, not limited to the one-dimensional case and are indeed of greatest interest in higher-dimensional spaces.

3. Ergodic phenomena

In this section we shall be concerned with iteration of functions and transformations, more particularly with the asymptotic properties of the sequence of iterated images of points. The great advances in ergodic theory of the last few decades have clarified the mathematical basis of statistical mechanics to a considerable if not to a complete extent. Roughly speaking, the analogues of the laws of large numbers in the theory of probabilities do now exist in the form of ergodic theorems. The more detailed analysis of analogues of the Gauss-Liapounoff-type theorems is by far less complete. We should mention here parenthetically, that often it is important to deal with transformations of noncompact spaces, e.g., the entire Euclidean space, into themselves. Certain theorems formulated originally for the compact case can be generalized under suitable formulation for such cases. So, for example, the Kronecker-Weyl theorem on the existence of ergodic means for rotations in n-dimensional space can be generalized, to some extent, as follows.

Let L be an *arbitrary* linear transformation of the Euclidean n-space into itself. Let C be any cone of directions in space. For almost every point p the sequence of iterated images $L^n(p)$ has a sojourn time in C. In other words, the ergodic limit of angles exists for almost all initial points p.

If the transformation

$$T\begin{cases} x_1' = f_1(x_1, \ldots, x_n) \\ \vdots \\ x_n' = f_n(x_1, \ldots, x_n) \end{cases}$$

of Euclidean n-space E^n to itself is linear:

$$x_1' = \sum_j x_i a_{ij} \qquad A = (a_{ij})$$

with all coefficients $a_{ij} > 0$, it is well known (Frobenius-Perron) that there exists a unique positive characteristic root r and a unique unit invariant vector $\bar{v} = (\bar{x}_1, \ldots, \bar{x}_n)$ with positive components such that $\bar{v}A = r\bar{v}$. Moreover, for every vector $v = (x_1, \ldots, x_n)$ with positive components, the sequence of points on the unit sphere:

$$vA^n/|vA^n|, \qquad n = 1, 2, 3, \ldots,$$

converges to $\bar{v}$. These facts establish the existence of "steady-state" distributions in many problems involving the multiplication and diffusion of particles.

A simple result of this sort cannot be hoped for in case the transformation T is nonlinear. Such transformations occur naturally in various physical problems involving interaction between the multiplying and diffusing particles. For example, if one takes account of the depletion of the medium in a multiplicative process, the equation for the moments of the probability flow is nonlinear. Also, if there are particles of, say, two different types, and the multiplication of each type depends upon the number of both types present, the corresponding transformation is also nonlinear.

To illustrate the simplest type of question that arises in such cases, suppose that the transformation T has the form

$$f_i(x_1, \ldots, x_n) = l_i(x_1, \ldots, x_n) + q_i(x_1, \ldots, x_n), \qquad i = 1, \ldots, n,$$

where the l_i are linear and the q_i pure quadratic forms in the variables $x_1, \ldots, x_n$.

The problem of interest concerns the asymptotic behavior of the sequence of directions of $T^\nu(x)$ generated by iteration of T on an initial vector x. The Frobenius theorem may perhaps generalize to the following, which we state as a conjecture.

1. Given a cone C of directions issuing from the origin and "almost any" vector $x = (x_1, \ldots, x_n)$, the "time of sojourn" of the iterates $T^\nu(x)$ in C exists.

2. For a given cone C the "sojourn time" of $T^\nu(x)$ in C may depend upon x but there are in this case only a finite number of values of such times.

The latter conjecture, if true, would mean that the space E^n splits into a finite number of disjoint subsets $S_1, \ldots, S_m$, all vectors x in S_i having the same sojourn time t for $T^\nu(x)$ in C.

The considerable variety of physical problems which can be cast in the form of a study of such transformations provides an interest in the investigation of their iterative properties. Thus the equation $\dot{x} = f(x)$ where x is a vector becomes

$$x^{\nu+1} = x^\nu + f(x^\nu) = g(x^\nu)$$

in difference form, which in turn leads to the transformation

$$x' = g(x)$$

and its iterations.

In similar fashion, a partial differential equation of the form

$$\partial u/\partial t = F(u, u_x, u_y, u_{xx}, u_{xy}, u_{yy}, \ldots)$$

when written in difference form on a finite mesh can be regarded from exactly the same point of view, the function $u(x, y)$ being considered as a vector

$$[u(x_1, y_1), \ldots, u(x_n, y_n)] = [u_1, \ldots, u_n] = u$$

the function u' at time $t + 1$ being given in terms of u at time t by an equation of form

$$u' = G(u)$$

The problems proposed on quadratic transformations in spaces of two or more dimensions seem rather difficult. It is of interest to consider in more detail the one-dimensional case.

4. *The Frobenius transform*

Let $y = f(x)$ be a measurable nonsingular transformation of the unit interval into itself. O. W. Rechard [1] has studied the transformation T_f of $L^1(0, 1)$ into itself defined by

$$\int_A T_f \xi \, dx = \int_{f^{-1}(A)} \xi \, dx$$

For reasons that shall appear later, this might be called the Frobenius-Perron transform corresponding to f. It can be shown that the transformation T_f has a nonnegative invariant function $\mu(x)$ such that

$$\lim_{n \to \infty} m[f^{-n}(N)] = 0$$

(where N is the subset of $(0, 1)$ on which μ vanishes) if and only if the set functions $m[f^{-n}(A)]$ are uniformly absolutely continuous with respect to the Lebesgue measure m. Under these circumstances, for every interval (a, b) on $(0, 1)$ and for almost every x, χ denoting the characteristic function,

$$\lim_{n\to\infty} \frac{1}{n+1} \sum_{j=0}^{n} \chi_{(a\ b)} [f^j(x)]$$

exists, and (if f is metrically transitive) is equal to $\int_a^b \mu\, dx$.

For example, if $g(x)$ is the broken-line function of Section 2, the transformation T_g is defined by

$$T_g \xi(x) = \tfrac{1}{2}\xi\left(\frac{x}{2}\right) + \tfrac{1}{2}\xi(1 - \tfrac{1}{2}x)$$

and the function $\mu(x)$ is identically 1. From this it follows that if $f(x)$ is the parabolic function $f(x) = 4x(1 - x)$, for which the transformation T_f is

$$T_f \xi(x) = 1/4\sqrt{(1-x)}\{\xi[\tfrac{1}{2}(1-\sqrt{(1-x)})] + \xi[\tfrac{1}{2}(1+\sqrt{(1-x)})]\}$$

then the corresponding function $\mu(x)$ is given by

$$\mu(x) = d/dx\,[2/\pi \sin^{-1}\sqrt{x}] = 1/\pi\sqrt{x}\sqrt{1-x}$$

These remarks suggest the following question. If a transformation $f(x)$ of the unit interval into itself is defined by a sufficiently "simple" function (e.g., a broken line function or a polynomial) whose graph does not cross the line $y = x$ with a slope in absolute value less than 1, does the corresponding F.-P. transform have a nontrivial invariant function? It is not even known if this is true for every transformation of the form

$$f(x) = 2x, \qquad 0 \leqq x \leqq \tfrac{1}{2}$$

$$f(x) = (2 - a) + 2(a - 1)x, \qquad \tfrac{1}{2} \leqq x \leqq 1$$

where $0 < a < \frac{1}{2}$. (For $a = \frac{1}{2}$ this is easily shown to be the case.)

The F.-P. transform corresponding to $f(x)$ can be thought of as the continuous analogue of the following transformation defined on the space of step functions on (0. 1). Let the unit interval be divided into n equal nonoverlapping subintervals $I_1, I_2, \ldots, I_n$ and define a_{ij} as that fraction of interval j which is mapped into interval i by $f(x)$. That is,

$$a_{ij} = m[I_i \cdot f^{-1}(I_j)]/m(I_i)$$

If now

$$\sigma_n(x) = \sum_{i=1}^{n} C_i \chi_I (x)$$

is a step function defined on the intervals I_j, then we define the transformed function

$$T_n \sigma_n(x) = \sum_{i=1}^{n} C_i' \chi_{I_i}(x)$$

where

$$C_i' = \sum_{j=1}^{n} a_{ij} C_j$$

That is the vector of coefficients (C_i') is just the result of operating on the vector (C_i) with the matrix a_{ij}.

If A and B are two subsets of (0, 1) each consisting of a finite number of intervals I_j and if $f^{-1}(A) = B$, then

$$\int_A T_n \sigma \, dx = \int_B \sigma \, dx$$

The matrix a_{ij} has its largest eigenvalue equal to 1 and a corresponding nonnegative eigenvector $(\bar{C}_i)$ which can be regarded as defining a step function $\bar{\sigma}_n(x)$ that is invariant under the transformation T_n. We conjecture that if the F.-P. transformation T_f has a nonnegative invariant function $\mu(x)$, then the invariant step functions $\bar{\sigma}_n(x)$ converge to $\mu(x)$ in $L'(0, 1)$ as n, the number of subdivisions of (0, 1), becomes infinite.

Under fairly weak restrictions on the matrix (a_{ij}) (e.g., if some power $(a_{ij})^k$ contains only positive elements), the theorem of Frobenius-Perron asserts that the invariant step function $\bar{\sigma}_n(x)$ is the limit, as j becomes infinite, of the sequence of iterates $T_n^j \sigma_n(x)$, where $\sigma_n(x)$ is any nonnegative step function not identically zero. Does a similar result hold for the general continuous F.-P. transform? That is, if $\xi(x) \geqq 0$ is not identically zero, does the sequence of iterates $T_f^j \xi(x)$ converge in $L'(0, 1)$ to $\mu(x)$? Computational evidence for the case $f(x) = 4x(1 - x)$ and $\xi(x) \equiv 1$ suggests that this is the case.

5. *Functions of two variables*

The following conjecture of the author has been proved by Zahorski [1]: for every function $f(x)$ continuous on the unit interval (0, 1), there exists a function $g(x)$ analytic on (0, 1) and a perfect set C on this interval such that $f(x) \equiv g(x)$ for all x of C. Is the analogue true in the plane if the functions $f(x, y)$ and $g(x, y)$ are continuous and analytic, respectively, on the unit square, and if the set C is required to be a direct product of two perfect sets?

Let $f(x_1, \ldots, x_n)$ be a real-valued continuous function defined on the "unit cube" $0 \leq x_i \leq 1$. Does there exist an arc in the cube on which the function is constant? *

Consider a continuous function $f(x, y)$ of two real variables which is associative

$$f(x, f(y, z)) = f(f(x, y), z)$$

for example, $x + y$, xy, $(x^2 + y^2)^{\frac{1}{2}}$, etc. What further condition on f guarantees that there exist a finite number of such functions g_i such that every associative continuous function f is conjugate to one of these, in the sense that for some $L(z)$

$$f(x, y) = L^{-1}(g_i(L(x), L(y)))$$

Thus $f(x, y) = xy$ and $g(x, y) = x + y$ are conjugate under $L(z) = \log z$ for positive x, y.

Obviously, the class of associative functions of two variables, if one merely requires continuity is very large — is it ever true, perhaps, that every continuous transformation of a plane $T: x' = f(x, y)$, $y' = g(x, y)$ can be obtained by composing a finite number of transformations T_i where f_i and g_i would be associative? Compare the wonderful recent results of Kolmogoroff [1].

6. *Measure-preserving transformations*

Does there exist a square integrable function $f(x)$ and a measure preserving transformation $T(x)$, $-\infty < x < \infty$, such that the sequence of functions $\{f(T^n(x)); n = 1, 2, 3, \ldots\}$ forms a complete orthogonal set in Hilbert space? (Banach)

* *Note added in proof*: A negative answer follows from recent constructions of R. H. Bing.

Let the real numbers $x = \sum_1^\infty a_i 2^{-i}$ $(a_i = 0 \text{ or } 1)$ on $(0, 1)$ be represented by the sequences

$$\sigma(x) = \{\ldots, a_6, a_4, a_2; a_1, a_3, a_5, \ldots\} \equiv \{\ldots, b_{-3}, b_{-2}, b_{-1}; b_0, b_1, \ldots\}$$

The transformation

$$T(x) = \sigma^{-1}\{\ldots, b_{-3}, b_{-2}; b_{-1}, b_0, b_1, \ldots\}$$

(right shift of one place) is known to be measure preserving. Is it true that T and its iterate T^2 are not conjugate under any measurable transformation $H(x)$ (i.e., $T \neq HT^2H^{-1}$)? More strongly, is it true that T has no measurable square root $S(x)$: $T(x) = S(S(x))$?

7. *Relative measure*

Does there exist, for every set A of measure zero(say on the interval), a countably additive measure function m_A under which at least all Borel subsets of A are measurable, and which has the properties

1. $m_A(A) = 1$; $m_A(p) = 0$, p a point of A,

2. for $A \supset B \supset C$, $m_A(C) = m_A(B) \cdot m_B(C)$.

In other words we desire a *class* of measure functions with a possibility of relativising it "uniformly."

Suppose that the basic space is the circumference E of the unit circle, and $m(X)$ the Lebesgue measure, which can be equivalently defined as

$$m(X) = \lim_{N\to\infty} (N+1)^{-1} \sum_{\nu=0}^{N} \chi_X(T^\nu(x))$$

for almost all x, where χ_X is the characteristic function of the subset X of E and T is any irrational rotation. For B taken as a Borel subset of a set A of measure zero, when does the limit

$$\lim_{N\to\infty} \sum_{\nu=0}^{N} \chi_B(T^\nu(x)) \Big/ \sum_{\nu=0}^{N} \chi_A(T^\nu(x))$$

exist, and for which subclass does this provide a suitable $m_A(B)$ function?

Another suggestion, due to A. L. Shields, is to consider

$$m_A(B) = \lim_{\varepsilon=0} m(B_\varepsilon)/m(A_\varepsilon)$$

where m is the ordinary Lebesgue measure and B_ε is the set of points whose distance from B is less than ε (similarly for A_ε).

(It is clear *a priori* that we shall not establish in this way an m_A for *all* Borel sets A of measure zero with the desired properties — in fact, for every A dense in the interval the measure would coincide with the measure of its closure. It would be of interest, however, to construct a class of measure functions with the above properties *1* and *2* for a sufficiently large class of the sets A.)

8. Vitali-Lebesgue and Laplace-Liapounoff theorems

A classical result (the "Vitali-Lebesgue theorem") in the theory of functions of a real variable is that, in a set Z of positive measure $m(Z)$, almost every point p of Z has density 1, that is to say, if I_n is any sequence of intervals all with midpoint p, whose lengths $m(I_n) \to 0$ as n becomes infinite, then

$$\lim_{n\to\infty} m(I_n \cdot Z)/m(I_n) = 1$$

(Compare Saks [1].)

One could try to strengthen this theorem by proving a definite *rate* of this convergence to 1. For example, knowing that

$$d_n \equiv m(I_n \cdot Z) - m(I_n) = 0(m(I_n))$$

can one assert more; e.g., is it true that for every $\varepsilon > 0$ and almost every p,

$$d_n = 0(m(I_n))^{\frac{3}{2}-\varepsilon}?$$

One can obviously restrict the set Z to the class of G_δ-sets. If Z is an open set, then $d_n = 0$ for all n sufficiently large. For sets which are both G_δ- and F_σ-sets, such a strengthening of the density theorem might well be possible, even to the extent of estimating, for any k, the measure of the set of those points of Z for which

$$\lim_{n\to\infty} d_n/[m(I_n)]^{\frac{3}{2}} < k$$

A similar investigation is possible in the ergodic theory. Let E be a measure space (say, the Euclidean cube), $T(p)$ a measure-preserving, metrically-transitive transformation of E onto itself.

The ergodic theorem states that for almost all p, and for every set A of positive measure $m(A)$,

$$\lim_{N\to\infty} (N+1)^{-1} \sum_{\nu=0}^{N} \chi_A(T^\nu(p)) = m(A)$$

$\chi_A(p)$ being the characteristic function of A (Chapter VI, Sections 2, 4, 7) that is,

$$d_N = \sum_{\nu=0}^{N} \chi_A(T^\nu(p)) - m(A)\cdot(N+1) = o(N+1)$$

What can one say about $d_N/(N+1)^{\frac{1}{2}+\varepsilon}$? This limit is certainly zero for *some* measure-preserving transformations of the interval. For example, if $T(p)$ is the "shift" transformation of Section 6 of the present chapter, the "central limit theorem" of Laplace-Liapounoff applied to the "Bernoulli" case of the additive probability theory states that

$$d_N = 0(N+1)^{\frac{1}{2}+\varepsilon}$$

for every $\varepsilon > 0$. Can one assert the above relation for almost all $T(p)$ of the type postulated? (Compare Feller [1].)

9. A problem in the calculus of variations

Suppose two segments are given in the plane, each of length one. One is asked to move the first segment continuously, without changing its length to make it coincide at the end of the motion with the second given interval in such a way that the sum of the lengths of the two paths described by the end points should be a minimum. What is the general rule for this minimum motion? It is clear from the Euler-Lagrange equations of the variational problem that, locally, the motion will be a composition of rotations and translations. (The problem could be stated for two such intervals given in the 3-dimensional space.) One could require alternately that instead of the sum, the square root of the sum of the squares of the lengths described by the end-points should be minimum.

More generally, one could pose an analogous problem of the "most economical" motion given a geometrical object A and another B congruent to it and requiring the motion from A to B to be such that a sum or integral of the lengths of paths described by individual points be minimum. This bears a certain relation to the problem of Monge of "déblais et remblais," but differs from it in that we require here rigidity of A through the course of the motion. One motivation behind the consideration of such questions is that in certain problems of mechanics of continua, e.g., in hydrodynamics the motions that are most prevalent, are singled out by extremal principles not unlike the above; but of course operating in a space of infinitely many dimensions.

10. A problem on formal integration

Let $f_1(x), \ldots, f_n(x)$ be arbitrary continuous functions. Does there exist a rational function $R(x)$ constructed from the f_i by rational operations such that the "indefinite integral"

$$\int R(x)dx$$

is not again such a rational combination of the f_i and of functions obtained by superposing the f_i? (Mazur and Ulam).

We mention this rather special question as a small example of more general and interesting problems involving the algebraic properties of finite "formal analytic" algorithms. Compare the paper by L. Bieberbach [1] also of Ritt [1] and, more specifically for the problems of the above type, Kaczmarz-Turowicz [1].

11. Geometical properties of the set of all solutions of certain equations

The class of all solutions of a *linear* differential equation forms a linear manifold in function space. We think here of functions satisfying the equation and given boundary conditions as points in the space of all such continuous and differentiable functions. What can one say about the geometric properties of the set of solutions of a differential equation which is *quadratic* in the unknown function and its derivatives? If the equation is of the type $Q(y, y') = 0$, where Q is a positive-definite quadratic form, the set of solutions has the property: given any k solutions, no other solution lies inside the simplex formed in function space with these k solutions as vertices. In other words, the manifold of solutions lies on an intersection of convex ("ellipsoidal") surfaces. Can one assert that the manifold M of solutions of an algebraic differential equation is formed by the intersection of (possibly infinitely many) cylinders, each of which is erected over a finite dimensional algebraic manifold A_i? That is to say

$$M = (A_1 \times E_1) \cdot (A_2 \times E_2) \ldots (A_i \times E_i) \ldots$$

where the A_i are as above, E_i are linear, infinitely dimensional hyperspaces in the function space.

CHAPTER VII

Physical Systems

1. Generating functions and multiplicative systems

Let us consider a system of particles of t distinct kinds such that a particle of type i, upon transformation, has a given probability $p_1(i; j_1, \ldots, j_t)$ of producing $j_1 + \ldots + j_t$ new particles, j_i of type i. The probability of a particular population $(j_1, \ldots, j_t)$ in the kth generation of progeny from a single particle of type i is given by the coefficient $p_k(i; j_1, \ldots, j_t)$ of the product $x_1^{j_1} \ldots x_t^{j_t}$ in the kth iterate of the generating transformation $x' = G(x)$:

$$x_i' = g_i(x_1, \ldots, x_n) = \sum_j p_1(i; j_1, \ldots, j_t) x_1^{j_1}, \ldots, x_t^{j_t}$$

This theorem on iteration of generating functions allows one to calculate the first moments of the distributions by multiplication of the matrices whose terms are the first partial derivatives of the g_i evaluated at $x_1 = x_2 = \ldots = x_t = 1$. Higher moments can be computed also, but the expressions become increasingly complicated.

Unfortunately, it is very difficult to obtain precise information about the behavior of these coefficients except in the simplest cases. If only one type of particle is involved, one can explicitly study the iterates of a generating function of the type

$$g(x) = (ax + b)/(cx + d)$$

where the a, b, c, d are chosen so that the coefficients of the power series for $g(x)$ are nonnegative and have sum unity. The iterates are easy to compute since iteration of $g(x)$ leads to functions of the same form. Analogously the transformation

$$x_i' = (\sum a_{ij}x_j + b_i)/(\sum c_{ij}x_j + d_i)$$

may serve for systems of t types of particles. The question now arises whether there exist groups (or semi-groups) of specific transformations in the t-dimensional space with *more* parameters than are available for the mappings of the above type which, when developed in power series, can be considered as generating transformations (i.e., coefficients of g_i nonnegative and with sum $= 1$)? This would allow a greater variety of transformations whose iterates could be obtained in closed form.

The expected number of particles of type j in generation k from one particle of type i is given by the number in the ith row and jth column of the kth power J^k of the Jacobian of $G(x)$ at $x = (1, 1, \ldots, 1)$. The moment matrix J, when positive, has a unique positive eigenvector v of norm 1 (a theorem of Frobenius-Perron) and for "supercritical" systems, it may be shown that "almost all" genealogies terminate in death or approach, ratio-wise, the vector v, in the sense of a natural measure defined in the space of genealogies. This statement constitutes an analogue of the strong law of large numbers for the "case of Bernoulli" for multiplicative processes (cf. Everett, Ulam [2, 3]). These results, however, represent only a first step in the theory of such processes. It would be important to establish the analogue of the central limit theorem. What are the asymptotic properties of such a process if the basic probabilities $p_1(i; j_1, \ldots, j_t)$ are not constant in time, but change in a specified way either explicitly in time or dependent upon the existing population? If the limit of the product of the Jacobians of the generating transformations $G_1, G_2, \ldots$ exists, does the population approach the corresponding vector v or die out with probability 1?

The reader will find several problems on multiplicative systems in the papers referred to above. These are concerned with the iteration of generating transformations given by polynomials or power series with nonnegative coefficients in n variables and the number of particles of each type present in the kth generation. The problem of total progeny from the first to the kth generation and of systems with source may be so studied. The behavior of the coefficients of the kth iterate of such transformations has not been determined. (See also Bellman and Harris [1].)

1a. Examples of mathematical problems suggested by biological schemata

The combinatorial complexities and analytical questions suggested by problems of genetics and by problems of structure of organic materials present features of purely mathematical interest. The well-known work of Volterra [1] on the struggle for survival and the subsequent work of W. Feller [2] dealing with certain systems of quadratic total differential equations contained important results on special nonlinear systems.

We shall mention briefly some related problems, leading also to a system of infinitely many nonlinear differential equations also suggested by biological situations — of course treated in an extremely simplified and schematized way.

Imagine a system of N particles which reproduce in discrete units of time (generations). In the simplest version assume that the reproduction is asexual. Each of these particles possesses an index k denoting the number of its "characteristics." This number may increase in time due to mutations occurring at random at a fixed rate in the population. We assume an advantage in acquiring additional characteristics considered to be improvements leading to higher probability for survival of an individual. Specifically, there is a probability $\alpha \ll 1$ for each individual to acquire an additional improvement in the course of one generation — α^2 being the probability of acquiring in one generation two improvements, etc. Another constant β defines for an individual the differential advantage in its survival. In the simplest scheme one may assume that the differential advantage for survival is proportional to the number of these improvements—that is to say, if one particle has an index k and the other one $k+j$, then the relative chance of survival of the richer individual over the poorer one is proportional to j. In a numerical treatment of the problem one may assume the population to be always normalized to a constant number N. The first problem concerns the number x_i of the particles with i advantages as a function of time in its dependence on the two constants α and β. A simple system of equation would be:

$$\Delta x_k = -\alpha x_k - \alpha^2 x_k - \frac{x_k}{N} \sum_{j=k}^{l} (j - k) \beta x_j ,$$

$$\Delta x_{k+1} = -\alpha x_{k+1} - \alpha^2 x_{k+1} + \alpha x_k + \beta x_{k+1} - \frac{x_{k+1}}{N} \sum_{j=k}^{l} (j - k) \beta x_j ,$$

$$\Delta x_i = -\alpha x_i - \alpha^2 x_i + \alpha x_{i-1} + \alpha^2 x_{i-2} + (i-k) \beta x_i - \frac{x_i}{N} \sum_{j=k}^{l} (j-k) \beta x_j ,$$

$$i = k + 2, \ldots, l - 2,$$

$$\Delta x_{l-1} = -\alpha x_{l-1} + \alpha x_{l-2} + \alpha^2 x_{l-3} + (l-1-k) \beta x_{l-1} - \frac{x_{l-1}}{N} \sum_{j=k}^{l} (j-k) \beta x_j ,$$

$$\Delta x_l = \alpha x_{l-1} + \alpha^2 x_{l-2} + (l - k) \beta x_l - \frac{x_l}{N} \sum_{j=k}^{l} (j - k) \beta x_j$$

Here the species k is the first one of any importance that is present in the generation and l is the last. One may assume that in each generation the k is determined by making it equal to the index of the first number in the sequence, $\{x_i\}$, which is equal to or greater than 1 and the l is made equal to 2 plus the index of the last number of the sequence, which is ≥ 1. The numerical investigation made by C. Luehr and the author led to the following results. The solution of the system seems to approach a steady state in the following sense: the average index $\bar{\imath}$ of the population existing in one generation increases linearly with time, and the distribution of the i around the $\bar{\imath}$ appears to approximate the Gaussian. The parameters of this normal distribution depend in a simple fashion on the constants α and β.

A problem next in complexity, but still extremely simplified compared to the real biological situations involves bisexual reproduction — that is to say, production of particles by pairs of particles. The equations will now be *essentially* nonlinear. One should assume that the advantageous characteristics acquired by mutations can be transmitted from either parent to the offspring in the next generation. We may assume again that the total population is constant by normalizing it and, in the simplest case, assume only two kinds of "genes". The advantage for survival (or in the increase of the number of offspring) depends, say, on the sum of the number of the improvement genes of each kind. If the offspring can acquire independently the genes from each parent, obtaining certainly those that are present in both and each gene which is present in one parent only with the probability equal to $\frac{1}{2}$, the equations could be

$$x'_{ij} = y_{ij}/\sum y_{ij}$$

where

$$y_{ij} = -\alpha x_{ij} + \alpha/2(x_{i,j-1} + x_{i-1,j}) + \beta[i + j - \min(i + j)]x_{ij} + 1/N \sum\sum \gamma_i^{kl} \gamma_j^{mn} x_{km} x_{ln}$$

where the choice of

$$\gamma_i^{kl} = 1/2^{(k-l)} \binom{(k-l)}{i - \min(k, l)}$$

corresponds to our rule on the inheritance of the extra genes. A numerical study of this system was undertaken by P. Stein and the author. Again a steady state distribution seems to establish itself with the rather curious property that only a few species with adjoining indices co-exist in any one generation. The speed with which the average number of "improvements" increases is constant.

This mathematical formulation is still very naive and too simple as compared to the biological reality. The number of kinds of genes (or phenotypes) is much greater than 2. Also one should study such systems with more realistic rules, i.e., the distinction between dominant and recessive genes and the Mendelian properties. The mathematical properties of solutions of such systems seem to be akin to those of the phenomena encountered in the study of nonlinear systems describing a vibrating string, etc., mentioned in Chapter VII, Section 8.

P. Stein succeeded in proving the following:

Consider the quadratic transformation in n-space given by the quadratic terms alone:

$$x'_{ij} = \sum\sum \gamma_i^{kp} \gamma_j^{mn} x_{km} x_{pn}$$

assume

$$x_{ij} \geqq 0, \qquad \sum x_{ij} = 1.$$

About the γ's assume only:

$$\gamma_i^{kl} > 0 \text{ only for } k \leqq i \leqq l; \quad \sum_{i=1}^{N} \gamma_i^{kl} = 1 \text{ and } \sum_{i=k}^{l} i\, \gamma_i^{kl} = (k + l)/2.$$

Under these assumptions the only fixed points of the transformation are of the form

$$x_{jj} = 1/(1+\tau)^2, \; x_{j+1,j} = \tau/(1+\tau)^2, \; x_{j+1,j+1} = \tau^2/(1+\tau)^2$$

all other x_{ij} being 0, τ is a parameter, $0 \leqq \tau \leqq \infty$. The iteration of the transformation, starting from any point, converges to one of these fixed points; the value of τ being computable from the initial conditions.

A still different class of problems, leading to a study of nonlinear (quadratic) transformations and their iterations originates from the following schema: imagine a large number of individuals (or particles) present in a given generation. Suppose these combine in pairs and produce, in the next generation, new particles, parents dying after procreating the new ones. Suppose the original particles are each one of N different types. A rule is now given for the type i ($i = 1, 2, \ldots, N$) produced by the mating of individuals of type j and k. In a random mating of particles the expected value of the fraction x_i' of particles of a given type in the next generation will be a quadratic function of the two fractions x_j and x_k. The equations would be

$$x_i' = \sum_{k,\, l=1}^{N} \gamma_i^{kl} x_k x_l \qquad i = 1, \ldots, N$$

where, if we assume that each pair produces exactly two new particles and specifically the rule of defining the type is that the γ's are either 0's or 1's, we may insist that for any index i not all γ_i^{kl} vanish and the system of equations really specializes to a form where each term in the product $(x_1 + x_2 + \ldots + x_N)^2$ will appear in exactly one row of the set of equations. For further simplification we may assume that the cross-products appear with the factor 2 (commutativity). So as an example we could have with $N = 4$,

$$\begin{aligned} x_1' &= x_1^2 + x_2^2 + x_4^2 + 2x_1x_4 + 2x_2x_4 + 2x_3x_4 \\ x_2' &= 2x_1x_3 + 2x_2x_3 \\ x_3' &= 2x_1x_2 \\ x_4' &= x_3^2 \end{aligned}$$

A study of all such transformations was made by P. Stein and the author in case $N = 3$. There are 97 nonequivalent possible "genetic" rules of this kind and all 97 corresponding transformations of this type were studied with regard to the properties of iterated sequences of each transformation. In some cases, starting with an arbitrary initial distribution (nondegenerate) one converges to a fixed point; that is to say, the ratios of the numbers of individuals of each type stabilize. In other cases the points seem to approach an oscillation between a finite number of fixed ratios. In every case, for $N = 3$ the first means, in time, of the ratios x_i exist for almost every point, it appears. The ergodic and asymptotic properties of the iterates of such transformations for $N > 3$ are unknown in general (see P. Stein and the author, [2]).

2. *Infinities in physics*

The simplest problems involving an actual infinity of particles in distributions of matter appear already in classical mechanics. A discussion of these will permit us to introduce more general schemes which may possibly be useful in future physical theories.

Strictly speaking, one has to consider a true infinity in the distribution of matter in all problems of the physics of continua. In the classical treatment, as usually given in textbooks of hydrodynamics and field theory, this is, however, not really essential, and in most theories serves merely as a convenient limiting model of *finite* systems enabling one to use the algorithms of the calculus. The usual introduction of the continuum leaves much to be discussed and examined critically. The derivation of the equations of motion for fluids, for example, runs somewhat as follows. One imagines a very large number N of particles, say with equal masses, constituting a net approximating the continuum which is to be studied. The forces between these particles are assumed to be given, and one writes the Lagrange equations for the motion of the N particles. The finite system of ordinary differential equations "becomes" in the limit $N = \infty$ one or several *partial* differential equations. The Newtonian laws of conservation of energy and momentum are seemingly correctly formulated for the limiting case of the continuum. There appears at once, however, at least one possible objection to the unrestricted validity of this formulation. For the very fact that the limiting equations imply tacitly the continuity and differentiability of the functions describing the motion of the continuum seems to impose various *constraints* on the possible motions of the approximating finite systems. Indeed, at any stage of the limiting process, it is quite conceivable for two neighboring particles to be moving in opposite directions with a relative velocity which need not tend to zero as N becomes infinite, whereas the continuity imposed on the solution of the limiting continuum excludes such a situation. There are, therefore, constraints on the class of possible motions which are not explicitly recognized. This means that a viscosity or other type of constraint must be introduced initially, singling out the "smooth" motions from the totality of all possible ones. In some cases, therefore, the usual differential equations of hydrodynamics may constitute a misleading description of the physical process.

On the other hand, the numerical solution of such a system of partial differential equations involves the use of a model of finitely many points approximating the continuum. The corresponding finite difference scheme must be carefully designed to insure not only that the distances between neighboring points are sufficiently small, but that various numerical stability conditions, e.g., so-called "Courant conditions," hold. This necessity shows again a number of implicit assumptions about the finite model approximating the mechanical system. The question whether the limit of the solutions of the approximating equations is in fact the solution of the limiting equation is, in the general case, open. The statement is probably false in the most general case.

Indeed, it may be that, in some future physical theories, the Euclidean continuum presently used as the exclusive model for distributions of matter will cease to be the *sole* convenient model for reality. It seems possible that in some cases spaces with the topology of the *Cantor* (perfect, nowhere dense) sets might serve to represent distributions of matter or energy.

3. *Motion of infinite systems, randomly distributed* *

We shall propose a few very simple mathematical questions illustrating the problems which will arise in the study of systems of this sort. They involve infinite assemblies of mass points with interactions assumed to exist between them. Such assemblies, while not finite, will *not* correspond to the continua which are presently employed in physical theories.

Suppose that we distribute a set of equal masses m on a line, on the integer points $0, \pm 1, \pm 2, \ldots$ by a probability scheme, placing a mass $m = 1$ on each point n *with probability* $\frac{1}{2}$ and leaving the point vacant with this probability. Since there is an obvious one-one correspondence between all possible initial distributions and real numbers on $(0, 1)$ in dyadic expansion, we can have a measure, e.g., the ordinary Lebesgue measure, in the set of distributions, and in what follows, the phrase "almost all" is understood in the sense of this measure. Let us further assume that between every two mass points there exists an attractive force inversely proportional to the square of their distance. Obviously the total force on each particle is well defined, the series of inverse squares of integers converging absolutely. We shall postulate also that colliding masses remain forever together, forming a single particle of mass equal to the sum of their masses, and that momentum is conserved under collision.

For times $t > 0$, the behavior of the system is described by an infinite system of Newtonian equations. We might remark parenthetically, that the various formulations of the principles of mechanics, all equivalent for finite systems, become in our case quite distinct. The total mass of our system being infinite, one has to use reformulations of the usual variational principles, or even the Lagrangian equations, to arrive at unequivocal statements.

The questions that arise concern the asymptotic behavior of such systems after long times. They can be put in this form. What is the measure of the set of initial distributions which will behave asymptotically in a specified fashion? Some such questions have

* The next three sections follow the author's article [2].

been answered. Many others suggest themselves (cf. Metropolis, Ulam [1]).

The situation becomes perhaps more "physically" interesting in two and three dimensions. The problem is still mathematically well defined. One can state that for almost every initial distribution it is true that, for every constituent mass point, the net force vector *exists* if the component forces are summed over successive spherical shells about the point; that is to say, the limit of these forces exists for almost all initial distributions. However, collision must now be understood in the sense of gravitational capture, i.e., the "colliding" points remain within some specified distance of each other for all time. Is it true that the series of forces on all masses of a distribution remain weakly convergent for all $t > 0$ if this condition obtains initially? The initial average density, in the obvious sense, of our system of particles, randomly distributed as they are at $t = 0$ is $= \frac{1}{2}$. Is it true that almost all distributions retain this density for all time?

In the case of one dimension there will be a tendency toward the formation of successively larger condensations. Is the same true in higher dimensions if by a condensation we understand a subsystem of points whose mutual distances all remain forever bounded? Will almost every distribution show a tendency toward the formation of "galaxies", "super-galaxies," etc.?

What force laws $F(r)$, or equivalently, what potential functions $V(r)$ have the property that, for almost every initial configuration, all forces remain well defined for all time under the ensuing motions, calculated from Newton's equations and assuming our conventions for collision?

Analogous but more difficult problems arise if we deal with countable systems, again randomly distributed initially but not restricted to the set of lattice points. The following general properties of our one-dimensional initially randomly distributed infinite systems are established very easily:

1. The masses appearing in the course of time will be unbounded. In other words, for almost every initial condition of our system there will exist for every M a time t such that a mass greater than M will appear after this time.

2. There will always exist single particles. In other words, for almost every system and for every t there will exist in the system points with unit mass.

3. The asymptotic density of our system remains constant and equal to the original density. We define the asymptotic density as follows. Consider the totality of particles contained in an interval from $-N$ to $+N$ and denote by M_N the total mass of all particles in this interval. The $\lim_{N=\infty} M_N/2N$ shall be called the asymptotic density if it exists. With our initial masses equal to 1, and the random placing of these masses on integer points, this limit (from Bernoulli's theorem) is equal to $\frac{1}{2}$. It is easy to see that this limit will exist and be equal to $\frac{1}{2}$ for all t. This is simply due to the fact that, given any t, the displacement of each particle will be bounded. If we take a sufficiently large interval, the flux across its ends will constitute an arbitrarily small fraction of the total number of particles and our assertion follows.

4. Arbitrarily large "holes" will appear in our system; that is, for almost every system and for all d there will exist a time t so that there will be infinitely many mass points separated by intervals larger than d. Moreover, for all greater times these long empty intervals will continue to exist.

These assertions are easy to prove in one dimension. In two or more dimensions, we shall *not* have, in general, collision between point masses and we would again have to define captures, that is to say, formation of double or multiple systems. The corresponding theorems on the existence of stable or semi-stable captures seem much harder to prove. An easier way to deal with an analogue of our system would be to give each point a finite size and then consider certain collisions as completely inelastic and leading to formations of larger masses. Property *2* and property *3* should then be easy to prove.

More interesting are the quantitative properties of such systems. For example, it would be interesting to calculate, even for systems in one dimension, the average mass of particles in our infinite collection at a given time t, and to determine the distribution of masses as a function of time. Another interesting question concerns the distribution of velocities of our particles as a function of time. (To define a meaningful average velocity one would have to introduce a cut-off in the distance between two particles, which approach each other, just before a collision; in the system consisting of mathematical points, these velocities become arbitrarily large during the collision process.) If we define a cut-off distance then the average velocity of our particles will have a meaning for all t and the question arises: what is the "temperature" of the system as a function of time?

In order to study these and other similar questions, a series of experiments were performed on a computing machine by John Pasta and the author. An attempt was made to imitate the infinite system by a finite one composed of a great number of masses placed on points of a regular subdivision of an interval with a decision, for placing or not placing mass points in successive positions, made by "throwing a die." In order to "approximate" an infinite system somewhat realistically, one has to imagine the two end points of the interval on which the points are located as coinciding; that is to say, we have a finite system of points on the circumference of a circle or a periodic structure. This attenuates the end effects. It is clear that such a finite system will imitate an infinite one only for a limited time. Given a finite system, it is certain that it will ultimately collapse to a single point, whereas in the infinite case, we saw that the asymptotic density will remain constant for all times. Therefore, in interpreting the results of a calculation made for a finite system, one has to carry them only up to a value T of time when the system still contains "many" points. In order to make a rigorous analysis, for the study of any given functional of the distribution of distances, masses, velocities, etc., of our system, one would have to give *a priori* inequalities for this T as a function of N and a $\delta > 0$; that is to say, with only a finite system of N points, we restrict T so that the functional of the system in which one is interested, computed up to time T, will differ in value by less than δ from the value of this functional for an infinite system (i.e., for almost all infinite systems). Many finite systems were computed (that is to say we started with many distributions of mass points given by our initial random procedure and in each case the subsequent motions were calculated with the hope of obtaining heuristic results on some functionals of such systems). The number of points initially taken was of the order of 1,000. Among the quantities calculated were the distribution of masses at any time t, the distribution of distances, etc. We shall indicate here merely very briefly some of the qualitative facts about them.

(a) The average mass of a particle appears to increase linearly with time.

(b) There is a suspicion, at least, that if one considers the distribution of masses of particles existing at time t in the units of the average mass at that time, this distribution tends to a fixed function. In other words a steady state may establish itself.

(c) A quantity which was called hierarchy was studied. This is defined as follows by induction. The original particles have by definition a hierarchy rank zero. When two particles of rank m and n collide, they form a particle whose hierarchy rank is equal to the greater of the two numbers m, n if $m \neq n$. In case $m = n$, the rank of the new particle is $m + 1$. This index gives an idea of the degree or hierarchy of conglomeration as distinguished from a mere increase of mass by accretion. The average hierarchy was increasing more slowly than the average mass, but presumably tends to infinity for an infinite system.

(d) The average kinetic energy was studied as a function of time. We have used in the computation a cut-off in the distance of approach in order to eliminate the arbitrarily high velocity just before collapse. The shape of this function is not known, but it is obvious that this average energy increases initially and then starts decreasing again, which, of course, is due to the fact that our system ultimately will end up as one big particle at rest. Presumably, our finite systems only imitate the infinite one up to the time when this average energy stops increasing. Nothing conclusive as yet, therefore, can be said about the change of "temperature" in time.

It should be pointed out here that the type of distribution (à la Bernoulli) which we have introduced in our numerical work could equally well be chosen differently. For example, we could assume that there is a fixed probability $\alpha \cdot dx$ of finding a unit mass in the interval dx. That is to say, the initial randomness could be chosen à la Poisson. It is easy to see that the convergence of the force on each particle of the system would be equally valid for almost all initial conditions with this set-up.

One could postulate a finite interval with infinitely many points of various masses distributed in this interval at time $t = 0$ by a given random process and then discuss the ensuing motions.

One could assume, of course, that not only the initial positions of our points on the line or in space are given at random, but also that there exist, at time $t = 0$, initial velocities of each of these points, given in a random fashion, say with a Maxwellian distribution.

4. Infinite systems in equilibrium

One of the qualitative differences in the behavior of finite and of infinite systems is that a system which has infinitely many mass points may exist in a state of static equilibrium, i.e., at rest, even when the forces between any two points are attractive. For example, assuming that any two points attract each other, the resultant force on each point may be still equal to 0: One can find a countably infinite set of point masses $m_1, m_2, \ldots$ and a set of initial positions $x_1, x_2, \ldots$ on the unit interval such that (a) $\sum_{i=1}^{\infty} m_i = 1$, (b) the masses attract each other according to the inverse square law, (c) the whole system is in static equilibrium, that is to say, the net force on each mass point exists and is zero. This equilibrium will not be stable, that is arbitrarily small displacements of such initial positions may lead to motions of the system which will make it collapse or, in any case, lead to configurations which with time increasing to infinity will differ more and more from the initial position.

It is easy to find distributions of such mass points with attracting forces like the above so that the initial motions will be expanding!

It would be of interest to find two "truly" three-dimensional infinite systems of points with total mass finite and so located that the net force on each point would be zero. One would like such systems to be of more than one dimension in the following sense. The set of all possible directions, i.e., angles between pairs of points should be dense on the circumference unit circle or — in a three-dimensional distribution of points — the set of directions between pairs of points should be dense on a unit sphere.

Can systems of the above sort be found which would be even dense on an interval or in a region of the plane or space?

5. *Random Cantor sets*

We have discussed an infinite but countable system of material points subject to forces acting on each of them. We shall now outline some models of mass distributions which combine a discrete character with certain properties of a continuum. One way to establish a rather simple distribution of this sort would be through the following process: Imagine a point with a mass equal to unity, located in the middle of the interval (0, 1). This mass point can now, either, with a probability p_1, remain forever in its original position, or, with a probability $p_2 = 1 - p_1$, split into two parts, each of mass $\frac{1}{2}$, which will be located in the positions $\frac{1}{4}$ and $\frac{3}{4}$, respectively. If the latter eventuality has occurred, we shall assume again that each of the two masses can, independently, just as before, either stay what it is, or each can independently split into two masses (equal $\frac{1}{4}$ each) which will be located at $\frac{1}{8}$ and $\frac{3}{8}$ for the first point, or $\frac{5}{8}$ and $\frac{7}{8}$ for the second point. This process is to continue indefinitely. We imagine that each of the points can split into two equal ones which will then be located to the left and right of it, with a probability p_2, or stay "dead" forever. We have thus a branching or multiplicative process which will define a possible distribution of masses on dyadically rational points of the interval. The process is defined by the two constants p_1, p_2. If p_2 should equal 1 and $p_1 = 0$, we would have a certainty of splitting every time and the process would lead to all rational binaries, each having, in the limit, mass zero. The closure of our set would be the full continuum of real numbers between zero and 1. If p_2 is less than 1, we shall get as a result of our continuing branching process a countable set of points whose closure will be, *with probability* 1, a nowhere dense set consisting of some isolated points and a perfect nowhere dense set of the kind defined by Cantor. There are three cases to distinguish, a subcritical, just critical, and supercritical system. In our simple set-up, these correspond to the cases $p_2 < \frac{1}{2}, = \frac{1}{2}, > \frac{1}{2}$, respectively. In the last case there is a finite probability for the process never ending and as a result, in addition to having a finite collection of points which are isolated, we shall obtain a perfect set on the interval as a closure of the "unending" part of the process. We have to make the sense in which we speak of a result "of one such process" more precise. What is meant, of course, is that one considers all possible outcomes of such a branching process. There exists a measure in the space of all possible branching processes defined in a rather natural way (cf. Everett, Ulam [3]). When we speak of the process leading with probability 1 to a set with given properties, we mean that the subset of the set of all processes with these properties has measure 1 in the space of all possible processes.

One can look upon the sets of points obtained by the above construction as describing "virtual positions of a physical object," or consider the space itself as being a collection of virtual symbols generated in such a way. These then will not, in general, form Euclidean continua. Neither will they consist of a discrete set of points. It is obvious that an analogous procedure can be effected in spaces of dimensions higher than one. One could, for example, perform our branching independently in all three dimensions, or one could imagine the following single process: A particle with mass 1, located at the center of a unit sphere splits with probability p_2 into two particles, each with mass $\frac{1}{2}$ and located on the opposite ends of an interval with length α_1. The direction of this interval can be obtained by a random process, say with isotropic distribution in space. Each of the ends with mass equal $\frac{1}{2}$ can again be subject to the same splitting possibility, say, with the same probability p_2 independently, and then split into two particles, each with mass equal $\frac{1}{4}$, located at the ends of an isotropically chosen interval of length α_2. If the process continues indefinitely, we shall obtain a three-dimensional analogue of the sets above. (It might be of interest to add here that it is by no means easy to determine the topological character of the resulting set. It is known since Antoine [1] that some perfect nowhere dense sets in 3-space are equivalent to sets located in one dimension under a homeomorphism of the whole three-dimensional space and others are not. The question which of the two is more likely under our process is not immediately answered.)

The above special way of constructing a space of symbols to correspond to a model of a physical situation is perhaps the simplest one of its kind and, so to say, applies to the "configuration space" of a particle. An analogous construction could be thought of as proceeding in the phase space. Not merely the positions, but the momenta or velocities of a particle could be generated by a process like the above, leading to a Cantor set structure of all "possible" values of the physical quantities in which we are interested. In the sequel we restrict ourselves to the configuration space alone. In addition, we shall confine ourselves here to classical, that is to say, nonrelativistic and nonquantum theoretical features of such models. Any pretended attempt to take the implications of such constructions for physical models more seriously would, of course, have to involve such constructions in Minkowski-Lorentz spaces.

We mentioned above some of the obvious properties of sets obtained by our branching process, e.g., that the sets will be, with probability 1, nowhere dense. To obtain more precise information about the nature of these sets seems difficult. Again exploratory numerical work seemed of some value and a series of computations on an electronic machine was undertaken as follows. Starting with the original point in the middle of the interval, the process was continued by the use of random numbers. That is to say, a great number of finite sets was determined, the process being stopped each time after a certain number of "generations." Given the probabilities p_1, p_2, one may obtain a number N of sets, produced by the splitting process, sufficient for a statistical study. If the number of generations k for the splitting process is kept constant, we shall obtain a variable number of points in the set. One can then compute the average value of any given functional of such a set. (The integration in the space of all possible branching processes is replaced by averaging on the N sets actually produced. One can easily justify such averaging as an approximation to integration for various functionals). The functionals studied were as follows.

1. Given a set of mass points at generation time k, one can compute its moment on inertia I. (The center of gravity of each system, as is obvious from its definition, is located in the original position at $\frac{1}{2}$, since our splitting preserves the center of mass.) If we average the value of the moment of inertia over all N sets which we have manufactured, we obtain an approximate value of the average moment of inertia of the infinite set.

2. Imagine that any two points of our set attract each other with a force proportional to the masses and inversely to the square of the distance. There arises the question: What is the value of the "gravitational self-energy"? Here again, if we compute this quantity for each of our N sets and take the average, we shall get an approximation to the average or expected value of this quantity in our infinite process. This value exists in the three-dimensional case.

In addition to obtaining the average, one can get a fair idea of the distribution of the value of this self-energy. Of course, in order to ascertain it with any accuracy, a very great number N of sets would have to be manufactured.

3. One could also ask about the mutual attraction of two systems of the above sort with n and m points located on an interval of length 1, but separated from each other, again assuming that any element of one set attracts any element of the other set with a force proportional to the product of the masses and inversely with the square of the distance.

The motivation for a calculation of quantities like self-energy is that it is *a priori* clear that the values of these will be less than the values for corresponding Euclidean continuum models and may have finite values in certain systems of our sort when these latter diverge. One may try to calculate a mass splitting process more general than the one outlined above. Namely, the probabilities for splitting which we have assumed to be constant from generation to generation and from point to point could be assumed to depend on the existence of another object of the above sort in proximity to our given system (cf. Chapter VIII Section 7). Indeed, if we assumed that a function $V(x)$ is given and the probability of multiplication is proportional to this function, we obtain a distribution u as a result of our splitting process which will obey an equation of the type of the Schrödinger equation

$$\Delta u + (E - V)u = 0$$

The process, as defined above, has the unsatisfactory feature that it still leaves mathematical points with finite masses. If one wants to insist on all mathematical points having masses zero, the following iteration of our procedure should be considered: Let L_ω denote the first passage to the limit of our process as defined above. We shall now iterate it in this fashion: each of the points which remain with finite mass will again be subject to a splitting procedure, say with the same probabilities, into two equal masses but this time located on the opposite ends of an interval shorter than the one in the first process by a fixed ratio R. For example, let $R = \frac{1}{100}$, then if it happened that the first mass point located at $\frac{1}{2}$ has not split during the first process, it will have a probability p_2 of splitting into two masses equal to $\frac{1}{2}$, but located at positions $(\frac{1}{2}) - (\frac{1}{100}) \cdot (\frac{1}{4})$ and $(\frac{1}{2}) + (\frac{1}{100}) \cdot (\frac{1}{4})$. Again, for example, if the point located at $\frac{3}{4}$ has not split in the first process, let it have probability p_2 of splitting into two masses each equal to $\frac{1}{4}$, and located at $(\frac{3}{4}) - (\frac{1}{100}) \cdot (\frac{1}{8})$ and $(\frac{3}{4}) + (\frac{1}{100}) \cdot (\frac{1}{8})$. If this continued, we should have a second limiting set $L_{2\omega}$. If, at the end of this procedure, some points still have finite masses, we continue, assuming now another still smaller ratio R for the splitting distance and repeat this splitting, obtaining $L_{3\omega}$ and so on. Now with probability 1, a sequence of these processes will lead to a Cantor-set of points so that at ordinal ω^2 all points will have mass zero and we shall obtain a distribution of density without finite values at any point.

Another generalization seems strongly suggested. In the processes described, the objects which are obtained by our branching are still algebraically of the nature of real numbers. It would seem that this is too special. The formalism of the new quantum theory would suggest considering each of the points "manufactured" as having spinlike properties. That is to say, there should be several *kinds*, $x_1, x_2, \ldots, x_n$ of these points. This is, in fact, already the case for the branching process as we described it in three dimensions.

In each of the processes of the above type one may consider a measure in the space of all their possible outcomes. This measure is defined in a natural way by defining it first for the set of all possible outcomes which have a specified appearance up to the kth generation for $k = 1, 2, 3, \ldots$. These special sets correspond to elementary intervals in the Lebesgue measure on the interval. Their measure is the probability of the specified special set of occurrences up to the kth generation. One may then extend it, in the usual way, to all sets of the Borel field over these elementary sets. This construction has been discussed for our simplest process L_ω and for an n-dimensional process, that is to say, particles of n types. Can it be generalized to our process of the type L_{ω^2}? Our set of all possible outcomes is, after closure, a continuum in which its own measure can be defined in a natural way. This allows one to integrate functionals of such a set: for example, of the points attracting each other according to a given law, one may define gravitational self-energy, etc.; then one can integrate the value of such a function of the set of all possible genealogies or outcomes in the sense of measure in this space. This way one obtains an average or expected value of such a quantity.

We repeat that mathematical studies of models of the kind suggested above may perhaps be of interest since the development of physical theories during the last few decades suggests a possibility of a continuing process of "atomization." The alternating vogues for a "field theory" and for the "elementary particles" points of view present, at a given time, either a topologically Euclidean continuum, the primitive mathematical entities being functions of continuous variables, or else a notion of fundamental particles the "insides" of which are not further analyzed. The interpretation of these ultimately small units of space evolve through stages: the atom becomes a nucleus surrounded by electrons, the nucleus in turn exhibits its inner components of nucleons, while at the present moment the protons and neutrons may be losing their right to a status of "particles" by exhibiting a definite substructure. All this, so to say, in the small — whereas, in the direction of the distribution of the physical universe in the large, such iterative processes also seems to exist: stars appear in clusters, the clusters surround galaxies. There exist clusters of galaxies, i.e., super-galaxies and perhaps one might see an indication of an infinite hierarchy at the other end of the scale.

One might say that neither a quantization of fields nor a relativistic quantization of space-time arrests this tendency of the successive models of physical reality to replace elementary particles by systems of more elementary ones.

It may therefore be interesting to imagine such processes continuing infinitely and in particular to consider the cases where the subdivision of mass (or energy) goes on forever without leading necessarily in the limit to a real number system or a Euclidean continuum in which the field is defined. On the contrary, in general these limits will lead to schemas of mass distribution which will be Cantor-set like and have the topology of the p-adic rather than the real numbers. One has to repeat that models which could even remotely claim a physical interest would have to be constructed in space-time rather than in the ordinary space and in the phase space and not in configuration space alone. It seems, also, that a quantum theoretical viewpoint would have to be adopted implying in particular that the physical interaction between two elements at a given stage of the process is of a "shorter range" than that between elements of different stages.

One might note that such a geometrization of physics, if it

ever takes place, would not consist of merely generalizing the Euclidean character of the distance in the large which is the content of Riemannian geometry. Since physical phenomena seem to radically change their properties in the small, it is clear that a differentiable metric will never bring out this feature. A more radical change, even that of local topology, as indicated above, would be necessary. The increasing recent evidence for the frequency of transmutation in properties of what at each stage was considered an elementary particle, and the multiplicative character of the phenomena in the small may make it amusing, at least for a mathematician, to consider models of this sort.

6. *Dynamical flow in phase space*

A dynamical system of n mass points may be represented, in the well-known fashion, by a single point in the $6n$-dimensional space. The totality of all possible initial positions and momenta will define a set of points in the $6n$-dimensional space — i.e., in the phase space — of the generalized coordinates and momenta. The change in time of the positions and velocities of the system defines then a measure-preserving flow of the phase space into itself. Mathematical work during the last few decades has brought a rigorous description of some properties of such transformations; in particular the ergodic theorems of von Neumann and G. D. Birkhoff (cf. Hopf [1], Khinchin [1]) provide a rigorous mathematical foundation for the ideas of statistical mechanics. One also knows that, among all possible measure-preserving continuous flows of a manifold into itself, the ergodic ones, i.e., those which are metrically transitive, form in a certain sense the general case (J. C. Oxtoby and S. Ulam [1]). These results show that almost all continuous measure-preserving flows possess the property that the time averages are equal to space averages, but this has not yet been proved in the general case of actual dynamical flows, that is to say, flows defined by differential equations with given Hamiltonians. In addition, there is a need for further information still of a qualitative kind about general properties of dynamical flows in phase space. One such property, first defined and studied by Poincaré, is that of "mixing". This involves the following property, broadly speaking: given any region A in phase space and any other region B, after a sufficiently long time the measure of the image of the region A which is contained in the region B will have approximately the value $m(A)$ multiplied by $m(B)$.

A quantity of great physical interest is the *rate* at which mixing proceeds. This problem is, of course, of interest in an actual hydrodynamical flow, of a fluid in three-dimensional space. Imagine that the initial appearance of the flow is quite regular, say almost laminar, but with a small irregularity superimposed initially as a perturbation. As time goes on, the motion may become more and more irregular and after sufficiently long time very complicated and turbulent.

If one should consider the actual fluid approximated by a large number k of points and their phase space, then the region B of the phase space whose points correspond to highly irregular velocity distributions of the given three-dimensional fluid undoubtedly occupies a very large proportion of the total volume of the entire phase space. It is important for the theory of turbulence to know something about the rate at which the set A of points, each of which corresponds to smooth or regular velocity distributions of the actual fluid tends to penetrate into the much larger region B of the phase space corresponding to turbulent motions. This question is of course a special case of the general problem: how to estimate for a dynamical system the rate of transition in the flow of phase space from one region into another.

The mathematical treatment of ergodic properties of mechanical systems is largely measure-theoretic. The definition of mixing is also of that nature. It would be interesting, in order to provide an additional description of the behavior of such systems, to introduce *metric* notions. The notion of measure is natural in phase space — Liouville's theorem refers to such measures established for any Lagrangian system of coordinates. The topology of the phase space is also given quite naturally by these coordinates. When it comes to defining a metric, that is to say distance between any two points of phase space, no such unique definition seems apparent. In order to treat both the coordinates and the momenta on a comparable basis one must reduce, say, the latter dimensionally to quantities in units of length since distance will involve additively both coordinates and momenta. If this were done, then the notion of mixing could be investigated, in addition to its measure-theoretic behavior, from a more geometric point of view. One could, for example, demand that, for a general flow, two fixed points at a given initial distance traverse space in such a manner that their distance as a function of time, averaged in time, tends to the average distance between *any* two points of phase space. For the latter to be defined, the phase space has to be bounded (compact). More strongly yet, one could define a "metric mixing" by requiring the analogue of the above property for turbulence on k-tuplets of points.*

Another illustration of the type of problem involving the estimation of the rates is the following: Consider a problem of three attracting bodies, say with equal masses, and with given total energy — the initial conditions being such that the kinetic energy is roughly equal to the potential energy of the system. For certain special initial conditions, the three points may, for all time, remain in a bounded portion of the configuration space. With other initial conditions one of the points may, after a certain time, start escaping from the remaining two (in the sense, for example, that its distance from the center of mass of the other two points will increase without limit). The first question is, given

* An analogue of the above property of mixing has been investigated in the finite case: Instead of a continuous space, we have a finite set of points. One may consider a transformation, i.e., a permutation of such a set of points. One may then consider, e.g., the average distance in the positions of two particular points. For a random permutation this turns out to be asymptotically equal to $n/3$ when n is the number of points in the space, for the obvious distance $|i - j|$ between any two points i, j of the set. The notion of a metric in a phase space would be useful for "geometrization" of various other physical properties of the flow. A useful metric will have to depend on the particular Hamiltonian describing the given system.

the total energy of the system, what is the volume of phase space corresponding to conditions which will guarantee its boundedness in space for all time. In case of unstable systems, what is their average lifetime, that is to say, time after which the escape in the above sense will take place? In the language of the phase space one is interested in the rate at which the volume, occupied by points corresponding to bounded configurations, goes into the volume corresponding to systems in which one of the three bodies is escaping from the other two.

Since hardly any quantitative work exists on problems of rates in the above sense, numerical work on computing machines may be of heuristic value in suggesting the kind of theorems which one might try to prove. The usual arguments on "relaxation" times are based on the size of volumes in the phase space alone and are in general lacking in rigor. In the last chapter we shall discuss some heuristic possibilities now open through computations on electronic machines

7. Some problems on electromagnetic fields

The mathematical investigations of Poincaré, Birkhoff, and their followers, of general qualitative and asymptotic properties of motions of dynamical systems and the corresponding ergodic theories are now in need of generalization to systems of infinitely many degrees of freedom. Recent developments in the new field of magneto-hydrodynamics — a study of the motion and behavior of ionized matter in electromagnetic fields — stimulate corresponding investigations on a more complicated topological level. In particular, a theory is needed to describe the topological properties of families of magnetic lines of force in space; in the most elementary case, these may be imagined as due to steady electric currents on a fixed system of curves (wires).

There are interesting problems arising in this connection. To start with the simplest: consider a single closed curve and a steady electric current stowing through it. Does there exist such a curve with at least one single line of magnetic force g dense in some region of space? (It is easy to find systems of wires such that magnetic lines will be, in general, dense on surfaces; statements sometimes made to the effect that a magnetic line is either closed or else goes from infinity to infinity are obviously incorrect. For example, if the currents are on the two curves: the z-axis and the circumference of the unit circle in the x, y plane, the lines of force will be dense on tori, except for a countable set of surfaces.)

It would be interesting to describe the system of magnetic lines of force due to a current flowing on a knotted (infinitely thin) wire. In particular, suppose the current flows through a "clover-leaf" knot. Does the system of lines of magnetic force in space surrounding the knot reflect topologically the "knottedness" of the curve? Such systems of curves may exhibit considerable topological complexity even when generated by currents flowing on straight lines, as shown by calculations on the properties of lines of force due to currents flowing on the three straight lines $x = 1$, $y = 0$; $y = 1$, $z = 0$; $z = 1$, $y = 1$.

It would be of interest to know the ergodic properties of a line of force (ergodic with respect to a particular parametrization); e.g., the behavior of the sequence of points one unit of length apart on a line of force. For example, are such sequences uniformly dense in some regions of space?

Answers to questions of this sort would, of course, give merely preliminary material towards a theory which should generalize the existing work on dynamical systems of n particles. Clearly, systems involving infinitely many degrees of freedom lead to consideration of phase spaces of infinitely many dimensions. In the absence of a definition of invariant measure (invariant under the flow defined by the Hamiltonians of the problem) the first attempts would be to approximate the infinitely many variables by systems with a finite number and then attempt a passage to the limit for the functionals in question. The additional difficulties due to the fact that Hamiltonians involve forces dependent on velocities (e.g., Lorentz forces) require, in the finite approximation, a study of nonholonomic systems.

8. Nonlinear problems

In some previous sections we have mentioned a few questions referring to nonlinear functional equations. A short discussion will be given below of a specific problem of some physical interest. This problem was investigated through computations on an electronic machine and was meant to be the first of a systematic sequence of such problems, increasing in complexity, involving nonlinear equations. This work was planned by E. Fermi, J. Pasta, and the writer [1] — a preliminary account of the motivation for this program and the part of it already performed will be given below.

The first problem involved the case where the nonlinear terms were small compared to the main, linear ones and could be treated as perturbations for the initial range of the parameters. The first such system studied was that of a vibrating string with fixed ends, with forces between elements which contained in addition to the usual elastic linear terms, nonlinear ones, for example, terms quadratic in the displacements. The linear problem has the well-known periodic solutions; the presence of the additional terms provides, in time, a "mixing" of the states which the linear system would possess. The plan was to follow the motion for a long time and to obtain the *speed* with which such a system attained a statistical equilibrium, a state in which modes of vibration of all kinds are present. The calculations involved an approximation to the continuous string, by replacing it with a finite number of particles, and were performed during the summer of 1953 on an electronic computer in Los Alamos (the "Maniac," one of the first such machines built.)

This problem was to serve as the very simplest one. The ultimate aim was to discuss problems with more independent variables in the hope of obtaining material which would suggest some general features of behavior of systems with an infinite number of degrees of freedom, with nonlinear interaction terms as they occur between the oscillators in quantum theory or between the degrees of freedom in an electromagnetic field or a meson field. The mathematical possibilities concerning what might be called *quasi-states* (in the sense of Chapter VI, Section 3) were discussed with Fermi in this connection.

For the numerical work the continuum is replaced by a finite number of points (at most 64 in our actual computation) so that the partial differential equations defining the motion are replaced by a finite number of total differential equations. We have, therefore, a dynamical system of 64 particles. If x_i denotes the displacement of the ith point from its original position and α denotes the coefficient of a quadratic term in the force between the neighboring mass points (β that of the cubic term in other problems), the equations were

$$(1) \quad \ddot{x}_i = (x_{i+1} + x_{i-1} - 2x_i) + \alpha[(x_{i+1} - x_i)^2 - (x_i - x_{i-1})^2], \qquad i = 1, 2, \ldots, 64$$

or

$$(2) \quad \ddot{x}_i = (x_{i+1} + x_{i-1} - 2x_i) + \beta[(x_{i+1} - x_i)^3 - (x_i - x_{i-1})^3], \qquad i = 1, 2, \ldots, 64$$

The coefficients α and β were chosen so that, even at the maximum displacement time, the nonlinear part of the force was small compared to the linear term (e.g., of the order of one tenth of it). If we let the number of particles become infinite in the limit, we would obtain a partial differential equation containing, in addition to the terms in the usual wave equation, nonlinear terms of a complicated nature.

Another case studied later was:

$$\ddot{x}_i = A(x_{i+1} - x_i) + B(x_i - x_{i-1}) + C$$

where the parameters A, B, C are not constant but assumed different values depending on whether or not the quantities in parentheses were less than or greater than a certain value fixed in advance. This prescription amounts to assuming the force as a broken linear function of the displacement. This broken linear function may imitate to some extent a cubic dependence.

The solution of the corresponding linear problem is a periodic vibration of the string. If the initial position of the string is, say, a single sine wave, the string will oscillate in this mode indefinitely.

Starting with the string in a simple configuration, for example, in the first mode (or in other problems, starting with a combination of a few low modes), the purpose of our computations was to see how, due to nonlinear forces perturbing the periodic linear solution, the string would assume more and more complicated shapes, and, for t tending to infinity, how the total energy of the string would come to be distributed in all the Fourier modes. In order to observe this, the shape of the string, that is to say, x as a function of i and the total potential plus kinetic energy were analyzed periodically in Fourier series. This amounts to a Lagrangian change of variables: instead of the original $\dot{x}_i$ and x_i, $i = 1, 2, \ldots, 64$, we may introduce $\dot{a}_k$ and a_k, $k = 1, \ldots, 64$, where

$$a_k = \sum x_i \sin \frac{ik\pi}{64} \tag{4}$$

The sum of kinetic and potential energies in the problem with a quadratic force is

$$E_{\text{kin}} + E_{\text{pot}} = \tfrac{1}{2} \sum_i (\dot{x}_i^2 + (x_{i+1} - x_i)^2 + (x_i - x_{i-1})^2)$$

or

$$E_{\text{kin}} + E_{\text{pot}} = \sum_k \tfrac{1}{2} \dot{a}_k^2 + 2a_k^2 \sin^2 \frac{\pi k}{128} \tag{5}$$

if we neglect the contributions to potential energy from the quadratic terms in the force which amounted in our case to at most a few per cent of the total.

In each problem reported here, the system was started with zero velocities at time $t = 0$. The length of a computational time cycle used varied somewhat from problem to problem. What corresponded in the linear problem to a full simple period of the motion was divided into a large number of time cycles (up to 500) in the computation. This is necessary for accuracy. Each problem ran through many "would-be periods" of the linear problem, so the number of time cycles in each computation ran to many thousands. That is to say, the number of swings of the string was of the order of several hundred, if by a swing we understand the period of the initial configuration in the corresponding linear problem. The distribution of energy in the Fourier modes (5) was noted after every few hundred of the computation cycles. The accuracy of the numerical work was checked by the constancy of the quantity representing the total energy. In some cases, for checking purposes, the corresponding linear problems were run, and these behaved correctly within one per cent or so, even after 10,000 or more cycles.

The calculation of the motion was performed in the x_i variables; after every few hundred cycles the quantities referring to the a_k variables were computed by the above formulae. It should be noted here that the calculation of the motion could be performed directly in the a_k and $\dot{a}_k$. The formulae, however, become unwieldy and the computation would take longer time. The computation in the a_k variables could have been more instructive for the purpose of observing directly the interaction between the a_k's.

One should now say here that the results of our computations showed features which were, from the beginning, surprising. Instead of a gradual, continuous flow of energy from the first mode to the higher modes, all of the problems show an entirely different behavior. Starting in one problem with a quadratic force and a pure sine wave as the initial position of the string, we indeed observed initially a gradual increase of energy in the higher modes as predicted e.g., by Rayleigh in a perturbation analysis. In the first problem, mode 2 starts increasing first, followed by mode 3, and so on. Later on, however, this gradual sharing of energy among successive modes ceases. Instead, it is one *or* the other mode that predominates. For example, mode 2 decides, as it were, to increase rather rapidly at the cost of all other modes and becomes predominant. At one time, it has more energy than all the others put together! Then mode 3 undertakes this role. It is only the first few modes which exchange energy among themselves, and they do this in a rather regular fashion. Finally, at a later time, mode 1 comes back to within one per cent of its initial value so that the system seems to be almost periodic. All our problems have at least this one feature in common. Instead of a gradual increase of all the higher modes, the energy is exchanged, essentially, among only a certain few. It is, therefore, very hard to observe the rate of "thermalization" or mixing in our problem which was the initial purpose of the calculation.

Looking at the problem from the point of view of statistical mechanics, the situation could be described as follows: the phase space required by a point representing our entire system has a great number of dimensions. Only a very small part of its volume is represented by the regions where only one or a few out of all possible Fourier modes have divided among themselves the bulk of the total energy. If our system, with nonlinear forces acting between the neighboring points, should behave as a good example of a transformation of the phase space which is ergodic or metrically transitive, then the trajectory of almost every point should be everywhere dense in the whole phase space. With overwhelming probability this should also be true of the point which at time $t = 0$ represents our initial configuration, and this point should spend most of its time in regions corresponding to the equipartition of energy among various degrees of freedom. From the results, this seems hardly the case. The ergodic sojourn times in certain subsets of the phase space may show a tendency to approach limits as guaranteed by the ergodic theorem. These limits, however, do not seem to correspond to equipartition even in the time average. Certainly, there seems to be very little, if any, tendency towards equipartition of energy among all degrees of freedom at any given time. In other words, these systems certainly do not show much mixing. Rather, the behavior of such physical systems suggests the existence of "quasi-states" in quadratic problems, discussed mathematically in Chapter VI, section 3.

CHAPTER VIII

Computing Machines as a Heuristic Aid

1. Introduction

In recent years a considerable amount of interesting heuristic mathematical work has been made possible by the development of fast electronic computing machines. Much more of this, primarily in mathematical physics, but also in combinatorial analysis and number theory can be expected in the near future. A knowledge of the behavior of special cases has always been useful and a background of experimental results has played an important role, notably in number theory; this was emphasized by Gauss himself. Means of greatly enlarging this background are now within reach and at a cost in time spent in such experimentations which has been enormously reduced in comparison with hand calculations. In the following sections we shall give a few examples of work already performed together with a rather arbitrary illustrative selection of possible future studies.

It might be thought that the interesting questions of combinatorial analysis lead immediately to numbers so large that memory limitations prohibit a useful study of such problems even on the present fast computing machines. Such is not always the case. Some problems have this general character: the number of elementary operations (e.g., additions and multiplications) may indeed be very large, i.e., of the order of many tens of millions, but the "memory" involved is only moderate, of the orders of hundreds to tens of thousands of positions. Indeed the rather widespread "fear" of enormously large numbers appearing in the study of combinatorial problems is often a misapprehension. While it is true that the fundamental combinatorial functions like 2^n or $n!$ increase rapidly with n, it seems that in many cases the formulae expressing the asymptotic behavior of certain interesting combinatorial functions are not only suggested but already rather well illustrated for small n. For example, the density of primes less than ten thousand gives a fair picture of the asymptotic density. Speaking loosely, many asymptotic formulae become valid within a small error for moderate values of n.

In such cases this may be due to a combinatorial fact which we shall vaguely describe as follows: Let Z be a subset of the set of all integers such that in its definition at most l operations of Boolean algebra and elementary arithmetical operations are used; the number of quantifiers and the operations of formation of direct product used in the definition is p. Let $f(N)$ be the number of integers in Z not exceeding N and suppose that $r = \lim_{n\to\infty} f(N)/N$ exists. In certain cases, given $\varepsilon > 0$, an estimate on $N(\varepsilon)$ could perhaps be given such that $|(f(N)/N)-r| < \varepsilon$ for $N > N(\varepsilon)$, *a priori* in terms of ε, l, and p. This will depend most sensitively upon p. Such an estimate cannot be given *a priori,* valid for *all* general recursive functions. This follows from the results of Gödel. In many practical cases, however, when p is small, one might expect some result like $N(\varepsilon) \sim (l^p/\varepsilon)$.

2. Some combinatorial examples

We shall mention at first a few questions of enumeration.

Given the set E of integers from 1 to n and two permutations S_1 and S_2 of this set selected at random, what is the expected number σ of elements in the group generated by these two permutations?

Analogously, let T_1 and T_2 be two random transformations of the set E into itself; what is the expected number τ of elements in the generated semi-group? Is $\sigma > \tau$?

These numbers of course do not exceed $n!$ and n^n, respectively. What is the relation between the sizes of τ and σ? If we postulate three or more permutations and the same number of transformations, does the relation between τ and σ reverse?

In any case, one can investigate such questions empirically for small n, on the machine. This can be done by using only a moderate number, say 100, of randomly selected pairs of permutations and even though 100 is very small compared to $(n!)^2$ even in cases where n equals, say, 7; the average values of σ and τ could be approximated well with a good probability from such sampling.

To give another example of such a sampling or "Monte Carlo" approach to combinatorial problems, one could mention the problem of the walk of the knight on the chess board. This consists of finding a succession of moves of the knight which would cover each position on the board exactly once. Euler found many solutions, but no general procedure for generating such walks is known even for the case of the ordinary 8×8 board. Certain practical recipes, e.g., the one (given by Euler) of moving into places which are already almost surrounded, seem to be successful. It is quite obvious that a computing machine can perform attempts to find solutions by trial and error even embodying rules like the above and also more complicated ones which involve retracting the last $1, 2, \ldots, k$ moves in cases when blind alleys develop. The memory requirements in such a problem are small and it would appear that the advantage of the speed of the machine is considerable in situations of this sort where the advantage of a visual survey available to the human brain is very small. The number of attempted walks of the knight which can be performed on the machine in the course of a few hours is of the order of 10^6. A result of such statistics would be the proportion of successes in trials involving given rules. M. Wells has obtained results on the Los Alamos computing machines for this problem, comparing the 8×8 case with other $n \times n$ cases.

Heuristic results can be obtained on combinatorial aspects of an algebra of quantifiers in models of projective algebras (cf. Chapter I, Section 5). If E is an infinite set, it is known that starting with two sets A and B in $E \times E$ one can obtain by the operations of projective algebra an infinity of new sets. If E is finite, consisting of n points, the question arises of finding a basis for the class of all subsets of $E \times E$. Of course the n^2 individual points form such a basis. But what is the minimal number of sets of a basis? The projective algebras generated by two fixed randomly selected subsets A and B of $E \times E$ are of some interest. What is the expected number of elements in the projective algebra thus generated? This could be investigated on a computing machine for all small n, perhaps up to 7 or so. The obvious symmetry and conditions of the problem reduce the number of cases to be investigated from $(2^{(n^2)})^2$ to something like 2^{2n} for small values of n.

Still another example: We have mentioned before the problem of approximating a $1-1$ transformation T: $x' = f(x, y), y' = g(x, y)$ of a product space into itself by a composition of transformations of the two types U: *(1)* $x' = x$, $y' = k(x, y)$ and *(2)* $x' = y$, $y' = x$. The finite analogue would be to have the space E consist of a small number n of points, say $n = 8$ and try to obtain all transformations of the space $E \times E$ by composition of the transformations of the above type. A similar question for the three-dimensional space, i.e., E^3, involving transformations of the type indicated in Chapter 4, Section 2a could be settled by work on computing machines for small values of n.

The combinatorial illustrative lemma of Chapter 2, Secton 1 has been verified by P. Kelly for $n \leqq 5$ by an examination of all cases. Probably a fast computer could examine all cases for $n = 8$ in a relatively short time.

The question whether the isometry of two spaces A^2 and B^2 implies the isometry of A and B where the latter are finite sets, with a metric such that the distance assumes only the values 1 or 2, has also been studied by Kelly [1] who verified it for small n. This problem also lends itself to investigation for somewhat larger n by the electronic computer.

3. Some experiments on finite games

As is well known, a game between two players can be considered as taking place according to the following schema: A combinatorial graph is defined starting with one vertex and branching successively into a set of segments. That is to say, we have n_1 segments issuing from the initial point corresponding to the choice of any one of n_1 possible moves by the first player. Then from the end of *each* of these segments we may assume there exist n_2 new segments (without loss of generality we may assume that this number is the same at the end of every initial segment). The choice among these is left to the second player. After that we have n_3 vertices issuing from the ends of these, and so on. In a finite game the game terminates after k moves. In the set of $N = n_1 \cdot n_2 \ldots n_k$ final vertices a subset W of these is defined as the winning positions. There might be given another subset D of points corresponding to the drawn games. Most of the games like checkers, chess, etc. can be thus formulated, although one should remember that what we consider as a position of the game may be a rather complicated structure involving the sequence of all previous positions, etc.

A winning strategy (for the first player) means the existence of an integer i_1 so that for every choice of i_2 there will exist a choice i_3 so that for every choice of i_4 and so on, the element constructed belongs to the set W. On the computing machines in Los Alamos some extremely simple cases of games of this sort were studied (P. Stein and S. Ulam [1], and W. Walden [1]). These experiments concerned comparisons between players of different strength, e.g., the computation of the winning chances for a player employing a perfect strategy versus one who "sees" only a limited number of moves ahead, when the set W was given at random.

Recently the writer has tried to find similar game formulations in some other mathematical situations. So, for example, the two players A and B, instead of selecting intervals as in the example above, could construct together, in turn, two permutations of the set of N integers. A selects n_1, B selects n_2, etc., up to n_N, obtaining a permutation E_1. After that they similarly construct a permutation E_2. If E_1 and E_2 generate the group S_N of all permutations on N integers it is the player A who wins, otherwise B wins.

To give a "topological" example, suppose a cube is subdivided into a great many smaller cubes and one of their vertices is given as a starting point. Two players, A and B, play as follows: The first player selects an edge of a small cube issuing from the starting point. The second player has to continue by selecting *another* edge whose beginning coincides with the end of the segment just given, otherwise arbitrary. Together the two players establish a path. This path must intersect itself at some time. If the closed curve thus obtained for the first time is knotted, let us say that the first player wins. In other words, A tries to close the curve so that it will have a knot in it; the second player tries to unwind the arc before it closes. Obviously some knowledge of topology is desired to plan a good strategy in this game! To study a game of this sort, preliminary explorations would be helpful on the computing machine which could try out tentative strategies.

The examples given are intended merely as arbitrary illustrations of the general possibility of converting special combinatorial studies to game situations. The desirable corresponding theorems would assert the existence of winning strategies. The writer found it rather amusing to consider how one can "gamize" various mathematical situations (or perhaps the verb should be "paizise" from the Greek word *παιζειν*, to play).

4. Lucky numbers

The sieve of Eratosthenes yields the primes among the natural integers. One can think of variations in the definitions of sieves. The following procedure was investigated numerically on the electronic computing machine in Los Alamos (Gardiner *et al.* [1]). In the sequence of all integers we strike out every second one, that is to say, all the even numbers. The first number remaining (apart from 1, which will not be counted) is 3. We shall now strike out every *third* integer, counting only the remaining ones, that is to say, this time we will strike out the integers 5, 11, 17, etc. In the remaining sequence the first number not used before is 7. Therefore we shall strike out every seventh number, counting among the remaining ones again. This will eliminate 19, etc. We proceed in this manner *ad infinitum.* The numbers that remain after the sequence of these procedures we might call, for example, lucky numbers. They are: 1, 3, 7, 9, 13, etc.

It turns out that many asymptotic properties of the prime number sequence are shared by the lucky numbers. Thus, for example, their asymptotic density is $1/ln(n)$. The numbers of twin primes and of twin luckies exhibit a remarkable similarity up to the integer $n = 100{,}000$, the range which we have investigated on the machine. The number of adjacent primes differing by 4 or by 6, 8, etc. is, in this range, very similar to the corresponding number of adjacent luckies. It also happens that within the range investigated every even number is a sum of two lucky numbers. The lucky numbers of the form $4n+1$ and $4n+3$ seem to be equally distributed, etc.

Another sieve would be a random one, that is to say, we retain the integer n with the probability $1/ln(n)$. In the class of sequences thus obtained one could try to prove theorems for almost every such sequence — statements about their distribution in detail such as representation of integers by sums of these, distribution of gaps between adjacent ones, etc. (See Erdös, Jabotinsky [1]; Chowla, Hawkins, forthcoming papers.)

It would perhaps not be without interest for number theory to randomize similarly procedures leading to primes in the quadratic fields. Analogously one could randomize the sequences of squares of integers or cubes of integers, etc., and consider the analogue of Waring's theorem. An experimental study on computing machines would quickly suggest statements likely to be true.

5. Remarks on computations in mathematical physics

The next few pages will deal with a few simple examples of physical problems which can be usefully studied through computations on modern machines. It is obvious that computational work on a large scale is immensely helpful in testing out theories on physical models; problems in general relativity, for example, lead to equations so complex that in most cases it will be numerical work which will test the validity of the models. Problems of constitution of the stars with assumed models of energy generation can be solved in special cases by such computations. One could name problems in atomic and molecular physics, etc., etc., where numerical checks of existing theories will be increasingly available. The mathematical complexity of theoretical physics has been steadily increasing during the last decades. This is so, not only because of its increased scope in embracing more complicated systems, but also because of changes in the very foundations of physics: the entities once treated as "points" (e.g., the molecules, the atoms, the nucleons themselves exhibit now a combinatorial substructure). In a number of problems one considers interactions between a considerable number of points (or fields with many degrees of freedom), that is, one deals with the n-body problem for sizeable values of n, but not large enough so that they can be handled by statistical or thermodynamical methods. There is little hope of obtaining solutions with the method of classical analysis alone, so that heuristic investigations involving an examination of many special cases seems indicated for the present.

The small selection of calculations described in the sequel is intended to illustrate some of the instances where understanding of the problems might be furthered by computation of special cases; the electronic computing machines enable one to make, as it were, "experiments in theory" by computations, either of the classical or of the "Monte Carlo" type computations (Section 8 contains a short account of the latter).

6. Examples from electromagnetism

Suppose a system of steady currents is given flowing through one or several curves fixed in space. One is interested in the field of the magnetic lines of force due to these curves. The calculation of these lines involves a numerical integration of a simple system of ordinary differential equations. The force field, at every point, is determined by curvilinear integrals (the law of Biot-Savart). We mentioned before some questions concerning the behavior in the large of such systems of curves — their topological and ergodic properties, for instance, in Chapter VII, Section 7, the problem of the field due to a simple cloverleaf knot. This can be realized for computational purposes by a space polygon with six sides. To begin with, simpler configurations were calculated on the electronic computer in Los Alamos. The first system studied was one given by currents flowing through two infinite straight lines, one the y-axis, the other wire parallel to the x-axis and intersecting the z-axis at the point $(0, 0, d)$. Starting with an arbitrary point in space, the magnetic line of force is traced out by computing the position of a sequence of points in the direction of the force at the preceding point. Taking the points sufficiently close and correcting for the changes in the field, one obtains an idea of the properties of the lines of force (tangent everywhere to the force vector). In order to obtain information about some qualitative properties of such lines in the large, the machine computed the number of times a line of force winds around each of the given straight lines. This is done by computing the Gauss integral. Some lines of force continue looping one of the wires; some others loop them both at once. Also the machine will compute the number of times a line of force crosses the surface of a given sphere. Other simple topological invariants, e.g., the Kronecker index of a mapping of directions onto a given sphere can be calculated and printed as a result of the computation. The next system involved three straight wires all skew to each other. These computations were considered as a preparation for the computation of a field due to the six-sided knotted polygon. The computation of such integral-valued invariants will not be affected by the accumulation of errors, unavoidably present in a numerical computation (see, e.g., a paper by Borsuk and the author [2] on ε-invariance of mappings). Roughly speaking, this is due to the fact that the computation may be so arranged that the field of error vectors is rotation free in the large.

7. *The Schrödinger equation*

The time independent Schrödinger equation

$$\Delta\psi + (E - V)\psi = 0$$

can be put into the form

$$\partial u/\partial t = \Delta u - Vu$$

through a substitution $u = \psi e^{-Et}$ by introducing a "time" parameter t. Both the lowest eigenvalue E and its characteristic function $\psi(x, y, z)$ can be obtained approximately by the following calculation: We imagine a population of fictitious point particles which multiply at the point x, y, z according to the given function $V(x, y, z)$ and which diffuse randomly through space, that is to say, perform a random walk as indicated in the equation by the Laplacian. The "game" played on the machine is then to process a great number of particles, each of which diffuses in space and at the same time multiplies its weight according to the given potential function. Processing a great number of the particles one obtains a population which asymptotically will tend to one possessing a density ψe^{Et}. The boundary conditions can be properly handled; the usual conditions at infinity are automatically fulfilled in such a model, and if they are prescribed on finite surfaces they may be satisfied by introducing suitable rules, e.g., absorption of particles on such surfaces. The procedure will give the lowest eigenvalue and its characteristic function. To obtain the successive eigenvalues and their characteristic functions becomes increasingly difficult by such methods. For example, to obtain the second characteristic function one would have to start with a population consisting of two kinds of particles ("black" and "red") corresponding to the positive and negative values of the function. One would start with a distribution orthogonal to the first eigenfunction (assuming it already determined) and then allow this population to undergo a diffusion-multiplication process with a suitable convention of mutual annihilation of the black and red particles whenever they appear at the same point (really a zone in space). The asymptotic population would then represent our second eigenfunction.

Such a statistical approach possesses an advantage in problems where a greater number of space dimensions is involved and the given potential function is complicated so that the usual methods of approximation become prohibitively lengthy. The main point, however, is as follows. One may only be interested in the value of certain given functionals $F_i(\psi)$, $i = 1, \ldots, k$, of the unknown solution ψ. In a classical approach one has to "know" ψ itself or at least know it at a sufficient number of points. Through a statistical approach like the above such functionals might be directly computed through sampling on those values of ψ which are obtained in the calculation. This is a general feature of the so-called Monte Carlo method, which we shall describe very briefly in the next section.

8. Monte Carlo methods

The sampling procedures in question might be said to consist in the "physical" production of models of combinatorial situations in mathematical problems, or of the playing of games with distributions of "particles" produced on computing machines, which distributions will constitute approximate solutions of physical problems.

Perhaps the simplest example is given by the problem of evaluation of a definite multiple integral. Suppose we want to find the volume of a region R contained in the unit cube in n dimensions defined by a given number of inequalities, that is to say, evaluating

$$\iint_{(R)} \cdots \int dx_1 \ldots dx_n ,$$

where the region R is defined by inequalities

$$\varphi_1(x_1 \ldots x_n) < 0,\ \varphi_2(x_1 \ldots x_n) < 0, \ldots, \varphi_k(x_1 \ldots x_n) < 0$$

The elementary numerical procedures would consist essentially of counting the lattice points of a subdivision of a cube in the space of n dimensions containing R and ascertaining the proportion of those satisfying the given inequalities. This is impossible in practice when n is moderately large, sometimes even when n is greater than 3, and the subdivision on each axis involves something like 100 points. The obvious statistical procedure would be instead to take a number of points *at random* in the n-dimensional cube, with uniform probability in this cube and count among those selected the points satisfying the given inequalities. Obviously for a precision of a per cent or so one needs to take only a number of points (of the order of 10^4) much smaller than the total number of all lattice points in a 100^n set. This is clear in case we know *a priori* that the volume of the region in question is not very small compared to that of the containing cube. In such a case special procedures are indicated. To illustrate them we shall take two examples of problems suggesting less simple-minded sampling.

Suppose one is interested in calculating the probability of a successful outcome of a solitaire (game of cards for one) where we assume that skill plays no role: a pure game of chance. In cases where the game has a very small probability of success most actual plays will end in failure and only an upper limit would be obtained for this probability. To get some idea of a lower limit > 0, suppose one obtains (still obeying the rules of the game) in a significant proportion α of tries, a situation where only rather few, let us say 10, cards remain to be played. It might be useful to try, in such cases, various permutations of the remaining 10 cards and continue the play from there. By examining a large number of such 10 factorial permutations one could obtain the number β expressing the chance that starting with one of the previous positions we obtain a win. A reasonable guess for the chance of success from the beginning without "cheating" would then be of the order of $\alpha \cdot \beta$. The idea is to decompose the problem into two or more stages in order to economize in the number of experiments performed for our sampling.

A similar problem is encountered in the transmission of particles (neutrons or gamma rays) traversing a series of materials serving as a shield for these particles. One is interested in cases where only an extremely small fraction of such particles which scatter at random and possibly produce other particles of this type, manage to traverse the whole shield. Again it is clear that in trying models of such random walks on the machine one could waste most of the experiments. The obvious thing to do is again to decompose the geometry of the shield or, more generally, the history of the process, into two or more stages, ascertain the proportion of traversals of the first stage; then starting with a typical situation after the traversal, try again the random walks from there on, etc.

An analogue of the Laplace-Liapounoff theorem is needed for estimates of errors in multiplicative and diffusive processes.

The computing machines are well suited for experiments of the above sort on the qualitative behavior of solutions of linear partial differential equations. Broadly speaking, one tries to change them to a form of integro-differential equation describing a diffusive and branching process, and uses the digital computer as an analogy machine as it were. Equations which are quadratic or of higher order in the unknown functions and their derivatives lead to a more complicated Monte Carlo procedure. As an example let us consider as a simple generalization of the Schrödinger equation discussed above, a bilinear system of two partial differential equations:

$$\partial u_1/\partial t = \alpha_1 \Delta u_1 + V_1(u_2)u_1,$$

$$\partial u_2/\partial t = \alpha_2 \Delta u_2 + V_2(u_1)u_2$$

u_1 and u_2 being unknown functions of the coordinates x, y, z and the time t; α_1, α_2, say, two given constants; V_1, V_2 given functions for simplicity linear in u_1 and u_2 and also involving the independent variables x, y, z. One would like to know the asymptotic form of u_1 and u_2 (for large values of t). This problem may be looked upon as a generalization of the diffusion model of the Schrödinger equation. It would correspond to a model of a system of two particles with the *potential* function depending only on x, y, z replaced by functions of the distributions. Such a system of linked particles presents, somewhat in the spirit of a field theory, a nonlinear problem. In general there will not exist eigenfunctions. The separation into a time independent system will not be possible. Instead one hopes for an almost periodical behavior, or at least that the unknown functions are summable by first means in time. One could again try a numerical approach in the study of such systems through a Monte Carlo procedure. In this case the fictitious particles whose density is to correspond to u_1 and u_2 are to diffuse and multiply, but this time not according to a given function V of coordinates but rather proportionally to a given function of the density of the particles of the other type. It is therefore necessary to operate with a frequent census of the particles of each type to ascertain the value of the "potential."

In practice an iterative procedure is necessary with the hope of converging to a self-consistent solution.

Prima facie sampling methods like the above work for problems involving distributions of nonnegative real numbers, specifically for Markoff chains involving iteration of matrices with nonnegative coefficients. Let us indicate, however, a way to study "experimentally" the behavior of matrices with arbitrary real terms, or even of more general ones. This possibility rests on the fact that real numbers may be considered as matrices with all terms nonnegative, by starting with the two matrices

$$\begin{pmatrix} 1 & 0 \\ 0 & 1 \end{pmatrix} \quad \text{and} \quad \begin{pmatrix} 0 & 1 \\ 1 & 0 \end{pmatrix}$$

representing the numbers $+1$ and -1, respectively. This correspondence obviously preserves both additions and multiplications. Any system described by an $n \times n$ matrix of real numbers can be interpreted probabilistically using $2n \times 2n$ matrices with nonnegative terms. The sampling would then involve two kinds of particles ("black" and "red") with the transformation rules given by the matrices above. Again in this case, however, a frequent census of the result is necessary since the final results depend on the *difference* between the numbers of black and red particles at each point. One can realize stochastic models for matrices with complex terms by having 4 kinds of particles, or more general algebras over real numbers through a suitable number of *kinds* of particles. The Dirac equation could be studied in such a way (cf. Everett, Ulam [5]).

9. Hydrodynamical problems

Numerical work is essential in many problems of hydrodynamics of compressible flow since at the present time one cannot obtain solutions in closed form. Analytic work alone is unable, in general, to throw light on the asymptotic behavior of solutions or even on their simpler general properties.

In problems involving more than one space dimension, one encounters serious difficulties in planning numerical work. The differential equations of Euler and Lagrange can be replaced, of course, by approximate systems of difference equations, the time variable, too, can be dealt with by a discrete succession of steps in time. The practical difficulties of executing such computations are considerable: if the problem involves large displacements of the fluid or gas (large compared to the initial dimensions of the system) and if the distortions of shape are sizeable, the initial network of Lagrangian points will change its structure in time and very often cease to provide an adequate basis for an approximate representation of the differential expressions which one has to compute (e.g., to evaluate density and pressure gradients). It would seem that the Eulerian system of fixed divisions of space would be more suitable and indeed in many problems this is so. However, other difficulties arise in this treatment; for example, if the fluid moves to vacuous regions — or in problems involving two or more fluids — it is very hard to obtain an idea about the position of the boundary between the two fluids. This is so since in the Eulerian treatment the dependent variables are density and the velocity and from these alone it is impossible to distinguish between a region which is completely occupied by a rarified fluid and one in which half of the region contains a fluid at normal density. In the Eulerian treatment there will be a fictitious computational diffusion from such regions and one loses sight of the position of the true boundaries.

In practice, in problems of this sort, one has to proceed in time through discrete periods called cycles such that the relative displacements in these times are small or moderate, then refine or change the spatial subdivision of the continuum and shorten the time step, a process which may not converge beyond a finite value of time.

Calculations can be considered that are of less "orthodox" variety. Broadly speaking, at least two different ways to approach numerically the problems dealing with the dynamical behavior of continua suggest themselves. In what follows a brief account will be given of some work in this direction, performed by J. Pasta and the author, on computing machines. This work was followed by additional computational experiments of this sort. F. Harlow [1], of Los Alamos Scientific Laboratory, is mainly responsible for the evolution of such methods.

One could try to base the hydrodynamical computations on the kinetic model of the gas or fluid. Assuming as a model of physical reality the Boltzmann integral differential equation, one may calculate the properties in the large of a motion of N "points" — atoms or molecules which have given velocity distributions. These are subject to fluctuations and yield the macroscopic motions only as statistical averages. It is well known, e.g., that the Navier-Stokes equations can be derived from this model. The question is whether it is computationally feasible to obtain the macroscopic quantities like density, pressure, and velocity as functions of time on this basis. The answer, it seems, is in the negative for the time being. If one takes the Boltzmann equations literally and considers the individual points of calculation as representing atoms, then to obtain the average velocity of a point of the gas one would require, say, k particles per cell in a space resolution in which we are interested. Let the number of cells in the computation be l. If on a linear scale the number of intervals is d, $l = d^r$, where r is the number of dimensions of the problem (for problems in 1, 2, or 3 space dimensions without special symmetries, $r = 1, 2, 3$), the total number of points in the calculation would be equal to kd^r. In a statistical study if one wants a mean error of the order of 1 per cent, then $k \sim 10^4$. If the linear resolution is to be of the order of, say, 5 per cent, then d is 20. Even in problems in one space dimension one would have to consider 200,000 points. This is much too high for present computers and makes it clear that the "points" of the calculation should be thought of as representing not individual atoms, but rather large aggregates of them. The dynamical behavior of such a point has to be schematized so as to represent a statistical average of a great number of atoms. The numerical work will not use the Boltzmann equations, but simpler equations which are its consequences. The implicit assumption in such a setup is that our conglomeration of atoms remains coherent during the entire course of the problem. In other words, the small globules of the fluid do not become too much distended and distorted. One should now define the density ρ of the fluid, and the pressure through the equation of state. In this outline we may specialize, say, to either an isothermal or adiabatic process, that is to say, pressure $p = f(\rho)$. The pressure gradients which are available through gradients of density will be calculated by estimating these gradients through the instantaneous appearance of our system of points, representing, we state again, the positions of centers of mass of fluid globules. The problem is then one of finding a rule or recipe to estimate the density at a point of space, given only a finite system of points (rather widely separated in practice). In the limiting case of a very large number of points one could simply count the number of points in a square or cube of a fixed mesh and this number will be proportional to the density. In practice we are limited to a moderate number of points for the whole fluid, the question is how to estimate this density in a most reliable way. Consider the case of two dimensions. One can think of the points located initially in a regular array, for example, on the vertices of a rectangular division, or better, vertices of a triangular subdivision of space. Of course, after some cycles of the computation the geometry of the system will change and one could, for example, estimate the density as follows: enclose each point by triangles with the closest points as vertices. The smallest (in area) triangle in whose interior a given point is located would give, through the ratio of its area to that of the original triangle at least an idea of the change in density at that point.

This procedure suffers from several drawbacks. They are due to the question of computational stability of our calculation. The selection of the nearest points leads to discontinuity in time of the area. In the equations of motion for each of our points we need the gradients of the density. And in this crude way of calculating the densities themselves the computation of *differences* in fixed directions in space is not straightforward. The nearby points may not be sufficiently close and these gradients may be very inaccurately estimated. In the case of points on the boundary of the fluid one would need special prescriptions. These errors would accumulate very seriously with time.

Another way that suggests itself is to introduce numerically the forces due to pressure gradients. One could imagine repulsive forces acting between any pair of our points. These would, in simple cases, depend upon the distance alone (the forces should derive from potentials if we assume scalar pressure, i.e., no viscosity or tensor forces, etc.). The form of the potential will, of course, depend upon the equation of state. In a one-dimensional problem the nature of this correspondence is clear. It suffices to have forces between neighboring points only. The continuity of motion guarantees the permanence in time of the relation of neighborhood. The situation is, however, completely different in one or more space dimensions. The neighbors of a given point will change in the course of time. If one tried to calculate the resultant force on a point due to *all* points in the problem, not merely the neighboring points, the computational work increases — it will grow with the square of the number of points considered. Therefore a "cut-off" for the force is necessary at some distance so that we need not calculate it if the distance between two points exceeds a certain constant. The pressure gradient would then be given directly as a resultant of all the forces acting on a point due to its neighbors and thus depends on the actual positions without reference to the initial configuration, *"forgotten" by the system.* Let us summarize briefly such a computational scheme. The particles represent small parts of the fluid. The forces due to pressure gradients are introduced directly by imagining that neighboring or "close" points repel each other. The dependence of this force on the distance between points is so chosen that in the limiting case of very many points, it would represent correctly the equation of state. That this is possible, in principle, is clear *a priori*: the density is inversely proportional, in the limit of a very large number of points, to the square (in two dimensions) or cube (in three dimensions) of the average distance between them. The pressure is a function of density, and this being a function of the distances, we obtain an analogue of the equation of state by choosing a suitable distance dependence of the force. The Lagrangian particles are at

$$P_1\colon (x_1, y_1, z_1),\ P_2\colon (x_2, y_2, z_2), \ldots, P_N = (x_N, y_N, z_N)$$

The forces (repulsive) between any two are given by

$$F_{i,j} = F(d(P_i, P_j))$$

where $d(P_i, P_j)$ is the distance between the two points.

The average value of d at a point of the fluid is, in the limit of $N \to \infty$, a function of the local density: $\bar{d} \cong \rho^{-\frac{1}{3}}$ for three dimensions.

The pressure p is a function of ρ alone in, say, isothermal or adiabatic problems. The pressure gradients are thus replaced by forces acting directly on point masses.

There is so far no general theory and the convergence of such finite approximations to the hydrodynamical equations remains to be proved, but even more important than that would be an estimate of the speed of convergence.

In order to test some of these proposals on actual problems, some numerical computations were performed on an electronic machine, the "Maniac" in Los Alamos and on its prototype at the Institute for Advanced Study in Princeton. (Cf. Pasta and Ulam [1]). The main purpose of these calculations was to test the feasibility of such numerical schemes rather than to obtain quantitative precision of the results. One of the problems concerned the study of the motion of a heavy fluid on top of a lighter one, both contained in a cube — with an initial irregularity present at their boundary. The problem involved following the development of this unstable situation — known as Taylor's instability — for motions in the large — beyond the infinitesimal stage — leading to gradual mixing of the two fluids.

With the comparatively crude network of points, representing the fluid, one could hardly expect that the details of the motion would be correctly represented; it was hoped that the behavior of a few functionals of the motion would be more accurately depicted.

One of the functionals which was computed as a function of time was the total kinetic energy of the particles divided into two parts: the kinetic energy of the horizontal and vertical motions separately. Even though the motion itself was very irregular, these quantities seemed to present rather smooth functions of time. In the case of a stable configuration with the lighter fluid on top and the heavy fluid on the bottom, there would be only a periodic interchange of kinetic and potential energies. In the unstable case, there should be an increase of kinetic energy which persists for a considerable time. The hope is that with many more calculations one could try to guess from such numerical results the form of an empirical law for the increase of this quantity as a function of initial parameters.

Another quantity of interest is the spectrum of angular momentum which may be defined, for example, in the following way. We draw around each particle a circle of fixed radius and calculate the angular momentum in each such region. We then take the sum of the absolute values or the sum of the squares of these quantities. This gives us an over-all measure of angular momentum on a scale given by the radius. Varying the radius, these numbers may then be studied as a function of both time and the radius. It is interesting to consider the rate at which the large scale angular momentum is transferred to small scale eddies — also, the spectrum in the asymptotic state, if it exists — all this, of course, as a function of the parameters of the problem: the ratio of the densities of the two fluids and the external force. It is clear that very many numerical experiments would have to be performed before one would trust such pragmatic "laws" for the time behavior of mixing or increase in vorticity.

Another reason for selection of this problem was an interest in the degree of spatial mixing of two fluids, starting from an unstable equilibrium. To measure quantitatively such mixing in configuration space, a functional was adopted similar to the one just mentioned for the spectrum of angular momentum. At a given time t, a circle is drawn around each point P. In each circle, we look at the ratio r of the number of heavy particles to the total number of particles in this region. We take the quantity $\mu(P, t) = 4r(1 - r)$. This gives an index of mixing of the fluid in the circle, being equal to zero if only one fluid is present and equal to 1, if particles of both fluids are equally present in it. We average this quantity over all the circles. Initially this average is very close to zero, the only region where it differs from zero being around the interface. As the mixing proceeds, our average measure of mixing increases with time. Again one might expect that after a sufficient number of numerical experiments, one would obtain an idea of the time behavior of this "mixing functional," or at least of the time T taken for the over-all mixing (which is initially close to zero) to become of the order of $\frac{1}{2}$ or $1/e$. This time, for dimensional reasons, must depend *inter alia* on $\sqrt{(L/a)}$ where L is the depth of the vessel and a the acceleration of the lighter fluid into the heavier one.

A comparison with experiments actually performed in Los Alamos in the problem of Taylor's instability for the case of a heavy gas on top of a light gas with an irregularity in the interface between them showed configurations not unlike those developed in the calculation. Of course, if one wanted to put credence in calculations like the above, the form of the proper force law imitating the real equation of state would have to be carefully chosen. The actual behavior, in time, of our functionals of the motion which we have calculated was very smooth despite the complicated nature and increasing "turbulence" of the motion itself.

A series of calculations was performed by F. Harlow and M. Evans [1], using a somewhat different method but still employing instead of the "classical" difference equation a finite approximation to the motion of the continuum by calculating motions of points representing globules of the fluid. The results obtained agreed in elementary cases with the analytic results and in other cases agreed quite well with experiments.

10. Synergesis

In addition to, and in some aspects quite beyond the above type of use of electronic computing machines one can conceive more general possibilities for large scale *experimentation* on problems of pure mathematics or in exploration of tentative ideas in physical theories. The program which will be outlined here requires certain equipment additional to the existing electronic machines and certain changes in their operation. These are not yet available but will soon exist since engineering work in this direction is in progress.

Speaking very broadly, the present machines operate on a set of given instructions (a flow diagram and a code) which once given make the machine proceed autonomously in solving problems of mathematical analysis or physics. The course of the operations is completely prescribed and the limited flexibility of the machine consists, roughly speaking, in taking one or the other course of computation depending on the value of numbers just computed. These so called *decisions* made by the machine in practice, so far, involve only a limited and prescribed set of changes in the logical course which is given.

One could conceive of a more general plan. Instead of using the machine as a robot or, as it were, as a player piano whose tunes are written in advance, the machine could be kept in constant communication with an intelligent operator who changes even the logical nature of the problem during the course of a computation, at will, after evaluating the results which the machine provides. Of course such possibilities exist already, but to a *very limited* extent. We shall try to indicate by some examples how a much more intimate relation could provide useful results. It is obvious for one thing that a computing machine can very quickly provide examples of geometrical or combinatorial situations which can be studied by a mathematician. Thus it obviously can play a role analogous to but vastly more efficient and helpful than scratch paper and pencil work which a person studying a problem uses to provide himself with a visual or a memory aid. Certainly the machine could perform quickly some of the drudgery of elementary algebraic of analytic calculations. In a search for examples or counter-examples the machine could calculate and then display on a screen elements of geometrical objects envisaged by the working mathematician and provide a responsive "scratch paper" if not an audience on which one could try one's ideas.

Obviously for activities of this sort, a rapid access to the machine is necessary and to work a computer in such a fashion would require a way to change its program more substantially than by just changing some constants of the computation in a short time. Also, the machine has to provide a quick illustration and display of the computed quantities and figures. In other words, the problem is of constructing ways of informing the machine of the desired course of computation and conversely of having the machine communicate the results obtained.

One rather wide field of application of a display technique will be in the study of properties of functions of several variables. In many problems it is important to find the critical values of a function of several real variables $f(x_1, x_2, \ldots, x_n)$, the function f being given analytically, e.g., as a polynomial in elementary functions or quotients of such polynomials. It is well known that the usual procedure employed to find, numerically say, local minima or maxima of a single function requires a search which is extremely time consuming. If the number of independent variables is large, for example 5 or more, no really efficient method of finding all the critical points has been proposed. Imagine now that the values of the function can be quickly computed on a grid of points which forms, say, a two-dimensional section in the given space and the resulting function of two variables, i.e., a surface displayed as in a projection, in perspective, on a tube. (A code for calculating axonometric projections is readily made.) The eye quickly notices the region or the regions where minima are likely to be assumed. By a quick change of scale and a magnification, that is to say, a subdivision of the region in question into a greater number of points, a computing machine can act as a microscope of arbitrary power. All we are saying is that in the search for critical points one can utilize, instead of the blind recipes embodied in a search code, the visual perception of the brain which is still much quicker than any now-known automatic code for "recognition."

By critical points we mean, for example, all of the points of space where all the first partial derivatives of the given function vanish. Beyond that, the recognition of other properties of a function (say, of 2 variables, for example), the finding of valleys, ridges, etc., immediately striking the human eye, would be, at the present time, almost impossible to code in a reasonable time for an automatic search by the machine.

Moreover, for functions of 3 or 4 variables one should also have a quick way to instruct a machine to select a desired two-dimensional section, to establish on it a set of independent grid points, to compute the value of functions on these points and display it in perspective on a screen. More important still will be the ability to change the scale of the independent and dependent variables by a general linear transformation. If T is a transformation, one wants to "see" also the transformations of the form HTH^{-1}, where H is a given one-one change of coordinates — i.e., the conjugate transformation. In this still elementary type of problem, it would be convenient to have a quick way of displaying the Fourier transform of the function. Also a display of the functions defined by the derivatives and gradients of the given functions for final verification of the character of interesting points would be convenient. The next step should involve a study of implicit functions, i.e., surfaces in n-dimensional space. Given a function $f(x_1, \ldots, x_n) = 0$, one would like to display the appearance of all desired two-dimensional sections through the n-dimensional surface. An idea of the three-dimensional sections could be obtained by moving two-dimensional sections continuously on the screen. Obviously there are technical problems, e.g., of persistence of image on the screen.

The next aim, still more ambitious, would be to provide for a series of "experiences" with problems computed on the machine so that its operator would acquire, after some practice, a *feeling* for the four-dimensional space as a result of such experimentation. Imagine, for example, that we consider the problem, in three-dimensions, of threading a given solid through a given closed space curve, an exercise which involves "*trying*" since no simple criteria about projections seem to be sufficient to decide whether or not one can push a given solid through a given curve. The physical process of effecting such motion could be imitated by the machine by making it compute the successive positions of the solid following given manual instructions (to be quickly transmitted numerically) about the rotations and translations in the three-dimensional space. The contact between the two sets would be tested after each trial displacement.

An analogous problem is: to thread a four-dimensional solid through a closed two-dimensional surface in E, something for which no tactile or visual experience exists. It is possible that a certain facility for such tasks could be developed, to some extent at least, by continued experimentation on machines which would compute the course of motion of the given solid and display to the operator the appearance of three-dimensional sections and at least inform the operator of collisions between the given surface and the object which has to be disjoint with it. The above is perhaps an arbitrary and already very complicated example of what should be a systematic approach to acquiring, by practice, a feeling for four-dimensional geometry. The problem above is only partially topological, and principally metrical. A systematic attempt to familiarize oneself with the purely *topological* properties of complexes in four dimensions would, however, be more difficult to formulate and to code.

But even in two dimensions, if one studies, instead of single real valued functions, *transformations* defined on the plane into itself interesting heuristic methods seem to be possible. For a given transformation $T(p)$ of the plane we may want to study the properties of the iterates of this transformation, that is to say, the asymptotic properties of the sequence of points $T^k(p)$, to find its fixed points, i.e., $p = T(p)$, involution points, i.e., points p such that $T^k(p) = p$, $k > 1$, and invariant subsets S of the plane, i.e., sets such that $T(S) \subset S$, etc. If T is given analytically, it is a simple matter to compute $T(p)$ for a great number of points p. We can then imagine a way to connect the given points to their images by arrows on a display tube. Here, as in all previous problems, the question of which is the most instructive method of display is not easily answered. Perhaps one could select the given points on some curve, e.g., on a circle, and then display the curve consisting of the image of this circle. Suppose we want to find the fixed point of a transformation. The proposal would be to start with a circle, look at its image, then decide upon a suitable motion of the given circle so that its image will fall essentially inside or essentially outside itself. Then from Brouwer's fixed point theorem we would know that the search for the fixed point is narrowed to the interior of the curve. There should then be a rapid manual method of "steering" the initial curve whose image is displayed, in any desired direction, of changing its size, and possibly the eccentricity if we start with an ellipse, etc.

All this applies, mutatis mutandis, in the n-dimensional space, where we would of course be able to display only various plane sections, or, perhaps better, a 2-dimensional section of an initially given sphere and its image. One transformation for which such a study was undertaken is the one discussed in Chapter II, Section 3. This transformation contains in itself all the information about the general algebraic equation of the nth order. There are many still unsolved questions concerning its fixed points, invariant manifolds, etc. John Ackley has coded the "Whirlwind" machine at the Lincoln Laboratories of M.I.T., in Cambridge, Massachusetts, for a preliminary study of this sort.

Computations on a machine are particularly well suited for the display of asymptotic properties of iterates. It is obvious that in order to tabulate only the interesting points, it is preferable to have a visual evaluation of the iterated properties of the many points rather than print out an enormous number of numerical values. The latter have to be evaluated by a person in order, one by one, and it is in general very hard to guess in advance what properties of the sequence one wants to look at — it seems difficult to foresee and put in a code ahead of time the desired criteria for selecting significant special cases.

Another example of such interplay between the operations performed on the machine and displayed visually, and the decisions made by the operator upon evaluation of the results: such cooperation might be useful in finding solutions to differential equations with boundary values. In the simplest cases, those of the ordinary differential equations, say of the 2nd order, with prescribed boundary conditions at 2 points, one could satisfy the conditions at the first given point, then by integrating on the machine presumably follow the trend of the solution before it is computed at the second boundary and by sensing the trend, so to say, modify the arbitrary parameter assumed at the 1st point, e.g., the value of the 1st derivative. It is clear that a saving in time can be obtained by an intelligent intervention quite beyond any automatic coding of obvious changes.

But even in the study of systems of equations of the 1st order, by exhibiting the vector field on a screen (in 3 or a higher number of dimensions showing the vector field in projections) one could guess the location of singular points "at a glance", and also perhaps form an idea of the behavior of solutions in the large. To give an example: suppose the problem concerns the behavior of the magnetic lines of force in space due to steady currents flowing on a given curve or a system of curves as mentioned in Section 6 of the present chapter. C. Luehr and the writer considered solution of the problem by computations on an electronic machine in two cases. *1.* In three-dimensional space a current flows on two infinite lines; one of these is the x-axis, the other the line $x = 0$, $z = 1$. *2.* The current flows through a closed curve forming a clover-leaf knot. The problem was to study the qualitative topological features of the field of all the lines of force in space. The general ergodic properties of the fields of lines of force are not understood — they are of practical interest now in magneto-hydrodynamics and in astronomy.

The computations were performed on the "Maniac" in Los Alamos. This work is still proceeding, but without a good mechanism of display, the progress in evaluating the properties of the lines of force is slow and the evaluation of the printed data laborious. A visual display would certainly permit one to diminish the number of trials, i.e., the initial points through which one computes the lines of force.

In the study of partial differential equations the advantage of a quick survey of results would, it seems, be even greater. As an example of what we have in mind, the following investigation should be mentioned. A rather deep connection appears to exist between the behavior of solutions of the Hamilton-Jacobi partial differential equation

$$(\partial s/\partial x)^2 + (\partial s/\partial y)^2 + (\partial s/\partial z)^2 + V(x, y, z) = \partial s/\partial t$$

on one hand, and the problem of the random walk, with a variable step proportional to the given function V, described by a diffusion equation. The conjecture, proved so far only in the case of one independent space variable and still open for two or more dimensions, concerns the following problem: Given a function $V(x_1, x_2, \ldots, x_n)$ consider, at time t, the front of the Hamilton-Jacobi wave $S(x^0, x) = t$ originating from any point x^0. This surface is obtainable by the Huyghens construction of envelopes of spheres. On the other hand, consider the crest of the probability function $w(x_1, x_2, \ldots, t)$ corresponding to a random walk from x^0 with a step whose variable length is proportional at the position $(x_1, x_2, \ldots, x_n)$ to the value of the given function V (cf. Everett, Ulam [6]).

By the *crest* of the probability function w we mean the surface $t = f(x, x^0)$ defined by $\partial w/\partial t = 0$. The conjecture states that f and S are simply related; for $n = 1$ one shows $f \sim S^2$ under suitable conditions. Apart from this, we should like to state here the belief that the field of curves or sheaves of surfaces corresponding to the equation $S = t$, where S is the Hamilton-Jacobi function, if suitably displayed on a screen, would help form ideas about the behavior of solutions of a dynamic problem for a great variety of initial conditions all at once.

In problems of hydrodynamics it is again obvious that for qualitative studies, e.g., of the progress of mixing of two fluids, a display of the position of the interface between the two fluids or gases — starting from an unstable configuration — would permit good choices of initial conditions and significant values of parameter. (One of these parameters is the value of the initial acceleration of the lighter fluid into the heavier one.) Here again, in absence of *a priori* knowledge of the character of the motion, it is very hard to code in advance automatic criteria for the selection of such parameters.

It is in the study of games and in the actual playing of games that *synergesis*, i.e., the continuing collaboration between the machine and its operator should prove of immediate value. Recently, some progress has been achieved in coding for electronic computers to play games like checkers (Samuels, IBM Research Lab., on the 704) and chess (Cf. Kister *et al.* [1]). While in the simpler game of checkers the existing code allows the machine to play a very good opening and middle game, in the game of chess the quality of the play, as coded so far, is very rudimentary. No doubt progress will be made during the next few years, but the writer does not believe that chess games of master quality will be achieved in the near future!

An intermediate, and in a way less ambitious, program would be to have codes prepared for assisting a human chess player to explore on a machine certain *special* sequences of moves quite far ahead (i.e., for perhaps 6–8 moves) and then to flash the tentative positions on a screen. This, when coupled with calculations of various evaluation functions generalizing those already conceived would possibly improve the game!

To play a "fair game", the two opponents would have to have at their disposal two identical machines! In other words, it seems that a computing machine could act as a "second" of the player, helping to analyze the positions. To the writer at least, an intelligent program of this sort seems nearer at hand than a completely automatic program enabling the machine to play a high quality game. Going beyond the existing games, it is perhaps permissible to speculate on the invention of entirely new games to be played between two players, each provided with a computer.

All the examples mentioned above deal with methods to facilitate solution of given problems by some general heuristic methods — either of Monte Carlo type, or consisting of specific numerical work. Beyond this it seems that the computing machines will soon be able, at least in elementary cases, to prove theorems in the customary mathematical formalisms. The interesting work of N. Rochester and his associates (Gelernter, Roth, and others) makes it appear plausible that an automatic code will soon be available for *proving* theorems in certain elementary domains (in Euclid's geometry, for example). It will probably take a long time until an analogous program will be realized for more extensive mathematical disciplines. Here again it could be pointed out that an intermediate program of collaboration between a human operator and a machine seems easier of accomplishment! In fact, one may start by practicing this by proving theorems in Euclidean geometry about triangles, etc.; or certainly, if one wanted to automatize proofs in projective geometry, a *display* of the geometric constructions envisaged by a human operator plus automatic searching by the machine for the routine syllogisms might provide a quicker way to find formal proofs.

BIBLIOGRAPHY

ALEXANDER, J. W.
[1] Some Problems in Topology, *Verhandlungen des Internationalen Math. Kong.*, Zürich, 1932, pp. 249–257.
ALEXANDROFF, P. and HOPF, H.
[1] Topologie I, *Die Grundlehren der Math. Wissenschaften*, Vol. XLV, J. Springer, Berlin, 1935.
ANDERSON, R. D.
[1] The Algebraic Simplicity of Certain Groups of Homeomorphisms, to appear in *Proc. Am. Math. Soc.*
ANTOINE, L.
[1] Sur l'homéomorphe de deux figures et leurs voisinages, *J. Math., Ser.* 8, 4 (1921), pp. 221–325.
ARNOLD, V. I.
[1] *Doklady Akad Nauk U.S.S.R.*, 114, No. 4 (1957).
AUERBACH, H.
[1] Sur un problème de M. Ulam concernant l'équilibre des corps flottants, *Studia Math.*, 7 (1938), pp. 121–142.

BELLMAN, R. and HARRIS, T.
[1] On Age-Dependent Binary Branching Processes, *Ann. of Math.* 55 (2) (1952), pp. 280–295.
BIEBERBACH, L.
[1] Operationsbereiche von Funktionen, *Verhandlungen des Internationalen Math. Kong.*, Zürich, 1932, pp. 162–172.
BING, R. H.
[1] The Cartesian Product of a Certain Nonmanifold and a Line is *E*. *Bull. Am. Math. Soc.* 64 (1958), pp. 82–84.
BIRKHOFF, G.
[1] *Lattice Theory*, Rev. Ed., Am. Math. Soc. Colloq. Pub. XXV, 1948.
BORSUK, K.
[1] Drei Sätze über die n-dimensionale Euklidische Sphäre, *Fund. Math.*, 20 (1933), pp. 177–190.
BORSUK, K. and ULAM, S. M.
[1] On Symmetric Products of Topological Spaces, *Bull. Am. Math. Soc.*, 37 (1931), pp. 875–882.
[2] Über gewisse Invarianten der ε-Abbildungen, *Math. Ann.*, 108 (1933), pp. 311–318.
BOTT, R.
[1] On Symmetric Products and Steenrod Squares, *Ann. of Math.*, 57 (2) (1953), pp. 579–590.

CANTOR, G.
[1] *Contributions to the Founding of the Theory of Transfinite Numbers*, Dover Publications, New York.

CASHWELL, E. D. and EVERETT, C. J.
[1] *A Practical Manual on the Monte Carlo Method for Random Walk Problems*, LA–2120 Office of Technical Services, U. S. Dept. of Commerce, Washington, D. C.
COHEN, L. W. and ULAM, S. M.
[1] On the Algebra of Systems of Vectors and Some Problems in Kinematics, *Bull. Am. Math. Soc.*, 50 (1944), Abstract 38, p. 61.

EGGLESTON, H. G.
[1] A property of Plane Homeomorphisms, *Fund. Math.*, 42 (1955), pp. 61–74.
ERDÖS, P.
[1] On a Lemma of Littlewood and Offord, *Bull. Am. Math. Soc.*, 51 (1945), pp. 898–902.
ERDÖS, P. and JABOTINSKY, E.
[1] On Sequences of Integers Generated by a Sieving Process, *Koninkl. Nederl. Akad. van Wetenschappen, Amsterdam Proceedings, Ser. A*, 61 (1958), pp. 115–128.
ERDÖS, P. and KAKUTANI, S.
[1] On Non-Denumerable Graphs, *Bull. Am. Math. Soc.* 49 (1943), pp. 457–461.
EVERETT, C. J. and ULAM, S. M.
[1] Projective Algebra I, *Am. J. Math.*, 68 (1946), pp. 77–88.
[2] Multiplicative Systems in Several Variables, I, AECD–2164 (1948); II, AECD–2165 (1948); III, AECD–2532 (1948), Technical Information Branch, Oak Ridge, Tenn.
[3] Multiplicative Systems I, *Proc. Nat. Acad. Sci.*, 34 (1948), pp. 403–405.
[4] On the Problem of Determination of Mathematical Structures by Their Endomorphisms, *Bull. Am. Math. Soc.*, 54 (1948), Abstract 285t, p. 646.
[5] On an Application of a Correspondence Between Matrices Over Real Algebras and Matrices of Positive Real Numbers, *Bull. Am. Math. Soc.*, 56 (1950), p. 63.
[6] Random Walk and the Hamilton-Jacobi Equation, *Bull. Am. Math. Soc.*, 56 (1950), Abstract 97t, p. 63.

FELLER, W.
[1] The Fundamental Limit Theorems in Probability, *Bull. Am. Math. Soc.*, 51 (1945), pp. 800–832.
[2] Diffusion Processes in Genetics, *Proceedings of the Second Berkeley Symposium on Mathematical Statistics and Probability*, Univ. of California Press, Berkeley, 1951, pp. 227–246.
FERMI, E., PASTA, J. and ULAM, S. M.
[1] Studies of Nonlinear Problems, LA-1940, Office of Technical Services, U. S. Dept. of Commerce, Washington, D.C.
FOX, R. H.
[1] On a Problem of S. Ulam Concerning Cartesian Products, *Fund. Math.*, 34 (1947), pp. 278–287.

GALE, D. and STEWART, F. M.
[1] *Infinite Games with Perfect Information*, Contributions to the Theory of Games, Vol. II, Princeton, N. J., 1953.

GARDINER, V., LAZARUS, R., METROPOLIS, N. and ULAM, S. M.
[1] On Certain Sequences of Integers Defined by Sieves, *Math. Mag.*, 29 (1956), pp. 117–122.
GÖDEL, K.
[1] Über formal unentscheidbare Sätze der Principia Mathematica und verwandte Systeme I, *Monatshefte Math. Phys.*, 38 (1931), pp. 173–198.
[2] *The Consistency of the Axiom of Choice and the Generalized Continuum Hypothesis with the Axioms of Set Theory*, Ann. of Math. Studies, No. 3, Princeton Univ. Press., Princeton, 1940.
HALMOS, P. R.
[1] Algebraic Logic I, *Compositio Math.*, 12 (1955), pp. 217–249; II, *Fundamenta Math.*, XLIII (1956), pp. 255–325.
HARLOW, F. H. and EVANS, M. W.
[1] The Particle-in-Cell Method for Hydrodynamic Calculations, LA–2139, Office of Technical Services, U. S. Dept. of Commerce, Washington, D. C.
HILBERT, D.
[1] Mathematische Probleme, *Nach. Akad. Wiss. Göttingen*, 1900, pp. 253–297.
HOPF, E.
[1] *Ergodentheorie*, Ergebnisse der Math., Chelsea Publishing Co., New York, 1948.
HUREWICZ, W. and WALLMAN, H.
[1] *Dimension Theory*. Princeton Math. Ser. No. 4, Princeton Univ. Press, Princeton, 1941.
HYERS, D. H.
[1] On the Stability of the Linear Functional Equation, *Proc. Nat. Acad. Sci. U. S.*, 27 (1941), pp. 222–224.
HYERS, D. H. and ULAM, S. M.
[1] On Approximate Isometries, *Bull. Am. Math. Soc.*, 51 (1945), pp. 288–292.
[2] Approximate Isometries of the Space of Continuous Functions, *Ann. of Math.*, 48 (2) (1947), pp. 285–289.
[3] Approximately Convex Functions, *Proc. Am. Math. Soc.*, 3 (1952), pp. 821–828.
[4] On the Stability of Differential Expressions, *Math. Mag.*, 28 (1954), pp. 49–64.

KACZMARZ, S. and TUROWICZ, A.
[1] Sur l'irrationalité des intégrales indéfinies, *Studia Math.*, 8 (1939), pp. 129–134.
KELLY, P. F.
[1] On Isometries of Square Sets, *Bull. Am. Math. Soc.*, 51 (1945), pp. 960–963.
KHINCHIN, A. Y.
[1] *Mathematical Foundations of Statistical Mechanics*, Dover Publications, New York, 1940.
[2] *Three Pearls of Number Theory*, Graylock Press, Rochester, New York, 1952, Chap. I.
KISTER, J., STEIN, P., ULAM, S., WALDEN, W., and WELLS, M.
[1] Experiments in Chess, Jour. Ass'n. Computing Machinery, 4 (1957), pp. 174–177.

Kolmogoroff, A. N.
[1] On the Representation of Continuous Functions of Several Variables in the Form of Superposition of Continuous Functions of One Variable and Addition, *Doklady Akad. Nauk S.S.S.R.*, 108, No. 2 (1956).
Kuratowski, C.
[1] *Topology I*, Monografje Matematyczne, Warsaw, 1933.
[2] Les ensembles projectifs et l'induction transfinie, *Fund. Math.*, 27 (1936), pp. 269–276.
[3] Les suites transfinies d'ensembles et les ensembles projectifs, *Fund. Math.* 28 (1937), pp. 186–196.
Kuratowski, C. and Ulam, S. M.
[1] Quelques propriétés topologiques du produit combinatoire, *Fund. Math.*, 19 (1932), pp. 247–251.
[2] Sur un coefficient lié aux transformations continues d'ensembles, *Fund. Math.*, 20 (1933), pp. 244–253.

Lévy, P.
[1] Trois théorèmes de calcul des probabilités, *Compt. rend.*, 238 (1954), pp. 2283–2286.
Lusin, N.
[1] *Leçons sur les Ensembles Analytiques*, Gauthier-Villars, Paris, 1930.

McKinsey, J. C. C.
[1] On the Representation of Projective Algebras, *Am. J. Math.*, 70 (1948), pp. 375–384.
MacMahon, P. A.
[1] *Combinatory Analysis*, Vol. I, II, Cambridge Univ. Press, New York, 1916.
Metropolis, N. and Ulam, S. M.
[1] On the Motion of Systems of Mass Points Randomly Distributed on the Infinite Line, *Bull. Am. Math. Soc.*, 55 (1949), Abstract 299t, p. 670.
Moise, E. E.
[1] Affine Structures in 3-Manifolds, VIII. Invariance of the Knot-Types; Local Tame Imbedding, *Ann. of Math.*, 49 (1954), pp. 159–170 and references to I–VII.
Montgomery, D. and Zippin, L.
[1] *Topological Transformation Groups*, Interscience Publishers, New York–London, 1955.
Morse, M.
[1] *The Calculus of Variations in the Large*, Am. Math. Soc. Colloq. Pub. XVIII, New York, 1934.
Mycielski, J., Swierczkowski, S. and Zieba, A.
[1] On Infinite Positional Games, *Bull. acad. polon. sci.*, 4 (1956), pp. 485–488.
Mycielski, J. and Zieba, A.
[1] On Infinite Games, *Bull. acad. polon. sci.*, 3 (1955), pp. 133–136.

Netto E.
[1] *Lehrbuch der Kombinatorik*, 2nd Ed., Chelsea Publishing Co., New York.

OXTOBY, J. C. and ULAM, S. M.
[1] Measure Preserving Homeomorphisms and Metric Transitivity, *Annals of Math.* 42 (1941), pp. 874–920.

PASTA, J. and ULAM, S.
[1] Heuristic Numerical Work in Some Problems of Hydrodynamics, *Math. Tables and Other Aids to Computation*, XIII (1959), pp. 1–12.

PÓLYA, G. and SZEGÖ, G.
[1] *Aufgabe und Lehrsätze aus der Analysis*, Vol. I, II, Dover Publications, New York.

PONTRJAGIN, L.
[1] *Topological Groups*, Princeton Math. Ser. No. 2, Princeton Univ. Press, Princeton, 1939.

RECHARD, O. W.
[1] Invariant Measures for Many-One Transformations, *Duke Math. J.*, 23 (1956), pp. 477–488.

RITT, J. F.
[1] *Integration in Finite Terms*, Columbia Univ. Press, New York, 1948.

ROTH, K. F.
[1] On Certain Sets of Integers, *J. London Math. Soc.*, 28 (1953), pp. 104–109.

SAKS, S.
[1] *Theory of the Integral*, Monografje Matematyczne VII, Stechert and Co., New York, 1937.

SCHREIER, J.
[1] Über die Drehungsgruppe im Hilbertschen Raum, *Studia Math.*, 5 (1934), pp. 107–110.

SCHREIER, J. and ULAM, S. M.
[1] Über die Permutationsgruppe der natürlichen Zahlenfolge, *Studia Math.*, IV (1933), pp. 134–141.
[2] Über topologische Abbildungen der euklidischen Sphären, *Fund. Math.*, 23 (1934), pp. 102–118.
[3] Eine Bemerkung uber die Gruppe der topologischen Abbildungen der Kreislinie auf sich selbst, *Studia Math.*, 5 (1935), pp. 155–159.
[4] Sur le nombre des générateurs d'un group topologique compact et convexe, *Fund. Math.*, 24 (1935), pp. 302–304.
[5] Über die Automorphismen der Permutationsgruppe der natürlichen Zahlenfolge, *Fund. Math.* 24 (1935), pp. 158–260.

SHAPIRO, H. N.
[1] Note on a Problem in Number Theory, *Bull. Am. Math. Soc.*, 54 (1948), pp. 890–893.

SIERPINSKI, W.
[1] *Hypothèse du Continu*, 2nd Ed., Monografje Mathematyczne Tom IV, Warsaw, Chelsea Publishing Co., New York, 1934.
[2] *General Topology*, Univ. of Toronto Press, Toronto, 1952.

SPERNER, E.
[1] Ein Satz über Untermengen einer endlichen Menge, *Math. Z.*, 27 (1928), pp. 544–548.

STEIN, P. R. and ULAM, S. M.
[1] A Study of Certain Combinatorial Problems Through Experiments on Computing Machines, *Proceedings of the High Speed Computer Conference*, Louisiana State University, Baton Rouge, 1955.
[2] *Quadratic Transformations*, Part I, Los Alamos Scientific Laboratory Report No. 2305, Office of Technical Services, U.S. Department of Commerce, Washington 25, D.C.

ULAM, S. M.
[1] Zur Masstheorie in der allgemeinen Mengenlehre, *Fund. Math.*, 16 (1930), pp. 141–150.
[2] Infinite Models in Physics, Applied Probability, *Proceedings of Symposia in Applied Mathematics*, Vol. VII, McGraw-Hill Book Co., New York, 1957.

ULAM, S. M. and VON NEUMANN, J.
[1] On the Group of Homeomorphisms of the Surface of the Sphere (abstract), *Bull. Am. Math. Soc.*, 53 (1947), p. 506.

VOLTERRA, V.
[1] *Leçons sur la Théorie Mathématique de la Lutte pour la Vie*, Gauthier-Villars, Paris, 1931.

WALDEN, W.
[1] A Study of Simple Games Through Experiments on Computing Machines, *Contributions to the Theory of Games*, Vol. III, Annals of Math. Studies No. 39, Princeton, 1957.

WEIL, A.
[1] *L'intégration dans les groups topologiques et ses applications*, Actualités Scientifiques et Industrielles 869, Hermann et Cie., Paris, 1938.

WEYL, H.
[1] Mathematics and Logic, *Am. Math. Monthly*, 53 (1946), pp. 2–13.

ZAHORSKI, Z.
[1] Sur l'ensemble des points singuliers d'une fonction d'une variable réelle admettant les dérivés de tous les ordres, *Fund. Math.*, 34 (1947), pp. 183–245; Supplément au mémoire, "Sur l'ensemble . . .", *Fund. Math.*, 36 (1949), pp. 319–320.

Commentaries

Commentaries to Part I

[1] Remark on the generalised Bernstein's theorem (pp. 3–5, this volume)

For numerous references to the literature on equivalence by finite decompositions, see Banach,[1] pp. 114–8, and comments ibidem, pp. 324–7. For the most important developments, see Tarski,[4] Theorems 11.28, 16.28, and 16.9. Let us add the following references: Morse,[1] and for the reader interested in the theorems on cardinal numbers which do not need the axiom of choice, since their proofs usually are constructions of suitable countably infinite decompositions, Bradford.[1] Estimates for the minimal number of parts necessary for a decomposition which establishes the theorem $3M = 3N \rightarrow M = N$, or many other results of that type contained in the literature quoted above, are not known.
J. Mycielski

[2] Concerning functions of sets (pp. 6–8)

This paper is the first published proof of the existence of a prime nonprincipal ideal in the Boolean algebra of all subsets of an infinite set. Independently and about the same time, Tarski[1] obtained this result as a corollary of his algebraic treatment of conditions necessary and sufficient for the existence of a measure (equivalence of the existence of a measure to the nonexistence of paradoxical decompositions) which he developed later in Tarski.[2,3,4]

The existence of a prime ideal in every Boolean algebra is the key lemma of the Stone representation theorem; see, e.g., Kuratowski and Mostowski.[1] Concerning related theorems and the role of the axiom of choice in Ulam's proof see the papers of Sierpiński,[1] Gladkii,[1] Halpern,[1] Feferman,[1] and Mycielski;[3] in these papers more references are given.

Solovay[2] announced (discovered independently by David Pincus, to appear) that the existence of a real-valued nonnegative measure μ over all subsets of a countable set N such that $\mu(N) = 1$ and $\mu(\{p\}) = 0$ for every $p \in N$ implies the existence of a set of reals without the property of Baire. Hence, by Solovay's result,[2] this cannot be proved without using the axiom of choice for some uncountable family of sets.
J. Mycielski

[3] Zur Masstheorie in der allgemeinen Mengenlehre (pp. 9–19)

This paper extended the earlier results of Banach and Kuratowski (Banach,[1] pp. 182–6 and 200–3 and comments ibidem, pp. 333–7 and 340–5) by showing:

(1) If there is a real-valued measurable cardinal, then either $2^{\aleph_0}$ is such or the smallest such cardinal is 01-measurable.

(2) The smallest real-valued measurable cardinal is weakly inaccessible.

(3) The smallest 01-measurable cardinal is strongly inaccessible.

Let μ denote the smallest 01-measurable cardinal and θ_1 the smallest uncountable strongly inaccessible cardinal. (3), which has been independently proved by Tarski, opened the problem of whether $\mu = \theta_1$. This problem was solved by Hanf and Tarski in 1960. See Keisler and Tarski.[1] It turned out that μ (if it exists) is much larger than θ_1; in fact the set of cardinals less than μ is closed under all known "inductive constructions" of cardinals (e.g., let θ_τ be the first cardinal in the sequence $\aleph_0 = \theta_0, \theta_1, \ldots, \theta_\xi, \ldots$ of all strongly inaccessible cardinals such that $\overline{\overline{\tau}} = \theta_\tau$ then $\theta_\tau < \mu$). The concept of inductive construction of cardinals is not altogether clear, but various exact concepts of that sort have been formulated (restricted classes of definitions "from inside") and such closure property of μ has been proved for them (see Keisler and Tarski,[1] where earlier references are given, Hanf and Scott,[1] and Vaught[2]).

Solovay[3] and Jensen and Fremlin independently proved an extension of (2) similar to the extension of (3) of Hanf and Tarski;[1] namely, the set of cardinals less than the first real-valued measurable cardinal is closed with respect to all known "inductive constructions" related to the notion of weak inaccessibility in the same way that the other constructions were related to strong inaccessibility (m^+ takes place of 2^m).

Let M denote the axiom "there exists a 01-measurable cardinal μ." Work following the discovery of Hanf and Tarski has shown the great deductive power of M (see Scott,[1] Gaifman,[1] Rowbottom,[1] Silver,[1] Solovay,[1] and Martin[1]). Among the most interesting consequences of M are the following: (i) All uncountable sets of real numbers of class $\sum_2^1$ ($= PCA$) have perfect subsets (Solovay[1]). (ii) All sets of the class $\sum_2^1 \cup \prod_2^1$ are Lebesgue measurable and have the property of Baire (Solovay, not yet published). (iii) Infinite positional games defined by sets of the class $\sum_1^1 \cup \prod_1^1$ are determined (Martin[1]). (iv) The set of constructible real numbers is countable (Rowbottom[1]). (v) For every $\alpha > 0$ the first ω_α constructible sets constitute an elementary substructure of the structure constituted by the first $\omega_{\alpha+1}$ constructible sets (Gaifman,[1] Silver[1]). Although it came as a surprise that such conclusions could be derived from *M*, the statements (i)–(iii) were always expected, one could say wished for, since the notions involved in them were invented. But these propositions were hopeless as Gödel proved that (i)–(v) are all independent of (i.e., their negations are consistent with) the usual axioms of the Zermelo-Fraenkel set theory *ZF*. Thus, everybody working in the foundations of mathematics is sensitive to the question of whether *M* is a consistent and admissible axiom.

The question of consistency of *M* cannot be settled unless *M* is inconsistent in *ZF*. Indeed, all known mathematical theories *T* "understood" better than $ZF + M$ are such that the consistency of *T* can be proved in $ZF + M$. Hence, by a theorem of Gödel, the consistency of $ZF + M$ is not provable in *T*. Solovay[3] has proved that $ZF + M$ is consistent if $ZF +$ an axiom claiming the existence of a real-valued measurable cardinal (which is apparently weaker than *M*) is consistent.

Concerning the question of admissibility of *M*, this commentator is convinced that the answer is yes; i.e., that *M*, and hence all the above-mentioned consequences of *M*, are in the best style of classical mathematics. It seems that the only justification of classical mathematics, normally formalized in *ZF*, is that it works and is beautiful to the human mind (both these features are so rich in content and so interrelated—e.g., the practical applicability of mathematics and the "interpreted character" of *ZF* is a part of them—that we cannot attempt to describe them here) and $ZF + M$ has the same properties. There are other new and powerful axioms which also satisfy the above criteria, e.g., the axiom *C* of existence of compact cardinals (see, e.g., Mycielski[3]) or the axiom *D* which claims that every infinite positional game, defined by a set of reals constructible from the set of all reals, is determined. The axiom *C* implies *M* in *ZF*. The consistency of $ZF + M$ and even stronger theories was proved in $ZF + D$ by Solovay (unpublished). Such new axioms yield solutions to old problems, like (i)–(v) stated here, and, strictly speaking, they are the only essentially new developments of classical mathematics. Of course we are ready to abandon most classical mathematics if the axiom of infinity of *ZF* is proved to be inconsistent; similarly we will abandon *M* (or *C* or *D*) if it is disproved.

It is known that *M* (if consistent) is not powerful enough to imply that all Δ^1_3 $(= \Sigma^1_3 \cap \Pi^1_3)$ sets of reals are Lebesgue measurable or have the

property of Baire and that M is consistent with the continuum hypothesis (see Silver[2]). Thus stronger axioms "of infinity" than M are needed if we want to prove by means of them, e.g., measurability of all projective sets and other important consequences of D. For such problems C does not seem to work better than M.
J. Mycielski

[6] (with K. Kuratowski) Quelques propriétés topologiques du produit combinatoire (pp. 32–36)

This paper brought the, today classical, analog of the theorem of Fubini with the concept of measure 0 replaced by first category.
J. Mycielski

[7] (with S. Mazur) Sur les transformations isométriques d'espaces vectoriels, normés (pp. 37–39)

Here Mazur and Ulam show that the metric structure of a normed space X already determines the linear structure. If X is strictly convex (i.e., $||x + y|| = ||x|| + ||y||$ holds only for vectors x, y on the same ray from the origin), this is an obvious fact, since in this case for every $x, y \in \mathrm{X}$, $(x + y)/2 = z$ is the only vector which satisfies $||z - x|| = ||z - y|| = ||x - y||/2$. Mazur and Ulam give here a simple but ingenious construction of the algebraic middle point of a segment in terms of distances in the general case. The result of Mazur and Ulam was the starting point for several other investigations. Let X and Y be normed spaces and let F be an isometry from X into Y such that $\mathrm{F}(\theta) = \theta$. If Y is strictly convex or if F is onto (by the result of the present paper) then F is linear. Simple examples show that in general F need not be linear. However, Figiel [1] showed (under the obvious assumption that Y is the closed linear span of FX) that there is a linear operator T from Y into X such that $||\mathrm{T}|| \leqslant 1$ and $\mathrm{TF}x = x$ for every $x \in \mathrm{X}$. This clearly generalizes the result of Mazur and Ulam. (See also Holsztyński,[1] and Charzyński.[1])

Another direction in which the result of Mazur and Ulam may be generalized is this: Assume that X and Y are Banach spaces and F a map from X onto Y such that for every $x_1, x_2 \in \mathrm{X}$, $||x_1 - x_2|| \leqslant ||\mathrm{F}x_1 - \mathrm{F}x_2|| \leqslant \lambda||x_1 - x_2||$ for some constant λ. If $\lambda = 1$ the result of Mazur and Ulam shows that F must be affine (i.e., a linear map followed by a translation). This clearly is no longer true if $\lambda > 1$. However, the following question is still open. Does the existence of F imply that there is a linear isomorphism T from X onto Y? Several recent results (Lindenstrauss[1] and Enflo[1,2]) show that the answer may

be positive, especially if X and Y are reflexive. Much work has still to be done in order to clarify the situation.
J. Lindenstrauss

[11] (with K. Kuratowski) Sur un coefficient lié aux transformations continues d'ensembles (pp. 45–54)

Here Kuratowski and Ulam study continuous mappings between two compact sets in a metric space. The result on the unicoherence of countable products of unicoherent metric continua [p. 248] was first extended to countable products of unicoherent metric, connected and locally connected spaces in Eilenberg,[1] and then to arbitrary products of arbitrary unicoherent, connected and locally connected spaces in Ganea.[1] The question [p. 252] of whether two quasi-homeomorphic closed manifolds are necessarily homeomorphic is still unanswered; however, it follows from Eilenberg[1] and Ganea[1] that they have the same homotopy type. The question [p. 252] of whether the fixed-point property is invariant under quasi-homeomorphism has been answered by Borsuk:[1] the fixed-point property is not invariant for arbitrary continua, but is invariant for two absolute neighborhood retracts.
T. Ganea

[12] (with J. Schreier) Über die Permutationsgruppe der natürlichen Zahlenfolge (pp. 55–62)

For other studies of groups of permutations of infinite sets see McKenzie.[1,2] See also Banach,[1] p. 250, and comment ibidem, p. 356.

McKenzie noticed that Theorem 2 of this paper can be improved as follows. There are two permutations ϕ and ψ of a countable set N which generate all the permutations of N which move only finitely many points. In fact, let $N = Z \cup \{p\}$, where Z is the set of integers and $p \notin Z$. Put $\phi(x) = x + 1$ for $x \in Z$, $\phi(p) = p$, $\psi(0) = p$, $\psi(p) = 0$ and $\psi(x) = x$ for $x \in Z$, $x \neq 0$. It is clear that ϕ and ψ generate every transposition of two elements of N and hence have the desired property.

Topological questions for groups of permutation were also studied by Gaughan.[1] Swierczkowski[1] has proved that the alternating group A cannot be isomorphically imbedded in any locally compact connected group.
J. Mycielski

[13] (with K. Borsuk) Über gewisse Invarianten der ϵ-Abbildungen (pp. 63–70)

Here Borsuk and Ulam study the concept of ϵ-map, i.e., a map f: $X \to Y$ of a compact metric space such that diam $f^{-1}(y) < \epsilon$ for every $y \in f(X)$. This concept is due to Alexandroff.[1] Hurewicz and Wallman[1] give modern proofs of the main applications of ϵ-maps to dimension theory: 1) if dim $X \leqslant n$ then, for every $\epsilon > 0$, the set of ϵ-maps of X into the $(2n + 1)$-dimensional cube is dense in the space of all such maps; 2) dim $X \leqslant n$ if, and only if, for every $\epsilon > 0$ there is an ϵ-map of X into an n-dimensional polytope. Generalizations of ϵ-maps and of the latter result to nonmetric noncompact spaces may be found in Dowker.[1] In connection with his work on ϵ-maps, Ulam raised the following question [Problem 21, The Scottish Book]: does there exist for every $\epsilon > 0$ an ϵ-map of the 2-dimensional disk onto the 2-dimensional torus? The question was negatively answered simultaneously by Fort[1] and Ganea.[2] In the latter paper it is proved that an n-dimensional ANR (absolute neighborhood retract) which may be mapped by arbitrary ϵ-maps onto n-dimensional closed manifolds (depending on ϵ) has the homotopy type of such a manifold, its separation properties by closed subsets are similar to those of closed n-manifolds and, if $n = 2$, the ANR is necessarily homeomorphic to a closed surface. When $n = 3$, ANR is no longer necessarily homeomorphic to a closed 3-manifold (see Ganea[1]). However, for arbitrary n, such an ANR is necessarily a generalized n-manifold in the sense of Wilder. See Mardesic and Segal.[1]

Call a topological property α global if for every compactum $A \in \alpha$ there exists an $\epsilon > 0$ such that if f: $A \to f(A)$ is a continuous map such that the diameter of the set $f^{-1}(y)$ is less than ϵ for every point $y \in f(A)$, then $f(A) \in \alpha$. It is shown in this paper that the existence of an essential map into S^n is a global property. Some applications of this theorem are given. These results belong to earliest results concerning theory of global properties. For other results concerning similar questions see Kuratowski and Ulam [11], Borsuk,[1] Eilenberg,[1,2] Granas, [1,2] and Holsztyński and Iliadis.[1]
K. Borsuk, T. Ganea

[15] (with J. Schreier) Eine Bemerkung über die Gruppe der topologischen Abbildungen der Kreislinie auf sich selbst (pp. 74–78)

Here Schreier and Ulam study groups of homeomorphisms. For other papers concerned with such groups one could mention Fine and Schweigert,[1] Ander-

son,[1] Whittaker,[1] Kirby,[1] and Lee.[1] See also [17], [24], and comments to these papers.
J. Mycielski

[16] (with Z. Łomnicki) Sur la théorie de la mesure dans les espaces combinatoires et son application au calcul des probabilitiés I. Variables indépendantes (pp. 79–120)

This is a contribution to the measure-theoretic formulation of the theory of probability. Other contributions have been made by Borel, A. Łomnicki, Steinhaus, and Kolmogorov [p. 237]. This paper contains ideas and results previously announced in 1932 and 1933, before Kolmogorov's.[1] The paper is devoted only to independent random variables. It was to be the first of a planned series on applications of Cartesian products in probability theory. The authors announced [p. 239] publication of a further paper, devoted to Markov chains. However it has not appeared. In this paper a sequence of independent random variables is meant as the direct product of a sequence of probability spaces (E_j, M_j, m_j), where $E_j =$ real line [see p. 265], while in the Kolmogorov scheme a sequence of independent random variables is understood as a sequence of (real) functions measurable with respect to the same probability space (E, M, m) and satisfying the condition of independence. The two approaches, while not being identical, are in fact equivalent: there exist some theorems which express explicitly this equivalence (see, e.g., Loève,[1] propositions A and B). Nevertheless the approach of Łomnicki and Ulam stresses the great role of Cartesian products in probability theory. Łomnicki and Ulam were the first authors to have formulated the general theorems on the existence of the direct product of finitely many abstract measures: Theorem 1 [p. 245], and of infinitely many such measures: Theorem 2 [p. 252]. These theorems have been announced without proofs in Ulam's communication during the Congress of Zurich (1932) [8].

Saks has given a proof of Theorem 1 of this paper which is different from the proof given by the authors. The proof of Theorem 2 given by authors is incomplete. Sparre Andersen and Jessen[1] write: "in the proof of lemma 4 the number N is chosen twice." Let us add that Sparre Andersen and Jessen[1] noted that two other proofs by Sparre Andersen and by Doob are also incomplete. Correct proofs of Theorem 2 have been given by von Neumann[1] and, the same proof, independently by Kakutani.[1] At present, the proof of Theorem 2 can be found in many monographs and textbooks. Theorem 2 can be generalized in different ways. A theorem on independent fields of sets formulated by Marczewski and proved in 1940 by Banach[1] (p. 275–90) contains Theorem 2 as a special case (see also Marczewski[1]). On the other hand, Banach's theorem can be also proved by the aid of Theorem 2 (Sikorski[1] and Sherman[1]). Another generalization of Theorem 2 consists in passing from product spaces to projective limits (Prokhorov[1]). The Daniell-Kolmogorov theorem (see, e.g., Halmos,[3] pp. 212, 214) on the non-direct product of measures (or its generalizations; e.g., for compact measures, see Marczewski[1] and Ryll-Nardzewski[1]) permits one to derive Theorem 2 from Theorem 1 in the case of probability measures on the field of Borel subsets of the real line (or, more generally, for all compact measures). It seems that theorems on product measures in such form, i.e., without Theorem 2 in the whole generality, are sufficient for the purposes of probability theory (see, e.g., Neveu,[1] pp. 72, 78).

At present, the theory of content, i.e., of finitely additive nonnegative set functions, is rarely applied in the probability theory and is used rather as a tool in the theory of measure, i.e., of σ-additive set functions. Thus, Theorems 3 and 4 [p. 256] are not so important as Theorems 1 and 2. In the nine probabilistic examples given by the authors [pp. 257–263], finite additivity is used only twice [Example 2, p. 257 and Example 9, p. 262], namely, in the case where the considered spaces are countable and the definition of an appropriate σ-additive probability measure is impossible. It seems that also at present the possibility of applying "content" in probability theory is not quite clear. There exists another mathematical scheme for the description of some situations for which the definition of a σ-additive probability measure is impossible. The theory of abstract conditional probability of Rényi[1] permits one to describe, among other cases, the uniform probability distribution on the real line or on the set of integers.

Certain corrections should be noted. On p. 238 instead of "Voir Theorem 4" read "Voir Theorem 3." On p. 244 the condition which defines the general (not necessarily direct) product of measures must be formulated in another way, e.g.: if $X_{j_0} \subset E_{j_0}$ for a certain $j_0(1 \leqslant j_0 < n)$ and $X_j = E_j$ for all $j \neq j_0(1 \leqslant j \leqslant n)$, then $m(X_1 x \ldots x X_n) = m(X_{j_0})$. On p. 251 instead of $M(F)$ read $m(F)$. On p. 254 the proof of lemma 4 is not complete.
Z. Łomnicki

[17] (with J. Schreier) Über topologische Abbildungen der euklidischen Sphären (pp. 121–137)

Related sharper results for the special case of the closed interval [0, 1] were

obtained by Jarník and Knichal.[1] See also [12], [15], [18], [19] and comments to these papers.
J. Mycielski

[18] (with H. Auerbach) Sur le nombre de générateurs d'un groupe semi-simple (pp. 138–140)

M. Kuranishi[1] has improved Theorem II of this paper. Namely, he proved that every connected semisimple Lie group G has a subgroup dense in G which is free with two free generators. See also [17], [19] and comments to these papers.
J. Mycielski

[19] (with J. Schreier) Sur le nombre des générateurs d'un groupe topologique compact et connexe (pp. 141–143)

Put

$$D = \{(a, b): a, b \in G, \text{ and } a, b \text{ generate a dense subgroup of } G\}$$
$$F = \{(a, b): a, b \in G, \text{ and } a, b \text{ generate a free nonabelian subgroup of } G\}$$

Kuranishi[1] has proved that if G is a perfect (i.e., the commutator of G equals G) connected Lie group, then D is an open set in G^2. Results of Balcerzyk and Mycielski[1] and Mycielski[1] imply that if G is a connected nonsolvable locally compact group, then F is residual (comeagre) in G^2. Hence, by the result of this paper, if G is a nonabelian connected metrisable compact group then $D \cap F$ is residual in G^2. This result and that of Mycielski[2] yield also that for such G there exists a perfect (i.e., nonempty closed and without isolated points) set $P \subseteq G$ such that P is a set of free generators of a free subgroup of G and every couple $a, b \in P, a \neq b$ generates a subgroup dense in G. See also Mycielski[4] for a stronger statement. It is not clear whether the last facts concerning nonabelian connected metric compact groups are true more generally for all nonsolvable (or only semisimple) connected metric locally compact (or even Lie) groups. One should also consult two papers of Schreier.[1,2] See also [12], [17], [18] and comments to these papers.
J. Mycielski

[21] (with J. Schreier) Über die Automorphismen der Permutationsgruppe der natürlichen Zahlenfolge (pp. 148–150)

Related results were proved by Schreier.[3] Freudenthal [1] proved that all the automorphisms (not only the continuous ones) of the group of rotations of R^3 around the origin, and of some other Lie groups, are inner. Whittaker[1] proved the same for groups of homeomorphisms of compact manifolds with or without boundary.

The problem at the end of the paper is still open. See also [12] and comment.

J. Mycielski

[22] (with J. C. Oxtoby) On the equivalence of any set of first category to a set of measure zero (pp. 151–156)

From a letter of J. C. Oxtoby to the editors: "It should be pointed out that the main result, at least as regards the existence of such a transformation of the square, was proved earlier by Brouwer.[1] (I did not come upon this paper until long afterward.) However, our results regarding category, and the method, appear to have been new. The only subsequent work in this direction that I know of is the remark that in the case of the square the transformation can be effected by a product homeomorphism, that is, one of the form $f \times g$, where f and g are automorphisms of the sides. I included this result in my mimeographed notes on measure and category (1957), which I am currently engaged in rewriting for possible publication by Springer." The notes have since appeared in print (Oxtoby[3]).

In the same letter (dated May 27, 1970) Professor Oxtoby answers a question of Mycielski by proving a theorem in a sense dual to the result of this paper (the terminology is that of Kuratowski[1]): "Theorem. Let X be a complete separable metric space. Let μ be a σ-finite nonatomic Borel measure in X with $\mu(X) > 0$. Given any set $A \subset X$ with $\mu(A) = 0$ there exists a Borel automorphism f of X of class 2, 2 such that $\mu(f(E)) = \mu(E)$ for every Borel set $E \subset X$ and such that $f(A)$ is of first category."

"Proof. We can add to A a countable dense set and a nullset of power c and still cover A with a G_δ nullset. Hence we may assume that A is itself a dense G_δ nullset of power c. The complement of A contains a set B homeomorphic to $N =$ the set of irrational numbers. Since $N = N \times N$, B can be chosen so that $\mu(B) = 0$. Then A and B are disjoint G_δ nullsets of power c. By Kuratowski,[1] p. 358, Remarque 2°, there exists a homeomorphism g of class 1, 1 of A onto B. Define

$$f = \begin{cases} g & \text{on } A \\ g^{-1} & \text{on } B \\ \text{identity} & \text{on } C = X - (A \cup B) \end{cases}$$

Then f is a 1—1 map of X onto itself and the restriction of f or f^{-1} to A, B, or C is of class 1. Since A, B, and C are of additive class 2, it follows (Kuratowski,[1] p. 284, IV 1) that f is a Borel automorphism of class 2, 2. We have $\mu(f(E)) = \mu(f(E) \cap C) = \mu(E \cap C) = \mu(E)$ for every Borel set E, and $f(A) = B \subset X - A$ is of first category."
J. Mycielski

[23] (with J. C. Oxtoby) On the existence of a measure invariant under a transformation (pp. 157–163)

An alternative proof of Theorem 1, not dependent on the use of generalized limits, can be obtained by extending T to a transformation on a compact superspace and then applying the Riesz representation theorem. See Oxtoby.[1] The paper referred to in footnote 3 was never published, but a proof of the result in question is given in Oxtoby.[2] The result can also be derived from a theorem of von Neumann.[3]
J. Oxtoby

[24] (with J. C. Oxtoby) Measure-preserving homeomorphisms and metrical transitivity (pp. 164–210)

Oxtoby and Ulam establish the result that metrical transitivity is the "general case" in the sense of category for certain large classes of transformations. Much subsequent work has been stimulated by this paper. Mention is made of Halmos,[1,2] Tulcea,[1] Rohlin,[1] Katok and Stepin,[1] and Oxtoby.[1] The simple characterization of measure topologically equivalent to Lebesgue measure obtained in Theorem 2 has proved useful, and new applications are still being found. See, e.g., Krickeberg.[1] The problem referred to in footnote 42 was subsequently solved by H. Anzai, and independently by L. W. Green (unpublished). If a one-parameter group is ergodic, then so are all but countably many of its members. It should be noted that a proof of Theorem 9, in the case $n = 2$, was given earlier by Brouwer.[1] In connection with Theorem 11, P. Erdös remarked that the point p may be unique. For instance, let T be the automorphism of $E = (0, 1)$ that takes x into x^2, let $\{n_k\}$ be a sequence of positive integers such that k/n_k is not convergent, and take $C =$

$\{0\} \cup \bigcup_{k=1}^{\infty} T^{2n_k} [1/4, 1/2]$. Then 0 is the only point of E whose images fall in C with a definite frequency.
J. Oxtoby

Many related developments have taken place since this paper was written, especially in the direction of differential equations and differentiable ergodic (or metrically transitive) flows on manifolds. These answer, at least partially, some of the questions stated in the introduction of this paper. One must emphasize that the differentiable case seems to be quite a different problem from that of the continuous case discussed by these authors. Recently, Anosov has written that A. B. Katok has shown that on every compact manifold of dimension $\geqslant 4$, there is a smooth ergodic dynamical system. Anosov[1] (see also Arnold and Avez[1] or Sternberg[1]), has defined a class of differentiable dynamical systems, now called Anosov flows. Anosov showed that they are ergodic, and that measure-preserving perturbations of them are also ergodic. Geodesic flows on unit tangent bundles of compact Riemannian manifolds of negative curvature are Anosov flows, and thus this result includes and extends the work of Hopf and Hedlund and Hopf on ergodicity of geodesic flows. By showing that certain flows are Anosov flows, one obtains examples of differentiable ergodic flows with good stability properties. This class includes suspensions of hyperbolic toral automorphisms. New examples of differentiable ergodic flows on suspensions of nilmanifolds were found in this way in Smale,[1] "infra nilmanifold" examples in Shub[1] and mixed nilpotent, semi simple examples in Tomter.[1] The conjecture of Birkhoff that ergodicity is a generic property has been shown to be false by Kolmogorov, Arnold, and Moser. See Arnold and Avez[1] or Sternberg[1] for an account of this theory which contrasts sharply with the results on the analogous topological problem of Oxtoby and Ulam. Perhaps a highlight of this work is the successful systematic introduction of the notion of genericity ("in the general case") into the study of dynamical systems. Today, this notion is basic for the study of dynamical systems, using the idea of Baire set or category, in the same way as done in this paper. See Smale[1] for a review of this.
S. Smale

[25] What is measure? (pp. 211–216)

This is Ulam's first expository paper. The role of the continuum hypothesis, which is not mentioned here, was emphasized in a later account of the problem of measure given by the author in 1967 in a filmed lecture entitled "Measure and set theory," in the M. A. A. Individual Lecture Series. A fuller

discussion of the possibility of nonmeasurable projective sets is contained in the author's book [69]. It should be noted that the uniqueness of Lebesgue or Haar measure, mentioned on page 601, is limited to measurable sets. There exist many distinct invariant extensions of these measures. Concerning the result announced in Ref. 5, see Banach,[1] p. 291.
J. Oxtoby

[27] (with D. H. Hyers) On approximate isometries (pp. 217–221), and [33] (with D. H. Hyers) Approximate isometries of the space of continuous functions (pp. 243–246)

Hyers and Ulam study a "stability" problem for isometry. By an ϵ-isometry of one metric space into another is meant a transformation which changes distances by less than ϵ, where ϵ is some positive number. Given an ϵ-isometry T the question arises whether a true isometry U exists which approximates T within $k\epsilon$ where k is some positive constant depending only on the metric spaces. The first paper answers the question in the affirmative for an ϵ-isometry mapping a Hilbert space onto itself. The same question was also answered in the affirmative for ϵ-isometries between two spaces of continuous functions on compact metric spaces in the second paper. Bourgin[1] has answered the question in the affirmative in the case of an ϵ-isometry of E_1 onto E_2 where E_1 is any Banach space and E_2 is a uniformly convex space satisfying certain additional conditions. However, there are so far no results for the general case where E_1 and E_2 are both arbitrary Banach spaces.
D. Hyers

[28] (with C. J. Everett) On ordered groups (pp. 222–230)

Everett and Ulam obtain for ordered groups results previously known only in the commutative case. Iwasawa[2] and Ogasawara[1] have shown that a conditionally complete lattice group is commutative. Hence, Theorem 1 implies that integrally closed ordered groups are necessarily commutative. In view of Lemma 2, this settles G. Birkhoff's Problem 3 referred to in the Introduction: Archimedean lattice groups are commutative. Obviously, Theorems 3 and 4 have lost their significance. That the converse of Lemma 1 is false may also be shown by the example of the two-dimensional vector space over the reals where $(x, y) > 0$ means that both $x > 0$ and $y > 0$. When this paper was written it was not known what orders free groups can admit. Since then it has been shown independently by several authors that free groups can be

linearly ordered; see, e.g., Iwasawa[1] or Neumann.[1] Moreover, they can carry no lattice-group structure other than linear order.
W. Fuchs

[30] (with C. J. Everett) Projective algebra I. (pp. 231–242)

This paper is important on two counts. First, it served as one of the major way-stations in the development of comprehensive algebraic versions of logic, such as cylindric and polyadic algebras. These latter algebras give a simpler way of treating quantifiers than projective algebras. However, projective algebras remain the simplest apparatus for an algebraic analysis of the usual projection operation of classical topology. Thus, secondly, the paper is also important in its own right. The theorem on completions (Section 3) has only recently been extended to other algebraic forms of logic. The representation theorem of Section 4 has remained a model for similar representation theorems for other two-dimensional algebraic logics. This representation theorem was extended to arbitrary projective algebras by McKinsey. See Henkin, Monk, and Tarski[1] and McKinsey.[1]
D. Monk

[33] (with D. H. Hyers) Approximate isometries of the space of continuous functions (pp. 243–246)

See commentary to [27].

[46] (with D. H. Hyers) Approximately convex functions (pp. 251–258)

Hyers and Ulam study a stability problem for convexity. A function f defined on a convex subset S of an n-dimensional Euclidean space E_n is called ϵ-convex if $f(hx + (1 - h)y) \leqslant hf(x) + (1 - h)f(y) + \epsilon$ for all x and y in S and for $0 \leqslant h \leqslant 1$. It is shown that if f is an ϵ-convex function defined on a convex open set G of E_n, then there is a convex function ϕ defined on G such that $|f(x) - \phi(x)| \leqslant k_n\epsilon$ for all $x \in G$ where $k_n = (n^2 + 3n)/(4n + 4)$. In the case of semicontinuous ϵ-convex functions, Green[1] has improved this result by obtaining a smaller value of k_n which is shown to be the best possible for $n \geqslant 3$. Green[1] has also studied a similar problem for approximately subharmonic functions.
D. Hyers

[51] (with D. H. Hyers) On the stability of differential expressions (pp. 259–264)

Here Ulam and Hyers study the stability problem for differential expressions. The following basic result is proved. Let f be a real function of a real variable having an nth derivative in a neighborhood N of a point $x = a$. If $f^{(n)}(a) = 0$ and $f^{(n)}(x)$ changes sign at $x = a$, then for each $\epsilon > 0$ there exists $\delta > 0$ such that for each function g having an nth derivative in N and satisfying $|g(x) - f(x)| < \delta$ in N, there exists a point $x = b$ such that $g^{(n)}(b) = 0$ and $|b - a| < \epsilon$. A similar result is proved for the differential expression $f'(x) - F(x, f(x))$. No analogous results are known for functionals such as those of the calculus of variations.
D. Hyers

[86] (with C. J. Everett) On some possibilities of generalizing the Lorentz group in the special relativity theory (pp. 265–287)

The Lorentz group generalization worked out is quite original; in the first six years after its publication no particular echo has been recorded in the physical literature; but it could occur in the future, especially if the possibility of recovering the classical group, as a particular approximation, will be proved. The great role played by the Lorentz group in physics stands upon the law of "Lorentz invariance," according to which the set V of all physical states of any given system observed in a frame E has to be mapped into itself when the system is observed from another frame E' in uniform motion with respect to E. In quantum physics this law becomes particularly meaningful since V acquires the structure of a normed vector space with an orthonormal basis and the Lorentz group determines an isomorphism with a group of linear transformation on V. Section 8 contains important steps to examine the possibility of satisfying the axioms of quantum mechanics in the framework of a vector space V over the p-adic field. The space-time notion loses its usual meaning for dynamical systems smaller, or evolving faster, than the atomic nuclei (this is the limit of 10^{-13} centimeter and 10^{-24} second referred to in Section 1); such a notion exists inasmuch as there exists a physical theory inherently consistent and able to fit the experimental behavior of these systems. The serious difficulties encountered by modern methods of theoretical physics ("quantum field theory") raise interest about the possibilities of building a dynamical theory on new space-time schemes like the one proposed in this paper.

Let us remember that the adopted definition of Lorentz transformations ($T\Gamma \subset \Gamma$) refers to the independence of light velocity from the frame of reference; it is a very-well-founded experimental fact, at least if space and time have their classical meaning. Of course the distinction between "proper" and "improper" transformations has nothing to do with the use of the same words made in physical literature, where "proper" excludes space or time reflections ($X' = -X$, $t' = -t$) which on the contrary are not included. The most curious result seems to be the recovery of an absolute meaning of length and time intervals, independent of the reference frame, as in the Galilean picture (Section 5). The procedure used makes explicit use of the norm $|X| = \max(\nu(x_i);\ i = 1, \ldots, n)$ introduced in the position-space $P = K^n$, where $n \geqslant 2$ is then assumed (Section 3). For a better understanding it might be useful to compare with the case $n = 1$, namely, one space coordinate $x \in K$, one time cordinate $t \in K$. As a guide, let us sketch a few points. Clearly the properties $N1$—5 reduce to $V1$—5 if $n = 1$ and a rotation becomes multiplication by $r \in K$, $\nu(r) = 1$; Theorem 2.1 and the statement (c) of Theorem 2.2 trivially hold. Writing $T = \begin{pmatrix} a & b \\ c & d \end{pmatrix}$ of order 2 over $K = Q_p$, the Lemma of Section 3 is no longer true: actually $\nu(c) > \nu(d)$ would imply $\nu(cx + d) = \nu(c) > 0$ if $\nu(x) = 1$, hence $cx + d \neq 0$. Similarly the Lorentz transformation $\tilde{T} = \begin{pmatrix} p & 1 \\ 1 & p \end{pmatrix}$ ($\tilde{T}\Gamma \subset \Gamma$ easily follows from V5) violates Theorem 3.1. Thus the characterization of improper transformations is different in the case $n = 1$; e.g., $\tilde{T}$ maps the axis of the light cone outside the cone $\left(\tilde{T}\begin{pmatrix} 0 \\ t \end{pmatrix} = \begin{pmatrix} t \\ pt \end{pmatrix},\ \nu(t) > \nu(pt)\right)$, a situation without analogy when $n \geqslant 2$. Nevertheless Theorem 3.2 can be paraphrased for $n = 1$ if the "proper" transformations are defined by $\nu(a) = \nu(d)$, $\nu(b) < \nu(d)$, $\nu(c) < \nu(d)$; this is enough to get the result of Section 5, i.e., the absence of length-contraction and time-dilatation, as far as proper transformations are concerned. In the classical case T reads $a = d = (1 - v^2)^{-1/2}$, $b = c = v(1 - v^2)^{-1/2}$, $v(< 1)$ being the relative velocity of the two frames (light velocity $= 1$): the valuation here used gives $\nu(1 - v^2) = 1$, which, roughly speaking, accounts for the Galilean behavior.

E. Beltrametti

[91] (with G. H. Meisters) On visual hulls of sets (pp. 288–90), and
[93] (with W. A. Beyer) Note on the visual hull of a set
(pp. 291–296)

Ulam, Meisters, and Beyer discuss the concept of the visual hull of a set, an

idea slightly analogous to the convex hull of a set. An additional reference is a much longer paper by Larman and Mani.[1]
W. Beyer

[97] (with J. Mycielski) On the pairing process and the notion of genealogical distance (pp. 304–311)

Theorems 1 and 2 were improved by Kahane and Marr.[1] These authors relaxed the condition that each couple produces exactly two children keeping only the condition that the size of each generation is n. Marr proved that

$$\int_R d_r(a, b) d\mu = 2(n-1),$$

if a and b belong to the same generation A_i and $a \neq b$, and he and Kahane got an upper estimate of $\int_R \rho_r(a, b) d\mu$. *Errata:* page 229, line 14, for B_j put A_j, lines 14–21, for $\neq$ put $\not\supset$, line 26, for $p^{s(t)}$ put $p_0^{s(t)}$.
J. Mycielski

Commentaries to Part II

[38] (with N. C. Metropolis) The Monte Carlo method (pp. 319–325, this volume)

The modern Monte Carlo school seems to have originated at Los Alamos after the Second World War in work on nuclear reactions. Some of the contributors were Everett, Fermi, Metropolis, R. D. Richtmyer, Ulam, and von Neumann. The method itself under other names can be traced back to much earlier times. But the appearance of the stored-program electronic computer made large scale sampling experiments practical. This paper is an important, early summary of the statistical approach to the study of differential equations and other analytical equations occurring in natural sciences. For an overview of Monte Carlo methods, see Halton.[1]
W. Beyer

[48] Random processes and transformations (pp. 326–337)

Ulam presents "a general point of view and specific problems connected with the relation between descriptions of physical phenomena by random processes and the theory of probabilities on the one hand and the deterministic descriptions by methods of classical analysis in mathematical physics on the other." Part of what Ulam presents forms the foundations of the theory

of Monte Carlo calculations. The random ergodic theorem announced in Ref. 10 was actually proved in an earlier paper of Pitt,[1] a paper which went unnoticed for much too long a time. For a fuller account, see Kakutani.[2]
J. Oxtoby, W. Beyer

[50] (with N. C. Metropolis) A property of randomness of an arithmetical function (pp. 338–339)

The exact distribution here approximated by the Poisson is, of course, the binomial distribution: $\lambda_i/n = \binom{n}{i}(n-1)^{n-i}/n^n$. The authors were led to experiment on the "middle square" function because it had been suggested by von Neumann about 1946 as a source of numbers for use in simulating random behavior. (See the remarks by Hammer,[1] Forsythe,[1] and von Neumann[2] where another aspect of Ulam's interest in random-number generation is mentioned.) Tests such as the one described here tend to show that the "middle square" function has the properties of random mapping. In a sequel to this paper, Metropolis[1] gave further empirical results about the "middle square" function, in particular the number of "trees." (The graphs are not really trees; they contain a single cycle.) Kruskal [1] showed that the average number of "trees" in the decomposition of a random mapping is asymptotically $\frac{1}{2}[\log n(2n) + \gamma]$, confirming the conjecture made in the present paper. For surveys of more recent work on random mappings, see Riordan[1] and Knuth;[1] the latter book discusses improved forms of random number generation which have made the original middle-square method obsolete.
D. Knuth

[53] (with V. L. Gardiner, R. Lazarus, and N. C. Metropolis) On certain sequences of integers defined by sieves (pp. 340–345)

The present paper constitutes the first detailed study of a sieve-process different in kind from the classical sieve of Eratosthenes. Not surprisingly, it aroused considerable interest and motivated some important theoretical work on the asymptotic density of sequences produced by analogous sieves. In the light of our present-day knowledge, the following comments seem appropriate.

1. Lucky numbers do indeed have prime-like distributions, but only in the first approximation. In 1957, Hawkins and Briggs[1] showed that the nth lucky number l_n is given asymptotically by

(1) $l_n \sim n \log n$;

this was improved by Chowla, as quoted in Hawkins and Briggs,[1] to

(2) $l_n = n \log n + \frac{n}{2} (\log \log n)^2 + o[n(\log \log n)^2]$.

This expression should be compared with the analogous result for the nth prime:

(3) $p_n = n \log n + n \log \log n + o(n \log \log n)$.

Thus, the luckies are somewhat sparser than the primes. An actual count performed (1969) on the Los Alamos Scientific Laboratory's Maniac II computer yields the comparisons shown in Table I between $L(n)$, the number of luckies $\leqslant n$, and $\pi(n)$ (the count for $\pi(n)$ includes 1, but not 2, in the list of primes). Among the 244,388 luckies less than 3,750,000 there are 26,700 primes. Furthermore, for all $n > 11$ in the range we have $l_n > p_n$, as conjectured by Ulam.

Subsequent to the work of Hawkins, Briggs, and Chowla, Briggs,[1] and

Table I

$n \times 10^{-5}$	$L(n)$	$\pi(n)$
2	16448	17984
4	30981	33860
6	44935	49098
8	58544	63951
10	71918	78498
12	85074	92938
14	98091	107126
16	110987	121127
18	123740	135072
20	136412	148933
22	148999	162662
24	161525	176302
26	173958	189880
28	186335	203362
30	198665	216816
32	210899	230209
34	223132	243539
36	235278	256726
37.5	244388	266717

Wunderlich and Briggs[1] considered analogous but more general sieves. They found that the resulting sequences are always asymptotically sparser than the primes; that is, that the second term in the analogue of (2) is always greater than $n \log \log n$ in order of magnitude.

2. There seems to have been little work on the additive properties of the lucky sequence. For example, the conjecture that there is an infinite number of lucky "twins" $l_n = l_{n-1} + 2$ has not been examined theoretically; it is probably just as difficult to prove as the corresponding conjecture for primes. Table II gives the counts, for all luckies $\leqslant 3.75 \times 10^6$, of the number of gaps of size 2, 4, . . . , 30 between successive luckies and compares them with the corresponding counts for the primes.

Both sequences show marked peaks at gaps which are multiples of 6, but the lucky gap distribution appears somewhat more "regular" in the sense that the ratio of the number of gaps for the form $6M + 2$ to the number of the form $6M + 4$ is closer to unity (cf. the Hardy-Littlewood [1] predictions for the prime sequence). At the present time nothing further can be said about this.

3. The Goldbach conjecture—that every even number is the sum of two prime numbers (in at least one way)—has been verified for all even numbers $2n$ in the range $6 \leqslant 2n \leqslant 10^8$. See Stein and Stein.[1] Owing to technical difficulties (see Section 5 of this commentary) it has not yet been possible to verify the analogue of Goldbach's conjecture for luckies over this

Table II

Gap	No. of gaps (luckies)	No. of gaps (primes)
2	23799	25418
4	23941	25180
6	33981	42427
8	15595	17787
10	15586	22637
12	25433	26821
14	9592	14271
16	9581	10051
18	17389	17433
20	7977	8539
22	8003	7582
24	11041	10277
26	4061	4570
28	4075	4889
30	7673	7756

same range. Over the much smaller range $2 \leqslant 2n \leqslant 3.75 \times 10^6$, however, the "Lucky-Goldbach" conjecture is indeed true (Maniac II, 1969). For the very modest range $2 \leqslant 2n \leqslant 2 \times 10^5$ there exists (see Stein and Stein[1]) a tabulation of the number of solutions $\lambda(2n)$ of the equation $2n = l_i + l_j$, $l_i \leqslant l_j$ (l_i, l_j luckies). In the same reference a corresponding tabulation is made for the prime case; i.e., there is given a list of the number of solutions $\nu(2n)$ of the equation $2n = p_i + p_j$, $p_i \leqslant p_j$ (p_i, p_j primes). The "Goldbach curve" for primes, i.e., the plot of $\nu(2n)$ against $2n$, is less regular than that for luckies. It does, however, have the advantage that its detailed structure is quite accurately predictable (over the range considered) by means of a probabilistic formula based on the number of primes in certain residue classes (and hence depending ultimately on the value of $\phi(n)$, the Euler function). In Stein and Stein[2] an attempt was made to derive an analogous formula for $\lambda(2n)$, but the results were disappointing. As yet no one has proposed any sieve-independent characterization of lucky numbers; the nearest analogue to $\phi(n)$ seems to be the number of integers $\leqslant n$ remaining when one sieves with a *fixed* number of luckies. This function, however, does not yield precise enough information to enable one to predict the fluctuations in $\lambda(2n)$.

4. The strong presumption that the Goldbach conjecture is true for both primes and luckies has motivated some extensive numerical research into the additive properties of other sequences, both deterministic and random. For random sequences with a prescribed density distribution, a relatively complete theoretical treatment is possible. For example, it has been proved (see Everett and Stein[1]) that, for "almost all" sequences of randomly generated, positive, odd integers with prime-like distribution, the Goldbach conjecture is true from some point on. (Here we mean, of course, the "weaker" Goldbach conjecture: that, with at most a *finite* number of exceptions, every even number is the sum of two elements of the sequence.) This sheds no light on the truth of the conjecture for particular sequences (such as luckies or primes). It does suggest, however, that the detailed properties of these sequences—in principle derivable from an analysis of the corresponding sieves—are largely irrelevant to the truth or falsehood of the Goldbach conjecture (the same can be said of the "twin" conjecture, which is indeed true for almost all random sequences with suitable density distributions. See Everett and Stein.[1] As an example of this, consider the observation that both the primes and the luckies appear to be much more numerous than is necessary to form a binary (additive) basis for the evens; indeed, it is known (see Halberstam and Roth[1]) that very sparse sequences

can form such a basis. To illustrate this superabundance of primes and luckies (as well as that of other sequences), a simple algorithm was devised. See Stein and Stein.[1] When this algorithm is applied to the lucky sequence, it yields a set of 6096 luckies which constitutes a basis for all even numbers $2 \leqslant 2n \leqslant 3.75 \times 10^6$, i.e., for all even numbers lying within the range of our most extensive table of luckies. Over this range, the corresponding number for the primes is 6211. Thus at most 2.5% of the luckies are required (note that the sets produced by this algorithm are sparse but not necessarily "minimal").

One might think that the property of forming a basis depends on the presence of the first few members of the sequence, but experiments indicate that this is not so. Choose a suitable finite sequence $\{a_j\}_{1 \leqslant j \leqslant n}$ and remove the first $k \ll n$ members. When the algorithm is applied to this sequence $\{a_{k+1}, a_{k+2}, \ldots, a_n\}$ one expects it to produce a basis for all evens $N_0 \leqslant 2n \leqslant N$, where $N \cong a_n - a_{k+1}$ and N_0 is "close to" $2a_{k+1}$. For example, if one starts the lucky sequence with $l_{16} = 67$ one can construct a basis with 6090 elements for all evens in the range $256 \leqslant 2n \leqslant 3.75 \times 10^6$ (for the prime sequence beginning with $p_{18} = 67$, one can construct a basis with 6168 elements for all evens $150 \leqslant 2n \leqslant 3.75 \times 10^6$). In the case of primes, this type of experiment has been carried much further; for example, if one removes the first 2885 primes, so that the sequence begins with $p_{2886} = 26263$, the algorithm generates a basis of 10491 primes for all evens in the range $53638 \leqslant 2n \leqslant 10^7$.

5. Experimental studies of the lucky number sequence are much less advanced than the corresponding studies on primes. The reason for this is that the preparation of an extensive list of luckies is quite laborious. Despite several ingenious suggestions (see Wunderlich[1]), it would be a major undertaking to generate all the luckies $\leqslant 10^8$; for primes, on the other hand, this can be done today in less than half an hour.

Despite the relative inefficiency of the lucky sieve, it does have associated with it several "checking" algorithms, the prime analogues of which are at best cumbersome. For the convenience of the reader we list three of these; they are all, of course, simply rearrangements of the basic sieve definitions.

(*a*) To calculate the *k*th lucky, knowing all of the luckies $\leq k$ (this is given, in a different notation, in Hawkins and Briggs[1]):

1. Let l_{m-1} be the greatest lucky $\leq k$.
2. Set $R_m = k$ initially.
3. The general iteration step is

$$j = \left[\frac{R_m}{l_{m-1} - 1}\right] \quad \text{if} \quad l_{m-1} - 1 \nmid R_m,$$

$$j = \left[\frac{R_m}{l_{m-1} - 1}\right] - 1 \quad \text{if} \quad l_{m-1} - 1 \mid R_m$$

(here $[x]$ means, as usual, the greatest integer $\leq x$), with

$$R_{m-1} = R_m + j.$$

4. Iterate on m, ending up with

$$l_k = 2R_2 - 1.$$

(*b*) To find the nth number $S_n(k)$ sieved out by the kth lucky l_k:

1. Set $R_k = nl_k$ initially.
2. The general iteration step is then identical with (*a*)3 above (changing m to k).
3. Iterating on k, we end up with

$$S_n(k) = 2R_2 - 1.$$

(*c*) To test whether n (odd) is a lucky:

1. Set $R_2 = \dfrac{n+1}{2}$ initially.
2. The general iteration step is:

   if $l_k \mid R_k$, i.e., if $\dfrac{R_k}{l_k} = m$, then n is not a lucky (in fact, it is the mth number sieved out by l_k); if $l_k \nmid R_k$, then set

$$R_{k+1} = R_k - \left[\frac{R_k}{l_k}\right].$$

3. Iterate on k until either $l_k \mid R_k$ (see preceding step) or until $l_k > R_k$. In the latter case, n is the R_kth lucky.

We note, in connection with (*a*), that it is not so easy to find the kth prime algorithmically (short of carrying out the full sieve process). What *is* easy to find is $\pi(n)$, the number of primes $\leqslant \sqrt{n}$. Correspondingly, it is $L(n)$, the number of luckies $\leqslant n$, which is difficult to check.

P. Stein

[55] (with J. M. Kister, P. R. Stein, W. Walden, and M. B. Wells) Experiments in chess (pp. 346–349)

This paper appeared at a time when computer chess was in its infancy. Prior to the games of "6 × 6 chess" played by Maniac I against human opponents, some other computer chess games had been reported; these games, however, were not physically played by an extant computer, but rather were "simulated" by human players following a specific but unimplemented program. Thus the "experiments" in question have at least the distinction of priority. The paper was accorded the further honor of being translated into Russian and published in the very heartland of chess—the Soviet Union.

Even in 1956 the precepts for computer play enunciated by the authors were in no sense new; the important thing was to demonstrate that they worked in practice. Ten years after these papers appeared the art (or science) of programming a computer to play (8 × 8) chess had advanced to such a degree that there was in existence at least one program capable of winning 80% of its games against "nontournament" players (this, at least, is the claim). This program, and the state of the art in general, is described in Greenblatt, Eastlake, and Crocker.[1]

It must be said that much of the progress is due to the great advances in computer technology during the past fifteen years. The "new ideas" incorporated into the M.I.T. program seem quite natural, perhaps even obvious, given that there actually exist technical methods of implementing them.

The opinion voiced in this paper to the effect that no machine will be coded to play *master* chess during the next 20 years, will probably stand. It would be rash, however, to be dogmatic in one's pronouncements without having seen the book by former World Champion Mikhail Botvinnik.[1]
P. Stein

[78] Combinatorial analysis in infinite sets and some physical theories (pp. 388–400)

Ulam discusses relations between infinite constructions in set theory and infinite constructions in physical theories. Some of the mathematical problems mentioned in this paper have already been solved. Let *SH* denote Souslin's hypothesis and *CH* Cantor's hypothesis ($2^{\aleph_0} = \aleph_1$). All four statements *SH* and *CH*, *SH* and non-*CH*, non-*SH* and *CH*, and non-*SH* and non-*CH* are now known to be consistent with the usual axioms of set theory *ZF* including the axiom of choice. For *SH* and *CH*, see R. B. Jensen (unpublished). For *SH* and non-*CH*, see Solovay and Tennenbaum.[1] For non-*SH* and *CH*,

see Jensen.[1] For non-SH and non-CH, see Jech.[1] The problem on the infinite positional game defined on pp. 346 and 347 is partly solved; e.g., player A has a winning strategy if and only if E is at most countable. (Galvin, Mycielski, and Solovay, to appear.) But it is not known if there are any sets E such that player B has a winning strategy.
J. Mycielski

[81] (with P. R. Stein) Non-linear transformation studies on electronic computers (pp. 401–484)

There appears to be no subsequent work on broken-linear transformations in two dimensions (Chap. VI of this paper). Even in one dimension there remains a host of intriguing problems. As an example, consider the following. Take the class of concave, symmetric (about $x = 1/2$) continuous functions $f(x)$ mapping the unit interval onto itself. We have

$$f(0) = f(1) = 0; \quad f\left(\frac{1}{2}\right) = 1; \quad \left|\frac{df}{dx}\right|_{0,1} > 1; \quad \left|\frac{df}{dx}\right|_{1/2} = 0.$$

Further, restrict $f(x)$ so that at the fixed point $x_f = f(x_f)$, $x_f \neq 0$, we have

$$\left|\frac{df}{dx}\right|_{x_f} > 1.$$

Problem: Give a constructive characterization of the subclass which has no finite limit sets (i.e., no attractive period of finite order).

An interesting type of one-dimensional "pathology" is discussed in Metropolis, Stein, and Stein.[1] This paper considers a class of essentially one-dimensional nonlinear transformations depending continuously on a single parameter α. These transformations have a countably infinite set of "families" of finite limit sets, each family corresponding to a certain range of α. Moreover, these families can be characterized as to order (in α) and number of points by employing purely combinatorial methods (cf. the comments at the beginning of Chapter VI of this paper). One should also compare Metropolis, Stein, and Stein.[2]

This paper is open to the criticism that it fails to indicate promising lines for the theoretical study of the nonlinear phenomena which Ulam and Stein discuss. As shown by the references quoted above, the appropriate tools for such a study are in the process of development, but one is a long way from being able to cope with the apparently pathological behavior exhibited

by even the simple class of transformations treated in this paper. At the present writing, therefore, it seems clear that the development of adequate theoretical methods will have to be preceded by further numerical experimentation, some of which will differ only in precision and level of sophistication from that exemplified by this paper and that of Metropolis, Stein, and Stein.[2]
P. Stein

[83] (with M. L. Stein and M. B. Wells) A visual display of some properties of the distribution of primes (pp. 485–489)

Ulam conceived this paper while doodling, during an uninteresting lecture. The simple idea of plotting the prime bit-pattern in spiral fashion, in order to display the density of primes in certain quadratic forms, illustrates the usefulness of the computer as a heuristic tool.
M. Wells

Bibliography for the Commentaries

P. Alexandroff

1. *Untersuchungen über Gestalt und Lage abgeschlossener Mengen beliebiger Dimension,* Ann. Math. ser. 2, vol. 30 (1928), pp. 101–87

E. Sparre Andersen and B. Jensen

1. *Some limit theorems on integrals in an abstract set,* Danske Vid. Selsk. Math-Fys. Medd. vol. 22, no. 14 (1946)

R. Anderson

1. *On homeomorphism as products of conjugates of a given homeomorphism and its inverse,* in Topology of 3–manifolds and related topics, Prentice-Hall, 1962, pp. 231–4

D. Anosov

1. *Geodesic flows on closed Riemannian manifolds with negative curvature,* Proc. Steklov Inst. Math. vol. 90 (1967). Amer. Math. Soc. Translation, Providence, R.I., 1969

V. Arnold and A. Avez

1. *Problèmes Ergodiques de la Mécanique Classique,* Gauthier-Villars, Paris, 1966

S. Balcerzyk and J. Mycielski

1. *On the existence of free subgroups in topological groups,* Fund. Math., vol. 44 (1957), pp. 303–8

S. Banach

1. *Oeuvres,* vol. I, Inst. Math. Acad. Polon. Sci., 1967

K. Borsuk

1. *Über stetige Abbildungen der euklidischen Räume,* Fund. Math., vol. 21 (1933), pp. 236–43
2. *Sur un problème de MM. Kuratowski et Ulam,* Fund. Math., vol. 31 (1938), pp. 154–9

M. Botvinnik

1. *Computers, Chess and Longrange Planning,* Springer Verlag, 1970

D. Bourgin

1. *Approximate isometries,* Bull. Amer. Math. Soc., vol. 52 (1946), pp. 704–14

R. E. Bradford

1. *Cardinal addition and the axiom of choice,* Ann. Math. Logic, vol. 3 (1971), pp. 111–96

W. Briggs

1. *Prime-like sequences generated by a sieve process,* Duke Math. J., vol. 30 (1963), pp. 297–312

L. Brouwer

1. *Lebesguesches Mass und Analysis Situs,* Math. Ann., vol. 79 (1919), pp. 212–22

Z. Charzyński

1. *Sur les transformations isométriques des espaces du type (F),* Studia Math., vol. 13 (1953), pp. 94–121

C. Dowker

1. *Mapping theorems for non-compact spaces,* Amer. J. Math., vol. 69 (1947), pp. 200–42

S. Eilenberg

1. *Transformations continues en circonférence et la topologie du plan,* Fund. Math., vol. 26 (1936), pp. 61–112
2. *Sur les transformations à petites tranches,* Fund. Math., vol. 30 (1938), pp. 92–5

P. Enflo

1. *On the nonexistence of uniform homeomorphisms between L_p-spaces,* Ark. Mat., vol. 8 (1970), pp. 103–5
2. *On a problem of Smirnov,* Ark. Mat., vol. 8 (1970), pp. 107–9

C. Everett and P. Stein

1. *On random sequences of integers,* Bull. Amer. Math. Soc., vol. 76 (1970), pp. 349–51

S. Feferman

1. *Some applications of the notions of forcing and generic sets,* Fund. Math., vol. 56 (1965), pp. 325–45

T. Figiel

1. *On nonlinear isometric embeddings of normed linear spaces,* Bull. Acad. Polon. Sci. Ser. Math. Astronom. Phys., vol. 16 (1968), pp. 185–8

N. J. Fine and G. E. Schweigert

1. *On the group of homeomorphism of an arc,* Ann. Math., ser. 2, vol. 62 (1955), pp. 237–53

G. Forsythe

1. *Generation and testing of random digits at the National Bureau of Standards, Los Angeles,* Monte Carlo Method, 1949, U.S. Nat'l. Bur. Standards, Appl. Math. Ser., vol. 12, pp. 34–5

M. Fort

1. *ε-mappings of a disk onto a torus,* Bull. Acad. Polon. Sci. Ser. Sci. Math. Astronom. Phys., vol. 7 (1959), pp. 51–4

H. Freudenthal

1. *Die Topologie der Lieschen Gruppen als algebraisches Phänomen I,* Ann. Math., Ser. 2, vol. 42 (1941), pp. 1051–74

H. Gaifman

1. *Measurable cardinals and constructible sets,* Notices Amer. Math. Soc., vol. 11 (1964), p. 771

T. Ganea

1. *Covering spaces and Cartesian products,* Ann. Soc. Polon. Math., vol. 25 (1952), pp. 30–42
2. *On ε-maps onto manifolds,* Fund. Math., vol. 47 (1959), pp. 35–44

E. Gaughan

1. *Topological group structures of infinite symmetric groups,* Proc. Nat. Acad. Sci. U.S.A., vol. 58 (1967), pp. 907–10

A. Gladkiĭ

1. *Effectively unbounded additive set-functions* (Russian), Kolomen Ped. Inst. Učen. Zap. Ser. Fiz.-Mat., vol. 2 (1958), pp. 111–6

A. Granas

1. *Über einer Satz von K. Borsuk,* Bull. Acad. Polon. Sci. Ser. Sci. Math. Astronom. Phys., vol. 5 (1957), pp. 959–62
2. *On continuous mappings of open sets in Banach spaces,* Bull. Acad. Polon. Sci. Ser. Sci. Math. Astronom. Phys., vol. 6 (1958), pp. 25–9

J. Green

1. *Approximately convex functions,* Duke Math. J., vol. 19 (1952), pp. 499–504

R. Greenblatt, D. Eastlake III, and S. Crocker

1. *The Greenblatt chess program,* AFIPS Proc. Fall Joint Computer Conference, vol. 31 (1967), pp. 801–10

H. Halberstam and K. Roth

1. *Sequences,* vol. 1, chapter I, Oxford Univ. Press, 1966

P. Halmos

1. *Approximation theorems for measure preserving transformations,* Trans. Amer. Math. Soc., vol. 55 (1944), pp. 1–18

2. *In general a measure preserving transformation is mixing,* Ann. Math., ser. 2, vol. 45 (1944), pp. 784–92

3. *Measure Theory,* Van Nostrand, 1950, pp. 212 and 214

J. Halpern

1. *The independence of the axiom of choice from the Boolean prime ideal theorem,* Fund. Math., vol. 55 (1964), pp. 57–66

J. Halton

1. *A retrospective and prospective survey of the Monte Carlo method,* SIAM Rev., vol. 12 (1970), pp. 1–63

P. Hammer

1. *The mid-square method of generating digits,* Monte Carlo Method 1949, U.S. Nat'l. Bur. Standards, Appl. Math. Ser., vol. 12, p. 33

W. Hanf and D. Scott

1. *Classifying inaccessible cardinals,* Notices Amer. Math. Soc., vol. 8 (1961), p. 445

G. Hardy and J. Littlewood

1. *Some problems of 'Partitio Numerorum': III. On the expression of a number as a sum of primes,* Acta Math., vol. 44 (1922), pp. 1–70

D. Hawkins and W. Briggs

1. *The lucky number theorem,* Math. Mag., vol. 31 (1957/58), pp. 277–80

L. Henkin, D. Monk, and A. Tarski

1. *Cylindric Algebras, Part I,* North-Holland, 1971

W. Holsztyński

1. *Linearization of isometric embeddings of Banach spaces. Metric envelopes,* Bull. Acad. Polon. Sci. Ser. Sci. Math. Astronom. Phys., vol. 16 (1968), pp. 189–93

W. Holsztyński and S. Iliadia

1. *Approximation of multi-valued by single-valued mappings and some applications,* Bull. Acad. Polon. Sci. Ser. Sci. Math. Astronom. Phys., vol. 16 (1968), pp. 765–9

W. Hurewicz and H. Wallman

1. *Dimension Theory,* Princeton Univ. Press, 1941

K. Iwasawa

1. *On the structure of conditionally complete lattice-groups,* Japan J. Math., vol. 18 (1943), pp. 777–89

2. *On linearly ordered groups,* J. Math. Soc. Japan, vol. 1 (1948), pp. 1–9

V. Jarník and V. Knichal

1. *Sur l'approximation des fonctions continues par les superpositions de deux fonctions,* Fund. Math., vol. 24 (1935), pp. 206–8

T. Jech

1. *Trees,* J. Symbolic Logic, vol. 36 (1971), pp. 1–14

R. Jensen

1. *The fine structure of the constructible hierarchy,* Ann. Math. Logic, vol. 4 (1972), pp. 229–308, esp. Theorem 6.2

B. Jessen

1. *Abstract theory of measure and integration,* Mat. Tidskr., ser. B, vol. 4 (1939), pp. 7–21

J. Kahane and R. Marr

1. *On a class of stochastic pairing processes and the Mycielski-Ulam notion of genealogical distance,* J. Combinatorial Theory, ser. A, vol. 13 (1972), pp. 383–400

S. Kakutani

1. *Notes on infinite product measure spaces. I.,* Proc. Imp. Acad. Tokyo, vol. 19 (1943), pp. 148–51

2. *Random ergodic theorems and Markoff processes with a stable distribution,* Proceedings of the Second Berkeley Symposium on Mathematical Statistics and Probability, 1951, pp. 247–61

A. Katok and A. Stepin

1. *Metric properties of measure preserving homeomorphisms,* Uspehi Mat. Nauk., vol. 25 (1970)

H. Keisler and A. Tarski

1. *From accessible to inaccessible cardinals,* Fund. Math., vol. 53 (1964), pp. 225–308

Lord Kelvin

1. *Nineteenth century clouds over the dynamical theory of heat and light,* Phil. Mag. 6th Ser., vol. 2 (1901), pp. 1–40

R. Kirby

1. *Stable homeomorphisms and the annulus conjecture,* Ann. Math., vol. 89 (1969), pp. 575–82

D. E. Knuth

1. *Seminumerical Algorithms,* Addison-Wesley, 1969, pp. 3–8, 453–5

A. Kolmogorov
1. *Grundbegriffe der Wahrscheinlichkeitsrechnung,* Springer, Berlin, 1933

K. Krickeberg
1. *Ein Isomorphiesatz über topologische Massräume,* Math. Nachr., vol. 37 (1968), pp. 59–66

M. Kruskal
1. *The expected number of components under a random mapping function,* Amer. Math. Monthly, vol. 61 (1954), pp. 392–7

M. Kuranishi
1. *On everywhere dense imbedding of free groups in Lie groups,* Nagoya Math. J., vol. 2 (1951), pp. 63–71, Corollary on page 71

K. Kuratowski
1. *Topology, Vol. I,* Polish Scientific Publishers, 1966

K. Kuratowski and A. Mostowski
1. *Set Theory,* North-Holland, 1968

D. Larman and P. Mani
1. *On visual hulls,* Pac. J. Math., vol. 32 (1970), pp. 157–71

Y. Lee
1. *Homeomorphisms on manifolds,* Kyungpook Math. J., vol. 7 (1967), pp. 31–6

J. Lindenstrauss
1. *On nonlinear projections in Banach spaces,* Mich. J. Math., vol. 11 (1964), pp. 263–87

M. Loève
1. *Probability Theory,* Van Nostrand, 1963, pp. 230–1

E. Marczewski
1. *Indépendance d'ensembles et prolongement de mesures (résultats et problèmes),* Colloq. Math., vol. 1 (1948), pp. 122–32
2. *On compact measures,* Fund. Math., vol. 40 (1953), pp. 113–24

S. Mardesic and J. Segal
1. *ε-mappings and generalized manifolds,* Michigan Math. J., vol. 14 (1967), pp. 171–82, 423–6

D. A. Martin
1. *Measurable cardinals and analytic games,* Fund. Math., vol. 66 (1970), pp. 287–91

R. McKenzie

1. *On elementary types of symmetric groups,* Algebra Universalis, vol. 1 (1971), pp. 13–20
2. *A note on subgroups of infinite symmetric groups,* Indag. Math., vol. 33 (1971), pp. 53–8

J. McKinsey

1. *On the representation of projective algebras,* Amer. J. Math., vol. 70 (1948), pp. 375–84

N. Metropolis

1. *Phase shifts-middle squares-wave equation,* Symposium on Monte Carlo Methods, Wiley, 1956

N. Metropolis, M. Stein, and P. Stein

1. *Stable states of a nonlinear transformation,* Numer. Math., vol. 10 (1967), pp. 1–19
2. *On finite limit sets for transformations on the unit interval,* to appear in J. Combinatorial Theory, ser. A

A. P. Morse

1. *Squares are normal,* Fund. Math., vol. 36 (1949), pp. 35–9

J. Mycielski

1. *On the extension of equalities in connected topological groups,* Fund. Math., vol. 44 (1957), pp. 300–2
2. *Independent sets in topological algebras,* Fund. Math., vol. 55 (1964), pp. 139–47
3. *Two remarks on Tychonoff's product theorem,* Bull. Acad. Polon. Sci. Ser. Sci. Math. Astronom. Phys., vol. 12 (1964), pp. 439–41
4. *Almost all functions are independent,* Fund. Math. (1973)

B. Neumann

1. *On ordered groups,* Amer. J. Math., vol. 71 (1949), pp. 1–18

J. Neveu

1. *Bases Mathématiques du Calcul des Probabilitiés,* Masson et Cie, 1964

T. Ogasawara

1. *Commutativity of Archimedean semi-ordered groups,* J. Sci. Hiroshima Univ., ser. A, vol. 12 (1943), pp. 249–54

J. Oxtoby

1. *Ergodic sets,* Bull. Amer. Math. Soc., vol. 58 (1952), pp. 116–36, propositions (2.1), (7.1), and (7.2)

2. *Homeomorphic measures in metric spaces,* Proc. Amer. Math. Soc., vol. 24 (1970), pp. 419–23
3. *Measure and Category,* Springer-Verlag, 1971

H. Pitt

1. *Some generalizations of the ergodic theorem,* Proc. Camb. Philos. Soc., vol. 38 (1942), pp. 325–43

Yu. Prokhorov

1. *Convergence of random processes and limit theorems in probability theory,* Theor. Probability Appl. (English Translation), vol. 1 (1956), pp. 157–214

A. Rényi

1. *On a new axiomatic theory of probability,* Acta Math. Acad. Sci. Hung., vol. 6 (1955), pp. 285–335

J. Riordan

1. *Enumeration of linear graphs for mappings of finite sets,* Ann. Math. Stat., vol. 33 (1962), pp. 178–85

V. Rohlin

1. *A "general" measure-preserving transformation is not mixing,* Dokl. Akad. Nauk SSR, vol. 60 (1948), pp. 349–51

F. Rowbottom

1. *Some strong axioms of infinity incompatible with the axiom of constructibility,* Ann. Math. Logic, vol. 3 (1971), pp. 1–44

C. Ryll-Nardzewski

1. *On quasi-compact measures,* Fund. Math., vol. 40 (1953), pp. 125–30

J. Schreier

1. *Über die Drehungsgruppe in Hilbertschen Raum,* Studia Math., vol. 5 (1934), pp. 107–10
2. *Eine Bemerkung über Erzeugende in kompakten Gruppen,* Fund. Math., vol. 25 (1935), pp. 198–9
3. *Über Abbildungen einer abstrakten Menge auf ihre Teilmengen,* Fund. Math., vol. 28 (1937), pp. 261–64

D. Scott

1. *Measurable cardinals and constructible sets,* Bull. Acad. Polon. Sci. Ser. Sci. Math. Astronom. Phys., vol. 9 (1961), pp. 521–24

S. Sherman

1. *On denumerably independent families of Borel fields,* Amer. J. Math., vol. 72 (1950), pp. 612–14

M. Shub

1. *Endomorphisms of compact differentiable manifolds,* Amer. J. Math., vol. 91 (1969), pp. 175–99

W. Sierpiński

1. *Fonctions additives non complètement additives et fonctions non mesurables,* Fund. Math., vol. 30 (1938), pp. 96–9

R. Sikorski

1. *Independent fields and Cartesian products,* Studia Math., vol. 11 (1950), pp. 171–84

J. Silver

1. *Some applications of model theory in set theory,* Ann. Math. Logic, vol. 3 (1971), pp. 45–110
2. *Measurable cardinals and Δ_3^1 well-ordering,* Ann. Math., ser. 2, vol. 94 (1971), pp. 414–16

S. Smale

1. *Differentiable dynamical systems,* Bull. Amer. Math. Soc., vol. 73 (1967), pp. 747–817

R. Solovay

1. *On the cardinality of Σ_2^1 sets of reals,* Foundations of Mathematics Symposium, Papers commemorating the sixtieth birthday of Kurt Gödel, Springer, 1969, pp. 58–73
2. *A model of set theory in which every set of reals is Lebesgue measurable,* Ann. Math., vol. 92 (1970), pp. 1–56
3. *Real-valued measurable cardinals,* Proc. Symp. in Pure Math., vol. 13, Part I, Amer. Math. Soc., 1971, pp. 397–428

R. Solovay and S. Tennenbaum

1. *Iterated Cohen extension and Souslin's problem,* Ann. Math., vol. 94 (1971), pp. 201–45

M. Stein and P. Stein

1. *Tables of the number of binary decompositions of all even number $0 < 2n < 200{,}000$ into prime numbers and lucky numbers,* Los Alamos Scientific Laboratory, LA-3106, 1964

2. *Experimental results on additive 2-bases,* Math. Comp., vol. 19 (1965), pp. 427–34

S. Sternberg

1. *Celestial Mechanics, Part II,* Benjamin, New York, 1969

S. Swierczkowski

1. *Finite subgroups of locally compact groups,* Colloq. Math., vol. 21 (1970), pp. 53–4

A. Tarski

1. *Sur les fonctions additives dans les classes abstraites et leur applications au problème de la mesure,* C. R. Soc. Sci. Lett. Varsovie, C. III, vol. 22 (1929), pp. 114–7
2. *Une contribution à la théorie de la mesure,* Fund. Math., vol. 15 (1930), pp. 42–50
3. *Algebraische Fassung des Massproblems,* Fund. Math., vol. 31 (1938), pp. 47–66
4. *Cardinal Algebras,* Oxford Univ. Press, 1949

P. Tomter

1. *Anosov flows on infra-homogeneous spaces,* Amer. Math. Soc. Proc. Symp. in Pure Math., vol. 14 (1970), pp. 299–328

A. Tulcea

1. *On the category of certain classes of transformations in ergodic theory,* Trans. Amer. Math. Soc., vol. 114 (1965), pp. 261–79

R. Vaught

1. *Indescribable cardinals,* Notices Amer. Math. Soc., vol. 10 (1963), p. 126

J. von Neumann

1. *Einige Sätze über messbare Abbildungen,* Ann. Math., vol. 33 (1932), pp. 574–86; lemma, p. 577
2. *Various techniques used in connection with random digits,* Monte Carlo Method, 1949, U.S. Nat'l. Bur. Standards, Appl. Math. Ser., vol. 12, pp. 36–38
3. *Functional Operators, Vol. 1,* Princeton (1950), chapter 10, sec. 4

J. Whittaker

1. *On isomorphic groups and homeomorphic spaces,* Ann. Math., vol. 78 (1963), pp. 74–91

M. Wunderlich

1. *Sieving procedures on a digital computer,* J. Assn. Computing Mach., vol. 14 (1967), pp. 10–19

M. Wunderlich and W. Briggs

1. *Second and third term approximations of sieve-generated sequences,* Illinois J. Math., vol. 10 (1966), pp. 694–700